# Lecture Notes in Electrical Engineering

## Volume 888

The book series *Lecture Notes in Electrical Engineering* (LNEE) publishes the latest developments in Electrical Engineering - quickly, informally and in high quality. While original research reported in proceedings and monographs has traditionally formed the core of LNEE, we also encourage authors to submit books devoted to supporting student education and professional training in the various fields and applications areas of electrical engineering. The series cover classical and emerging topics concerning:

- Communication Engineering, Information Theory and Networks
- Electronics Engineering and Microelectronics
- Signal, Image and Speech Processing
- Wireless and Mobile Communication
- Circuits and Systems
- Energy Systems, Power Electronics and Electrical Machines
- Electro-optical Engineering
- Instrumentation Engineering
- Avionics Engineering
- Control Systems
- Internet-of-Things and Cybersecurity
- Biomedical Devices, MEMS and NEMS

For general information about this book series, comments or suggestions, please contact leontina.dicecco@springer.com.

To submit a proposal or request further information, please contact the Publishing Editor in your country:

**China**

Jasmine Dou, Editor (jasmine.dou@springer.com)

**India, Japan, Rest of Asia**

Swati Meherishi, Editorial Director (Swati.Meherishi@springer.com)

**Southeast Asia, Australia, New Zealand**

Ramesh Nath Premnath, Editor (ramesh.premnath@springernature.com)

**USA, Canada:**

Michael Luby, Senior Editor (michael.luby@springer.com)

**All other Countries:**

Leontina Di Cecco, Senior Editor (leontina.dicecco@springer.com)

**** This series is indexed by EI Compendex and Scopus databases. ****

More information about this series at https://link.springer.com/bookseries/7818

Deepak Gupta · Rajat Subhra Goswami ·
Subhasish Banerjee · M. Tanveer ·
Ram Bilas Pachori
Editors

# Pattern Recognition and Data Analysis with Applications

Springer

*Editors*
Deepak Gupta
Department of Computer Science
and Engineering
National Institute of Technology Arunachal
Pradesh
Jote, Arunachal Pradesh, India

Rajat Subhra Goswami
Department of Computer Science
and Engineering
National Institute of Technology Arunachal
Pradesh
Jote, Arunachal Pradesh, India

Subhasish Banerjee
Department of Computer Science
and Engineering
National Institute of Technology Arunachal
Pradesh
Jote, Arunachal Pradesh, India

M. Tanveer
Department of Mathematics
Indian Institute of Technology Indore
Indore, Madhya Pradesh, India

Ram Bilas Pachori
Department of Electrical Engineering
Indian Institute of Technology Indore
Indore, Madhya Pradesh, India

ISSN 1876-1100 ISSN 1876-1119 (electronic)
Lecture Notes in Electrical Engineering
ISBN 978-981-19-1519-2 ISBN 978-981-19-1520-8 (eBook)
https://doi.org/10.1007/978-981-19-1520-8



This Springer imprint is published by the registered company Springer Nature Singapore Pte Ltd.
The registered company address is: 152 Beach Road, #21-01/04 Gateway East, Singapore 189721, Singapore

# Preface

The book coverage is concerned with the latest advancements in the areas of machine learning, computer vision and pattern recognition, computational learning theory, big data analytics, network intelligence, signal processing and their applications in real world. The topics covered in machine learning involve feature extraction, variants of support vector machine (SVM), extreme learning machine (ELM), artificial neural network (ANN) and other areas in machine learning. The mathematical analysis of computer vision and pattern recognition involves the use of Geometric techniques, scene understanding and modelling from video, 3D object recognition, localization and tracking, medical image analysis and so on. Computational learning theory involves different kinds of learning like incremental, online, reinforcement, manifold, multi-task, semi-supervised, etc. Further, it covers the real-time challenges involved while processing big data analytics and stream processing with the integration of smart data computing services and interconnectivity. Additionally, it covers the recent developments to network intelligence for analysing the network information and thereby adapting the algorithms dynamically to improve the efficiency. Lastly, it includes the progress in signal processing to process the normal and abnormal categories of real-world signals, for instance signals generated from IoT devices, smart systems, speech, videos, etc., and involves biomedical signal processing: electrocardiogram (ECG), electroencephalogram (EEG), magnetoencephalography (MEG), electromyogram (EMG), etc.


Jote, India — Deepak Gupta
Jote, India — Rajat Subhra Goswami
Jote, India — Subhasish Banerjee
Indore, India — M. Tanveer
Indore, India — Ram Bilas Pachori

# Contents

# About the Editors

**Dr. Deepak Gupta** is Assistant Professor at the Department of Computer Science & Engineering of National Institute of Technology Arunachal Pradesh. He received the Ph.D. degree in Computer Science & Engineering from the Jawaharlal Nehru University, New Delhi, India. His research interests include support vector machines, ELM, RVFL, KRR and other machine learning techniques. He has published over 60 referred journal and conference papers of international repute. His publications have around 862 citations with an h-index of 15 and i10-index of 29 (Google Scholar, 21/08/2022). He is currently the member of an editorial review Board member of Applied Intelligence. He is the recipient of the 2017 SERB-Early Career Research Award in Engineering Sciences which is the prestigious award of INDIA at early career level. He is a senior member of IEEE and currently an active member of many scientific societies like IEEE SMC, IEEE CIS, CSI and many more. He has served as a reviewer of many scientific journals and various national and international conferences. He is the General Chair of upcoming 3rd International Conference on Machine Intelligence and Signal Processing (MISP-2021) and associated with other conferences like IEEE SSCI, IEEE SMC, IJCNN, BDA 2021 etc. He has supervised 3 PhD students and currently 3 PhD students are enrolled under him. He is currently the Principal Investigator (PI) or Co-PI of 02 major research projects funded by the Science & Engineering Research Board (SERB), Government of India.

**Dr. Rajat Subhra Goswami** received his B.Tech. in Information Technology in 2005 from West Bengal University of Technology, West Bengal. He received his M.E. in Multimedia Development from Jadavpur University, West Bengal, in 2009 and then joined in Bengal Institute of Technology Shantiniketan, Bolpur, West Bengal, as Assistant Professor in CSE department. He became Assistant Professor of CSE department at National Institute of Technology, Arunachal Pradesh, Government of India, in 2011. He received Ph.D. in Computer Science and Engineering from National Institute of Technology Arunachal Pradesh in 2015. Currently, he is working as Associate Professor in the department of Computer Science and Engineering in National Institute of Technology, Arunachal Pradesh, Government of India. He has more than 10 years of experience as a teacher. Cryptography, big data and machine learning are his research areas. In separate national/international journals/conferences, he has written over 50 research papers. Two Ph.D. scholars were awarded under his supervision, and four scholars are now working in separate fields. He is a life member of Indian Science Congress Association and Cryptology Research Society of India.

**Dr. Subhasish Banerjee** received his Ph.D. in Computer Science and Engineering from National Institute of Technology, Arunachal Pradesh, in 2016 and M.Tech. degree in Computer Application from Indian Institute of Technology (ISM), Dhanbad, India, in 2012. Currently, he is working as Assistant Professor in the Department of Computer Science and Engineering in National Institute of Technology, Arunachal Pradesh. His research activities are mainly focused on cryptography, networking and information security. He is the author or co-author of more than 20 papers in international refereed journals and more than 20 paper contributions in referred international conference.

**M. Tanveer** is Associate Professor and Ramanujan Fellow at the Discipline of Mathematics of the Indian Institute of Technology Indore. Prior to that, he worked as a Postdoctoral Research Fellow at the Rolls-Royce@NTU Corporate Lab of the Nanyang Technological University, Singapore. During 2012 to 2015, he was an Assistant Professor at the Department of Computer Science and Engineering of the LNM Institute of Information Technology (LNMIIT), Jaipur. He received the Ph.D. degree in Computer Science from the Jawaharlal Nehru University, New Delhi, India. Prior to that, he received the M.Phil degree in Mathematics from Aligarh Muslim University, Aligarh, India. His research interests include support vector machines, optimization, machine learning, deep learning, applications to Alzheimer's disease and dementias. He has published over 100 referred journal papers of international repute. His publications have over 2500 citations with h index 28 (Google Scholar, July 2022). Recently, he has been listed in the world's top 2% scientists in the study carried out by Stanford University, USA. He has served on review boards for more than 100 scientific journals and served for scientific committees of various national and international conferences. He is the recipient of the 2017 SERB-Early Career Research Award in Engineering Sciences and the only recipient of 2016 DST-Ramanujan Fellowship in Mathematical Sciences which are the prestigious awards of INDIA at early career level. He is currently the Associate Editor - IEEE Transactions on Neural Networks and Learning Systems (Feb. 2022 - ), Associate Editor - Pattern Recognition, Elsevier (Nov 2021 - ), Action Editor - Neural Networks, Elsevier (Jan 2022 - ), Board of Editors - Engineering Applications of Artificial Intelligence, Elsevier (Jan 2022 - ), Associate Editor - Neurocomputing, Elsevier (Jan 2022 - ), Associate Editor - Cognitive Computation, Springer (Jan. 2022 - ), Editorial Board - Applied Soft Computing, Elsevier (Jan 2022 - ), International Journal of Machine Learning and Cybernetics, Springer (July 2021 - ), Associate Editor - Frontiers in Applied Mathematics and Statistics (Aug 2020 - ), Editorial Review Board - Applied Intelligence, Springer. He is/was Guest Editor in Special Issues of several journals including ACM Transactions of Multimedia

(TOMM), Applied Soft Computing, Elsevier, IEEE Journal of Biomedical Health and Informatics, IEEE Transactions on Emerging Topics in Computational Intelligence, Multimedia Tools and Applications, Springer and Annals of Operations Research, Springer. He has also co-edited one book in Springer on machine intelligence and signal analysis. He has organized many international/national conferences/symposium/workshop as General Chair/Organizing Chair/Coordinator, and delivered talks as Keynote/Plenary/invited speaker in many international conferences and Symposiums. He has organized several special sessions in top-ranked conferences including WCCI, IJCNN, IEEE SMC, IEEE SSCI, ICONIP. Amongst other distinguished, international conference chairing roles, he is the General Chair for 29th International Conference on Neural Information Processing (ICONIP2022) (the world's largest and top technical event in Computational Intelligence). Tanveer is currently the Principal Investigator (PI) or Co-PI of 11 major research projects funded by Government of India including Department of Science and Technology (DST), Science & Engineering Research Board (SERB) and Council of Scientific & Industrial Research (CSIR), MHRD-SPARC, ICMR.

**Prof. Ram Bilas Pachori** received the B.E. degree with honours in Electronics and Communication Engineering from Rajiv Gandhi Technological University, Bhopal, India, in 2001, the M.Tech. and Ph.D. degrees in Electrical Engineering from Indian Institute of Technology (IIT) Kanpur, Kanpur, India, in 2003 and 2008, respectively. He worked as Postdoctoral Fellow at Charles Delaunay Institute, University of Technology of Troyes, Troyes, France, during 2007–2008. He served as Assistant Professor at Communication Research Center, International Institute of Information Technology, Hyderabad, India, during 2008–2009. He served as Assistant Professor at Department of Electrical Engineering, IIT Indore, Indore, India, during 2009–2013. He worked as Associate Professor at Department of Electrical Engineering, IIT Indore, Indore, India, during 2013–2017 where presently he has been working as Professor since 2017. He is also Associated Faculty with Department of Biosciences and Biomedical Engineering and Center for Advanced Electronics at IIT Indore. He was Visiting Professor at School of Medicine, Faculty of Health and Medical Sciences, Taylor's University, Subang Jaya, Malaysia, during 2018–2019. He worked as Visiting Scholar at Intelligent Systems Research Center, Ulster University, Northern Ireland, UK, during December 2014. He is Associate Editor of *Electronics Letters, Biomedical Signal Processing and Control journal* and Editor of *IETE Technical Review Journal.* He is Senior Member of IEEE and Fellow of IETE and IET. He has supervised 12 Ph.D., 20 M.Tech. and 37 B.Tech. students for their theses and projects. He has more than 210 publications which include journal papers (126), conference papers (66), books (04) and book chapters (16). His publications have around 7500 citations with h index of 46 (Google Scholar, January 2021). He has been listed in the top h index scientists in the area of computer science and electronics by Guide2Research website. He has been listed in the world's top 2% scientists in the study carried out at Stanford University, USA. He has served on review boards for more than 100 scientific journals and served for scientific committees of various national and international conferences. He has delivered more than 135 talks in various conferences, workshops, short term courses and institutes. His research interests are in the areas of signal and image processing, biomedical signal processing, non-stationary signal processing, speech signal processing, brain–computer interfacing, machine learning and artificial intelligence in health care.

# Revolutions in Infant Fingerprint Recognition—A Survey

**Shilpa Chaman**

**Abstract** Fingerprint-based biometric recognition systems are routinely used in a numerous forensic laboratories, government and civilian applications because of the two basic traits: (1) Permanency: Fingerprint basic pattern of ridges and valleys remains invariant over time and (2) Uniqueness: Each individual has unique fingerprint. Even the fingerprints of identical twins are distinct from one another. Although fingerprint recognition is now a mature field but still accurate and reliable, infant recognition system is still a complex and intricate pattern recognition problem which yet has many aspects unexplored. An infant recognition system is urgently needed to stop the suffering of millions of infants due to malnutrition and vaccine-related diseases. Infant authentication will not only effectively provide nutritional supplements but also would assist in accurate and reliable identification of missing or abducted children and prevent baby swapping. The basic problem that has afflicted the infant health care system is infant matching or the ability to accurately link health records for the same infant across multiple hospitals and clinics. With a unique infant fingerprint ID, their previous health records can be accessed to provide better health care, financial services and government benefits throughout their lifetime. This paper has tried to explored child dermatoglyphics ('Derma' means skin in Greek, and 'Glyphic' depicts carvings), the scientific study of fingerprints of infants which not only includes an overview of existing techniques for infant fingerprint sensing, extraction and matching of features, but also the challenges and bench marking advancements made through deep convolutional neural networks. Finally, the paper also provides an insightful analysis of the challenges faced during database collection of infant fingerprints and future directions to improvise the existing infant fingerprint system by incorporating the age progression model of fingerprints.

**Keywords** Dermatoglyphics · Residual dense network · Crossing number · Infant fingerprint · Convolutional neural network

S. Chaman (✉)
St. Francis Institute of Technology, Mumbai, India
e-mail: shilpachaman@gmail.com



D. Gupta et al. (eds.), *Pattern Recognition and Data Analysis with Applications*,
Lecture Notes in Electrical Engineering 888,
https://doi.org/10.1007/978-981-19-1520-8_1

# 1 Introduction

Fingerprints are those distinctive and immutable impressions that are made by the series of dark lines called ridges and white areas between the ridges called valleys, on the epidermis of our finger [1]. This formation of fingerprints is not only decided by a genetic code in DNA, but also by the pressure on the fingers from the baby touching, its surrounding amniotic liquid and the position of fetus in the womb. Due to this reason, it is found that even identical twins do not have same fingerprints. The formation of pattern of fingerprints is almost complete by the time a fetus is 6 months old [2], and this configuration stays with an individual throughout his life. This makes fingerprint a unique biometric.

Although fingerprint recognition is a mature field, but still infants all over the world are deprived of their unique verifiable digital fingerprint ID. Due to lack of proper identification, infants, especially from poor and less developed countries, are not able to get proper vaccination, medical care and food supplements. According to World Health Organization (WHO) [3], the global vaccination coverage of DTP3 course (class of combination vaccines against three infectious diseases: diphtheria,tetanus and pertussis), from 2015 to 2018, was only 85%. UNICEF [4] declares that over 1.5 million children dies every year due to lack of proper vaccination. The key factors behind this is paucity of official ID for infants, which makes the tracking of vaccine schedules extremely difficult.

Overcrowding in hospitals and insufficient maternity wards often cause swapping of babies after birth. It is important to identify newborns and link their identities to their mothers in order to solve this issue. A proper identification would also help in investigation of missing or abducted child. However, paper records cannot be used to identify infants since they could be misplaced or stolen. So the most unique, socially acceptable [5], accurate, stable and practical means of infant recognition is fingerprint biometric recognition. This would not only help them to get a distinct national identification, but could also be used for their passport verification.

Other biometrics has the following shortcomings especially when we are dealing with infants:

- Face recognition [6] is challenging due to the variation in facial characteristics with age, gross head reflexes, pose and expression variations.
- Iris recognition [7] is difficult as the infant has to look at the camera with eyes open. This becomes also practically infeasible if the infant is sleeping or crying. Also, parents are concerned as it requires infrared illumination into the child's eye for iris capture.
- Footprint recognition [8] is difficult as the child feet must be properly cleaned to get footprinted, which can make the infant cry or become uncomfortable.
- Palmprint recognition [9] is tough as it requires to open an infant's fists to properly capture the palmprint, whereas capturing a fingerprint requires only opening a finger.

## 2 Fingerprint Recognition System

Sir Francis Galton [10] conducted extensive research on fingerprint recognition of infants. He inference that the fingerprint recognition is practicable for a child above 2.5 years of age. European Commission-Joint Research Center (JRC) mentioned in their technical report [11] that automatic fingerprint recognition can be done for children if the time gap is less than 4.5 years between the two captured fingerprints, provided the images are of sufficient quality.

Depending on the application background, an infant fingerprint biometric device may be used as an authentication or verification system. An infant biometric recognition (identification or verification) system requires to compare a registered or enrolled infant fingerprint sample against a newly captured biometric sample. The enrollment and identification processes are used by an identification method, whereas the enrollment and verification processes are used by a verification system, as described below:

**Infant Enrollment** In the infant enrollment system, the infant's extracted feature set is saved in the database as an infant fingerprint template. This template can be generated from either a single fingerprint or by including multiple samples. Enrollment is a basic step of registration before a biometric can be used for recognition, i.e., verification or identification.

**Infant Verification** One to One: The infant verification system is used to claim infant's identity by comparing its fingerprint against enrolled sample template saved in the database. The submitted claim or identity can be rejected or accepted using a one-to-one comparison method. . For example, one can identify whether an infant had taken vaccination or not.

**Infant Identification** One to Many: The infant identification system uses one-to-many comparisons to identify an infant by scanning the entire enrolled infant prototype database for a match and returning the enrollment reference's identifier if the match is successful. For example, one can identify whether a child is swapped or not. It can also be used to find the identity of a stolen or an abducted child over a large time lapse.

The modules mentioned in the following section are used in the infant enrollment, verification and identification processes required in the fingerprint recognition. Modules include: (a) fingerprint sensor; (b) feature extraction; (c) matching and decision-making; and (d) infant database.

## 3 Fingerprint Sensor Module

The acquisition of fingerprint image can be done by placing the fingertip of infant on the fingerprint sensor module, which produces an image of the finger impression, whose resolution is expressed in dots per inch (dpi) or pixels per inch. Image acquisition by fingerprint sensor is the most crucial step in the process of infant fingerprint recognition, and the loss of information at this stage is barely recoverable. This implicates that the fingerprint sensor designed for infants must have the following desirable characteristics:

- Resolution: Due to the softness of the skin, we may get poor contrast between some of the ridges, and the relative positions of the minutia are fairly variable. Therefore, in order to make infant fingerprints recognizable, the optical fingerprint reader must have resolution greater than 1000 ppi, so that it can easily acquire intricate details in infant fingerprints such as ridge ends and bifurcations.
- Compact and Ergonomic: In order to gently place the tiny infant fingers on the reader to capture the image, it should be compact and designed accordingly so that very short and delicate fingers of infant must be conveniently captured.
- Faster capture, lower distortion and motion blur: When an infant's finger is placed on a glass platen, the grip reflex causes the finger to respond in an unpredictable manner, so the reader must be fast to capture and should overcome any type of distortion due to motion blur or improper placement of finger.
- Wet or oily finger: For newborns, the outer layer of skin can peel during the first few weeks. After birth, the skin's integrity can also change. Infants can also develop the habit of sucking their fingers, which changes the texture of their skin. Such textural changes must be captured properly.

A fingerprint acquisition can be done by a sensor in either online or offline mode as explained below:

**Offline Sensing**: Offline system is usually partially automatic, and it is done by first smearing black ink on infant's finger and then scanning the acquired inked fingerprint card. This mode is not suitable for oily and soft baby skin as ridges get easily deformed when they come in contact with paper and ink.

**Online Sensing**: In a fully automatic online system, fingerprint is captured using a live scanner, and here, immediate recognition is done.

## 3.1 Optical Sensors

The technique used in most of the optical sensors is called frustrated total internal reflection (FTIR) [12]. The infant finger is put on the top of a glass prism, and when the light enters the prism, it is reflected at the valleys and absorbed at the ridges. The ridges can be differentiated from the valleys due to the lack of reflection. The light which is reflected from the prism is focused via a lens over a CCD or CMOS image sensor. They are quite robust and need less maintenance.

To capture the soft friction ridges of infants, Jain et al. [13] developed a custom made 1270 ppi fingerprint reader for infants, but it worked well only for infants older than 6 months. The recognition accuracy was further improvised by Engelsma et al. [14], for infants below 6 months of age by another fingerprint reader of 1900 ppi., which can capture minute friction ridge pattern, minutiae and pores using high fidelity infant fingerprint images.

### 3.2 *Solid-state Sensors*

Solid-state sensor consists of a 2D-pixel array of micro-cells on a silicon chip. Each micro-cell is touch-based sensor, and when the user touches its surface, it reads the fingerprint pixel. There are mainly four types of solid-state sensors: capacitive, piezoelectric, thermal and electric field. The major drawback of the capacitive sensors is that its cost rises dramatically with the sensing area dimensions.

Skin conditions of infant's finger are usually wet or oily where capacitive sensors produce significant error rates as compared to optical sensors. Marcialis et al. [15] has assessed experimentally that recognition accuracy obtained by optical sensors is 5.78 times greater than the accuracy obtained using a capacitive sensor.

### 3.3 *Ultrasound Sensors*

Echography is the basic principle behind ultrasound sensors. Its transmitter sends short acoustic pulses and the echo signal, which has obtained the depth image of the fingerprint and its ridge structure, is received by it [16]. Because it photographs the subsurface of the finger, this approach may not be influenced by dirt and oil accumulations or thin gloves on the finger. Still its usage is limited because ultrasound sensors are quite bulky, time consuming and expensive. In spite of producing good quality images, this technology is not frequently used for infant fingerprint recognition due to parental concerns.

To overcome the drawback of current fingerprint sensors related to sensing of wet/dry fingers of infants or distortions caused by motion blur or inadequate pressure, new sensing techniques like multi-spectral imaging [17] and 3D touchless acquisition [18] need to be explored further.

## 4 Feature Extraction Module

A fingerprint image has series of ridges (dark lines) and valleys or furrows (bright lines) as shown in Fig. 1a. Ridge ending or termination occurs when a fingerprint's ridges abruptly come to an end, and ridge bifurcation occurs when they divide into two. When analyzing the fingerprint pattern at different scales, different sorts of features emerge at different levels, such as (a) global level, (b) local level and (c) very fine level.

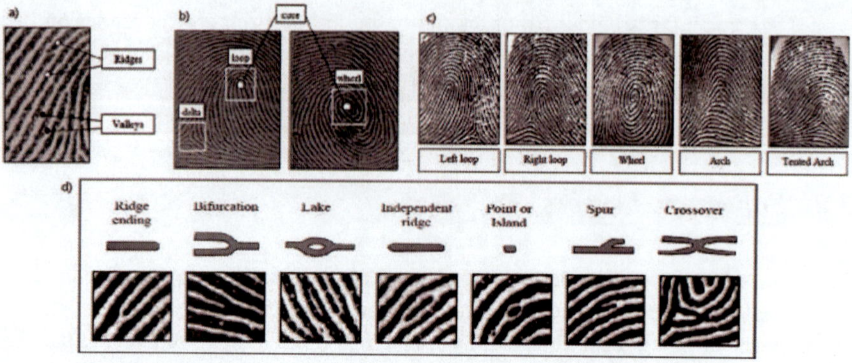


**Fig. 1** Fingerprint images showing **a** ridges and valleys patterns; **b** white boxes indicate singular regions, and circles indicate core points; **c** five major classes of fingerprints defined by Henry [12, 19]; **d** categorizes of minutiae 'Galton Details' [20]

- Global level: Fingerprint patterns show regions known as singular regions where ridge lines exhibit distinctive shapes, which may be broadly classified into three types, viz., (a) Whorl, (b) Loop and (c) Delta, as shown in Figs. 1b and characterized by O, ∩, △ shapes, respectively. Whorl fingerprint is a mix of two loops or one whorl with two deltas, whereas left loop and right loop normally have one loop and one delta. Arches are a wave-like pattern which can be plain arches or tented arches which rise to a sharper point. Such five major classes of fingerprint defined by Henry [19] are shown in Fig. 1c. For simplifying search and retrieval, matching algorithms use singular points like core which is the north most center point of the inner edge line point of loops and whorl, for assigning a distinct class to the fingerprint pattern. For fingerprints of the arch class, it is tough to define the core. As a result, the place of greatest ridge line curvature is designated as the core. Unfortunately, when imaged by an FTIR scanner, the newborn fingerprint may flatten on the surface and the ridges may fuse, resulting in feature fusion. It is difficult to precisely find the core point in all of the fingerprint images since newborn fingerprint patterns vary so much. Therefore, for accurate matching additional features like external fingerprint shape, orientation and frequency may also be recognized at the global level [12].
- Local level: Discontinuities of the ridges of fingerprint are often called minutiae. In 1892, Sir Francis Galton [20] did substantial research, and on scientific basis, he categorized minutiae and inferred that they remain permanent over an individual's lifespan. They are also called 'Galton Details' in his honor as shown in Fig. 1d. Although he mainly categorized them into seven different types, but the two most distinguished minutiae are: ridge endings and ridge bifurcations. When a ridge ends abruptly, that point is marked as a ridge ending, whereas the point where a ridge diverges into two branches is called ridge bifurcation. These minutiae points are very important features at local level and are highly salient as their distribution is quite peculiar. This minutiae extraction is a challenging task for infant fingerprints whose inter-ridge spacing is usually of 4–5 pixels or may be less than 1 pixel.

- Very fine level: Very fine features like intra-ridge width, ridge-shape, ridge-curvature, ridge-edge contours, dots and incipient ridges can also be extracted at very fine level. Another important fine level detail which is extracted at high resolution of about 1000 dpi is sweat pores details, whose positions and shapes are considered highly distinctive to aid infant fingerprint recognition systems where minutiae are not always available, even at resolutions over 1500 ppi. Nguyen et al. [21] suggested a pore-based matcher and pore extraction approach to improve the recognition accuracy for fingerprint identification.

The main task of feature extraction module is to convert input fingerprint image into its respective template. To accurately extract features for template formation, a fingerprint image has to undergo following important stages as described below: (a) ridge orientation and frequency; (b) segmentation; (c) singularity detection; (d) enhancement; (e) minutiae extraction.

### 4.1 Local Ridge Orientation and Frequency

Ridge orientation is an intrinsic property of a fingerprint image which is determined by the angle $\theta xy$ that it makes with the horizontal axis. It helps to make proper distinction between different types of minutiae points. Ratha et al. [22] proposed a novel adaptive ridge flow orientation technique for robust feature extraction. The local ridge frequency is another important feature which is defined at a point $(xo; yo)$. Hong et al. [23] counted the average number of pixels between two consecutive peaks of gray levels along the direction normal to the local ridge orientation $\theta xy$ and evaluated the local ridge frequency.

### 4.2 Segmentation

The fingerprint must be separated from the noisy background area, through the segmentation task. Local or global thresholding does not effectively isolate fingerprint images from background as they are striated patterns. To selectively isolate fingerprint area, Maio et al. [24] proposed a robust technique which discriminated foreground and background by using the average magnitude of the gradient in each image block, as the gradient response is high in the fingerprint area and small in the background.

### 4.3 Singularity Detection

For simplifying search and retrieval, matching algorithms uses singular points like core, deltas and loops. The fingerprint orientation image is used in the majority of the methodologies proposed in the literature for singularity detection. The most well-known approach by Kawagoe et al. [25] was Poincare' index. For singularity detection, a variety of methods have been suggested, which can be roughly classified as follows: (1) orientation image's local characteristics-based methods, (2) partitioning-based methods, (3) core detection and fingerprint registration-based methods [12].

### 4.4 *Enhancement*

The quality of image obtained by infant fingerprint is usually low due to motion blur, incorrect finger pressure, wet or dry fingers of infants. We must improve image quality because the efficiency of minutiae extraction methods and fingerprint recognition procedures is strongly based on it. As a result, to boost the clarity and sharpness of the friction ridge pattern, we will need to include a fingerprint enhancement module.

Hong et al. [23] proposed an image enhancement technique for fingerprint images which included steps like: (a) pixel-wise normalization of fingerprint image; (b) local orientation estimation using least mean square orientation algorithm; (c) local frequency estimation from x signature; (d) region mask estimation and (e) filtering. Zhang et al. [26] proposed image enhancement by using deep residual dense network (RDN) for image super-resolution in which they extracted successive features from all the convolutional layers.

### 4.5 *Minutiae Extraction*

Minutiae extraction is a challenging task especially when we are dealing with infant fingerprint images. Most of the matching algorithms for fingerprint recognition are based on the successful implementation of this stage. Following are the main steps required for minutiae extraction as shown in Fig. 2:

- Binarization: After enhancement, the fingerprint grayscale is transformed into a binary image.
- Thinning: To reduce the thickness of ridge line to almost one pixel, after binarization, a thinning is done.
- Minutiae Extraction: To extract minutiae points, the image is scanned pixel-wise and crossing number (CN) is used to determine minutiae points. The CN of a pixel in a binarized fingerprint image is defined as half of the sum of differences between pairs of adjacent pixels in the eight neighborhood. Depending upon the values of CN, the category of minutiae can be determined.

Few authors suggested direct method of minutiae extraction from the input grayscale images, without undergoing the process of binarization or thinning because they consider that binarization may introduce information loss and may not provide

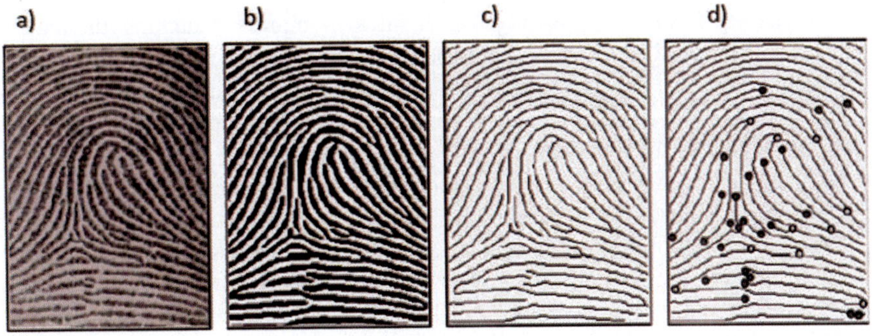


**Fig. 2** Steps for minutiae extraction. **a** input grayscale image; **b** enhancement and binarization; **c** thinning; **d** minutiae points extraction [24, 27]

good results in case of low quality images as in case of infant fingerprints. Also, the process of thinning the binarized image may introduce numerous amount of spurious minutiae. Maio et al. [24] suggested a direct minutiae extraction method in which they tracked the ridge lines in the grayscale image, according to the local orientation of the ridge pattern. David et al. [12] proposed a minutiae filtering stage after the extraction of minutiae points in order to remove the spurious minutiae detected in highly corrupted regions or introduced by the process of thinning. It is also found that by using deep networks [28], better quality minutiae extraction is feasible as compared to conventional approaches.

## 5 Matching and Decision-Making Module

During the fingerprint verification or identification stage, the feature set extracted by the feature extraction module is compared against the feature set which was previously stored in template(s) during the enrollment process, and a similarity or match score is generated. The matching module also confines a decision-making module, whose task is to take the decision according to the match scores. It can either validate a claimed identity or provide a ranking of the enrolled identities in order to recognize an individual. Algorithms for fingerprint matching can be classified in three ways :

- Correlation-based matching: In this matching algorithm, the similarity between the template and input fingerprint image is found using correlation (or cross-correlation) between their corresponding pixels. To compensate for brightness and contrast variations in images, a zero-mean normalized cross-correlation method was suggested by Crouzil et al. [29].
- Minutiae-based matching: In this matching process, the extracted minutiae from the template and input fingerprint image are stored as feature vector. They denote attributes like co-ordinates of minutiae in the fingerprint image, their orientation

and type like ridge ending or ridge bifurcation. In minutiae matching, the matching algorithm finds the spatial difference and orientation difference between the template and the input minutiae feature vectors, i.e., it must be minimum for same type of minutiae, to result in the maximal number of minutiae pairings. Some authors [30] have proposed global minutiae matching to determine the distinctiveness of fingerprint and to reduce the computational complexity. On the contrary, some authors [31] suggested local minutiae matching to overcome the effect of translations, rotations and scale changes in the matched images. But for infant fingerprint matching , more robust matching algorithm is required which should incorporate the advantages of both local and global minutiae matching [32].

- Non-minutiae feature-based matching: To improve the system accuracy and robustness, especially in cases where minutiae extraction is tough like poor quality fingerprint images of infants, or the sensor fingerprint area is small, in such cases, non-minutiae feature-based matching would be preferable. Features that can be used in non-minutiae-based matching are textual information (global or local), singularity information (type, number and position), attributes of ridge line (local orientation, frequency, shape), sweat pores details and many more. On the basis of these extracted features, non-minutiae-based methods do the matching of template and input fingerprint image.

## 6 Infant Fingerprint Age-Progression

Gottschlich et al. [33] suggested an isotropic growth model to model the effect of child growth on the isotropic rescaling of fingerprint images based on growth chart for stature, in order to achieve better accuracy of fingerprint matching systems. Another infant fingerprint age model was proposed by Perciozzi et al. [34], where they showed that for infant ages starting at one year old, after applying a growth factor interpolation method, their fingerprints can be up-scaled to adult size, without a significant loss of accuracy. A minutiae aging model was also proposed by Engelsma et al. [14], where they scale the minutiae sets using bi-cubic interpolation. In a similar manner, they aged the enhanced fingerprint images prior to passing them to the fingerprint matching module.

## 7 Infant Database Module

The system database is the repository of infant's biometric information. During the enrollment process, a template is created from the feature set extracted from the raw biometric image sample, and it is stored in the system database along with some biographic information such as infant's name, date of birth, address characterizing the infant. Based on the fingerprints, an infant fingerprint recognition system must effectively be able to show its ability to recognize (verify or identify) a child even after months of its initial enrollment. Therefore, a longitudinal fingerprint dataset needs to be maintained, which contains fingerprint images of the same infant over time at successive intervals for evaluation.

## 8 Discussion and Challenges

Although fingerprint recognition is a mature field, but still infant fingerprint recognition needs to address following challenges:

### *8.1 Fingerprint Matching*

Fingerprint matching is a challenging problem especially when it comes to matching of fingerprints of infants, as the quality of fingerprint images may not be good enough for robust feature extraction. In such a case, they may show large intra-class variation and small inter-class variation. Large intra-class variations may be attributed to factors like variable pressure, displaced or rotated fingers, partial overlap during fingerprint acquisition from infants. Also, changing skin condition and noise may cause errors in feature extraction of infant fingerprints. To solve these issues, feature-level fusion of both the local and global details [35] can be done to not only improve the matching algorithm performance but also to reduce the effect of noise by cancelation in combination process and to increase the fingerprint area for matching. CNN-based texture matcher [14] can also further improvise the matching algorithm.

### *8.2 Growth Model for Infant Fingerprint*

The analysis of growth of fingerprints with age has already been explored by many researchers, but it needs a more deeper analysis because the stature grows faster than the hand in early childhood and the respective scaling factors may also change with age. Thus, true body heights may also be considered for improvisation [33].

### *8.3 Database Collection*

Collection of dataset is the most challenging task due to following reasons :

- Cooperation of an infant's parents : The child's biometric data needs to regularly collected over a long period of time which requires the cooperation of an infant's parents in returning to the clinic multiple times for participation in the study, and there is a high risk that some of them would withdraw during the course of the study.

- Difficulty in data collection for children: It requires working with uncooperative infants who may become hungry or agitated during the data collection. Rahmun et al. [36], based on experimental results for fingerprints of 300 children, mentioned that it is difficult to acquire fingerprints of children below 12 years of age.
- Quality of data: Even if a significant number of infants would still be available, the minimum observation time needs to be large enough in order to have significant measurements (above noise) to draw robust conclusions. Infants fingerprints due to motion blur, wet/dry fingers may have low contrast parts, i.e., parts of the image where no distinction can be made between ridge lines.
- Availability of dataset: Regarding pre-existing dataset, only few resources exist [5, 14, 34], but still data subjects children needs to be tracked over a sufficiently long period of time (0–12 years) for the complete data analysis. Also, the biggest problems that the research community faces is that large datasets of fingerprint images acquired in real operational conditions are, rightly so, secured under data protection regulations that severely restrict the access to these data.

## 9 Conclusion

This survey has also emphasized the prospective benefits of infant fingerprint recognition system as a suitable biometric solution for child recognition in applications such as vaccination, tracking the missing/abducted infant, providing child health care, passport verification and national identification to provide government benefits throughout their lifetime. Thorough and integrated in-depth assessment about infant fingerprint has led to the conclusion that although children also posses unique fingerprints, but still automatic fingerprint recognition for children on real scenarios needs some improvisation, related to feature extraction and matching, since the recognition rates obtained with this technology for children are not similar to those reached for adults.

## References

1. Pankanti, S., Prabhakar, S., Jain, A.K.: On the individuality of fingerprints. IEEE Trans Pattern Analysis Mach Intell **24**(8), 1010–1025 (2002). https://doi.org/10.1109/TPAMI.2002.1023799
2. Babler, W.: Embryologic development of epidermal ridges and their configurations. Birth Defects original article series **27**(2), 95–112 (1991)
3. World Health Organization: Progress and challenges with achieving universal immunization coverage. https://www.who.int/immunization/monitoringsurveillance/who-immuniz.pdf (2018). Accessed 15-Feb-2019

4. United Nations Children's Fund: Immunization programme. https://www.unicef.org/immunization. Online; Accessed 15-Feb-2021
5. Jain, A.K., Arora, S.S., Cao, K., Best-Rowden, L., Bhatnagar, A.: Fingerprint recognition of young children. IEEE Trans Inf For Secur **12**(7), 1501–1514 (2017). https://doi.org/10.1109/TIFS.2016.2639346
6. Weingaertner, D., Bellon, O.R.P., Silva, L., Cat, M.N.: Newborn's biometric identification: Can it be done? In: VISAPP, vol. 1, pp. 200–205 (2008). https://doi.org/10.5220/0001093302000205
7. Beck, H.C., Ezon, I., Flom, L., Pitchford, C., Park, L.: Iris recognition technology in newborns. Invest. Ophthalmol. Vis. Sci. **49**(13), 2265 (2008)
8. Liu, E.: Infant footprint recognition. In: Proceedings of the IEEE International Conference on Computer Vision, pp. 1653–1660 (2017). https://doi.org/10.1109/ICCV.2017.183
9. Lemes, R.P., Bellon, O.R.P., Silva, L., Jain, A.K.: Biometric recognition of newborns: Identification using palmprints. In: 2011 International Joint Conference on Biometrics (IJCB), pp. 1–6. IEEE (2011). https://doi.org/10.1109/IJCB.2011.6117475
10. Galton, F.: Finger Prints of Young Children. British Association for the Advancement of Science (1899)
11. Fingerprint Recognition for Children: Technical report. Institute for the Protection and Security of the Citizen (2013). https://doi.org/10.2788/3086
12. Maltoni, D., Maio, D., Jain, A.K., Prabhakar, S.: Handbook of fingerprint recognition. Springer. New York (2008). https://doi.org/10.1007/b97303
13. Jain, A.K., Arora, S.S., Best-Rowden, L., Cao, K., Sudhish, P.S., Bhatnagar, A., Koda, Y.: Giving infants an identity: fingerprint sensing and recognition. ICTD, Ann Arbor, MI, 3–6 June 2016. https://doi.org/10.1145/2909609.2909612
14. Engelsma, J.J., Deb, D., Jain, A.K., Sudhish, P.S., Bhatnagar, A.: Infant-prints: fingerprints for reducing infant mortality. CVPR Workshop on CV4GC, 2019. https://doi.org/10.03624v1
15. Marcialis, G.L., Roli, F.: Fingerprint verification by decision-level fusion of optical and capacitive sensors. In: Proceedings Biometric Authentication, ECCV 2004 International Workshop, BioAW 2004, Prague, Czech Republic, May 15, 2004. https://doi.org/10.1007/978-3-540-25976-3_28
16. Schneider, J.K., Wobschall, D.C.: Live scan fingerprint imagery using high resolution C-scan ultrasonography. In: 25th Annual IEEE International Carnahan Conference on Security Technology, Oct 1991. https://doi.org/10.1109/CCST.1991.202196
17. Nixon, K.A., Rowe, R.K.: Multispectral fingerprint imaging for spoof detection. In: Proceedings of the SPIE Conference on Biometric Technology for Human Identification, vol. 5779, pp. 214–225, Orlando (2005). https://doi.org/10.1117/12.606643
18. Chen, Y., Parziale, G., Diaz-Santana, E., Jain, A.K.: 3D touchless finger-prints: compatibility with legacy rolled images. In: Proceedings of the Biometric Symposium, Biometric Consortium Conference, Baltimore (2006). https://doi.org/10.1109/BCC.2006.4341621
19. Lee, H.C., Gaensslen, R.E.: Advances in fingerprint technology. Elsevier Publishing, New York, 2nd edn. https://doi.org/10.1201/9781420041347
20. Galton, F.: Fingerprints. Macmillan, London (1892)
21. Nguyen, D., Jain, A.K.: End-to-end pore extraction and matching in latent fingerprints: going beyond minutiae (2019). arXiv:1905.11472
22. Ratha, N.K., Chen, S.Y., Jain, A.K.: Adaptive flow orientation-based feature extraction in fingerprint images. Pattern Recogn. **28**(11), 1657–1672 (1995). https://doi.org/10.1016/0031-3203(95)00039-3
23. Hong, L., Wan, Y., Jain, A.K.: Fingerprint image enhancement: algorithms and performance evaluation. IEEE Trans. Pattern Anal. Mach. Intell. **20**(8), 777–789 (1998). https://doi.org/10.1109/34.709565
24. Maio, D., Maltoni, D.: Direct gray-scale minutiae detection in fingerprints. IEEE Trans. Patt. Anal. Mach. Intell. **19**(1) (1997). https://doi.org/10.1109/34.566808
25. Kawagoe, M., Tojo, A.: Fingerprint pattern classification. Patt. Recogn. **17**, 295–303 (1984). https://doi.org/10.1016/0031-3203(84)90079-7

26. Zhang, Y., Tian, Y., Kong, Y., Zhong, B., Fu, Y.: Residual dense network for image super-resolution. In: Proceedings of the IEEE Conference on Computer Vision and Pattern Recognition, pp. 2472–2481 (2018)
27. Jain, A.K.: Flynn, Ross: Handbook of Biometrics. Springer. New York (2008). https://doi.org/10.1007/978-0-387-71041-9
28. Tang, Y., Gao, F., Feng, J., Liu, Y.: Fingernet: an unified deep network for fingerprint minutiae extraction. In: 2017 IEEE International Joint Conference on Biometrics (IJCB). IEEE (2017). https://doi.org/10.1109/BTAS.2017.8272688
29. Crouzil, A., Massip-Pailhes, L., Castan, S.: A new correlation criterion based on gradient fields similarity. In: Proceedings of 13th International Conference on Pattern Recognition, pp. 632–636 (1996). https://doi.org/10.1109/ICPR.1996.546101
30. Ratha, N.K., Karu, K., Chen, S., Jain, A.K.: A real-time matching system for large fingerprint databases. IEEE Trans. Patt. Anal. Mach. Intell. **18**(8), 799–813 (1996). https://doi.org/10.1109/34.531800
31. Chen, X., Tian, J., Yang, X., Zhang, Y.: An algorithm for distorted fingerprint matching based on local triangle feature set. IEEE Trans Inf For Secur **1**(2), 169–177 (2006). https://doi.org/10.1109/TIFS.2006.873605
32. Jiang, X., Yau, W.-Y.: Fingerprint minutiae matching based on the local and global structures. In: Proceedings 15th International Conference on Pattern Recognition, Sept 2000. https://doi.org/10.1109/ICPR.2000.906252
33. Gottschlich, C., Hotz, T., Lorenz, R., Bernhardt, S., Hantschel, M., Munk, A.: Modeling the growth of fingerprints improves matching for adolescents. IEEE Trans. Inf. For. Secur. **6**(3), 1165–1169 (2011). https://doi.org/10.1109/TIFS.2011.2143406
34. Preciozzi, Javier: Garella, Guillermo, Camacho, Vanina, Franzoni, Francesco, Di Martino, Luis, Carbajal, Guillermo, Fernandez, Alicia: Fingerprint biometrics from newborn to adult: a study from a national identity database system. IEEE Trans. Biomet. Behav. Identity Sci. **2**(1), 68–79 (2020). https://doi.org/10.1109/TBIOM.2019.2962188
35. Jain, A.K., Prabhakar, S., Hong, L., Pankanti, S.: FingerCode: a filterbank for fingerprint representation and matching. In: IEEE Computer Society Conference on Computer Vision and Pattern Recognition, vol 2, p. 193. https://doi.org/10.1109/CVPR.1999.784628
36. Rahmun, F., Bausinger, O.: Best practice fingerprint enrolment standards European visa information system (2010). https://bit.ly/2VcyyQN

# A Review of High Utility Itemset Mining for Transactional Database

**Eduardus Hardika Sandy Atmaja and Kavita Sonawane**

**Abstract** High utility itemset mining (HUIM) is an expansion of frequent itemset mining (FIM). Both of them are techniques to find interesting patterns from the database. The interesting patterns found by FIM are based on frequently appeared items. This approach is not that efficient to identify the desired patterns, as it considers only existence or nonexistence of items in database and ignores utility. However, the patterns are more meaningful for the user if the utility is considered. The utility can be quantity, profit, cost, risk, or other factors based on user interest. HUIM is another approach to find interesting patterns by considering utility of items along with the frequency. It uses minimum utility threshold to determine if an itemset is high utility itemset (HUI) or not. There are several challenges to implement utility from traditional pattern mining to HUIM. Lately, there are many research contributions that proposed different algorithms to solve these issues. This review work explores various HUIM techniques with detailed analysis of different strategies like apriori, tree based, utility lists based, and hybrid. These strategies are used to implement various HUIM techniques in order to achieve the effectiveness in pattern mining. The observations and analytical findings based on this detailed review done with respect to various parameters can be recommended and used for further research in the pattern mining.

**Keywords** Frequent itemset · High utility itemset · Apriori · Tree based · Utility list based · Hybrid

E. H. S. Atmaja (✉) · K. Sonawane
St. Francis Institute of Technology, Mumbai, India
e-mail: eduardus@student.sfit.ac.in

K. Sonawane
e-mail: kavitasonawane@sfit.ac.in

E. H. S. Atmaja
Sanata Dharma University, Yogyakarta, Indonesia



D. Gupta et al. (eds.), *Pattern Recognition and Data Analysis with Applications*, Lecture Notes in Electrical Engineering 888,
https://doi.org/10.1007/978-981-19-1520-8_2

## 1 Introduction

FIM is a technique to find interesting patterns by finding items which frequently appear together in the transaction [1]. It has been experimented by many researchers [2–6]. These researches are the basic techniques to solve FIM problem. The general problem is solved by finding interesting patterns using frequency of items in transaction. Minimum threshold of support is applied to decide whether or not an itemset is frequent. Every frequent itemsets that has confidence higher than minimum threshold of confidence becomes association rule. For example, a rule: Pencil, Eraser → Pen [Support = 10%, confidence = 90%]. It means that 10% customers bought pencil, eraser, and pen together in one transaction. The value 90% for confidence interprets that customers who bought pencil and eraser together also bought pen. This information can be used by the manager to set up marketing strategies by putting the items together in the display to increase sales. FIM only considers frequency to discover interesting patterns, but in reality there are many factors that affect meaningfulness of the patterns. Utility is an important factor that can make the generated patterns more meaningful. It can be quantity, profit, cost, risk, or other factors based on user interest. FIM does not consider these factors; as a result, there are many items with high utility that cannot be detected. HUIM is then introduced to solve the problem. It finds interesting patterns not only by considering frequency but also the utility. Minimum threshold of utility is applied to determine whether an itemset is HUI or not. With this approach, the undiscovered items which have high utility can be found.

The main challenge being faced by researchers in HUIM is that how to discover interesting patterns effectively and efficiently. Effective means that the interesting patterns truly represent the real-life conditions, for example, patterns should have high profit and correlation when they are appearing together in the transaction. Efficient means that the algorithms should produce interesting patterns by consuming less time and memory. The main efficiency problem is that the number of item combinations may be huge. To solve that problem, many researches have been proposed such as apriori based [7, 8], tree based [9–17], utility list based [18–34], and hybrid-based algorithm [35, 36]. Moreover, there are several HUIM variations such as high average utility itemset mining (HAUIM) [16, 37], HUIM in incremental databases [17, 38], HUIM in sequential database (HUSPM) [18–20, 33, 34], HUIM in regular occurrence [31], close HUIM [27, 29, 30], and correlated HUIM [26, 28]. These kinds of HUIM are introduced to make the results more meaningful.

In the last phase of pattern mining, once the HUIs are generated, post-processing steps can be applied to generate association rules leading to final validation of identified patterns. To evaluate performances of various strategies and techniques, researchers have used different interestingness measures. These measures have a significant role to decide whether a rule is interesting or not. It also can be used to rank the rules [1, 26, 28, 39]. With this interestingness measures, we can get patterns based on the user's potential interest. Along with this, performance of HUIM systems can be evaluated in terms of memory consumption and execution time. However, with the existing challenges, HUIM is found to be a very interesting area to explore. With this objective in mind, this review work tried to explore major HUIM algorithms along with their algorithmic details and performance comparisons to mine HUI.

## 2 Literature Review

The objective of this review is to explore, analyze, and compare various research contributions in HUIM. This paper presents the detailed study and analysis of different strategies with respect to working principle, its experimentation, and the performance observations for each algorithm. All the major findings are summarized along with advantages and disadvantages into six different sections (Point 2.1–2.6) as follows.

### *2.1 Apriori-Based Algorithm*

**Apriori** [2] is an algorithm that introduces downward closure property on FIM. It means that if an itemset is frequent, then every subset must be frequent, and if an itemset is infrequent, then every superset is infrequent. The pseudocode of apriori is given in Algorithm 1. The function in line 3 has an important role to generate candidate k-itemsets by pairing between itemsets from (k-1)-itemsets, and then the unnecessary candidates are removed by line 9. The main advantage of apriori is that it can generate all significant patterns, but it needs a lot of time to generate candidate itemset. Apriori also needs database scan for every iteration, and it makes the algorithm inefficient. The candidate generation also consumes much memory to save the candidates.

**Algorithm 1**: Apriori

Input : database D, minimum support threshold minSup
Output : frequent itemsets $F_k$

```
1 F1={i|i∈I ∧ i.count ≥ minSup}//frequent 1-itemsets
2 for(k=2; Fk-1≠∅; k++){
3  Ck=apriori_gen(Fk-1);//generate candidate k-itemsets
4  for each transaction t ∈ D {
5   Ct=subset(Ck, t);//identify all candidates
6   for each candidate c ∈Ct
7     c.count++;//increment support count
8  }
9  Fk={c ∈Ck|c.count ≥minSup}//frequent k-itemsets
10 }
11 return ∪Fk;//return all frequent itemsets
```

**UMining** and **UMining_H** [7] work similar to the origin apriori to mine HUI, but there are three steps modified, viz. first in the pruning step, utility upper bound in

Eq. (1) is used instead of calculating actual utility which is similar process to line 4–7 in Algorithm 1. Second, it uses utility values instead of frequent values. Third, in the generating step, since it does not assure downward closure property, the line 3 in Algorithm 1 cannot be adapted. The k-itemsets are created by combining (k-1)-itemsets with list items $I$ scanned from database. UMining_H uses same framework as UMining, but it uses Eq. (2) instead of Eq. (1) to calculate the upper bound. It also may prune some HUIs although it produces correct HUIs. Both algorithms do not apply downward closure property so it produces a large number of candidate itemset.

$$b\left(S^k\right) = \frac{\sum S^{k-1} \in C^{k-1} u\left(S^{k-1}\right)}{\left|C^{k-1}\right| - 1} \tag{1}$$

$$b'\left(S^k\right) = \frac{s_{min}}{\left|C^{k-1}\right| - 1} \times \sum_{S^{k-1} \in C^{k-1}} \frac{u\left(S^{k-1}\right)}{s\left(S^{k-1}\right)} \tag{2}$$

where $S^k$ is an itemset of $k$, $C^k$ is candidate itemset of $k$, $s$ is support, and $\left|C^{k-1}\right|$ is cardinality of $C^{k-1}$.

Two phases [8] has similar concept to downward closure property in apriori, namely transaction weighted downward closure (TWDC) property. It means that whenever an itemset is not HUI, every superset is not HUI. This concept reduces the search space because all supersets from low transaction weighted utilization (TWU) itemsets can be cut down. Although in the first phase there is a TWU pruning to save execution time, in the second phase it needs more time to extra database scan.

### 2.2 *Tree-Based Algorithm*

To reduce search space in apriori-based algorithm, tree structure algorithm based on pattern growth approach is proposed [12, 13]. **UP-Growth** [12] and **UP-Growth +** [13] are extension of FP-Growth [4] in FIM. It uses a compressed tree, namely UP-Tree to store crucial information about the utility. Figure 1a shows UP-Tree representation. There are four proposed strategies called discarding global unpromising items (DGU), discarding global node utilities (DGN), discarding local unpromising items (DLU), and decreasing local node utilities (DLN). These strategies are used to reduce candidate itemset. UP-Growth + added two new strategies are called discarding local unpromising items and their estimated node utilities (DNU) and decreasing local node utilities for the nodes of local UP-Tree by estimated utilities of descendant nodes (DNN).

**CHUI-Mine** [14] is an algorithm that can dynamically prune the tree during tree construction by reducing count of items. It uses tree structure similar to UP-Growth. When the count of an item is zero, then nodes having that item and its descendants can be removed. The pruned nodes are moved to the buffer and ready to mine using

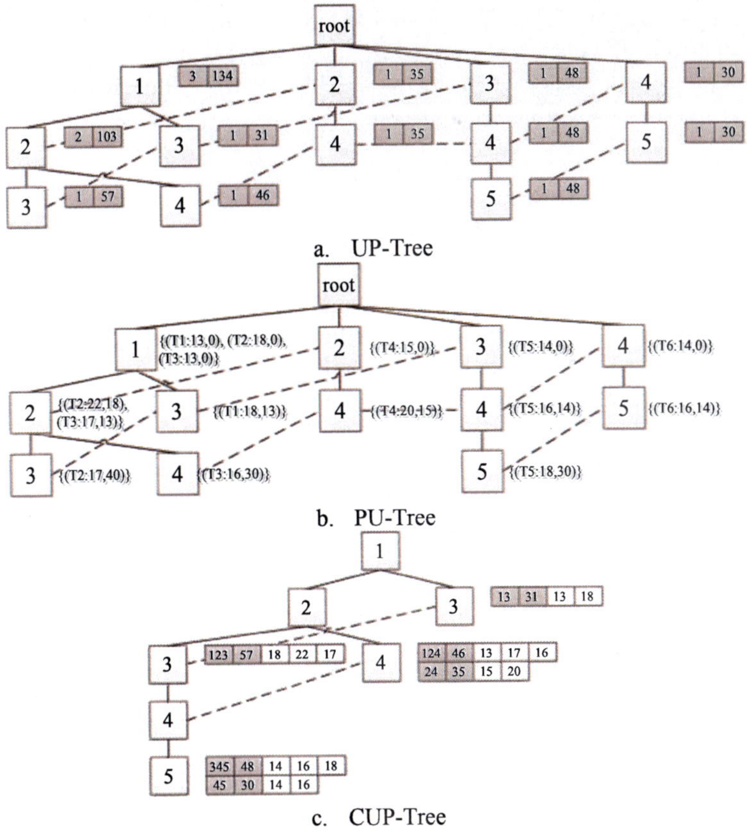


**Fig. 1** Tree representations

pattern growth approach in concurrent time without waiting for the tree insertion process finished. **MIP** [15] is also based on pattern growth approach. PUN-List is the new data structure. PUN-list which is node in PU-Tree contains list of transactions (using vertical data format). MIP works in PU-Tree with depth first search approach to explore and mine HUI. Figure 1b shows PU-Tree representation.

**CTU-Mine** [9] is an extension of CT-PRO [5] in FIM. Compressed transaction utility tree (CTU-Tree) is the proposed data structure which is very compact and efficient because it maps transactions into an ascending sequence of integers without generating more branch. CTU-Tree is similar to CUP-Tree used by CTU-PRO, but it sorts the items into TWU ascending order. It also does not have link node for the prefix item, and it can be mined using top-down approach. **CTU-PRO** [10] and **CTU-PROL** [11] are variants of CTU-Mine. CTU-PRO has new structure, namely compressed utility pattern tree (CUP-Tree). Figure 1c shows CUP-Tree representation. CUP-Tree has link node to the prefix item, so it can be mined quickly using bottom-up approach. CTU-PROL is similar to CTU-PRO, but it mines LocalCUP-Tree separately in concurrent time. CTU-PROL divides transactions into several parallel projections and generates its LocalCUP-Tree from GlobalCUP-Tree. Table 1 shows tree-based algorithms comparison.

### 2.3 *Utility List-Based Algorithm*

Utility list-based algorithm is proposed to improve performance of tree-based algorithms. **HUI-Miner** [21] is the former of utility list inspired by ECLAT [3] in FIM. The utility list is built to produce k-itemset similar to apriori, but there is no candidate generation. It uses enumeration tree to extend the search space. HUI-Miner prunes the search space by applying minimum utility threshold. **FHM** [22] proposed a novel structure, namely estimated utility co-occurrence structure (EUCS). It is a matrix in triangular shape that consists of items with its co-occurrence with other items. With EUCS, candidate itemset can be easily found without joining any item from utility list. Estimated utility co-occurrence pruning (EUCP) is the proposed strategy to prune candidate itemset from EUCS. **IMHUP** [23] is proposed to improve HUI-Miner and FHM. Indexed utility list (IU-List) is proposed to maintain the database efficiently. Two new techniques called reducing upper-bound utilities in IU-Lists (RUI) by decreasing upper-bound utilities in IU-List and combining HUI without creating IU-List (CHI) are applied.

**HMiner** [24] has two new data structures called compact utility list (CUL) and virtual hyperlink. It is used to store and determine duplicate transactions. The algorithm also uses three strategies to accelerate the mining process, viz. initial TWU computation and 1-itemset CUL generation, search tree exploration, and k-itemset CUL construction. It also uses several pruning properties such as U-Prune and EUCS to decrease the search space and mine HUI efficiently.

**ULB-Miner** [25] has a new structure called utility list buffer. It is used for storing the potential HUI by temporarily inserting the data in the data segment. StartPos and EndPos are two pointers that express start and end index of data segments. It helps to access the data quickly. Whenever an itemset is not needed anymore, it can be replaced by other candidate to maintain efficient memory management.

**SPHUI-Miner** [32] has a new data format called high utility-reduced transaction pattern list (HUI-RTPL) which consumes small memory. It has two data structures: selective database projection utility list (SPU-List) to reduce the scanning process and maintain information of the database and Tail-Count list which helps to prune the search space efficiently. It has two new upper bounds (TUP and PU) that also help to reduce the search space effectively. Table 2 shows utility list-based algorithms comparison.

**Table 1** Tree-based algorithms comparison

| Algorithm | Key principle | Parameters: execution time, memory usage | |
|---|---|---|---|
| | | Advantages | Disadvantages |
| UP-Growth | An extension of FP-Growth to mine HUI with compressed tree structure called UP-Tree and four new strategies: DGU, DGN, DLU, and DLN | Faster algorithm: The four strategies support in reducing the overestimation of utilities and candidates | – Dataset with distinct items, the tree grows bigger<br>– Leads to high memory consumption<br>– High traversal and mining time |
| UP-Growth+ | An extension of UP-Growth with two new strategies: DNU and DNN | Two additional strategies are effectively helping to reduce overestimation utilities and candidates and performs better than UP-Growth | The disadvantage is same as UP-Growth because it has same tree structure |
| CHUI-Mine | A pattern growth approach by dynamically pruning and mining the tree during the tree construction | – It reduces both: the number of candidates and the search space<br>– The dynamic pruning helps to reduce the memory consumption<br>– Concurrent mining makes it faster | If the tree structure grows widely and deeply, in some cases it makes the algorithm slower |
| MIP | A pattern growth approach with PU-Tree and PUN-List data structures | – PUN-List avoids costly and repeated utility computation<br>– PUN-List generates candidates efficiently | In sparse dataset, it consumes more memory because the PU-Tree is grown bigger |
| CTU-Mine | An extension of CT-PRO to mine HUI with compact tree structure, namely CTU-Tree | It can mine complete HUI and perform good in dense datasets | It overestimates the potential HUI, then it increases the memory consumption and computational process |
| CTU-PRO | A variant of CTU-Mine with more compact tree structure, namely CUP-Tree | – It has compact tree structure that can reduce the database size<br>– It is also efficient in sparse and relatively dense datasets | – It needs more time to do the local mining due to several global tree scanning<br>– It needs more memory consumption during local mining |
| CTU-PROL | A parallel version of CTU-PRO by using paralel projection to the transactions | – It uses parallel projection effectively and handles very large database<br>– Concurrent mining makes it faster | – It is slower for high utility threshold<br>– Parallel processing of data leads to more memory consumption |

**Table 2** Utility list-based algorithm comparison

| Algorithm | Key Principle | Parameters: execution time, memory usage | |
|---|---|---|---|
| | | Advantages | Disadvantages |
| HUI-Miner | The first utility list-based algorithm | It reduces execution time by avoiding candidate generation and utility calculation | It needs more time to join k-itemset among utility lists |
| FHM | An extension of HUI-Miner with a new pruning mechanishm, namely EUCP | EUCP helps to reduce execution time by reducing join operations | The performance decreases for dense dataset |
| IMHUP | Indexed utility list-(IU-List) based algorithm with two strategies, namely RUI and CHI | – It reduces execution time by reducing join operations and search space<br>– It reduces memory usage by reducing utility list construction | The upper-bound utility should be tightened |
| HMiner | A utility list algorithm with two new data structures, namely CUL and virtual hyperlink | The new data structures reduce memory usage | It consumes more memory for creating CUL for every itemset in sparse dataset |
| ULB-Miner | Another utility list algorithm that uses utility list buffer to reuse memory whenever possible | – It reduces execution time by accessing and mining the data structure quickly<br>– The utility list buffer consumes less memory | More distinct items lead to high memory usage |
| SPHUI-Miner | A projection utility list based with a new data formatm namely HUI-RTPL and two new data structures namely SPU-List and Tail-Count | – The data structure reduces memory usage<br>– It reduces execution time by reducing database scan and search space | The memory consumption increases when the minimum utility is decreased |

## 2.4 Hybrid-Based Algorithm

It is possible to combine tree and utility list-based algorithm [35, 36]. **mHUI-Miner** [35] is an algorithm that combines HUI-Miner and IHUP-Tree [17]. IHUP-Tree is used to avoid expanding items that do not appear in database. It makes the mining faster. The tree does not contain utility information, and it is used only to escort the mining and extension process. It needs low memory consumption, because there is no calculation and storing of utility in the tree. The concept of utility list from HUI-Miner is used to maintain information about utility of the items. These two strategies work together in the mining process, the tree helps to traverse and expand the search space, and the utility list is used to prune the candidate based on minimum utility. These two strategies make the algorithm more efficient, but it is weak on dense dataset.

The other hybrid-based algorithm is **UFH** [36] that combines UP-Growth+ and FHM. UP-Growth+ is used to construct and mine the tree. FHM is called after UP-Growth+ builds the conditional pattern base. The utility list is built based on conditional pattern base. Then FHM is called to mine the utility list. It means that FHM works with this utility list in local tree to mine HUI. The hybrid framework performs better because in UP-Growth+ there is a transaction merging process to reduce the memory consumption, and it also provides actual utility calculation to prune the tree efficiently. Then, FHM mines the utility list efficiently because the size of utility list is reduced by UP-Growth+ .

## *2.5 Other Variations*

The other HUIM problem is that it may produce long itemsets which gets an inconsistent predicted profit opposing the actual value. This condition happens if the count of distinct items is huge and the transactions contain many items. To overcome this challenge, HAUIM [16, 37] is proposed. It considers the average utility (consider both length and utility) of itemsets to decrease itemsets with unreasonable estimated profit. The earlier concept of HUIM assumes that transactional databases are static, especially the utilities. In real life, the utilities may change over time. For example, mask is cheaper before a pandemic, the price is increased because of high demand. This issue may produce inaccurate results on real datasets. Moreover, transactional databases also can be manipulated such as additions, deletions, and modifications. These kinds of problem can be solved by HUIM in incremental databases [17, 38]. Sequence dataset is the other problem in HUIM. Different from the usual dataset, sequence dataset maintains the order of the item that cannot be reordered. For example, DNA sequence cannot be reordered because it represents the important information about someone's DNA. To overcome the problem, HUSPM which maintains the important sequence of items is proposed [18–20, 33, 34]. Regular occurrence of items in HUIM [31] is also interesting to investigate. It can be used to investigate the occurrence behavior of itemsets with their utility values. For example, in the retail dataset, we can explore regular purchases items which have high profit. Close and maximal HUI are compact representation of HUIM [27, 29, 30]. It helps to decrease the number of candidate itemset. Close HUI means that if an itemset is HUI, then its supersets do not have the same frequency. Maximal HUI means that if an itemset is HUI, then its supersets are not HUI. It can prune redundant patterns efficiently.

### 2.6 Interestingness Measures

Interestingness measure is very essential in data mining. It can be used in HUIM either for pruning or ranking the patterns based on user-specific preferences [39]. A HUI may has low correlation to each items since there is only utility threshold calculation to prune the candidates. Some of the HUIM algorithms use interestingness measures to prune the candidate during the mining process [26, 28]. In [26], all confidence and bond measures are used, and in [28], Kulczynski measure is used to prune weakly correlated candidate itemsets. There are another interestingness measures such as $X^2$, lift, jaccard, cosine, and max confidence [1]. Interestingness measure can be used to rank the generated patterns by sorting the measure value either in ascending or descending order. The ranking represents patterns from the most interesting to the less interesting. This may provide more meaningful patterns because it has high correlation that represents the real condition.

### 2.7 Overall Observations and Analytical Key Findings

Based on review of high utility itemset algorithms above, the overall observations and analytical key findings can be described as follows:

i. Utility threshold plays significant role in HUIM algorithms as selection of this threshold directly impacts the search space, memory utilization, and the processing time.
ii. Changes in utility threshold values are based on the transactional databases as per the buying selling properties and strategies being applied in the real-time retail store transactions.
iii. Many researchers have contributed to handle these dynamics mentioned in point ii. This leads to trade-off among various parameters while trying to address the problems associated with the changing pattern in the utilities.
iv. So it still remains a challenge to handle such dynamics and comes to the completed and generalized solution for HUIM.
v. Observations based on contribution from various researchers indicating that the positive finding of various algorithms can be combined. The process of determining the threshold can be automated and can be generalized to gain interesting patterns leading to high profit.
vi. This should be experimented with the new real-time application areas, where the utilities are changing dynamically and also huge transactions are being generated.
vii. This study leads to the suggestion that there is need of changing not only the algorithm but also the strategy of applying the various key solutions in distributed manner to handle different problems with dedicated key solutions in parallel fashion in order to improve the overall performance.

## 3 Conclusion

HUIM is found to be effective over FIM as it considers the utility of every item and not just the frequency. This leads to benefits in terms of desired pattern generation along with the effective association rules. This mining approach is investigated by many researchers using several data structures such as array, tree, utility list, and hybrid based. Each data structures has its own advantages and disadvantages or constraints with respect to the processing and handling of different datasets. Based on the review of literature in this work, we found that most of the research done had aimed to improve the existing algorithms and also to address the issues so the HUI can be mined efficiently. Various challenges identified in this area are running time, memory consumption, and generation of desired patterns and association rules. We can delineate that there are basically four strategies named apriori, tree based, utility list based, and hybrid approaches. Most of the papers have discussed how to create pruning techniques to reduce the search space. Some algorithms applied more than one pruning mechanisms to eliminate irrelevant candidate. The pruning techniques greatly help the main algorithm to reduce the candidate itemset. We can state that a qualified pruning technique is also necessary to efficiently produce the HUIs.

Overall study of this research survey is opening the door toward new research directions along with the existing techniques (1) using efficient data structure such as tree and/or utility list based, (2) using efficient mining and pruning strategies, (3) working on utility thresholding and its impact in order to improve, (4) application of parallel programming/processing strategies, and (5) effective use of interestingness measures either to prune or rank the pattern that may help in final decision making or recommendations.

## References

1. Han, J., Kamber, M., Pei, J.: Data mining: concepts and techniques, 3rd edn. Morgan Kaufmann, Waltham (2012)
2. Agrawal, R., Srikant, R.: Fast algorithms for mining association rules. In: 20th International Conference on Very Large Data Bases, pp. 487–499. Morgan Kaufmann, San Francisco (1994)
3. Zaki, M.J.: Scalable algorithms for association mining. IEEE Trans. Knowl. Data Eng. **12**(3), 372–390 (2000). https://doi.org/10.1109/69.846291
4. Han, J., Pei, J., Yin, Y.: Mining frequent patterns without candidate generation. In: 2000 ACM SIGMOD International Conference on Management of Data, pp. 1–12. Association for Computing Machinery, New York (2000). https://doi.org/10.1145/335191.335372
5. Sucahyo, Y.G., Gopalan, R.P.: CT-PRO: abottom-up non recursive frequent itemset mining algorithm using compressed fp-tree data structure. In: IEEE ICDM Workshop on Frequent Itemset Mining Implementations. (2004)
6. Aryabarzana, N., Bidgoli, B.M., Teshnehlab, M.: negFIN: an efficient algorithm for fast mining frequent itemsets. Expert Syst. Appl. **105**, 129–143 (2018). https://doi.org/10.1016/j.eswa.2018.03.041
7. Yao, H., Hamilton, H.J.: Mining itemset utilities from transaction databases. Data Knowl. Eng. **59**(3), 603–626 (2006). https://doi.org/10.1016/j.datak.2005.10.004

8. Liu, Y., Liao, W., Choudhary, A.: A two-phase algorithm for fast discovery of high utility itemsets. In: 9th Pacific-Asia Conference on Knowledge Discovery and Data Mining, pp. 689–695. Springer, Berlin (2005). https://doi.org/10.1007/11430919_79
9. Erwin, A., Gopalan, R.P., Achuthan, N.R.: CTU-Mine: an efficient high utility itemset mining algorithm using the pattern growth approach. In: 7th IEEE International Conference on Computer and Information Technology, pp. 71–76. IEEE, Fukushima (2007). https://doi.org/10.1109/CIT.2007.120
10. Erwin, A., Gopalan, R.P., Achuthan, N.R.: A bottom-up projection based algorithm for mining high utility itemsets. In: 2nd International Workshop on Integrating Artificial Intelligence and Data Mining, pp. 3–11. Australian Computer Society, Australia (2007)
11. Erwin, A., Gopalan, R.P., Achuthan, N.R.: Efficient mining of high utility itemsets from large datasets. In: 12th Pacific-Asia Conferences on Knowledge Discovery and Data Mining, pp. 554–561. Springer, Berlin (2008). https://doi.org/10.1007/978-3-540-68125-0_50
12. Tseng, V.S., Wu, C.W., Shie, B.E., Yu, P.S.: UP-Growth: an efficient algorithm for high utility itemset mining. In: 16th ACM SIGKDD Interntional Conference on Knowledge Discovery and Data Mining, pp. 253–262. Association for Computing Machinery, New York (2010). https://doi.org/10.1145/1835804.1835839
13. Tseng, V.S., Shie, B.E., Wu, C.W., Yu, P.S.: Efficient algorithms for mining high utility itemsets from transactional databases. IEEE Trans. Knowl. Data Eng. **25**(8), 1772–1786 (2013). https://doi.org/10.1109/TKDE.2012.59
14. Song, W., Liu, Y., Li, J.: Mining high utility itemsets by dynamically pruning the tree structure. Appl. Intell. **40**, 29–43 (2014). https://doi.org/10.1007/s10489-013-0443-7
15. Deng, Z.H.: An efficient structure for fast mining high utility itemset. Appl. Intell. **48**, 3161–3177 (2018). https://doi.org/10.1007/s10489-017-1130-x
16. Yildirim, I., Celik, M.: An efficient tree-based algorithm for mining high average-utility itemset. IEEE Access **7**, 144245–144263 (2019). https://doi.org/10.1109/ACCESS.2019.2945840
17. Ahmed, C.F., Tanbeer, S.K., Jeong, B.S., Lee, Y.K.: Efficient tree structures for high utility pattern mining in incremental databases. IEEE Trans. Knowl. Data Eng. **21**(12), 1708–1721 (2009). https://doi.org/10.1109/TKDE.2009.46
18. Yin, J., Zheng, Z., Cao, L.: USpan: An efficient algorithm for mining high utility sequential patterns. In: 18th ACM SIGKDD International Conference on Knowledge Discovery and Data Mining, pp. 660–668. Association for Computing Machinery, New York (2012). https://doi.org/10.1145/2339530.2339636
19. Gan, W., Lin, J.C.W., Zhang, J., Chao, H.C., Fujita, H., Yu, S.: ProUM: projection-based utility mining on sequence data. Inf. Sci. Inf. Comput. Sci. Intell. Syst. Appl. J. **513**, 222–240 (2020). https://doi.org/10.1016/j.ins.2019.10.033
20. Gan, W., Lin, J.C.W., Zhang, J., Viger, P.F., Chao, H.C., Yu, P.S.: Fast utility mining on sequence data. IEEE Trans. Cybern. **51**(2), 487–500 (2020). https://doi.org/10.1109/TCYB.2020.2970176
21. Liu, M., Qu, J.: Mining high utility itemsets without candidate generation. In: 21st ACM International Conferene on Information and Knowledge Management, pp. 55–64. Association for Computing Machinery, New York (2012). https://doi.org/10.1145/2396761.2396773
22. Viger, P.F., Wu, C.W., Zida, S., Tseng, V.S.: FHM: faster high-utility itemset mining using estimated utility co-occurrence pruning. In: 21st International symposium on Methodologies for Intelligent Systems, pp. 83–92. Springer, Cham (2014). https://doi.org/10.1007/978-3-319-08326-1_9
23. Ryang, H., Yun, U.: Indexed list-based high utility pattern mining with utility upper-bound reduction and pattern combination techniques. Knowl. Inf. Syst. Int. J. **51**, 627–659 (2017). https://doi.org/10.1007/s10115-016-0989-x
24. Krishnamoorthy, S.: HMiner: efficiently mining high utility itemsets. Expert Syst. Appl. **90**, 168–183 (2017). https://doi.org/10.1016/j.eswa.2017.08.028
25. Duong, Q.H., Viger, P.F., Ramampiaro, H., Norvag, K., Dam, T.L.: Efficient high utility itemset mining using buffered utility-lists. Appl. Intell. **48**, 1859–1877 (2018). https://doi.org/10.1007/s10489-017-1057-2

26. Viger, P.F., Zhang, Y., Lin, J.C.W., Dinh, D.T., Le, H.B.: Mining correlated high-utility itemsets using various measures. Logic J. Interest Group Pure Appl Logics (IGPL) **28**(1), 19–32 (2018). https://doi.org/10.1093/jigpal/jzz068
27. Wu, C.W., Viger, P.F., Gu, J.Y., Tseng, V.S.: Mining compact high utility itemsets without candidate generation. In: High-Utility Pattern Mining: Theory, Algorithms and Applications, pp. 279–302. Springer, Cham (2019). https://doi.org/10.1007/978-3-030-04921-8_11
28. Vo, B., Nguyen, L.V., Vu, V.V., Lam, M.T.H., Duong, T.T.M., Manh, L.T., Nguyen, T.T.T., Nguyen, L.T.T., Hong, T.P.: Mining correlated high utility itemsets in one phase. IEEE Access **8**, 90465–90477 (2020). https://doi.org/10.1109/ACCESS.2020.2994059
29. Wei, T., Wang, B., Zhang, Y., Hu, K., Yao, Y., Liu, H.: FCHUIM: efficient frequent and closed high-utility itemsets mining. IEEE Access **8**, 109928–109939 (2020). https://doi.org/10.1109/ACCESS.2020.3001975
30. Vo, B., Nguyen, L.T.T., Bui, N., Nguyen, T.D.D., Huynh, V.N., Hong, T.P.: An efficient method for mining closed potential high-utility itemsets. IEEE Access **8**, 31813–31822 (2020). https://doi.org/10.1109/ACCESS.2020.2974104
31. Amphawan, K., Lenca, P., Jitpattanakul, A., Surarerks, A.: Mining high utility itemsets with regular occurrence. J. ICT Res. Appl. **10**(2), 153–176 (2016). https://doi.org/10.5614/itbj.ict.res.appl.2016.10.2.5
32. Bai, A., Deshpande, P.S., Dhabu, M.: Selective database projections based approach for mining high-utility itemsets. IEEE Access **6**, 14389–14409 (2018). https://doi.org/10.1109/ACCESS.2017.2788083
33. Lin, J.C.W., Li, Y., Viger, P.F., Djenouri, Y., Zhang, J.: Efficient chain structure for high-utility sequential pattern mining. IEEE Access **8**, 40714–40722 (2020). https://doi.org/10.1109/ACCESS.2020.2976662
34. Viger, P.F., Li, J., Lin, J.C.W., Chi, T.T., Kiran, R.U.: Mining cost-effective patterns in event logs. Knowl. Based Syst. **191**, 1–25 (2020). https://doi.org/10.1016/j.knosys.2019.105241
35. Peng, A.Y., Koh, Y.S., Riddle, P.: mHUIMiner: a fast high utility itemset mining algorithm for sparse datasets. In: 21st Pacific-Asia Conference on Knowledge Discovery and Data Mining, pp. 196–207. Springer, Cham (2017). https://doi.org/10.1007/978-3-319-57529-2_16
36. Dawar, S., Goyal, V., Bera, D.: A hybrid framework for mining high-utility itemsets in a sparse transaction database. Appl. Intell. **47**, 809–827 (2017). https://doi.org/10.1007/s10489-017-0932-1
37. Wu, J.M.T., Lin, J.C.W., Pirouz, M., Viger, P.F.: TUB-HAUPM: tighter upper bound for mining high average-utility patterns. IEEE Access **6**, 18655–18669 (2018). https://doi.org/10.1109/ACCESS.2018.2820740
38. Vo, B., Nguyen, L.T.T., Nguyen, T.D.D., Viger, P.F., Yun, U.: A multi-core approach to efficiently mining high-utility itemsets in dynamic profit databases. IEEE Access **8**, 85890–85899 (2020). https://doi.org/10.1109/ACCESS.2020.2992729
39. Geng, L., Hamilton, H.J.: Interestingness measures for data mining: A Survey. Assoc. Comput. Mach. (ACM) Comput. Surv. **38**(3), 9 (2006). https://doi.org/10.1145/1132960.1132963

# A Cross-Sectional Study on Distributed Mutual Exclusion Algorithms for Ad Hoc Networks

Ashish Singh Parihar and Swarnendu Kumar Chakraborty

**Abstract** The mutual exclusion problem has been substantially studied in the distributed systems. Various solutions have been proposed to achieve distributed mutual exclusion till date. These solutions are exposed to different network topologies as static and dynamic ones. On a broader categorical note, ad hoc networks are the best-suited representation of dynamic networks in which node mobility is highly unpredictable. Nowadays, wireless communication is everywhere; hence, the adaptability of ad hoc networks is getting increased day by day. Through this research article, we present a cross-sectional study on various existing distributed mutual exclusion algorithms imposed on ad hoc networks and their associated network variants including their performance metrics and fault-tolerant capabilities.

**Keywords** Distributed system · Distributed mutual exclusion · Flying ad hoc network · Resource allocation problem · Unmanned aerial vehicles

## 1 Introduction

Distributed system [1] is a cluster of independent autonomous nodes that appear as a single unit of coherent system to its corresponding end users. These nodes collaborate with each other in a way to solve a particular issue that is beyond the processing capacity of a single node. Since no shared memory or logical clock present in these systems, message passing is the only way to establish communication between the nodes. The concurrent access to a shared resource [2] by more than one process

A. S. Parihar (✉) · S. K. Chakraborty
Department of Computer Science and Engineering, National Institute of Technology (NIT), Jote, Arunachal Pradesh, India
e-mail: ashish.phd20@nitap.ac.in; ashish.parihar@kiet.edu

S. K. Chakraborty
e-mail: swarnendu@nitap.ac.in

A. S. Parihar
Department of Computer Science, KIET Group of Institutions, Delhi-NCR, Ghaziabad, Uttar Pradesh, India

 
D. Gupta et al. (eds.), *Pattern Recognition and Data Analysis with Applications*, Lecture Notes in Electrical Engineering 888,
https://doi.org/10.1007/978-981-19-1520-8_3

running on different processors needs extra attention to avoid any inconsistent state, i.e., access should be in a mutually exclusive manner. Any process can access the shared resource through its special code segment, namely as critical section (CS). Mutual exclusion ensures to allow only one process at a time to its CS and such solutions to maintain mutual exclusion in a distributed system known as distributed mutual exclusion (DME) algorithms. The prime classifications of DME algorithms are—token based and non-token based. A unique token circulates within the system, and a process holding that token at any time is allowed to invoke its CS in token-based DME algorithms [3–5, 28–30]. Whereas in non-token-based DME algorithms, a process interested to invoke its CS requires permission from a special set of processes in the system [6]. Network topology plays a vital role while designing any DME algorithm, especially on its performance measure and fault-tolerant capabilities. On a side, a tremendous level of work has already been imposed on static network topologies and remains a keen interest on dynamic network topologies on the other side.

As a major wireless classification [7], ad hoc networks are in trending and highly adaptable due to its dynamic node mobility model in the current era. Mobile ad hoc network (MANET) [8], vehicular ad hoc network (VANET) [9] and flying ad hoc network (FANET) [10, 11] are the various variants of the ad hoc network. Through this research article, we present a cross-sectional study on various existing distributed mutual exclusion algorithms imposed on ad hoc networks and their associated network variants including their performance metrics and fault-tolerant capabilities.

The rest of this article is organized as: In Sect. 2, we discuss ad hoc network and their variants. Section 3 presents the literature survey. In Sect. 4, a comparative study has been shown among various existed DME algorithms on ad hoc networks. Section 5 contains open challenges and future research directions. Finally in Sect. 6, we have concluded our work.

## 2 Ad Hoc Network

In the current era of rapid wireless technology advances, ad hoc networks are the most adaptable network topology in existence. The node mobility model is the main key feature of ad hoc network that makes this more appropriate as compared to others. MANETs are the first main classification of ad hoc networks which extended to VANET and FANET later. Mobile technology is one of the classic examples of MANET in which the underlying network is based on wireless communication. In VANET, sensors and devices are placed on vehicles to make them enable for communication while driving. An intelligent transportation system implements VANET as its background technology.

FANETs [11] are the most trending network topology that exists till date. Unmanned aerial vehicle (UAV) is the core component of FANET having a capability to fly in air without any human personnel. Due to the flying capacity of UAV,

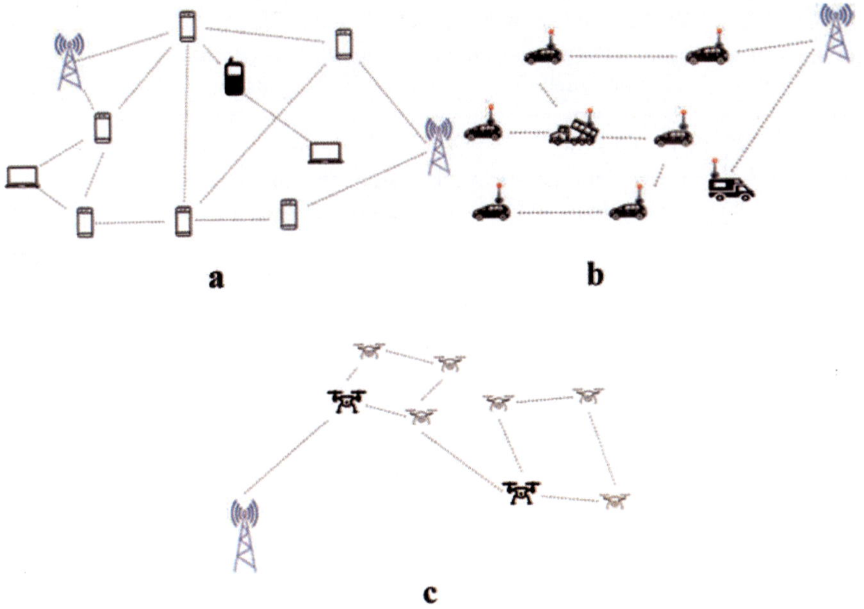


**Fig. 1** **a** MANET, **b** VANET, **c** FANET

it has a huge potential in terms of its functionality in various domains like, military operations [12], goods delivery [13], and forest monitoring [14] etc. Whereas the demand and applicability of FANET are getting increased in various fields, the collaboration and communication among the UAVs in FANET are complex due to its random node movement and unpredictable nature [15]. A sample visualization of MANET, VANET, and FANET can be seen in Fig. 1.

## 3 Literature Survey

The concept of DME in ad hoc networks was firstly proposed by Walter et al. [16] as reverse link (RL) protocol. They claimed their solution to fault-tolerant and provided the directed acyclic graph (DAG) as an underlying network topology. Communication is restricted only to neighbor nodes within the system. Message aging is done from an event-based strategy through RL protocol. Better results have been drawn as compared to existed static DME algorithms in terms of message complexity per CS. Self-stabilization [17] enables the protocol to handle trivial faults automatically, and based on that, Chen and Welch [18] proposed a self-stabilizing algorithm for MANET through virtual rings. Constraint to their solution is the pre-assumed node mobility model. Another self-stabilizing algorithm was proposed by Baala et al. [19] to handle node failures. Wu et al. [20] presented another DME algorithm as MUTEX based on the non-token strategy and uses a look-ahead [21] technique to reduce the number of message exchanges during a race condition within the system.

Sharma et al. [22] provided the first DME in VANET using a dual token strategy for various classifications of the cluster as inter and intra. Through their simulation, they claimed their algorithm to behave on the inversely proportional concept in between the number of the nodes in the system and message exchanges. Various intelligent transport systems [23–25] have been introduced in VANET as a DME implementation.

Further to FANET, the only existed DME algorithms are proposed by Khanna et al. [4, 5]. Initially, Khanna et al. [4] introduced token-based DME algorithm as mobile resource mutual exclusion (MRME) on a FANET with resource mounted on a UAV and rest of the nodes competes for their CS to invoke. They broadly categorized various states of nodes in terms of their token status and determined to allocate the token based on resource occupied UAV communication range with overall message complexity to O(N). Another work of Khanna et al. [5] supports local mutual exclusion to support non-neighboring nodes to invoke their CS simultaneously. They incorporate fuzzy logic to elect the leader in the system by considering various quality parameters of a node. Token loss handling is also one of the supported features in their solution.

## 4 Comparative Study

In this section, we firstly present a comparative analysis between all existed DME algorithms on ad hoc networks by comparing their performance metrics and capabilities to handle various faults. Table 1 comprises a detailed comparison for the same and also, few observations have been made on the basis of that as: Walter et al. [16] have lower messages exchange as compared to others in MANET but only supports links failures. Fault-tolerant to node failures has been effectively handled through Baala et al. [19] and Wu et al. [20] but costs to transmission message exchange. An application-oriented approach has been implemented in VANET via DME as of [23, 24]. Finally, in FANET, Khanna et al. [4, 5] introduced an effective approach to handle DME through UAVs and shown better results as compared to their older version.

Graph plots have been shown from Figs. 2, 3, 4 and 5 concluding the average number of messages exchange and synchronization delay to various proposed algorithms. Whereas from Figs. 2, 3 and 4, the average number of messages exchange with number of nodes is drawn respectively to MANET, VANET, and FANET based DME algorithms, and average synchronization delay is presented in Fig. 5 for all existing DME solutions.

**Table 1** Existing DME algorithms for ad hoc networks (Refer Appendix 1 for various notations)

| Algorithm | Publisher | Message Complexity (Worst Case) | Synchronization delay | Fault-tolerant capability | | |
|---|---|---|---|---|---|---|
| | | | | Token failure | Node failure | Link failure |
| *MANET* | | | | | | |
| [16] | WN (Kluwer Academic) | ~O (log N) | $1/€_r$ | ✘ | ✘ | ✓ |
| [18] | IWDA (ACM) | $u_b$ | μ | – | – | – |
| [19] | JPDC (Elsevier) | – | – | ✘ | ✓ | ✘ |
| [20] | PMC (Elsevier) | 3 N/2 | O(1) | ✘ | ✓ | ✓ |
| *VANET* | | | | | | |
| [22] | CCIS (Springer) | O(1/N) | T | ✘ | ✘ | ✘ |
| [23] | TPDS (IEEE) | – | – | – | – | – |
| [24] | JOS (Springer) | – | – | – | – | – |
| [25] | IEEE Access | – | – | – | – | – |
| *FANET* | | | | | | |
| [4] | CEE (Elsevier) | *O(N)* | T | ✓ | ✘ | ✘ |
| [5] | CC (Elsevier) | < O(N) | T | ✓ | ✘ | ✘ |

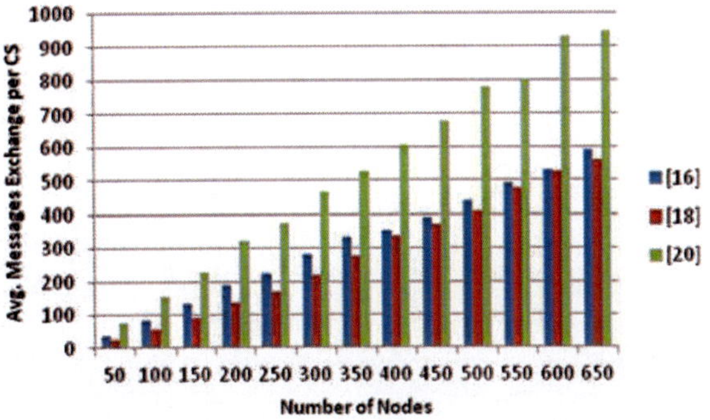


**Fig. 2** Average messages exchange per CS in MANET

## 5 Open Challenges and Future Research Directions

Despite the different applicability and implementation of DME in ad hoc networks, various significant challenges remain to be addressed in this domain. Here, in this section, we incorporate the possibilities of such area which are untouched and also discuss the challenges while merging those concepts in our possibilities.

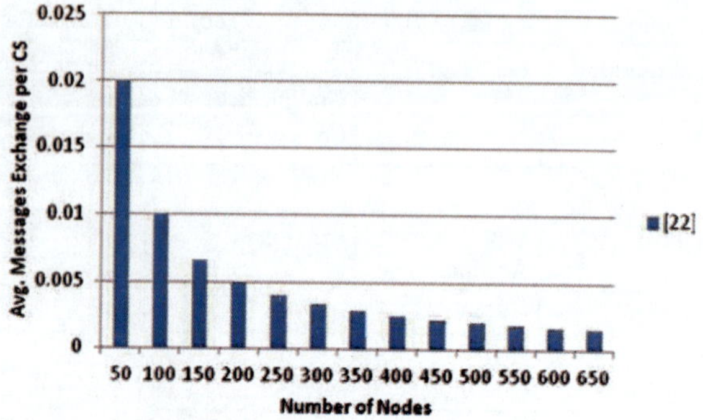


**Fig. 3** Average messages exchange per CS in VANET

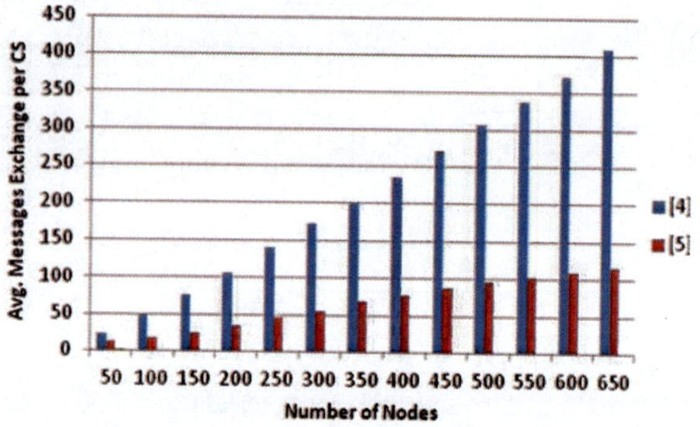


**Fig. 4** Average messages exchange per CS in FANET

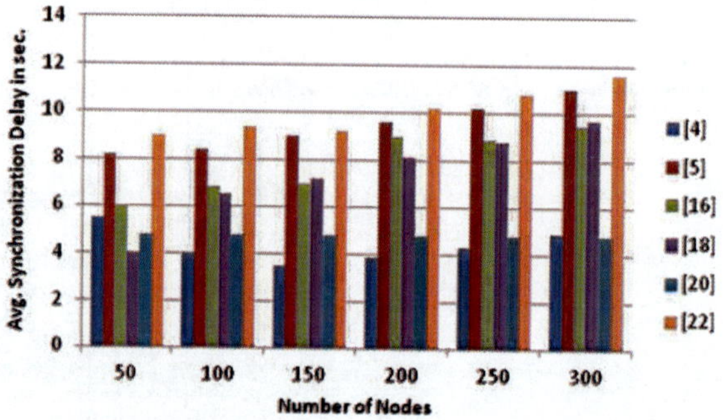


**Fig. 5** Average synchronization delay on all studied ad hoc network variants

As a prime classification of ad hoc network, MANET, VANET and FANET are in existence and due to a trend in the evaluation of wireless communication in the past few years, the adaptability of these networks is getting high. Major works as a part of different DME variants have already been applied on MANET and VANET (*discussed in* Sect. 3) but remain an unexplored area in FANET.

### 5.1 *Challenges*

Node mobility model in FANET is highly unpredictable due to their fly in the air. Srivastava and Prakash [11] mentioned various movement models in their work, but still, a pre-determination of such movement during the transmission is complex because of an open space. Assurance of packet/message delivery through routing protocol might be guaranteed but in an unexpected time bound [15]. Sustainability issues with UAVs are also one of the factors to be considered on as it depends on various parameters like power consumption, transmission range, and speed etc.

### 5.2 *Group Mutual Exclusion and Self-stabilizing Algorithms*

The concept of group mutual exclusion (GME) was firstly introduced by Joung [26] in which processes belonging to different groups in the system can invoke their CS simultaneously and enhance the overall throughput. Such involvement of processes to access the shared resource is still remains to be explored in FANET. A correlation might be established in between the groups of the system and clusters in FANET as a further approach to increase the throughput and lower the number of messages exchanged. Self-stabilization [17] also be another aspect to be think on in FANET where the solution develops a self-correction algorithm in itself to resolve the trivial faults in the system.

### 5.3 *Machine Learning/Blockchain Approach*

Machine learning (ML) involvement in ad hoc network to achieve DME is still an unexposed area in all regions. ML can insight and explore various aspects like identifying node movement patterns, packet travel through hops, and prediction of best suited DME algorithm etc. ML helps to build a model that represents the best-case possible solution based on the historical data available. Nowadays, blockchain is one of the most powerful technologies to ensure tamper-proof communication while message transactions. We recently approach this technology in one of the previous work [27] and again, this might be helpful while designing any secure DME algorithms for dynamic network for the purpose of data protection.

## 6 Conclusion

Through this article, we present a cross-sectional study on various existing distributed mutual exclusion algorithms imposed on ad hoc networks and their associated network variants including their performance metrics and fault-tolerant capabilities. Open challenges and direction to future research have also been discussed as a part of this study. Initially, we began our discussion on DME and various variants of ad hoc networks as MANET, VANET and FANET. Further, we reviewed different existing DME algorithms on these variants and found certain areas to be unexplored in FANET which are addressed accordingly along with the challenges. Finally, we hope that the work carried through this article might be useful and considered as a quick reference to get the insights of distributed mutual exclusion in ad hoc networks.

## Appendix 1

$N$ = *Number of nodes in network.*

$€_r$ = *arrival rate of Poisson process.*

$\mu$ = *Maximum delay in between node-to-node communication.*

$u_b$ = *Upper bound limit of messages generated by a node.*

$T$ = *Propagation time of a message.*

## References


1. Kshemkalyani, Singhal, M.: Distributed Computing: Principles, Algorithms, and Systems. Cambridge University Press (2008)
2. Dijkstra, E.W.: Solution of a problem in concurrent programming control. Commun. ACM **8**, 569 (1965). https://doi.org/10.1145/365559.365617
3. Parihar, A.S., Chakraborty, S.K.: Token-based approach in distributed mutual exclusion algorithms: a review and direction to future research. J. Supercomput. **77**, 14305–14355 (2021). https://doi.org/10.1007/s11227-021-03802-8
4. Khanna, A., Rodrigues, J.J.P.C., Gupta, N., Swaroop, A., Gupta, D., Saleem, K., de Albuquerque, V.H.C.: A mutual exclusion algorithm for flying Ad Hoc networks. Comput. Electr. Eng. **76**, 82–93 (2019). https://doi.org/10.1016/j.compeleceng.2019.03.005
5. Khanna, A., Rodrigues, J.J.P.C., Gupta, N., Swaroop, A., Gupta, D.: Local mutual exclusion algorithm using fuzzy logic for flying Ad hoc networks. Comput. Commun. **156**, 101–111 (2020). https://doi.org/10.1016/j.comcom.2020.03.036
6. Saxena, P.C., Rai, J.: A survey of permission-based distributed mutual exclusion algorithms. Comput. Stand. Interfaces **25**(2), 159–181 (2003). https://doi.org/10.1016/S0920-5489(02)00105-8
7. Benchaïba, M., Bouabdallah, A., Badache, N., Ahmed-Nacer, M.: Distributed mutual exclusion algorithms in mobile ad hoc networks. ACM SIGOPS Operating Syst. Rev. **38**(1), 74–89 (2004). https://doi.org/10.1145/974104.974111
8. Ismail, D.P.I.I., Ja'afar, M.H.F.: Mobile ad hoc network overview. Asia-Pac. Conf. Appl. Electromagnet. (2007). https://doi.org/10.1109/apace.2007.4603864
9. Jain, M., Saxena, R.: Overview of VANET: Requirements and its routing protocols. In: 2017 International Conference on Communication and Signal Processing (ICCSP) (2017). https://doi.org/10.1109/iccsp.2017.8286742

10. Bekmezci, İ, Sahingoz, O.K., Temel, Ş: Flying Ad-Hoc networks (FANETs): a survey. Ad Hoc Netw. **11**(3), 1254–1270 (2013). https://doi.org/10.1016/j.adhoc.2012.12.004
11. Srivastava, A., Prakash, J.: Future FANET with application and enabling techniques: Anatomization and sustainability issues. Comput. Sci. Rev. **39**:100359. https://doi.org/10.1016/j.cosrev.2020.100359, ISSN 1574-0137
12. Cevik, P., Kocaman, I., Akgul, A.S., et al.: The small and silent force multiplier: a swarm UAV electronic attack. J. Intell. Robot. Syst. **70**(1–4), 595–608 (2013). https://doi.org/10.1007/s10846-012-9698-1
13. Kerr, S.: UAE to develop fleet of drones to deliver public services. The Financ. Times World News. Retrieved **12** (2014)
14. Barrado, C., Messeguer, R., L´opez, J., Pastor, E., Santamaria, E., Royo, P.: Wildfire monitoring using a mixed air-ground mobile network. IEEE Pervasive Comput. **9**(4), 24–32 (2010), https://doi.org/10.1109/MPRV.2010.54
15. Sang, Q., Wu, H., Xing, L., Xie, P.: Review and comparison of emerging routing protocols in flying Ad Hoc networks. Symmetry **12**, 971 (2020). https://doi.org/10.3390/sym12060971
16. Walter, J.E., Welch, J.L., Vaidya, N.H.: A mutual exclusion algorithm for Ad Hoc mobile networks. Wireless Netw. **7**, 585–600 (2001). https://doi.org/10.1023/A:1012363200403
17. Dijkstra, E.: Self stabilization in spite of distributed control. Comm. ACM **17**, 643–644 (1974). https://doi.org/10.1145/361179.361202
18. Chen, Y., Welch, J.L.: Self-stabilizing mutual exclusion using tokens in mobile ad hoc networks. In: Proceedings of the 6th International Workshop on Discrete Algorithms and Methods for Mobile Computing and Communications—DIALM '02 (2002). https://doi.org/10.1145/570810.570815
19. Baala, H., Flauzac, O., Gaber, J., Bui, M., El-Ghazawi, T.: A self-stabilizing distributed algorithm for spanning tree construction in wireless ad hoc networks. J. Parallel Distrib. Comput. **63**(1), 97–104 (2003). ISSN 0743-7315, https://doi.org/10.1016/S0743-7315(02)00028-X
20. Wu, W., Cao, J., Yang, J.: A fault tolerant mutual exclusion algorithm for mobile ad hoc networks. Pervasive Mob. Comput. **4**(1), 139–160 (2008). ISSN 1574-1192, https://doi.org/10.1016/j.pmcj.2007.08.001
21. Singhal, M., Manivannan, D.: A distributed mutual exclusion algorithm for mobile computing environments. In: Proceeding of ICIIS'97, IEEE Computer Society, pp. 557–561 (1997)
22. Sharma, B., Bhatia, R.S., Singh, A.K.: An O(1/n) protocol for supporting distributed mutual exclusion in vehicular Ad Hoc networks. In: Nagamalai, D., Renault, E., Dhanuskodi, M. (eds.) Advances in Parallel Distributed Computing. PDCTA 2011. Communications in Computer and Information Science, vol 203. Springer, Berlin. /https://doi.org/10.1007/978-3-642-24037-9_14
23. Wu, W., Zhang, J., Luo, A., Cao, J.: Distributed mutual exclusion algorithms for intersection traffic control. IEEE Trans. Parallel Distrib. Syst. **26**(1), 65–74 (2015). https://doi.org/10.1109/TPDS.2013.2297097
24. Lim, J., Jeong, Y.S., Park, D.S., et al.: An efficient distributed mutual exclusion algorithm for intersection traffic control. J. Supercomput. **74**, 1090–1107 (2018). https://doi.org/10.1007/s11227-016-1799-3
25. Shehu, H.A., Sharif, M.H., Ramadan, R.A.: Distributed mutual exclusion algorithms for intersection traffic problems. IEEE Access **8**, 138277–138296 (2020). https://doi.org/10.1109/ACCESS.2020.3012573
26. Joung, Y.-J.: Asynchronous group mutual exclusion. Distrib. Comput. **13**, 189–206 (2000)
27. Parihar, A.S., Prasad, D., Gautam, A.S., Chakraborty, S.K.: Proposed end-to-end automated E-voting through blockchain technology to increase voter's turnout. In: Prateek, M., Singh, T.P., Choudhury, T., Pandey, H.M., Gia, Nhu N. (eds.) Proceedings of International Conference on Machine Intelligence and Data Science Applications. Algorithms for Intelligent Systems. Springer, Singapore (2021). https://doi.org/10.1007/978-981-33-4087-9_5

28. Parihar, A.S., Chakraborty, S.K.: Handling of resource allocation in flying ad hoc network through dynamic graph modeling. Multimedia Tools Appl. (2022). https://doi.org/10.1007/s11042-022-11950-z
29. Parihar, A.S., Chakraborty, S.K.: A simple R-UAV permission-based distributed mutual exclusion in FANET. Wireless Netw. (2022). https://doi.org/10.1007/s11276-022-02889-y
30. Parihar, A.S., Chakraborty, S.K.: A new resource-sharing protocol in the light of a token-based strategy for distributed systems. Int. J. Comput. Sci. Eng. In Press (2022)

# Estimation of Electromagnetic Pollution Index of Macrocell

**N. Padmavathy, M. C. Chinniah, and K. Ravi Varma**

**Abstract** The exponential increase of the usage of mobile handsets and other electronic gadgets has significant impact on the global atmospheric warming of a cell and is estimated as an electromagnetic pollution index. The evidence of electromagnetic pollution index measures using the FS model in the literature, but the authors generated a mathematical model that was erroneous. This research has focused on deriving correct model to measure the pollution index of macrocell using the free space propagation model considering macrocell simulation parameters like frequency, cell size, power, and radius of macrocell. The results show that high transmission power, frequency, and number of users have a significant impact on the environmental electromagnetic pollution index. The resulting electromagnetic pollution index values are harmful, i.e., greater than 150 Wh, and an ideal strategy for small electromagnetic pollution index (say, less than 150 Wh) has been proposed.

**Keywords** Electromagnetic pollution index · Electromagnetic radiation · FS propagation · Global warming · Green mobile communication · Line of sight · Macrocell · Polluted area · Polluting energy · Transmission power

## 1 Introduction First Section

Mobile communication technology has evolved very rapidly from the age of pagers to smart phones at the same time also include the integration of many heterogenous devices. Furthermore, an enormous scattering of BS on the roofs and neighborhood can similarly be irrefutably seen. In India, by the year 2023, it is estimated that almost 500.9 million mobile users would exists as seen in Fig. 1. As per the survey, India had the world's largest Internet population at about 483 million users in 2018. Due

N. Padmavathy (✉) · K. Ravi Varma
Department of Electronics and Communication Engineering, Vishnu Institute of Technology, Bhimavaram, Andhra Pradesh, India
e-mail: padmavathy.n@vishnu.edu.in

M. C. Chinniah
B. V. Raju Institute of Technology, Vishnupur, Narsapur, Medak District, Telangana, India

 
D. Gupta et al. (eds.), *Pattern Recognition and Data Analysis with Applications*, Lecture Notes in Electrical Engineering 888,
https://doi.org/10.1007/978-981-19-1520-8_4

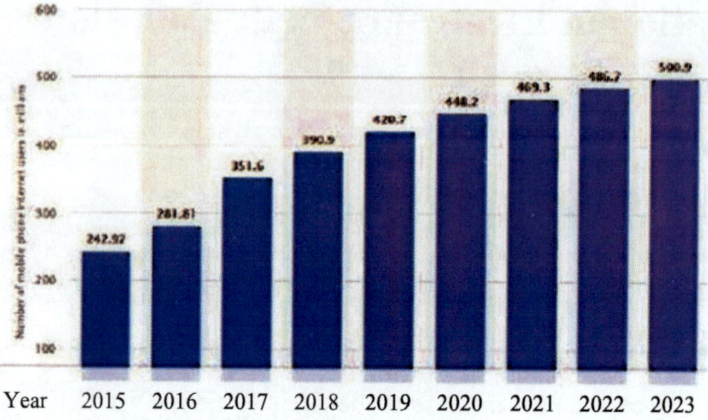


**Fig. 1** Number of mobile phone users in India since 2015 [1]

to lack of policy making on deployment of the cell phone tower in India, hence, people living in rural/urban/sparsely populated regions are severely affected by the radiations.

Because of the existence of EM, the BSs emit RF radiation as a type of non-ionizing radiation in the EM range, according to [2]. The Kyoto Protocol [2] is being used by the Indian government, as well as many other countries throughout the world, to diminish the energy usage. Moreover, India has been keen in reducing the intensity of carbon by 20–25% between 2005 and 2020 and is poised to achieve 35% reduction by 2029 [1]. With the Paris agreement, India has taken three quantitative climate change goals.

As per the Paris agreement [2] signed by India with Finland in November 2020, India had taken three strategic decisions on carbon reduction—1: is reduction in the emissions intensity of GDP by 33 to 35% by 2030 from 2005 level; 2: is achieving about 40% cumulative electric power installed capacity from non-fossil fuel-based energy resources by 2030; 3: is creating an additional carbon sink of 2.5 to 3 billion tons of carbon dioxide equivalent through additional forest and tree cover by 2030.

## 2 Literature Review

The increasing demand, constant development, and the speedy growth for continuous production of latest and advanced communication devices have a major result on the worldwide surroundings in terms of energy consumption, life-loss because of radiation effects, biological changes, and disappearance of living species, etc. Several reports have reported the decline of the sparrow population, colony collapse disorder (slow disappearance of bees) occurs due to deadly effects of the EM radiation. The most predominant previously mentioned issues have persuaded the researchers to examine for the promotion of green communication [3–5]. Grzegorz, 2015, clearly

**Table 1** Air quality index threshold and its counter effects

| Air quality index | Particulate matter |
|---|---|
| 301–500 | Hazardous |
| 201–300 | Very unhealthy |
| 151–200 | Unhealthy |
| 101–150 | Unhealthy for Sensitive groups |
| 51–100 | Moderate |
| 0–50 | Good |

emphasized that there is no scientific evidence of the effect of EM radiations on living organisms but as a hypotheses it is also highlighted that there are literature that shows evidence of the effect of EM radiation on the pineal gland (especially its hormone melatonin) causing sleep disorders, lower mood, reduced concentration, depression, and growth of cancer [6]. Worldwide Commission on Non-Ionizing Radiation Protection (ICNIRP) defined rules that particularly overlooked the potential long haul organic impacts like expanded danger of diseases like cancer [7].

Table 1 provides the effect of pollution index on the existence of the living beings. It is very much necessary to limit the radiation in order to reduce the adheres effects on health. For a healthy living, the quality index needs to be <50 Wh, and as the EM pollution increases beyond 300 Wh, its hazardous. While between 101 and 299 Wh, sensitive people developed breathing discomfort, and respiratory discomfort (asthma), and other serious health issues.

Table 2 provides details of the ICNIRP radiation norms for setting mobile towers in India. The international EMF radiation norms [8] for mobile towers vary between 0.001 W/m$^2$ and 12 W/m$^2$ (both values inclusive) at a frequency of 1800 MHz. RF radiation from cell phones was designated as 'possibly carcinogenic to humans (group 2B)' by the World Health Organization (WHO) [9] in early 2011. The panel on climate, horticulture, neighborhood, and local issues summarizes the natural impacts on verdure [10].

The European Council report [11] reported that 'EM fields from mobile telephony appear to have more or less potentially harmful, non-thermal, biological effects on plants, insects, and animals [12], as well as the human body when exposed to levels that are below the official threshold values.' Henceforth, the public need to take the simplest measures to remain far away from such natural impacts. As per the recommendations reported by GoI, the electromagnetic radiations are to be considered as a pollutant.

**Table 2** ICNIRP radiation norms for mobile towers in India from 01.09.2012Frequency (MHz) Power density levels (ICNIRP radiation norms)—W/m29000.4518000.921001.05

| Frequency (MHz) | Power density levels (ICNIRP radiation norms)—W/m$^2$ |
|---|---|
| 900 | 0.45 |
| 1800 | 0.9 |
| 2100 | 1.05 |

To summarize, EM radiations have prolonged impact on the human health, flora, and fauna. Literature also addresses that the EM radiation effects on the above mentioned have been documented but no statistics on this has been established so far. The work on studying the effect of EM radiation is still in its nascent stage, and there is no study done so far in the proposed area. However, a few literature survey show that few researchers have conducted experimental and comprehensive study on impact of cell size on the EM pollution. This paper proposes an approach that helps in understanding how the radiation can be reduced. The main objective is to estimate the EPI in line of sight (LOS) for a macrocell. The subsequent paragraphs would lead the readers to gain knowledge on estimation of EPI.

The radio frequency radiation radiated by BS and MSs is mostly measured as EM pollution index (EPI). The EPI is mathematically given as seen in (1).

$$\begin{aligned} \text{EPI} &\triangleq \text{Normalized polluted area} * \text{polluting energy} \\ \text{EPI} &\triangleq \text{PA}_{\text{norm}} * \text{PE} \end{aligned} \tag{1}$$

The threshold range from 4.7 $\mu$W/m$^2$ to 170 $\mu$W/m$^2$ [2, 4] is recommended for the normalized area of *PoP* [13].

$$PA_{\text{norm}} = \left(\frac{\text{Total area of all PoP's}}{\text{Area of the macrocell}}\right) = \left(\frac{A_1}{A}\right) \tag{2}$$

The electromagnetic pollution (EMP) can be reduced using a cell size-based approach [14]. The authors mainly focused on reducing the RF radiation and green-house gases by increasing the system capacity. A prediction model theory for multi-system BS and MS considering several antenna parameters [15] has been proposed to calculate electromagnetic radiation (ER). The results of the past investigations are summed up and there's a necessity to supervise EMP, while the terms EMP and EM have been utilized widely used broadly for quite a while [16], thus gives off an impression of being no definition or model of such contamination with regards to versatile correspondences. Such definition is essential to gauge, monitor, and oversee EMP as pointed in [17, 18].

## 3 EMP and EPI in Mobile Communications

The FS propagation model is employed to work out the signal strength of the received signal when the TR separation is a clear, unobstructed LOS path. Despite the actual fact that EM signals once passing through remote channels experiences blurring because of different deterrents, when compared to the signals traveling directly as line of sight. The EPI computation for FS model utilizing the simulation parameters like frequency (*f*), cell size (*j*), macrocell power (*P*), and cell radius (*D*) are given beneath.

### 3.1 Estimate of EPI for Free Space Model

The EPI can be calculated using (1) and (2) to get (3)

$$\text{EPI} \triangleq \left(\frac{\text{Total area of all PoP's}}{\text{Area of the Macrocell}}\right) * (\text{sum of energy in all PoP's}) \tag{3}$$

As a rule, amount of space of all PoPs (4) is resolved as an element of region because of base handset station and amount of *j* mobiles PoP territory.

$$A_1 = \alpha_{\text{BTS}} + \sum a_j \tag{4}$$

Sum of energy in all PoPs can be determined (5)

$$PE = \sum \int_0^T P_r \text{dt} \tag{5}$$

(6) is got after substituting (4) and (5) in (3);

$$\text{EPI} = \frac{\left[\alpha_{\text{BTS}} + \sum \text{a}_\text{j}\right]}{\text{A}} * \sum \int_0^\text{T} \text{P}_\text{r}\text{dt} \tag{6}$$

Power density for the free space propagation

$$s = \frac{P_t G_t}{4\pi} \tag{7}$$

Received power.

$$P_r = \frac{P_1 G_1}{4\pi} \tag{8}$$

The minimum received power for the FS model is (9).

$$P_{\text{min}} = \frac{P_1 G_1 G_r}{R_{\text{max}}^2} * \left(\frac{\lambda}{4\pi}\right)^2 \tag{9}$$

The transmitter power can be estimated as shown in (10).

$$P_t = \frac{P_{\text{min}} * R_{\text{max}}^2}{G_t G_r} * \left(\frac{4\pi}{\lambda}\right)^2 \tag{10}$$

Substitute (10) in (8), to get received power as shown in (11).

$$P_r = \frac{P_{min} * R_{max}^2}{G_r} * \left(\frac{4\pi}{\lambda}\right)^2 \tag{11}$$

Let Gr = 1; and substitute the receiver gain, the received power (12) is.

$$P_r = \frac{P_{min} * R_{max}^2}{G_r} * \left(\frac{4\pi}{\lambda}\right)^2 \tag{12}$$

Substitute (12) in (6), (13) to (16) is derived as

$$\mathrm{EPI} = \frac{[\alpha_{BTS} + \sum a_j]}{A} * \sum \int_0^T P_{min} * R_{max}^2 * \left(\frac{4\pi}{\lambda^2}\right) dt \tag{13}$$

$$\mathrm{EPI} = \left(\frac{[\alpha_{BTS} + \sum a_j]}{A}\right) * \left(P_{min} * R_{max}^2 * \left(\frac{4\pi}{\lambda^2}\right)\right) * \sum \int_0^T dt \tag{14}$$

$$\mathrm{EPI} = \left(\frac{[\alpha_{BTS} + \sum a_j]}{A}\right) * \left(P_{min} * R_{max}^2 * \left(\frac{4\pi}{\lambda^2}\right)\right) * \sum [T] \tag{15}$$

$$\mathrm{EPI} = \left(\frac{[\alpha_{BTS} + \sum a_j]}{A}\right) * \left(P_{min} * R_{max}^2 * \left(\frac{4\pi}{\lambda^2}\right)\right) * \left[\sum_1^{uv} \tau_i + \sum_1^{j} T_j\right] \tag{16}$$

Finally, (16) is the EPI equation for FS model considering macrocell.

ABCDEF of Fig. 2 represents a hexagon with a side length of '*a*', and the hexagon area can be calculated using (17).

$$\text{Area of cell(A)} = \frac{3\sqrt{3}}{2} a^2 \tag{17}$$

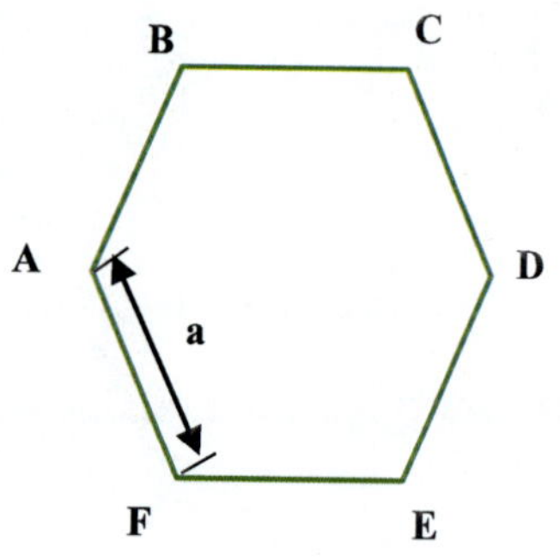


**Fig. 2** Hexagon with side length of 'a'

- Area of the PoP due to base transceiver station:

If the area of the cell is converted into six PoPs, then each PoP is consisting of one base transceiver, then the area of the PoP due to BTS is given as

$$\alpha_{\text{BTS}} = \frac{A}{6} \tag{18}$$

Consider a hexagonal shaped PoP as shown in Fig. 3. In this PoP, the $j$ mobiles are distributed randomly. The range of the mobile radiation is between 1 and 10 mts. Considering the sum of all mobiles area, i.e., consider a mobile which is radiating at '$a$' mts range.

*Case (1): Using right angled triangle (See* Fig. 4a*)*

$$\text{The area of the triangle} = 1/2 * \text{base} * \text{height}(18)$$
$$= 1/2 * \text{opposite length} * \text{adjacent length}$$

The hexagon can be divided into 12 equal number of right angled triangles, then the area of the mobile (which is hexagonal shape) with radiation of '$a$' mts is given by

$$= 12 * 1/2 * \text{opposite length} * \text{adjacent length}$$

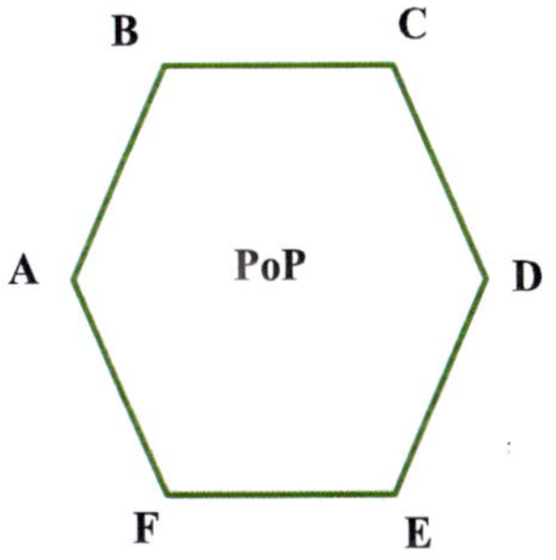


**Fig. 3** PoP representation or shape of POP

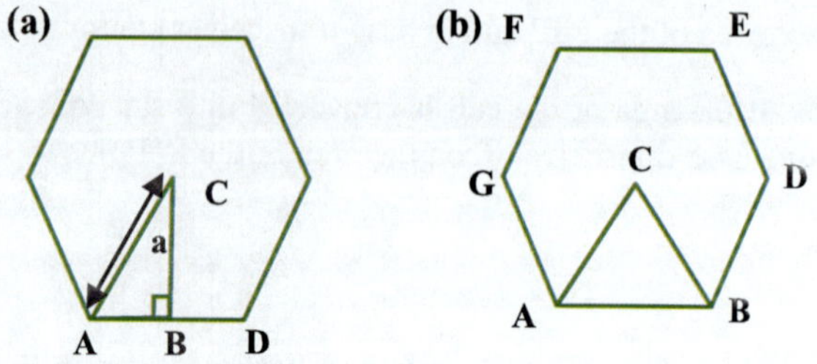


**Fig. 4** Dividing the hexagon as a **a** right angled triangle **b** equilateral triangle to find area

$$= 6 * \text{opposite length} * \text{adjacent length}$$

*Case (2): Using equilateral triangle (See* Fig. 4b*).*
The ΔABC is equilateral triangle, for equilateral triangle all sides are of equal length. So that the side of the hexagon is '*a*' mts, then the area of the hexagon is given as defined in (17) or (19)

$$\text{Areaofequilateraltriangle} = \frac{\sqrt{3}}{4}\text{a}^2 \tag{19}$$

In the hexagon, there are 6 equilateral triangles, and hence, the area of the mobile is as given in (20).

$$\text{Area of equilateral triangle} = 6\frac{\sqrt{3}}{4}\text{a}^2 \tag{20}$$

The maximum radius of the PoP is calculated (21) with the knowledge of $P_{\min}$, the minimum power (7) required at the maximum distance and area of the PoP.

$$a^2 = \frac{2A}{3\sqrt{3}}$$

$$a = \sqrt{\frac{2A}{3\sqrt{3}}} \tag{21}$$

Using (21), the maximum distance of the PoP can be determined. Here, area of the PoP is nothing but area of the base transceiver station.

$$a = \sqrt{\frac{2\alpha_{\mathrm{BTS}}}{3\sqrt{3}}} \tag{22}$$

- $\sum_{1}^{M} T_j$(average transmit time for $j$ mobiles)

$$T = \frac{d}{s} \tag{23}$$

$s$ is the speed which is found using (24)

$$s = \frac{c}{\sqrt{\varepsilon_r}} \tag{24}$$

where $\varepsilon_r$ is the dielectric constant, the electromagnetic waves are traveling through the air medium. So for area medium, the dielectric constant is '1'.

$$s = \frac{c}{\sqrt{1}}$$

$$s = c$$

i.e., speed of the electromagnetic waves is equal to that of light speed, i.e., $3*10^8$ m/s. Then, the time period is calculated for all the mobiles at different distances and all are summed together.

- $\sum_{1}^{uv} \tau_i$ (mean transmit time for $i_{\text{th}}$ channel of BTS)

Using Erlang B model, the average transmit time for $i$th channel of BTS is calculated as given in (25).

$$A_0 = \frac{Q_L * \tau_L}{60} \text{erlangs} \tag{25}$$

From (25), the transmit time (26) is

$$\tau_i = \frac{A_0 * 60}{Q_i} \tag{26}$$

The blocking probability is the number of blocked calls per hour and can be calculated using (27) for a $m$ number of indistinguishable equal assets like workers, phone lines, and so on.

$$B = \frac{\frac{A_0^m}{m!}}{\sum_{i=0}^{m}\left(\frac{A_0^i}{i!}\right)} \tag{27}$$

Erlang B table helps in finding the offered load by taking $B = 2\%$ and varying the $u$. The number of users '$j$' (28) can be calculated with the knowledge of total traffic in cell to the per user traffic ($Au$)

$$j = \frac{A_0}{A_u} \tag{28}$$

*Au* can be determined using (29)

$$\text{Au} = \text{request rate} * \text{holding time} \tag{29}$$

The total number of calls or maximum calls per hour in a cell 'Qi' is calculated using (30)

$$\text{Au} = \text{request rate} * \text{number of users} \tag{30}$$

# 4 Algorithm

An algorithm has been developed to estimate the EPI of a macro cell considering a FS model. The proposed algorithm considers several metrics like number of cells; radius of the cell; frequency; and the power to calculate the EPI of the macro cell. The simulation results are estimated using MATLAB 2018a run on windows 10 at speed of 1.80 GHz and explained briefly in the subsequent sessions.

*Step* 1: Initialize the input parameters like cell size (j), frequency (f), power (P), and cell radius (D)

*Step* 2: Find the area of cell (A) using (17),and also find the area of the base station (*aBT* ) using (18)

*Step* 3: Using (20), calculate the area due to mobiles in a cell. Repeat for mobiles having different cell ranges ($\sum aj$)

*Step* 4: Using (21), calculate the side length (a) or radius of the PoP. Also, the average transmit time for j mobiles is calculated using (23) at different distances in a PoP.

*Step*5:Using Erlang B model (i.e., (26), the mean transmit time for i channels of BTS is calculated. The parameters like offered load (A) is calculated using Erlang B model (26) (27), (28) and (29) or by using Erlang B table. Finally, the maximum call per hour in a cell (Qi) is calculated using (30).

*Step* 6: Calculate the mean transmit time for each and every channel of BTS by using the *Step* 5.

*Step* 7: Calculate the minimum power using (9)

*Step* 8: Estimate the EPI of a macrocell under FS propagation using (16).

## 5 Illustration

Macrocell type has been considered as a case study for the approximation of EPI under FS propagation. The essential simulation inputs utilized for the EPI estimation are provided in Table 3. To have a healthy environment, the optimum parameters like $f = 70$ MHz, $j = 500$, $D = 10{,}000$ m, and $P = 60$ W have been considered to evaluate EPI.

The EPI of the macrocell in FS propagation has been simulated considering the few simulation parameter as variable throughout the simulation, while maintaining other parameters a constant. Every mobile phone (transceiver) emits radiations called as pollution due to the influence of considered simulation parameters like $P$, $f$, and $D$ which increases with increasing cell size (i.e., number of users in a macrocell). From Fig. 5, it can be understood that with increasing the cell radius, the pollution index decreases drastically. For example (see red line), when cell radius is 20 km, the EPI is 285.52 Wh and at 40 km its 285.42 Wh with same number of users in the defined cell radius. With less number of users in large area, the pollution index falls down by 5%. If the number of users increases in large area, then each mobile device would radiate energy which causes pollution.

Thereby, increasing the sum of the radiations discharged by mobile phones that result in increase in EPI of the macrocell as seen in Fig. 5 (see black line). That is assuming 100 users in an area of 10 km releases EPI < 100 Wh radiation and 1000 users in same area would release 600 Wh radiation almost 83.33% increase in EPI.

From the simulated results considering LOS condition (FS propagation), it may be concluded that with sizable amount of mobile users (say 1000 users) on a multi-channel macrocell with operating frequency of 75 MHz; transmitting power 100 W over a cell radius of 40 km; the radiated pollution lies between 285.4 and 561.7 Wh. These indices indicate that all species are compelled to avoid physical activities outside and confine themselves just to indoor. To have a non-hazardous environment, it is advisable to have a macrocell design considering typical transmitting power of 60 W with its operating frequency of 70 MHz over a cell radius of 10 km with EPI as $67 \geq \text{EPI} \leq 125$ Wh, *i.e.*, pollution index reduces by 75% when 67 Wh and reduces by 56% if 125 Wh. For example, consider the cell radius of 10 km, the measured EPI is 285.9 Wh, and as the radius has been increased to 40 km, the measured EPI is 285.4 Wh, which shows a negligible small variations in the pollution index.

**Table 3** Simulation parameters

| Parameter | Specification |
|---|---|
| Cell type propagation model | Macrocell FS model |
| Indoor/Outdoor applications | Outdoor |
| Number of users ($j$) | 100–1000 |
| Maximum output power ($P$) | 40–100 W |
| Maximum cell radius ($D$) | 10–40 km |
| Frequency ($f$) | 60–75 MHz |

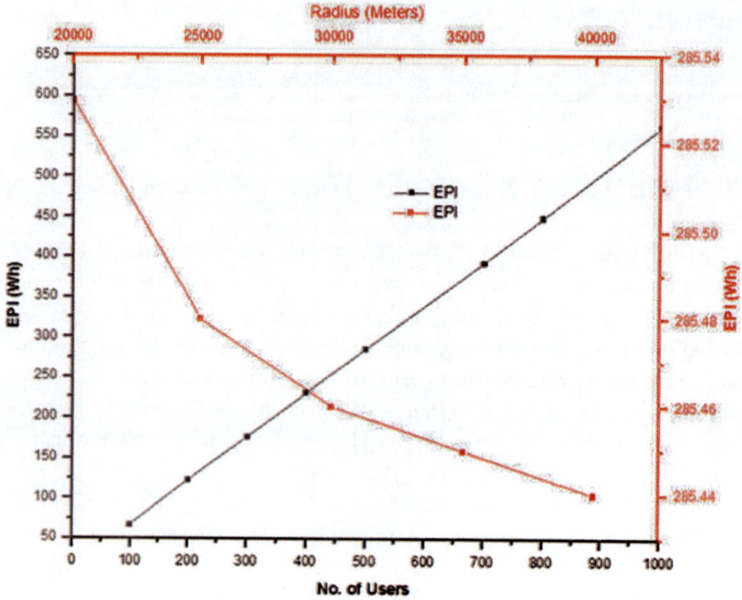


**Fig. 5** Effect of number of users (FS Model) and radius of macrocell on EPI

The transmitter converts the electrical signals to radio waves, and these signals are transmitted over longer distances by boosting the transmission power. Due to higher transmission power utilization, these RF signals (due to the base station and mobile phones) radiate energy as EM radiation that cause an increase in pollution. Accordingly, if power increases inevitably, the EPI increases as seen in Fig. 6 (see red line). Whereas, the power and frequency are directly proportional to each other. Generally, with increasing the transmitting frequency higher radiations are produced within the macrocell (see Fig. 6, black line). Therefore, this radiation rise leads to higher EPI.

Assuming a frequency of 60 MHz and frequency of 75 MHz, the difference in EPI between these two levels resulted in 37% radiation which is hazardous to human life and existence of birds. When the frequency increases, then there will be significant effect on EPI of the macro cell. Based on EPI (16), the $PA_{norm}$ depends on the ratio of A1 to A as seen in (2). The formula to find A1 is defined in (4). According to (2), as A increases, there is a fall in $PA_{norm}$ which leads to small EPI change and vice-versa, as observed in Fig. 6.

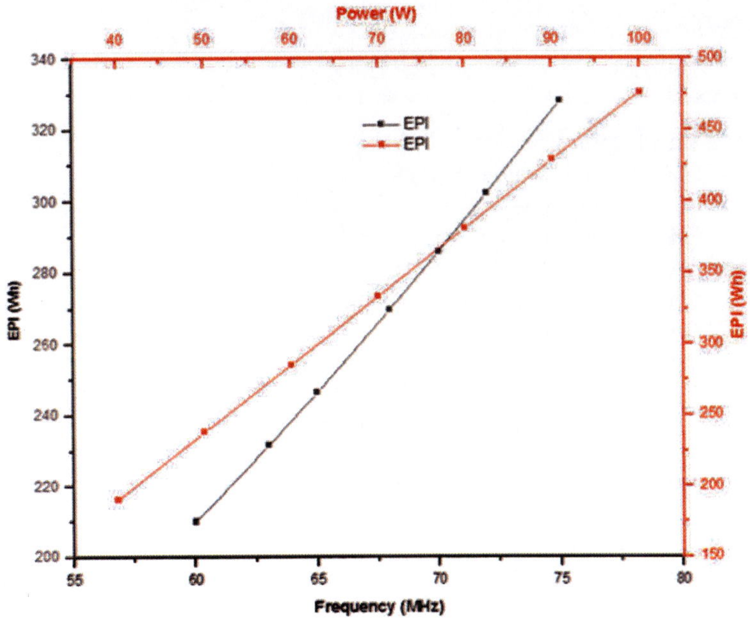


**Fig. 6** Effect of frequency and power (FS model) of macrocell on EPI

## 6 Conclusion

The use of the communication revolution (4G and 5G) and furthermore, the use of electronic gadgets dramatically incremented the electromagnetic radiations. This radiation presents electromagnetic contamination when the constraints of radiation have exceeded the edge esteems (see Table 2). Electromagnetic pollution significantly affects the climate, people, and living organic entities, and so on. It is crucial to think about the EPI estimation while developing a transceiver system. Henceforth, this research is focused on the EPI estimation of a macrocell utilizing FS model. In this paper, considering the typical power of 60 W with its operating frequency of 70 MHz over a phone range of 10 km for a set size of clients (say 100–200) in a macrocell prompts to a healthy environment. This examination infers that restricting the quantity of clients gives a decent solid healthy climate to macrocell range working in FS environment. Consequently, the fundamental boundary for a sound climate of a macrocell, an architect needs to consider enhanced qualities like, communicating power (60 W); frequency (70 MHz); cell span (10 km), and number of clients (restricted to 100–200 clients).

## Appendix

### Notations

| | |
|---|---|
| $\lambda$ | Wavelength |
| $\alpha_{\mathrm{BTS}}$ | Coverage area due to the BTS |
| $\tau_i$ | Mean calling time |
| $\sum a_j$ | Area of the PoP due to $j$ mobiles |
| $A$ | Area of the cell (radius) |
| $a$ | Side of a hexagon in meters |
| $A_0$ | Offered load (total traffic) |
| $A_1$ | Sum of the area of all packets of pollution |
| $B$ | Blocking probability |
| $d$ | Distance of the mobile from the BTS |
| $D$ | Cell Radius |
| $f$ | Frequency |
| $G_r$ | Receiving antenna gain |
| $G_t$ | Transmitting antenna gain |
| $j$ | Cell Size (number of clients/users/mobile nodes) |
| $P$ | Power in Wh |
| $PA_{\mathrm{norm}}$ | Normalized polluted area |
| PE | Polluting energy |
| $P_{\mathrm{min}}$ | Minimum power required at the maximum distance |
| $Pr$ | Receiver power |
| $P_t$ | Transmitter power |
| $Q_i$ | Maximum calls/hour/cell |
| $R_{\mathrm{max}}$ | Maximum distance covered by the BTS |
| $s$ | Speed of light in $3*10^8$ m/sec$^2$ |
| $T$ | Time period |
| $T_j$ | Mean transit time for $j$ mobiles |
| $u$ | Maximum number of frequency channels/cell |
| $u$ | Number of channels/base transceivers |
| $v$ | Number of base transceivers |
| $\varepsilon_r$ | Dielectric constant |

## References

1. https://economictimes.indiatimes.com/news/economy/policy/india-to-achieve-target-of-reducing-35-pc-emissions-intensity-before-2030-javadekar/articleshow/79430592.cms?Utmsource=contentofinterest&utm_medium=text&utm_campaign=cppst
2. Sarma, J.S.: Telecom regulatory authority of India, consultation paper on green. Telecommunications **3**, 1–52 (2011)

3. Fragopoulou, Y., Grigoriev, O., Johansson, L.H., Margaritis, L., Morgan, E., Richter. C. Sage.: Scientific panel on electromagnetic field health risks: Consensus points, recommendations, and rationales. scientific meeting: Seletun, Norway, Rev. Environ. Health **25**(4), 1–11 (2010)
4. Padmavathy, N., Sasi, K.S.: Green communication: An emerging telecommunication technology—Its research challenges. Techn. Appl. Springer **750**, 76–84 (2017)
5. Ravi Varma, K., Padmavathy, N.: Electromagnetic pollution index estimation of green mobile communication. In: 2nd IEEE International Conference on Intelligent Computing and Control Technologies (ICICICT 2019), pp. 59–63 (2019)
6. Ahlbom, U., Bergqvist, J., Bernhardt, H., Cesarini, J.P., Court, L.A., Grandolfo, M., Matthes, R.: Guidelines for limiting exposure to time-varying electric, magnetic, and electromagnetic fields (up to 300 GHz). Health Phys. **74**(4), 494–522 (1998)
7. Grzegorz, R., Bogdan, L., Arkadiusz, G., Andrzej, K., Marek, K., Janusz, P., Kazimierz, J., Piotr, T., Jacek, J., Dominik, A., Aukasz, S., Dawid, G.: The influence of electromagnetic pollution on living organisms: Historical trends and forecasting changes. BioMed Res. Int. 1–19 (2015)
8. TRAI.: Information Paper on Effects of Electromagnetic Field Radiation From Mobile Towers and Handsets, Information Paper No. 01/2014, 1–39 (2014)
9. Franco, M., Buonaguro.: WHO press release, IARC classifies radio frequency electromagnetic fields aspossibly carcinogenic to humans. **208**, 1–6 (2011)
10. Huss. J.: Committee on the environment, agriculture and local and regional affairs, the potential dangers of electromagnetic fields and their effect on the environment. Doc. 12608, Resolution 1815, 1–46 (2011)
11. Ulrich, W.: Bees, birds and mankind—Destroying nature by Electrosmog. A brochure series by the competence initiative for the protection of humanity. Environ. Democracy 1–40 (2009)
12. Technical Report.: Report on Possible Impacts of Communication Towers on Wildlife Including Birds and Bees. Ministry of Environment and Forests, Government of India, pp. 1–22 (2011). http://www.moef.nic.in/downloads/publicinformation/final_mobile_towers_report.pdf
13. Venkatapathy, P., Jena, J., Jandhyala, A.: Electromagnetic pollution index—A key at tribute of green mobile communications. Inst. Electri. Electron. Eng. Green Technol. Conf. 1–4 (2012)
14. Neeraj, K., Pandey, C.: A review on reducing radio frequency pollution effecting by cell size. In: Special Issue: Proceedings of 2nd International Conference on Emerging Trends in Engineering and Management (ICETEM), pp. 112–115 (2013)
15. Yang, J., Lei, W., Xianli, L., Jie, W., Yongjin, C., Cunzhen, P.: A prediction model for electromagnetic radiation of multi- system base station. In: Proceedings of the 2nd International Conference on Computer Science and Electronics Engineering (ICCSEE), pp. 3012–3015 (2013)
16. Avadhanulu, J.V.: Integrated Communication and IT infrastructure (ICITI) for the socio-economic development of auroville bioregion. In: Proceedings of the 32nd Asia-Pacific advanced network meeting, New Delhi, vol. 32, pp. 172–184 (2011)
17. Genc, O., Bayrak, M., Yaldiz, E.: Analysis of the effects of GSM bands to the electromagnetic pollution in the RF spectrum. Progress Electromagnet Res PIER **101**, 17–32 (2010)
18. Ismail, F., Hashim, W.: Predicting radio frequency radiation from mobile communication base stations. Int. J. Comput. Commun. Eng. **2**(4), 482–486 (2013)

# Prediction of Train Delay System in Indian Railways Using Machine Learning Techniques: Survey

Ajay Singh, D. Rajesh Kumar, and Rahul Kumar Sharma

**Abstract** Railway system all over the world faces many issues in detecting train delay. Train delay is the major issue in railway networks throughout the world. There are so many factors like bad weather, exogenous data, railway asset condition, infrastructure conflicts, human errors, etc., which are the main causes for delay in train arrival. According to the (TOI) Time of India newspaper, in India, several millions of people depend on the train for their travel, which increases year by year. But in India, most of the train does not run as per the schedule because of few railway tracks and poor signaling. This states that traveler gets delayed to their destination. One of the difficulties of predicting train delay is the unpredictability and uncertainty of times in railway traffic. This paper aims to survey many challenges and issues faced by train delay prediction systems in India, USA, China, German, Dutch, etc. Several methods and algorithms related to the prediction system survey know their efficiency with various kinds of datasets regarding that particular region. This survey is much efficient for the upcoming researcher to research this recent topic of train delay prediction system.

**Keywords** Support vector regression · Fuzzy logic · Fuzzy petri nets · Big data · Machine learning

---

A. Singh (✉)
Computer Science Department, Galgotias University, Greater Noida, UP, India
e-mail: ajay.singh_phd19@galgotiasuniversity.edu.in

D. R. Kumar
School of Computer Science and Engineering, Galgotias University, Greater Noida, UP, India
e-mail: d.rajeshkumar@galgotiasuniversity.edu.in

R. K. Sharma
Computer Science Department, G L Bajaj Institute of Technology and Management, Greater Noida, UP, India



D. Gupta et al. (eds.), *Pattern Recognition and Data Analysis with Applications*, Lecture Notes in Electrical Engineering 888,
https://doi.org/10.1007/978-981-19-1520-8_5

## 1 Introduction

Train delay is one of the common parameters used for predicting timetables and resolving infrastructure issues. Delay date is most prominent for dispatching of train and also for the operation of railway traffic. Sometimes train delays by hour-hours or day-day with the same day, and irregularity of train delays makes the operators tedious for planning railway operation in the short time.

Initially, the reason for train delay is external stochastic disturbance [1]. The initial delay created within an observed network is known as the original delay. Buffer times among train are less than the length of initial disturbance, and the delay is transmitted to other trains. Primary delay of one train delay affects trains arrival time and develops secondary delay. It is very tedious to compute secondary delay because it is based on primary delay. The reason for the primary delay is a technical failure, delayed boarding and alighting time of passengers, lower than planned running speeds, and bad weather condition [2]. With the use of statistical analysis, the primary train delay can be estimated from current empirical data. Similarly, the use of the analytical stochastic model can predict secondary train delay.

Accurate determination of train delay is the most prominent need for anticipative and proactive real-time control for railway traffic [3]. The traffic controller is responsible for detecting train's arrival time within their location to control the possibility of timetable realization. Accurate evaluation of departure and arrival time is most prominent for avoiding or minimizing delay propagation, maintaining connections, and offering consistent passenger data. One of the difficulties of predicting train delay is the unpredictability and uncertainty of times in railway traffic. The model for controlling real-time traffic control mainly concentrates on overcoming the combinatorial difficulty of train rescheduling [4, 5], delay management [6], crew rescheduling, and rolling stock [7, 8]. The developed methods can resolve complex examples in real time.

In current years, the uncertainty of train time has become a common problem for resolving the issues in railway traffic [9, 10] uses implementable solutions. The uncertainty is denoted by realization's probability distribution. However, current methods consider that train delay is due to a fixed probability distribution. They do not consider real-time data on train position, and delay may cause by equivalent parameters.

The ability to manage risk becomes necessary for transportation operations to enhance service quality and meet passenger's expectation. The delay because of unpredictable factors like facility failures, human error, and bad weather affects railway operation [11]. Those factors result in train delay and train cancelation. These issue constraints the quality of services provided by the railway.

Related to China railway [12], train's average punctuality at the destination is below 90%, whereas delay of five minutes is acceptable [13]. In Norway, the best routes have punctual arrival of 94%, whereas the worst route has an arrival rate of 8% [14]. High-speed railway [15] is the most famous transportation mode for the people of China. Recently, more than 2660 high-speed trains are being operated in China every day. Numerous disruptions affect the operation of the train, which ultimately affects the people who depend on the train for daily purposes. Disruption means divergence from the plan. The sources of disruption are infrastructure breakdown, bad weather, human error, etc. All these disruptions affect the reach of the train to the destination location. This leads to the prediction and monitoring of the train delay prediction system [16–19].

In USA, the railroads have unpredictable runtime [20] because of infrastructure constraints from the network's topology, including single track and sliding [21]. Single track requests for the human dispatcher to segment complex dispatching and passing movements. This is one of the causes for train delay. Whereas in multi-track routes, they run with predictability and regularity and accordingly allow the train to attain their runtime [22]. In addition, US network has a heterogeneous train mix related to physical characteristics and train type. Freight train varies from passenger train, both trains vary with priority and length, in some location both trains will take halt, all these factors affect the predictability of delay experienced by train, so there is need to predict the train delay which becomes necessary [23, 24].

Similarly, [25] Indian railway is the 4th largest network in the world, which has more than 20 thousand trains [26]. In India, most of the trains cannot reach the destination on time because of poor signaling, high congestion, and more counts of trains. In country like India, the majority of the population depend on the railway. Every day, nearly 5.2% of people use the train. Frequent delay by train is quite common in India, where people face lots of problems and inconvenience if predicted in advance will help people using the train to plan their journey regarding their work. In some cases, people cannot book train reservations from their place to destination, so people go for break in the journey. The major issue with the break journey is that if people found their first train, they would ultimately miss the second train. These issues can be overcome only by prediction of train delay which is most prominent and necessary. Train delay prediction provides flexibility to reschedule the journey.

## 2 Empirical Study

The main objective of this survey is to give way for the growth of basic knowledge about the prediction of train delay systems and act as the basement for future work within this field. Here, the researcher discusses various algorithms and tools that are most suitable for determining train delay prediction systems in better way.

## 2.1 Models for Train Delay Prediction System

In the Serbian Railway network, [1] two methods deal with the primary delay of the train. The first model was developed with no historical data on train delays. Instead, it uses data related to the length of train delays and frequency occurring of trains. For this, fuzzy logic is utilized to compute train delay with experience, knowledge, and expertise of railway person who is the main incharge for regulating railway traffic. The fuzzy petri net model's fuzzy system's parameters are determined from data collected from timetable information and personnel interviews to forecast time delays.

**Fuzzy petri nets**. This utilizes fuzzy logic instead of Boolean logic [27]. The concept of fuzziness is applied in petri net by using the mechanism of fuzzy reasoning over petri nets structure.

**Petri nets (PN)**. It is a mathematical tool utilized to simulate and analyze simultaneous systems [28]. PN's theory depends on bipartite graphs. PN also describes discrete distributed systems (DDS). The distributed system is designed with a bipartite directed graph with two sets of nodes they are: Set of place denotes system object or state, whereas set of events represents system's dynamics. Petri net consists of five tuples, $PN = (Q, K, J, W, N)$, where

- $Q = \{Q_1, Q_2, \ldots..Q_n\}$ denotes finite set of places.
- $K = \{K_1, K_2, \ldots..K_n\}$ denotes finite set of events.
- $J$: an input function, $(Q \times K) \rightarrow M$, where $M$ denotes non-negative integer number, where $j(q \times k)$ is number of arcs from place $q$ to transition $k$.
- $W$: denotes output function, $(K \times Q) \rightarrow M$, where $w(k \times q)$ denotes count of arcs from k transition to q place.
- $N$: $Q \rightarrow \{0, 1, 2, 3...\}$ is primary marking assigning to $q$ place with non-negative integer $t$ which means marking place $q$ with $t$ tokens.

Similarly, high-level petri net is modeled with parameters like: HLP = ($Q$, $K$, $C$; Type, Pre, Post, $N_0$), where

- $C$ denotes the non-empty finite set of the non-empty domain where every element of $C$ is known as type.
- $Q \cup K \rightarrow C$ is function utilized to assign types to places and find transition modes.
- Pre, post: TRANS $\rightarrow$ $\mu$ PLACE are pre- and post-mapping with:

$$\text{TRANS} = \{(k, n) | k \epsilon K, n \epsilon \text{Type}(k)\} \tag{1}$$

$$\text{PLACE} = \{(q, g) | q, g \epsilon \text{Type}(q)\} \tag{2}$$

$N_0 \epsilon \mu$ PLACE is multiset called as primary asking of net.

PN and HLPN have similar computational power, but HLPN has more modeling power because of their good structuring facilities. HLPN can deal with complex datasets because every token is assigned with their own colors. Token colors are utilized to build functions and logic expressions. HLPN provides various tools for describing, developing, and analyzing difficult railway system including:

- HLPN efficiently verifies parallel systems by evaluating standards and safety rules for operation of train and analysis of timetable.
- HLPN uses graphical illustration which is simple to understand.
- HLPN can be modified easily due to its modularity.

**Fuzzy logic in fuzzy petri net.** Designing a fuzzy system with HPLN or PN guesses that element is redefined in such way that fuzzy data can be introduced. For the particular feature of fuzzy system, functional and structural elements are defined. The parameters for fuzzy logic system vary for various cases. Similarly, the second model depends on adaptive network fuzzy inference system (ANFIS), which supports the system based on train delay's historical data. The delay is utilized for training the neuro-fuzzy ANFIS system. Once the result is evaluated, then using the fuzzy petri net, ANFIS model is replicated. Here, data is gathered from a real-time system to train and test novel neuro-fuzzy models to compute train delay and launch connection among train parameters and equivalent delays. In the future, fuzzy petri net is enhanced by adding modules regarding train routes to manage the rising conflicts using fuzzy logic to design plans for train dispatch in traffic control. Fuzzy petri net model with train conflict model integrates to resolve conflicts rising in train route similar to the real-time system of train managing conflicts.

**Stochastic train delay prediction in large network.** The stochastic model is utilized for delay forecast and propagation of arrival and departure of the train, which is also suitable for other public transport types [29]. The stochastic model is fairly real, and it is formulated with event graph which designs train schedule and waiting conditions among planned move possibilities. It consists of the train's common waiting policies, profiles regarding driving time on travel arcs based on departure time, and the buffer time of train driving and train stop. On travel arcs, discrete distribution of travel is selected arbitrarily, which is used to test various scenarios, mainly with constraints of the systems. A fundamental property of this method is it uses dynamic updates regarding new delay data. For provided incoming data from an external source, the data is used with the whole network. The event graph is plotted, which is a kind of acyclic graph. Hence, delay propagation is performed in the topological order of events. There is a need to propagate once primary distribution over the event graph starts. Then, new forecast and efficient status data is propagated in the forward cone regarding the equivalent event that is part of the network that can reach it. Here, two kinds of distribution are utilized like one-point distribution used for previously realized events, and another distribution is the arbitrary discrete distribution used for events that lie in the future. Stochastic delay prediction is quietly costly. Experiments are performed on the German train network, and the waiting rule among connecting trains is less than 14 s which is considered to propagate for entire discrete distribution for whole traffic day. Here, prediction is compared with two various days like weekend and midweek with four sets of waiting rules among connecting trains. This work is performed with artificial distribution for train delay prediction. In the future, artificial distribution is replaced by empirical distribution for gathered statistical data for several months.

### 2.2 *Big Data Analytics*

Here, [30] data-driven train delay prediction system (TDPS) is built for larger railway networks using big data, statistical tools, and learning algorithms. This system is mainly built to give useful data regarding traffic management and train dispatching using various techniques and tools. Data is gathered from railway data system regarding historical train movement. Train delay is caused due to various issues like natural disasters or human errors. So in order to predict the train delay, TDPS is built. Here, single train profile is used for predicting time delay. The aim is to detect train delay of that particular time in fixed time in the future $k = k+$ at proceeding checkpoints. Here, most recent historical data is taken into account for forecast. From previous observation, train delay determination issues are mapped to classical time-varying multivariate regression issues.

In conventional framework, set of data $C_m = \{(y_1x_1),(y_2x_2)\ldots\ldots(y_nx_n)\}$ with $y_j \in Y \in \mathbb{R}^c$ and $x_i \in X \in \mathbb{R}$ is taken from automation system. The aim is to discover unknown mode $\wp$: $Y \to X$ through model $\mathcal{M}$: $Y \to X$ is selected by algorithm $\text{Å}_{\mathcal{H}}$ id distinct by set of hyperparameters ($\mathcal{H}$). The accuracy of $\mathcal{M}$ model denotes $\wp$ unknown system can be estimated with reference to various measures of accuracy. Here, issues with train delay prediction are mapped with dynamic multivariate regression model. Let us take train of interest as $K_t$, which is at checkpoint $D_j^{K_t}$ with $j \in \{0,1,\ldots,m_d\}c$. Accordingly, Y input space consists of

- Current day of week
- Boolean value represents whether current day is working day or holiday.
- Train delay, dwell time, and running time for $K_t$ for $k \in [k_0 - \delta^-, k_0]$
- Train delay, dwell time, and running time for other train $K_u$ with $u \neq t$ which is running over same section of railway network during day for $k \in [k_0 - \delta^-, k_0]$.

Regarding to output space $X$, it consists of $D_i^{K_t}$ with $i \in \{j + 1, j + 2,\ldots,m_d\}$ where $k_0 + \delta^+$ is equal to NT of $K_t$ for every $D_i^{K_t}$. At last, $C_m$ has been developed by exploring historical dataset consisting of all data gathered during day in $[c_0 - \Delta^-, c_0]$. Which can easily solve issues with dynamism? Rail traffic management system (RMS) is created to manage inherent difficulties of rail service by combined and holistic view of operational performance, efficiently assuring high-level train operation by giving accurate TDPS to TMS to enhance management of traffic and train dispatching. Here, experiments were conducted on real-time data from RFI. For purpose of validation, RFI provides rights to use six months of data of Italian Railway Network. In the future, exogenous data is used as external resource which affects operation of railway dispatching.

## 2.3 Support Vector Regression

Machine learning concept is utilized for prediction of arrival train delay in Serbian Railway [31]. Support vector machine (SVM) is utilized to analyze train delay and compared it with artificial neural network (ANN). Here, experts opinion is considered as method for fetching data which is related to train arrival delays. For fetching data, interviews are conducted with experts regarding train delay; among them, seven factors are chosen which affect train arrival delay in Serbian Railway. The input variables are:

- Train category for passengers.
- Schedule time for arrival of train at station.
- Influence of infrastructure said by expert
- Percentage of journey finished by distance wise.
- Traveled distance.
- Traveled time.
- Headway.

Figure. 1 shows correlation among train delays and input variables. Let training data be $\{(Y_1, x_1), \ldots..(Y_m, x_m)\}$, where $Y_j$, $x_i$ represent input data and target data, the aim of SVR is predict function $f(Y)$ that $(b)$ has at most $\in$ divergence from target data, and $(a)$ is flat. Here, $\in$-SVR is applied with radial basic function.

$$F(Y, U) = \sum_{i=1}^{m} U_i \exp(-\gamma ||Y - Y_i||^2) \tag{3}$$

where

- $\gamma$ denotes parameter.
- $Y_i$ denotes input vector of training data.
- $U$ denotes unknown parameter which is calculated to reduce the function.

$$Min \frac{1}{2} ||\mathrm{U}||^2 + D \cdot \sum_{j=1}^{m} \max(|x_j - f(Y_j, U)| - \epsilon, 0) \tag{4}$$

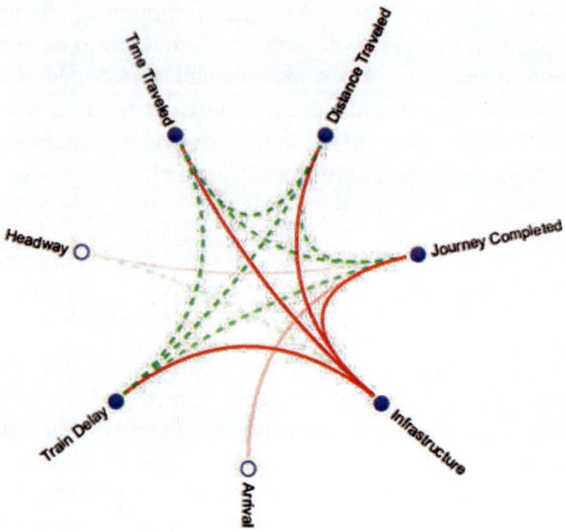


**Fig. 1** Green dashed and red solid links denote positive and negative correlation, respectively, [31]

$U \in \mathbb{R}$

where parameter $D > 0$ controls tradeoff among flatness of $f(\cdot)$ and quantity up to which deviation greater $>\epsilon$ is tolerated. The dual of this optimization issue is resolved with use of technique called convex programming.

Parameter selection and data processing are most prominent for best SVR performance. Primarily, entire variables are scaled linearly with range [0, 1]. Second, $\epsilon$—SVR consists of parameters like $\gamma$, $D$, and $\epsilon$. The values of these parameters are computed to know strength of SVM model. This is performed through combination of grid search method and cross-validation. Here for cross-validation, fivefold is taken and do grid search over the proceeding parameters: $\gamma = 2^b$, $b \in \{2,\ldots,8\}$; $a \in \{-4,\ldots 4\}$; $D = 1.1^d\ (\max_j x_j - \min_j x_j)$, $d \in (-22,\ldots-12\}$.

The generalization power of two model (SVM and ANN) is validated by comparing them by training and testing them on multiple randomly chosen subsets of data. A statistical comparison is performed to validate the performance of model on datasets. Support vector machine with LIBSVM [32] outperforms the ANN by achieving high average rate on testing.

### *2.4 Train Delay Prediction Based on Machine Learning Techniques*

In India, most of the people for their journey depend on the train [33]. So, there is a need to predict the delay time of the train to schedule the procedure works. Here for prediction, previous data regarding train delay is integrated with weather report data to determine train delay. Here, four various machine learning methods are utilized

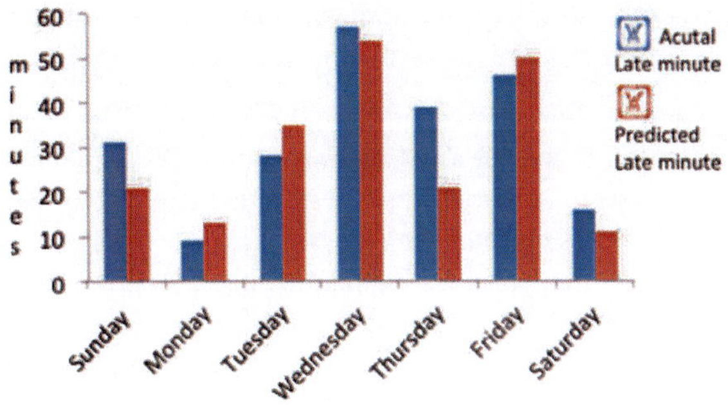


**Fig. 2** Actual late minute versus predicted late minute [33]

for prediction of train delay. Here, Indian train detail is gathered from Indian Railway API. Train delay data is manually gathered from Indian Railway website [34], and weather data is gathered from the open weather map API. Particular region data regarding train delay is combined with weather data collected from that region. This combination serves as a single file. There is close contact among past delays in the specified region and climatic conditions of that region. So hereby Fig. 2, it is clear that past delay and weather data are most prominent for the train delay prediction system. Using the scikit-learn Python library, data is segmented to test data and training data. Seventy percentage data is used for training the model, and 30% of data is used for the purpose of testing. K-fold cross-validation is utilized for estimating various models of machine learning, where k denotes integer. The models of machine learning are as follows:

**Linear Regression (LR) Model.** It uses approximate real-time values that depend on sequence variables. Here, relationship is established among dependent and independent variables by best fit line. This line is called as regression line and is denoted by linear equation

$$Y = M * X + C \tag{5}$$

where

- $Y =$ denotes dependent variable.
- $X =$ denotes independent variable.
- $M =$ represents slope, and C represents intersect.

**Gradient Boosting Regression (GBR) Model.** This is a boosting algorithm utilized to deal with lot of data to detect with higher prediction. This integrates various base estimators in a way to enhance robustness over single estimator. It integrates numerous average or weak predictor to develop strong predictor. This is most suitable for projects based on data science.

**Decision Tree Regression.** It observes object features and trains the model in tree structure to determine data in the future to develop continuous meaningful output, where continuous output means the output is not discrete which means not denoted by discrete value but with known set of the numbers.

**Random Forest Regression**. This is utilized for classification issues and regression types of issues. RF is commonly trained through the bagging method. RF regression is an optimized and convenient model for decision tree. RF consists of additional randomness by searching the best feature while splitting the tree. This results in low bias and high variance results in a good model.

The researcher computes the value for mean absolute error (MAE), root mean square error (RMSE), and R2 for both unknown train and known train acquired from test data shown in Table 1. These values for the journey were computed between actual late minutes and determined late minutes. It is noted from Table 1 below that RF gives better results compared to the other three methods.

Train delay for every day related to detected delay and actual delay is illustrated in below Fig. 2.

From above Table 1, it is clear that 90.01% of accuracy is obtained by LR model, 91.68% of accuracy is obtained by GBR model, 93.71% of accuracy is obtained by DT model, and to the highest of 95.36% accuracy is obtained by RFR model. RFR had best average value for R2 for nearly 8 min as depicted in Fig. 3.

In the future, other deep learning models or algorithms are investigated to predict train delay and to evaluate with the huge dataset to determine their effectiveness. Additionally, some factors like the count of railway track, route information, kind of rail engine, etc., are used for strong prediction to enhance prediction performance.

**Table 1** Comparison of machine learning models [33]

| Machine learning models | $R^2$ | RMSE | MAE | Accuracy (%) |
|---|---|---|---|---|
| RFR | 0.87 | 81.73 | 49.28 | 95.36 |
| DT | 0.69 | 102.47 | 74.31 | 93.71 |
| GBR | 0.51 | 98.27 | 54.29 | 91.68 |
| LR | 0.53 | 80.07 | 49.86 | 90.01 |
| RFR | 0.87 | 81.73 | 49.28 | 95.36 |

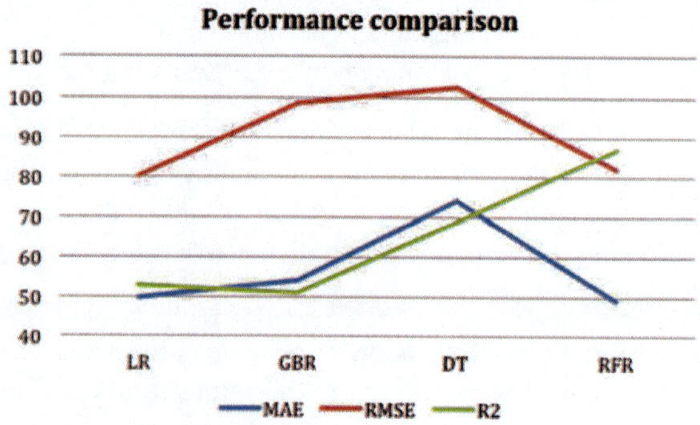


**Fig. 3** Performance line chart [33]

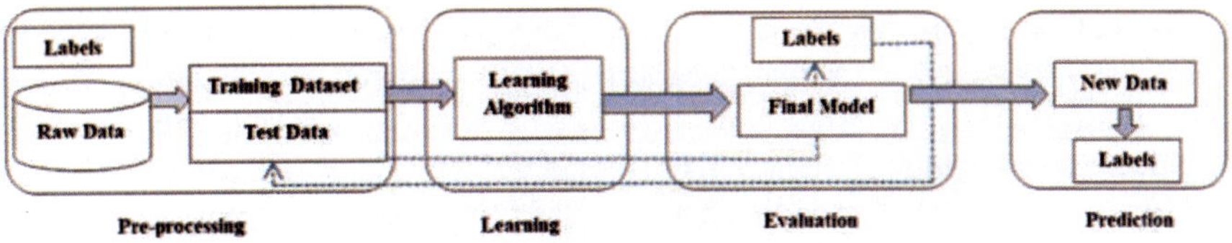


**Fig. 4** Machine learning-based train delay prediction system [33]

**Multivariate Regression**. Here, more than one input variable are utilized for the estimation of the target [35]. Model with two input variables is shown as

$$X = G_0 + G_1.y_1 + G_2.y_2 \tag{6}$$

Common equation for this model with m input variable is shown as:

$$X = G_0 + G_1.y_1 \ldots\ldots\ldots\ldots + G_m.y_m \tag{7}$$

where $G$ denotes regression coefficient, and m denotes count of predictors (Fig. 4).

**Neural Network**. It works by organizing neurons in layer to create expected output [35]. First layer is input layer, whereas final layer is output layer. The layers in between first and last layer is known as hidden layer. Each neuron has activation function. Network parameters are weight and biases of every layer. The aim of NN is to study parameters of network such that detected output is similar as ground truth. Backpropagation feedforward network is utilized for train delay prediction, and it does better prediction with few error.

**Kernel Regression**. For prediction of train delay, two ensemble-based models like context-aware RF and kernel regression models are used. Ensemble model utilizes dissimilar set of models, statistical and simulations dependent to develop prediction for train delay. Context-aware RF is used for network traffic states like current headway, stretch conflicts exogenous weather, and work zone data, whereas kernel regression captures dynamics of train delay. This model is widely used for delay prediction of wide passenger service network. Here, prediction system is used on big data, weather and exogenous data in Germany.

The main concept of KR is to preserve reference catalog for every movement of train. The forecast is developed by sum of weight of the weight of reference catalog, where weight is calculated by measuring similarities among reference set and train of interest. This method is also used to for bus movements [36]. For the case of train movement, reference set is defined for every single-ordered set of station that is $\{T_1, T_2, T_3, \ldots\ldots\ldots\ldots T_m\}$. Every trajectory in reference set N is represented by arrival departure pair for ordered set of station, that is, $c^n = \{c_1^n(C), c_2^n(B), , c_2^n(C), c_3^n(B), \ldots\ldots c_M^n(B)\}$. Arriving at start station and also departure at end station are ignored. For partial trajectory, c = $\{c_1(C), c_2(B), c_2(C), c_3(B), \ldots\ldots\ldots c_L(B)\}$ of train that has progress till L stop, now forecast is made for downstream station $\widehat{c_{L+g}}$.

Trajectory y, x similarity is measured using Gaussian kernel:

$$\text{Kern}(y, x) = \exp(-||y - x||^2/a) \tag{8}$$

where a is bandwidth parameter which is utilized to control the weight spread in reference set. From forecasting perspective, recent delay observation is most prominent than older observation. To account this, u as window parameter is utilized to restrict two trajectory comparisons. To normalize kernel argument, $\sigma_j r^2$ empirical variance at every station departure/arrival is employed. With respect to delay in arrival departure, the kernel weight is calculated as:

$$\text{Kern}(\text{c}, c^n) = \exp(-\frac{1}{a}\sum_{j=u}^{L}\sum_{r\in[B,C]}\frac{(c_j(r) - c_j^n(r))^2}{\sigma_j(r)^2}) \tag{9}$$

Forecast delay is calculated by:

$$\widehat{c_{L+g}}(r) = c_L + \frac{\sum_{n\in N}\text{kern}(c, c_n)(c_{L+g}(r) - c_L^n)}{\sum_{n\in N}\text{kern}(c, c_n)} \tag{10}$$

Delay was denoted in relative or absolute terms. Three various mechanisms are tested to denote trajectory as illustrated in Fig. 5.

The first part depends on travel time, second part depends on delays, third part depends on additional delay that is delay ensue since before stop. Kernel depends on additional delays, and delays were significantly outperformed those depend on travel time. The procedure depends on additional delay states that reference set was needed to recalculate every time by itself which was observed. This computational was too expensive, and final employed model depends on delays.

There are few practical concerns necessary to address. The kernel model can be utilized only with operational train. Kernel is tested for non-operational train with the dispatch time in the future, but the performance is not much better. There are several kernel reference catalogs for every service. For infrequent train, catalog gives low quality for forecast. Threshold-dependent heuristic was utilized to discard forecast produced by reference catalogs. Additionally, two variants of model are tested, one to

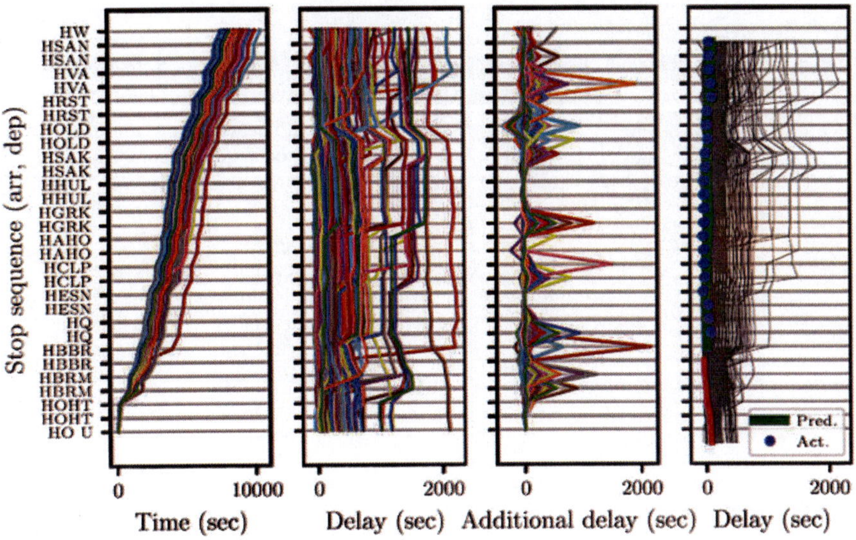


**Fig. 5** Kernel plot for sample 383 trajectories

compute similarity of trajectory depending on departure/arrival message, and another one for intermediate passing message, which is produced at control points. Addition of extra data does not contribute to accuracy of forecast. This is partially because of higher empirical variance noted at intermediate control points (Table 2).

## 3 Conclusion and Future

Railway system all over world faces many issues in detecting train delay. Train delay is the foremost issues in railway network throughout the world. There are some many factors like bad weather, exogenous data, railway asset condition, infrastructure conflicts, human errors, etc. which is one of the main causes for delay in train arrival. Here, study is made on several paper related to train delay prediction system with various methods and algorithm with different kinds of datasets related to their particular region. This prediction is much useful for the passenger to schedule their task by knowing their delay time of train. This survey supports researcher to make research on train delay prediction system with various methods and algorithms.

**Table 2** Train delay prediction models

| Ref. No | Tools/methodology | Dataset | Conclusion | Research gap or future work |
|---|---|---|---|---|
| [2] | Effective delay propagation (EDP) algorithm depends on time event graph | Dutch National Railway timetable | EDP's efficiency enables the system to use in real-time application | Automatic computation of optimal dispatching decision |
| [37] | ANN | Dataset: Iranian Railway | For evaluation, multimodal logistic regression and decision tree are used, and it achieves 90% accuracy | It is further enhanced by using metaheuristic methods like hybrid or genetic algorithm |
| [38] | Model based on ANN, multi-layered perceptron, GA-BPNN model | Large quantity of historical data from observed organization | Provide good enough solution for conflicts | Particular circumstances are not noticed in BPNN model for extraction of data. This problem should be investigated in the future. The proposed system takes more time |
| [15] | Zero truncated negative binominal distribution model | Recorded data from HSR in China | Reduce occurrence of disruption | Reduce delayed traffic before and during disruption |
| [39] | N-Markov model, RFR, N-OMLMPF algorithm | Indian railway | Predict late minutes at inline station | For enhancing accuracy of prediction rate, parameters like railway asset condition to be included |
| [40] | Extreme learning machine (ELM), shallow and deep ELM | Historical data regarding train delay | Prediction system achieves better accuracy | Most prominent parameter weather data is missing |
| [41] | Machine learning model | 3-month dataset related to weather report, train schedule, and train delay is taken | The model is useful for passengers, railway operators by giving most accurate prediction result | The model would be executed with recent data in the future |

(continued)

**Table 2** (continued)

| Ref. No | Tools/methodology | Dataset | Conclusion | Research gap or future work |
|---|---|---|---|---|
| [35] | XGBOOST algorithm and SVR algorithm | Real-time operational data of 2018 | Model exhibits sound applicability over period of time depending on SVR and XGBOOST | Delay duration for every train in PD sequence is not found, so it is the future work to determine |

## References


1. Milinković, S., Marković, M. Vesković, S., Ivić, M., Pavlović, N.: A fuzzy petri net model to estimate train delays. In: Simulation modelling practice and theory, vol. 33, pp. 144–157, (2013). https://doi.org/10.1016/j.simpat.2012.12.005
2. Goverde, R.M.: A delay propagation algorithm for large-scale railway traffic networks. In: Transportation Research Part C: Emerging Technologies, vol. 18, no. 3, pp. 269–287, (2010). https://doi.org/10.1016/j.trc.2010.01.002
3. Corman, F., Kecman, P.: Stochastic prediction of train delays in real-time using Bayesian networks. In: Transportation Research Part C: Emerging Technologies, vol. 95, pp. 599–615, (2018). https://doi.org/10.1016/j.trc.2018.08.003
4. Corman, F., D'Ariano, A., Pacciarelli, D., Pranzo, M.: Dispatching and coordination in multi-area railway traffic management. In: Computers & Operations Research, vol. 44, pp. 146–160 (2014). https://doi.org/10.1016/j.cor.2013.11.011
5. Meng, L., Zhou, X.: Simultaneous train rerouting and rescheduling on an N-track network: A model reformulation with network-based cumulative flow variables. In: Transportation Research Part B: Methodological, vol. 67, pp. 208–234 (2014). https://doi.org/10.1016/j.trb.2014.05.005
6. Dollevoet, T., Huisman, D., Kroon, L., Schmidt, M., Schöbel, A.: Delay management including capacities of stations. Transp. Sci. **49**(2), 185–203 (2015). https://doi.org/10.1287/trsc.2013.0506
7. Nielsen, L.K., Kroon, L., Maróti, G.: A rolling horizon approach for disruption management of railway rolling stock. Eur. J. Oper. Res. **220**(2), 496–509 (2012). https://doi.org/10.1016/j.ejor.2012.01.037
8. Potthoff, D., Huisman, D., Desaulniers, G.: Column generation with dynamic duty selection for railway crew rescheduling. Transp. Sci. **44**(4), 493–505 (2010). https://doi.org/10.1287/trsc.1100.0322
9. F. Corman, L. Meng.: A review of online dynamic models and algorithms for railway traffic management. In: IEEE Transactions on Intelligent Transportation Systems, vol. 16, no. 3, pp. 1274–1284 (2014). doi:https://doi.org/10.1109/TITS.2014.2358392
10. Quaglietta, E., Corman, F., Goverde, R.M.: Stability analysis of railway dispatching plans in a stochastic and dynamic environment. J. Rail Transp. Planning Manage. 3(4), 137–149 (2013). https://doi.org/10.1016/j.jrtpm.2013.10.009
11. C. Wen, J. Li, Q. Peng, B. Li, J. Ren.: Predicting high-speed train operation conflicts using workflow nets and triangular fuzzy numbers. In: Proceedings of the Institution of Mechanical Engineers, Part F: Journal of Rail and Rapid Transit, vol. 229, no. 3, pp. 268–279 (2015). https://doi.org/10.1177/0954409713509978
12. Huang, P., Wen, C., Fu, L., Peng, Q., Li, Z.: A hybrid model to improve the train running time prediction ability during high-speed railway disruptions. Saf. Sci. **122**, 104510 (2020). https://doi.org/10.1016/j.ssci.2019.104510

13. Lessan, J., Fu, L., Wen, C., Huang, P., Jiang, C.: Stochastic model of train running time and arrival delay: a case study of Wuhan–Guangzhou high-speed rail. Transp. Res. Rec. **2672**(10), 215–223 (2018). https://doi.org/10.1177/0361198118780830
14. Harris, N.G., Mjøsund, C.S., Haugland, H.: Improving railway performance in Norway. J. Rail Trans. Planning Manage. **3**(4), 172–180 (2013). https://doi.org/10.1016/j.jrtpm.2014.02.002
15. Xu, P., Corman, F., Peng, Q.: Analyzing railway disruptions and their impact on delayed traffic in Chinese high-speed railway. In: IFAC-PapersOnLine, vol. 49, no. 3, pp. 84–89 (2016). https://doi.org/10.1016/j.ifacol.2016.07.015
16. Goverde, R.M., Meng, L.: Advanced monitoring and management information of railway operations. J. Rail Trans. Plann. Manag. 1(2), 69–79 (2011). https://doi.org/10.1016/j.jrtpm.2012.05.001
17. Corman, F., D'ariano, A.: Assessment of advanced dispatching measures for recovering disrupted railway traffic situations. In: Transportation Research Record, vol. 2289, no. 1, pp. 1–9 (2012). https://doi.org/10.3141/2289-01
18. Veelenturf, L.P., Kidd, M.P., Cacchiani, V., Kroon, L.G., Toth, P.: A railway timetable rescheduling approach for handling large-scale disruptions. Transp. Sci. **50**(3), 841–862 (2016). https://doi.org/10.1287/trsc.2015.0618
19. Meng, L., Zhou, X.: Robust single-track train dispatching model under a dynamic and stochastic environment: A scenario-based rolling horizon solution approach. Transp. Res. Part B: Methodological **45**(7), 1080–1102 (2011). https://doi.org/10.1016/j.trb.2011.05.001
20. Barbour, W., Samal, C., Kuppa, S., Dubey, A., Work, D.B.: On the data-driven prediction of arrival times for freight trains on US railroads. In: 21st International Conference on Intelligent Transportation Systems (ITSC). pp. 2289–2296. IEEE (2018). https://doi.org/10.1109/ITSC.2018.8569406
21. Murali, P., Dessouky, M., Ordóñez, F., Palmer, K.: A delay estimation technique for single and double-track railroads. Transp. Res. Part E: Logistics Transp. Rev. **46**(4), 483–495 (2010). https://doi.org/10.1016/j.tre.2009.04.016
22. Sogin, S.L., Lai, Y.-C., Dick, C.T., Barkan, C.P.: Comparison of capacity of single-and double-track rail lines. Transp. Res. Rec. **2374**(1), 111–118 (2013). https://doi.org/10.3141/2374-13
23. Ghofrani, F., He, Q., Goverde, R.M., Liu, X.: Recent applications of big data analytics in railway transportation systems: A survey. Transp. Res. Part C Emerg. Technol. **90**, 226–246 (2018). https://doi.org/10.1016/j.trc.2018.03.010
24. Barbour, W., Mori, J.C. M., Kuppa, S., Work, D.B.: Prediction of arrival times of freight traffic on US railroads using support vector regression. Transp. Res. Part C Emerg. Technol. **93**, 211–227 (2018). https://doi.org/10.1016/J.TRC.2018.05.019
25. Arshad, M., Ahmed, M.: Prediction of Train Delay in Indian Railways through Machine Learning Techniques (2019). https://doi.org/10.26438/ijcse/v7i2.405411s
26. George, S.A., Rangaraj, N.: A performance benchmarking study of Indian Railway zones. Benchmarking: Int. J. (2008). https://doi.org/10.1108/14635770810903178
27. Chen, W.-L., Kan, C.-D., Lin, C.-H., Chen, T.: A rule-based decision-making diagnosis system to evaluate arteriovenous shunt stenosis for hemodialysis treatment of patients using fuzzy petri nets. IEEE J. Biomed. Health Inf. **18**(2), 703–713 (2013). https://doi.org/10.1109/JBHI.2013.2279595
28. Henry, M.H., Layer, R.M., Zaret, D.R.: Coupled Petri nets for computer network risk analysis. Int. J. Crit. Infrastruct. Prot. **3**(2), 67–75 (2010). https://doi.org/10.1016/j.ijcip.2010.05.002
29. Berger, A., Gebhardt, A., Müller-Hannemann, M., Ostrowski, M.: Stochastic delay prediction in large train networks. In: 11th Workshop on Algorithmic Approaches for Transportation Modelling, Optimization, and Systems, 2011: Schloss Dagstuhl-Leibniz-Zentrum fuer Informatik. https://doi.org/10.4230/OASIcs.ATMOS.2011.100
30. Oneto, L. et al.: Train delay prediction systems: a big data analytics perspective. Big data Res. **11**, 54–64 (2018). https://doi.org/10.1016/j.bdr.2017.05.002
31. Marković, N., Milinković, S., Tikhonov, K.S., Schonfeld, P.: Analyzing passenger train arrival delays with support vector regression. Transp. Res. Part C: Emerg. Technol. **56**, 251–262 (2015). https://doi.org/10.1016/j.trc.2015.04.004

32. Chang, C.-C., Lin, C.-J.: LIBSVM: A library for support vector machines. In: ACM Transactions on Intelligent Systems and Technology (TIST), vol. 2, no. 3, pp. 1–27 (2011). https://doi.org/10.1145/1961189.1961199
33. Arshad, M., Ahmed, M.: Train delay estimation in Indian Railways by including weather factors through machine learning techniques. In: Recent Advances in Computer Science and Communications, vol. 12, pp. 1–00 (2019). https://doi.org/10.2174/26662558136661909120095739
34. https://runningstatus.in/
35. Wen, C., Mou, W., Huang, P., Li, Z.: A predictive model of train delays on a railway line. J. Forecast. **39**(3), 470–488 (2020). https://doi.org/10.1002/for.2639
36. Andres, M., Nair, R.: A predictive-control framework to address bus bunching. Transp. Res. Part B: Methodological, **104**, 123–148 (2017). https://doi.org/10.1016/j.trb.2017.06.013
37. Yaghini, M., Khoshraftar, M.M., Seyedabadi, M.: Railway passenger train delay prediction via neural network model. J. Adv. Transp. **47**(3), 355–368 (2013). https://doi.org/10.1002/atr.193
38. Hu, J.: Application of artificial neuron network in analysis of Railway delays. Open J. Soc. Sci. **4**(11), 59 (2016). https://doi.org/10.4236/jss.2016.411005
39. Gaurav, R., Srivastava, B.: Estimating train delays in a large rail network using a zero shot markov model. In: 21st International Conference on Intelligent Transportation Systems (ITSC), 2018, pp. 1221–1226. IEEE. https://doi.org/10.1109/ITSC.2018.8570014
40. Satyakrishna, J., Sagar, R.: Train delay prediction systems using big data analytics. Int. J. Innovative Res. Comput. Commun. Eng. **6**(3) (2018)
41. Wang, P., Zhang, Q.-P.: Train delay analysis and prediction based on big data fusion. Transp. Saf. Environ. **1**(1), 79–88 (2019). https://doi.org/10.1093/tse/tdy001

# Valence of Emotion Recognition Using EEG

**Avinash L. Tandle**

**Abstract** Affective computing requires a sound algorithm that can distinguish, evaluate, process and simulate human affects. This article proposes frontal theta asymmetry models which quantify the valence of evoked emotion due to musical stimulus using features frontal theta power asymmetry of the participant, and Appreciator and Non-Appreciator are the self-responses given to the stimulus as labels. Each model classified using SVM and validated the performances using various performance metrics of machine learning algorithms. The proposed models total emotion valence index theta (TEVI$\theta$) and emotion valence index frontal F78 (EVIF78$\theta$) perform uniformly outstanding accuracy of model TEVI, EVIF78 89.66% and 88.22%, respectively. Findings uncover the connection of neuronal and self-responses to the musical stimulus of subjects. Appreciator and Non-Appreciator of melodic boost have a novel pattern on fronto anterior regions. The outcome of study can be implicated for various engineering and clinical application.

**Keywords** EEG · Emotion · SVM

## 1 Introduction

Affective computing is an assortment of computational methods that use machine learning on biological data to predict the emotional and psychological response of humans [1]. Affective computing has enormous applications in the fields of engineering, health, entertainment, human interaction and marketing [1]. Affective computing by measuring the neuronal activity of evoked emotions accurate rather than measuring physiological responses due to stimulus [1]. EEG is the most suitable modality because it directly maps the neuronal activity outstanding temporal resolution 1 msec and good spatial resolution 10 mm, and EEG is an appropriate to carry stimulus-based experiment as it is [2]

A. L. Tandle (✉)
MPSTME Mumbai Campus, NMIMS University, Mumbai, India
e-mail: avinash.tandle@nmims.edu
URL: http://www.nmims.edu/



D. Gupta et al. (eds.), *Pattern Recognition and Data Analysis with Applications*, Lecture Notes in Electrical Engineering 888,
https://doi.org/10.1007/978-981-19-1520-8_6

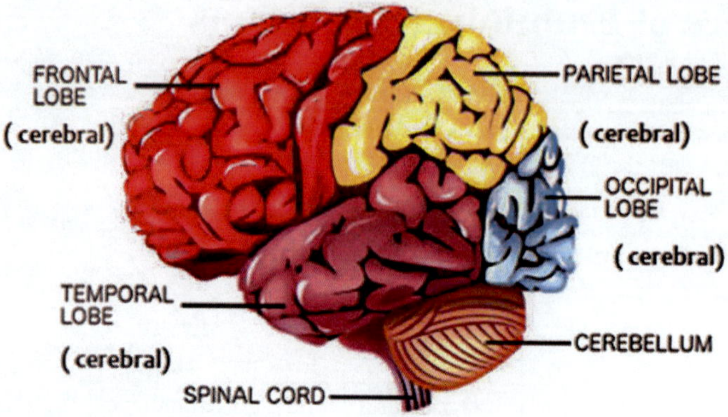


**Fig. 1** Functional organization of the human brain adopted from [5]

- Non-ionizing
- Simple to operate, handy
- Silent
- No fear of close place
- Relatively low cost
- Stimulation experimentation can be easily design
- Easy to design HCI applications

Prior to understanding EEG signals, it is required to comprehend the functional organization of the brain. Cerebrum, cerebellum and brain stem are the three parts of the human brain refer Fig. 1. The human cerebellum divided into four part such as frontal lobe, parietal lobe, occipital lobe and temporal lobe; each lobe associated with some mental functions such as frontal lobe associated with emotion processing, critical thinking, temporal lobe associated with hearing and memory, occipital lobe take cares of vision related tasks and parietal lobe associated with taste and pain [3, 4]. The adult brain has average 100 billions neurons [6]. Each neuron communicates with other neurons to process the stimulus by means of electrical and chemical signal. The electrical communication of neuron generates the oscillation this oscillation called brain wave or EEG signal. The frequency ranges of the signal of these signal are 0.5–100 Hz, while amplitude ranges 10–100$\mu$V[7]. Refer Table 1 for functional and electrical characteristics of these waves.

## 2 Related Literature

The present article is addressing the research gaps mentioned in the literature review article refer article [8] for detail literature review.

For the pleasant musical stimulus, left frontal EEG activity raised, and for unpleasant musical stimulus, right frontal EEG activity raised, and the author also found EEG asymmetry distinguishes intensity of emotions [9], for the various musical stimuli like

**Table 1** Electrical characters tics of brain waves

| Brain wave | Freq. (Hz)/Amp. (μV) | Mental function |
|---|---|---|
| $\delta$ | 0–4<br>10-=100 | • Oblivion<br>• During a profound dreamless sleep |
| $\theta$ | 4–8<br>10–50 | • Subconscious mind<br>• Focused attention<br>• Emotion processing |
| $\alpha$ | 8–12<br>5–25 | • Calm mental state |
| $\beta$ | 12–30<br>0.1–1 | • Hyper focused cognition |
| $\gamma$ | 30–99<br>$\ll 0.1$ | • Hyper brain activity<br>• Memory processing |

classical music, jazz, rock-pop and natural sounds. The author discovered positive emotional attributions were associated with an incriminating left temporal activation, negative emotion by increasing the right fronto-temporal cortex. The author also discovered the activation was more in female compared to male [10]. Charming and upsetting emotions were evoked by happy and sad musical stimuli; author discovered happy music was related with increase in frontal mid-line $\theta$ power [11]. The author explored the association of EEG signal and music evoked emotion responses using four musical excerpt [12]. Author investigated frontal theta asymmetry using stimulus *Raag Bhairavi* [13].

## 3 Methods

The experimental approach adopted for study as shown in Fig. 2. *Raag Bhairavi* instrumental classical music is unfamiliar music, unfamiliar stimulus and most appropriate for the building of an emotion classification and identification system [14]. In related literature, various of emotions are considered for emotion classification. Some emotions get overlapped with other emotion when higher number of emotion considered [15]. The experiment started with ethics committee permission from Dr. R. N Cooper Municipal General Hospital. The participants selected are normal right-handed subject; mostly, clinical and engineering staff normalcy and handedness [16] are confirmed by clinical supervisors. Clinical supervision satisfies major recommendation of author [17]. The participant selection, stimulus selection, duration of stimulus, EEG recording and the classification of Appreciator and Non-Appreciator of subjects on the basis of self-responses to the stimulus as mentioned reference [13, 18]. Refer [8, 13, 18] for experimental protocol experimental recording, artifact removal and feature extraction steps.

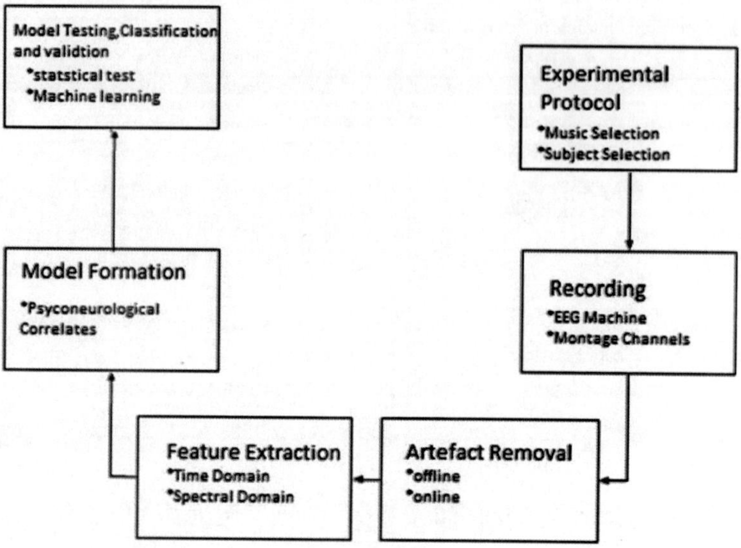


**Fig. 2** Experimental approach

## 4 Model Formation

The models are formed by finding stated in the literature survey and approach adopted [8] using frontal electrodes Fp1, F7 and F3, while Fp2, F8 and F4 on the left and right hemisphere in a referential montage taking A1 and A2 as reference electrodes, respectively, as mentioned in reference [13, 18].

$x(n)$, $n = 0, 1, 2, 3 \ldots. N$ are the sampled values of filtered EEG from left hemisphere and right hemisphere during listening to musical stimulus for 10 min. Refer Eq.(1) for Fourier transform of spectral theta band. Equations 2 and 3 represent total frontal theta power(TF$\theta$) left and right hemispheres Eqs. 4, 5, 6 and 7 represent four models of emotion valence index of frontal theta asymmetry [19]. Using equations, theta power for all subjects is computed.

$$x(\theta) = \sum_{i=1}^{n} x(n)e^{\frac{-2\pi\theta n}{N}} \tag{1}$$

$$TF\theta PL = \theta P(Fp1A1) + \theta P(F7A1) + \theta P(F3A1) \tag{2}$$

$$TF\theta PR = \theta P(Fp2A2) + \theta P(F8A2) + \theta P(F4A2) \tag{3}$$

$$TEVI\theta = [\log_{10}(TF\theta PL) - (\log_{10}(TF\theta PR))] \tag{4}$$

$$EVIFp1p2\theta = [\log_{10}\theta PF_{p1A1} - \log_{10}\theta PF_{p2A2}] \tag{5}$$

$$EVIF78\theta = [\log_{10}\theta PF_{7A1} - \log_{10}\theta PF_{8A2}] \tag{6}$$

$$EVIF34\theta = [\log_{10}\theta PF_{3A1} - \log_{10}\theta PF_{4A2}] \tag{7}$$

## 5 Machine Learning Classifier

For the evoked emotion classification, supervised SVM algorithm is most suitable. The kernel tricks in SVM transform the data to find the optimal boundary to detect possible output. Nonlinear kernel tricks can incarceration more intricate relationship between data points [20]. The following feature makes it more suitable to test and validate the model

- Prediction speed is very high
- High training speed
- Great accuracy
- Results are interpretable
- Suitable for small dataset

All models are classified using SVM exerting linear, radial, polynomial and sigmoid kernels described above. For classification dependent variables, frontal theta activity on the left and right hemisphere and feature vector Appreciator and Non-Appreciator of musical stimulus grouped from the Likert scale of self-responses of all participants converetd as Appreciator and Non-Appreciator refer Figs. 3, 4, 5 and 6 for classification of SVM using various kernels for models TAEVI$\theta$, EVIF78$\theta$, EVIF34$\theta$ and EVIFp12$\theta$.

## 6 Model Performance

Model can be assessed using various metrics for example true positive rate, specificity, precision, false discovery rate, $F1$ score, Mathew correlation coefficient (MCC), Youden index, ROC; many metrics are biased metrics such sensitivity, specificity accuracy [8, 21]. ROC is best metric as it gives trade-off between sensitivity and specificity; ROC is best metric to build the model [22–24]. Equation (8) represents formula of confusion matrix, whereas Eqs. (9)–(19) represent true positive rate (TPR), true negative rate (TNR), precision (PPV), negative prediction Value (NPV), false positive rate (FPV), false discovery rate (FDR), F1 score, Mathew correlation coefficient (MCC), accuracy, and Youden index (YI) or markedness, respectively.

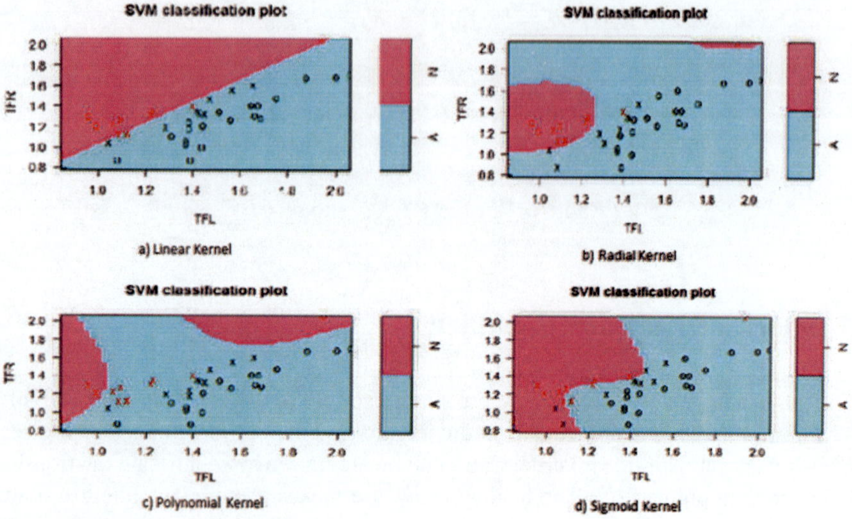


**Fig. 3** TAEVI$\theta$ classification using SVM

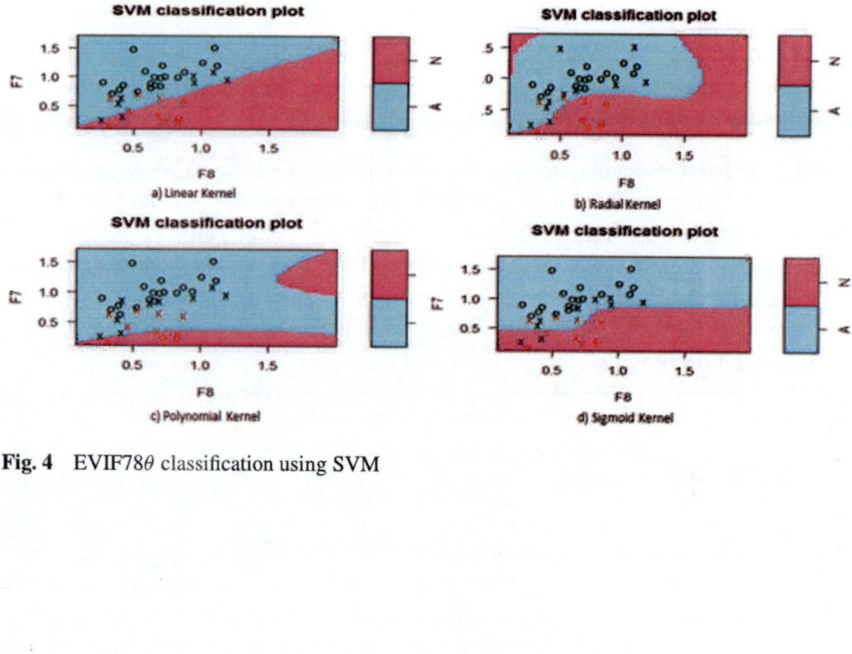


**Fig. 4** EVIF78$\theta$ classification using SVM

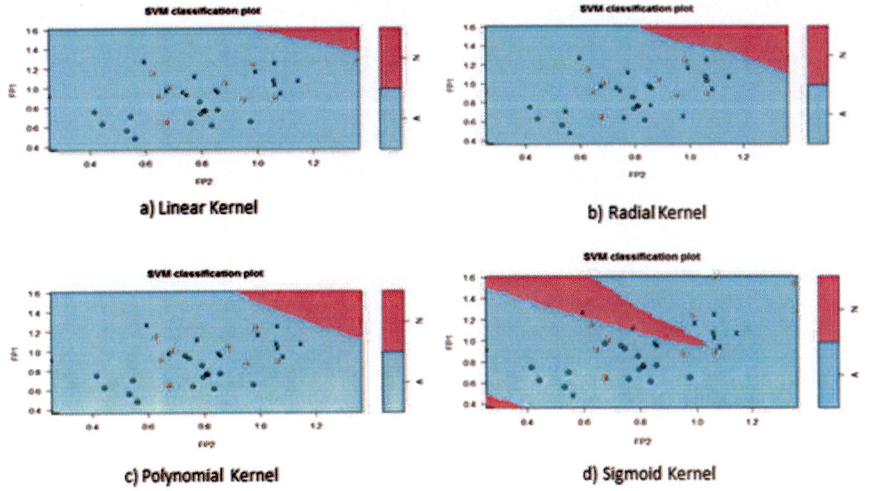


**Fig. 5** EVIF34$\theta$ classification using SVM

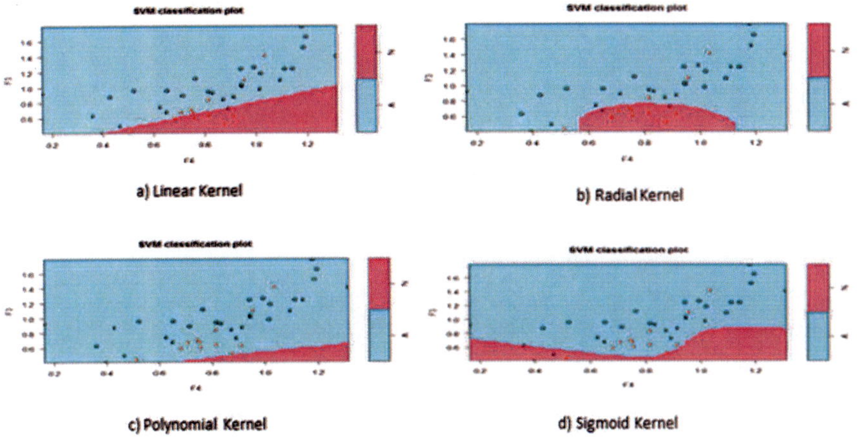


**Fig. 6** EVIFp12$\theta$ classification using SVM

$$Cp = \begin{bmatrix} \text{Tp Fp} \\ \text{Fn Tn} \end{bmatrix} \tag{8}$$

$$\text{TPR} = \frac{\text{Tp}}{\text{Tp} + \text{Fn}} \tag{9}$$

$$\text{TNR} = \frac{\text{Tn}}{\text{Tn}} + \text{Fp} \tag{10}$$

$$\text{PPV} = \frac{\text{Tp}}{\text{Tp} + \text{Fp}} \tag{11}$$

$$\text{NPV} = \frac{\text{Tn}}{\text{Tn}} + \text{Fn} \tag{12}$$

$$\text{FPR} = \frac{\text{Fp}}{\text{Fp} + \text{Tn}} \tag{13}$$

$$\text{FDR} = \frac{\text{Fp}}{\text{Fp} + \text{Tp}} \tag{14}$$

$$F1 = 2\frac{[\text{PPV}] \times [\text{TPR}]}{[\text{PPV}] + [\text{TPR}]} \tag{15}$$

$$M = \frac{(\text{Tp} \times \text{Tn}) - (\text{Fp} \times \text{Fn})}{\sqrt{(\text{Tp} + \text{Fp})(\text{Tp} + \text{Fn})(\text{Tn} + \text{Fp})(\text{Tn} + \text{Fn})}} \tag{16}$$

$$\text{Accuracy} = \frac{\text{Tp} + \text{Tn}}{\text{Tp} + \text{Tn} + \text{Fp} + \text{Fn}} \tag{17}$$

$$Y\text{-Index} = \text{TPR} + \text{TNR} - 1 \tag{18}$$

$$\text{Markedness} = \text{PPV} + \text{NPV} - 1 \tag{19}$$

## 7 Discussions and Implications

The emotional valence index is plotted in Fig. 7 and Table 1 of Ref. [18] for all 41 subjects from the figure. Subjects 1 and 2 are Appreciators as per self-report (Likert scale 4) and per EVI of both $0.8442\,\mu V^2$/Hz and $0.3518\mu V^2$/Hz, respectively For Subjects 3 and 4 are Non-Appreciators with EVI $-0.3469\mu V^2$/Hz and -0.0218 $\mu V^2$/Hz. The EVI of most of the Appreciator is positive, while for Non-Appreciator is negative. Results of SVM classifiers for radial, linear, polynomial and sigmoid kernels presented in table classification using all model perform well in radial kernel model TEVI$\theta$ and EVIF78$\theta$ perform uniformly outstanding, whereas model EVIF34$\theta$ perform moderately and a model EVIFp12$\theta$ designated poor performance. ROC of three models shown in figure area under the curve (AUC) for three models TEVI$\theta$, EVIF78$\theta$ EVIFp12$\theta$ and EVIF34$\theta$ are 89.66%, 88.22%, 75% and 78.74%, respectively. Model TEVI$\theta$,EVIF78$\theta$ performing substantially identically (Table 2). Our findings are consistent with results in the literature. For the Appreciator of music, theta power is high on the left frontal hemisphere, and for Non-Appreciator, theta power right frontal region is increased. The formed valence index for Appreciator is positive, while for Non-Appreciator is negative. The valence index is not correlating with the Likert scale but corresponding with formed groups, i.e., Appreciator and Non-Appreciator. Findings uncover the connection of neuronal and self-reactions of subjects Appreciator and Non-Appreciator of melodic boost that has a novel pattern. Positive emotions evoked because of subjects preferring prompts approach, while negative emotions evoked due aversions of jolt drives evasion [25]. In the present research, the evoked emotion due to Indian instrumental traditional music recommends is more noteworthy on terminal F7-F8. This result will help in evaluating evoked emotion utilizing fewer electrode by shaping a versatile gadget computational speed will increment due to the decreased number of electrode, and accuracy of proposed system [19] will improve due to fewer artifacts which the primary requirements of various clinical and engineering applications.

**Implication Engineering** The outcome of study reveals the correlation of neuronal responses and evoked emotion. The finding also leads to electrode reduced system

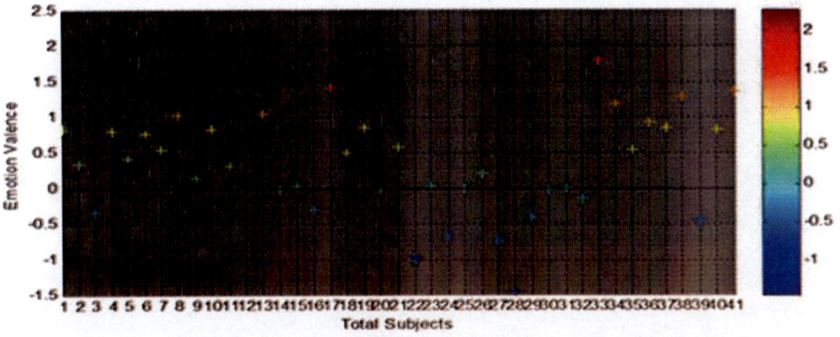


**Fig. 7** Valence of emotion total frontal $\theta$ asymmetry

**Table 2** Performance of models using various evaluating attributes

| Model | Ker. | TPR | TNR. | PPV. | NPV | PPV | FDR | FNR | FPR | Acc. | F1 | MCC | YI | Marked |
|---|---|---|---|---|---|---|---|---|---|---|---|---|---|---|
| TAEVI$\theta$ | Rad | 97 | 83 | 93 | 91 | 93 | 6 | 3 | 16 | 93 | 95 | 0.86 | 0.79 | 0.84 |
| | Lin | 100 | 75 | 91 | 100 | 91 | 9 | 0 | 25 | 93 | 95 | 0.82 | 0.75 | 0.91 |
| | Pol | 97 | 33 | 78 | 80 | 78 | 22 | 3 | 67 | 78 | 86 | 0.46 | 0.30 | 0.57 |
| | Sig | 83 | 75 | 88 | 64 | 88 | 11 | 17 | 25 | 80 | 85 | 0.69 | 0.58 | 0.53 |
| EVIF78$\theta$ | Rad | 97 | 83 | 93 | 91 | 93 | 7 | 3 | 17 | 93 | 95 | 0.86 | 0.80 | 0.84 |
| | Lin | 90 | 67 | 87 | 73 | 87 | 13 | 10 | 33 | 83 | 88 | 0.67 | 0.56 | 0.59 |
| | Pol | 100 | 58 | 85 | 100 | 85 | 15 | 0 | 42 | 88 | 92 | 0.70 | 0.58 | 0.85 |
| | Sig | 90 | 67 | 87 | 73 | 87 | 13 | 10 | 33 | 83 | 88 | 0.68 | 0.56 | 0.59 |
| EVIFp12$\theta$ | Rad | 100 | 17 | 74 | 100 | 74 | 26 | 0 | 83 | 76 | 85 | 0.35 | 0.16 | 0.74 |
| | Lin | 100 | 17 | 74 | 100 | 74 | 26 | 0 | 83 | 76 | 85 | 0.35 | 0.16 | 0.74 |
| | Pol | 100 | 25 | 76 | 100 | 76 | 24 | 0 | 75 | 78 | 87 | 0.43 | 0.25 | 0.76 |
| | Sig | 86 | 8 | 69 | 20 | 69 | 31 | 14 | 92 | 63 | 77 | 0.13 | 0.06 | 0.12 |
| EVIF34$\theta$ | Rad | 93 | 67 | 87 | 80 | 87 | 12 | 7 | 33 | 85 | 90 | 0.70 | 0.59 | 0.67 |
| | Lin | 93 | 50 | 82 | 75 | 82 | 19 | 7 | 50 | 80.5 | 87 | 0.57 | 0.43 | 0.57 |
| | Pol | 100 | 0 | 70 | 0 | 70 | 29 | 0 | 100 | 70.7 | 83 | 0 | 0 | 0 |
| | Sig | 90 | 17 | 72 | 40 | 72 | 28 | 10 | 83 | 80 | 68 | 0.13 | 0.06 | 0.12 |

All values are rounded off, Ker.—kernel; Rad.—Radial; Lin—Linear; Pol.—Polynomial; Sig.—Sigmoid such a system suable for neuromarketing application. In the current neuromarketing research, for the most part, uses event-related potential ERP [26]. ERPs are all around befitted to examine issues about the rate of neural activity and less very much suited to inquire about inquiries concerning the area of such occasion [26].

**Implication Clinical** In the subjects of mental depression, focused attention got impaired. Musically stimulated frontal theta asymmetry can be used to quantify the depression. Frontal theta asymmetry biomarker of depression [27] as frontal alpha [28]. The present system of depression diagnosis is entirely qualitative and qualitative system which is error-prone. Music-evoked quantitative depression diagnosis system will be the best solution.

## 8 Future Work

EOG artifact is the most challenging artifact to remove from EEG; an automated EOG artifact removal algorithm should be investigated. This study only focuses mainly on mental attentiveness function, during music listening to music psychological process of memory processing and takes this needs to investigate by formulating a psychoneurological hypothesis. Non-supervised machine learning algorithm should be used to discover the neuronal reason for many psychological processes that take place during music processing. For musical stimulus how the brain of the subjects with dementia, Alzheimer and other affective disorder needs to investigate by creative psychoneurological models testing and validating using various machine learning algorithms. Non-supervised machine learning algorithm should be used to discover biological reason of many psychological processes that take place during music processing.

## 9 Conclusion

Models EVIF$\theta$F78, EVIF$\theta$F34 and EVIF$\theta$Fp12 perform best in the radial kernel of SVM indicating nonlinearity data of on that region, whereas TAEVI$\theta$ equally performs well in linear kernel reporting on the overall frontal region data which is linear interprets emotion modulated with stimulus on the frontal region. The outcome of the model-based study narrows down to the specific location (EVIF$\theta$F78) of the frontal region which signifies in electrode reduction, due to electrode reduction complexity, artifact, will be drastically reduced, whereas computation speed will increase making more reliable system for quantifying evoked emotion for various engineering and clinical applications.

## References


1. Picard, R.W.: Affective computing. M.I.T Media Laboratory Perceptual Computing Section Technical Report No. 321
2. Lystad, R.P., Pollard, H.: Functional neuroimaging: a brief overview and feasibility for use in chiropractic research. J. Canad. Chiropractic Association **53**(1), 59–72. 0008-3194 (2009)
3. Gray, H.: Gray's Anatomy. Random House, The Classic Collector's Edition. New York (1988)
4. Chen, P.: Principles of biological science (2011)
5. Alotaiby, T., El-Samie, F.E.A., Alshebeili, S.A., Ahmad, I.: A review of channel selection algorithms for EEG signal processing. J. Adv. Signal Process. **1** (2015). https://doi.org/10.1186/s13634-015-0251-9
6. Patel, N.D.: An EEG-based dual-channel imaginary motion classification for brain computer interface. Master of Engineering Science, Thesis, Lamar University (2011)
7. Kamel, N., Malik, A.S.: EEG/ERP Analysis Methods and Applications. CRC Press (2014)
8. Tandle, A.L., Joshi, M.S., Dharmadhikari, A.S., Jaiswal, S.V.: Mental state and emotion detection from musically stimulated EEG. Brain Inf. **5**(2), 14 (2018). https://doi.org/10.1186/s40708-018-0092-z
9. Schmidt, L., Laurel, J.: Frontal brain electrical activity (EEG) distinguishes valence and intensity of musical emotions. Cognit. Emot. **15**(4), 487–500 (2001). https://doi.org/10.1080/02699930126048
10. Altenmüller, E., Schürmann, K., Lim, V.K., Parlitz, D.: Hits to the left, flops to the right: different emotions during listening to music are reflected in cortical lateralisation patterns. Neuropsychologia **40**(13), 2242–56 (2002). https://doi.org/10.1016/s0028-3932(02)00107-0
11. Sammler, D., Grigutsch, M., Fritz. T., Koelsch, S.: Music and emotion: electrophysiological correlates of the processing of pleasant and unpleasant music. Psychophysiology **44**(2), 293–304 (2007). https://doi.org/10.1111/j.1469-8986.2007.00497
12. Lin, Y.P., Wang, C.H., Wu, T.L., Jeng, S.K., Chen, J.H.: Support vector machine for EEG signal classification during listening to emotional music., In: Proceedings of the 2008 IEEE 10th Workshop on Multimedia Signal Processing **15**(4), 127–130 (2008). https://doi.org/10.1109/MMSP.2008.4665061
13. Tandle, A., Jog, N., Dharmadhikari, A., Jaiswal, S., Sawant, V.: Study of valence of musical emotions and its laterality evoked by instrumental Indian classical music : an EEG study International Conference on Communication and Signal Processing (ICCSP), pp. 276–280 (2016). https://doi.org/10.1109/ICCSP.2016.7754149
14. Thammasan, N., Moriyama, K.: Familiarity effects in EEG-based emotion recognition. Brain Informatics **4**(1), 39–50 (2017). https://doi.org/10.1007/s40708-016-0051-5
15. Jatupaiboon, N., Pan-ngum, S., Israsena, P.: Real-time EEG-based happiness detection system. Hindawi Publishing Corp. Sci. World J. (2013). https://doi.org/10.1155/2013/618649
16. Oldfield, R.C.: The assessment and analysis of handedness: the Edinburgh inventory. Neuropsychologia **9**(1), 97–113 (1971)
17. Brouwer, A.M. ,Zander, T.O., van Erp, J.B.F., Korteling, J.E., Bronkhorst, A.W.: Using neurophysiological signals that reflect cognitive or affective state: six recommendations to avoid common pitfalls. Front. Neuroscince **9**, 136 (2015). https://doi.org/10.3389/fnins.2015.00136
18. Tandle, A., Jog, N., Dharmadhikari, A., Jaiswal, S..: Estimation of valence of emotion from musically stimulated EEG using frontal theta asymmetry. In: 12th International Conference on Natural Computation, Fuzzy Systems and Knowledge Discovery (ICNCFSKD) (2016). https://doi.org/10.1109/FSKD.2016.7603152
19. Tandle, A., Dikshant, S., Seema, S.: Methods of neuromarketing and implication of the frontal theta asymmetry induced due to musical stimulus as choice modeling. Procedia Comput. Sci. **132**, 55–67 (2018). https://doi.org/10.1016/j.procs.2018.05.059
20. Home Page,https://community.alteryx.com/t5/Data-Science-Blog/Why-use-SVM/ba-p/138440. Last accessed Sept 2018
21. Powers, D.M.W.: Evaluation: from precision, recall and F-measure to ROC. Informedness, markedness and correlation. J. Mach. Learn. Technol. **2**(1), 37–63 (2011)

22. Fawcett, T.: An introduction to ROC analysis. Pattern Recogn. Lett. **27**(8), 861–874 (2006). https://doi.org/10.1016/j.patrec.2005.10.010
23. Home Page, https://wikipedia.org/wiki/Precision. Last accessed Sept 2018
24. Home Page, https://towardsdatascience.com/choosing-the-right-metric-for-evaluating-machine-learning-models-part-2-86d5649a5428 Last accessed Sept 2018
25. Davidson, R.J. (eds.): The Asymmetrical Brain. MIT Press, Cambridge, pp. 565–615
26. Luck, S.: An introduction event related potential Techniques (2005)
27. Dharmadhikari, A., Tandle, A., Jaiswal, S., Sawant, V., Vahia, V., Jog, N.: Frontal theta asymmetry as a biomarker of depression. East Asian Arch Psychiatry **28**, 17–22 (2018). https://doi.org/10.12809/eaap181705
28. Fachner, J., Gold, C., Ala-ruona, E., Punkanen, M., Erkkilä, J.: Depression and music therapy treatment—clinical validity and reliability of EEG alpha asymmetry and frontal midline theta: three case studies EEG assessment. Music Therapy, ICMPC (2010), vol. 11, 11–18

# A Deep Learning-Based Approach for an Automated Brain Tumor Segmentation in MR Images

**Puranam Revanth Kumar, Amogh Katti, Sachi Nandan Mohanty, and Surender Nath Senapati**

**Abstract** Brain tumor classification is done by biopsy, which is not normally performed due to conclusive brain surgery without invasive interventions, improving technologies and machine learning can help radiologists detect tumors. MRIs are a commonly used imaging technique for the study of these tumors, but the vast volume of knowledge given by MRI prevents manual breakup over an acceptable time span, reducing the use of accurate quantitative calculations in clinical practice. This paper introduces an approach to program division based on convolution neural networks (CNN). The proposed work uses a 300 MR image dataset from Kaggle with 70% training and 30% testing. In addition to having a positive impact on overfitting, Kernel $3 \times 3$ enables the construction of a deeper architecture, provided the lower number of weights in the network. The application of force standardization was also discussed here as a pre-preparation step, which has proven extremely viable in MRI images for brain tumor division, despite its lack of regularity in CNN division strategy and data extension. Detection of accurate cancer cells in high-density areas that are impure is difficult. The extraction and identification of tumor from brain MRI scans are conducted using the MATLAB tool. The accuracy was 92.50% with good generalization capabilities and good speed of execution, and for medical diagnostic radiologists, the new developed CNN architecture will be an essential decision-making tool.

P. Revanth Kumar (✉)
Department of Electronics and Communication, IcfaiTech (Faculty of Science and Technology), Hyderabad, India
e-mail: revanth123451.rk@gmail.com

A. Katti
Department of Computer Science and Engineering, IcfaiTech (Faculty of Science and Technology), Hyderabad, India

S. Nandan Mohanty
Department of Computer and Information Technology, College of Engineering, Pune, India

S. Nath Senapati
Department of Radiation Oncology, Acharya Harihar Regional Cancer Centre, Cuttack, India

 
D. Gupta et al. (eds.), *Pattern Recognition and Data Analysis with Applications*, Lecture Notes in Electrical Engineering 888,
https://doi.org/10.1007/978-981-19-1520-8_7

**Keywords** Brain tumor segmentation · Convolutional neural network · Deep learning · MR imaging

## 1 Introduction

The unwanted cell development produces a lump in the brain called brain tumors are formed by the growth of uncontrollable cells in an unregulated manner in the brain [1]. Early diagnosis of cancer can delay death, even though not always. The tumor may be benign, pre-carcinoma, or malignant in comparison with cancer. Benign tumors differ since they are not typically distributed and surgically removed to other organ and tissue [2]. Magnetic resonance imaging (MRI) is widely employed in the diagnosis of brain tumor patients, but it is not practical to make effective use of such imaging methods due to a lack of sufficient means to handle vast volumes of data generated by the image acquisition system. Gliomas, meningiomas, and pituitary tumors comprise all primary brain tumors [3].

Tumors that develop from components of the brain other than nerve cells and blood vessels are commonly referred to as glioma. The membranes around the brain and around the central nervous system are, on the other hand, a source of meningiomas, and lumps in cranium are hypo-physical tumors [3–6]. The only difference is that, among the three tumors, meningiomas are frequently benign and are the most commonly malignant. Unlike meningiomas, which develop slowly, pituitary tumors, though benign, can induce various medical problems [5, 6]. Given the foregoing, distinguishing between all three types of cancers is a critical step in the clinical evaluation process and the final successful treatment of patients.

A number of problems can arise with MRI images [7, 8] such as the intensity of homogeneity or the variation of power ranges between related arrangements and scanners. We use preprocessing processes such as tilt field and power standardization to expel these disservices in the MRI images. MRI is one of the most widely employed ways of effective total representation [9] in tumor therapy.

The rest of the paper is planned as follows: in Sect. 2—literature survey, in Sect. 3—we give information about the methodology, in Sects. 4 and 5—the implementations, performances, and experimental evaluation and investigation of this work followed by the conclusion with recommendations for future work.

## 2 Background and Related Work

### *2.1 Background*

In this section, CNN-based classification methods are presented briefly. The convolution network, which consists of alternating convolution, pooling layers, and several completely connected layers, is a commonly used method for biologic image segmentation. These methods have been loosely segregated into two categories of analysis and decision-making. The first is patched, the latter end to end. The network input is a patch with fixed and odd sizes, and the output is the center pixel class. A sliding window divides the images into patches based on the specification. The network is often trained by means of patches extracted from an image collection, with the same way of dividing the truth of the ground. This style is simple and easy to understand for image segmentation. And through patch-based techniques, the data imbalance can be overcome quickly.

In [10], DCNN was proposed for multimodal image segmentation and three architectures. The patches for these three architectures are 28 × 28, 12 × 12, and 5 × 5, and the kernels for these architectures are 5 × 5, 3 × 3, and 2 × 2, respectively. There are 2, 2, and 1 layers of convolutions. It shows that the DCNN is able to effectively segment brain tumors. In the meantime, it shows that the patch size and the size of the convolution filter affect the results if the brain tumor is patch-based. There are few layers in each of the three structures, the deeper the DCNN, the greater the characteristics. Consequently, [11] is adding extra layers by reducing filter size and pooling layers. The size of the input is 33 × 33 compared with [10], and the size of the kernel is 3 × 3. This network is deeper than the previous network and has enhanced efficiency. In addition, the entire patched segmentation process was introduced in [10].

Havaei et al. [12] proposed a two-way architecture that incorporates the features of various size networks. This study consists of two paths of different sizes of convolution kernels. The characteristics obtained from two directions are concatenated, moving through the convolution layer and the layer of SoftMax. In addition, some cascade architectures based on a two-way architecture and combining different input patches are also available. The input patches in [14, 15] are extracted from the axial, coronary, and sagittal sections of the view. To produce the final results, all three analyses will then be combined. This often takes advantage of the presented contextual details. In addition, each layer's phase size is set to 1. This avoids redundant measurements and during research helps the section of images to be cut by slice. In the end-to-end methods, the success of net training and image segmentation at the image stage was achieved with the combination of completely convoluted or de-convoluted networks. In comparison to the patch-based technique, the end-to-end method will reduce numerous redundant measurements. Most approaches to the segmentation of brain tumors are mainly based on FCN [16], SegNet [17], and U-Net [18]. These approaches are essentially in the form of a devolution. The U-Net has been implemented by [19] to automatically segment brain tumors. A novel 3D segmentation approach based on cross-modality was suggested by [18].

## 2.2 Related Work

The segmentation technique for the brain tumor does not exist perfectly, but several groundbreaking solutions for the automatic segmentation of the tumor are constantly being implemented. The distinction between strength, form, location, and frontiers of the brain tissue varies between individuals and is a major challenge for automated segmentation. The innovative result of deep learning is a signatory to these problems in the segmentation and classification of images. This section will address several profound methods of learning in the automated segmentation of brain tumors used in the processing of MRI data.

In the neural network architecture, small overlapping filters like $3 \times 3$ are used to preserve the larger CNN depth and to learn more about the inputs in every network learning layer. [20] proposed one of those blocks consisting of three blocks of convolution (11 layers deep) and six layers with a $3 \times 3$ filter, two layers of max, followed by three layers fully connected. The authors used prepossession to balance all images' sensitivity by normalizing and filtering sound through calculation of the default deviation and the mean intensity value of all training pictures before network training. The suggested model for the entire tumor was 88% accurate, for the core tumor 83% correct, and for the active tumor 77% correct based on BraTS data.

A new neural network for automatic brain tumors segmentation was proposed in MRI images [21] in three-dimensional confounding. It takes a lot of computing time, but 3D visualization makes it easier of tumor development for radiologists. The $3 \times 3 \times 3$ convolution filters, as well as the batch standardization layers, ReLu, and 3D max pooling layers, are used to reduce the size of feature maps. The 3D inputs are arranged in a 4D volume and are visible in height, width, image channels, and number of modes in four dimensions. In the MRI dataset, 87, 77, and 73% for the whole area of the tumor, the main tumor, and the active zone have been accurate.

The automatic segmentation of brain tumors by teaching separately all methods and integrating post-processing SegNet production has been suggested by [22]. First, standardization and bias field correction inputs are important for eliminating unnecessary errors that boost segmentation efficiency. SegNet is used for the independent training of the four related MRI modes. There are a couple of encoders and decoders in the architecture (down sample) (up sampling). The encoder uses 13 convolution layers with $3 \times 3$ filters, batch standardization layers, ReLU, and max pools with two filters. The decoder also has 13 convolutional layer to complement the corresponding encoder. High-dimensional features extracted from the decoder are fed to the SoftMax layers to classify each pixel type independently. The segmentation technique obtained a precision of 85% for the entire tumor, 81% for the central tumor, and 79% for the enhancement of the tumor.

The patch-based CNN methodology utilizes CNN's intrinsic functionality for the detection of patterns and also performs extremely exact division within the MRI and implemented a cascaded two-way CNN model, which extracted at the same time a big $37 \times 37$ patch size and a minor $19 \times 19$ patch size [23]. The architecture has a variety of learning criteria in order to discourage overfitting, the maxout, and dropout layers used in architecture that may contribute to overfitting during school. The model consists of seven convolution layers of varying filter sizes, allowing the CNN model to understand the features of various sizes and to use the ReLU. The network was trained to end the initial CNN output and the second CNN entry by cascading. In addition to the 3D slicer image package for bias field correction, similar to artifacts that had previously been available, this led to improved segmentation results.

## 3 Methodology

An overview of the proposed system approach is given in Fig. 1. Preprocessing, CNN classification, and post-processing are three main stages.

### 3.1 Preprocessing

The images of MRI are subject to predisposing to field distortion and homogenization. As of today, due to field homogeneity in attractive reverberation images, there are some formulas for correcting uniform power. X-ray pictures are changed by the inclination field bending. This makes the force of similar tissues to fluctuate over the picture. To adjust it we connected the N4ITK technique [2]. This is not enough to guarantee the tissue's dispersion by intimidation. As there will be a range of power, even if the same patient's MRI is taken at different times in the same scanner, the results will be the same. We are thus using the methodology of force standardization in this sense. It makes the patient's challenge and ability more comparatively cross-cutting and receiving. The N3 (nonparametric non-coherence uniformity) algorithm is the influential force of the homogeneity correction technique [11]. It is iterative and seeks a smooth field of propagation that increases the strong repetition of tissue diffusion control content. It is a properly programmed methodology. What is more, it

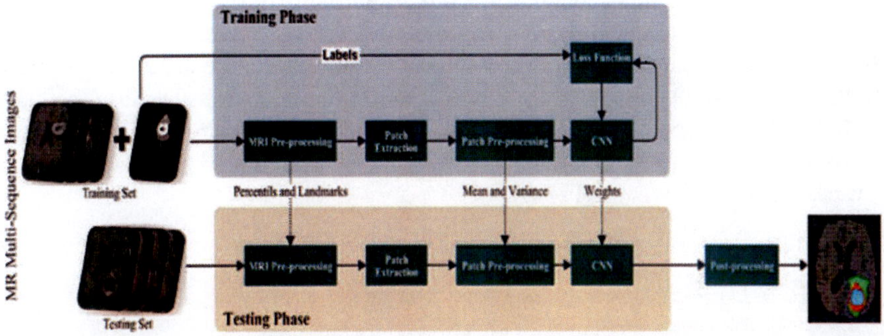


**Fig. 1** Overview of the proposed method

appears not to require an earlier display of tissue for submission. You may refer to any image of the MRI. The MRI's power scales are implemented by force standardization. Without an established scale of force, it would be impossible to summarize the relative action of various types of tissue across different volumes in the view of the tumor. The test is based on the standardization of the regulation of ensemble methodology proposed by [23] is used here for assessment. The stage is being designed in two phases, and the changes are being organized. The principal motivation behind the planning phase is to find the default scale restrictions and to prepare histograms for the candidate volume for the standard histogram scale. A more equal histogram is obtained for each category. The mean power calculation and the standard deviation are indicated in the planning arrangement. These characteristics are used to standardize test patches. Force standardization involves the scheduling of stage and the organization of transition. We discover the standard scale parameters during our preparation stage, and when adjusting, the histogram organization of the data picture is mapped to the default scale [intrigue power (IOI)].

### 3.2 *Convolution Neural Network*

In CNN, we have different layers. Using convolution layers [13, 14] allows maps to be caused by convoluting a flag or image with parts. The unit in the part defines the relation to the previous layer by the weights of the artifacts on these points. Piece weights square measure modified in the middle of getting ready stage by back propagation to optimize the safe properties of the results. Because all units of analog part maps share these square elements, the convolutional layers are less weight-to-order and less coordinate than those of thick FC layers. Moreover, an analogous feature of the invariability of location perception is separately recognized since a constant variable is convoluted over a complete picture. In the absence of a few convolution layers, the distorted highlights are often typical of a deeper depth. For example, the update of the primary layers shows the square measure massed as subjects, bits, or queries in the required layers [15]. In CNN, some steps are initialized, enabled, pooled, regularized, raised data, and loss function. To accomplish convergence, the initialization system sets up data augmentation: To increase the size of the set preparation and decrease overfitting, this can be achieved. We have restricted the extension of the descriptions to turning operations because the class of the repair is specified by the focal voxel. A few developers even recommend image interpretations [16], but this may be achieved by adding the incorrect class to a work-around for division (Fig. 2).

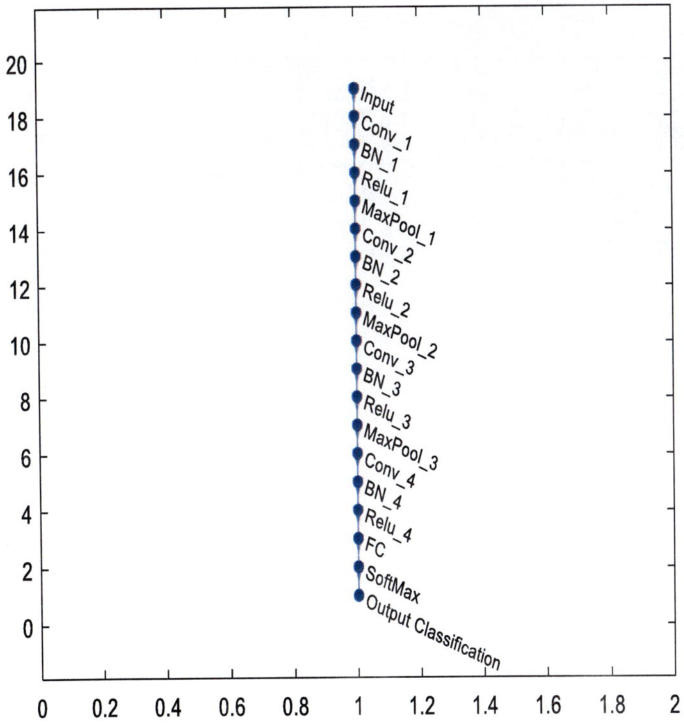


**Fig. 2** Proposed system architecture of CNN model for classification of brain tumor

## 3.3 *Pooling*

This combines the features of the included maps that are spatially close together. This combination of possibly repetitive highlights reduces the computational heap of the following stages by making the representation, for example, inconsequential subtle components, more minimized, and invariant to small image shifts. The use of full pooling or regular pooling [15] is more frequently demonstrated when used together. Loss functionality: The loss feature should be minimized during the preparation process.

## 3.4 *Post-processing*

To find out about the tumors, some of the tiny clusters left after the CNN phase will be reprocessed. We need to exclude the clusters acquired during the CNN segmentation to establish these constraints. They will then see the results.

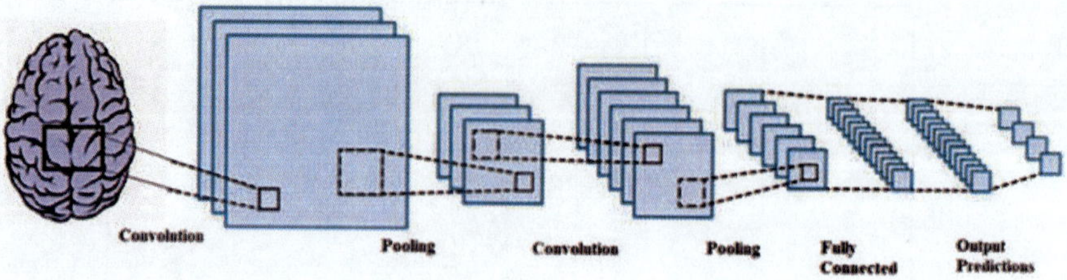


**Fig. 3** CNN architecture

## 3.5 *Rectified Linear Unit*

ReLU function is used in deep neural networks. Recently, it has been shown that the convergence of Tanh functions has improved six times. Mathematically, rectified liner units are defined as follows (Fig. 3):

$$f(x) = \max(0, x)$$

$$\text{If } (x) = 0,\ f(x) = \max(0, x) + \alpha \min(0, x) \text{ and } x \geq 0,\ f(x) = x$$

$$f(x) = \max(0, x) + \alpha \min(0, x)$$

## 3.6 *Loss Function*

It is the function to be minimized during training:

$$H = -\sum_{j \in \text{voxels}} \sum_{k \in \text{voxels}} c_{j,k} \log\left(\widehat{c_{j,k}}\right) \quad (1)$$

The probabilistic predictions (after the SoftMax) are represented by $\hat{c}$, while the target is represented by $c$.

# 4 Results and Discussion

Even small details are significant in the field of biomedical imaging, as incorrect interpretation can lead to a blunder in the diagnosis. So, in order to identify brain tumors, we introduce a deep learning method that integrates residual relations along with parametric ReLU with a value of $\alpha = 0.01$. In our proposed model, we used cross-channel standardization to standardize image data. Unlike the other networks in this paper, our proposed model provides better performance with validation accuracy 92.50%. In addition, the importance of the measurement metrics reinforces the increased performance of our network. Higher precision is desirable for segmentation to be treated as a good percentage.

The development of the network using either the tumor region or any other input is more easily done, but also requires sorting approaches or an expert devoted to categorizing those components. To our interpretation, Urban G [21] shows with a precision of 87% the best results for literature using segmented image parts as inputs. The $3 \times 3$ convergence filter decreases the map size, the batch normalization levels, the ReLU, and the percentage of the 3D maximum pooling layers. It is observed that CNN's proposed architectures perform better with 92.50% accuracy in performance measurements than other popular deep learning models.

For the calculation of CNN computation time, training parameters are important. The training parameters should also be set equal for all models, and the same dataset should be used.

CNN designs take longer for learning, but provide more classification efficiency due to the increase in layers and training configurations. The network can be used as a classification until a network is trained, and it takes only a few seconds to segment the image with a trained model. However, it can take hours for healthcare professionals to manually segment tumors. The proposed image classification techniques are accurate, quick, and can be used with low-cost data. It will allow physicians to identify brain cancers easily and reliably, which will save the lives of many people (Figs. 4 and 5).

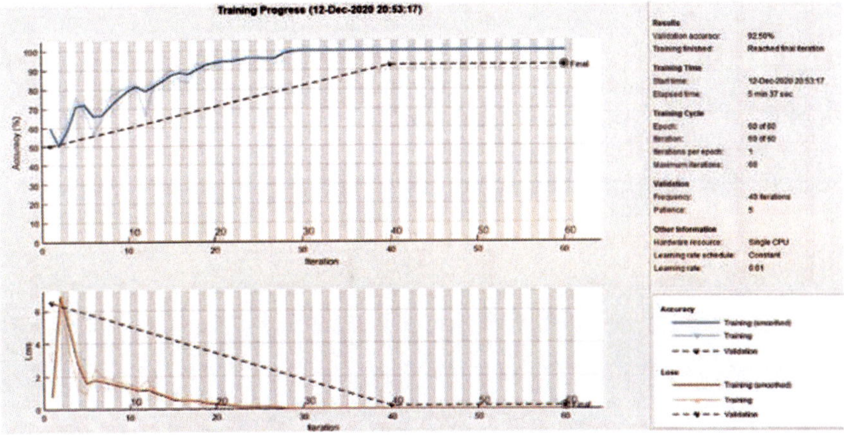


Fig. 4 Performance evaluation of proposed CNN model

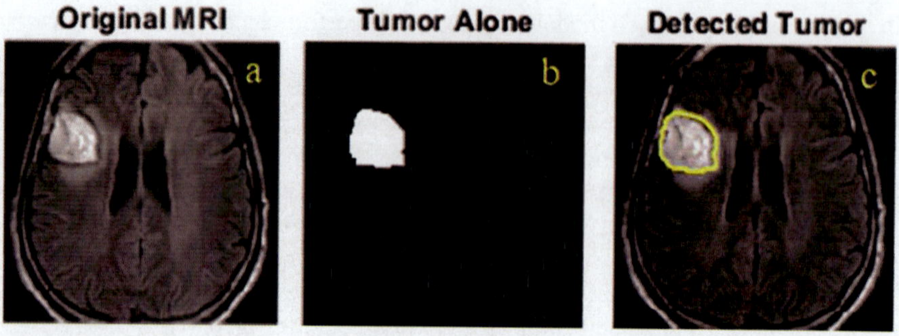


**Fig. 5** Segmentation results when the network was trained on the Kaggle dataset and evaluated on the same dataset, the following results were obtained: **a** original image, **b** segmentation mask, and **c** detected tumor

## 5 Conclusion

In the proposed works, we have introduced an architecture of a neural network that can identify brain tumors with greater accuracy and false prediction information to be used by radiation specialists in order to properly diagnose biomedical imaging. Not only does our proposed network classify the tumor of the brain in MR images, it also preserves tiny contours and limits. The network's findings are particularly good when the already trained network is fed with different datasets. The network can be inferred that the issue of reshaping the network with multiple datasets is avoided. Owing to the addition of residual and parametric RELU connections, the proposed network is more flexible in terms of layers and more balanced compared with other systems. In closing, the significant increase in the precision measurements obtained by the proposed network showed that the tumor capacity of the network was best classified over other network architectures.

The proposed CNN models will in the future be enhanced by the use of the hybrid CNN model for segmentation of various filter sizes, all MRI image modalities will be included in the tumor segment, and the classification result will be improved by an increased mini-batch scale between 64 and 128 and a maxi-epoch between 60 and 80 (or 120).

## References

1. World Health Organization Cancer. https://www.who.int/news-room/fact-sheets/detail/cancer. Last accessed 12 Mar 2021
2. Priya, V.V., Shobarani.: An efficient segmentation approach for brain tumor detection in MRI. Indian. J. Sci. Technol. **9**(19), 1–6 (2016)
3. Abler, D., Rockne, R.C., Büchler, P.: Evaluating the effect of tissue anisotropy on brain tumor growth using a mechanically coupled reaction–diffusion model. In: New Developments on

Computational Methods and Imaging in Biomechanics and Biomedical Engineering, vol. 33, pp. 37–48 (2019)
4. DeAngelis, L.M.: Brain tumors. New Engl. J. Med. **344**, 114–123 (2001)
5. Louis, D.N., Perry, A., Reifenberger, G., VonDeimling, A., Figarella-Branger, M., Cavenee, W.K., Ohgaki, H., Wiestler, O.D., Kleihues, P., Ellison, D.W.: The 2016 World Health Organization classification of tumors of the central nervous system: a summary. Acta Neuropathol. **131**(6), 803–820 (2016)
6. Kumar, P.R., Sarkar, A., Mohanty, S.N., Kumar, P.P.: Segmentation of white blood cells using image segmentation algorithms. In: 5th International Conference on Computing, Communication and Security (ICCCS), pp. 1–4. IIT, Patna (2020)
7. Prastawa, M., Bullitt, E., Ho, S., Gerig, G.: A brain tumor segmentation framework based on outlier detection. Med. Image Anal. **8**(3), 275–283 (2004)
8. Menze, B.H., Leemput, K.V., Lashkari, D., Weber, M.A., Ayache, N., Golland, P.: A generative model for brain tumor segmentation in multi-modal images. In: Medical Image Computing and Computer Assisted Intervention (MICCAI 2010), pp. 151–159. Springer, New York (2010)
9. Dan, C.C., Giusti, A., Gambardella, L.M., Schmidhuber, J.: Deep neural networks segment neuronal membranes in electron microscopy images. Adv. Neural. Inf. Process. Syst. **25**, 2852–2860 (2012)
10. Hoseini, F., Shahbahrami, A., Bayat, P.: An efficient implementation of deep convolutional neural networks for MRI segmentation. J. Dig. Imag. **2**, 1–10 (2018)
11. Pereira, S., Pinto, A., Alves, V., Silva, C.A.: Brain tumor segmentation using convolutional neural networks in MRI images. IEEE Trans. med. Imag. **35**(5), 1240–1251 (2016)
12. Havaei, M., Davy, A., Warde-Farley, D., Biard, A., Courville, A., Bengio, Y., Pal, C., Jodoin, P.M., Larochelle, H.: Brain tumor segmentation with deep neural networks. Med. Image. Anal. **35**, 18–31 (2015)
13. Crimi, A., Menze, B., Maier, O., Reyes, M., Winzeck, S., Handels, H.: Brainlesion: Glioma Multiple Sclerosis Stroke and Traumatic Brain Injuries, pp. 75–87. Springer International Publishing, Cham (2016)
14. Zhao, X., Wu, Y., Song, G., Li, Z., Zhang, Y., Fan, Y.: A deep learning model integrating FCNNs and CRFs for brain tumor segmentation. Med. Image. Anal. **43**, 98–111 (2017)
15. Kamnitsas, K., Ledig, C., Newcombe, V.F., Simpson, J.P., Kane, A.D., Menon, D.K., Rueckert, D., Glocker,B.: Efficient multi-scale 3D CNN with fully connected CRF for accurate brain lesion segmentation. Med. Image Anal. **36**, 61 (2016)
16. Badrinarayanan, V., Kendall, A., Cipolla, R.: SegNet: a deep convolutional encoderldecoder architecture for image segmentation. IEEE Trans. Pattern Anal. Mach. Intell. **39**(12), 2481–2495 (2017)
17. Ronneberger, O., Fischer, P., Brox, T.: U-net: convolutional networks for biomedical image segmentation. In: International Conference on Medical Image Computing and Computer-Assisted Intervention, pp. 234–241 (2015)
18. Tseng, K.L., Lin, Y.L., Hsu, W., Huang, C.Y.: Joint sequence learning and cross–modality convolution for 3D biomedical segmentation, pp. 3739–3746 (2017)
19. Dong, H., Yang, G., Liu, F., Mo, Y., Guo, Y.: Automatic brain tumor detection and segmentation using u-net based fully convolutional networks. In: Conference on Medical Image Understanding and Analysis, pp. 506–517 (2017)
20. Zhuge, Y., Krauze, A.V., Ning, H., Cheng, J.Y., Arora, B.C., Camphausen, K., Miller, R.W.: Brain tumor segmentation using holistically-nested neural networks in MRI images. Med. Phys. **44**(10), 5234–5243 (2017)
21. Urban,G., Bendszus, M., Hamprecht, F., Kleesiek, J.: Multi-modal brain tumor segmentation using deep convolutional neural networks. In: MICCAI BraTS (Brain Tumor Segmentation) Challenge. Proceedings, Winning Contribution, pp. 31–35 (2014)
22. Alqazzaz, S., Sun, X., Yang, X., Nokes, L.: Automated brain tumor segmentation on multimodal MR image using SegNet. Comput. Visual Media **5**(2), 209–219 (2019)
23. Hussain, S., Anwar, S.M., Majid, M.: Brain tumor segmentation using cascaded deep convolutional neural network. In: 39th Annual International Conference of the IEEE Engineering in Medicine and Biology Society (EMBC), pp. 1998–2001 (2017)

# MZI-Based Electro-optic Reversible XNOR/XOR Derived from Modified Fredkin Gate

**Shashank Awasthi, Sanjeev Kumar Metya, and Alak Majumder**

**Abstract** Arithmetic logic unit (ALU) is the heart of any computational logic module, and most of the operations in it are supported by XNOR/XOR gates, which also play a vital role even in encryption/decryption circuits. Thus, efficient and reliable operation of XNOR/XOR using emerging technologies has become a point of interest. The recent advancement of IC technology finds itself in a spot of bother due to the excessive heat dissipation, which is needed to be addressed, and reversible logic (RL) has emerged as a potential candidate. This paper unveils the exploration of reversible XNOR/XOR logic using a single cell of $4 \times 4$ MFG. Electro-optic Mach–Zehnder interferometer (MZI) is considered to realize the proposed logics under beam propagation method using OptiBPM tool. The power model of output ports is calculated and verified through MATLAB simulation.

**Keywords** Reversible logic · XNOR/XOR · Electro-optic effect · Modified Fredkin gate · Mach–Zehnder interferometer

## 1 Introduction

In the era of nanotechnology, where researchers are finding it hard to preserve the heat dissipation in a conventional CMOS technology, it is a high time to think over reversible logic (RL) [1]. This is validated by [2], which shows that increasing gate count leads to an increase in power density and on-chip device temperature. Landauer [3] explained that a bit computation in irreversible logic corresponds to a minimum of kTln2 ($2.8 \times 10^{-21}$ J) amount of energy dissipation where $k$ stands for Boltzmann constant and $T$ refers temperature in kelvin. This small energy may culminate a bigger value when a large number of transistors are integrated on a single die following the Moore's law. To resolve this issue, the concept of RL came into existence to preserve both information and heat dissipation and hence to find applications in quantum dot

S. Awasthi (✉) · S. Kumar Metya · A. Majumder
Integrated Circuit and System (I-CAS) Lab, Department of Electronics and Communication Engineering, National Institute of Technology, Yupia, Arunachal Pradesh 791112, India
e-mail: shashank.1801@gmail.com

 
D. Gupta et al. (eds.), *Pattern Recognition and Data Analysis with Applications*, Lecture Notes in Electrical Engineering 888,
https://doi.org/10.1007/978-981-19-1520-8_8

cellular automata, quantum and optical computing [4, 5]. A few prominent reversible gates (RG) are, namely the Fredkin gate, Peres gate and Feynman gate, etc. which are used to emulate the fundamental basic logic gates. To implement these RGs, several approaches have been rigorously studied in the literature [6–8]. One of the most vital logical operations are XOR/XNOR, which is employed to configure major building blocks of arithmetic circuits present in computer ALUs. Also, XOR logic finds application in designing encryption/decryption circuits such as stream cipher, e.g., set-top box of cable channels, wireless handsets, etc. From security point of view, using XOR/XNOR or any other bit-wise operators makes no difference as it actually depends upon the randomness of key stream generation. However, as single XOR/XNOR on chip can be utilized in both encryption and decryption, it saves silicon space comparing to other [9]. The reversibility of such blocks depends on the fact whether the unit design cell is reversible which preserves the bits during computation and gives away least heat dissipation. Photons being ultrafast are unquestionably superior to realize RG rather than the existing CMOS technology. Various optical switches have been exercised to observe the functionality of RG [10, 11], and lithium niobate-based MZI having Pockels effect has emerged to outperform other optical switches due to its better switching speed and inherent property of higher refractive index. In this article, we have studied the logical implications and mathematical power model of 4 × 4 MFG [11, 12] and explore the reversible XNOR/XOR operation, which is validated in terms of many influencing device factors.

The organization of the paper is as follows: Sect. 2 deals with the overview of MFG and its logical implementation of XNOR/XOR, which is followed by the mathematical modeling and its MATLAB simulation in Sect. 3. The simulation results through OptiBPM are shown in Sect. 4, whereas Sect. 5 contributes the single unit MZI and system level analysis followed by Sect. 6 that concludes the paper.

## 2 Overview of Modified Fredkin Gate

Figure 1 shows a 4 × 4 universal RG [11] famously called as MFG and having 16 distinct combinations of operation ensuring the principle of reversibility. The electro-optic implementation of MFG is done using 8 MZIs, thus defining an optical cost of 8. The Boolean expression of MFG is given by;

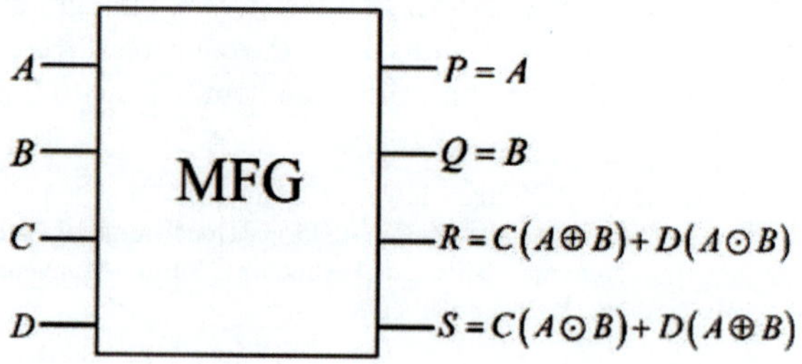


**Fig. 1** Block diagram for MFG

$$\begin{aligned} P &= A \\ Q &= B \\ R &= C(A \oplus B) + D(A \odot B) \\ S &= C(A \odot B) + D(A \oplus B) \end{aligned} \tag{1}$$

From Eq. (1), it is evident that an ancilla input at each of '$C$' and '$D$' guides MFG to output XNOR and XOR logic at port '$R$' and '$S$'. With $C = 0$ and $D = 1$, the equation for '$R$' and '$S$' can be re-written as given in Eq. (2) and accordingly the logical behavior may be seen at Table 1, which conveys that the MFG swaps the ancilla inputs at the output '$R$' and '$S$', only when the inputs $A$ and $B$ are equal.

$$\begin{aligned} R &= A \odot B \\ S &= A \oplus B \end{aligned} \tag{2}$$

**Table 1** Truth table for MFG (blue color defines XNOR/XOR operation)

| Inputs | | Ancilla | | Garbage | | Outputs | |
|---|---|---|---|---|---|---|---|
| *A* | *B* | *C* | *D* | *P* | *Q* | *R* (XNOR) | *S* (XOR) |
| 0 | 0 | 0 | 0 | 0 | 0 | 0 | 0 |
| 0 | 0 | 0 | 1 | 0 | 0 | 1 | 0 |
| 0 | 0 | 1 | 0 | 0 | 0 | 0 | 1 |
| 0 | 0 | 1 | 1 | 0 | 0 | 1 | 1 |
| 0 | 1 | 0 | 0 | 0 | 1 | 0 | 0 |
| 0 | 1 | 0 | 1 | 0 | 1 | 0 | 1 |
| 0 | 1 | 1 | 0 | 0 | 1 | 1 | 0 |
| 0 | 1 | 1 | 1 | 0 | 1 | 1 | 1 |
| 1 | 0 | 0 | 0 | 1 | 0 | 0 | 0 |
| 1 | 0 | 0 | 1 | 1 | 0 | 0 | 1 |
| 1 | 0 | 1 | 0 | 1 | 0 | 1 | 0 |
| 1 | 0 | 1 | 1 | 1 | 0 | 1 | 1 |
| 1 | 1 | 0 | 0 | 1 | 1 | 0 | 0 |
| 1 | 1 | 0 | 1 | 1 | 1 | 1 | 0 |
| 1 | 1 | 1 | 0 | 1 | 1 | 0 | 1 |
| 1 | 1 | 1 | 1 | 1 | 1 | 1 | 1 |

## 3 XNOR/XOR BPM Layout

OptiBPM, being a tool run under beam propagation method (BPM), is used to design and implement the MFG-based XNOR/XOR with titanium (Ti)-diffused lithium niobate (LN) MZI and monitors the optical field propagated through the waveguide. Figure 2 shows the BPM layout of the configuration with 8 MZIs. The input signals (also referred to as control signal (CS)) are fed to MZIs in the following order: Input *A* is fed to MZI1, input *B* is given to MZI2, MZI7 and MZI8, whereas ancilla input '$C = 0$' is fed to MZI3 and MZI6, and finally, the input '$D = 1$' is driving MZI4 and MZI5. CS plays a vital role in guiding the MZI to either behave as a bar switch or cross switch with CS = logic '1' (= 6.8 V) and CS = logic '0' (= 0 V), respectively. It is very crucial to smartly consider the propagation direction of the design due to anisotropic property of LN. In this paper, it has been considered to exercise *z* cut wafer propagated in *y* direction [11]. The various parameters that are used in designing the MZI and electro-optic XNOR/XOR are given in Table 2 [11].

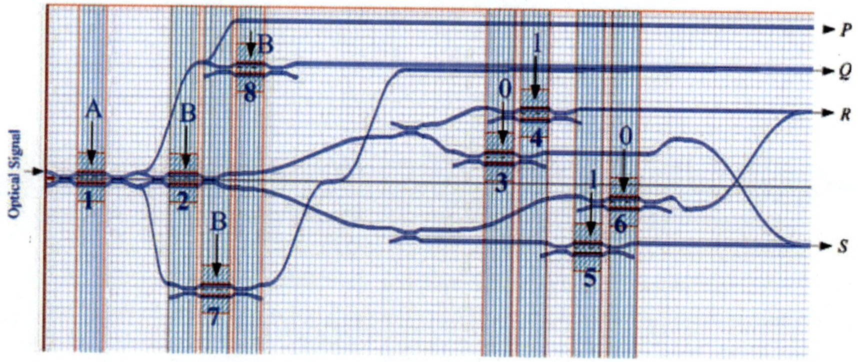


**Fig. 2** BPM layout for XNOR/XOR logic from MFG

**Table 2** MZI designing parameters

| Parameters | Value |
|---|---|
| Length of a MZI (μm) | 33,000 |
| Thickness (μm) | 8 |
| Wavelength (μm) | 1.3 |
| Substrate thickness (μm) | 10 |
| Cladding thickness (μm) | 2 |
| Refractive index (Ti:LN) | 2.2 |
| Buffer refractive index | 1.47 |
| Air refractive index | 1 |

## 4 Simulation Results and Mathematical Modeling

The simulation results of the MFG-based reversible XNOR/XOR logic are presented in Fig. 3, where the laser light is provided to the top input port of MZI1, and based on the CS of respective MZIs, the light travels through the desired path to yield the expected output as briefed in the following case. The combination AB = 00 makes CS of both MZI1 and MZI2 to be logic '0' thereby steering them to act as a cross switch. Hence, MZI1 receives the light from top input port and emerges out from lower output port to hit the lower input port of MZI2 only to come out through the upper output port of it. This now travels through the upper 3-dB coupler to hit MZI3 and MZI4 top input port, which are guided by ancilla input '0' and '1', respectively, as mentioned in previous section. Because MZI3 is a cross switch, the light is terminated at the lower output port and MZI4 allows the light to travel through its bar port. Accordingly, the final output is yielded as $P = 0$, $Q = 0$, $R = 1$ (XNOR operation) and $S = 0$ (XOR operation), which may be verified from Fig. 3a. In the similar way, the other cases of Figs. 3b, c and d may be studied.

To verify the simulation results and the truth table presented in Table 1, the mathematical model of the power at the output ports is carried out (refer Eq. 3–6) following the relations presented in [11]. The power models are executed on MATLAB to observe the logical implications as shown in Fig. 4, where the first row to last row represents the output combinations of MFG-based XNOR/XOR module for input conditions of 00–11, respectively. Considering the coordinates (row, column) (2,3), (3,3) and (4,4) in Fig. 4, the output wave is noted to reach around 0.5 V only to roll down to 0 V and the same is considered as logic '0'.

$$P_P = \sin^2\left(\frac{\Delta\phi_1}{2}\right) \tag{3}$$

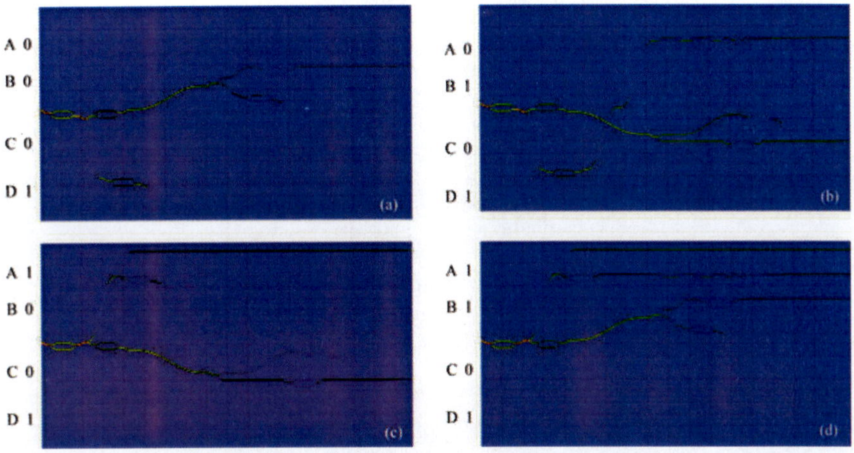


Fig. 3 BPM simulation results of XNOR/XOR logic

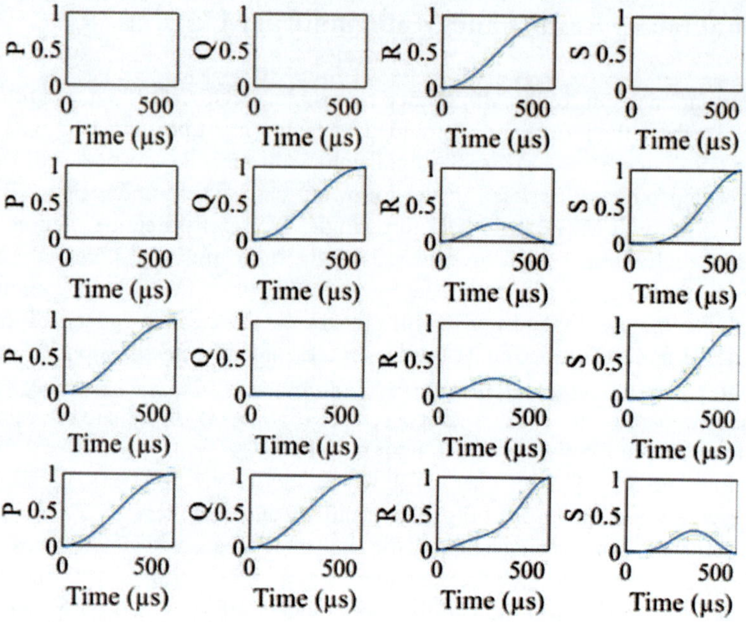


**Fig. 4** MATLAB simulation results

$$P_Q = \sin^2\left(\frac{\Delta\phi_1}{2}\right)\sin^2\left(\frac{\Delta\phi_7}{2}\right) + \cos^2\left(\frac{\Delta\phi_1}{2}\right)\sin^2\left(\frac{\Delta\phi_8}{2}\right) \tag{4}$$

$$\begin{aligned} P_R = &\left\{\sin^2\left(\frac{\Delta\phi_1}{2}\right)\sin^2\left(\frac{\Delta\phi_2}{2}\right) + \cos^2\left(\frac{\Delta\phi_1}{2}\right) + \cos^2\left(\frac{\Delta\phi_2}{2}\right)\right\} \\ &\sin^2\left(\frac{\Delta\phi_4}{2}\right) + \left\{\begin{array}{l} \sin^2\left(\frac{\Delta\phi_1}{2}\right)\cos^2\left(\frac{\Delta\phi_2}{2}\right) \\ +\cos^2\left(\frac{\Delta\phi_1}{2}\right)\sin^2\left(\frac{\Delta\phi_2}{2}\right) \end{array}\right\}\sin^2\left(\frac{\Delta\phi_6}{2}\right) \end{aligned} \tag{5}$$

$$\begin{aligned} P_S = &\left\{\sin^2\left(\frac{\Delta\phi_1}{2}\right)\cos^2\left(\frac{\Delta\phi_2}{2}\right) + \cos^2\left(\frac{\Delta\phi_1}{2}\right)\sin^2\left(\frac{\Delta\phi_2}{2}\right)\right\} \\ &\sin^2\left(\frac{\Delta\phi_5}{2}\right) + \left\{\begin{array}{l} \sin^2\left(\frac{\Delta\phi_1}{2}\right)\sin^2\left(\frac{\Delta\phi_2}{2}\right) \\ +\cos^2\left(\frac{\Delta\phi_1}{2}\right)\cos^2\left(\frac{\Delta\phi_2}{2}\right) \end{array}\right\}\sin^2\left(\frac{\Delta\phi_3}{2}\right) \end{aligned} \tag{6}$$

# 5 Study of Performance Metrics

The unit cell of the design is a Ti-diffused LN-based MZI, which is an anisotropic material, and hence, its performance gets altered as a function of refractive index of the material. Referring [13], we know how any alteration in various parameters like electrode voltage and wavelength affects the perfect execution of an MZI. The coupling ratio and extinction ratio are studied in [13] as a function of wavelength and horizontal diffusion length. In this paper, the execution of MZI is explored in terms of insertion loss and excess loss as a function of wavelength ($\lambda$) and electrode voltage (V).

## 5.1 Insertion Loss

Insertion loss (IL) refers to the loss of light between input and output ports of a waveguide [14] and is expressed by the following relation:

$$\text{IL} = -10\log_{10}\frac{P_{\text{OUT}}}{P_{\text{IN}}} \tag{7}$$

where $P_{\text{OUT}}$ and $P_{\text{IN}}$ are the power available at the output and input ports, respectively.

Evidently, Eq. 7 corresponds to the fact that a least possible IL is expected for the proper functioning of the MZI as a switch. A greater IL incurs a cost in terms of light presence at the desired output port, which may have an impact on other MZIs in the system. The IL due to the variation in switching (or electrode) voltage (SV) is depicted at Fig. 5a while keeping the wavelength fixed at 1.3 μm and diffusion stripe thickness at 0.051 μm. With the CS = logic '1', the graph records a minimum IL of 0.02 dB at 6.8 V when the major portion of light passes to the expected bar port of MZI. If the SV gets varied extensively, there will be significant increase in the IL slope, which may give away malfunction for a system-level application. On

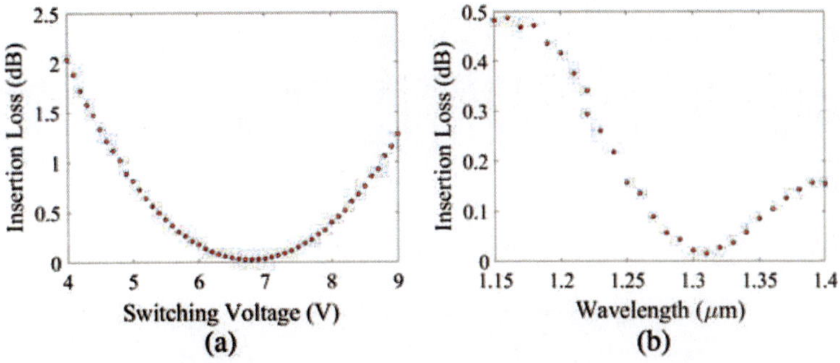


**Fig. 5** Insertion loss in a single unit MZI w.r.t. **a** switching voltage and **b** wavelength

the other end, an optical waveguide follows the principle of optical window [10] for efficient transferring of information through the light source. Out of the four windows, the second window with $\lambda = 1.3\ \mu m$ is considered for this study due to minimum dispersion and attenuation of 0.5 dB/km. It is clearly perceived from Fig. 5b that beyond a small variation of 1.3 µm, the MZI functionality may fall due to increased IL.

## 5.2 Excess Loss

Any light source while traveling through a waveguide may lose some light in terms of intensity and amplitude. Thus, the amount of light that is wasted in addition to the light that is coupled in two or more output ports is termed as excess loss (EL). Mathematically, it is written as:

$$\text{EL} = -10\log_{10}\frac{P_1 + P_2 + \cdots + P_n}{P_{\text{in}}} \tag{8}$$

In a MZI, being a 2 × 2 coupler, the light is fed to any one of the input ports, and all the light is expected to be present at either output ports depending on the electrode voltage. Thus, Eq. (8) may be re-written as:

$$\text{EL} = -10\log_{10}\frac{P_{\text{out1}} + P_{\text{out2}}}{P_{\text{in}}} \tag{9}$$

Thus, ideally both the output ports must have all parts of light without leaking while traversing through the waveguide.

Figure 6a confirms our MZI to leak a tiny light thereby offering the 99.9% of input light intensity at the desired output of interferometer switch. As per Fig. 6b,

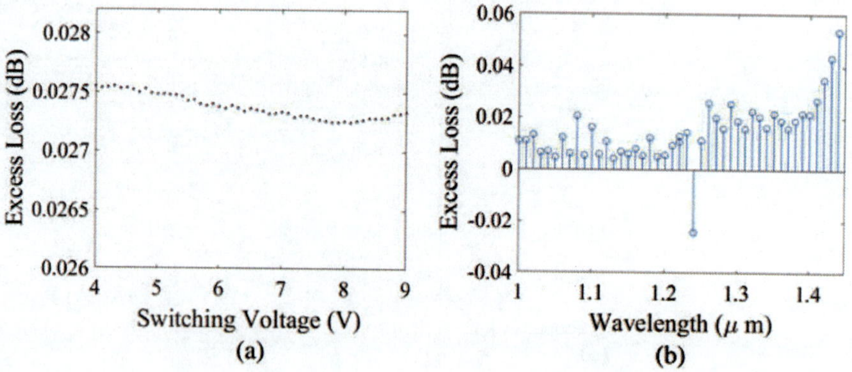


**Fig. 6** Excess loss in a single unit MZI w.r.t. **a** switching voltage and **b** wavelength

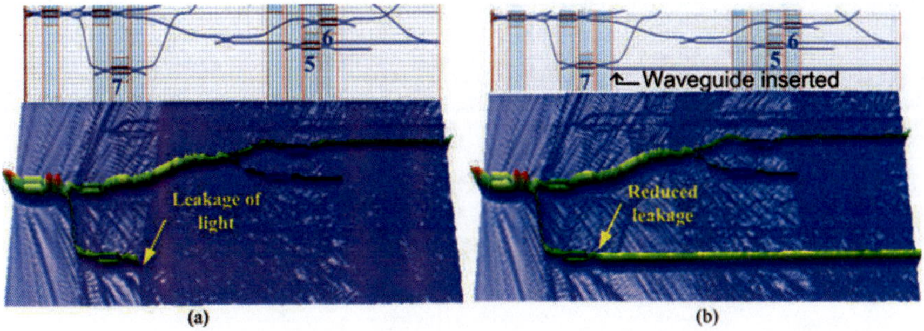


**Fig. 7** Leakage of light **a** before waveguide and **b** after waveguide insertion

even though the wavelength ranging from $1.1 < \lambda$ ($\mu$m) $< 1.2$ offers negligible EL, the tolerance of $\lambda$ is very small to maintain light within second optical window and that small variation in wavelength does not affect the performance of MZI.

## *5.3 System-level Analysis*

To validate the performance of a single MZI, a random MZI is considered from the system of MFG-based XNOR/XOR for an input combination of '00'. This means, CS to MZI1, MZI2, MZI7 and MZI8 is set to logic '0', and thus, they behave as cross switches. If the undesired port of a MZI in a certain system remains floating, it invites massive interference which may lead to complete functional failure. For example, if we consider MZI7 to be studied in the proposed XNOR/XOR design, and it is evident from Fig. 7a that the cross port of it is not connected to any other MZIs thereby resulting to an excessive leakage of light. This problem may be solved by placing an optical waveguide terminator [15] or a waveguide of certain length at the end of the floating MZI port as exercised and shown in Fig. 7b to reduce the amount of leakage.

As the input combination is fixed at '00', the light traverses to the port of XNOR logic (refer Table 1) through MZI4, which is taken into account for system level analysis. EL and IL as a function of switching voltage and wavelength are studied and found to offer optimum result around 6.8 V and 1.3 $\mu$m only as shown in Fig. 8, which confirms a decent tolerance of both the influencing factors to maintain the functionality of the design. It is to note that the tolerance of $\lambda$ in Figs. 8c and d has to be within the second optical window to have the minimum IL and EL.

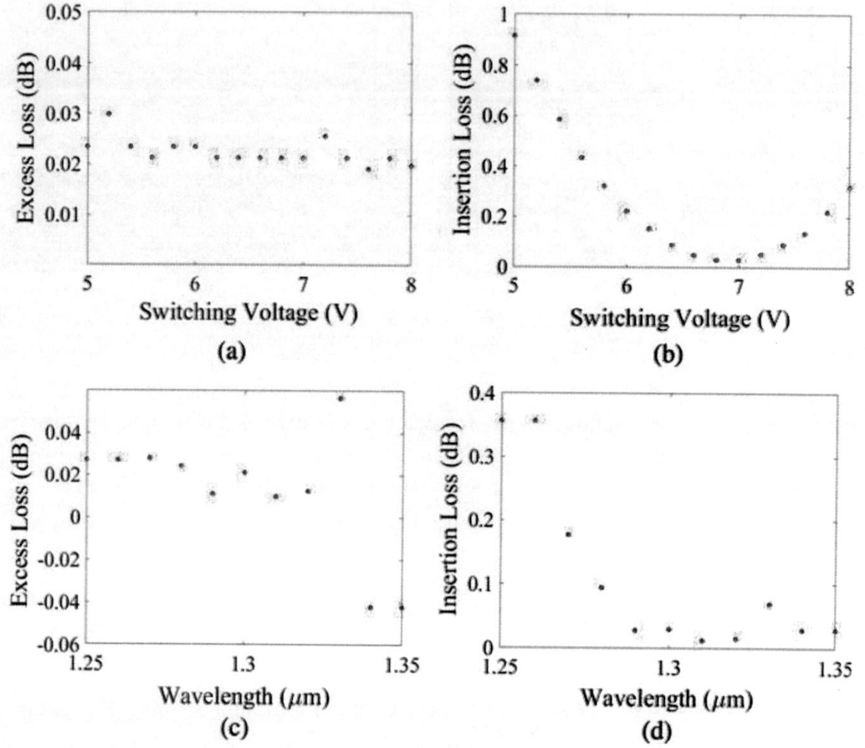


**Fig. 8** Study of MZI4 **a** EL versus SV, **b** IL versus SV, **c** EL versus λ and **d** IL versus λ

## 6 Conclusion

Electro-optic (EO) or Pockels effect is most widely used method in observing arithmetical/logical operations (ALOs) using various optical switches due to its nonlinear electro-optic property. Out of many ALOs based on RG, modified Fredkin gate-based reversible XNOR/XOR is presented with its EO design of Mach–Zehnder interferometer switch and using titanium-diffused lithium niobate (Ti: $LiNbO_3$) material. $LiNbO_3$ is superior to other EO materials such as $BaTiO_3$, GaAs, $LiTaO_3$ due to its best electro-optic tensor coefficient and minimum voltage requirement. The simulation is carried out in OptiBPM tool, and the power equations have been verified through MATLAB simulation to verify the truth table. The proper functionality of MZI unit cell and the XNOR/XOR system is analyzed in terms of insertion loss and excess loss as a function of switching voltage (SV) and wavelength (λ) to achieve the optimum results at the SV = 6.8 *V* and λ = 1.3 μm. The future scope of the proposed design can be considered in optical memory and online security systems for encryption and decryption process to make the online system ultrafast.

**Acknowledgements** The work is carried out under Visvesvaraya PhD Scheme and SMDP-C2SD, MeitY—Govt. of India.

## References

1. Bennett, C.H.: Logical reversibility of computation. IBM J. Res. Dev. **17**, 525–532 (1973)
2. Pop, E., Sinha, S., Goodson, K.E.: Heat generation and transport in nanometer-scale transistors. Proc. IEEE **94**(8), 1587–1601 (2006)
3. Landauer, R.: Irreversibility and heat generation in the computing process. IBM. J. Res. Dev. **5**(3), 183–191 (1961)
4. Tarpadhar, C., Chattopadhyay, T., Roy, J.N.: Mach Zehnder interferometer based all-optical reversible logic gate. Opt. Laser. Technol. **42**(2), 249–259 (2010)
5. Thapliyal, H., Ranganathan, N.: Reversible logic-based concurrently testable latches for molecular QCA. IEEE Trans. Nanotechnol. **9**(1), 62–69 (2010)
6. Kang. M.S., Heo, J., Choi, S.G., Moon, S., Han, S.W.: In: Optical Fredkin gate assisted by quantum dot within optical cavity under vacuum noise and sideband leakage. Sci. Rep. **10**(1), 1–13 (2020)
7. Swathi, M., Rudra, B.: Implementation of reversible logic gates with quantum gates. In: 2021 IEEE 11th annual computing and communication workshop and conference (CCWC), pp. 1557–1563 (2021)
8. Abbas, M.N., Abdulnabi, S.H.: Plasmonic reversible logic gates. J. Nanophoton. **14**(1), 016003 (2020)
9. Zhou,H.: A humble theory and application for logic encryption. IACR Cryptol. ePrint Arch., **696** (2017)
10. Singh, K., Kaur, G.: Interferometric architectures based all-optical logic design methods and their implementation. Opt. Laser. Technol. **69**, 122–132 (2015)
11. Awasthi, S., Biswas, A., Metya, S.K., Majumder, A.: Optical configuration of modified Fredkin gate using lithium-niobate-based Mach-Zehnder interferometer. Appl. Opt. **59**(23), 7083–7091 (2020)
12. Picton, P.: Modified Fredkin gates in logic design. Microelectron. J. **25**(6), 437–441 (1994)
13. Awasthi, S., Biswas, A., Metya, S.K., Majumder, A.: Electro-optic reversible toffoli gate with optimal count of LiNbO3 Mach-Zehnder interferometers. In: 2020 IEEE nordic circuits and systems conference (NorCAS), Oslo, Norway, pp. 1–7 (2020)
14. Khare, R.P.: Fiber Optics and Optoelectronics. Oxford University Prress, USA (2013)
15. Duminas, P.: Optical waveguide termination having a doped, light-absorbing slab. U patent no. WO 2017/193740 Al (2017)

# Secured Remote Access of Cloud-Based Learning Management System (LMS) Using VPN

**Paramita Chatterjee, Rajesh Bose, Subhasish Banerjee, and Sandip Roy**

**Abstract** Globally, education system is in changing process. The new tendency is everywhere, from traditional classroom systems to digitalization systems. Cloud-based learning management systems (LMS) will drive the educational sector forward in the future years because they will provide end users with a flexible, easy-to-use, secure, and cost-effective learning process. Especially when the economy is in a slump due to global pandemic situation, cloud-based LMS model is the appropriate and most trusted learning model in global educational sector. It can be used through web in remote places with multiple users on same platform. Hence, the first thing which required is secured access of said LMS model. This security based on some protocols like Hypertext Transfer protocol Secure (HTTPS) which is maximum considered presently. It is more effective tool than the other security protocols like (Hypertext Transfer Protocol) HTTP. Connecting cloud-based LMS securely in remote locations using virtual private network (VPN) technology and secure socket layer (SSL) and Citrix access gateway product, which ensures security from authorizations to application-level protection and client-side security, improves the usability of this LMS among remote users of various educational institutions in a global pandemic situation. The goal of this work is to propose an SSL VPN design that uses the Citrix access gateway product (which includes Windows and UNIX-based systems and web-enabled applications) to ensure safe data transfer of a user-friendly cloud-based LMS in remote locations.

**Keywords** Secure remote access · Cloud security · Virtual private network (VPN) · Secure socket layer (SSL) · Hypertext Transfer Protocol Secure (HTTPS) · Citrix access gateway · LMS security

---

P. Chatterjee (✉) · R. Bose · S. Roy
Department of Computational Science, Brainware University, Barasat, Kolkata, India
e-mail: pc11april@gmail.com

S. Banerjee
National Institute of Technology, Itanagar, Arunachal Pradesh, India
e-mail: subhasish@nitap.ac.in



D. Gupta et al. (eds.), *Pattern Recognition and Data Analysis with Applications*,
Lecture Notes in Electrical Engineering 888,
https://doi.org/10.1007/978-981-19-1520-8_9

## 1 Introduction

Nowadays, the worldwide educational sector, whether it is schools, colleges, or any educational service provider, uses LMS to give competitive service to their end users, allowing them to gather and refresh their knowledge at any time and from any location. However, because there are various remote users using same, all service providers face huge issues in terms of security platform through cloud-based network access. Here comes the role of virtual private network (VPN) which gives a private network increasing power so that it can act as internet among public network. It is allowing host computer to communicate through data exchanging among shared network with appropriate function, security and maintaining different internet policies in private network. The choice of getting exact VPN in action depends on user's mode of exercise. VPN is having mainly two functions: remote or distant approach for end users and site-to-site approach for two different geographically positioned which able to connect with a Wide Area Network (WAN). IPSec, SSL are two most trusted VPN technologies being used in today's network system. These two are having their own capacities and challenges. Secure socket layer (SSL) VPN is a growing technology that makes remote access VPN strengthen and by using which a user can float a web browser to access any remote VPN connections from an internet to enable location. Another use of VPN can be to conceal actual information regarding remote connections that are difficult to measure. The main challenges regarding appropriate choice of VPN are depending on its security, functionality, and traffic measurement ability. The more perfect VPN one can use the more technical sound access of remote network or internet facility users can enjoy. As a result, users can use the LMS more effectively. In this paper, it has shown and discussed the parameters of selecting appropriate VPN based on its effective performance.

The research work is having following sections: Sect. 2 presents a brief knowledge of secured remote access. Section 3 described related works till date by extensively going through related articles and researches. Section 4 said about the proposed works and methodology with optimum explanations in details. Section 5 briefly described result analyses. Lastly, the conclusion and way forward described in Sect. 6.

## 2 Secured Remote Access

Remote access [1] is a special power given to any individual to control any computer or network from any geographical location with the help of network connection. It is a system where physical presence of a user does not require. A user can reach any computer, file, or any related documents whenever it is needed by remote access facilities. It is actually helping the service provider or LMS users of any concern situated at different locations to get connected so that they can have well informed, and by this way, it also helps in increasing usability and understanding between service provider and users. It is actually an industry where different technical support teams solve the LMS user's problems from any remote locations whether it is within or outside of organization. Remote access VPN connection is the most trusted way in this system as it is prudent for secured encrypted communication over internet network. Another main motive of using VPN technology is to give a secure platform to remotely access the organizational branches at different locations so that the users can access other corporate facilities and resources. Actually, it is used to manage related to coalition of network connection, software and hardware applications.

## 2.1 *Virtual Private Network (VPN)*

Virtual private network (VPN) is a program that allows for safe and encrypted connections across public networks such as the internet. It encrypts data in subway protocols at the dispatch end and decrypts it at the receiving end, including the network address, allowing for more secure online operations. It is also applied to furnish remote access to software hosted on exclusive network. One should have the authentication process to use the VPN app like password, biometric, or security token. VPN apps mostly use through mobile devices for giving more secured data protection. Users should not, however, confuse private browsing, which does not require any encryption and is an optional feature. In essence, a VPN establishes an end-to-end connection that is only accessible by authorized users. To continue, this VPN client (software application) should be installed on every point, either in local or in the cloud. The execution may be influenced by a variety of factors such as the user's internet speed, the quality of services provided by the service provider, and so on. There are now a few protocols that must be followed in order to obtain an accurate rating of security in connected systems. Here are a few protocols from a long list of VPN protocols: IP security (IPsec), secure socket layer (SSL), and transplant layer security (TLS), point-to-point tunneling protocols (PPTP), Layer 2 tunneling protocols (L2TP) open VPN.

Benefits of remote access VPN are it allows the remote users to access to LMS network from any geographical location that can be useful for service provider and end user who is highly mobile or situated remotely. In other way, remote users can take lead of using public Wi-Fi connection as data are in encrypted version which is more secure to access. So, while planning to use remote access VPN, a controller always have to have enough number of VPN software license with network bandwidth which can minimal the latency for remote users.

## 2.2 *SSL VPN*

A secure socket layer (SSL) VPN [2] is used in normal browser to get encrypted data from HTTP traffic through SSL protocol. It also protects real-time protocol (RTP) traffic. A remote user can reach to the LMS network and web applications through SSL VPN. SSL VPN is having mainly three protocols. Handshake Protocols

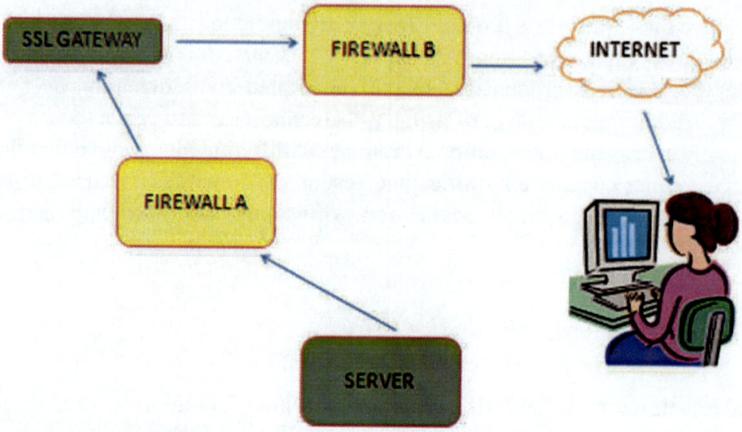


**Fig. 1** Architecture of SSL VPN [2]

negotiate the encryption algorithm and authorize the server to the users. Record Protocol helps client and server both to share key to encrypt data. In alert protocol, user can get to know the error, if any. Figure 1 is describing the architecture of SSL VPN, and its functionalities are mentioned below.

SSL VPN gateway is a common gateway of all VPN connections of the organization. It also helps to initiate internal application server of all connections which is protected by Firewall both inside and outside. Outside Firewall gives permission to connect with a machine and any internet with SSL VPN gateway. In the first step, an end user is connecting, after authentic log-in, to an organization's gateway with a list of given applications. Simultaneously, internal server is connected through inside Firewall by SSL gateway. As a result, a user gets a summarized feedback by SSL VPN gateway, and SSL VPN tunnel is created between end user's device and SS VPN gateway.

Table 1 is describing the comparison between IPSec and SSL VPN (Table 2).

## 2.3 *Citrix Access Gateway/NetScaler Gateway*

Citrix authorizes [3] its end users to supply dependent, less risk access of services associated with the cloud (SaaS) or other environment. It is helping in restricting the user's access to the internet by web filtering and web isolation too. Citrix recommends three way-out to its end users to carry out risk-free access with secured mechanism (single sign-on) for the applications, information, and network. These three options are as below:

**Table 1** Comparison between IPSec and SSL VPN

| Parameters | IPSec VPN | SSL VPN |
|---|---|---|
| Gateway location | Implemented on the Firewall | Behind the Firewall |
| Security/control | Broad access creates security concerns | More granular controls require more management |
| Application | Can support all IP base applications | Best for browser-based application like email, file sharing, etc. |
| Network layer as it works | Operates at layer 3 | Operates at layer 4–7 |
| endpoints | Require host-based clients | Browser based with optional thin client |
| Connectivity | Connect entire remote host network | Connect specific applications and services |
| Complexity | More | Less |

**Table 2** Depicts the strengths and weaknesses of SSL VPN

| Strength of SSL VPN | Weakness of SSL VPN |
|---|---|
| Support remote access<br>No special software is required<br>Browser-based session<br>Granular-level control<br>Specific application gives more security<br>Simple configuration | Restricted users<br>Vendor specific installation<br>Not appropriate for remote site require always on-link |

- Citrix gateway
- Citrix gateway service
- Citrix access control

Citrix access gateway is a network mechanism which securely transfers any inquiry anywhere with policy-based smart access control. It is actually less risk involved and capacity to long distance access to applications which helps in giving answers to user's problems. Citrix gateway furnished improved functionalities for communication to Citrix virtual apps, Citrix virtual desktops applications from remote access. In this environment, users can get needed explanations as per as secured, completely trusted performance through Citrix gateway only. For high-latency network like home and public Wi-Fi, Citrix HDX enlightened data transport, a protocol developed on UDP, helps in faster the traffic over high latency which is only supported by Citrix gateway. It is very important to use appropriate VPN to access to cloud-based LMS server network. Citrix gateway provides the actual solution to this as full SSL VPN. This is useful for both service provider and user in respect to access the application on cloud environment. As it is a browser based, users can reach to all web in a cloud any time and from anywhere through Wi-Fi access. They can automatically connect on LMS network through internet. Citrix also merges with third-party providers (authentication) which relates RADIOUS, LDAP, TACACS, and diameter-based mechanisms.

### 2.4 Security Aspect

The users cannot measure that security level of their traffic flow in public cloud, neither they prevent or create security protocol to prevent. The security factor increases as the users are working remotely, using multiple devices with multiple operating systems and multiple networks. The following risk factors are:

- Security related to authentication
- Virus and threats from user's computer to LMS network
- Split tunneling
- Host identify verification
- Secure desktop
- Man-in-the-middle attack

### 2.5 Optimal Usability

To give more advanced security and execution on the network, SSL VPN is the current trend than traditional VPN IP security. On the other hand, data protection and LMS application tunneling are the smart process rather than data back up and network tunneling. Presence of service-level guarantee (SLA) is important to stabilized internet latency. As per benefit is concerned, SSL VPN is far ahead than other VPN [4] technologies. This provides low-risk platform between remote users and private network which are using SSL protocol, its successor transport layer security. Apart from security, SSL VPN gives user-friendly interface. The ultimate aim of SSL VPN is to give protected and trustworthy cloud connection to the service providers so that end users can have a secured, fast, and user-friendly access too.

## 3 Related Work

After going through few related studies, it has been found that the researches are developing based on VPN cloud-based remote access. Viewpoints of some papers are given below:

In [4], it describes the development of the cloud related to VPN measuring the security risk. In [2], it describes the comparison between IPSec and SSL VPN. In [3], author tries to focus on Citrix for accessing any application from any geographical location irrespective of any device. Like TOR (multiple VPN services) [5], Hotspot Shield, and other services are going with unique fingerprints. In [6], Yamada et al.

discussed a technique for encrypted traffic. In [7], it has found different applications of VPN in context to Android usability, and in [8], it described its risk. In [9, 10], author described about the VPN clients and respective VPN server connection. In [11, 12], it described the activity of different malicious entity in network traffic when the connection is established. In [13], author discussed about TLS interception related with the certificates trusted locally for effective work of VPN services. In [14], it described risky situations related with sensitive data. In [15], it described the related with the VPN services for information security. Authors [16] proposed an approach of man-in-the-middle attack in VPN network traffic and forwarded a technique using public key infrastructure (PKI) involving massive key management. In this paper, it is considered that the raw data are available without encryption for analyzing and finding out unencrypted VPN traffic with the help of application layer proxy. In paper [17], it created new model called virtual private network as a service (VPNaaS) to know the requirement of VPN service in discrete architecture with the help of cloud. The paper [18] described use of different VPN to know the connection between cloud computing and cloud services and their different software implementations. In paper [19], author discussed about the unique method called round trip time (RTT) to restrict the international traffic. In paper [20], it discussed about VPN-based app and its privacy, security, protocols, and traffic. In [21], author discussed about the traditional security measures of VPN and its modern applications. In the paper [22], it is described as case study of remote real-time data analysis for maintaining work flow. In paper [23], it discussed about new digital fabrication and introduced a new service called fabrication as a service (FaaS) in cloud. In paper [24], it has discussed that the remote use of cloud network in pandemic situation as the education system started new mode of study which is online study.

## 4 Proposed Works and Methodology

In this proposed work, it is trying to carry out two numbers of Citrix access gateway box in a datacenter. Figure 2 will show the conceptual network diagram of the datacenter, where the proposed box is in DMZ zone of the datacenter. The author is trying to publish the LMS server URL through this box, and users can access the LMS security.

Other networking products that interact with access gateway include the LMS server load balancer, Firewalls, routers, and IEEE 802.11 wireless devices. When installing an access gateway in any network infrastructure, there is no need to make any changes to the existing hardware or internal network. To get more secured, the access gateway is placed inside internal demilitarized zone (DMZ). Access gateway is introduced in following networks, at the time of installing in the DMZ, (A) private network, and (B) public network (routable IP address). Actually, private network is internal network, internet defined as public network. This access gateway may be used as partition local area networks internally to access control and security. The

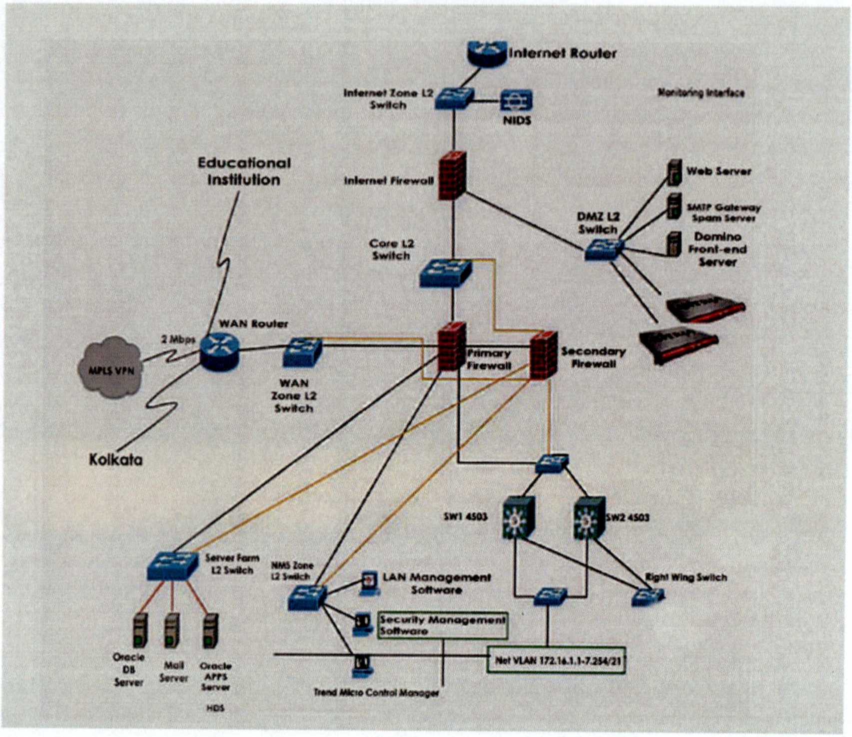


**Fig. 2** Conceptual network diagram—secured remote access of cloud-based learning management system

partition can be created among: (A) wired and wireless networks (B) data and voice networks.

Figure 3 is describing the methodology;

All the steps of methodology are mentioned in Fig. 3 above described as below in subsections.

## 4.1 *Installing the Access Gateway*

Identify the Internet Protocol (IP) addresses we need to configure the appliance. These are included in Fig. 4:

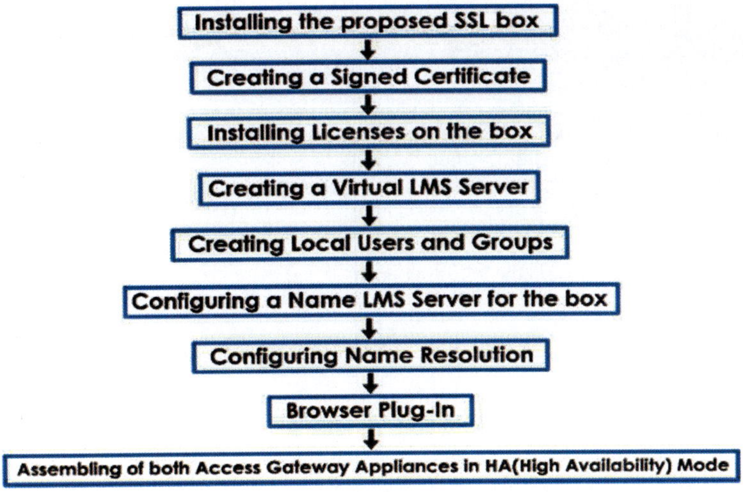


**Fig. 3** Methodologies for implementation of secured remote access of cloud-based LMS

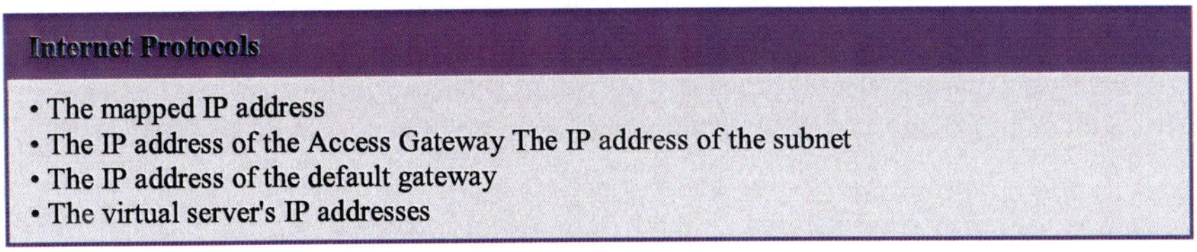


**Fig. 4** Internet Protocol

## *4.2 Creating a Signed Certificate*

A certificate authority (CA) sign server certificate should be assembled on the access gateway. If there is no CA-authorized certificate, we can generate this certificate request that is sent to the CA for signing. The steps for creating and installing a CA-signed certificate on the access gateway are:

- Private key creation
- Request for certificate signing
- Getting the signed certificate
- Binding this private key and certificate to the virtual server

## *4.3 Installing License on the Access Gateway*

Licensing processed in the following order in Fig. 5.

**Licensing Order**

1) Getting License Authorization Code (LAC) in email.
2) Assembling Access Gateway with host name.
3) Allocate Access Gateway licenses from MyCitrix.com. Using host name binding licenses to appliance in time of allocation process.
4) Generate this license entitlement and download this license file.
5) Install the license file on this Access Gateway.

**Fig. 5** Licensing processed in the following order

Introduced this license on access gateway in directory named; /nsconfig/license directory.

This license defines huge number of concurrent users which able to log on to this appliance. User licenses are locked to the appliance host name or FQDN. The host name needs to be changed in three places:

On the access gateway using the command line interface

- In the rc.conf file
- In the hosts file

The rc.conf and hosts file must always be in the /nsconfig directory. If these two files are not in the directory, they must be created using a text editor, such as Vi.

## 4.4 Creating Virtual Server

The virtual server is the key logical component of the access gateway. A virtual server created using configuration utility or command line interface is depicted in Fig. 6.

**Creating Virtual server using configuration utility**

- In the left pane of Configuration Utility, click SSL VPN.
- Select SSL VPN policy manager from the right pane.
- Select Virtual Servers from the Configured Policies / Resources menu.
- Select Create new virtual server from the Related Tasks menu.
- In Name, type the virtual server's name.
- Type the IP address in the IP address field. In Port, type port number.
- Select a certificate from the Available section of the Certificates tab, then click Add. If the certificate is signed by a CA, select Add as CA, then Create, then Close.
- At the top of the page in Configuration Utility, click Save.

**Fig. 6** Command line interfaces

## 4.5 Creating Local Users and Groups

Locally created authentication, users are created for assembling into groups which are situated at access gateway. Followed by the application takes place for session policies, authorization, and other activities like creating bookmarks, specifying applications, and specifying IP address of file shares, servers to which generally user has access.

## 4.6 Creating Name Server for Access Gateway

Name server can be added directly to the access gateway. This name server is then used by the access gateway to resolve DNS queries. We can add this name server by specifying just the IP address. This simplifies the process of configuring a DNS server for the access gateway. Name servers are bound to the access gateway globally.

## 4.7 Configuring Name Resolution

To allow for proper client operation through the access gateway, a valid DNS name server must be configured. Configure name of DNS service IP address and port number that DNS service listens on. There are two steps:

- Configure DNS service
- Binding DNS service to the virtual server

## 4.8 Browser Plugin

The access gateway provides two kinds of plugins that are automatically downloaded and executed on client devices after users are successfully authenticated.

## 4.9 Assembling of Both Access Gateway Appliances in High Availability (HA) Mode

To assemble primary or secondary access gateway in the high availability pair.

(i) Log on to either the primary or the secondary access gateway as an administrator using the default user name nsroot.
(ii) Set the password for the administrator account and RPC node to the same value. The password must be the same on both appliances.

(iii) Make sure the following entries are specified properly in the ns.conf file:
(a) IP address of the access gateway (b) ID and IP address of the access gateway
(iv) On the secondary access gateway, synchronize the certificates, licenses, startup scripts, and other configuration files with those on the primary access gateway.
(v) Modification of the present system IP address.
(vi) Disable necessary interfaces on primary and secondary access gateway, which not connected or being used for traffic.
(vii) Disable monitoring for any interfaces whose failure do not cause a high availability mode failover.
(viii) At a command prompt, type: save.
(ix) Connect the primary or secondary access gateway back to the network.

## 5 Result Analyses

By typing web address in browser, user connects to access gateway. A log on page appears, and user needs to provide prior given id and password. In case of configuration of external authentication of servers, access gateway communicates with server and defers whole process to it. If on board authentication is in use, user authentication is performed locally on the access gateway.

After successful authentication of user, initiation of access gateway tunnel starts. Permission of authenticated browser plugin being installed (after download) by access gateway is initiated. Alternatively, on user's device, secure access client is activated. In addition, access gateway installs an encrypted, per-session cookie on the user's computer which contains the user's authentication credentials. In the case of the Java applet plugin [18], the plugin begins with a list of resource Internet Protocol addresses (preconfigured), port numbers.

If there are any client-based security policies exist, access gateway checks. If there are, it runs those checks on client access. Verification of security on client device is related required steps like (1) security-related operating system updates, (2) antivirus protection, and (3) perfectly configuration of firewall. Access gateway either blocks user (for log on) or places them in a quarantine group action, when the client is unable to do security check. If log on fails, necessary updates or packages required download and install. We can also configure pre-authentication policies that check the client device before the user logs on and is authenticated. If the users do not pass the pre-authentication scan, they are not allowed to log on. If authentication is successful, the access gateway logs on client. The tunnel is now established, and all data of internal network traverse authenticated and encrypted tunnel.

After establishment of session, users are instructed. They can choose resources for accessing to an access gateway portal page. Users also see the plugin's secure remote session window in the lower right corner of their screen. This window remains on the users' desktop as long as the session is active.

Each time the client accesses an internal resource, the access gateway checks to see whether any application name-based configurations or time-of-day-based configurations are in place for this resource. If there are name-based configurations in place, the access gateway verifies that the user's client application is allowed to access the requested resource. If time-of-day-based configurations are in place, the access gateway verifies that the user is logging on during a permitted time of day.

Access gateway displays an error message to the client in case of failure to check. If both checks are feasible, some activities are done by access gateway:

(i) Finding of requested resources.
(ii) Establishment of secured connection between client and needed resources.

If there are both access gateway implementation configured as (1) high availability pair and (2) pair experiences a failover, all user sessions established with the primary access gateway are transferred to the secondary access gateway. Users do not have to log on again, although they might need to restart any applications they were using. The client may end this active session and may tick logout button in secure remote session window or in the secure access client, or by closing the secure remote session window. After logging off the session, client no longer has access to internal resources, and encrypted session cookie is also removed.

## 6 Conclusions and Future Scope

In present educational system, globally the cloud-based LMS model is now largely accepted as medium of new e-educational system. While talking of e-learning, users whether student or teacher or educational service provider can be situated remotely and access from anywhere or any geographical locations. Here is most important factor secured and user-friendly access to the given system as multiple users using same platform. And hence, the usability of appropriate VPN is utmost important. Not only the data security or secured access, but network traffic is also need to take care of while using the internet. SSL VPN and Citrix, all are having their role in network security, data encryption, and secured access of remote LMS users. To prevent internal as well as external threats, especially when users are more tending toward using mobile devices, One log-in or Single Sign-on (SSO) Log-in is getting importance among future security analysts so that one can access easy and reliable to remote data. For all cloud LMS applications, One Log-in access is getting popular as the trusted experience platform for.

(i) Easy, low-risk access
(ii) Enable security policies
(iii) Check VPN activities to lower the risk

In this paper, it has tried to suggest an optimum solution of using appropriate VPN, and in the future research work, it will be based on configuring cloud-end and user-end security policies and utilize proper authentication method based on distributed architecture and distributed database.

## References


1. Roy, S., Bose, R., Sarddar, D.: Self-servicing energy efficient routing strategy for smart forest. Braz. J. Sci. Technol. **3**(13), (2016)
2. Chawla, B.K., Gupta, O.P., Sawhney, B.K.: A review on IPsec and SSL VPN. Int. J. Sci. Eng. Res. **5**(11), 21–24 (2014)
3. Bose, R., Roy, S., Sarddar, D.: A billboard manager based model that offers dual features supporting cloud operating system and managing cloud data storage. Int. J. Hybrid. Inf. Technol. **8**(6), 229–236 (2015)
4. Jakimoski, K., Bogoevski, V., Kochov, D.: Carrier-class VPN to cloud evolution. Int. J. Grid. Distrib. Comput. **8**(6), 41–48 (2015)
5. Vallina-Rodriguez, N., Amann, J., Kreibich, C., Weaver, N., Paxson, V.: A tangled mass: the android root certificate stores. In: Proceedings of the 10th ACM International on Conference on Emerging Networking Experiments and Technologies, pp. 141–148, ACM, Sydney, Australia (2014)
6. Song, Y., Hengartner, U.:Privacyguard: a VPN-based platform to detect information leakage on android devices. In: Proceedings of the 5th Annual ACM CCS Workshop on Security and Privacy in Smart Phones and Mobile Devices, pp. 15–26, ACM, Denver, CO, USA (2015)
7. Fahl, S., Harbach, M., Muders, T., Baumgartner, L., Freisleben, B., Smith M.: Why eve and mallory love android: an analysis of android ssl (in) security. In: Proceedings of the 2012 ACM Conference on Computer and Communications Security, pp. 50–61, ACM, Raleigh, NC, USA (2012)
8. Abimbola, A.A., Munoz, J.M., Buchanan, W.J.: Nethost sensor: investigating the capture of end-to-end encrypted intrusive data. Comput. Secur. **25**(6), 445–451 (2006)
9. Martin, R.: Snort—lightweight intrusion detection for networks. In: Proceedings of the 13th USENIX Conference on System Administration, LISA '99, pp. 229–238, USENIX Association, Seattle, WA, USA(1999)
10. Li, X., Karanvir, S.G., Cooper, G.H., Guzik, J, R.: Encrypted data inspection in a network environment. US Patent 9176838 (2013)
11. He, G., Xu, B., Zhu, H.: AppFA: a novel approach to detect malicious android applications on the network. Secur. Commun. Netw. **2854728**, 1–15 (2018)
12. Niu, W., Zhang, X., Yang, G.W., Zhu, J., Ren, Z.: Identifying APT malware domain based on mobile DNS logging. Math. Probl. Eng. **4916953**, 1–9 (2017)
13. Kajal, R., Saini, D., Grewal, K.: Virutal private network. Int. J. Adv. Res. Comput. Sci. Softw. Eng. **2**(10), 428–432 (2012)
14. Leavitt, N.: Anonymization technology takes a high profile. Computer. **42**(11), 15–18 (2009)
15. Zhang, Z., Chandel, S., Sun, J., Yan, S., Yu, Y., Zang, J.: VPN: a boon or trap? : A comparative study of MPLs, IPSec, and SSL virtual private networks. In: Proceedings of The 2018 2nd International Conference on Computing Methodologies and Communication (ICCMC). IEEE, pp. 510–515. Erode, India (2018)
16. Karuna Jyothi, K., Indira Reddy, B.: Study on virtual private network (VPN), VPN's protocols and security. Int. J. Sci. Res. Comput. Sci, Eng. Inf. Technol. **3**(5) (2018)
17. Bhat, A. Z., Dalal, K. A.S., Singh, A.V.:Virtual private network as a service-a need for discrete cloud architecture. In: Proceedings of the 5th International Conference on Reliability, Infocom Technology and Optimization (Trends and Future Directions) (ICRITO 2016). IEEE, pp. 526–532 (2016)

18. Huang, C., Smith, P., Sun, Z.: Secure network solutions for enterprise cloud services. cloud technology: concepts, methodologies. Tools Appl. IGI Global Chap. 68, 1464–1486 (2015)
19. Fujkawa, H., Damiani, E., Yamamoto, y.: Network virtualization by differentially switched VPN for stable business communication with offshore computers. J. Reliable. Intell. Environ. **2**, 119–130 (2016)
20. IKram, M., Radriguez, N. V., Senevirante S., Ali Kaafar, M, Paxson, V.: An analysis of the privacy and security risks of android VPN permission-enabled. In: Proceedings of the Internet Measurement Conference IMC'16. pp. 349–364 ACM (2016)
21. Kuldeep, K., Singh, V.V., Gupta. H.: A new approach for the security of VPN. In: Proceedings of the 2nd International Conference on Information and Communication Technology for Competitive Strategies ICTCS'16. Vol. 13, pp 1–5. ACM (2016)
22. Guo, Y., Mohamed, I., Abou-Sayed, O., et al.: Cloud computing and web application based remote real-time monitoring and data analysis : slurry injection case study, Onshore USA. J. Petrol. Explor. Prod. Technol. **9**, 1225–1235 (2019)
23. Cornetto, G., Mateos, J., Touhafi, A. et al.:Design ,simulation and testing of a cloud platform for sharing digital fabrication resources for education. J. Cloud Comput. **8**(12) (2019)
24. Favale, T., Soro, F., Trevisan, M., Drago, I., Mellia, M.: Campus traffic and e-learning during covid-19 pandemic. Comput. Netw. (2020)
25. Chakraborty, S., Bose, R., Roy, S., Sarddar, D.: Auditing deployed software licenses on cloud using a secure loopback protocol. Int. J. Recent. Technol. Eng. **8**(3), 1–5 (2019)
26. Bose, R., Chakraborty, S., Roy, S.: Explaining the workings principle of cloud-based multi-actor authentication architecture on banking sectors. In: Proceedings of Amity International Conference on Artificial Intelligence, pp. 764–768 (2019)
27. Mukhopadhyay, B., Bose, R., Roy, S.: A novel approach to load balancing and cloud computing security using ssl in iaas environment. Int. J. Adv. Trends. Comput. Sci. Eng. **9**(2), 2362–2370 (2020)
28. Sarddar, D., Roy, S., Bose, R.: An efficient edge servers selection in content delivery network using voronoi diagram. Int. J. Recent and Innov. Trends. Comput. Commun. **2**(8), 2326–2330 (2014)
29. Chatterjee, P., Bose, R., Roy, S.: A review on architecture of secured cloud based learning management system. J. Xidian. Univ. **14**(7), 365–376 (2020)
30. Roger, D., Mathewson, N., Paul, S.: TOR: the second generation onion router. Technical Report, Naval Research Laboratory, Washington, DC, USA, (2004)
31. Yamada, A., Miyake, Y., Takemori, K., Studer, A., Perrig, A.: Intrusion detection for encrypted web accesses. In: Proceedings of The 21st International Conference on Advanced Information Networking and Applications Workshops (AINAW'07), vol. 1, pp. 569–576, Niagara Falls, Ont., Canada (2007)
32. Weaver, N., Kreibich, C., Dam, M., Paxson ,V.: Here be web proxies. In: Proceedings of the International Conference on Passive and Active Network Measurement, pp. 183–192, Springer, Los Angeles, CA, USA (2014)
33. Reis, C., Gribble, S.D., Kohno, T., Weaver, N.C.: Detecting in-flight page changes with web tripwires. In: Proceedings of the 5th USENIX Symposium on Networked Systems Design and Implementation, **8**, pp. 31–44, San Francisco, CA, USA. (2008)
34. Vallina-Rodriguez, N., Sundaresan, S., Kreibich,C., Paxson, V.: Header enrichment or ISP enrichment? emerging privacy threats in mobile networks. In: Proceedings of the 2015 ACM SIGCOMM workshop on hot topics in middle boxes and network function virtualization, pp. 25–30, ACM, London, UK (2015)
35. Weaver, N., Kreibich, C., Paxson, V.: Redirecting DNS for ads and profit. In: Proceedings of the UNISEX workshop on free and open communications on the internet 2011. **2**, San Francisco, CA, USA (2011)

36. Ikram, M., Vallina-Rodriguez, N., Seneviratne, S., Kaafar, M.A., Paxson, V.: An analysis of the privacy and security risks of android VPN permission-enabled apps. In: Proceedings of the 2016 Internet Measurement Conference, pp. 349–364, ACM, Santa Monica, CA, USA (2016)
37. Sudin, S., Ahmad, R.B., Syed Idrus, S.Z.: A model of virus infection dynamics in mobile personal area network. J Telecommun. Electron. Comput. Eng. **10**(2–4), 197–201 (2018).
38. Goh, V. T., Zimmermann, J., Looi, M.:Towards intrusion detection for encrypted networks. In: Proceedings of the 2009 International Conference on Availability, Reliability and Security. IEEE, pp. 540–545, Fukuoka, Japan (2009)

# Surface EMG Signal Classification for Hand Gesture Recognition

**Priyanshu Raj, Binish Fatimah, and B. Sushma**

**Abstract** This paper develops a classification algorithm to recognize basic hand movements using surface electromyography (sEMG) signals. This can be used in various applications related to brain computer interface (BCI), in particular for orthotic exoskeletons. The algorithm is developed by decomposing the given sEMG signal into narrowband signals and computing features like mean, variance, skewness, kurtosis, and Renyi entropy from each of the sub-band signals so obtained. The performance of three popular signal decomposition algorithms namely variational mode decomposition, discrete wavelet transform, and empirical mode decomposition are compared for a publicly available dataset. The dataset includes six basic hand gestures namely lateral, palmar, cylindrical, hook, tip, and spherical. The average accuracy obtained for recognizing six hand gestures for five healthy subjects is 95.33% using variational mode decomposition, 97.78% using empirical mode decomposition, and 97.89% using discrete wavelet transform. The proposed work studies the efficacy of using two-channel sEMG signal for recognizing these hand movements.

**Keywords** Surface EMG · Variational mode decomposition · Machine learning · Discrete wavelet transform · Pattern recognition · Empirical mode decomposition

## 1 Introduction

Around ten million of the world population are reported as amputee, in which three million are armed amputee. Approximately half million amputees are reported in India, with more than 23,500 cases reported every year [22]. The majority of these

P. Raj · B. Fatimah (✉) · B. Sushma
Department of Electronics and Communication Engineering, CMR Institute of Technology, Bengaluru, India
e-mail: binish.f@cmrit.ac.in

P. Raj
e-mail: priy17ec@cmrit.ac.in

B. Sushma
e-mail: sushma.b@cmrit.ac.in



D. Gupta et al. (eds.), *Pattern Recognition and Data Analysis with Applications*, Lecture Notes in Electrical Engineering 888,
https://doi.org/10.1007/978-981-19-1520-8_10

cases belong to low income working age group, and these accidents affects their life tremendously. Significant advances in prosthetic limbs have been reported by the medical community and in the robotics area using electroencephalogram (EEG), electromyography (EMG), and surface electromyography (sEMG) signals. Exoskeleton prosthetic limbs can help people with amputee limbs to perform daily life activities such as basic hand gestures using myoelectric control systems.

Biomedical signals can capture vital information regarding the functioning of human body and are used extensively to diagnose various pathological conditions. Some of these signals can carry similar information, and the choice of the biomedical signal depends on the application in hand. EMG and EEG data acquisition are not as convenient and user-friendly as sEMG, which can effectively capture the required muscular information and therefore can be used in hand gesture detection. The sEMG signals are collected in a non-invasive manner and can capture the neuromuscular activity in the form of an electrical signal. The research and technological advances in the field of biomedical sensors and devices such as the Myo armbands have created an opportunity for researchers to explore these signals for a variety of applications related to brain computer interface devices. sEMG signals can be used to develop healthcare devices for assisting people with amputee limbs and for patients with neuro-degenerative diseases to help them in their daily activities. Also, depending on the extent of damage to amputee limb the sEMG-based assisting device can be manufactured with different degrees of freedom.

Authors in [4] used auto regression coefficients, Hjorth features, integral absolute value, mean absolute value, root mean square, and cepstral features to classify ten hand movements using myoelectric signals. An average accuracy of 92.3% was obtained with the multiclass support vector machines (SVMs) with radial basis function as the kernel. Vasanthi and Jayasree [35] computed various time domain features and compared the results obtained with machine learning algorithms, deep learning networks, ANN, and cascaded feed forward ANN. Here, support vector machine classifier gives the best result of 98.88%. Authors in [25] used support vector machine to classify fifteen hand gestures using sEMG signals collected using eight sensors. The best accuracy of 79.36% was obtained using radial basis function as the kernel.

Ahsan et al. [1] computed root mean square value, standard deviation, variance, mean absolute value, waveform length, zero-crossings, and slope sign change to train artificial neural network (ANN) to detect four hand movements collected from three subjects. ANN has been explored by various authors, such as in [17] neural network was trained with signals collected from a number of subjects to classify four hand gestures. ANN has also been used by Zhang et al. [39] to develop a real-time hand gesture identification algorithm. The algorithm classifies five hand gestures collected from twelve subjects with an average classification accuracy of 98.7%. In [28], sEMG signal has been used for hand movement recognition for a bionic hand.

Geng et al. [15] showed that the instantaneous values of high-density sEMG can be effectively used for hand movement recognition. The sEMG images of eight hand movements were used with deep convolutional network and an accuracy of 99% was obtained using majority voting over 40 frames. In [26], a hand gesture recognition algorithm has been proposed which is robust to different arm postures. In order to do so, the authors have collected EMG signals and signals from an accelerometer. Features such as the average value and the waveform durations are used to classify eight hand gestures based on the maximum likelihood estimation. Tunable Q-wavelet transform (TQWT) has been used in [23, 33] to decompose the sEMG signal. In [23], a TQWT-based filter bank was developed and Kraskov entropy was computed from each sub-band signal. Subasi and Qaisar in [33] used the mean absolute value, average power, standard deviation, skewness, kurtosis of the coefficients obtained from the sub-band signals, and the absolute mean value ratios of the neighbouring sub-band signals.

In [21], intrinsic mode functions (IMFs) are obtained using empirical mode decomposition (EMD) for four channel sEMG signals collected for seven hand gestures of thirty subjects. Deep convolutional network based on ResNet are then used with the first three IMFs to obtain the required identification. EMD is a popular choice for non-stationary signals such as biomedical signals. Authors in [27, 38] have also used EMD to decompose sEMG signals for hand movement classification. Sapsanis et al. in [27] used various time domain features such as the mean of the absolute values of signal, number of slope sign changes, waveform length, number of zero-crossings, and statistical features such as variance, kurtosis, and skewness. The algorithm was validated on a publicly available dataset and an average accuracy of 89.21% was obtained for classifying six hand movements of five subjects. Yan et al. [38] used autoregressive (AR) model parameters obtained for each IMF and classified four hand gestures using least squares support vector machines. In [37], variational mode decomposition (VMD) has been used to represent the sEMG signals as variational mode functions (VMFs). Composite permutation entropy index is computed from each of these VMFs, and machine learning algorithms are then employed to classify the hand gestures. The FDM has shown its efficacy in many applications such as detection of sleep apnoea events [12], modelling, audio signal processing [11], ECG and EEG signal analysis [10, 13, 14, 31].

In this work, we present the comparison of the performance of popular signal decomposition techniques including VMD, EMD, and DWT (discrete wavelet transform). Each sEMG signal is decomposed into multi-scale components and time based and statistical features including mean, variance, skewness, kurtosis, and Renyi entropy are computed for each sub-band signal. Different machine learning algorithms are then used to classify the feature space. A freely available dataset from UCI machine learning repository has been used in this work to test the hand gesture classification algorithms based on each decomposition scheme.

The paper is presented in five sections. Section 2 provides a detailed discussion on the dataset, and Sect. 3 presents the proposed algorithm. Simulation studies and conclusions are presented in Sects. 4 and 5, respectively.

## 2 Dataset

The dataset used here is acquired from the UC Irvine machine learning repository, under the name "sEMG for Basic Hand movements Data Set". It includes two databases, where the first contains sEMG signals collected from two male and three female participants. The subjects considered in the study does not have an amputee limb and thus can be treated as sample from healthy population. Each subject performs six hand movements namely tip (TI), spherical (SP), lateral (LA), palmar (PA), hook (HO), and cylindrical (CY). Each gesture is repeated 30 times. The sEMG signals in the dataset have been acquired using a two channel programming kernel of the National Instruments (NI) Labview. sEMG signals have been de-noised using frequency selective filters, and the signal obtained after processing lies between 15 and 500 Hz.

The second database includes the sEMG signal acquired over three days from one healthy male participant for six hand grasps. Each movement is conducted hundred times over three consecutive days. This database unlike the first can be used to test the time invariance property of the hand movement recognition algorithm.

## 3 Methodology

The machine learning-based algorithm developed in this paper consist of decomposing the de-noised sEMG signals using multi-scale decomposition techniques and extracting features from the sub-band signals so obtained, as shown in Fig. 2. Different machine learning algorithms are then trained using the feature set. In the dataset considered in this work, the sEMG signal has been collected using two channels, we could either take correlation of these channels as the single input to the proposed scheme as done in [23] or we can consider individual information which will give us a feature vector in a higher dimensional space as considered in this work. The sEMG signals are represented as multi-scale components using three algorithms including VMD, EMD, and DWT.

EMD was proposed by Huang in [19] as an adaptive time-frequency analysis algorithm for non-stationary and nonlinear signals. EMD decomposes the signal into finite multi-scale components termed as intrinsic mode functions (IMFs). The set of IMFs makes complete basis for the given signal and should fulfil two conditions, the number of extrema and the number of zero-crossings should be equal or their difference is not more than one. The second condition states that at any instant the mean value of the envelope defined by the local maxima and the envelope defined by the local minima is zero. EMD has been employed in umpteen signal processing applications like denoising, pattern recognition, neuroscience, financial time series prediction, ocean data and seismic data analysis, etc. [3, 7, 16, 18, 32]. EMD is not robust to noise and suffers from sifting issues, moreover it is based on an empirical algorithm not on mathematical equations. In order to overcome these limitations, various authors have presented variants of EMD [5, 36].

VMD decomposes signal into intrinsic modes known as variational mode functions. It was proposed in 2015 by [9] to improve the noise and sampling properties of EMD. Unlike EMD, it is a non-recursive adaptive algorithm to obtain VMFs concurrently such that backward error can be taken into account. VMD decomposes the given signal into finite number of narrowband signals such that the VMFs reconstructs the given signal exactly or in the leasts squares sense. The VMFs, $v_i(t)$, of a continuous time finite energy signal $x(t)$ are given as

$$x(t) = \sum_i v_i(t) = \sum_i A_i(t)\cos(\phi_i(t)) \tag{1}$$

where $A_i(t)$ is the instantaneous amplitude and $\phi_i(t)$ is the instantaneous phase of $v_i(t)$. Here, each $v_i(t)$ is sparse with specific properties. For more details, refer [9].

DWT decomposes the given signal into dyadic sub-band signals. Unlike EMD and VMD, DWT is not a signal adaptive algorithm. DWT has been used by various researchers in varied applications including denoising, feature extraction, image processing, etc. [8, 24, 29]. Researchers have used DWT for multi-scale modelling of various stationary and cyclostationary signal, and it has also been explored for non-stationary and nonlinear signals as well. If $\psi[n]$ is a wavelet with a support in $[-K/2, K/2]$, a discrete wavelet scaled by $a^j$, is expressed as

$$\psi_j[n] = \frac{1}{\sqrt{a^j}}\psi\left(\frac{n}{a^j}\right), \qquad 1 \le a^j \le NK^{-1} \tag{2}$$

The discrete scaling filter, $\phi_j[n]$ is defined as

$$\phi_j[n] = \frac{1}{\sqrt{a^j}}\phi\left(\frac{n}{a^j}\right) \tag{3}$$

DWT decomposes successively each approximation $a_j \in V_j$ into a coarser approximation $a_{j+1} \in V_{j+1}$, and the detailed coefficient $d_{j+1} \in W_{j+1}$. $\{\phi_{j,n}\}_{n\in\mathbb{Z}}$ and $\{\psi_{j,n}\}_{n\in\mathbb{Z}}$ are orthonormal bases of $V_j$ and $W_j$. The approximate coefficients and detailed coefficients of level $j+1$, represented as $a_{j+1}$ and $d_{j+1}$, respectively, are obtained using the following equation:

$$a_{j+1}[n] = a_j * h[2n] \tag{4}$$

$$d_{j+1}[n] = a_j * f[2n] \tag{5}$$

where "$*$" denotes convolution, $h[n]$ is the impulse response of low-pass filter, $H(z)$, and $f[n]$ is the impulse response of the high-pass filter, $F(z)$ as shown in Fig. 1.

The narrowband components obtained using EMD, VMD, or DWT are then used to compute the features. Considering the performance of various time domain and frequency domain features, we have chosen the following time domain statistical features for the problem addressed in this work (Fig. 2).

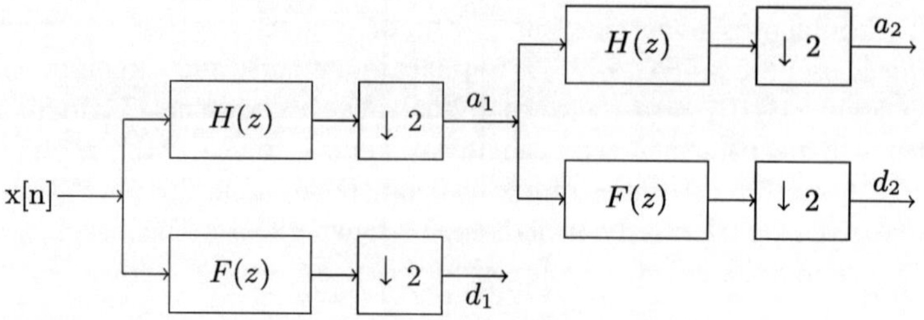


**Fig. 1** Block diagram of DWT

1. Mean value of the $k$th sub-band signal

$$\mu_k = \frac{1}{L}\sum_{i=1}^{L} s_k[i], \tag{6}$$

where $s_k[i]$ denotes the $k$th sub-band signal and $L$ is the length of the signal.

2. Variance of the $k$th sub-band signal

$$\sigma_k^2 = \frac{1}{L}\sum_{i=1}^{L} (s_k[i] - \mu_k)^2, \tag{7}$$

3. Skewness of the $k$th sub-band signal

$$\text{Skewness} = \sum_{i=1}^{L}\left(\frac{s_k[i] - \mu_k}{\sigma_k}\right)^3, \tag{8}$$

4. Kurtosis of the $k$th sub-band signal

$$\text{Kurtosis} = \sum_{i=1}^{L}\left(\frac{s_k[i] - \mu_k}{\sigma_k}\right)^4 \tag{9}$$

5. Renyi entropy of the $i$th sub-band signal

$$\text{Ent} = \frac{1}{1-\alpha} log_2\left(\sum_{i=1}^{L} p\left(s_k[i]\right)^{\alpha}\right), \tag{10}$$

where $p(s_k[i])$ is the discrete probability of $s_k[i]$, $\alpha$ is the order of the Renyi entropy, $\alpha \geq 0$ and $\alpha \neq 1$.

The feature vector, thus, obtained for both channels are used to train machine learning algorithms. Performance of machine learning-based recognition algorithms

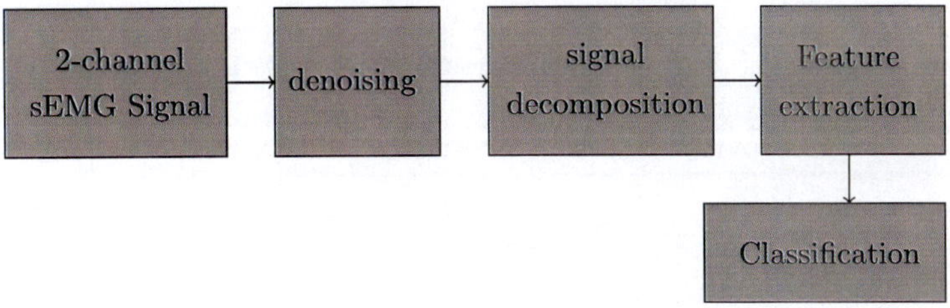


**Fig. 2** Proposed methodology

depend on the feature vector used and, also, on the machine learning algorithm selected. In the next section, to choose the best classifier, we will compare various machine learning classifiers using extracted feature set based on performance metrics used in classification algorithms.

## 4 Numerical Results

We now discuss the simulation results procured using the proposed algorithm. Table 1 presents the results obtained for classifying the six hand movements when the selected signal decomposition scheme is VMD. Here, the first three VMFs are used for feature extraction as increasing the number of VMFs did not improve the recognition rate. A 10-fold cross-validation scheme has been used in this work with different machine learning algorithms such as SVMs with linear, quadratic, cubic and Gaussian kernels, ensemble bagged trees (EBT), k-neighbouring neighbour (kNN), ensemble subspace discriminant (ESD), and ensemble subspace kNN (ESkNN). The best accuracy obtained for Sub#1 is 93.89% using SVM cubic, 96.11% for Sub#2 with ESD, 97.22% for Sub#3 with EBT, 94.44% for Sub#4 for SVM cubic, and 95.00% for Sub#5 for EBT. The simulations have been carried on MATLAB 2020b.

Results attained using the EMD algorithm are presented in Table 2. The best results as reported in the table are obtained using the first two IMFs. The best accuracy obtained for Sub#1 is 95.56% using SVM linear and quadratic, 97.78% for Sub#2 with linear discriminant, 97.78% for Sub#3 with EBT, 98.89% for Sub#4 for SVM quadratic, and 98.89% for Sub#5 for SVM quadratic and linear.

The results obtained using the DWT are shown in Table 3. Wavelet Symlets four have been used in the DWT. The best accuracy obtained for Sub#1 is 95.56% using EBT and ESD, 98.33% for Sub#2 with ESD, 98.33% for Sub#3 with EBT, 98.33% for Sub#4 for ESD, and 98.89% for Sub#5.

Table 4 presents the confusion matrix obtained when the second database is considered and EBT classifier is used. The classification accuracy of 83.2% is obtained in this case. The signals in the first database were acquired in a single session and therefore does not give an idea about the time variance property of the hand movement detection algorithm.

**Table 1** Performance comparison of several machine learning classifiers with 10-fold cross-validation for each subject using first three VMFs obtained with VMD

| Classifier | Accuracy (%) for five subjects | | | | |
|---|---|---|---|---|---|
| | Sub#1 | Sub#2 | Sub#3 | Sub#4 | Sub#5 |
| Linear discriminant | 93.33 | 95.56 | 89.44 | 84.44 | 90.00 |
| SVM linear | 92.78 | 96.11 | 92.78 | 92.22 | 93.33 |
| SVM quadratic | 93.33 | 93.33 | 94.44 | 93.89 | 94.44 |
| SVM cubic | **93.89** | 93.33 | 92.22 | **94.44** | 92.78 |
| SVM Gaussian | 90.56 | 94.44 | 90.00 | 91.67 | 92.22 |
| kNN | 88.33 | 92.78 | 80.56 | 76.67 | 92.78 |
| EBT | 90.56 | 93.33 | **97.22** | 92.22 | **95.00** |
| ESD | 91.67 | **96.11** | 92.22 | 87.22 | 88.89 |
| ESkNN | 82.78 | 89.44 | 87.22 | 86.11 | 93.33 |

The best results are bolded

**Table 2** Performance comparison of several machine learning classifiers with 10-fold cross-validation for each subject using first two IMFs obtained using EMD

| Classifier | Accuracy (%) for five subjects | | | | |
|---|---|---|---|---|---|
| | Sub#1 | Sub#2 | Sub#3 | Sub#4 | Sub#5 |
| Linear discriminant | 95.00 | **97.78** | 92.78 | 98.33 | **98.89** |
| SVM linear | **95.56** | 96.11 | 93.89 | 98.33 | **98.89** |
| SVM quadratic | **95.56** | 95.00 | 96.11 | **98.89** | **98.89** |
| SVM cubic | 95.00 | 94.44 | 93.33 | 98.33 | 98.33 |
| SVM Gaussian | 93.33 | 93.33 | 94.44 | 95.56 | 95.00 |
| kNN | 86.67 | 91.67 | 88.33 | 93.33 | 96.67 |
| EBT | 94.44 | 95.56 | **97.78** | 96.67 | 96.67 |
| ESD | 93.89 | 97.78 | 92.78 | 97.22 | 97.78 |
| ESkNN | 87.22 | 92.78 | 94.44 | 94.44 | 94.44 |

The best results are bolded

From Tables 1, 2 and 3, it is noted that for the chosen features and dataset, the performance of DWT is superior than VMD and EMD. Finally, we tabulate the results presented by various authors in the literature for the UCI dataset in Table 5. For the proposed framework, DWT performs better than VMD and EMD, however, the obtained accuracies are low compared to algorithms presented in [23, 30]. While [23] utilized TQWT based filter bank, [30] obtained better results using multichannel convolutional neural networks.

**Table 3** Performance comparison of machine learning classifiers when the decomposition scheme used is DWT

| Classifier | Accuracy (%) for five subjects | | | | |
|---|---|---|---|---|---|
| | Sub#1 | Sub#2 | Sub#3 | Sub#4 | Sub#5 |
| SVM linear | 93.33 | 95.56 | 95.56 | 97.78 | 97.22 |
| SVM quadratic | 93.89 | 96.67 | 95.56 | 97.22 | **98.89** |
| SVM cubic | 93.33 | 96.11 | 94.44 | 96.67 | 98.33 |
| SVM Gaussian | 91.11 | 95.00 | 94.44 | 95.66 | 96.11 |
| kNN | 81.11 | 91.11 | 85.66 | 90.00 | 96.67 |
| EBT | **95.56** | 96.11 | **98.33** | 97.22 | 97.78 |
| ESD | **95.56** | **98.33** | 96.11 | **98.33** | **98.89** |
| ESkNN | 89.44 | 93.89 | 90.00 | 93.89 | 94.44 |

The best results are bolded

**Table 4** Confusion matrix obtained for the second database using EBT classifier

| Predicted class → True class ↓ | Lateral | Tip | Spherical | Cylindrical | Palmar | Hook |
|---|---|---|---|---|---|---|
| Lateral | 220 | 13 | 0 | 2 | 53 | 12 |
| Tip | 15 | 224 | 0 | 5 | 29 | 7 |
| Spherical | 0 | 1 | 292 | 6 | 0 | 1 |
| Cylindrical | 1 | 11 | 2 | 268 | 0 | 18 |
| Palmar | 57 | 25 | 1 | 0 | 209 | 8 |
| Hook | 10 | 13 | 1 | 13 | 7 | 256 |

**Table 5** Performance comparison of the proposed algorithm with the existing hand movement recognition algorithms using common dataset

| Author | Mean CA (%) for five subjects | | | | | |
|---|---|---|---|---|---|---|
| | Sub#1 | Sub#2 | Sub#3 | Sub#4 | Sub#5 | Average |
| Sapsanis et al. [22] | 87.25 | 88.05 | 85.53 | 90.42 | 94.80 | 89.21 |
| Iqbal et al. [20] | 82.78 | 87.67 | 83.11 | 90 | 90 | 86.71 |
| Akben [2] | 93.04 | 86.66 | 97 | 99.23 | 97.66 | 94.72 |
| Too et al. [34] | – | – | – | – | – | 95.74 |
| Bergil et al. [6] | 90.90 | 94.83 | 97.83 | 94.85 | 96.37 | 94.96 |
| Sikder et al. [30] | 98.15 | 98.15 | 96.3 | 100 | 100 | 98.52 |
| Nishad et al. [23] | 98.33 | 97.78 | 99.44 | 98.89 | 98.83 | 98.55 |
| Proposed work (DWT) | 95.56 | 98.33 | 98.33 | 98.83 | 98.89 | 97.89 |

# 5 Conclusions

In this paper, the performance of VMD, EMD, and DWT algorithms is compared for sEMG signal classification application. The dataset used in the paper consists of sEMG signals collected from five healthy subjects for six most commonly used hand gestures. Each sEMG signal is first decomposed into multiple sub-band signals using VMD, EMD, or DWT algorithms. Time domain and statistical features are then computed for each narrowband constituents of the sEMG signal so obtained. The average accuracy reported by various machine learning algorithm is 95.33% with VMD, 97.78% with EMD, and 97.89% with DWT. The accuracy can be increased with deep learning and ANN.

# References


1. Ahsan, M.R., Ibrahimy, M.I., Khalifa, O.O.: Electromyography (EMG) signal based hand gesture recognition using artificial neural network (ANN). In: 2011 4th International Conference on Mechatronics (ICOM), pp. 1–6 (2011)
2. Akben, S.: Low-cost and easy-to-use grasp classification, using a simple 2-channel surface electromyography. Biomed. Res. **28**, 577–582 (2017)
3. Ali, H., Hariharan, M., Yaacob, S., Adom, A.H.: Facial emotion recognition using empirical mode decomposition. Expert Syst. Appl. **42**(3), 1261–1277 (2015)
4. Amamcherla, N., Turlapaty, A., Gokaraju, B.: A machine learning system for classification of emg signals to assist exoskeleton performance. In: 2018 IEEE Applied Imagery Pattern Recognition Workshop (AIPR), pp. 1–4 (2018)
5. Barbosh, M., Singh, P., Sadhu, A.: Empirical mode decomposition and its variants: a review with applications in structural health monitoring. Smart Mater. Struct. **29**(9), 093001 (2020)
6. Bergil, E., Oral, C., Ergul, E.: Efficient hand movement detection using k-means clustering and k-nearest neighbor algorithms. J. Med. Biol. Eng. 1–14 (2020)
7. Chacko, A., Ari, S.: Denoising of ECG signals using empirical mode decomposition based technique. In: IEEE International Conference on Advances in Engineering, Science and Management (ICAESM-2012), pp. 6–9 (2012)
8. Chakraborty, A., Banerjee, A.: An adaptive and automated image fusion algorithm based on DWT for real time applications. In: 2019 4th International Conference on Information Systems and Computer Networks (ISCON), pp. 677–682 (2019)
9. Dragomiretskiy, K., Zosso, D.: Variational mode decomposition. IEEE Trans. Signal Process. **62**(3), 531–544 (2014)
10. Fatimah, B., Javali, A., Ansar, H., Harshitha, B.G., Kumar, H.: Mental arithmetic task classification using Fourier decomposition method. In: 2020 International Conference on Communication and Signal Processing (ICCSP), pp. 0046–0050 (2020)
11. Fatimah, B., Preethi, A., Hrushikesh, V., Singh B.A., Kotion, H.R.: An automatic siren detection algorithm using Fourier decomposition method and MFCC. In: 2020 11th International Conference on Computing, Communication and Networking Technologies (ICCCNT), pp. 1–6 (2020)
12. Fatimah, B., Singh, P., Singhal, A., Pachori, R.B.: Detection of apnea events from ECG segments using Fourier decomposition method. Biomed. Signal Process. Control **61**, 102005 (2020)
13. Fatimah, B., Singh, P., Singhal, A., Pachori, R.B.: Hand movement recognition from SEMG signals using Fourier decomposition method. Biocybern. Biomed. Eng. (2021)

14. Fatimah, B., Singh, P., Singhal, A., Pramanick, D., Pranav, S., Pachori, R.B.: Efficient detection of myocardial infarction from single lead ECG signal. Biomed. Signal Process. Control **68**, 102678 (2021)
15. Geng, W., Du, Y., Jin, W., Wei, W., Hu, Y., Li, J.: Gesture recognition by instantaneous surface EMG images. Sci. Rep. **6**, 36571 (2016)
16. Han, J., van der Baan, M.: Empirical mode decomposition for seismic time-frequency analysis. GEOPHYSICS **78**(2), O9–O19 (2013)
17. Hasan, M.M., Rahaman, A., Shuvo, M.F., Ovi, M.A.S., Rahman, M.M.: Human hand gesture detection based on EMG signal using ANN. In: 2014 International Conference on Informatics, Electronics Vision (ICIEV), pp. 1–5 (2014)
18. Hong, L.: Decomposition and forecast for financial time series with high-frequency based on empirical mode decomposition. Energy Procedia **5**, 1333–1340 (2011), 2010 International Conference on Energy, Environment and Development—ICEED2010
19. Huang, N., Shen, Z., Long, S., Wu, M., Shih, H., Zheng, Q., Yen, N.C., Tung, C.C., Liu, H.: The empirical mode decomposition and the Hilbert spectrum for nonlinear and non-stationary time series analysis. Proc. R. Soc. Lond. Ser. A Math. Phys. Eng. Sci. **454**, 903–995 (1998)
20. Iqbal, O., Fattah, S.A., Zahin, S.: Hand movement recognition based on singular value decomposition of surface EMG signal. In: Proceedings of IEEE Region 10 Humanitarian Technology Conference, pp. 837–842 (2017)
21. Kisa, D.H., Ozdemir, M.A., Guren, O., Akan, A.: EMG based hand gesture classification using empirical mode decomposition time-series and deep learning. In: 2020 Medical Technologies Congress (TIPTEKNO), pp. 1–4 (2020)
22. Mishra, S., Nayak, S., Sahoo, P., Sharma, M., Equebal, A.: Impact of exoskeletal prosthesis on energy expenditure in female amputees during walking with two different level of amputation. Int. J. Health Sci. Res. **10**, 377–381 (2020)
23. Nishad, A., Upadhyay, A., Pachori, R., Acharya, U.R.: Automated classification of hand movements using tunable-Q wavelet transform based filter-bank with surface electromyogram signals. Future Gener. Comput. Syst. **93** (2018)
24. Pasti, L., Walczak, B., Massart, D., Reschiglian, P.: Optimization of signal denoising in discrete wavelet transform. Chemom. Intell. Lab. Syst. **48**(1), 21–34 (1999)
25. Pomboza-Junez, G., Terriza, J.H.: Hand gesture recognition based on SEMG signals using support vector machines. In: 2016 IEEE 6th International Conference on Consumer Electronics—Berlin (ICCE-Berlin), pp. 174–178 (2016)
26. Rhee, K., Shin, H.C.: Electromyogram-based hand gesture recognition robust to various arm postures. Int. J. Distrib. Sens. Netw. **14**(7), 1550147718790751 (2018)
27. Sapsanis, C., Georgoulas, G., Tzes, A., Lymberopoulos, D.: Improving EMG based classification of basic hand movements using EMD, vol. 2013, pp. 5754–5757 (2013)
28. Shi, W.T., Lyu, Z.J., Tang, S.T., Chia, T.L., Yang, C.Y.: A bionic hand controlled by hand gesture recognition based on surface EMG signals: a preliminary study. Biocybern. Biomed. Eng. **38**(1), 126–135 (2018)
29. Shirazi, F.A., Mahjoob, M.J.: Application of discrete wavelet transform (DWT) in combustion failure detection of IC engines. In: 2007 5th International Symposium on Image and Signal Processing and Analysis, pp. 482–486 (2007)
30. Sikder, N., Mohammad Arif, A.S., Nahid, A.: Heterogeneous hand guise classification based on surface electromyographic signals using multichannel convolutional neural network. In: 2019 22nd International Conference on Computer and Information Technology (ICCIT), pp. 1–6 (2019)
31. Singhal, A., Singh, P., Fatimah, B., Pachori, R.B.: An efficient removal of power-line interference and baseline wander from ECG signals by employing Fourier decomposition technique. Biomed. Signal Process. Control **57**, 101741 (2020)
32. Sonali, Singh, O., Sunkaria, R.K.: ECG signal denoising based on empirical mode decomposition and moving average filter. In: 2013 IEEE International Conference on Signal Processing, Computing and Control (ISPCC), pp. 1–6 (2013)

33. Subasi, A., Qaisar, S.: Surface EMG signal classification using TQWT, bagging and boosting for hand movement recognition. J. Ambient Intell. Humaniz. Comput. (2020)
34. Too, J., Abdullah, A.R., Saad, N.M.: Classification of hand movements based on discrete wavelet transform and enhanced feature extraction. Int. J. Adv. Comput. Sci. Appl. **10**(6), 83–89 (2019)
35. Vasanthi, S.M., Jayasree, T.: Performance evaluation of pattern recognition networks using electromyography signal and time-domain features for the classification of hand gestures. Proc. Inst. Mech. Eng. Part H J. Eng. Med. **234**(6), 639–648 (2020)
36. Wu, Z., Huang, N.E.: Ensemble empirical mode decomposition: a noise-assisted data analysis method. Adv. Adapt. Data Anal. **01**(01), 1–41 (2009)
37. Xiao, F., Yang, D., Lv, Z., Guo, X., Liu, Z., Wang, Y.: Classification of hand movements using variational mode decomposition and composite permutation entropy index with surface electromyogram signals. Future Gener. Comput. Syst. **110**, 1023–1036 (2020)
38. Yan, Z.G., Wang, Z., Ren, X.: Joint application of feature extraction based on EMD-AR strategy and multi-class classifier based on LS-SVM in EMG motion classification. J. Zhejiang Univ. Sci. A **8**, 1246–1255 (2007)
39. Zhang, Z., Yang, K., Qian, J., Zhang, L.: Real-time surface EMG pattern recognition for hand gestures based on an artificial neural network. Sensors **19**(14) (2019)

# Improved Energy Efficiency in Street Lighting: A Coverage-Based Approach

**Tanmoy Dey and Parag Kumar Guha Thakurta**

**Abstract** An approach to reduce the energy consumption of the street lighting system is proposed in this paper through the efficient deployment of sensor-based lights over a geographical area. The target area for installing the streetlights is divided into multiple Voronoi cells with non-overlapping coverage. Each such non-overlapping region contains at least one street light. These lights are equipped with different sensors for controlling the intensity of light depending on street conditions. An auto changeover technique has been used here to regulate the illuminance level of the lights depending on the road conditions across the different duration of a day. The lighting unit is equipped with solar photovoltaic cells, which convert solar energy to electricity to initiate the battery charging during the daytime. The streetlight can automatically switch to a utility power source if there is an insufficient charge in the battery. The simulation results show the effectiveness of the proposed approach over existing street lighting techniques.

**Keywords** WSN · Street light · Deployment · Target area · Coverage · Energy · Intensity control · Illuminance · Threshold

## 1 Introduction

Nowadays, wireless sensor networks (WSNs) have typical coverage applications [1] of an area under surveillance using sensor nodes. One of those is the efficient deployment of streetlights to obtain less energy consumption. A WSN-based smart lighting [2] can monitor an area by sensing the environment, communicating with neighboring nodes, and performing pre-determined operations based on the data collected from the environment. In case of installation, the streetlights are deployed

T. Dey (✉) · P. K. G. Thakurta
National Institute of Technology Durgapur, Durgapur, West Bengal, India
e-mail: tanmoydeynitd@gmail.com

P. K. G. Thakurta
e-mail: paragkumar.guhathakurta@cse.nitdgp.ac.in



D. Gupta et al. (eds.), *Pattern Recognition and Data Analysis with Applications*, Lecture Notes in Electrical Engineering 888,
https://doi.org/10.1007/978-981-19-1520-8_11

so that the target area will be fully covered, as said in [3], with either no or minimum overlapping of the coverage area of two neighboring sensors.

According to a survey in [4], there are more than 315 million streetlights in the entire world, which emits nearly 110 million tons of $CO_2$ per year, which is almost 5% of global $CO_2$ emissions. A 40% of electrical energy is getting wasted, which costs nearly 20 billion dollars [5]. Therefore, the design of a cost-effective utilization of streetlights and carbon emission reduction is the primary goal. To achieve this, an efficient light-emitting diode (LED) luminary with illumination level control [5] is the current need. India consumes about 20% of electric power for both street lighting systems and residential lighting [5]. Here, the streetlights take a significant part, while India faces a deficiency of electricity [6]. This paper highlights an efficient deployment of the streetlights with LED lights through a wireless sensor network interface for controlling the illuminance level of the light so that unnecessary electricity wastage can be reduced.

An approach to reduce the energy consumption of the street lighting system is proposed here through efficient deployment of the sensor-based lights. In order to install those lights competently, the target area is divided into multiple Voronoi cells with non-overlapping coverage. These lights are equipped with motion sensors for controlling light. In this arrangement, the proposed solar-powered LED streetlight is charged with adequate solar energy. An auto changeover technique is used such that the streetlight can automatically switch to a utility power supply when the battery has insufficient charge. The charging of the battery of the sensor node in the daytime is initiated via a Photovoltaic (PV) solar panel. In the dusk, the streetlight is automatically turned on with a minimum threshold intensity level ($I_{Th}$) with the help of a light-dependent resistor (LDR). Subsequently, the battery starts discharging. Until sunrise, if the street light detects any object movement, the intensity of light increases from such minimum threshold level to maximum intensity level for a preset time period. When this preset period expires, the intensity of light would gradually reduce to $I_{Th}$ level. In the meantime, if the sensor further perceives any movement, the intensity is increased again to the maximum. On the other hand, if no motion is detected in that preset time, the light would turn OFF automatically and remains in passive mode until either any movement is detected or any signal is received from neighboring lights. The simulation results represent the acceptability and effectiveness of the proposed method over other existing techniques in terms of various aspects of energy efficiency.

The remaining part of this paper has been structured as follows: Sect. 2 discusses a brief study of the related works. Section 3 presents the preliminaries and system model followed by the problem statement discussed in Sect. 4. Section 5 describes the proposed methodology. Various simulation results have been given in Sect. 6. Finally, the paper is concluded in Sect. 7.

## 2 Related Work

Many kinds of research have been done on monitoring an area toward energy efficiency using WSN. Dhivvya et al. [7] discussed the application of different types of wireless sensors to monitor energy waste in buildings. They presented a complete system for achieving sustainable urbanization through effective monitoring of energy resources. Priyadarshi and Gupta [3] discussed a technique to increase the area under sensor coverage through smaller mobility of nodes. However, it is not suitable for satisfying energy constraints, and redeployment of sensor nodes is very difficult here. Thet et al. [2] presented a design, implementation, and testing of an intelligent lighting system for better visibility comfort, high reliability, along with energy efficiency. Again, Jabbar et al. [8] proposed an efficient control operation of the street lighting system based on the availability of sunlight and the movement or motion detection by wireless communication support. However, this concept does not guarantee full coverage of an area.

In short, various energy-efficient street lighting systems are discussed concerning different coverage aspects. However still, there is scope to improve energy efficiency in this regard. So, the streetlight system proposed here introduces an auto changeover technique to change the light intensity as per requirement. As a result, the streetlight can automatically switch to a utility power supply when the battery has insufficient charge.

## 3 Preliminaries and System Model

### *3.1 Network Model*

In the proposed work, the WSN consists of physically distributed sensor nodes communicating with each other and the sink node or base station (BS) via wireless links, as shown in Fig. 1. Here, the BS can monitor and control each of these nodes. Each node includes an LED light and a passive infrared (PIR) sensor, which detects the appearance of an object in the visibility range of the light. Solar photovoltaic (PV) cell is considered as the source of power supply of the sensor node. Furthermore, a sensor node can be in one of the two modes: active and passive (power-saver) modes. In the active mode, the light attached to the node remains ON after sensing the presence of any moving object by the PIR sensor, and subsequently, the camera can capture the object present within its range. On the other hand, in passive mode, the lights remain OFF during daytime and at night until the sensor detects any moving object. However, the transceiver unit remains ON to receive signals from the neighboring active sensor nodes if requires.

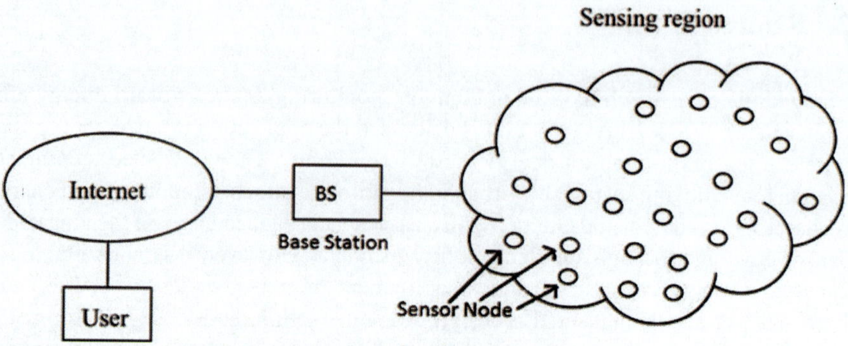


**Fig. 1** WSN containing several sensor modes deployed in a region

## 3.2 *Target Area Layout*

In the proposed work, the concept of the Voronoi diagram [9] is utilized for dividing the target area under surveillance into numerous Voronoi cells, as shown in Figs. 2 and 3. Here, we consider a set ($P$) of $n$ distinct points $p_1, p_2, ..., p_n$ in the target area, and these points are called Voronoi sites. Each Voronoi cell must obey the following characteristics.

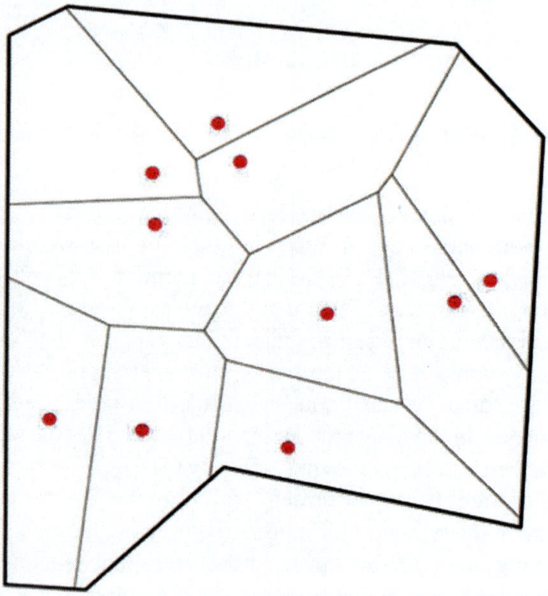

**Fig. 2** Voronoi diagram of a polygonal area

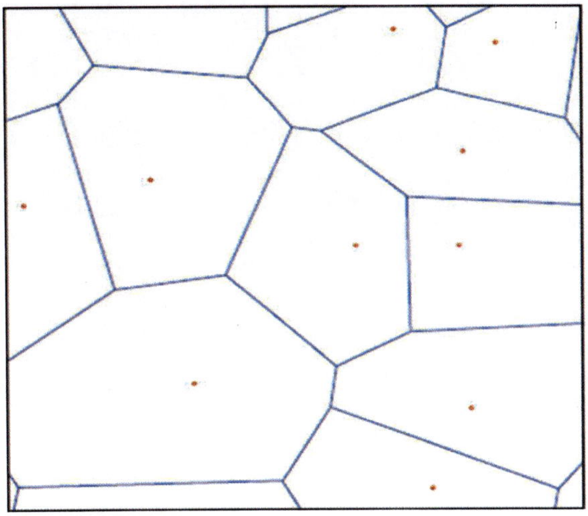

**Fig. 3** Voronoi diagram of a geographical area

- Every point $p_i \in P$ lies in precisely one Voronoi cell.
- If any point $q \notin P$ belongs to the same cell as $p_i$, then the Euclidian distance from point $p_i$ to point $q$ would be less than the same from point $p_j$ to point $q$, where $p_j$ is another point in $P$ [10].
- The shared edge of two adjacent Voronoi cells is the bisector line $\overline{p_i p_j}$ between the two points $p_i$ and $p_j$ in the corresponding two cells.

After dividing the target area into several Voronoi cells, the sensor nodes attached with the streetlights are to be deployed in such a way that the entire area of interest can be monitored through efficient coverage.

## 3.3 *Coverage Used*

The term 'coverage' is generally interpreted as how well a WSN can be used to monitor a target area of interest. In this work, three types of coverage such as area coverage, point coverage and barrier coverage are introduced next in the perspective of the proposed work.

### 3.3.1 Area Coverage

As shown in Fig. 4, area coverage is used to estimate the minimum number of sensor nodes attached to the lights so that every point in the target area is monitored by at least one sensor. When each point in the target area is observed by at least $k$ ($\geq 1$) active sensors every time, then the WSN is known as a k-coverage network, where $k$ is called the degree of coverage. The Voronoi diagram helps sensors to distribute the

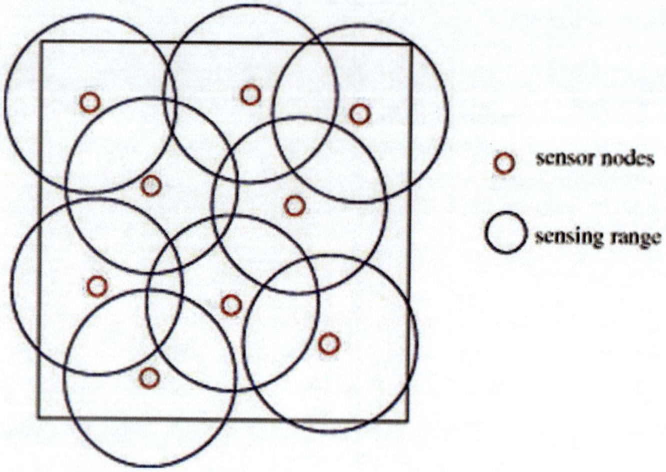


**Fig. 4** Area coverage

sensing task by partitioning the area in a meaningful way. Thus, the Voronoi cell of a sensor node $S$ is the subset of the area in which all points are closer to $S$ than any other sensor node.

### 3.3.2 Point Coverage

A point in the target area is said to be covered by a sensor if it lies within its sensing range. So, by point coverage, a set of given points (also termed as targets) in the target area can be monitored by at least one active sensor all the time, as shown in Fig. 5. Here we have three sensors viz. $S_1$, $S_2$, and $S_3$. $P_1$, $P_2$, ... $P_{13}$ are the points of interest. Each of these points is covered by at least one sensor.

### 3.3.3 Barrier Coverage

Barrier coverage aims to securely monitor the belt region [11] or boundary of the target area against the trespassers trying to enter through this critical area by providing the total coverage with the minimum number of sensor nodes [11]. Figure 6 shows the layout of the barrier coverage.

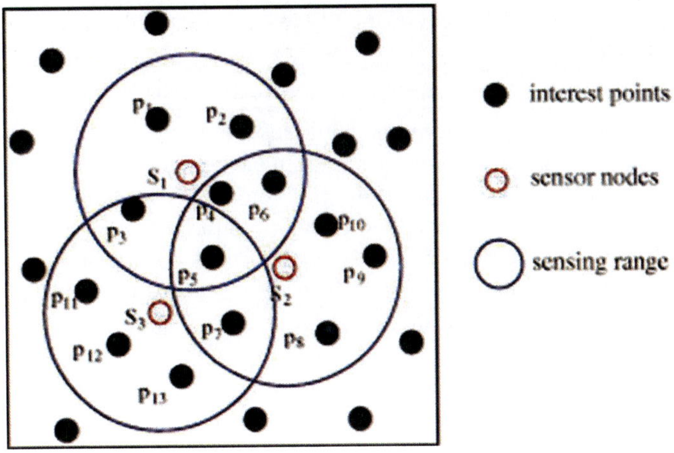


**Fig. 5** Point coverage

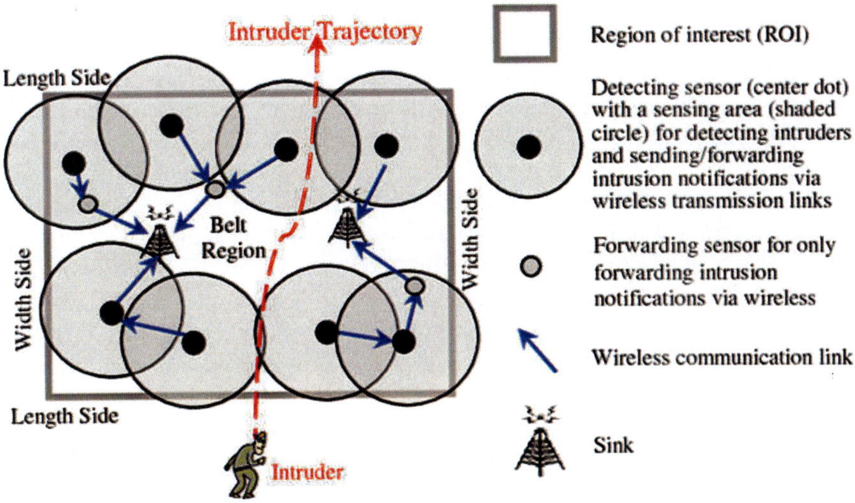


**Fig. 6** Barrier coverage

## 4 Problem Statement

A set of sensors, $S = \{S_1, S_2, \ldots, S_n\}$ in a two-dimensional (2-D) area A is given. Each sensor node ($S_i$) can monitor any point within its transmission range ($R_i$). The objective is to deploy each sensor $S_i$, $i = 1 \ldots n$, at coordinate ($x_i$, $y_i$) inside region A, such that the minimum number of sensor nodes can cover the entire target region. As a result, energy efficiency for the proposed work can be obtained to a great extent.

In order to obtain such an outcome, the following issues with respect to the target area are considered.

- The entire target area is divided into several Voronoi cells, as discussed earlier.
- The sensor node(s) of each Voronoi cell must cover the boundary of the target area so that the entry of any intruder through the boundary can be detected.
- Each node includes one solar-powered dimmable LED light, which can be controlled accordingly to reduce power consumption.
- If an active node detects any object, it sends a signal to the nearby passive nodes to activate and switch on the light. Otherwise, it remains off in passive mode to reduce the power consumption.

## 5 Proposed Methodology

In the proposed work, adaptive street lighting goes into dim during no movement detection and brightens for detecting movement by the sensor. In order to obtain such an outcome, different sensor nodes are connected with the microcontroller to perceive information from the environment. These sensors are motion sensor, light sensor, brightness sensor, and temperature sensor. To increase the lifetime of the battery, a solar panel has been attached to it so that in the daytime, the battery gets charged from the solar PV cell, and at night, the LED lamp uses that battery charge. When a vehicle or a passerby comes within the visibility range of light, the motion sensor detects the presence of the corresponding object and intimates the microcontroller [12]. The temperature sensor is receiving temperature data, and according to this information, the brightness sensor controls the intensity of LED light. Hence, the LED light gets on. If no object is detected, the light is dim with low intensity, and if there is any moving object in the visibility range of light, then the light is brightened again with high intensity. ZigBee network having a specific coverage range with improved data rate [13] is connected with the microcontroller, which collects the data from every sensor node and communicates it to the base station (BS). This working procedure of the proposed methodology is shown with a block diagram in Fig. 7.

The proposed lighting system does not require all the lights in active state throughout the whole day, as discussed earlier. It follows a dynamic arrangement of the active and passive modes depending on the street condition, as shown in Fig. 8. Since about half of the lights are in passive mode unless all the nodes detect an object in its sensing region, it reduces the power consumption of the lighting system to a great extend.

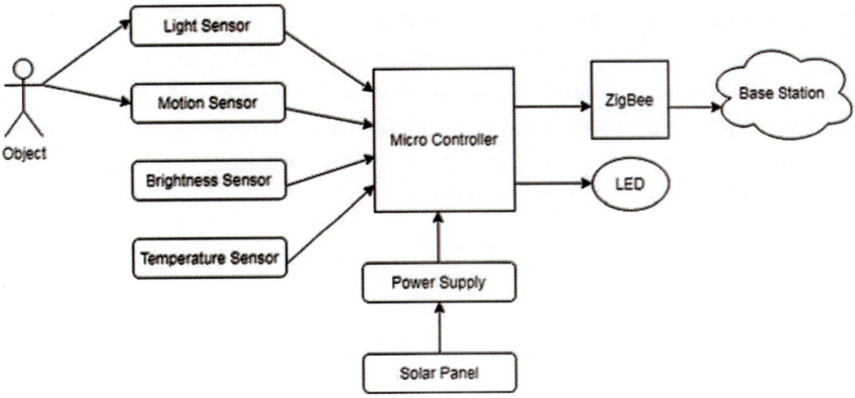


**Fig. 7** Block diagram of the proposed methodology

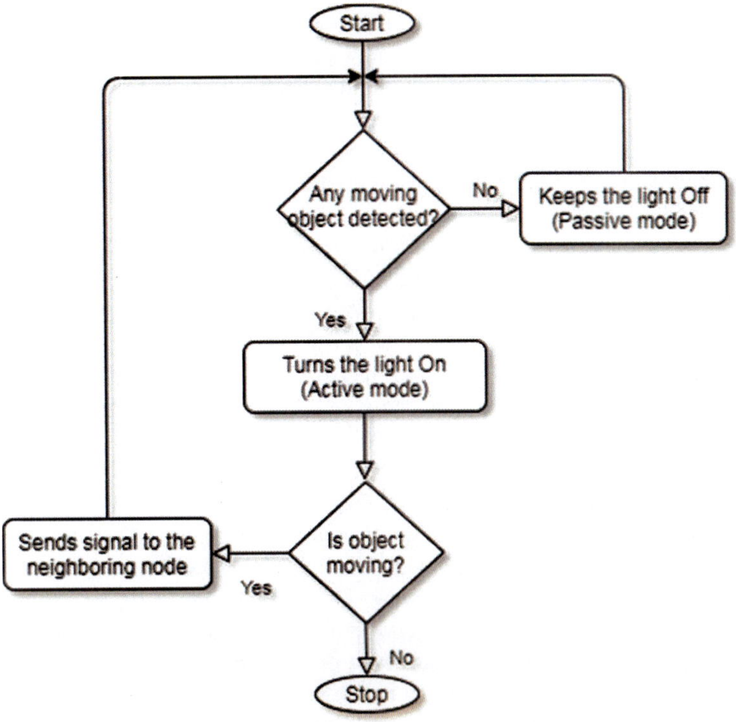


**Fig. 8** Active and passive modes in proposed work

## 6 Result and Performance Comparison

The streetlight with a dynamic brightness control strategy can reduce energy cost by almost 80% and substantially reduces maintenance time and cost. In order to measure the effectiveness of this proposed system, the energy consumption (*E*) of the streetlights by the proposed approach is determined by the following expression.

$$E = \sum W \times I \times H \tag{1}$$

where '*W*' represents the Wattage of the streetlight, '*I*' denotes the percentage of the intensity level of the streetlight at a particular time and '*H*' represents the number of hours during which the streetlight remains ON or active mode.

In order to compare the energy efficiency of the proposed work with respect to other existing techniques, let us consider, there are 20 smart LED street lights installed throughout the target area, where each light with the power consumption of 75 W and the typical duration for which the streetlights remain ON is 10 h. Here, three types of streetlights, such as High-Pressure Sodium (HPS) streetlights, traditional LED streetlights, and LED streetlights with automatic intensity control, are considered for the simulation works. We used MATLAB R2011a to do the comparative study among the results. The parameters used for this comparison are wattage of light, illuminance level and active time (in Hours).

Since our proposed lighting system controls the intensity level of lights as per requirements such as full intensity, threshold intensity, or OFF when not needed. Consider a particular light is in 100% intensity for 3 h, in threshold (say 30%) intensity for 4 h, and in passive mode with 0% intensity for 3 h. Thus, the energy consumption becomes 315 W per light as per (1). Furthermore, the average $CO_2$ emission for electricity [6] is estimated as $C = 475 \times \text{Total KWh}/1000$ Kg. Various results of the proposed system are shown in Figs. 9, 10 and 11. Here Fig. 9 shows how daily power consumption changes with the number of streetlights deployed. Figure 10 represents the daily electricity cost concerning the number of lights, and Fig. 11 shows the amount of $CO_2$ emission with respect to the number of lights. Since the electricity cost and amount of $CO_2$ emission are proportionally related to energy consumed, these three results show the same relational behavior. In addition, a comparative study between the proposed technique and existing approaches is highlighted in Table 1. It is to be noted here that the streetlight equipped with solar cells does not consume any electricity from the utility power supply; as a result, it works free of cost till the provided battery backup is consumed.

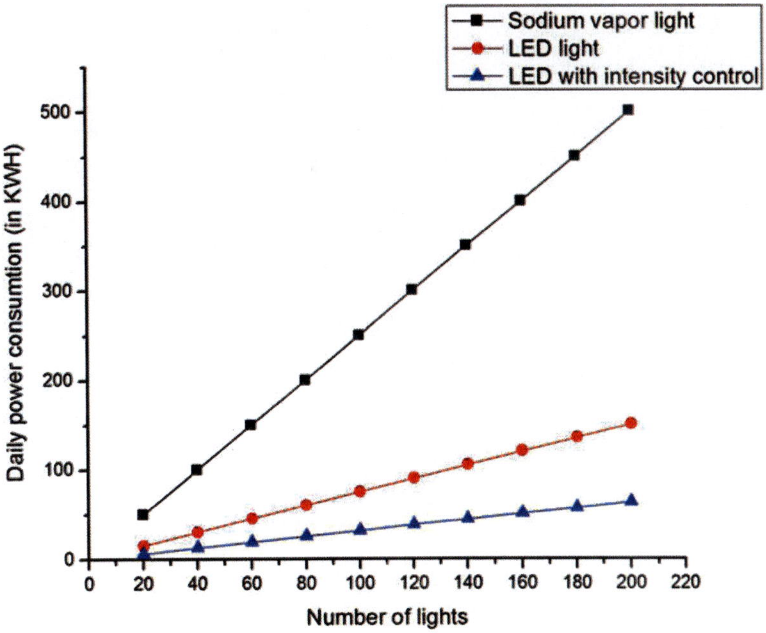


**Fig. 9** Number of lights versus daily power consumption

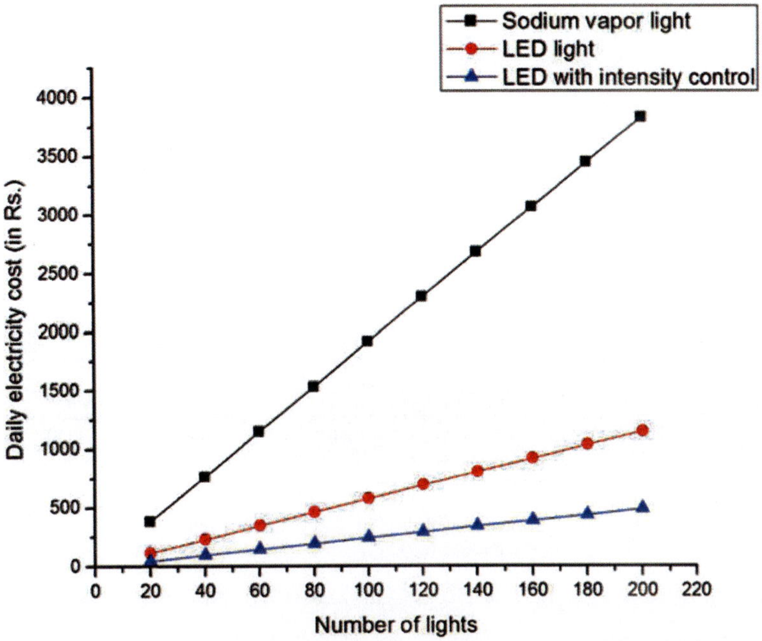


**Fig. 10** Number of lights versus daily electricity cost

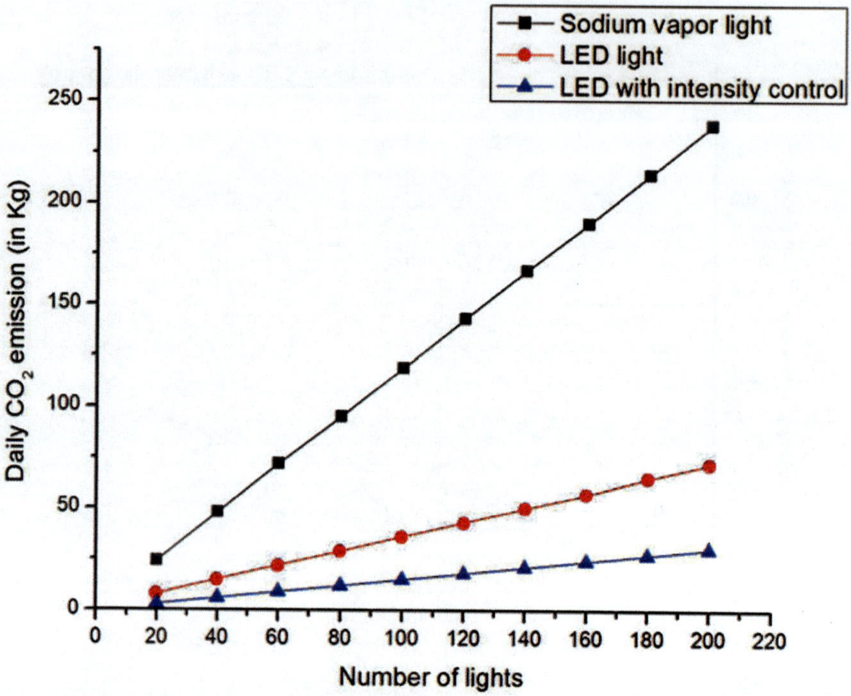


**Fig. 11** Number of lights versus daily $CO_2$ emission

**Table 1** Comparative study between the existing and proposed system

| | Sodium vapor light | Existing LED light | LED with diming control |
|---|---|---|---|
| Power (W) | 250 | 75 | 75 |
| Duration (h) | 10 | 10 | 10 |
| Yearly consumption (KWh) | 912.5 | 273.75 | 115 |
| Cost of electricity (Rs/KWh) | 7.65 | 7.65 | 7.65 |
| Yearly electricity cost (Rs/light) | 6980 | 2094 | 880 |
| Yearly $CO_2$ emission (kg) | 433 | 130 | 55 |

## 7 Conclusion

The aim of the proposed system in this paper is to find cost-effective and energy-efficient street lighting systems using WSNs. In order to reduce the electricity consumption by the streetlights, an efficient technique that facilitates controlled modules has been discussed here. The microcontroller is used to control different illuminance levels for street lighting conditions as desired. Furthermore, dimmable LED lighting in the proposed work can obtain an energy-saving performance. The proposed approach utilizes dimmable LED and wireless sensor technology in order to activate the street lighting within an area in a regulating way. Again, it reduced the electricity cost compared to existing lights, and the proposed approach has assured a significant reduction in $CO_2$ emission. The significance of considering coverage to obtain energy efficiency is discussed thoroughly. The performance of the proposed work using LED street lights with automatic intensity control can show an improvement over other existing lighting techniques. In the future, the proposed approach can be utilized to ensure fault tolerance at every point of the target area.

## References

1. Yarinezhad, R., Hashemi, S.N.: A sensor deployment approach for target coverage problem in wireless sensor networks. J. Ambient Intell. Human. Comput. (2020)
2. Thet, L.M., Kumar, A., Xavier, N., Panda, S.K.: A smart lighting system using wireless sensor actuator network. In: Intelligent Systems Conference (2017)
3. Priyadarshi, R., Gupta, B.: Coverage area enhancement in wireless sensor network. Springer-Verlag GmbH Germany, part of Springer Nature (2019)
4. Zhang, G., You, S., Ren, J., Li, D., Wang, L.: Local coverage optimization strategy based on voronoi for directional sensor networks. Sensor **16**, 2183 (2016)
5. Bhairi, Edake, Kangle, Madgundi, Bhosale.: Design and implementation of smart solar LED street light. In: International Conference on Trends in Electronics and Informatics (2017)
6. https://www.iea.org/reports/global-energy-co2-status-report-2019/emissions. Last accessed 30 June 2021
7. Dhivvya, J.P, Jayakrishnan, V.M, Thomas, E.K., Ramesh, M.V., Divya, P.: Towards energy conservation in campus using wireless sensor network. In: IEEE Global Humanitarian Technology Conference (2017)
8. Jabbar, Yuzaidi, Yan, Bustaman, Hashim, AlAriqi.: Smart and green street lighting system based on arduino and RF wireless module. In: 8th International Conference on Modeling Simulation and Applied Optimization (2019)
9. Souvaine, D., Horn, M., Weber, J.: Voronoi Diagram. Comp 163: Computational Geometry, Tufts University (2004)
10. Lee, S.K.: Search reverse nearest neighbor query on air. In: Fourth International Conference on Information Technology (2007)
11. Benahmed, T., Benahmed, K.: Optimal barrier coverage for critical area surveillance using wireless sensor networks. Int. J. Commun. Syst. (2019)

12. Imran, L.B., Rana, M., Latif, A., Farhan, M., Tariq, T.: Real-time simulation of smart lighting system in smart city. Int. J. Space-Based Situated Comput. **9**(2) (2019)
13. Allahham, A.A., Rahman, M.A.: A smart monitoring system for campus using Zigbee wireless sensor networks. IJSECS **4**(1), 1–14 (2018). ISSN: 2289-8522

# Cognitive IoT for Future City: Architecture, Security and Challenges

**Saikat Samanta, Achyuth Sarkar, and Aditi Sharma**

**Abstract** Internet of things has little capacity of its own. It has to be intellectual to reap the real benefits of IoT. The challenges in this area have prompted researchers to focus their attention on developing cognitive strategies for IoT use. Cognitive Computing can make IoT more advanced, smarter and immersive. This article reflects on the integration of IoT with Cognitive Artificial Intelligence. In addition to the presentation of the basic principle of Cognitive Computing, the paper touches on a variety of CIoT problems. There will be a special discussion on how automation has reached a new standard for CIoT. The difficulties in adopting CIoT, along with the social and ethical issues, were described and deeply addressed. In this sense, the market principles of CIoT and some potential implementations have been illustrated. In this article, we suggest a future city architecture using Cognitive Internet of Things. Finally, we recognize potential possible challenges and possibilities that could arise through the design of the planned architecture.

**Keywords** Smart city · Artificial intelligence · Machine learning · Encryption · Data analysis

## 1 Introduction

We are now at the edge of the intelligent universe, in which anything can be done automated and with or without human interference. Innovation is supposed to introduce us to the fantasy world, defined as the Internet of Things. IoT is a digital network,

S. Samanta (✉) · A. Sarkar
Department of Computer Science and Engineering, National Institute of Technology Arunachal Pradesh, Papum Pare, Arunachal Pradesh, India
e-mail: s.samanta.wb@gmail.com

A. Sarkar
e-mail: achyuth@nitap.ac.in

A. Sharma
Department of Computer Science and Engineering, School of Technology, Quantam University, Roorkee, Uttarakhanda, India



D. Gupta et al. (eds.), *Pattern Recognition and Data Analysis with Applications*, Lecture Notes in Electrical Engineering 888,
https://doi.org/10.1007/978-981-19-1520-8_12

**Table 1** Recent literature analyses

| Year | Reference | Technology approach | Focus |
|---|---|---|---|
| 2017 | [1] | Cognitive IoT architecture | The architecture model consists of a single central server that stores all data |
| 2018 | [2] | Cognitive computing | A examination of cognitive computing, including its evolution from information discovery to cognitive science and big data |
| 2018 | [3] | Cloud and fog computing | Issues with cloud-IoT coupling are discussed, accompanied by a comparison of fog and cloud computing |
| 2018 | [4] | Cognitive architecture | A smart home framework focused on IoT and ICT solutions with a focus on usability for a smarter lifestyle |
| 2018 | [5] | Security of IoT | IoT-CPS security challenges, risks, and solutions |
| 2019 | [6] | IoT challenges | IoT challenges and comprehend the interdependence of these challenges to aid in the growth of smart cities |

which enables everything to be connected to the real world through the Internet. We are supposed to make IoT smarter. We cannot truly understand the ability and vision of IoT without AI. AI will bring more progress in this field and autonomous actions to the smart world. Cognitive AI reaches beyond the regular intellect of the machine to cognitive thought and to the reasoning of solutions of problems such as human intelligence. Cognitive AI transforms an unintelligent computer into a humanoid intelligence which allowing computers to read, think and interact with people in natural language. The Cognitive IoT is an awareness that applies context and makes a decision and communicates it to people.

Table 1 summarizes some recent literature analyses and studies. First, we discuss the basics of IoT and CIoT's layout and terms. We introduce a new framework for CIoT-based architecture in this paper.

We present the latest research the architecture that illustrates various realms of the future city network, such as intelligent house, intelligent industrial automation and intelligent grid to construct several cognitive-based applications.

Our solution differs from others in that it uses a unique architecture in solving privacy issues and security concerns. The paper comes to a close with some final thoughts and plans for future studies on the subject.

## 2 Integrated Technologies

In this section, we introduce some basic knowledge regarding IoT, AI, Cognitive IoT and Cognitive AI.

## 2.1 IoT and AI

The Internet is a connectivity of different devices. The Internet vision was further expanded by IoT. Not only the machines are interconnected here but also they can be linked to all beings on earth. The fundamental aim was to automatically and systematically collect and exchange data. IoT devices can detect or read the world they are installed in or around them. The sensed data was collected and evaluated in order to gain knowledge. In reality, the essence of IoT is to make everyone and every time knowledge accessible through all barriers [7]. Without any specific external instruction IoT devices can do their function independently. They gather information individually and proactively share it with other network IoT devices.

IoT technologies have been found to be in operation of engineering, logistics, shipping, agriculture, hospitals, home, grid and transportation, etc. Advances in technology that have resulted allowed IoT to gather environmental information and communicate with the physics world at a low cost, cheap encoding and cheap bandwidth with pervasive wireless coverage and smart phones [8]. The distance between physical structures and the cyber universe has thus been minimized and the monitoring of these machines has been made easier. In brief, IoT has taken us, through universally connected computers, to the modern opportunities of an intelligent future for smart homes. As just a new phase towards automation, the beneficial features of IoT have been taken by enterprises and organizations to allow for centralized monitoring and control [9].

IoT-based automation can minimize the operating costs by linking and collaborating with each other relative to the manual process by automating monitoring and maintenance of isolated and autonomous devices [10]. From the point of view of computer science research, AI is a system that comprises several branches such as neural networks, deep learning, fuzzy systems, genetic algorithms, natural language processing and many others. However, many areas of research are different from AI, but IoT seems to be the product of a synthesis of several fields of study. AI will increase data processing precision, efficiency and speed. Human intelligence can be derived from the IoT data by applying intelligent algorithms [11]. This lowers the level of production and operating performance in the current framework plans which were otherwise unachievable before [10]. There are still several automation-related opportunities in IoTs and in AI to resolve issues and challenges. The current field of research highlights things to be trained and how well they communicate as expected. The research into how things can be improved and cleverer in the future.

## 2.2 Cognitive AI and Cognitive IoT

AI means the capacity of the computer to learn or understand. AI can execute only those functions that are described by strict guidelines that are coded by humans. To overcome the limitation, the AI people have embraced knowledge of the human mind and behavioural mechanism and encouraged it beyond the study of it. They named this amalgamation the natural continuation of the present AI for Cognitive AI. Cognitive AI plays an important part in bridging the human–machine gap. CIoT seeks to enhance efficiency and achieve IoT wisdom through Cognitive Computing's cooperative mechanisms [12].

Today's IoT is usually based on observing the surroundings and responding appropriately. IoT-connected systems usually make decisions based on pre-programmed versions. They should assume on the basis of accessible sensed data. However they are not fully autonomous devices and can make their choices depending on the immediate context. They will be able to sense, interpret and comprehend, very similar to human beings, as well as to assimilate this knowledge to an excerpt of actionable knowledge and functional patterns [13]. They must be aware of the context in which IoT is used and behave accordingly [14].

The IoT's aim is to erase the distinction between human and physical universes by individually communicating and exchanging knowledge amongst our surrounding objects with us. In order to reap the full advantages of IoT, we must use Cognitive Computing as an update to what we consider Cognitive IoT [15]. By reacting to the current state of the network, cognitive networks aim to attain optimum efficiency. CIoT will make the decision using cognitive networks by knowing the existing state of the network and analysing perceived knowledge [16].

## 3 CIoT Potential Application for Future City

AI has made devices smarter, more flexible, more robust, more conscious and more powerful. Over time, maybe in the very distant future AI will be completely converted into Cognitive AI. Although a majority of work is getting closer, the understanding of CIoT to the fullest degree is a ways away to go. In general cognitive devices will be the next big technologies that will have a huge effect on industry and the environment, health care, culture and life, make recommendations and transactions, etc. [17]. AI would also improve the productivity of current technology. Some IoT-based AI including Cognitive AI applications are described below (see Fig. 1).

### *3.1 Intelligent Living*

Home automation and intelligent environment are no more only imagination, but the different items we find in their everyday lives are smart. They will improve our ease of living by supporting our everyday lives in a variety of ways. For example, a home automation detects the presence of an object at house and, based mostly on background of the individual, appropriate resources are enabled. Smart life simplifies our lives and eliminates dependency Smart living does have a major role to play in

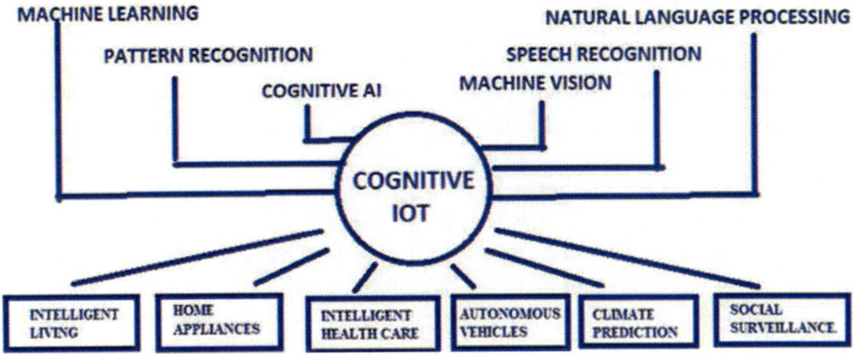


**Fig. 1** Cognitive IoT and its application

improving the lives of elderly people. However, a lot needs to be achieved in this respect, and more focus needs to be paid to it.

## 3.2 *Intelligent Health Care*

One of several greatest effects of CIoT is on healthcare services. Health technologies with intelligence may be used to track the state of health for ill people. Each detritions identified could be assessed in order to avoid a hazardous medical problem, saving people's lives in time health help attached to a human could be used to track heart rate, blood pressure, oxygen levels in the blood, blood sugar, tiredness or exhaustion, epilepsy, etc. In the event of an emergency, the monitoring system connected to the patient could warn to health services. IoT aims to track senior citizens living remote location from home. Early notice of an urgent health condition can avoid bad injuries from occurring [18].

## 3.3 *Home Appliances*

Cognitive IoT has made it possible for home appliances to be smart. The web application allows you to associate perception that allows the computer to understand the expectations of the user, the pattern of operation and their regular routine. The useful knowledge created by the system can be used as guidance for engineering and development to create smarter machines that can support people more intelligently.

### 3.4 Autonomous Vehicles

Autonomous cars, which tend to be a futuristic novel, are no longer the focus of imagination, but a rather advanced fact. The numerous functions that the driver would perform may be supplemented by CIoT sensors that imitate the driver's eyes and ears, and cognitive AIs that imitate the driver's intellect. To negotiate the passage, a car with awareness connects with others, thus reducing signalling by sound and light. Service assistance is immediately called for by the cognitive AI in the car if the sensors sense any malfunction. In the operation of the system, sometimes required steps such as halting the car and raising the alarm are applied for any significant phenomenon. In comparison, previous car data and traffic data observed may be used for traffic control, signalling, traffic redirection, etc. [13].

### 3.5 Social Surveillance

A significant IoT-centric application is social surveillance. The IoT monitors incidents or conditions and initiates effective protective steps for the safety of persons. Automated workplace temperature and light monitoring, depending on people's mood or tension. People mistakenly abandon their objects whilst in transport or are frequently left unsupervised. New sensors allow the IoT to "hear" sound input from groups of individuals [19]. Other social influence of CIoT is to help people interact/communicate with one another effectively by increasing group dynamics [20].

### 3.6 Climate Prediction

Modern IoT system can identify and measure the state of the area. Data obtained from sensors connected to smart technology, vehicles, houses, smart phones and social media networks offers enough information to reliably predict the climate per each area or region. Climate Firm, air pressure data, sensed by a large number of mobile phones, is gathered and analysed by cloud technologies for precise and instant weather forecasting [21]. Important information could be expected from the weather report in advance. It could help to recommend regular switches for early exits or to carry an umbrella when it is raining. Kids suffering from cold or asthma are frightened into taking care as temperatures are going to dip. This form of application includes complex atmospheric simulations that can be performed by integrating Cognitive Computation and IoT to atmospheric physics.

### 3.7 *Real-Time Monitoring*

A large complex network is often generated by connected IoT devices. Such systems constantly produce information that needs to be processed in real-time in order to take effective action at the appropriate time and place The AI device provides real-time data analytics to the IoT system [22]. Actual research is mostly about information collection in real-time, so that the machine can respond in the same time frame of the timeframe wherein the information or intervention request is made. Actual research is a time-critical method that relies on variables such as network latency, data information processing, pattern detection, historical data inference, data management and stream data retrieval.

## 4 Cognitive IoT Architecture for Future City

Our proposed design requires a minimum number of separate frameworks to be developed for various technologies. We use several data sets obtained from different sensors to serve various cognitive computing functions. The proposed structure is shown in Fig. 2. We would have easier, more real-time methods to support the complexity of data generated by smart communities efficiently. The proposed CIoT-based architecture consists of three layers, the future city application platform, data knowledge layer and the cognitive system layer. Every section is listed in depth below:

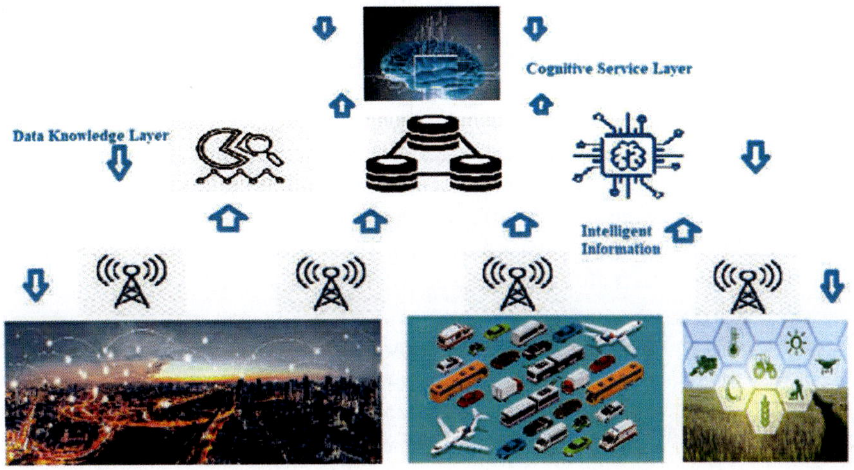


**Fig. 2** Proposed future city architecture

## 4.1 *Future City Application Platform*

Future city consists of intelligent houses, intelligent electricity, intelligent transport, intelligent agriculture and intelligent enterprises, etc. Both unstructured and structured data are generated here. Sensor data from intelligent residential buildings includes various human variables such as thoughts, sound, brain function and so on.

**Intelligent buildings** Intelligent buildings are consisting of multiple detectors that allow data from various sources to be gathered to optimize light systems, lifts, etc. Houses gather data such as feelings, the atmosphere that together offer several solutions, such as maximizing energy usage by heat treatment and lighting management, air product quality, elevators, based on user expectations to improve their functionality.

**Intelligent energy** Data from both the sensory and mental regions of the brain was combined and used to create the intelligent visualization method. Going to connect to cognitive tool for imaging and weather prediction, energy providers offer innovative safety technologies to power plants, like predicting safety threats in machines whilst increasing efficiency.

**Smart industry** Similar to the data obtained from sensors in the world and sensors in the brain industry output can be enhanced to incorporate human data such as workflow processes and qualitative information, efficient decisions can be taken in full detail and natural assets can be best controlled in industries.

## 4.2 *Data Knowledge Layer*

These solutions can be seen as a major symbiotic relationship of cognitive outcomes and big data. The amount of data produced by IoT-based sensor systems, virtual sensors and home apps, in the form of both data from multiple sources and data from multiple sources, is growing [23]. Research by Kambatla et al. [24] suggest that a large quantity of data is produced in big data. The cognitive process is affected by a variety of influences, including the smart city network, the IoT framework and the data access layer.

Importantly, cognitive processes have the capacity to evaluate, understand and recall an issue that might be cognitively important to an organization. The key characteristics of the neural method are awareness and learning capacity to change without reprogramming; the creation and review of theories based on the existing level of knowledge of the framework.

The method, in which certain complex improvements can be implemented by it, demonstrates the complexities of cognitive computation over the processing of massive data. Natural interaction can be found to have a profound impact on the synthesis of natural language and make the cognitive system particularly attractive.

Based on the analysis of Santos et al. [25], the analysis of the data set would make for a clearer understanding and resolution of a complicated set of issues because there are a number of information sources. Such data may be regarded as unstructured, organized, multimedia and textual information. It is considered that the power to make choices based on evidence is in the cognitive system. Value is considered in big data as an attribute that means useless data volume before the data is translated to the source of information.

### 4.3 Cognitive Service Layer

This level describes the mechanism by which the cognitive programming algorithm is constructed. We illustrate the steps that include preprocessing, data interpretation, cognitive feature extraction and deep learning.

Data preprocessing: There might be noisy or fragmented data in the raw data obtained from sources. The incorporation of this information would result in inefficient artificial intelligence-driven models based on cognitive computation. If it is inconsequential to the result, the lost data is omitted, or it may be substituted to prevent destroying data.

Data analysis: Analysis of data includes the collection of cognitive traits which are important for the preparation of our model. Multiple characteristics are chosen that represent diverse applications instead of establishing a different cognitive model with each specific use case in a future city.

Machine learning: Machine learning enables software systems based on the training obtained to include analytics or prediction. Machine learning has learned to think intelligently so that its learning habits can be formed on their own and further evaluated. Artificial intelligence is what this evolution is calling. Cognitive computing has taken another step further by combining the way people learn and teach computers to think with a far more logical way.

## 5 Overview of Security Issue and Discussion

In this section we will survey security related issue and proposed some precautions for that challenges. Our private and confidential data is gathered, preserved and exchanged with international agencies in relation to our everyday activities and our lives. As seen in Fig. 3, we address the open problems and security issue or this architecture. In addition, large volumes of such data are processed for potential analytics and information retrieval in distributed cloud computing systems. Data processing and usage-related technical advancements outperform the advancement of privacy, trust and protection frameworks. As such from several entry points in the IoT data network, attackers could intercept personal data.

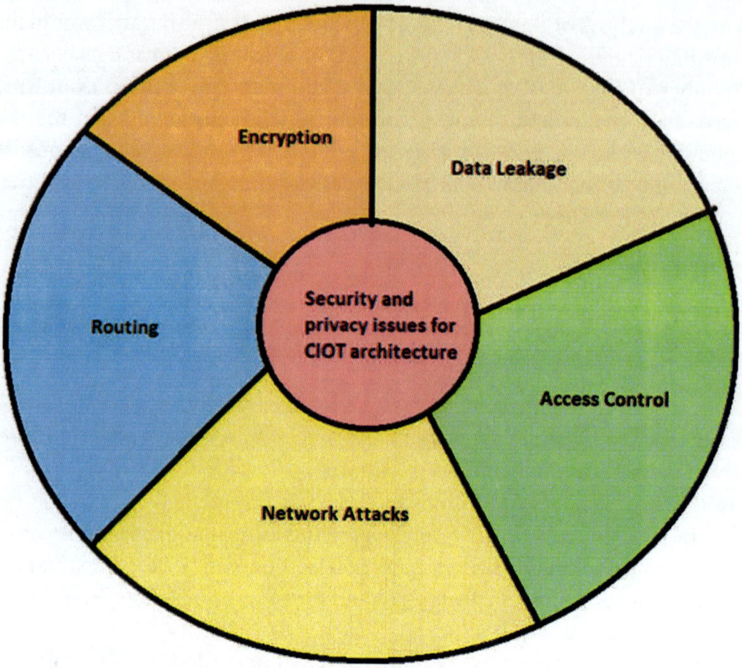


**Fig. 3** Security and privacy issue for CIoT architecture

A big task of cyber-security is data safety. Reasonable countermeasures are also required to reduce the possibility of leakage of information, which may have an effect on personal data protection, privacy and protection [26]. Some citizens claim and prefer that they are discreet and that the right of identity and monitoring in public spaces is still maintained [27].

The principle of protection by design requires that safety risks should be handled across all stages and across all interfaces. When attached to the real world, security concerns are more important because risks switch from accessing information to managing processes. We should secure records, computers, networks, servers and the cloud to improve the protection within IoT systems. In order to provide data security on the move, constructive vulnerability thwarting intelligence should be introduced.

Challenges of the existing CIoT architecture must be protected. We address some precaution for different challenges in the layer of CIoT architecture shown in Table 2.

**Table 2** Precaution for different challenges in the layer of CIoT architecture

| Layer | Challenges | Precaution |
|---|---|---|
| Future city application platform | Data leakage<br>Jamming<br>Trustworthiness | Policies for lost data in software<br>Data reduction<br>Regulation formulation that is flexible and consultative with consumers |
| Data knowledge layer | Identification of Information<br>Data accessibility<br>The integrity of data<br>Secondary application | Encryption is used to preserve records<br>Schemes for access management dependent on device hierarchy<br>Anonymization of data should be used<br>Data validation is used before use |
| Cognitive service layer | Attacks on the side channels<br>Service disruption | Often use authentication protocols<br>Using public networks to encrypt data |

# 6 Future Direction

Several recommendations can be made based on the analysis of this paper. Encryption techniques, authentication protocols, data anonymization techniques, and other approaches to avoid invalidated access to the CIoT network are all part of a major research field in the protection and privacy of CIoT in smart cities. Access monitoring and supervision, safe system discovery, spoofing protection and data leakage will all be made possible with blockchain, as long as end-to-end encryption is used.

Another area to focus on is the advancement of cost-effective storage strategies and low-power hardware. Decentralized networks have been suggested as the best options for increasing device efficiency from an implementation standpoint.

In addition, there is a lot of future work in the field of AI. This involves the development of data fusion techniques to make it possible to use heterogeneous data sources, as well as intelligent data reduction/feature selection approaches to eliminate redundant data. This would result in a faster response time and better implementation efficiency.

# 7 Conclusions

Cognitive IoT is providing various innovative and smart services for our community. This cognitive intelligence will make the IoT environment more sustainable to risks and attacks. The trust and privacy problems in IoT environments become more complicated because of the large number of interconnected devices. We have established similar protection concerns and cognitive IoT-related threats. We address the research in the areas of IoT, CIoT and smart city design. We introduce the future city architecture based on CIoT. In our article, we introduced a feasible framework for cognitive IoT-based systems. Proposed architecture explains various cognitive features which offer real-time solutions to future city platforms. The proposed CIoT architecture solves the new, dynamic and scaling problems that concern smart cities when processing large number of IoT files. Finally, we are addressing the problems and possibilities that emerge with our proposed architecture.

## References

1. Patra, M.K.: An architecture model for smart city using cognitive Internet of Things (CIoT). In: Proceedings of the 2017 2nd IEEE International Conference on Electrical, Computer and Communication Technologies, ICECCT 2017. Institute of Electrical and Electronics Engineers Inc. (2017). https://doi.org/10.1109/ICECCT.2017.8117893
2. Chen, M., Herrera, F., Hwang, K.: Cognitive computing: architecture, technologies and intelligent applications. IEEE Access **6**, 19774–19783 (2018). https://doi.org/10.1109/ACCESS.2018.2791469
3. Aazam, M., Zeadally, S., Harras, K.A.: Fog computing architecture, evaluation, and future research directions. IEEE Commun. Mag. **56**, 46–52 (2018). https://doi.org/10.1109/MCOM.2018.1700707
4. Baig, M.N., Himarish, M.N., Pranaya, Y.C., Ahmed, M.R.: Cognitive architecture based smart homes for smart cities. In: Proceedings of the 2nd International Conference on Trends in Electronics and Informatics, ICOEI 2018, pp. 461–465. Institute of Electrical and Electronics Engineers Inc. (2018). https://doi.org/10.1109/ICOEI.2018.8553774
5. Kim, N.Y., Rathore, S., Ryu, J.H., Ho Park, J., Park, J.H.: A survey on cyber physical system security for IoT: issues, challenges, threats, solutions (2018). https://doi.org/10.3745/JIPS.03.0105
6. Janssen, M., Luthra, S., Mangla, S., Rana, N.P., Dwivedi, Y.K.: Challenges for adopting and implementing IoT in smart cities: an integrated MICMAC-ISM approach. Internet Res. **29**, 1589–1616 (2019). https://doi.org/10.1108/INTR-06-2018-0252
7. Industrial Automation Industry Exploring and Implementing the Internet of Things: https://blog.isa.org/industrial-automation-industry-exploring-implementing-internet-of-things. Last accessed 02 July 2021
8. IoT: Implementation and Challenges | OpenMind. https://www.bbvaopenmind.com/en/technology/digital-world/iot-implementation-and-challenges/. Last accessed 02 July 2021
9. Pramanik, P.K.D., Pal, S., Choudhury, P.: Beyond automation: the cognitive IoT. Artificial intelligence brings sense to the internet of things. In: Lecture Notes on Data Engineering and Communications Technologies, pp. 1–37. Springer Science and Business Media Deutschland GmbH (2018). https://doi.org/10.1007/978-3-319-70688-7_1
10. The Automation Internet of Things | Automation World. https://www.automationworld.com/products/data/blog/13307756/the-automation-internet-of-things. Last accessed 02 July 2021
11. AI and the IoT: Are We Truly Prepared for What's Coming? | IT Business Edge. https://www.itbusinessedge.com/networking/ai-and-the-iot-are-we-truly-prepared-for-whats-coming/. Last accessed 02 July 2021
12. Cai, H., Xu, B., Jiang, L., Vasilakos, A.V.: IoT-based big data storage systems in cloud computing: perspectives and challenges. IEEE Internet Things J. **4**, 75–87 (2017). https://doi.org/10.1109/JIOT.2016.2619369

13. Sathi, A.: Introduction. In: Cognitive (Internet of) Things, pp. 1–12. Palgrave Macmillan US (2016). https://doi.org/10.1057/978-1-137-59466-2_1
14. Vodyaho, A.I., Osipov, V.Y., Zhukova, N.A., Chervontsev, M.A.: Cognitive technologies in monitoring management. Autom. Doc. Math. Linguist. **53**, 71–80 (2019). https://doi.org/10.3103/s0005105519020080
15. Choi, N., Kim, D., Lee, S.-J., Yi, Y.: Fog operating system for user-oriented IoT services: challenges and research directions (2016)
16. Zhang, M., Zhao, H., Zheng, R., Wu, Q., Wei, W.: Cognitive Internet of Things: concepts and application example. Undefined (2012)
17. Worldwide Big Data and Analytics Spending Guide. https://www.idc.com/tracker/showproductinfo.jsp?containerId=IDC_P33195. Last accessed 02 July 2021
18. Embedding Intelligence in the Internet of Things—THINK Blog. https://www.ibm.com/blogs/think/2016/02/embedding-intelligence-in-the-internet-of-things/. Last accessed 02 July 2021
19. The Cognitive Era Presents Opportunities For Enhanced Collaboration. https://www.forbes.com/sites/ibm/2015/12/14/the-cognitive-era-presents-opportunities-for-enhanced-collaboration/?sh=744440c01301. Last accessed 02 July 2021
20. Weathering Hurricane Season with Cognitive, IoT—THINK Blog. https://www.ibm.com/blogs/think/2016/05/weathering-hurricane-season-with-cognitive-iot/. Last accessed 02 July 2021
21. AI Could Be The Catalyst To Unleash The Power of IoT. https://www.oodlestechnologies.com/blogs/AI-Could-Be-The-Catalyst-To-Unleash-The-Power-of-IoT/. Last accessed 02 July 2021
22. Sicari, S., Rizzardi, A., Grieco, L.A., Coen-Porisini, A.: Security, privacy and trust in Internet of things: the road ahead (2015). https://doi.org/10.1016/j.comnet.2014.11.008
23. Talari, S., Shafie-Khah, M., Siano, P., Loia, V., Tommasetti, A., Catalão, J.P.S.: A review of smart cities based on the internet of things concept (2017). https://doi.org/10.3390/en10040421
24. Kambatla, K., Kollias, G., Kumar, V., Grama, A.: Trends in big data analytics. J. Parallel Distrib. Comput. **74**, 2561–2573 (2014). https://doi.org/10.1016/j.jpdc.2014.01.003
25. Santos, M.Y., Oliveira e Sá, J., Andrade, C., Vale Lima, F., Costa, E., Costa, C., Martinho, B., Galvão, J.: A big data system supporting Bosch Braga Industry 4.0 strategy. Int. J. Inf. Manage. **37**, 750–760 (2017). https://doi.org/10.1016/j.ijinfomgt.2017.07.012
26. Corea, F.: Introduction to data (2019). https://doi.org/10.1007/978-3-030-04468-8_1
27. Christidis, K., Devetsikiotis, M.: Blockchains and smart contracts for the Internet of Things (2016). https://doi.org/10.1109/ACCESS.2016.2566339

# A Heuristic Model for Friend Selection in Social Internet of Things

**Ashish Kumar, Sunil Kumar Singh, and Pawan Kumar Chaurasia**

**Abstract** The Social Internet of Things (SIoT) is emerging as a future of information dissemination in society. It is a new concept in the combination of social network with IoT, may enable interaction between humans and objects. Many times, finding desired services may lead to high computation and memory demands as the number of friends increase over time. Therefore, it is in need to judiciously select future friends by applying a few heuristics to minimize the computational load and memory. In this work, we have incorporated three different heuristic-based strategies to select the future friend and also to optimize the average degree, average path length, and average clustering coefficients. The performance analysis of the model depicts its effectiveness in terms of future friend selection.

**Keywords** Social Internet of Things · Internet of Things · Friend selection · Social networking

## 1 Introduction

Autonomous interaction between social networking and the Internet of Things is an emerging interdisciplinary research area [1] and it is being emerged as the Social Internet of Things (SIoT). The term social networking is used as Internet-based media platforms to remain connected with family, friends, colleagues, customers, and clients, etc. Social media platforms like Facebook, Twitter, Instagram, and many others are being used for social purposes, business purposes, or both [2, 3]. Figure 1, indicates the collaboration of social networking and the Internet of Things.

Internet of Things has emerged as a key concept for moving toward automation on social networking. It is defined as the network of physical Things/objects that are embedded with sensors, softwares, and other technologies for connecting and exchanging data with other devices. The term IoT was coined by Kevin Ashton in

A. Kumar · S. K. Singh (✉) · P. K. Chaurasia
Mahatma Gandhi Central University, Motihari, Bihar, India
e-mail: sunilsingh.jnu@gmail.com; sksingh@mgcub.ac.in



D. Gupta et al. (eds.), *Pattern Recognition and Data Analysis with Applications*, Lecture Notes in Electrical Engineering 888,
https://doi.org/10.1007/978-981-19-1520-8_13

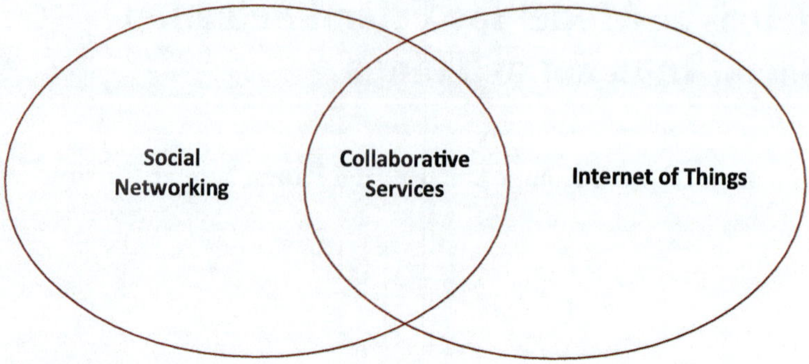


**Fig. 1** Social Internet of Things

1999 [4] to ease out the many application areas like healthcare, agriculture, smart cities, education, and social networking [5].

Appropriate friend selection in this new paradigm SIoT which is an integration of social networking, and the Internet of Things is not an easy task. The idea behind the selection of friends is to look for desired services, using its friendship. Friend selection always has been a key issue especially when we have limitations on the number of friends. Generally, friend selection depends on the following three concerns; first, what kind of people tend to have more friends? And second, what kinds of people are plausible to be chosen as a friend? And the last one, assortative personalities (the kind of people who have similar levels of openness) affect the friend selection [6].

If we look at the number of friends on Facebook, we cannot have more than 5000 friends on our friend list. Therefore, it is very difficult to decide, whom we should keep on our friend list, and to whom we should remove. Many social networking sites have their policies and constraints for their users. To deal with the better friend selection, a few important concepts are required to be highlighted which are as follows:

## *1.1 Social Network*

Nowadays most people using Internet-based social networking platforms to communicate with family, friends, colleagues, clients, or customers. Social networking has become a prominent footing for communication looking to engage people. The social network is made of nodes (People, Organizations, and Groups) that are tied by social links where these links may be directed, undirected, or multi-directed. In the social network, each node's position and links could also be individually analyzed and compared with those of other actors to analyze their comparative importance in the network.

## 1.2 Social Internet of Things

SIoT takes the scope of backing the emerging applications and networking services for the IoT more effectively and efficiently to overcome the meticulous issues of scalability and heterogeneity [7]. The term "Social Networking" roam across the online social networking platforms as well as social relations among people. In general, the communication of an object depends on its relationship with the other objects. Closely associated objects may communicate directly without any intermediation from neighbor objects. The smart objects communicate with their friends directly and by continuing this process they try to communicate with the devices which are not in friends.

SIoT enables the formation of enchanting applications that simplifies the interaction among humans and physical objects over the digital space thereby assisting the creation of a shared community that operates toward the improvement of society. The key research areas of SIoT are mainly Service Discovery and Service Composition.

Service Discovery is the basic ingredient that finds, what service can be retrieved from the objects similar to the persons looking for information and services on the social networks.

Service composition enables the communication between objects, somewhere the services are recognized by the service discovery component. The objects pair in SIoT, makes the service discovery process more prominent and scalable [7]. Figure 2, shows the IoT devices association with Social networks.

This work proposes a heuristic approach based on friendship selection strategies. The objectives of the work are as follows;

- To select a better future friend for meaningful information dissemination.
- To remove a friend from the friend list which is obsolete in the list.

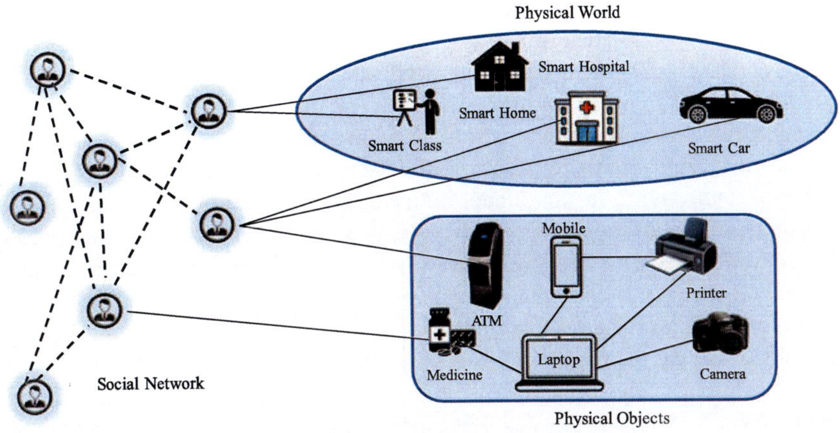


**Fig. 2** SIoT components

The organization of the work is as follows. Section 2 briefs the recent related works. Sections 3 and 4 describe the problem definition and proposed model. Section 5 states the working of the model with the help of an illustrative example. A performance study of the proposed model is done in Sect. 6 by its simulation in python along with the analysis of the results. Finally, the conclusion of the work is mentioned in Sect. 7.

## 2 Related Work

SIoT plays a vital role in the dissemination of information. There are a few similar works that have been done to select a future friend for exchanging the information. A friendship selection-based model is given by Nitti et al. [8]. In which, a few friend selection strategies have been applied to select the right friend. Also, they have analyzed the impact of strategies in terms of network navigability by measuring the giant components, the average degree of connections, local clustering, and average path length. Observation suggests that a few better strategies are still required to be explored.

A model [7] indicates that IoT devices can collect information from surroundings and abstracted as a service. The IoT devices along with social networking can build a social relationship to discover devices along with services viz; Social Internet of Things (SIoT). SIoT can provide reliable and trustworthy networking solutions for utilizing the social network of friends. It can play its significant role in service discovery, relationship management, service composition, and trust management, etc. Model [9] also classified the trust and friendliness-based approaches in the SIoT with important highlights of scalability, adaptability, and suitable network structures.

An advanced heuristic-based friend selection model given by Ramasamy and Arjunasamy [10] in SIoT environment. Finding the desired service in SIoT environment with the help of friends may lead to high computation and memory demands. Therefore, it is highly required to select a friend judiciously to minimize the computation load and also minimize the network navigability. In this model, after applying the heuristic, network navigability is slightly improved.

The model [11] also focused on wise friendship selection for the benefit of overall network connectivity. In this model, efficient, distributed, dynamic solutions are proposed to select the right friend, considering the overall network connectivity. The solution is based on a game-theoretic approach and shapely-value-based algorithm. Performance is analyzed in terms of the average number of hopes and also compared with the standard solutions where objects are not bounded with any limit in terms of having the number of friends. Arjunasamy and Rathi [12] proposed the relationship management-based heuristic to manage the objects which are directly communicating with the devices. Because cope up with the memory and computational capacity of each object is an important task. Results analysis of the work shows that the average degree and average path length of the network are reduced quite significantly.

A few other works also have been done to maintain the trust level among the devices [13] in SIoT environment. Even after considering the works done in the field of SIoT, still there is a scope to devise some new heuristic mechanism to improve the friendship selection for disseminating the information. The work done in the paper is aimed to propose an efficient heuristic to effectively select the future friends.

## 3 Problem Definition

Many of us using social media platforms for communicating with friends, family, colleagues, and other people all over the world. When the friend list of the node is reached the maximum timeline then, no new request can be accepted by the user. If a new request is important than existing friends but no room available then we have to remove some friends from the list to accept the new request. Various social media platforms having their own constraints to impose a limit on the maximum number of friends. For example, in Facebook maximum of 5000 friends can be connected to a node. The model is aimed to propose a heuristic-based approach to select the best future friends.

## 4 The Proposed Model

In the proposed model, we have applied three strategies for friendship selection and to remove a friend from the friend list if it is reached to its maximum limit. A friend which will be removed from the friend list will be having the least priority in the list and even it will have lower importance than the request arrives.

Figure 3, shows the flow of the model along with the applied strategies; Common neighbors', Jaccard Coefficient, and Adamic Adar index.

In Fig. 3, $X$ send request to $Y$, then $Y$ can accept the request only when if $Y_{\text{degree}} < N_{\max}$ and if it is not then applying the heuristic to find a friend the one which is having less importance to remove from the list. The meaning of the symbols used, are indicated in Table 1.

Strategies used in the model for friendship selection are as follows;

- **Strategy 1**: Many users send the request to $Y$ and $Y_{\text{degree}} < N_{\max}$ timeline accepts the new request based on the highest common Neighbors' otherwise if $Y_{\text{degree}} < N_{\max}$ then there is no room available for the new request, remove the friend from $Y's$ list based on the common neighbor value, then accept the new request.
- **Strategy 2**: When many users like $X$ send the request to $Y$ and $Y_{\text{degree}} < N_{\max}$ timeline accepts the new request based on the highest Jaccard Coefficient value otherwise when $Y_{\text{degree}} < N_{\max}$ then for accepting the request it is required to remove the friend from the $Y's$ list whose common neighbors is minimum and then accept the new request based on the highest Jaccard Coefficient value.

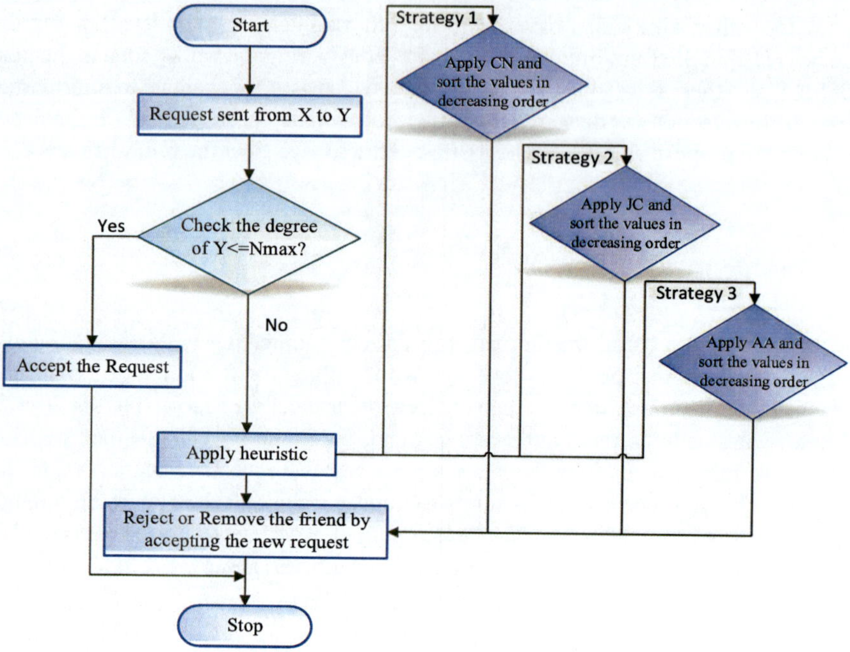


**Fig. 3** Flow chart of the proposed methodology

**Table 1** Notations

| Symbol | Description |
|---|---|
| $N$ | Number of nodes |
| $X, Y$ | Represents the nodes in the network |
| $C_i$ | Local clustering coefficient |
| $e_n$ | Number of linked set neighbors of n |
| $K_n$ | Number of neighbors of a node |
| $Y_{\text{degree}}$ | Number of nodes directly connected to $Y$ |
| $N_{\max}$ | Maximum permissible degree of any node |

- **Strategy 3**: In the last strategy, when many users like $X$ send the request to $Y$ and again if $Y_{\text{degree}} < N_{\max}$ timeline accepts the new request based on the highest Adamic Adar Index value. Otherwise, $Y_{\text{degree}} < N_{\max}$ then there is no room to accept the new request, then for accepting the new request, remove the friend from $Y's$ list and then after, accept the new request based on the highest Adamic Adar Index value.

The neighbor-based heuristics used for implementing the model are as follows;

I. *Common neighbors*
Common neighbors are the intersection operation of the two sets. The greater values give the chance of future friendship between them and vice versa [14].

$$\text{Common Neighbors } (X, Y) = |N(X) \cap N(Y)| \tag{1}$$

where $X$ and $Y$ are two nodes in the network, in which $N(X)$ indicates the number of nodes adjacent to $X$ and $N(Y)$ indicates the number of nodes adjacent to $Y$.

II. *Jaccard Coefficient*
It compares the number of common neighbors of node $X$ and $Y$ concerning the total neighbors of $X$ and $Y$. With help of Union operation, we can calculate the total number of neighbors in the set. The working of the metric JC depends on set overlapping and it is based on the ratio of intersection and union operations. It may suffer from division by zero when it contains the empty set in the denominator [15].

$$\text{Jaccard Coefficient } (X, Y) = \frac{|N(X) \cap N(Y)|}{|N(X) \cup N(Y)|} \tag{2}$$

III. *Adamic Adar Index*
It is used to find out the probability that a node $X$ is associated to a node $Y$ as the sum of the number of mutual neighbors. Also, it can determine the intersection of neighbor-sets of two nodes in the network or graph but emphasize the smaller overlap. The working of this metric depends on the commonality between the two problem nodes [16].

$$\text{Adamic Adar Index } (X, Y) = \sum_{u \in N(X) \cap N(Y)} \frac{1}{\log(|N(u)|)} \tag{3}$$

## 4.1 Flow Chart

The flow chart, to explain the working of applied strategies is given in Fig. 4. In the proposed model, JC, CN, or AA any one of the techniques can be used to find the suitable node which is required to remove for adding a new node into the list.

Figure 4, shows the detailed working of applied strategies in the proposed model. The objective of the flow chart is to show the process of adding a future friend by removing an obsolete friend from the list.

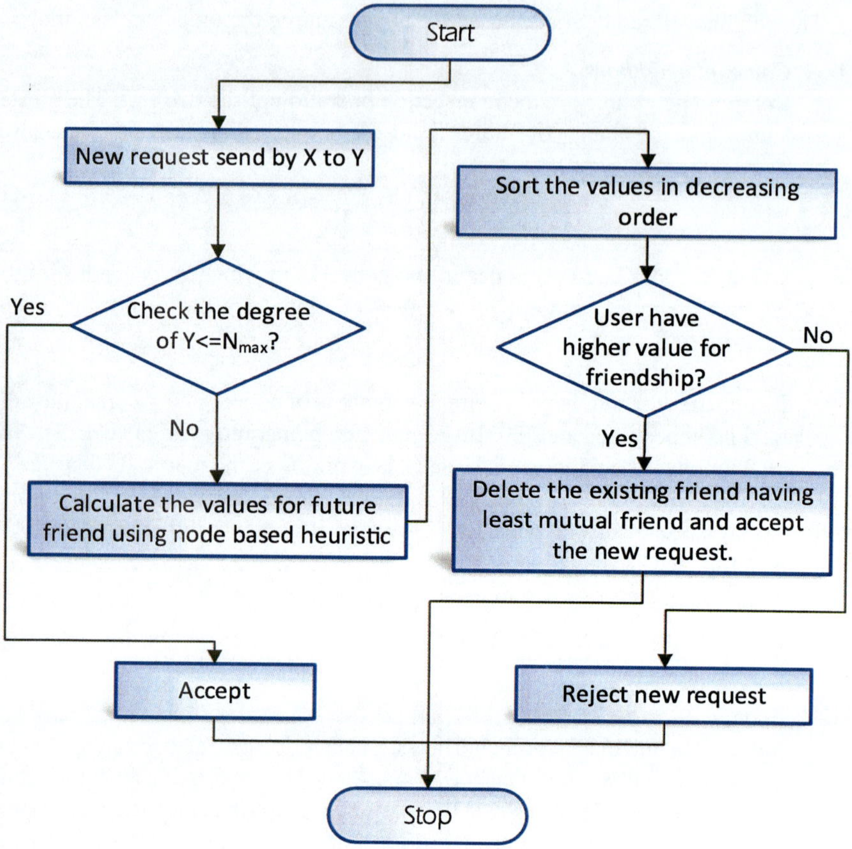


**Fig. 4** Flow chart

## 4.2 Algorithm

The algorithm of the proposed model is as follows which addresses the limitation of already existing models.

---

Algorithm: Future Friend Selection
Input: *X* sends a friend request to *Y*
Output: Friend request Accepted or Rejected
Steps: Initialize the degree of each node in the network

---

(continued)

(continued)

1. New request from node $X$ to node $Y$
2. $if Y_{degree} < N_{max}$ then
3. Compute the value of nodes by applying the CN, JC, AA-based heuristics
4. Accept the friend request based on a higher-value request
5. The request may be rejected if the value is too low for future requests.
6. ***else_if*** $Y_{degree} \geq N_{max}$ then
7. *if (X among the first* $N_{max}$) then
8. Select a friend $Z_{del}$ from the existing list with the least value and delete from the list
9. Create the space for a new friend and accept the request
10. *else*
11. Reject the request to add to the list
12. ***else***
13. ***end***

In the algorithm, initially start with a new friendship request, line number 2–5 is used to simply accept the request by checking the usefulness of the request with the help of JC, CN, AA values if receiving node $Y_{\text{degree}} < N_{\text{max}}$.

Line number 6–12 check the maximum limit of friends in a friend list if it is $Y_{\text{degree}} \geq N_{\text{max}}$, then based on calculated values check $X$ is in the top $N_{\text{max}}$ or not. If it is, then simply accept the request by deleting $Z_{\text{del}}$ which is having the least mutual friend among the list of nodes. Otherwise, simply reject the request.

In the end, line number 13 indicates the termination of the algorithm. Line numbers 1–13 will be repeated for every new request received.

## 4.3 Parameters

A few key parameters used in the proposed model for evaluation and explanation are as follows. A node refers to each object in the social network and edges are the friendship link between any two objects. The shortest distance between any two nodes in the network is averaged as the average path length. Degrees of the nodes in SIoT network are one of the important parameters to analyze its behavior. Average degree is defined as the ratio of the total number of degrees of each node in the network to the total number of nodes where the degree is defined as the number of connections to the nodes.

The average clustering coefficient [17] and the Giant component are the important parameters to analyze the network. The average clustering coefficient is used to measure the nodes' closeness. It is calculated for each node; the range is 0–1. The clustering coefficient can be different for the directed and undirected network [17] as indicated in Eqs. 4 and 5. Equation 6 represents the average clustering coefficient for the directed network.

For Undirected Network

$$CC_n = \frac{2e_n}{(k_n * (k_n - 1))} \quad (4)$$

For Directed Network

$$CC_n = \frac{e_n}{(k_n * (k_n - 1))} \quad (5)$$

Average Clustering Coefficient for directed Network

$$\overline{c} = \frac{1}{n}\sum_{i=0}^{n} C_i \quad (6)$$

The giant component is also one of the popular parameters that refer to the connected component in the network wherever a cluster of connected nodes is organized. It also refers to the higher is the network navigability of the network. The large complex networks often contain a giant component that holds a large percentage of all nodes.

## 5 An Illustrative Example

In this section, we have taken a scenario of the network to explain the working of the model. In which, BA model is taken with 15 nodes, $N_{\max} = 10$, and $m = 2$ (average overrun); Nodes 4, 1, 5, 10 want to become the friend of node 0.

So, in this model, we have calculated the number of common neighbors of the nodes 4, 1, 6, 10 with node 0 which is shown in Table 2.

From Table 2, it can be seen that the maximum common neighbors are between node 0 and node 4 which is 3. From Fig. 5, it can be seen that the degree of node 0 is 4 $(\text{Node } 0)_{\text{degree}} = 4$. We can see that $(\text{Node } 0)_{\text{degree}} < N_{\max}$ because $N_{\max} = 10$. Therefore Node 0 can first accept the request of the node which is having the highest common neighbors. In Fig. 5, we can see that (Node 4) is connected with (Node 0) and it is indicated by the thick brown line.

**Table 2** Nodes with common neighbors

| Requesting nodes | Receiver node | Common neighbors |
|---|---|---|
| Node 4 | Node 0 | 3 |
| Node 1 | Node 0 | 2 |
| Node 6 | Node 0 | 1 |
| Node 10 | Node 0 | 0 |

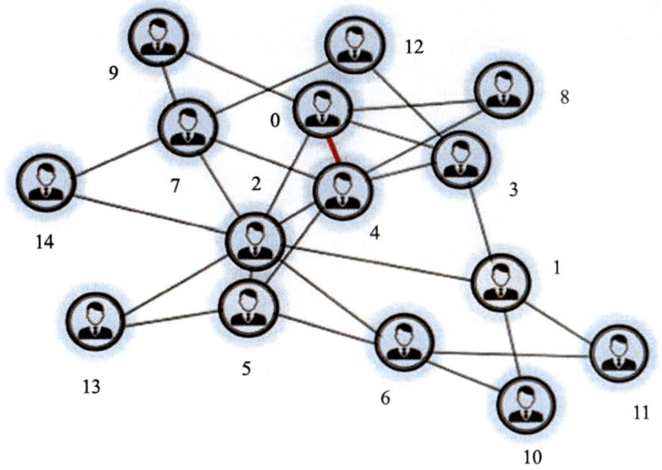


**Fig. 5** SIoT network

After reaching to $N_{max}$, a node cannot send the request to another node in the network. So our proposed model can suggest a future friend to the nodes which are having the highest node-based heuristic value. The request receiving node can remove a friend from the list which is having fewer mutual friends and can accept the request of a suggested friend.

## 6 Performance Analysis

To analyze the outcomes of the model, simulation is done in the python programming language using the "**NetworkX**" package. It is used for creation, manipulation, and to study the dynamics and functions of complex networks. The performance of the proposed model is observed on a Facebook dataset collected from Stanford Large Network Dataset collection [18] and BA model.

In this section, experiments are conducted to observe the performance of the model in terms of the following parameters; Network navigability, Average degree, Average clustering coefficients, Average path length.

### *6.1 Experiment 1*

This set of experiments are carried out to observe the network Navigability of the proposed model where navigability refers to the network in which we have a short path between all or almost all pair of nodes in the network.

**Table 3** Parameter of Facebook and BA model

| Parameters of network | Facebook | BA model |
|---|---|---|
| Nodes | 4039 | 1000 |
| Edges | 88,234 | 4975 |
| Network diameter | 8 | 5 |
| Average path length | 3.68 | 2.7 |
| Average clustering coefficient | 0.605 | 0.108 |
| Average degree | 43.6 | 9.99 |

Table 3, shows the values of various parameters applied on the BA model and Facebook dataset.

In this section, we have also observed the degree distribution of the BA model with the number of nodes are 1000 and $m = 5$ (averaged over 5 runs). Figure 6 shows the degree distribution of the BA model. Observation shows that higher degree node frequency is comparatively low.

We have also analyzed the Facebook dataset to observe the degree distribution, and it is observed that the nodes which are having low frequency are of high degree while high-frequency nodes' degree is comparatively low as shown in Fig. 7.

Figures 6 and 7 show the degree distribution of the Facebook and BA model through a log scale which is also known as the power-law function.

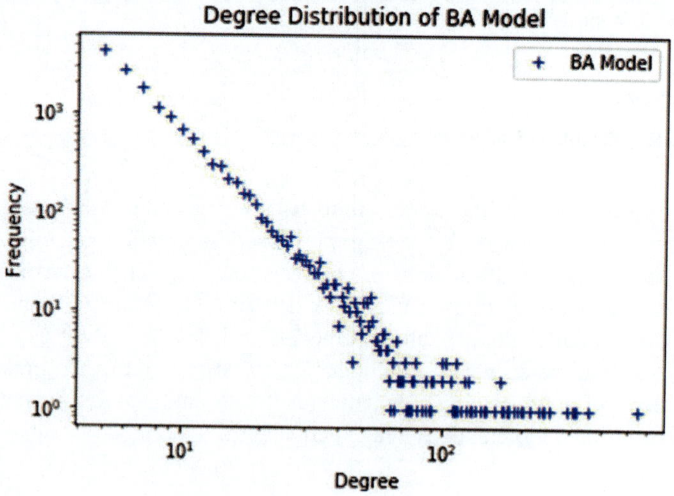


**Fig. 6** Degree distribution of BA model

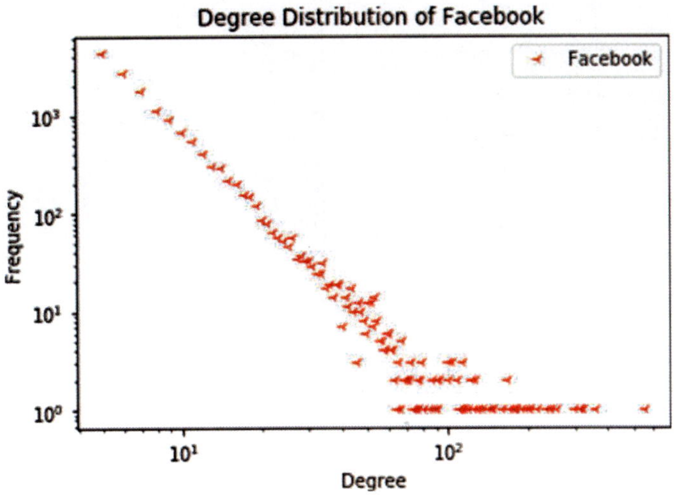


**Fig. 7** Degree distribution of Facebook

## 6.2 *Experiment 2*

This set of experiments are carried to observe the comparative analysis of all the three strategies used in the proposed model. The parameter used for comparative analysis is Average Degree, Average path length, and Average clustering coefficients.

The input parameters are as follows; the number of nodes is 1000, and the average number of edges is 4975.

Figure 8, shows the variation in the average degree of nodes on a varying number of $N_{\max}$ connections. It can be observed that after increasing the $N_{\max}$ average degree is also being increased.

Figure 9, indicates the variation in average path length on a varying number of $N_{\max}$ connections, and it is overserved that average path length is decreasing when

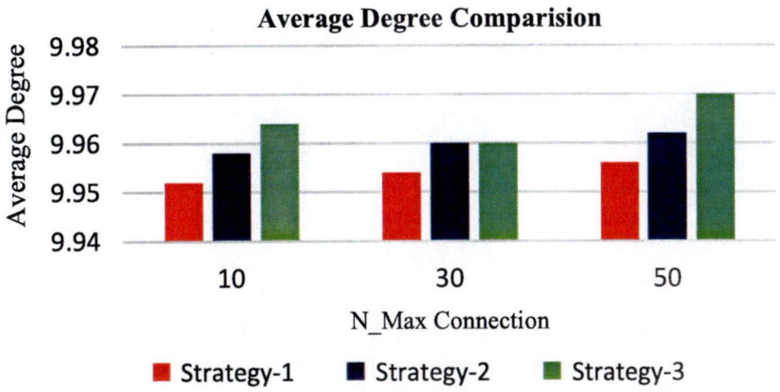


**Fig. 8** Comparison of average degree

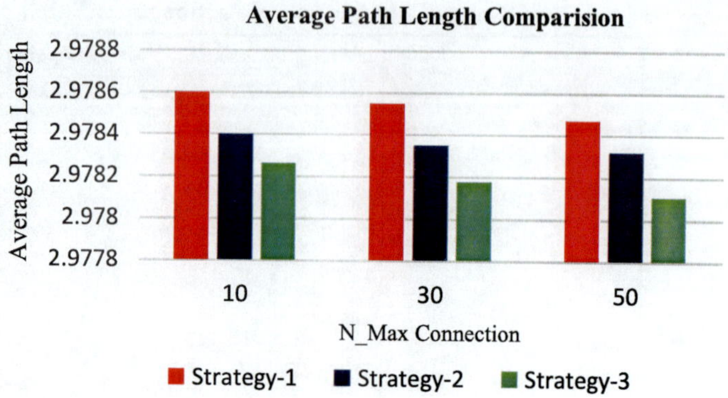


**Fig. 9** Comparison of average path length

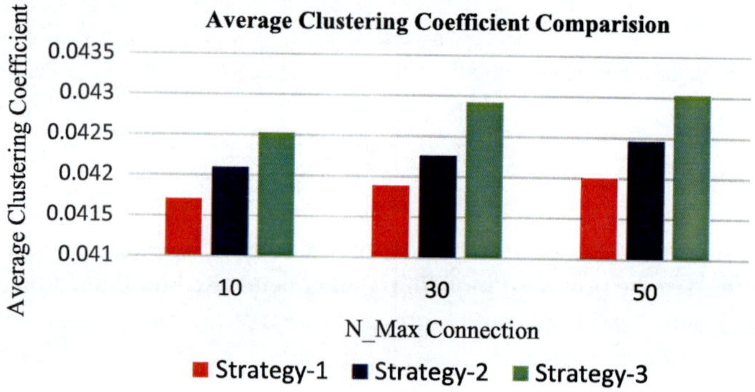


**Fig. 10** Comparison of average clustering coefficient

we increase the $N_{\max}$. It is lowest for all the three strategies when the value of $N_{\max}$ is 50.

Average clustering coefficients has been observed in Fig. 10 on varying $N_{\max}$. It is observed that the average clustering coefficient is highest when $N_{\max} = 50$. Overall observation derived from Figs. 8, 9, and 10 shows that Strategy-1 is performing better in comparison to other strategies.

## 7 Conclusion

In this work, we have applied the proposed heuristic strategies on the BA model and Facebook data set. The model is aimed to observe the performance in terms of minimizing the computational load and memory in SIoT. We have applied the strategies to select the better future friend and removing the obsolete friend from the list especially when the friend list is exhausted. An illustrative example confers the working of the model, in which we can add the future friend by removing the less important one.

Comparative analysis of the proposed strategies is also tested on varying $N_{\max}$ to observe the performance on the parameters' Average degree, average path length, and average clustering coefficients. Results conclude the effectiveness of the proposed model in SIoT environment. In future work, a few soft computing-based techniques can be applied to minimize the service discovery time by selecting the appropriate friend.

## References

1. Rho, S., Chen, Y.: Social Internet of Things: applications, architectures and protocols (2019)
2. Culnan, M.J., McHugh, P.J., Zubillaga, J.I.: How large US companies can use Twitter and other social media to gain business value. MIS Q. Exec. **9** (2010)
3. Roy, A., Maxwell, L., Carson, M.: How is social media being used by small and medium-sized enterprises? J. Bus. Behav. Sci. **26**, 127 (2014)
4. Chung, A.E., Jensen, R.E., Basch, E.M.: Leveraging emerging technologies and the "Internet of Things" to improve the quality of cancer care. J. Oncol. Pract. **12**, 863 (2016)
5. Weaver, A.C., Morrison, B.B.: Social networking. Computer (Long. Beach. Calif). **41**, 97–100 (2008)
6. Zhou, Y., Zhang, Z., Wang, K., Chen, S., Zhou, M., Zhang, J.: Personality and emerging adults' friend selection on social networking sites: a social network analysis perspective. PsyCh J. **10**, 62–75 (2021)
7. Roopa, M.S., Pattar, S., Buyya, R., Venugopal, K.R., Iyengar, S.S., Patnaik, L.M.: Social Internet of Things (SIoT): foundations, thrust areas, systematic review and future directions. Comput. Commun. **139**, 32–57 (2019)
8. Nitti, M., Atzori, L., Cvijikj, I.P.: Friendship selection in the social Internet of Things: challenges and possible strategies. IEEE Internet Things J. **2**, 240–247 (2014)
9. Amin, F., Ahmad, A., Sang Choi, G.: Towards trust and friendliness approaches in the social Internet of Things. Appl. Sci. **9**, 166 (2019)
10. Ramasamy, T., Arjunasamy, A.: Advanced heuristics for selecting friends in social Internet of Things. Wirel. Pers. Commun. (2017). https://doi.org/10.1007/s11277-017-4759-1
11. Militano, L., Nitti, M., Atzori, L., Iera, A.: Enhancing the navigability in a social network of smart objects: a shapley-value based approach. Comput. Netw. **103**, 1–14 (2016)
12. Arjunasamy, A., Rathi, S.: Relationship based heuristic for selecting friends in social Internet of Things. Wirel. Pers. Commun. **107**, 1537–1547 (2019). https://doi.org/10.1007/s11277-019-06344-8
13. Lin, Z., Dong, L.: Clarifying trust in social Internet of Things. IEEE Trans. Knowl. Data Eng. **30**, 234–248 (2017)
14. Aljubairy, A., Zhang, W.E., Sheng, Q.Z.: SIoTPredict : a framework for predicting relationships in the social internet. Springer International Publishing (2020). https://doi.org/10.1007/978-3-030-49435-3
15. Hwang, C., Yang, M., Hung, W.: New similarity measures of intuitionistic fuzzy sets based on the Jaccard index with its application to clustering. Int. J. Intell. Syst. **33**, 1672–1688 (2018)
16. Adamic, L.A., Adar, E.: Friends and neighbors on the web. Soc. Netw. **25**, 211–230 (2003)
17. Watts, D.J., Strogatz, S.H.: Collective dynamics of 'small-world' networks. Nature **393**, 440–442 (1998)
18. SNAP: Stanford Large Network Dataset Collection. https://snap.stanford.edu/data/index.html#socnets. Last accessed 31 March 2021

# A Fuzzy String Matching-Based Reduplication with Morphological Attributes

**Apurbalal Senapati, Arunendu Mondal, and Soumen Maji**

**Abstract** String matching is a common problem in the field of computer science, and it is a common operation in various language processing tasks. Several efficient algorithms have been developed for string matching problems like Knuth-Morris-Pratt (KMP) algorithm [1], Rabin-Karp's algorithm [2], matching using a finite-state machine, etc. But in natural language processing, the problem related to string matching is much complex and the conventional string matching algorithm does not fulfil their requirement. This paper presented one such issue related to the string matching on partial reduplication. In that context, a fuzzy-based string matching technique has been proposed. In this approach, the fuzzy membership is not only considered based on the character/symbol or sub-string matching rather some other grammatical information like morphological information, prosodic pattern, etc., are considered. The experiment is done on the Bengali dataset, and finally, the system is tested on real-life text to measure the accuracy.

**Keywords** Natural language processing · String matching · Fuzzy · Reduplication

## 1 Introduction

Reduplication is a linguistic phenomenon that exists in almost all human languages. It is defined as a multi-word or lexeme that repeats two identical or almost identical parts. In other words, it implies the repetition of a linguistic entity, and sometimes, it is treated as morphological and phonological processes [3]. Sometimes, the root word and its morphology are repeated to form a new word. For example, goody-goody, bye-bye, night-night, flip-flop, etc., are used in the English language. Sometimes, the repetitions change their semantics or meaning. In some languages [4] (like Bengali, Hindi, Bodo, etc.), the root word is with a slight change also takes part of reduplication. The exact semantic of reduplication needs to capture in

A. Senapati · A. Mondal · S. Maji (✉)
Central Institute of Technology Kokrajhar, Kokrajhar, Assam 783370, India
e-mail: s.maji@cit.ac.in



D. Gupta et al. (eds.), *Pattern Recognition and Data Analysis with Applications*, Lecture Notes in Electrical Engineering 888,
https://doi.org/10.1007/978-981-19-1520-8_14

**Fig. 1** Bengali–English Google translation that could not translate the reduplicated information (translation date: 27-03-2021)

various language processing applications like machine translation, anaphora resolution, information retrieval, etc. In the above example, the reduplication in bye-bye is the exact reduplication, but in the case of the flip-flop, it is not exact but a partial match and called the partial reduplication. The morphology of partial reduplication is relatively simple in English like European languages, but languages like South-Asian, African are complex and difficult to identify from the text.

Reduplication is relatively simple in English-like languages and is less frequently used in language. But in Indic language like Bengali, Hindi, it is used frequently in language, and their morphology is complex [4]. Hence, it needed special attention whilst it uses in various language processing applications.

The use of reduplication (partial) is shown in Fig. 1 in Bengali, and it also shows that its translation in English using the Bengali–English Google translation (translated on 27th March 2021). The result clearly shows that the system could not capture the reduplicated information and hence produce a wrong result. It clearly shows the importance of the study of reduplication.

### *1.1 Types of Reduplication*

Different types of reduplications are there and vary from language to language. Language is a dynamic entity and it evolves in time. For example, the etymological study shows that reduplicated features in English do not appear at all until the fifteenth century [5]. So in this study, first gives a brief outline of the English reduplication and next explained it in the Bengali language.

It is already mentioned that there are two types of reduplication, such as full and partial. But, based on their morphological formation, they can be categorized into different types and it varies upon the languages. In English, it is express in a different way, e.g. imitate sounds (bow-wow, ding-dong, etc.), suggests alternative patterns (ping-pong, flip-flop, etc.), sometimes are disparaging (wishy-washy, dilly-dally, etc.) or it intensifies in meaning (tip-top, teeny-weeny, etc.) [6]. Reduplication is not a frequent means of occurring lexemes in English, rather it is perhaps the most uncommon one. Most of the reduplicated words forms involve a similar morphological or phonological form that results in the rhyme of words [7]. Sometimes, two separate meaningful words combine the reduplication (examples culture-vulture, flower-power, etc.), or sometimes, one of the words is not meaningful, as in super-duper, or both, as in namby-pamby. In some cases, the results are a combination of two existing words, like culture-vulture and flower-power, but more usually, one of the elements is not meaningful, as in super-duper, or both, as in namby-pamby.

The reduplication is a complex morphology for South–Asian language. It is used frequently in the languages to satisfy the various pragmatic and linguistic reasons and purposes. From the structural aspect, there are six category of reduplication in the Bengali language [4], which are as follows with examples:

(i) Exact reduplication: repetition of the same word, and it looks like ww or w–w or 'w w', where 'w' is any token of the language.
Example, বছরবছর/bachharbachhar (every year), whereas the meaning of বছর/bachhar is a year. দিন-দিন/din-din (day by day), whereas the meaning of দিন/din is a day.
Note that in exact reduplication, each word has valid mining, and in most of the cases, it is different from their reduplicated meaning.

(ii) In this category, the duplicated word is the inflection of the first word.
Example, ধবধবে/dhabdhabe (pure white colour), whereas ধব/dhab and ধবে/dhabe are not a meaningful word. In the example টকটকে লাল/taktake lal (deep red colour), whereas the meaning of টক/tak is sour but টকে/take is not a meaningful word.
It shows that in this category, each individual word is not a valid word but the reduplicated form gives a valid meaning.

(iii) In this category, the first part is inflected, and then, the inflected one is repeated.
Example, ঘরেঘরে/ghareghare (in every house) whereas the meaning of ঘরে/ghare is in house. Similarly, কানেকানে/kanekane (secretly), whereas the meaning of কানে/kane is in ear.
The semantic behaviour of this category is almost similar to the category (i).

(iv) In this category, semantically two words or almost similar words are used to form the reduplication.
Example, চালচুলো/chalchulo (economically poor) where the meaning of চাল/chal is rice and চুলো/chulo means cooking burner. Similarly, চুরিচামারি/churichamari (robbery) where the meaning of চুরি/churi is theft, চামারি/chamari means illegal work.
Noted that, in this case, the semantic of each individual token is almost similar, and their reduplicated meaning is also same as of individual word.

(v) Sometimes, an eco-word is co-occurrence along with to generate the reduplicated word.
Example, জলটল/jaltal (water, beverage, etc.) where the meaning of জল/jal is water and টল/tal is the eco-word. Similarly, খাবারদাবার/khabardabar (varieties food) where the meaning of খাবার/khabar is food and দাবার/dabar is eco-word.
Generally, in these cases, the first token has a specific meaning, but with the eco-word, it changes its meaning. It is also noted that the composite meaning is almost similar to that of the first word but in plural form. But this rule does not applicable in all cases.

(vi) Sometimes, the repetition of onomatopoeic words behaves like reduplications. Examples, ছমছম/chhamchham (feeling of sound of silence), খলিখলি/khilkhil (sound of laugh), ঝনিঝনি/jhinjhin (jingling), etc.
These are also considered under reduplication, and in these cases, the semantic is related to sound of some real event or virtual feeling.

## 2 Previous Study on Reduplication

Most of the theories on reduplication are contributed by the linguistic people, and the study has started long back in various Indic languages [8–10]. After the advancement of the computer, it has drawn special attention and produced several works in computational aspects to explore the semantic. Senapati and Garain [4] have ruled the reduplicated pronoun in the application of anaphora resolution. They have also [4] tried to develop an algorithm to find the reduplication from the raw corpora. A semantic-based analysis of reduplication is carried out by Bandyopadhyay [11], whereas an extensive study is found from Dash [12]. Dolatian et al. [13] tried to model the reduplication with 2-way finite-state transducers. There are so many attempts to model reduplication using the finite-state machine [14–18].

## 3 Problem in Existing System and Proposed Solution

Most of the computational approaches are based on the morphological construction of the duplicated words. The morphological similarity implies that the similar or almost similar of words in terms of their use of characters, word length, and use of vowel modifiers in the words. But such similarities do not cover all types of reduplication. Especially for the partial reduplication identification, the classical string matching algorithms are not suitable, rather they need special treatment. Some researchers [4] tried to address by the heuristic approach, but they face high-false positive cases. To solve this problem, we have introduced a fuzzy-based string matching that produced a better result.

## 3.1 Fuzzy-Based String Matching

Fuzzy is a mathematical concept introduced by Zadeh [19] use to measure the quantity or qualitative of the vagueness of data. The literary meaning of fuzzy refers to the concept of vague or not. For example, a person of the age of 50 years. It cannot consider that person as an old man or a young man. Because that person has some features of an old man and some features of a young man. In such cases, it is difficult to define such a hard decision rather we need to define such activities in a fuzzy manner. Particularly, in this case, it is more suitable to say that person is an old man with weightage 0.6, and person is a young man with weightage 0.4.

Fuzzy string matching technique is the matching technique that matches a pattern approximately rather than exactly. In other word, the fuzzy matching is a matching technique that will find the degree of matches even when there is some mismatch in characters in the words. This is also termed approximate string matching. So, this concept can be import in the finding of the partial reduplication in the corpus. There is an in-built fuzzy Python library FuzzyWuzzy that is used in string matching. That library uses the Levenshtein distance to calculate the differences by counting the mismatching characters between the strings. For example, using the FuzzyWuzzy library, the matching score of the pair ('হাবি/habi', 'জাবি/jabi') and ('হাবি/habi', 'হারি/hari') is 67 in both the cases, but the pair ('হাবি/habi', 'জাবি/jabi') is a valid reduplication but ('হাবি/habi', 'হারি/hari') is not a reduplication at all. So this cannot be a matching criterion for partial reduplication because there involve some other linguistic features like morphology, suffix, prefix, rhythms, etc., and hence, we need to develop separate fuzzy string matching techniques incorporating all such linguistic features.

There are five morphological features which are used in the fuzzy string matching criteria. In computational viewpoint, the morphological features are defined as the (a) length of characters excluding the vowel modifiers, (b) length of characters in the vowel modifiers, (c) ordered sequence of the characters, excluding the vowel modifiers, (d) ordered sequence of characters of vowel modifiers, (e) position of mismatch.

The following partial reduplicated example illustrated the features with a concreate example.

Consider the partial reduplicated strings $s1$ = 'চাকরি' and $s2$ = 'বাকরি'.
Vowel modifier of string $s1$ =< া, ি > and of string $s2$ =< া, ি >
String excluding vowel modifier of $s1$ =< চ, ক, র > and for $s2$ =< ব, ক, র >

Whenever partial reduplication comes, it calculates the fuzzy similarity score based on the above morphological features. It calculates an aggregate score, and it is considered a reduplicated pair if its fuzzy score is exceeding a predefined threshold score.

How to assign the fuzzy score for each morphological feature is a critical issue and what should be the threshold value is also important. This is done by the survey of the existing reduplication in the Bengali literatures. Philosophically, it is assuming that when it matches all the morphological properties, its highest score should near 1 and none of these properties matches, then the score should be near zero, i.e. the range of membership value is [0, 1]. Since, there are five morphological features are used, so, for every matching feature/criteria, the fuzzy score is considered 1/5 = 0.2. For the position mismatch criteria (*e*), a score of 0.20 has been considered when mismatched at 1st position, the score will be 0.10 when a mismatch is in 2nd position, and so on and the threshold score has been set to 0.75.

## 4 Results and Discussion

Though the reduplication is a common phenomenon but the partial reduplication is relatively less frequent in the regular Bengali text. To test the performance of our system, five random news articles are selected from the AnandabazarPatrika, a leading Bengali newspaper, where 20 partial reduplications are injected randomly into the text and are considered as the text data. Some of such injected partial reduplication are {'খবরা খবর'/khobora-khabor (information), 'সামনা সামনি'/samna-samni (face to face), 'ঝাপটা ঝাপটি'/jhapta-jhapti (act like wrestling), 'আম আদমি'/aam-aadmi (common people), 'মার কাট'/mar-kat (bloodshed), …}.

The fuzzy-based string matching system retrieves the 18 from the 20 injected partial reduplications and fails to identify two cases. On the other hand, the system retrieves two more cases from the text which are not valid reduplication. Hence, the precision = 18/20 = 0.9 and recall = 18/20 = 0.9.

### 4.1 *Error Analysis*

The above results show that the system fails in two cases (i.e. false negative = 2) and besides, it identified two wrong instances (i.e. false positive = 2). To investigate these erroneous cases, the error analysis is performed, and details are represented in a confusion matrix in Table 1. The table also shows the false instances in false negative and false positive cells, respectively. In case of false negative ('আম আদমি'/aam-aadmi (common people), 'মার কাট'/mar-kat (bloodshed)), the morphological structure of these cases is different which is not explained above and is not incorporated in the fuzzy score calculation. So their total calculated fuzzy score goes below the threshold value and not treated as reduplicated. Similarly, in the case of false positive ('গড়িয়ে গিয়ে'/gariye-giye (after rolling), 'নিজেই নিজের'/nijei-nijer (by itself)), there is a similarity in morphological structure, and hence, the calculated fuzzy score goes above the threshold value and treated as reduplicated.

**Table 1** Confusion matrix

| $n$ = 20 | Detected reduplicated | Fail to detected reduplicated | Total |
|---|---|---|---|
| Actual reduplicated | True positive = 18 | False negative = 2 (‘আম আদমি’, ‘মার কাট’) | 20 |
| Not reduplicated | False positive = 2 (‘গড়িয়ে গিয়ে’, ‘নিজেই নিজেরে’) | | |
| Total | 20 | | |

## 5 Conclusion

The paper reported an ongoing work of fuzzy-based partial reduplication identification technique. Earlier, this problem is addressed by the heuristic and 2-way finite transducer. In comparison with existing systems from an algorithmic viewpoint, our approach is more sophisticated, robust, and scalable. With the fine-tuning of the fuzzy scoring scheme, the incorporation of unrevealed morphological features for better performance. This technique can be imported easily for other morphologically rich languages.

## References


1. Knuth, D., Pratt, M.: Fast pattern matching in strings. SIAM J. Comput. **6**(2), 323–350 (1977)
2. Karp, R.M., Rabin, M.O.: Efficient randomized pattern-matching algorithms. IBM J. Res. Dev. **31**(2), 249–260 (1987)
3. Rubino, C.: Reduplication. Max Planck Institute for Evolutionary Anthropology, Leipzig (2013)
4. Senapati, A., Garain, U.: A computational approach for corpus based analysis of reduplicated words in Bengali. In: International Conference on Intelligent Text Processing and Computational Linguistics, pp. 456–466. Springer, Cham (2015)
5. Millward, C.M., Hayes, M.: A Biography of the English Language. Nelson Education (2011)
6. Crystal, D.: The Cambridge Encyclopedia of the English Language. Ernst Klett Sprachen (2004)
7. Burridge, K.: Gift of the Gob: Morsels of English Language History. Harper Collins (2010)
8. Chattopadhyay, S.K.: Bhasa-Prakash Bangala Vyakaran, 3rd edn. Pupa publication (1992)
9. Chaudhuri, B.B.: Bangla Dhwanipratik: Swarup o Abhidhan (Bangla Sound Symbolism: Properties and Dictionary). Paschimbanga Bangla Academy, Kolkata (2010)
10. Thompson, H.R.: Bengali: A Comprehensive Grammar, pp. 663–672. Routledge publication (2010)
11. Bandyopadhyay, S.: Identification of reduplication in Bengali corpus and their semantic analysis: a rule-based approach. In: Proceedings of the Workshop on Multiword Expressions: From Theory to Applications (MWE 2010), pp. 72–75. Beijing (2010)
12. Dash, N.: A Descriptive Study of Bengali Words, pp. 225–251. CUP (2015)
13. Dolatian, H., Heinz, J.: Modeling reduplication with 2-way finite-state transducers. In: Proceedings of the Fifteenth Workshop on Computational Research in Phonetics, Phonology, and Morphology, pp. 66–77 (2018)

14. Walther, M.: Finite-state reduplication in one-level prosodic morphology. arXiv preprint arXiv:cs/0005025v1 (2000)
15. Beesley, K.R., Lauri, K.: Finite-State Morphology: Xerox Tools and Techniques. CSLI, Stanford (2003)
16. Cohen-Sygal, Y., Shuly, W.: Finite-state registered automata for non-concatenative morphology. Comput. Linguist. **32**(1), 49–82 (2006)
17. Hulden, M.: Finite-state machine construction methods and algorithms for phonology and morphology (2009)
18. Hulden, M., Shannon, T.B.: A simple formalism for capturing reduplication in finite-state morphology. In: Proceedings of the 2009 Conference on Finite-State Methods and Natural Language Processing: Post-proceedings of the 7th International Workshop FSMNLP 2008, pp. 207–214 (2009)
19. Zadeh, L.A.: Fuzzy sets. Inf. Control **8**(3), 338–353 (1965)

# Accelerating LOF Outlier Detection Approach

**Abhaya, Mohini Gupta, and Bidyut Kr. Patra**

**Abstract** Outliers are the deformities in the data that diverges from the normal behavior. Detection of outlier points is a crucial task as it leads to the extraction of the discordant observations in different domains. One of the most popular density-based outlier detection techniques is local outlier factor (LOF), and later many variants of this approach are also introduced. These techniques have more execution time as they calculate the outlier score for every data point. In this paper, we propose an approach that first detects the data points which have a high probability of being an outlier (i.e., probable outliers) based on $Z$-score and modified $Z$-score statistical techniques. Subsequently, we compute the anomaly score of only these probable outliers. Therefore, we avoid to calculate the outlier score of a substantial number of data points. We conducted experiments on synthetic dataset as well as on real-world datasets, and experimental results demonstrate that our proposed approaches outperform the popular outlier detection technique LOF and its variants.

**Keywords** Density-based outlier detection method · Local outlier factor · $Z$-score · Modified $Z$-score

## 1 Introduction

Outlier detection is investigating the abnormal activities that are generated by some uncertain mechanisms. These unusual activities behave differently across the domains. The discovery of these activities is significant in various applications like traffic monitoring, intrusion detection, credit card fraud analysis, medical field, etc. [1]. Many approaches are introduced to identify anomalies in literature [2–6]. These approaches can be categorized into three classes: (a) supervised outlier detection techniques, (b) semi-supervised outlier detection techniques, (c) unsupervised outlier detection techniques. Supervised techniques require a dataset that has labeled

Abhaya (✉) · M. Gupta · B. Kr. Patra
National Institute of Technology, Rourkela, Rourkela, India
e-mail: abhayasharma77@gmail.com



D. Gupta et al. (eds.), *Pattern Recognition and Data Analysis with Applications*,
Lecture Notes in Electrical Engineering 888,
https://doi.org/10.1007/978-981-19-1520-8_15

normal and outlier classes. Semi-supervised techniques require the dataset, which has labeled normal classes, and if an instance does not belong to that class, then considered as an outlier point. Unsupervised outlier detection techniques detect the anomalies in an unlabeled dataset. These techniques consider that most of the similar instances in the dataset are normal while remaining are outliers.

There are different distance and density-based approaches used for detecting the outliers. Distance-based techniques are based on threshold values. If the distance between an instance and its $k$th nearest neighbor is beyond the threshold limits, then that data object is considered as an outlier [6]. In the density-based outlier detection techniques, each data instance density is estimated and compared with respect to its neighbors. These approaches do not perform well in regions of the data that have varying densities. The various ways are used to obtain the anomaly score of the data object. Among all the proposed approaches, LOF is the most popular technique for detecting the outliers [2]. LOF is an efficient technique and is applied in various application domains like in the mining of outliers in large databases [7], sensor networks [8], streaming data [9], intrusion detection systems [10], etc. LOF computes the score that determines how a data point is isolated with respect to its neighborhood. Later, its other variants are also introduced like COF [3], LDOF [4], etc. In LOF, the anomaly score of each data instance is computed, which leads to higher execution time.

In this paper, we propose an approaches that reduce the execution time of LOF and its variants by detecting the probable outliers. We apply a clustering approach to the whole data, and the properties of each cluster are used to identify the probable outliers. We efficiently utilize the statistical technique $Z$-score and modified $Z$-score for identification of probable outlier points. Subsequently, we compute LOF of only probable outliers. The summary of our main contributions in this paper are

- $k$-means clustering approach is exploited to divide the dataset into number of clusters to reduce the computation time.
- Efficiently utilized a statistical $Z$-score method to identify the probable outlier points.
- To overcome the limitation of $Z$-score method, another statistical approach modified $Z$-score is employed to select the probable outlier points.
- To compare the performance of our proposed approaches to the popular outlier detection approach LOF, one performance evaluation metric is introduced in this paper.
- One synthetic dataset and four real-world datasets are used to evaluate our proposed approaches for detection of outlier points.

The rest of the paper is ordered as follows. Section 2 explains the related work on anomaly detection and LOF in detail. The proposed work is illustrated in Sect. 3. Experimental results and analysis are discussed in Sect. 4. We conclude the paper in Sect. 5.

## 2 Related Work

Most of the researches on outlier detection are based on density-based approaches. The concept behind the density-based approach is that it computes the relative density of an instance and compare it with respect to its neighbors. The point is considered as a normal point, if the density of data instance is close to its neighborhood points. If the density of an instance deviates significantly from its neighbors, it is regarded as an outlier. In these methods, the density of an object is measured locally to test whether the data object lies in the dense region or the sparse region. In all approaches, the anomaly score for each data instance is evaluated. The instance which has a higher anomaly score is considered as an outlier.

Breunig et al. introduced a concept named local outlier factor (LOF) to detect outliers [2]. It is one of the most popular density-based outlier detection approaches which is totally based on the statistics of $k$-nearest neighbors. In LOF, firstly, the reachability distance is computed, which measures how far an instance $p$ is from an instance $o$. The reachability distance *reach-dist* of point $p$ with respect to point $o$ is $reach\text{-}dist(p, o) = \max\{k\text{-}distance(o), d(p, o)\}$.

Where, *k-distance(p)* is the distance of instance $p$ from its $k$th nearest neighbor and $d(p, o)$ is the actual distance between $p$ and instance $o$.

The reachability distance concept is further use for another useful concept local reachability density. For each instance $o \in N_{MinPts}(p)$, where $N_{MinPts}(p)$ are the minimum number of nearest neighbors of $p$, the local reachability density $lrd_{MinPts}(p)$ is the inverse of the average of reachability distance of point $p$ to its $k$ nearest neighbors.

Finally, it computes a outlier score denoted as LOF for each point $p$, which is average of the ratio of local reachability density of a point to its nearest neighbors. The anomaly score of a point $p$ is

$$LOF_{MinPts}(p) = 1/\left(\frac{\sum_{o \in N_{MinPts(p)}} \frac{lrd_{MinPts}(o)}{lrd_{MinPts}(p)}}{|N_{MinPts(p)}|}\right) \tag{1}$$

The point $p$ is considered as an anomaly, if $LOF(p) >> 1$. LOF computes the anomaly score of each data instance, and $topN$ points with the highest values are declared as $topN$ outliers. Therefore, the number of computations will be huge, and it leads to high-execution time and also does not perform well for scattered datasets.

To improve the execution time, Goldstein et al. brought up the concept, which is inspired by an expectation-maximization algorithm [11]. It computes the local outlier factor (LOF) incrementally and runs faster than the standard method. In this algorithm, the dataset is randomly partitioned into the number of chunks, and for each point, the nearest neighbor is computed within the chunk. Based on the computed nearest neighbors, local reachability distance (LRD) and local outlier factor (LOF) are calculated for all the data points, the same as in Eq. 1. The point is considered as a normal point, if the LOF score is close to 1 and will not be used for further processing. For the remaining points, the process repeats until the $n$-outliers are detected. It takes less execution time than LOF. However, the outlier detection accuracy is not at par with the original LOF approach. Further, Poddar et al. proposed another approach that reduces the execution time [12]. In this approach, a metric DevToMean is used, which recognizes the normal data points, and computation of anomaly scores for these normal data points is avoided for further processing. They applied a clustering technique $k$-means for the partition of data points into different clusters, and then within a cluster, they computed DevToMean for each data point. Based on the concept that inlier points situated in the dense region and outliers in the sparse region, they consider the data points as a normal point if the DevToMean score is close to 1. For outlier points, the value of DevToMean is high ($>> 1$) means the distance of point with the mean is high or is close to 0 means near to the mean point. They avoid the outlier score computation for normal points; however, for the computation of selected probable outlier points, all the data instances are used. For the selection of probable outlier points, they applied 10 percentile rule on sorted DevToMean values.

In this paper, we propose an approaches that overcome the execution time issues of LOF and its variants and also improve the performance. We exploit two statistical approach such as $Z$-score and modified $Z$-score. Iglewicz and Hoaglin [13] introduced these statistical approaches for outlier labeling. $Z$-score is based on the normal distribution concept. Computation of $Z$-score is given in Eq. 2.

$$Z_i = \frac{(x_i - \bar{x})}{s} \tag{2}$$

where $x_i$ is the observation, $\bar{x}$ is the mean of all the observations and $s$ is standard deviation.

The mathematical concept median is used for modified $Z$-score calculation given in Eq. 3.

$$M_i = \frac{0.6745(x_i - \bar{x})}{\text{MAD}} \tag{3}$$

where $x_i$ is the observation, $\bar{x}$ is median of ordered observations and MAD is the median of the absolute deviation of each observation from the median $\bar{x}$.

## 3 Proposed Methodology

LOF is an outstanding outlier detection technique. It computes anomaly scores using the statistics of $k$-nearest neighbors. However, computing outlier scores for each data point is not required as the number of outlier points is very less compared to the normal points. In this paper, an algorithm is introduced that computes the outlier score of only those data points which have a higher probability of being an outlier. We termed those data points as "probable outliers". We avoid outlier score computation of normal points without removing it.

In our propose approach, we first divide the dataset into a number of chunks, and each cluster properties are exploited to find the probable outlier points. Here, we use the $k$-means clustering algorithm as it has linear time complexity. Let $\pi = \{C_1, C_2 \ldots C_k\}$ be the clustering obtained after applying $k$-means algorithm on the dataset $D$. We compute the distance of each instance $i$ in the cluster $C_j$ from its center $mC_j$ as $d(i, m_{C_j})$. After obtaining the distance of all the points, we find the mean and standard deviation of distances within the cluster $C_j$ denoted as $d_{\text{mean}_j}$ and $\sigma_j$, respectively. Subsequently, we compute the $Z$-score of each data instance $i$ by using

$$Z\text{-score}(i \in C_j) = \frac{d(i, m_{C_j}) - d_{\text{mean}_j}}{\sigma_j} \tag{4}$$

In our approach, $Z$-score plays a vital role in identifying the probable outliers across the clusters. It measures the deviation of the point from the mean. Within a cluster, if the $Z$-score value of a instance $i$ differs significantly from the rest of the points, then it can be treated as a probable outlier denoted by $prob\_outlier$. There are two possible scenarios associated with the $Z$-score: (i) If the cluster center $m_{C_j}$ lies in the dense region, very likely, the outliers will reside far away from the cluster center and (ii) if the center of the cluster $m_{C_j}$ will lie in the sparse region, an outlier may also lie into the sparse region close to $m_{C_j}$.

Here, first scenario is for the regular shaped clusters. The cluster center $m_{C_j}$, in that case, will lie in the dense region so the outliers will reside far away from the cluster center. In such a case, the $Z$-score will have a much greater value. So the data objects having a higher value could be the probable outliers. The second scenario is for an arbitrarily shaped cluster; the center of such cluster will lie in the sparse region. In this case, the value of the $Z$-score will be less for these data points. The instances having a lesser $Z$-score will also be considered as probable outlier points. Therefore, data points with very high and low $Z$-score values can be considered as probable outlier points.

After obtaining the $Z$-score of each instance $i$, we accommodate all the $Z$-scores and sort them. In our approach, we consider top-$N$ percentile and bottom-$N$ percentile points on the basis of $Z$-score as probable outlier points. Once we detect the probable outliers, the outlier score of only those points is computed. This leads to reduction in the execution time of the entire procedure. It can be noted that we compute the anomaly score of only probable outliers. Therefore, the execution time of LOF is reduced. The procedure is discussed in Algorithm 1.

**Algorithm 1** $Z$-score-based outlier detection

**Input:** *Dataset D, #clusters*
**Output:** *outlier-score*
*cluster* ← *k-meansClustering(D, #clusters)*
**for** *each* $C_j \in cluster$ **do**
  $m_{C_j} \leftarrow mean(C_j)$
  **for** *each* $i \in C_j$ **do**
    $d(i, m_{C_j}) \leftarrow dist(i, m_{C_j})$
  **end for**
**end for**
*Compute mean and standard deviation of distances for every cluster* $C_j$ *as* $d_{mean_j}$ *and* $\sigma_j$, *respectively.*
**for** *each point* $i \in C_j$ **do**
  *Compute* $Z\text{-}score(i)$
**end for**
*points* ← *sort(Z-score)*
*prob-outlier* ← *Filter Z-scores by applying N percentile rule*
**for** *each* $i \in$ *prob-outlier* **do**
  *Compute LOF of i considering entire D*
**end for**

In the case of a small dataset, $Z$-score-based approach does not perform well. To improve the performance of the proposed approach, we applied a statistical technique modified $Z$-score. In the modified $Z$-score based approach, after computation of the distance $d(i, m_{C_j})$ of each point $i$ from the center of each cluster in the $Z$-score based outlier detection approach, we ordered the distance and found out the median and MAD of each distance. We compute the modified $Z$-score of each data point based on this equation:

$$M_i = \frac{0.6745(d(i, m_{C_j}) - \bar{d})}{\text{MAD}} \tag{5}$$

where $\bar{d}$ is the median of ordered distance of each point and MAD is the median of the absolute deviation of the distance from the median ($\bar{d}$). After obtaining the modified $Z$-score of each point $i$, we combine all the scores and sort them. In this approach, we consider the top-$N$ percentile and bottom-$N$ percentile based on modified $Z$-score value as probable outlier points. After the detection of probable outlier points, we calculate the outlier score for only these points. In this way, we can reduce the outlier score computation time of normal points, which do not play any role in identifying any actual outlier point. We do not delete these data points, only we do not consider these points for further calculation. This leads to execution time reduction and also improves the performance of the approach. The procedure is deliberated in Algorithm 2.

**Algorithm 2** Modified Z-score-based outlier detection

**Input:** *Dataset D, #clusters*
**Output:** *outlier-score*
$cluster \leftarrow$ *k-meansClustering(D, #clusters)*
**for** $each\ C_j \in cluster$ **do**
  $m_{C_j} \leftarrow mean(C_j)$
  **for** $each\ i \in C_j$ **do**
    $d(i, m_{C_j}) \leftarrow dist(i, m_{C_j})$
  **end for**
**end for**
*Order the distance* $d(i, m_{C_j})$ *and find the median* $\bar{d}$
*Find the absolute deviation of each distance* $d(i, m_{c_j})$ *from the median* $\bar{d}$
**for** $each\ i \in C_j$ **do**
  *Compute modified Z-score(i)*
**end for**
$points \leftarrow sort$*(modified Z-score)*
*prob-outlier* $\leftarrow$ *Filter modified Z-scores by applying N percentile rule*
**for** $each\ i \in$ *prob-outlier* **do**
  *Compute LOF of i considering entire D*
**end for**

## 4 Experimental Results

The experiments are performed on synthetic dataset as well as on real-world datasets in order to evaluate the proposed algorithm. We compared execution time of our approaches with three existing techniques such as LOF [2], FastLOF [11], and DevToMean [12].

We introduced one evaluation measure *detection factor@n* to compare our proposed approach with the popular density-based approach LOF. We considered the outlier obtained by the LOF technique as ground truth and compared our both approaches Z-score-based outlier detection and modified Z-score-based outlier detection with this ground truth. Computation of *detection factor@n (DF@n)* is given in Eq. 6.

$$DF@n = \frac{|A \cap B|}{n} \tag{6}$$

where $A$ is the outlier set obtained by the LOF approach, $B$ is the outlier set obtained by our proposed approach, $|A \cap B|$ represents number of elements present in the set $(A \cap B)$ and $n$ is the number of injected outlier points in the dataset.

We also considered recall as an evaluation metric given in Eq. 5.

$$\text{Recall} = \frac{\text{TP}}{(\text{TP} + \text{FN})} \tag{7}$$

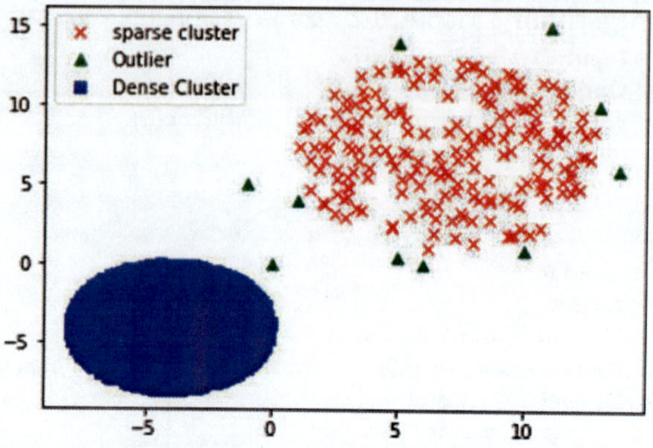


**Fig. 1** Non-uniform dataset

where true positive (TP) is when the observation is relevant and is predicted to be relevant, and false negative (FN) is when the observation is relevant but is predicted as irrelevant.

## *4.1 Dataset Description*

We designed one synthetic dataset named as non-uniform dataset. It has varying density, and 10 outlier points are injected over the feature space (Fig. 1). We also performed our experiments on four UCI machine learning real-world datasets. In the case of real-world datasets, we considered one class or group of classes as normal points, and we injected the data points from other classes and considered as outlier points. The details about the synthetic datasets and the real-world are illustrated in Table 1.

**Table 1** Detail description of datasets

| Dataset | Dimension | Size | #Injected outliers |
|---|---|---|---|
| Non-uniform | 02 | 6310 | 10 |
| Statlog (landsat satellite) | 36 | 5742 | 10 |
| Pendigits | 16 | 9952 | 15 |
| Mammography | 06 | 10,933 | 10 |
| Shuttle | 09 | 45,606 | 20 |

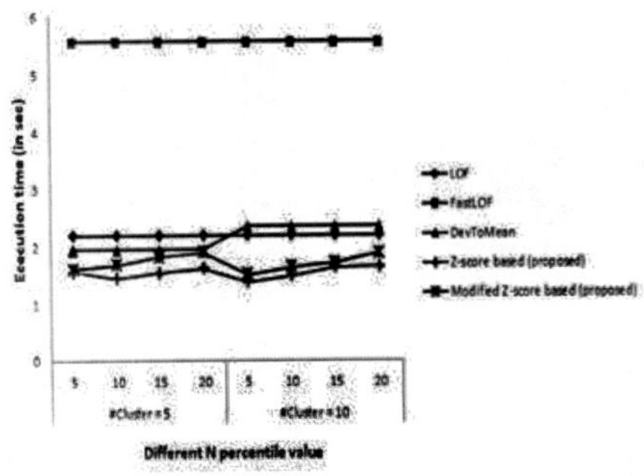


(a) Non-uniform dataset

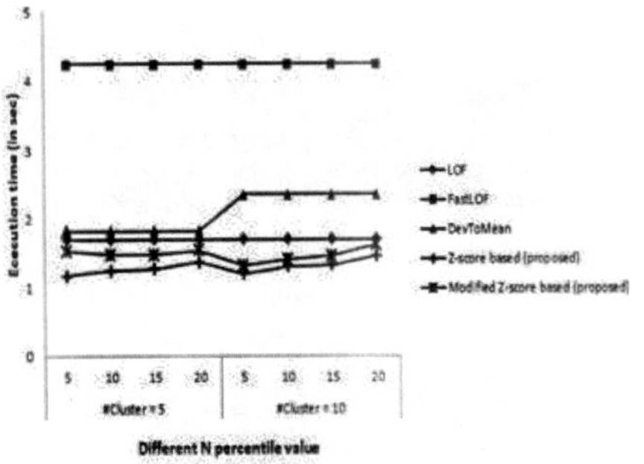


(b) Statlog dataset

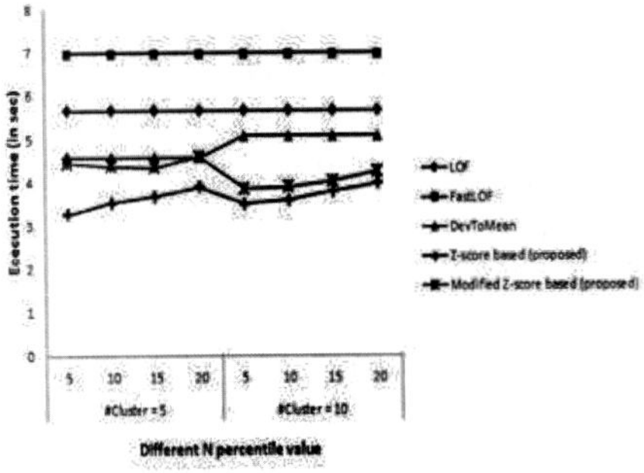


(c) Pendigits dataset

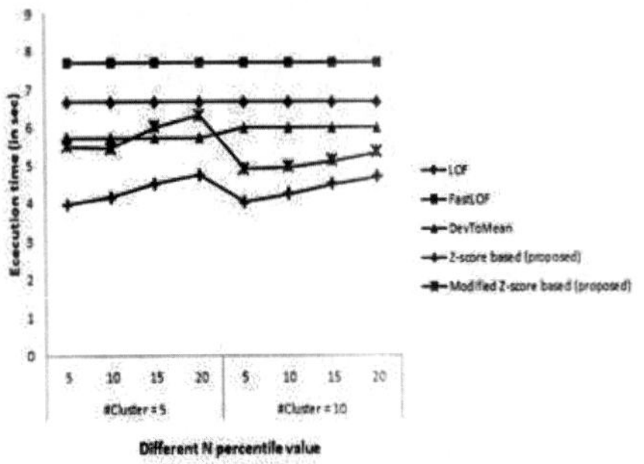


(d) Mammography dataset

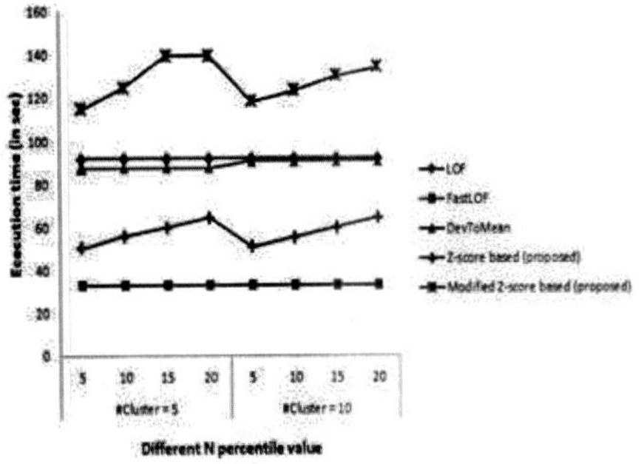


(e) Shuttle dataset

**Fig. 2** Execution time comparison of different dataset

## *4.2 Result Analysis*

The experiments are performed by considering the different number of clusters (5 and 10) as the parameter for our proposed approach and DevToMean existing approach. In the LOF approach, there is no concept of clustering. So, we keep the number of nearest neighbors constant ($knn = 20$) for this approach. Another existing approach FastLOF is based on the different number of chunks concept, where the size of the

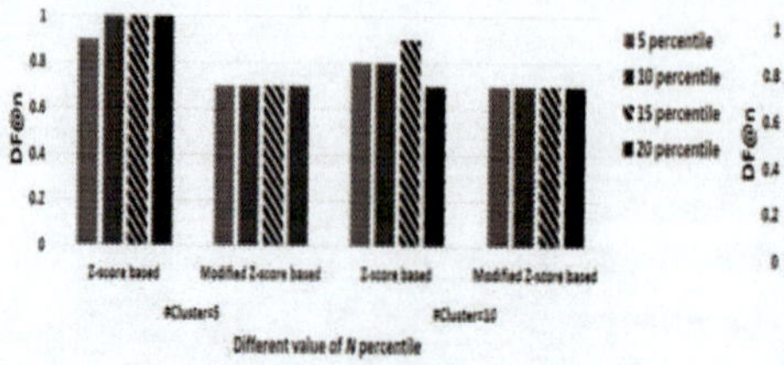


(a) Non-uniform dataset

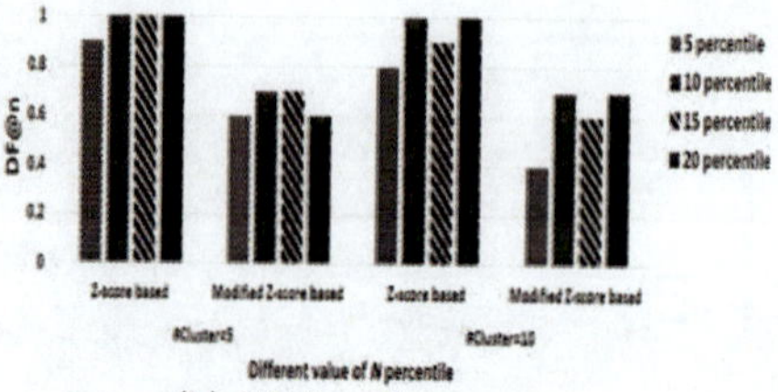


(b) Statlog dataset

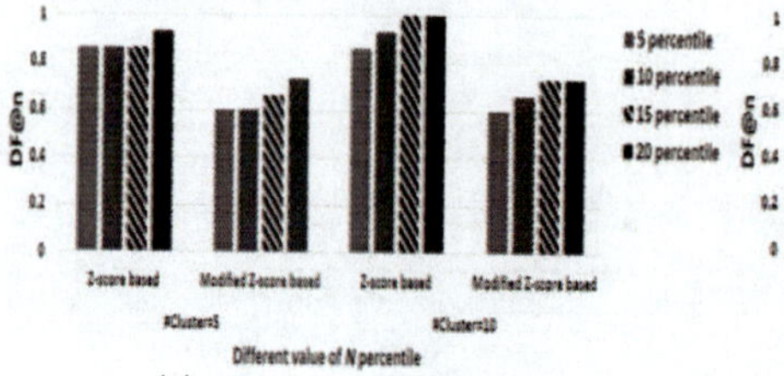


(c) Pendigits dataset

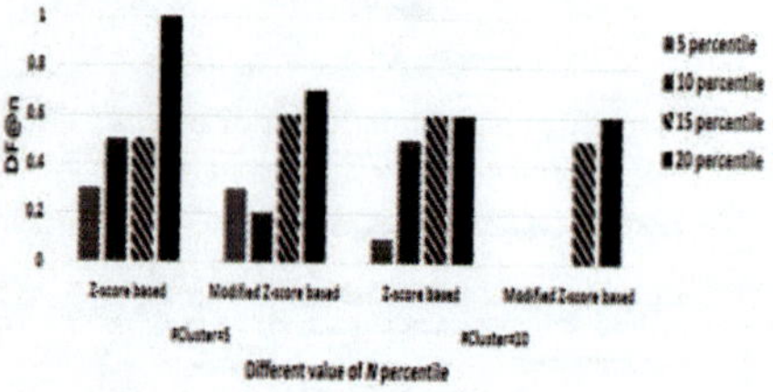


(d) Mammography dataset

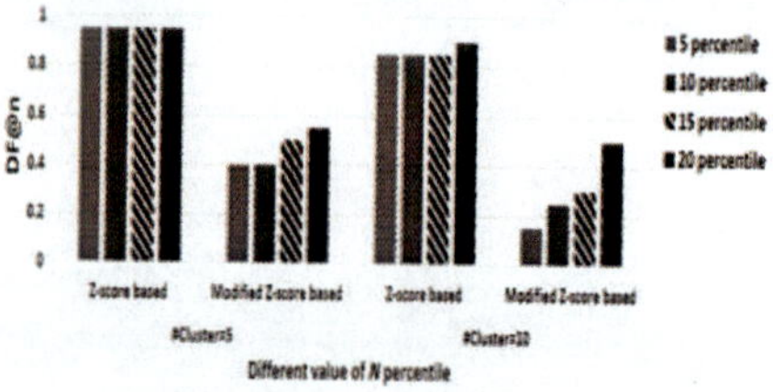


(e) Shuttle dataset

**Fig. 3** DF@n of different dataset

chunk is $\sqrt{D}$, $D$ is the size of the dataset. Also, we experimented with different values of $N$ percentile for the selection of probable outlier points from the dataset in our $Z$-score-based and modified $Z$-score-based outlier detection approach.

The execution time comparison with LOF and its variants on the synthetic non-uniform dataset is reported in the plot (Fig. 2a). Our proposed $Z$-score-based approach takes the least execution time than all the existing approaches. In the case of a modified $Z$-score, its execution time is lesser than the existing approach but more than the $Z$-score-based method. Statlog dataset comparison is reported in the plot (Fig. 2b). We can notice that the $Z$-score-based approach takes the least time than all the existing approaches as well as a modified $Z$-score-based approach in different numbers of the cluster and different $N$ percentile probable outlier data points. The execution time result of the pendigits dataset is shown in plot (Fig. 2c). In this dataset, the $Z$-score based approach takes the least time than all the existing approaches. Modified $Z$-score-based approach in different numbers of the cluster and different $N$ percentile probable outlier data points takes more execution time than $Z$-score based approach. The execution time result of the mammography dataset is reported in the plot (Fig. 2d). In this dataset, $Z$-score based approach takes the least time than

**Table 2** Recall comparison with #cluster=5 and 10

| Dataset | Approach and #cluster | | | | | | | |
|---|---|---|---|---|---|---|---|---|
| | Z-score-based (proposed) | | Modified Z-score-based (proposed) | | LOF | FastLOF | DevToMean | |
| | 5 | 10 | 5 | 10 | | | 5 | 10 |
| Non-uniform | 70.00 | 90.00 | **100.00** | **100.00** | 70.00 | 10.00 | 70.00 | 70.00 |
| Statlog | **60.00** | 50.00 | **60.00** | **60.00** | 50.00 | 10.00 | 30.00 | 30.00 |
| Pendigits | 60.00 | 60.00 | **93.33** | 86.66 | 60.00 | 6.66 | 66.66 | 66.66 |
| Mammography | 10.00 | 10.00 | 10.00 | 10.00 | 0.00 | 0.00 | **20.00** | 10.00 |
| Shuttle | **50.00** | 45.00 | 40.00 | 25.00 | 45.00 | 40.00 | 40.00 | 40.00 |

Bold represents the best performance

all the existing approaches as well as a modified Z-score-based approach. The result of shuttle dataset is reported in the plot (Fig. 2e). We can infer that FastLOF has the least execution time than all the other approaches. In this dataset, the modified Z-score has the higher execution time compared with the existing and Z-score-based approach.

We experimented with a different value of $N$ percentile for $DF@n$ measure computation. Figure 3a represents the comparison of our approaches Z-score-based and modified Z-score-based with the ground truth, which is the outlier score obtained by popular technique LOF on a non-uniform synthetic dataset. We can notice that our Z-score based approach is comparable with the popular LOF method. $DF@n$ results with statlog dataset is reported in Fig. 3b. Z-score-based approach with 5 number of clusters performance is very near to the LOF approach. Obtained $DF@n$ results on the pendigits dataset and shuttle dataset by Z-score based detection approach is also comparable with the approach LOF (Fig. 3c, e). In the mammography dataset (Fig. 3d), LOF hardly detects outlier point. As a result, during comparison of our performance with the LOF approach, $DF@n$ is degraded.

We used evaluation measure recall for performance calculation of all the techniques (proposed and existing). In the LOF and FastLOF approach, there is no concept of clustering. Table 2 depicted the recall of each approach on different datasets. We experimented with #cluster=5 and #cluster=10. Here, #cluster represents the number of cluster. We only represent the recall of the 10 percentile selection of probable outlier points. Recall in bold represents the best performance on each dataset. For #cluster = 5, modified Z-score achieves 100.00% recall, which is better than Z-score-based method and all the existing methods. In the case of Statlog dataset, performance of Z-score-based approach and modified Z-score based approach is equivalent and also better than all the other existing approach. LOF and FastLOF are not able to detect outlier points on the mammography dataset. In this dataset, performance of DevToMean approach is better than other approaches. On the shuttle dataset, LOF performs better than a modified Z-score-based approach. However, our other approach Z-score-based outperforms all the other techniques with 50.00% recall. When we experimented with #cluster=10, performance of FastLOF and LOF is poor for all the dataset. The popular technique LOF performs better on shuttle dataset. However, our proposed Z-score based approach is comparable with LOF on this data. For all the other datasets, modified Z-score performs better than other approaches (existing and Z-score-based) [14].

## References

1. Chandola, V., Banerjee, A., Kumar, V.: Anomaly detection: a survey. ACM Comput. Surv. CSUR **41**(3), 15 (2009)
2. Breunig, M.M., Kriegel, H.P., Ng, R.T., Sander, J.: LOF: identifying density-based local outliers. In: Proceedings of the 2000 ACM SIGMOD International Conference on Management of Data, pp. 93–104 (2000)

3. Tang, J., Chen, Z., Fu, A.W.C., Cheung, D.W.: Enhancing effectiveness of outlier detections for low density patterns. In: Pacific-Asia Conference on Knowledge Discovery and Data Mining, pp. 535–548. Springer, Berlin, Heidelberg (2002)
4. Zhang, K., Hutter, M., Jin, H.: A new local distance-based outlier detection approach for scattered real-world data. In: Pacific-Asia Conference on Knowledge Discovery and Data Mining, pp. 813–822 (2009)
5. Hubballi, N., Patra, B.K., Nandi, S.: NDoT: nearest neighbor distance based outlier detection technique. In: International Conference on Pattern Recognition and Machine Intelligence, pp. 36–42 (2011)
6. Knorr, E.M., Ng, R.T.: Algorithms for mining distance-based outliers in large datasets. VLDB **98**, 392–403 (1998)
7. Jin, W., Tung, A.K., Han, J.: Mining top-n local outliers in large databases. In: Proceedings of the Seventh ACM SIGKDD International Conference on Knowledge Discovery and Data Mining, pp. 293–298 (2001)
8. Salehi, M., Leckie, C., Bezdek, J.C., Vaithianathan, T.: Local outlier detection for data streams in sensor networks: revisiting the utility problem invited paper. In: 2015 IEEE Tenth International Conference on Intelligent Sensors, Sensor Networks and Information Processing (ISSNIP), pp. 1–6. IEEE (2015)
9. Salehi, M., Leckie, C., Bezdek, J.C., Vaithianathan, T., Zhang, X.: Fast memory efficient local outlier detection in data streams. IEEE Trans. Knowl. Data Eng. **28**(12), 3246–3260 (2016)
10. Alshawabkeh, M., Jang, B., Kaeli, D.: Accelerating the local outlier factor algorithm on a GPU for intrusion detection systems. In: Proceedings of the 3rd Workshop on General-Purpose Computation on Graphics Processing Units, pp. 104–110 (2010)
11. Goldstein, M.: FastLOF: an expectation-maximization based local outlier detection algorithm. In: Proceedings of the 21st International Conference on Pattern Recognition (ICPR2012), pp. 2282–2285. IEEE (2012)
12. Poddar, S., Patra, B.K.: Reduction in execution cost of k-nearest neighbor based outlier detection method. In: International Conference on Mathematics and Computing, pp. 53–60. Springer, Singapore (2018)
13. Iglewicz, B., Hoaglin, D.C.: How to Detect and Handle Outliers, vol. 16. ASQ Press (1993)
14. Jin, W., Tung, A.K., Han, J., Wang, W.: Ranking outliers using symmetric neighborhood relationship. In: Pacific-Asia Conference on Knowledge Discovery and Data Mining, pp. 577–593. Springer, Berlin, Heidelberg (2006)

# Evaluating Quality of Machine Translation System for Digaru-English with Automatic Metrics

**Rushanti Kri and Koj Sambyo**

**Abstract** The machine translation output requires evaluation to measure its relevancy to the reference for checking its quality. The evaluation of the output translations can be done manually by human translator, but it requires immense amount of time and labor cost. Therefore, the automatic evaluation metrics has been introduced which is widely used for machine translation evaluation task. In this paper, the translation quality of Digaru-English translation output of 10,500 sentence pair is measured using the automatic evaluation metrics. Further, the translation output of Digaru-English from the Moses SMT system has been evaluated by the BLEU, NIST, METEOR, and TER metric. Further, manual evaluation of Digaru-English has been examined and judged based on fluency and adequacy.

**Keywords** BLEU · NIST · TER · METEOR · Evaluation metrics

## 1 Introduction

Determining the quality of a translation has been quite difficult since the very beginning as there are many possible ways to interpret a translation with respect to its reference. The use of humans for evaluation has been stated as time-consuming and expensive. Most of the low-resourced languages like Digaru have relatively free word order, complex grammatical construction and are morphologically rich [11]. This property of a language makes evaluation even harder when compared to the English language which does not have declensions [3]. The evaluation approach was first proposed by Miller and Beeber-center in 1956 followed by Pfaffine in 1965. Evaluation was only performed by the human judges during the initial phase of machine translation evaluation which was particularly subjective [8]. In the recent years with the rapid increase in human and computer interaction through different languages, it has resulted in higher demand of efficient language processing tools which has piqued the interests of researchers in the field of natural language processing. For translation

R. Kri (✉) · K. Sambyo
Department of Computer Science and Engineering, NIT Arunachal Pradesh, Jote, India
e-mail: krirushanti@gmail.com



D. Gupta et al. (eds.), *Pattern Recognition and Data Analysis with Applications*,
Lecture Notes in Electrical Engineering 888,
https://doi.org/10.1007/978-981-19-1520-8_16

accuracy, computer needs to perform several analyzes on two large parallel aligned corpora. Advancements in the field of machine translation in the recent years have led to the analysis of parallel corpus with higher accuracy and almost perfect translation of a target sentence given its source sentence from the parallel corpora [2]. The automatic evaluation metrics impart assessment of the translation accuracy and its quality. Although the evaluation metrics provides us with relatively faster and cheaper evaluation of a translation quality without any requirement of the bilingual speakers, it still does not provide reliable scores for an individual sentence. For a much reliable output, the need of a large dataset is required [20].

This paper discusses the different type of automatic evaluation metrics like the BLEU, TER, NIST, METEOR, etc., and compares the score generated form these metrics for the Digaru-English corpus consisting of 10,500 parallel sentences. The translated output generated from the Moses SMT system has been evaluated on the basis of its fluency and adequacy to judge its quality [10].

## 2 Related Works

The parameters like fluency and adequacy determine the relativity of a translation with its reference for manual evaluation. Manual evaluation is a labor-intensive process and requires linguistic experts for both language and hence is cost effective. Therefore, several automatic evaluation metrics are being introduced based on scoring the output for judging the quality. The metric is the measurement that evaluates MT output to represent its quality. Quality scores should be assigned by the MT evaluation metric for higher correlation with the human judgment that is usually subjective in nature [14]. Several machine translation evaluations metric perform well with large amount of data; evaluation measures are relatively less reliable on shorter translation (Turian et al. 2003). Automatic metric must include five attributes for it to be acceptable, and this fives attributes include correlation, reliability, sensitivity, consistency, and generality [1]. The automatic evaluation came in light with the introduction of BLEU metric which is based on the n-gram match between the candidate and reference. BLEU is the standard metric till date as it highly correlates with the human judgment based on the average output of individual sentence error (Papineni et al. 2001). The NIST was then introduced which is also an automatic evaluation metric that calculate the matched n-grams and attaches heavier weight for rare word. The NIST was introduced to improve BLEU and uses arithmetic mean of the n-gram matches between candidate and reference translation, and a brevity penalty is also introduced (Doddington 2002). The TER metric has been designed to give very intuitive metric that requires less data compared to other automatic evaluation techniques and avoid labor-intensive work required in human evaluation (Snover 2006). The METEOR metric makes use of recall which is calculated at word level. Tuning of weights is required in METEOR to match human judgment [1]. Languages like German, French, etc., with English translation had been used for determining the correlation between automatic metrics and human evaluation where metric with higher correlation-coefficient determines translation quality of a system.

## 3 Automatic Evaluation

The automatic evaluation metric evaluates the output of a MT system to judge the quality of the translated output in context to its reference translations [12]. Quality of any automatic evaluation metric for the translated output must assign scores that correlates with the human evaluation of the same. The automatic evaluation metrics were developed to reduce higher cost, subjectivity, etc., of the human judgment [19]. There are several automatic machine translation evaluation metrics like the BLEU, NIST, METEOR, TER, etc., which gives effective results mostly based on the precision, recall, editing distance, and f-measure [5–7]

1. **Precision**: which gives us the correct words count in the machine translation output.
2. **Recall**: provides us with the total no. of words that are correct in the reference.
3. **Both Precision and Recall**: The combination of the precision and recall gives us the F1 score which can be calculated by $F1 = 2PR/(P + R)$.
4. **Levenshtein edit distance**: provides the total number of insertions, deletion, and the substitution needed to transform the machine translation output to its reference.

### *3.1 The BLEU Metric*

The bilingual evaluation understudy (BLEU) metric was developed to measure the relativity of SMT output with the human reference translation with the help of weighted match average. BLEU metric makes use of modified n-gram precision. If the reference translation matches with a candidate word, the reference translation is considered exhausted [15].

The BLEU metric requires brevity penalty to compensate for the difference in length of the candidate and reference translation and is calculated as

$$P_B = \begin{cases} 1, c > r \\ e\left(1^{-rc}\right), c \leq r \end{cases}$$

$r$ *and* c are the length of reference corpus and candidate translation, respectively.

**The BLEU Metric**

$$\text{BLEU} = P_B \exp\left(\sum_{n=0}^{N} w_n \log p_n\right)$$

Here, $w_n$ is positive weights summing to 1, $N$ is the maximum length of n-grams, and $p_n$ is the $n$-gram precision.

## 3.2 The NIST Metric

The NIST score is an evaluation metric (Doddington 2002) that formulates the n-gram to give us an insight to the n-gram output and the degree of information it can provide, the rarer a correct n-gram the higher will be the weight assigned to it.

The NIST score can be calculated by the use of below equation:

$$\text{NIST Score} = \sum_{n=1} \left\{ \sum_{\substack{\text{all } w_1, w_2, \ldots, w_n \\ \text{that co - occur}}} \text{info}(w_1, w_2, \ldots, w_n) / \sum_{\substack{\text{all } w_1, w_2, \ldots, w_n \\ \text{in sys output}}} \quad (1) \right\}.$$

$$\exp\left\{\beta \log^2\left[\min\left(\frac{L_{sys}}{\overline{L}_{ref}}, 1\right)\right]\right\}$$

In the above equation, $\overline{L}_{\text{ref}}$ represents the average no. of words in a reference translation which is the overall averaged of all the reference translation, $L_{\text{sys}}$ is the total no. of words scored based on the translation. The NIST score is stable in scoring and reliability. Similar to the BLEU score, NIST score can be generated for various language pairs, and it requires a source translation and one or more reference translations [16].

The NIST score for the Digaru-English translation output has been normalized under 0–1 by using min–max normalization:

$$Z = \frac{\mu - \alpha}{\beta - \alpha}$$

where $Z$ is the normalized value, $\mu$ is the actual value, $\alpha$ represents the minimum of $\mu$, and $\beta$ represents the maximum value of $\mu$.

## 3.3 The METEOR

Metric for evaluation of translation with explicit ordering (METEOR) is an automatic evaluation metric that generates translation score by computing the alignments with reference to its exact word, synonym, stemmed word matching, and paraphrase matching of the word and sentences [1]. These modules lay out alignment between the reference and the predicted translation [4].

Execution of the precision accompanied with recall gives higher relativity to human judgment at sentence level [13].

Let us consider few notations:

$\varphi$: Number of unigrams in reference translation.

$\omega$: Number of unigrams in predicted translation.

$\theta$: Number of identical unigrams between $\varphi$ and $\omega$.

The unigram precision $\rho$ can be obtained by

$$\rho = \frac{\theta}{\omega}$$

The unigram recall $\gamma$ can be obtained by

$$\gamma = \frac{\theta}{\varphi}$$

And, F-measure £ generated while computing METEOR score provides the mean of $\rho$ and $\gamma$ with the formula:

$$\pounds = \frac{2 * \rho * \gamma}{\rho + \gamma}$$

The use of higher order n-grams is not required as there is no need for explicit word-word matching of the n-grams. METEOR produces additional features such as the matched synonym, stemming, and exact word. Unlike the bilingual evaluation understudy (BLEU) and NIST metrics that depends only on precision, METEOR makes use of both precision and recall.

### 3.4 TER Metric

The translation edit rate (TER) which has been proposed by Matthew Snover and Bonnie Dorr 2006 calculates the minimum number of edits required to change a hypothesis in order to match it with one of the references. The TER score measures the hypothesis with its nearest reference [18]. TER score requires minimum number of editing which include the insertion, deletion, substitution of a word, and shifting of word sequences.

$$\text{TER} = \frac{\text{no. of edits}}{\text{avg. of reference words}}$$

**Table 1** Digaru-English MT system

| S. No. | Corpus description | Total no. of instances |
|---|---|---|
| 1 | Training set | 8590 |
| 2 | Testing set | 1000 |
| 3 | Tuning set | 910 |

Translation edit rate is commonly used when the machine translation output is produced for post-editing purpose since it helps in estimating the total amount of changes need to be done by the human translators to produce translation of human (gold) quality.

## 4 Human Evaluation

The manual evaluation technique which is also termed as the human evaluation is among the most used technique for judging the machine translation quality. The human evaluation requires an expert who is fluent in either one or both the language and aware of the linguistic features behind the translation. Therefore, a greater number of experts are requested to evaluate the translation, and final evaluations are made and justified statistically. Manual evaluation requires effort and is costly and very time-consuming process.

Human evaluation is usually done by rating translation on a fixed scale where the highest rating symbolizes higher quality translation and lowest rating indicates the opposite. Adequacy and fluency of a translation determine the quality of translation based on the level of meaning it preserve and how fluent is the translated output.

## 5 Corpus Description

The automatic evaluation has been performed on the output of Digaru-English corpus consisting of 10,500 parallel aligned sentence pairs. The Moses SMT system has been used to obtain the output of the Digaru-English corpus consisting of the training and testing sets [9] (Table 1).

## 6 Result and Analysis

The quality of Digaru-English translation output has been analyzed, and based on the scores, the performance of the automatic evaluation metric has been judged. The BLEU/NIST is based on the n-gram precision, and the output of 3 g from training and tuning has been evaluated and compared below.

**Table 2** BLEU score with and without tuning

| n-gram | Without tuning | With tuning |
|---|---|---|
| 1-g | 0.4596 | 0.4847 |
| 2-g | 0.1843 | 0.2252 |
| 3-g | 0.1232 | 0.1442 |

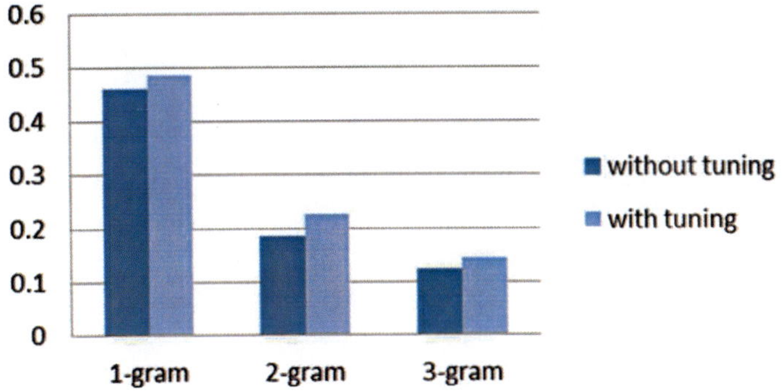


**Fig. 1** Comparison of BLEU score from Table 2 with and without tuning

## 6.1 *BLEU Score With and Without Tuning*

The BLEU score has been evaluated based on the precision and the degree of information a translation could provide. The Digaru-English corpus has been trained and tuned, respectively. After application of the tuning parameter, the score of the BLEU has increased its efficiency for the unigram, bigram, and the trigram values.

The BLEU score of 1, 2, and 3 g for the Digaru-English corpus with and without tuning is given in Table 2 and its corresponding comparative graph is given in Fig. 1 respectively.

## 6.2 *NIST Score With and Without Tuning*

The tuning parameter has refined the scores in both the BLEU and the NIST metric for evaluation of the Digaru-English corpus. The BLEU score of 1, 2, and 3 g for the Digaru-English corpus with and without tuning is given in Table 3 and Fig. 2.

**Table 3** NIST score with and without tuning

| n-gram | Without tuning | With tuning |
|---|---|---|
| 1-g | 2.5187 | 3.0054 |
| 2-g | 0.4940 | 0.6250 |
| 3-g | 0.0870 | 0.1057 |

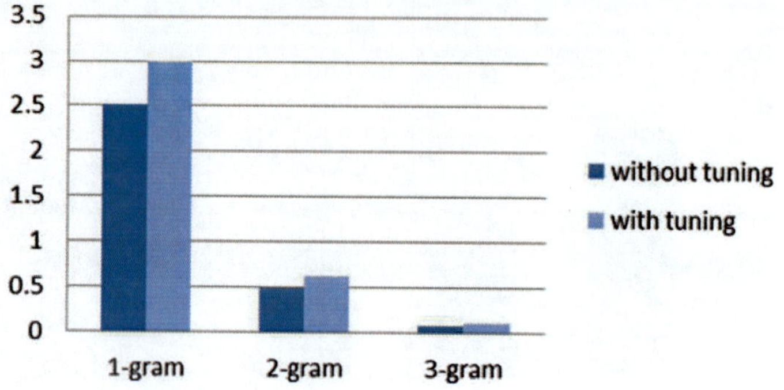


**Fig. 2** Comparison of NIST score from Table 3 with and without tuning

## 6.3 *Automatic Evaluation on Digaru-English Corpus*

The result of different evaluation metric with value ranging between 0 and 1 has been applied on the Digaru-English corpus to determine the quality of the translation output. Table 4 provides the output for different automatic evaluation metrics applied on the Digaru-English corpus.

Figure 3 depicts the comparative analysis performed on the Digaru-English corpus with different automatic evaluation metric where the TER metric has the highest value within 0–1 range followed by NIST, METEOR, and the lowest BLEU. Here, the TER metric calculates the no. of edits between the reference and the hypothesis, and the lower the result of TER in 0–1 range better will be the quality of the translation output.

The higher value of TER for Digaru-English corpus suggests more edit required in the Digaru-English translation output.

**Table 4** Comparative analysis of automatic metrics for Digaru-English

| No. | Automatic evaluation metrics | Output |
|---|---|---|
| 1 | BLEU | 0.1947 |
| 2 | NIST | 0.3056 |
| 3 | TER | 0.7983 |
| 4 | METEOR | 0.2432 |

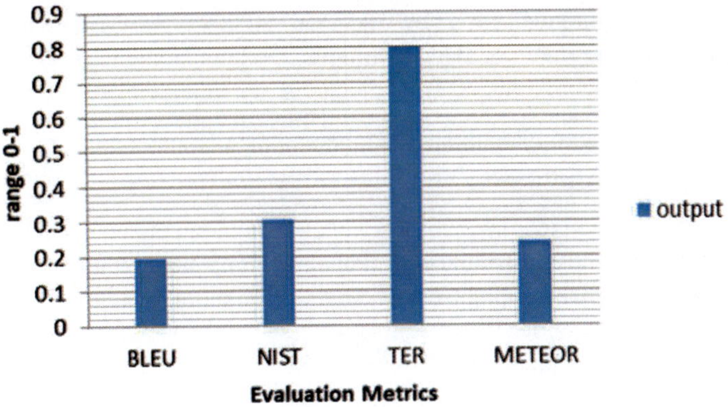


**Fig. 3** Comparative analysis on output of different automatic evaluation metrics performed on Digaru-English corpus

## 6.4 *Human Evaluation for Digaru-English*

To further examine the translated output of Digaru-English corpus, human evaluators who are native speaker having Digaru as their first language has been approached. The automatic evaluation metrics scored very low for Digaru-English translation quality and quality judgment failed in most of the important quality parameters. Here, we make use two parameter *adequacy* and *fluency* to judge the quality of our predicted translation. Few samples of the translated English sentences from the predicted outputs are examined with its reference translation and the input translation in Digaru of the same sentence pair for examining their translation quality from different perspective.

Table 5 shows the best, moderate, and worst translation output according to the human translators for Digaru-English machine translation output.

In Table 5, the prediction ranges from perfectly fluent and adequate translation to translation which are fluent and to translations that does not relate with the reference in any word. Since Digaru is a morphologically rich language [17], evaluating the translated output has been quite difficult by the automatic metrics since it requires large dataset for obtaining higher scores, therefore it can be concluded that the SMT

**Table 5** Human evaluation on Digaru-English translation

| S. No. | Source (Digaru) | Reference | Predicted | Quality parameter |
|---|---|---|---|---|
| 1 | a lai, haa nyu kasadi | Hey, I know you | Hey, I know you | Adequate and fluent |
| 2 | cho chow na | Start running | let us run | Fluent |
| 3 | haa naara | I am not well | I am sick | Partially adequate and fluent |
| 4 | tachung bari hanana | Get out of bed | bed come down | Neither adequate nor fluent |

system performed quite well for a new and small 10,500 Digaru-English corpus that gave several nearly perfect translation outputs.

## 7 Summary/Conclusion

This paper provides the performance level of several automatic evaluation metrics on a relatively small Digaru-English corpus. The n-gram score for 1, 2, and 3-g has been evaluated with tuning and without tuning where the scores with tuning parameter are slightly higher compared to the scores without tuning. Form the comparative analysis performed among the BLEU, NIST, METEOR, and TER metric, the TER has the highest score suggesting that the Digaru-English translation output requires a lot of edit to match the reference.

Even with the availability of several evaluation metrics to judge the quality of translation, there is a need of large dataset for the metrics to perform its finest. For a small corpus of 10,500 Digaru-English sentence pairs, the automatic evaluation metrics scored quite low. Therefore, the translation output has been evaluated by the human translator to check the translation quality based on fluency and adequacy.

## References


1. Banerjee, S., Alon, L.: METEOR: an automatic metric for MT evaluation with improved correlation with human judgments. In: Proceedings of the ACL Workshop on Intrinsic and Extrinsic Evaluation Measures for Machine Translation and/or Summarization, pp. 65–72 (2005)
2. Bhattacharyya, P.: Machine Translation. CRC Press, Boca Raton, USA. ISBN: 978-1-4398-8, xxv 234 (2015)
3. Dabre, R., Fabien, C., Sadao, K., Pushpak, B.: Leveraging small multilingual corpora for SMT using many pivot languages. In: Proceedings of the 2015 Conference of the North American Chapter of the Association for Computational Linguistics: Human Language Technologies, pp. 1192–1202 (2015)
4. Denkowski, M., Alon, L.: Meteor universal: language specific translation evaluation for any target language. In: Proceedings of the Ninth Workshop on Statistical Machine Translation, pp. 376–380 (2014)
5. Gupta, U., Deepak, G.: An improved regularization based Lagrangian asymmetric ν-twin support vector regression using pinball loss function. Appl. Intell. **49**(10), 3606–3627 (2019)
6. Gupta, U., Deepak, G.: Lagrangian twin-bounded support vector machine based on L2-norm. In: Recent Developments in Machine Learning and Data Analytics, pp. 431–444. Springer, Singapore (2019)
7. Hazarika, B., Deepak, G.: Density-weighted support vector machines for binary class imbalance learning. Neural Comput. Appl. 1–19 (2020)
8. Kalyani, A., Hemant, K., Shashi, P., Ajai, K.: Assessing the quality of MT systems for Hindi to English translation. arXiv preprint arXiv:1404.3992 (2014)
9. Koehn, P., Hieu, H., Alexandra, B., Chris, C., Marcello, F., Nicola, B., Brooke, C., et al.: Moses: open-source toolkit for statistical machine translation. In: Proceedings of the 45th Annual Meeting of the Association for Computational Linguistics Companion Volume Proceedings of the Demo and Poster Sessions, pp. 177–180 (2007)

10. Koehn, P.: Statistical Machine Translation. Cambridge University Press (2009)
11. Kri, R., Koj, S.: Phrase-based machine translation of Digaru-English. In: Electronic Systems and Intelligent Computing, pp. 983–992. Springer, Singapore (2020)
12. Lin, C., Eduard H.: Automatic evaluation of summaries using n-gram co-occurrence statistics. In: Proceedings of the 2003 Human Language Technology Conference of the North American Chapter of the Association for Computational Linguistics, pp. 150–157 (2003)
13. Melamed, I., Ryan, G., Joseph T.: Precision and recall of machine translation. In: Companion Volume of the Proceedings of HLT-NAACL 2003-Short Papers, pp. 61–63 (2003)
14. Nyodu, K., Koj, S.: Automatic identification of Arunachal language using K-nearest neighbor algorithm. In: 2018 International Conference on Advances in Computing, Communication Control and Networking (ICACCCN), pp. 213–216. IEEE (2018)
15. Papineni, K., Salim, R., Todd, W., Wei-Jing, Z.: Bleu: a method for automatic evaluation of machine translation. In: Proceedings of the 40th Annual Meeting of the Association for Computational Linguistics, pp. 311–318 (2002)
16. Przybocki, M., Kay, P., Sébastien, B., Gregory, S.: The NIST 2008 metrics for machine translation challenge—overview, methodology, metrics, and results. Mach. Transl. **23**(2), 71–103 (2009)
17. Roger, B.: A Dictionary of Tawra, a Language of Arunachal Pradesh, Orthographic Edition. McDonald Institute of Archaeological Research University of Cambridge
18. Snover, M., Bonnie, D., Richard, S., Linnea, M., John, M.: A study of translation edit rate with targeted human annotation. In: Proceedings of Association for Machine Translation in the Americas, vol. 200(6) (2006)
19. Strassel, M., Mark A., Kay P., Zhiyi, S., Kazuaki, M.: Linguistic resources and evaluation techniques for evaluation of cross-document automatic content extraction. In: LREC (2008)
20. Wołk, K., Danijel, K.: Comparison and adaptation of automatic evaluation metrics for quality assessment of re-speaking. arXiv preprint arXiv:1601.02789 (2016)
21. Turian, J. P., Shea, L., & Melamed, I. D.: Evaluation of Machine Translation and its Evaluation (2003).
22. Papineni, K., Roukos, S., Ward, T., & Zhu, W. J.: BLEU: a Method for Automatic Evaluation of Machine Translation (2002). https://doi.org/10.3115/1073083.1073135.
23. Doddington, G. Automatic evaluation of machine translation quality using n-gram co-occurrence statistics pp. 138–145 (2002). https://doi.org/10.3115/1289189.1289273.
24. Snover, M., Dorr, B., Schwartz, R., Micciulla, L., & Makhoul, J.: A study of translation edit rate with targeted human annotation (2006).

# PREP: Prerequisite Relationship Extraction Using Position-Biased Burst Analysis

**Aditya Limaye, Sujyoth S. Karkera, Hardik Khatri, and Vijay T. Raisinghani**

**Abstract** An intelligent tutoring system (ITS) provides personalized instructions and feedback to learners without requiring expert intervention. One of the components of an ITS is the domain modeling module where the system stores the dependencies between the concepts in the domain using a concept map. These concept dependencies are used by an ITS to determine how to sequence the teaching of concepts to a student and as a representation of student knowledge. This requires an expert to manually find concepts in the domain and enter the dependencies between concepts. We aim to automate the process of domain modeling by finding relationships among concepts in a domain. One such type of relationship is a prerequisite relationship which represents the learning order of a pair of concepts. Determining prerequisite relationships helps in developing an effective concept map for an ITS. These relationships can be extracted by finding patterns between occurrences of the concepts in the text. We propose a method "Prerequisite Relationship Extraction using Position-biased burst analysis" (PREP) based on burst analysis to determine prerequisite relationships between the concepts present in unstructured text using the order of occurrence of concepts. We have evaluated the proposed method using textbooks spanning multiple domains like data mining, geometry, precalculus, and physics to test its robustness. The proposed method results in as much as a 17% improvement in precision over existing methods.

**Keywords** Prerequisite relationships · Concept maps · Learning order

---

A. Limaye (✉) · S. S. Karkera · H. Khatri · V. T. Raisinghani
Department of Information Technology, Mukesh Patel School of Technology Management and Engineering, NMIMS, Mumbai, Maharashtra 400056, India
e-mail: aditya.limaye50@nmims.edu.in

S. S. Karkera
e-mail: sujyoth.karkera39@nmims.edu.in

H. Khatri
e-mail: hardik.khatri43@nmims.edu.in

V. T. Raisinghani
e-mail: vijay.raisinghani@nmims.edu



D. Gupta et al. (eds.), *Pattern Recognition and Data Analysis with Applications*, Lecture Notes in Electrical Engineering 888,
https://doi.org/10.1007/978-981-19-1520-8_17

# 1 Introduction

A concept map summarizes a domain graphically in terms of concepts as nodes and relations among concepts as edges connecting these nodes. Concept maps are used for evaluating student knowledge, capturing expert knowledge, curriculum planning, and mapping concept dependencies. Intelligent Tutoring Systems (ITSs) use prerequisite relationships to determine the learning order of concepts. A prerequisite relationship is a dependency relation between two concepts. It represents the concepts which must be known to or studied by a learner before approaching another concept. Since a large number of concepts are present in a domain, it becomes increasingly tedious for domain experts to manually enter the dependencies between concepts. We propose a method for automatically determining prerequisite relationships from unstructured text documents.

The existing methods that attempt to automate the extraction of relationships between concepts are co-occurrence-based methods [1, 2] and burst-based methods [3, 4]. According to the findings in [3], co-occurrence-based methods perform poorly compared to burst-based methods as high co-occurrence of concepts is not a necessary and sufficient condition to determine the strength of prerequisite relationships. Burst-based methods [3, 4] are based on Kleinberg's algorithm [5]. A burst of a concept is a set of consecutive sentences where a concept has an increased frequency of occurrence. The patterns between the bursts of a pair of concepts are used to determine the prerequisite relationship between the concepts. The drawback of this approach is that every burst is given equal importance. However, some bursts of a concept could be more important in finding prerequisite relationships between concepts than other bursts. The proposed method addresses this problem by giving precedence to the initial bursts of a concept over the other bursts.

We have evaluated and validated the results of our method and compared it with the Burst Analysis method [3]. The precision of the extracted prerequisite relationships is used to compare both the methods. We have validated our results using statistical hypothesis testing. We have tested the proposed method on English language textbooks across domains like precalculus, geometry, physics, data mining, and computer science which are used in the AL-CPL [6, 7] and PRET [8] datasets.

This paper is organized as follows: in Sect. 2, the related work is discussed; in Sect. 3, the proposed methodology is discussed; in Sect. 4, we describe the evaluation parameters; in Sect. 5, the results of the evaluation and the basis for validation of the proposed method are discussed; and in Sect. 6, we summarize our work and conclude this study.

# 2 Related Work

Wang et al. [9] uses Wikipedia as an external knowledge source along with the table-of-contents of a textbook to determine relationships between concepts. The method assumes that a Wikipedia article will exist for every concept mentioned in the book. However, this may not hold true for domains with less coverage in Wikipedia. The method given in [10] does not depend on any external knowledge sources. It extracts rule-based semantic relations between concepts from unstructured texts using domain-specific patterns. Relations determined by this method are very domain-specific and are only extracted if explicitly mentioned in the document. Such relations are also not suitable for finding concept dependencies. Concept dependencies are better represented by prerequisite relationships [3].

A prerequisite relationship is a dependency relation between two concepts. It represents which concept a learner must know or study before approaching another concept. ITSs could use prerequisite relationships to determine the learning order of concepts. The method given by Adorni et al. [3] is predicated on the idea that concepts have intervals with high frequency known as bursts. After detecting bursts for each concept, the method identifies patterns formed by the bursts of each pair of concepts to determine and quantify the prerequisite relationship between them.

Determining prerequisite relationships from unstructured text is a nascent field of research. Adorni et al. [3] provides state-of-the-art performance for prerequisite relationship extraction from unstructured text. However, the method has the following problems:

1. Every burst is given equal importance. However, certain bursts of a concept may be more important than other bursts. For example, the burst where a concept is defined is more important than a burst where a concept is mentioned in passing.
2. The direction of prerequisite relationships is determined by using only the order of first occurrence of the concepts.
3. The method does not consider the instances where a concept is being referred to by a pronoun. For example, a concept may be referred to by the pronoun "it" after it has been introduced.

Our work focuses on addressing problem "1" to improve prerequisite relationship extraction from unstructured text.

## 3 Proposed Methodology

We propose a new approach for generating concept maps from unstructured text. As discussed in the previous section, Burst Analysis [3] assumes that every burst of a concept is equally important for determining prerequisite relationships. However, this assumption fails to consider that the initial bursts of a particular concept could be more important as these bursts hold the formative stages of the concept; and the other concepts that are mentioned in these stages have a higher chance of being prerequisites. Therefore, better extraction of prerequisite relationships can be achieved by considering the initial bursts of a concept.

The proposed method is divided into the following phases: first, the bursts of all the concepts in the document are extracted. After extraction, patterns are identified

**Fig. 1** Bursts extracted for concept "Network Layer"

between the bursts. Using these patterns, the prerequisite relationship for a given pair of concepts is determined.

## 3.1 *Burst Extraction*

A text document $D$ and a list of concepts $C = \{c_1, c_2, \ldots, c_k\}$ appearing in $D$ are given as input to the method. $D$ is split into a list of sentences $S = \{s_1, s_2, \ldots, s_n\}$ where $s_n$ is the $n$th sentence of $D$.

The bursts for each concept $c_k$ are extracted using Kleinberg's algorithm [5]. A set of consecutive sentences where a concept has an increased frequency of occurrence is identified as a burst of that concept. Such sets of consecutive sentences are detected throughout the text. Finally, the start and end of each burst of $c_k$ in terms of sentence indices is stored in $B_{c_k} = \{[b_{1,\text{start}}, b_{1,\text{end}}], [b_{2,\text{start}}, b_{2,\text{end}}], \ldots, [b_{n,\text{start}}, b_{n,\text{end}}]\}$ where $b_{n,\text{start}}$ denotes the index of the sentence from which the $n$th burst starts and $b_{n,\text{end}}$ denotes the index of the sentence at which the $n$th burst ends.

For example, in Fig. 1, the bursts extracted for the concept "Network Layer" can be observed. The 1st burst of "Network Layer" starts at the 5th sentence and ends at the 11th sentence. Similarly, the further bursts are extracted from the text and are finally stored in $B_{\text{Network Layer}} = \{[5, 11], [30, 35], [45, 60], [65, 80]\}$.

## 3.2 *Pattern Identification*

Once the bursts for each concept in $C$ have been detected, patterns are identified among these bursts. For a given pair of bursts $B_{c_u}[i]$ (referring to the $i$th burst of concept $c_u$) and $B_{c_v}[j]$ (referring to the $j$th burst of concept $c_v$), a weight $W$ is given to the pattern. This weight is determined according to the degree to which the pattern implies a prerequisite relationship. Similar to Lee et al. [4] and Adorni et al. [3], burst patterns are identified according to the set of patterns given in Allen's interval algebra [11] along with a tolerance gap proportional to the burst length.

For example, in Fig. 2, the identified patterns between the bursts of the concepts "Network Layer" and "Routing" can be observed. The 2nd bursts of both the concepts *start* together. The 3rd burst of "Network Layer" *includes* the 3rd burst of "Routing". The 4th bursts of both the concepts *overlap* each other.

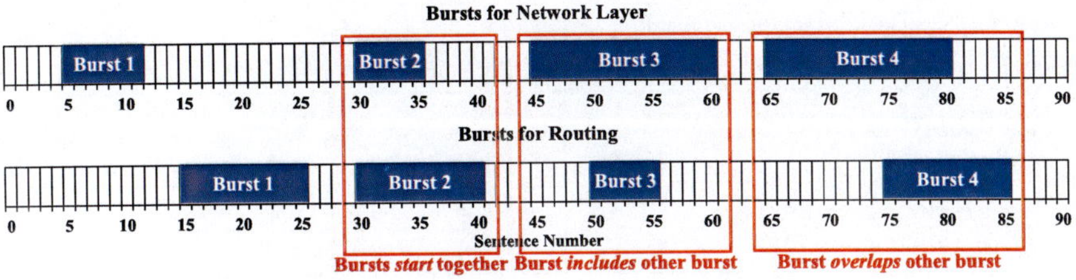


**Fig. 2** Identified patterns between bursts of "Network Layer" and "Routing"

### *3.3 Determining Prerequisite Relationship*

To determine the strength of prerequisite relationships, we use a formula based on the prerequisite relationship (PR) formula given by Adorni et al. [3] as shown in Eq. (1). The strength of the prerequisite relationship between every pair of concepts in $C$ is determined. These relationships are then used to generate a concept map.

$$PR_{X,Y} = \sum_i \left( W \frac{f(X, B_X[i]) \times |B_X|}{\sum_m |B_X[m]|} \frac{\sum_j \frac{f(Y, B_Y[j]) \times |B_Y|}{\min(i,j)}}{\sum_n |B_Y[n]|} \right) \tag{1}$$

The strength of the prerequisite relationship between a pair of concepts $(X, Y) \in C$ is determined by adding the pattern weights of the concepts and normalizing it by using the following factors:

- $f(X, B_X[i])$: the frequency of concept $X$ in the $i$th burst of concept $X$
- $|B_X|$: the number of bursts of concept $X$
- $\sum_m |B_X[m]|$: the total sentences in all the bursts of concept $X$
- $\min(i, j)$: the minimum of the burst indices $i$ (the index of the $i$th burst of concept $X$) and $j$ (the index of the $j$th of concept $Y$).

Concepts that are used to explain or define a concept have a higher chance of being prerequisites as they are utilized to explain that concept. Since it is not possible to determine which burst includes the definition or explanation of a concept, we assume that a concept is defined and explained in its initial bursts. Hence, we consider the inverse of the burst indices so that the patterns between the earlier bursts of concepts have a higher influence on the strengths of prerequisite relationships.

For example, if the bursts of concept $Y$ form patterns with the initial bursts of concept $X$, these patterns will have a higher influence in determining the strength of the prerequisite relationship between $X$ and $Y$.

**Table 1** Details of the textbooks used for evaluation

| Name | # Chapters | # Concepts | # Pairs | # Prerequisites |
|---|---|---|---|---|
| Geometry [14] | 12 | 89 | 1681 | 524 |
| Precalculus [15] | 13 | 223 | 2060 | 699 |
| Physics [16] | 30 | 152 | 1962 | 487 |
| Data Mining by Yang [17] | 8 | 120 | 826 | 292 |
| Data Mining by Aggarwal [18] | 20 | 120 | 826 | 292 |

# 4 Evaluation

This section describes the evaluations performed on PREP. We have tested the method on textbooks across different domains. We have used precision as a measure to compare the performance of PREP and Burst Analysis [3]. We have also performed hypothesis testing to validate the results.

## 4.1 *Dataset*

We have evaluated our algorithm on the PRET [6] and AL-CPL [7, 8] datasets.

PRET [6] is a dataset having prerequisite relations that have been extracted by four domain experts from a chapter in a computer science textbook [12]. The dataset [6] consists of 185 concepts and 526 prerequisite relationships.

The AL-CPL [7, 8] dataset is based on the Wiki concept map dataset [13] containing prerequisite relationships collected from textbooks on geometry, precalculus, physics, and data mining. The dataset contains prerequisite relationships for each domain. The details of the textbooks used for evaluation are given in Table 1.

## 4.2 *Evaluation Metrics*

We evaluated our algorithm on the datasets [6–8] mentioned in Sect. 4.1. We compared the relationships extracted by the algorithms with the relationships annotated by experts. We have computed the precision for the top 300 relationships extracted by both the methods. The formula for calculating the precision is given in Eq. (2).

$$\text{Precision} = \frac{\text{number of correct relationships extracted}}{\text{total number of relationships extracted}} \tag{2}$$

A relationship extracted by the algorithm is considered as a *correct relationship* if the same relationship has been annotated by experts.

We have performed hypothesis testing to validate our results by using one-tailed Z-test. One-tailed Z-test is used to validate whether the proposed method performs better than the Burst Analysis method [3].

### 4.3 Experimentation

We have tested the proposed method with varying input sizes to evaluate its robustness to different input sizes. The textbooks used for evaluation were divided into parts by considering factors like number of chapters, size of the book, and the memory size of the testing computer. The various inputs given to the method are as follows:

- Chapter-wise division: The contents of the textbook were given as input to the method chapter-wise. The aim of this division was to observe the extracted prerequisite relations pertaining to a single chapter.
- Part-wise division: The contents of the textbook were given as input part-wise with each part consisting of multiple chapters. The aim of this division was to extract prerequisite relations which featured inter-dependencies between different chapters within the divisions of the book.
- Sliding window of chapters: The contents of the textbook were given as input to the method as a sliding window of chapters. For example: the chapters 1, 2, and 3 were given as input in the first iteration; the chapters 3, 4, and 5 were given as input in the second iteration; and so on. The aim of this type of division was to find whether the intersecting chapters between two windows had an impact on the extraction of prerequisite relationships.

After dividing the textbook into parts (based on the type of division), we gave each section as input to the algorithm in an iterative manner. The extracted prerequisite relationships were scored as discussed in Sect. 3.3. We have considered the top 300 relationships extracted by both the methods for evaluation.

## 5 Results and Discussion

In this section, we discuss in detail the results of our evaluation of PREP. As mentioned in Sect. 4.2, we have evaluated our algorithm on textbooks belonging to multiple domains. However, due to shortage of space, we have only presented the results obtained on the precalculus textbook [15]. This section is divided as follows: in Sect. 5.1, we provide chapter-wise input of the textbook [15]; in Sect. 5.2, we provide the textbook [15] in one-third divisions as the input; in Sect. 5.3, we provide a sliding window of chapters as the input; in Sect. 5.4, we validate our findings by applying statistical hypothesis testing on the results; and in Sect. 5.5, we discuss the observations made on the evaluation results.

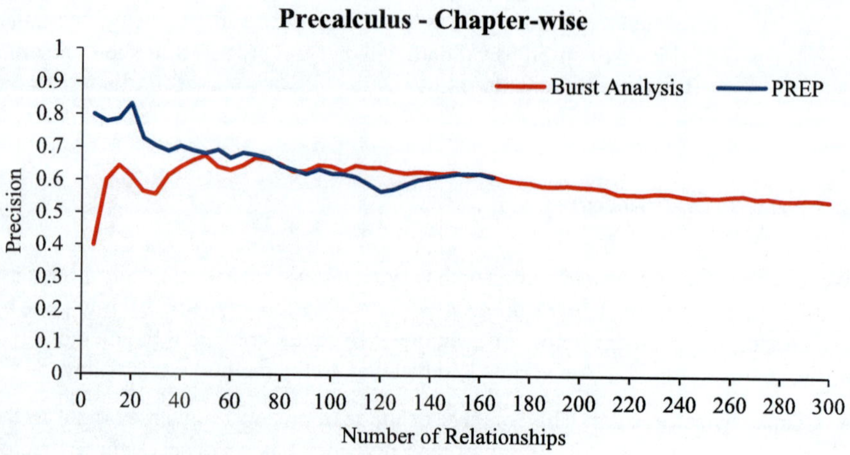


**Fig. 3** Comparison of precision; input given chapter-wise

## 5.1 *Chapter-Wise Division*

We tested PREP and the Burst Analysis method [3] by feeding them the textbook [15] chapter-wise. Figure 3 shows the precision of both the methods.

The peak precision of PREP is greater than that of the Burst Analysis method [3]. The average precision of PREP is 0.621 with an S.D. of 0.066, while that of Burst Analysis method [3] is 0.596 with an S.D. of 0.046. The PREP method yields an improvement of 4.19% over the precision of the Burst Analysis method [3].

## 5.2 *Part-Wise Division*

We tested PREP and the Burst Analysis method [3] on the textbook [15] by taking the input in three parts. Figure 4 shows the precision of both the methods.

The peak precision of PREP is greater than that of the Burst Analysis method [3]. The average precision of PREP is 0.724 with an S.D. of 0.083, while that of Burst Analysis is 0.617 with an S.D. of 0.047. PREP yields an improvement of 17.34% over the precision of the Burst Analysis method [3].

## 5.3 *Sliding Window*

We tested PREP and the Burst Analysis method [3] on the textbook [15] by using a sliding window of 3 chapters at a time. Figure 5 shows the precision of the methods.

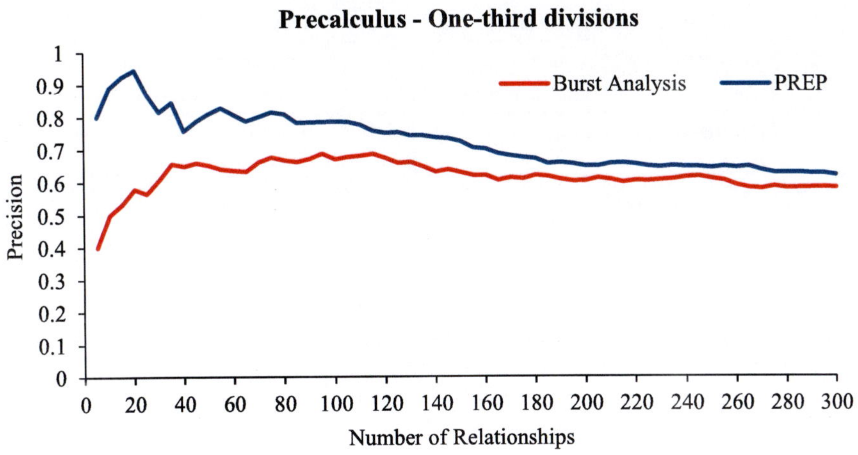


**Fig. 4** Comparison of precision; input divided into three parts

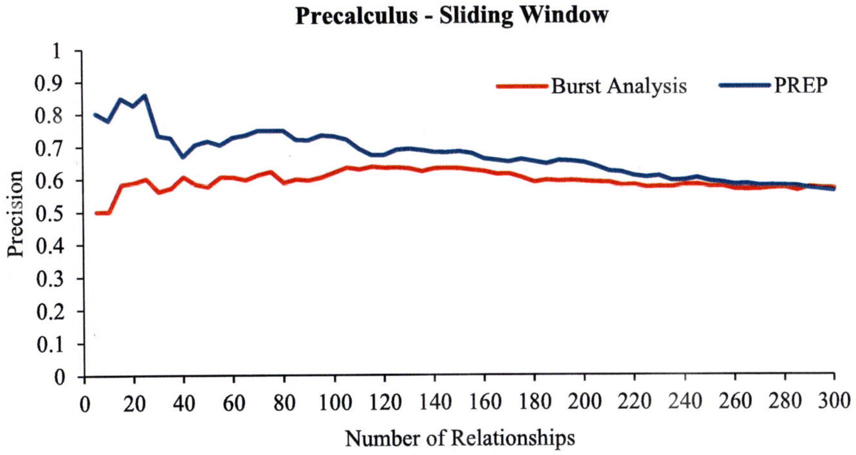


**Fig. 5** Comparison of precision; input given as a sliding window

The aim of using a sliding window of chapters as input was to observe the effect on precision of the transitive relationships present in the common chapters of consecutive windows.

The peak precision of PREP is greater than that of Burst Analysis method [3]. The average precision of PREP is 0.669 with an S.D. of 0.07 and that of Burst Analysis [3] is 0.592 with an S.D. of 0.02. The PREP method yields an improvement of 13.01% over the precision of the Burst Analysis method [3].

**Table 2** Results of the hypothesis testing

| S. No. | Type of input | Mean | | Z-stat | Z-critical | Null hypothesis |
|---|---|---|---|---|---|---|
| | | Burst analysis [3] | PREP | | | |
| 1 | Chapter-wise | 0.596 | 0.621 | 2.35 | 1.64 | Rejected |
| 2 | One-third | 0.617 | 0.724 | 8.71 | 1.64 | Rejected |
| 3 | Sliding window | 0.592 | 0.669 | 7.78 | 1.64 | Rejected |

## 5.4 *Hypothesis Testing*

We can observe that PREP yields better precision compared to the Burst Analysis method [3]. To further validate this claim, we have applied one-tailed Z-test to investigate the following hypotheses:

- $H_0$: PREP does not perform better than the Burst Analysis method [3]
- $H_1$: PREP performs better than the Burst Analysis method [3].

For one-tailed Z-test: If Z-stat is greater than Z-critical, then we reject the null hypothesis ($H_0$). Table 2 shows the results of the hypothesis testing.

For each input type (chapter-wise, one-third and sliding window), Z-stat is greater than Z-critical which leads to the conclusion that the null hypothesis ($H_0$) is rejected. Thus, we can claim that PREP performs better than the Burst Analysis method [3] in terms of precision for prerequisite relationship extraction.

## 5.5 *Observations*

In this section, we discuss the observations made during the evaluation of the proposed method.

The proposed method yields a 12.02%, 11.58%, 11.45%, and 8.7% improvement in precision for data mining [17], geometry [14], precalculus [15], and physics [16] textbooks, respectively, over the Burst Analysis method [3]. The increase in precision was observed across different input sizes (chapter-wise, part-wise, and sliding window) showcasing the robustness of the proposed method across multiple domains as well as different input sizes.

We observed that the size of the input in terms of the number of chapters is directly proportional to the precision of the extracted relationships. For instance, the average precision calculated for chapter-wise input of the precalculus textbook [15] is 0.621, whereas the average precision for one-third division is 0.724. This increase in precision could be caused due to an increased sample size for finding patterns between concept bursts. The increased sample size enables the algorithm to find relationships between concepts across multiple chapters.

## 6 Conclusion and Future Work

Building concept maps from unstructured text is a non-trivial task as there is a dearth of features through which prerequisite relationships can be determined. We have proposed a method to extract prerequisite relationships from unstructured text. The proposed method improves over the Burst Analysis method [3] by incorporating the position of bursts to calculate the strength of prerequisite relationships between concepts. Our method yields a 12.02%, 11.58%, 11.45%, and 8.7% improvement in precision for data mining [17], geometry [14], precalculus [15], and physics [16] textbooks, respectively, over the Burst Analysis method [3].

Since prerequisite relationship extraction is a nascent field of research, an adequate amount of research has not been conducted on the factors which may affect the accuracy of extracting prerequisite relationships from text.

One such factor that our proposed method does not consider is the writing style adopted by the author of the textbook. The approach of extracting prerequisite relationships could be varied depending on whether the author has used a bottom-up approach or a top-down approach for explaining the concepts while writing the textbook. Different writing styles could lead to different burst patterns between prerequisite concepts which would require different weights being assigned to patterns for proper identification of prerequisite relationships.

Automatic extraction of prerequisite relationships would help an ITS in generating learning orders for students which would be helpful in breaking down the learning of complex subjects. Moreover, it could also help students in making mind maps of textbooks while studying a subject. Our proposed method would serve as a foundation for further improvements in prerequisite relationship extraction from unstructured text.

## References

1. Gordon, J., Zhu, L., Galstyan, A., Natarajan, P., Burns, G.: Modeling concept dependencies in a scientific corpus. In: Proceedings of the 54th Annual Meeting of the Association for Computational Linguistics (Volume 1: Long Papers), vol. 1, pp. 866–875 (2016)
2. Liang, C., Wu, Z., Huang, W., Giles, C.L.: Measuring prerequisite relations among concepts. In: EMNLP, pp. 1668–1674 (2015)
3. Adorni, G., Alzetta, C., Koceva, F., Passalacqua, S., Torre, I.: Towards the identification of propaedeutic relations in textbooks. In: Isotani, S., Millán, E., Ogan, A., Hastings, P., McLaren, B., Luckin, R. (eds.) Artificial Intelligence in Education, pp. 1–13. Springer International Publishing, Cham (2019)
4. Lee, S., Park, Y., Yoon, W.C.: Burst analysis for automatic concept map creation with a single document. Expert Syst. Appl. **42** (2015)
5. Kleinberg, J.: Bursty and hierarchical structure in streams. Data Min. Knowl. Disc. **7**(4), 373–397 (2003)
6. Liang, C., Ye, J., Wang, S., Pursel, B., Giles, C.L.: Investigating active learning for concept prerequisite learning. In: 32nd AAAI Conference on Artificial Intelligence, AAAI 2018. pp. 7913–7919. AAAI Press (2018)

7. Liang, C., Ye, J., Zhao, H., Pursel, B., Giles, C.L.: Active learning of strict partial orders: a case study on concept prerequisite relations. In: Lynch, C.F., Merceron, A., Desmarais, M., Nkambou, R. (eds.) EDM 2019—Proceedings of the 12th International Conference on Educational Data Mining, pp. 348–353. International Educational Data Mining Society (2019)
8. Alzetta, C., Koceva, F., Passalacqua, S., Torre, I., Adorni, G.: PRET: prerequisite-enriched terminology. A case study on educational texts. In: Proceedings of the Fifth Italian Conference on Computational Linguistics (CLiC-it 2018), Torino, Italy, 10–12 Dec 2018. CEUR-WS.org (2018)
9. Wang, S., Liang, C., Wu, Z., Williams, K., Pursel, B., Brautigam, B., Saul, S., Williams, H., Bowen, K., Giles, C.L.: Concept hierarchy extraction from textbooks. In: Proceedings of the 2015 ACM Symposium on Document Engineering, pp. 147–156. ACM, Lausanne Switzerland (2015)
10. Kaushik, N., Chatterjee, N.: A practical approach for term and relationship extraction for automatic ontology creation from agricultural text. In: 2016 International Conference on Information Technology (ICIT), pp. 241–247 (2016)
11. Allen, J.F.: Maintaining knowledge about temporal intervals. Commun. ACM **26**, 832–843 (1983)
12. Brookshear, J.G.: Computer Science: An Overview. Addison-Wesley Publishing, Reading, MA (1997)
13. Wang, S., Ororbia, A., Wu, Z., Williams, K., Liang, C., Pursel, B., Giles, C.L.: Using prerequisites to extract concept maps from textbooks. In: Proceedings of the 25th ACM International on Conference on Information and Knowledge Management, pp. 317–326. ACM, Indianapolis Indiana USA (2016)
14. Greenberg, D., Jordan, L., Gloag, A., Cifarelli, V., Sconyers, J., Zahnerm, B.: CK-12 Basic Geometry. CK-12 Foundation (2012)
15. Stewart, J., Redlin, L., Watson, S., Panman, P.: Precalculus: Mathematics for Calculus. Cengage Learning, Boston, MA (2016)
16. Horner, M., Halliday, S., Blyth, S., Adams, R., Wheaton, S.: Textbooks for High School Students Studying the Sciences (2008)
17. Aggarwal, C.C.: Data Mining. Springer International Publishing, Cham (2015)
18. Yang, X.-S.: Introduction to Algorithms for Data Mining and Machine Learning. Academic Press, an imprint of Elsevier, London, United Kingdom; San Diego, CA (2019)

# Enhancing the Security and Performance of Cloud-Based Distance Education System

Anirudh Bishnoi, Anupma Sangwan, Anju, and Rishi Pal Singh

**Abstract** There are many methods that enable a learner to receive education online from distant locations, but online education is growing fast. There remains the issue of the security of digital content during transmission over the cloud. However, it was noted that the security of digital information has to be increased. The emphasis of the current study is on the safety and efficiency improvement of distant learning systems in the cloud. Several kinds of cloud computing research have been conducted to offer remote learning. Previous research found issues with data performance and security. There is a need for a mechanism that could transfer educational content from one place to another in less time securely. During data transmission, the education materials have to be protected and compressed. In order to decrease the size of the packet, research utilized a content replacement method. The proposed work has improved security with the use of the substitute mechanism Exclusive OR (XOR), since other types of encryption are time-consuming, such as the Advance encryption standard (AES) and the Data encryption standard (DES). The technology is safe and quick, since data is first compressed and encrypted on the sender's side before sending it. During packet transmission, proposed work has provided a solution to manage packet loss, errors, packet hijacking and many different attacks. Finally, data is decrypted and decompressed at the receiver end after receiving. The probability of cracking encrypted files also gets reduced as the data is encrypted after compression. The comparison of the proposed work is made to the previous Rivest–Shamir–Adleman (RSA), Deoxyribo Nucleic Acid (DNA) cryptography mechanism. The impact of attacks on packets in the case of the proposed work, RSA and DNA cryptography has been simulated. The simulation results confirm that the proposed work is more secure than previous RSA and DNA cryptography.

**Keywords** Cloud computing · Distance education system · Compression · Security · RSA · DNA

---

A. Bishnoi · A. Sangwan (✉) · Anju · R. Pal Singh
Guru Jambheshwar University of Science &Technology, Hisar (Haryana), India
e-mail: anulathwal@gmail.com



D. Gupta et al. (eds.), *Pattern Recognition and Data Analysis with Applications*, Lecture Notes in Electrical Engineering 888,
https://doi.org/10.1007/978-981-19-1520-8_18

# 1 Introduction

## *1.1 Cloud Computing*

Cloud computing provides services over a network that could be public or private. These clouds can be found in remote locations. It could be utilized in a wide area network (WAN) as well as in a local area network (LAN). The virtual private network could also make use of cloud computing. Many applications, such as email and web-based conferences, are hosted in the cloud. Platform independence is provided by cloud computing. The dominance of cloud computing in the education sector [1, 2] is increasing rapidly. Corona virus disease (COVID) has further accelerated the demand for cloud-based education systems. Students self-enroll for classes; staffs are available to provide learning; and administration personnel provide support through the use of a cloud-based education system. Students obtain their homework in the presence of an Internet connection [3]. Teachers can send study materials directly to the students. Additionally, teachers could hold live sessions on the cloud system to educate their students. It has been observed that cloud-based distance learning mechanisms are also used in industries for professional training [4]. The online clouds provide technical and competent information to the professionals. These clouds [5] host interactive and multimedia websites with high-quality content. This content takes time during transmission. Moreover, these contents need to be protected. However, several mechanisms are built to increase their security, but these security mechanisms reduce data communication performance. Thus, there remains a need for a system that could provide security for digital content hosted on the educational cloud [6, 7] without affecting the performance of data communication. Moreover, it has been observed that there remains the issue of packet dropping and hijacking.

## *1.2 Challenges and Issues in Distance Education Cloud Computing*

Existing studies have looked at the challenges of implementing cloud infrastructure in the education field [7] for faculty, employees, and students. Researchers have also looked at protection risks and danger classifications. The areas where cloud computing can affect education, are examined [8, 9]. A big concern that the education sector faces in developed countries is security monitoring. Moreover, students in underdeveloped countries [10] are deprived of appropriate teaching management [11]. Security issue involves the hacking and cracking activities by intruders.

The main issue with not implementing an online education cloud system is a lack of security. Clouds are currently used by educational institutions, but they are always concerned about safety issues. Previously, DNA [12] and RSA [13, 14] cryptography mechanisms were used in previous research. Issues related to safety are also faced by independent clouds which are different from above. Delivery of significant data by service providers in a secure manner is a difficult task. This data is transmitted by means of the Internet. Due to this, it has become necessary to consider data security in a cloud environment. The following are some of the security concerns raised by cloud computing:

- *Data integrity* refers to instances in which human mistakes are made when entering data into a system. Errors may occur during the transmission of data from one system to another. Hardware failures, such as crashing hard drives, may cause errors.
- *Data theft:* Cloud computing employs an external data server to perform flexible and cost-effective activities. As a result, there is a chance that information may be stolen from an external server.
- *Data loss* is considered as a significant cloud computing concern. If banking, business transactions, and Research and Development (R and D) ideas are all carried out online, unauthorized persons can collect data shared in the cloud.
- *Privacy concerns:* With cloud computing, the protection of user data is a top priority. Since many servers are external, the provider must ensure that the data is protected against unauthorized access.
- *Challenges at the user level:* It is essential for the user to ensure that there is no risk of data loss as a result of their own actions or the actions of other users sharing a shared cloud server.
- *Security issues at the supplier level:* Cloud is the ideal choice if high security is provided by the supplier.

This research has made an effort in order to provide better security with high per-romance in favor of a cloud computing environment for distance education. Performance has been increased by reducing the size of data by applying a content replacement mechanism where large words have been replaced by small words. In order to increase security, the cryptography technique has been employed. Cloud-based education systems are frequently used by students, teachers and professionals. The growing need for distance education is the motivation for the proposed work. The rapid development in the field of cloud computing is a source of motivation. There has been research that has provided solutions for remote learning and secure data transmission via the cloud. The performance and security issues found in existing research have motivated the proposed model. The proposed work addresses the security and performance issues associated with cloud-based distance learning. This study proposed a mechanism that will allow the integration of exclusive or and data replacement in order to enhance data transmission security and decreases the chances of error and packet dropping. In this way, the proposed work has contributed to reliable, secure, flexible and scalable approach to distance learning. The remainder of this paper has been organized in six sections. Section 1 has presented role of cloud in distance learning and motivation of research. Sections 2 and 3 explained existing research field of cloud computing in education system, cryptography security and data compression. Section 4 introduces the research methodology used for the proposed model. Process flow of proposed model has been discussed. Sections 5 presented simulation outputs. Section 6 discusses the research's findings and scope.

## 2 Related Work

Much research has already been carried out in the areas of cloud-based distance learning, encryption and data compression. But still, there is a need to do more work to introduce a mechanism that could provide distance learning features over the cloud in a more secure manner without affecting the performance.

### *2.1 Researches in Area of Cloud Computing in Education System*

Many researchers have considered the features of the present electronic learning (E-Learning) system [1] for distance learning. Some research represents the needs of cloud computing in online education [2, 5], and it has been discovered that cloud computing is a dynamically scalable system. It has the ability to provide Internet-based services. Due to technical enhancements, virtual technologies are playing an important role in online education. The research concentrated on the utilization of an online education system that is dependent on cloud computing environments. During this research, it has been observed that the diversity and importance of the data that has been used in education is increasing because of technology enhancement [3]. Research has presented the role of web technologies and their contributions to the distance learning system. Cloud computing ramifications and problems in the area of academia [4] have been addressed. Researchers have explained the security loopholes in cloud computing and the prevention mechanisms to restrict the attack on the cloud environment. The research presents e-learning approaches by utilizing cloud computing [5]. Distance Education Technologies has been presented as e-learning system along with cloud computing [6]. The focus of the study is on how cloud infrastructure services are used in e-learning to support e-learners. Some studies [7] focused on five areas: conceptual and pedagogical dimensions, applications available in the field of education, data and resource management, advantages and limitations of cloud computing in education, database management system (DBMS) incorporation into cloud-based services, and DBMS integration into cloud-based services. Some cloud computing models in distance learning [8] opted to provide a cost-effective web-based solution at any time and any place and focused on economical solutions for cloud-based distance education systems. Novel observations [9] on cloud computing in education are proposed to analyze education sectors that are making use of cloud computing as a service. Online cloud-based education for underprivileged students from underdeveloped countries [10] is supposed to be provided. The development of teaching resources to perform teaching management [11] has been introduced for innovative practice teaching.

## 2.2 Researches in Cryptographic Security

Many researchers have implemented DNA Cryptography [12] for cloud computing. The Huffman Algorithm has also been used in such research to perform data compression. These researchers have also used the concept of socket programming to build new mechanisms for securing data over the cloud. The RSA mechanism [13, 15] has been frequently used for cryptography to secure data in a cloud environment. The issue with such a mechanism is its time consumption during encryption operations. Several other researchers have also used different cyber security [14] mechanisms to protect the e-learning environment on the cloud platform. They also analyzed security issues in cloud-based e-learning [16]. Many researchers have discussed the needs, scope, issues, and challenges in cloud environments, considering cost and security [17–22]. Secure access and storage of data in cloud computing [23–25] has been provided in existing research. Some authors used digital signatures with Diffie Hellman key exchange [26], while others used the AES encryption algorithm to boost data protection in cloud storage.

## 2.3 Researches Data Compression

To increase performance, there is a need to introduce the concept of data compression. Because compressed content is quickly transferred over the network, the likelihood of error and packet loss has decreased. However, the Huffman algorithm [12] has been used in many types of research to reduce the size of content because it provides lossless data compression, but the time consumption by the Huffman algorithm is high. The data security model for cloud computing [27] has been introduced considering the requirement of data compression. Also, recent literature has addressed the successful review [17] of Cloud-based web platforms for interactive learning platform Systems Integration.

# 3 Problem Statement

The proposed work is focused on implementing distance education in a cloud environment. The major challenge is the network speed and size of data. The educational content needs to be secured and compressed during transmission. The proposed work would integrate the proposed mechanism into the educational module and test the performance and security of the proposed cloud-based education system and do a comparative study of traditional and the proposed work.

# 4 Methodology Used in Proposed Work

The proposed work has focused on the security of data using encryption mechanisms and compression of data using compression mechanisms. The educational content would be first compressed and then encrypted before transmission from the server to the client. As shown in the process flow, the education content has been considered as D. After compression, the content is converted to CD. Then an encryption mechanism is applied. After applying encryption data, CDE is transferred to the receiving client. Here, the contents are decrypted and decompressed to restore their actual educational content.

## *4.1 Internal Working of Proposed Work*

This section presents how the proposed work compresses the large-sized packet and how the security to the data is provided by applying the XOR-based encryption mechanism (Fig. 1).

## *4.2 Data Compression Using Replacement Mechanism*

The large-size content took a lot of time to travel over the network. Moreover, there are always chances of packet dropping and hijacking. Thus, in the proposed work, the content has been compressed before sending it. There should be less data loss during compression, so the proposed work has used a replacement mechanism to reduce the size of the data. The large-sized strings are replaced with small-sized strings that are mentioned in the replacement table. The replacement table contains the strings with their corresponding small-sized strings. The string in the data packet is replaced with its corresponding string only if that string is available in the replacement table. For example, if the data packet consists of a "Computer" string and the small string in the replacement table is "_c1_" corresponding to "computer", then the "Computer" word in the data packet would be replaced by "_c1_". In this way, all strings in the data packet will be replaced. As a result, the size of the data packet has been reduced. Moreover, this could be termed as first-level encryption where data is not the same as the original data.

## *4.3 XOR-Based Encryption*

It has been observed that traditional techniques such as AES and RSA take more time during data encryption due to the complexity of the algorithm. The proposed

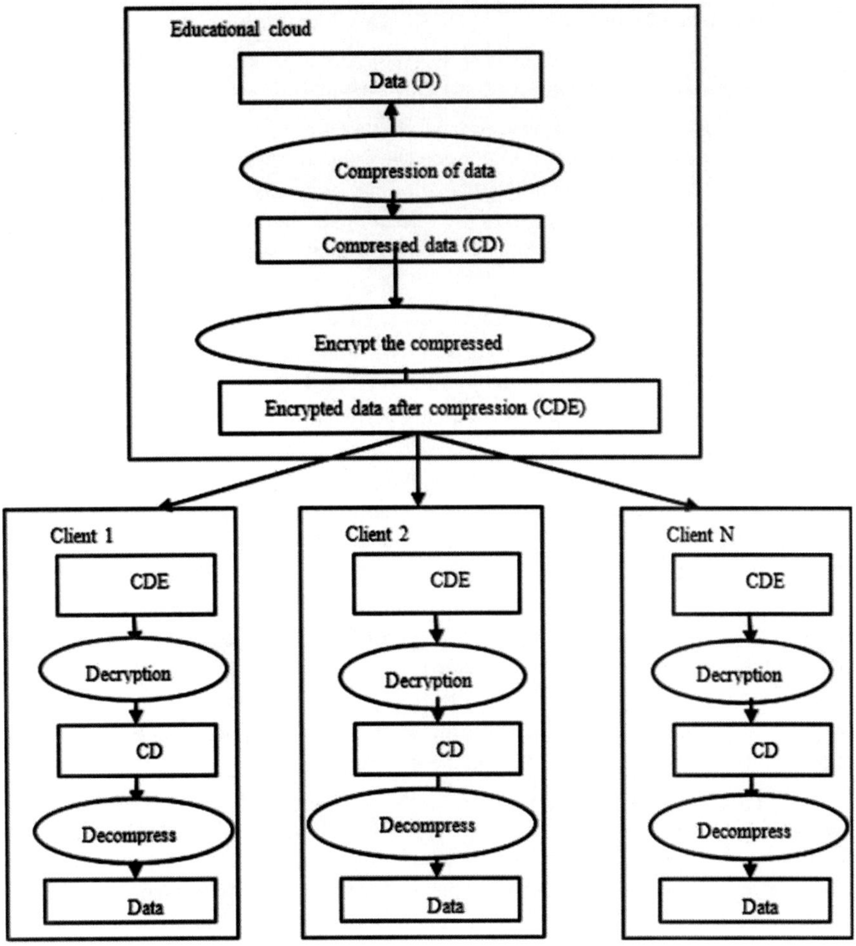


**Fig. 1** Process flow of proposed work

work applies an XOR operation after content replacement to encrypt the data. The content that is to be sent is processed using the XOR mechanism after compression. The XOR operation has been used to encrypt the content in the proposed work. The working of the XOR mechanism is explained in this section. Assume the user needs to send data "8" and the XOR token is "5". After XORing 8 and 5, the result is 13.

On the receiving end, the data is again processed by XOR mechanism to produce the actual information. XOR operations return a value of 1 if the bits are not the same and a value of 0 if the bits are the same. The content that would be processed by the XOR operation is sent to the receiver side. On the receiver end, during the decryption phase, data is again decoded by XOR operation and the real contents are restored by a replacement table.

# 5 Implementation

During the development of the network application, the sender and receiver module was developed in the "NetBeans IDE 6.1" environment. On the receiver end, the port number, file path and the token to decode the data are stated. The XOR mechanism has been used to encrypt data. The user-defined port was used to securely transmit data. Because port numbers 1–1023 are already reserved for existing protocols. The file name is set in the file path name text box to store the received content in a text file. On the receiver end, the user would specify the port number, file name and token to apply XOR operation on the incoming data. Then the user clicks on the enable upload option to enable the receiver. During sender implementation, the module works to transfer data. Here, the port, the path of the file, the Internet protocol address of the server and the token to encode the code are set. Here, the sender would set the port number that should be the same as that of the client-side port. Then the user specifies the file path and the name that is to be sent. The IP address of the receiver is set in the IP address box. In the token box, the sender would send the code to encrypt data using the XOR mechanism. During the implementation of the receiver, the path of the file, port number, and decoding code is set. A text file would be transmitted from the sender to the receiver. The large string content of this file would be replaced by small words and data would be encoded by applying the XOR application. During the implementation of server-side code, it is a must that the port number should be more than 1023. During the execution of the sender module, the port must be user-defined. Moreover, it must be the same as the port number used on the receiver side. To compress the size of packets to be transferred over the network, the data compressing mechanism has been used. The large-sized contents are replaced with small-sized ones in this case. The sender and receiver module has been developed on NetBeans platform using Java as a programming tool. During simulation, the time consumption in the case of previous work and in the case of proposed work is noted according to a different number of packets. Simulation work has been performed in a MATLAB environment.

## 5.1 *Simulation for Time/Error/Packet Size*

**Time Consumption**

Time taken has been simulated in the case of the proposed system in comparison to previous RSA; advanced RSA and DNA cryptography-based research are shown in Fig. 2. The proposed work makes use of an exclusive order during encryption and the compressed data has been encrypted. Previous research, on the other hand, has used RSA, a DNA mechanism that takes longer to encrypt data. Furthermore, previous research did not compress the data prior to transmission. Thus, the time consumption is evidently less as compared to others, due to the smaller size of the data packets.

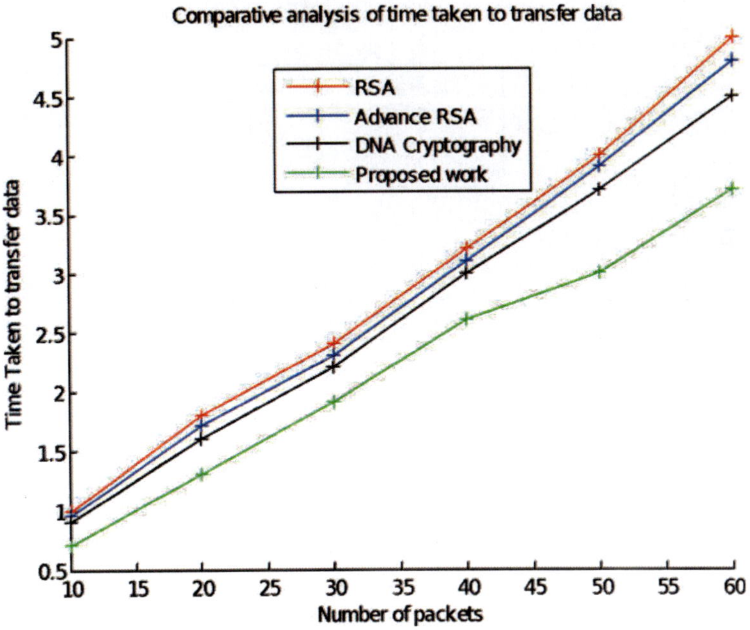


**Fig. 2** Comparison of time taken in RSA, advance RSA and DNA cryptography with proposed work during transmission of data

### Error Rate

There remain chances of errors during data transmission. However, if the packet size is reduced and the packet remains on the network for a shorter period of time, the likelihood of an error is reduced. Also, the length of the string is reduced using a replacement mechanism that decreases the chances of error. However, previous research using RSA and DNA cryptography [12, 13, 15] mechanisms did not reduce packet size. As a result, the current study has the potential to reduce the error rate. Figure 3 shows a comparative analysis of the error rate for previous RSA, Advance RSA, DNA cryptography and proposed work.

### Packet Size

The replacement technique used in the proposed work has reduced the content length, resulting in a smaller packet size. So, present research allows smaller data packets as compared to previous research. Previous studies using RSA and DNA [12, 13, 15] cryptography did not compress the data. Comparative analysis of packet size in the cases of RSA, Advance RSA, and DNA cryptography has been done with the proposed model (Fig. 4).

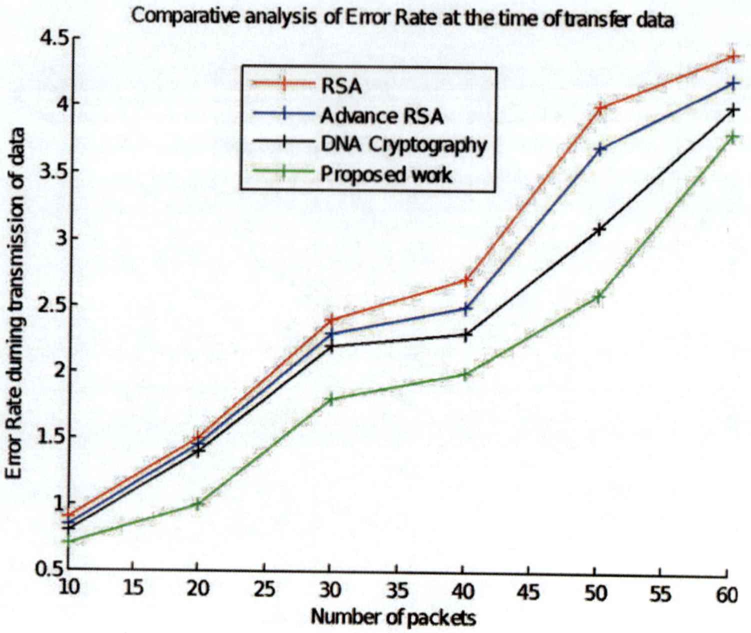


**Fig. 3** Comparison of error rates for RSA, advance RSA and DNA cryptography with the proposed work

## *5.2 Matlab Simulation for Comparative Analysis of Security*

This section presents the impact of the proposed work on security. In the case of the proposed work, the number of packets affected is less as the number of attacks increases. From previous research, it has been found that DNA cryptography [12] is better as compared to RSA [13] and advanced RSA [15]. But the proposed work is better than DNA cryptography. Based on the figures below, it is concluded that the affected packets are fewer in the proposed work than in the RSA and DNA-based cryptography approaches.

**Man-in-the-Middle**

Its impact on the packet in the case of RSA, Advance RSA, and DNA cryptography and proposed work in the case of these attacks are shown below in Fig. 5.

**Brute Force Attack**

A brute force attack involves guessing login information via trial and error. Encryption keys and a hidden web page are also used. A comparative analysis of this attack is shown in Fig. 6.

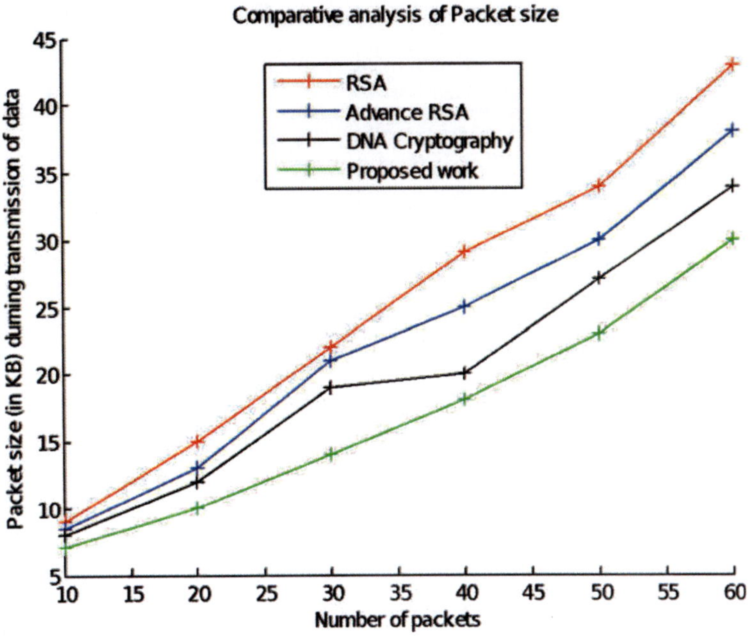


**Fig. 4** Comparison of packet size for RSA, Advance RSA and DNA cryptography with the proposed work

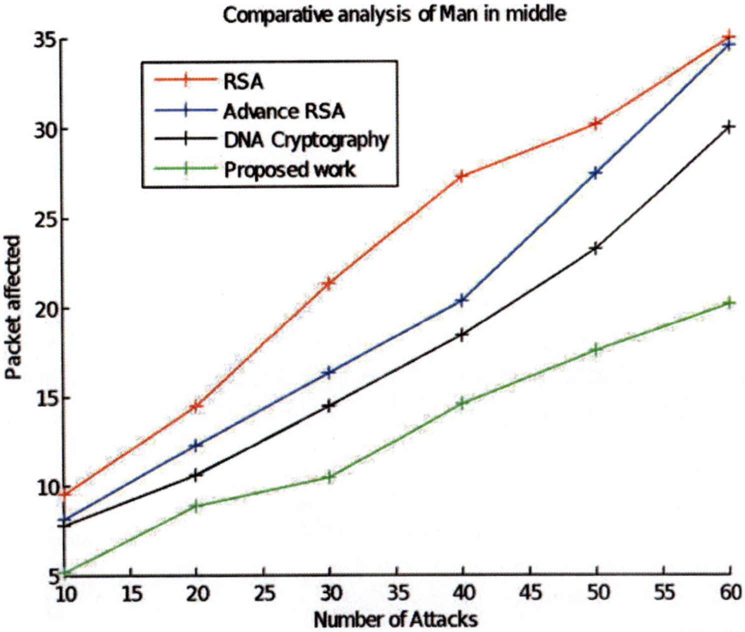


**Fig. 5** Comparative analysis in case of attack Man-in-the-Middle

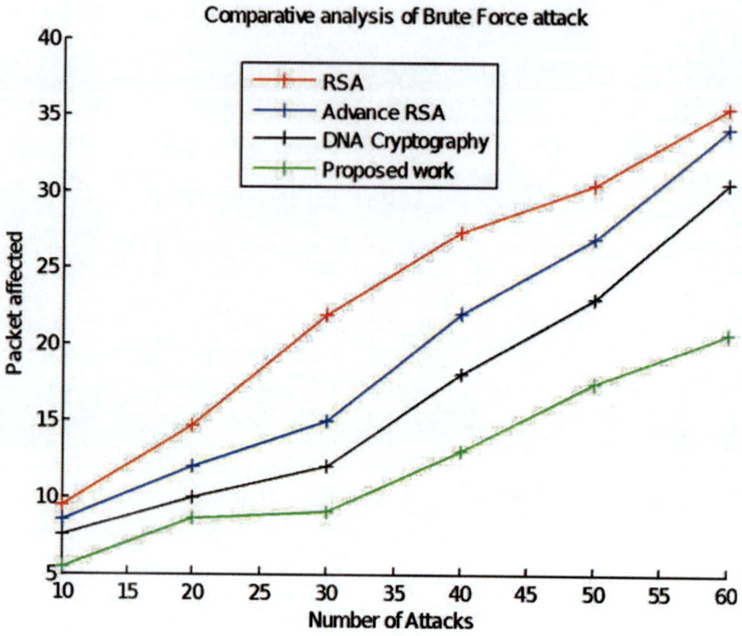


**Fig. 6** Comparative analysis in case of Brute force attack

**Denial-of-Service**

A Denial-of-Service (DoS) attack is a kind of cyber-attack that attempts to prevent people from accessing a computer or network resource. Due to reduced size of packet and less time taken during transmission over the network, the probability of Denial-of-Service has been reduced. Thus, the impact of Denial-of-Service is less in case of the proposed work. Figure 7 depicts a comparison of Denial-of-Service attacks.

**Traffic Hijacking**

If the data is left on the network for an extended period of time, the likelihood of traffic hijacking increases. But the proposed work has reduced this probability. However, previous research ignored the factors that influence traffic hijackings. Figure 8 presents the comparative analysis of traffic hijacking in the case of RSA, Advance RSA, DNA cryptography and proposed work.

**Access Violation**

Proposed work is making use of user-defined port and security keys for exclusive or getting modified each and every time in different sessions. Thus, access violation issues have been resolved in the proposed work. Figure 9 presents the comparative analysis of access violations in the case of RSA, Advance RSA, DNA cryptography and proposed work.

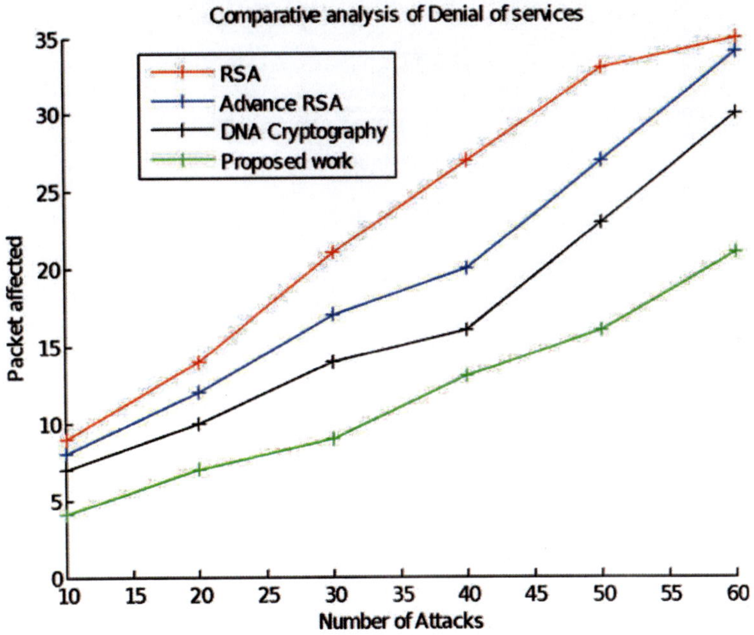


**Fig. 7** Comparative analysis in case of Denial-of-Service

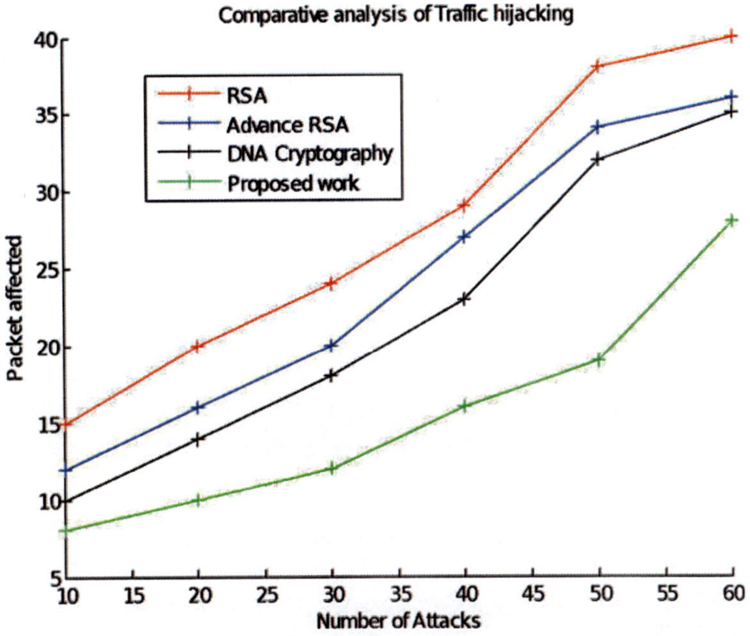


**Fig. 8** Comparative analysis in case of traffic hijacking

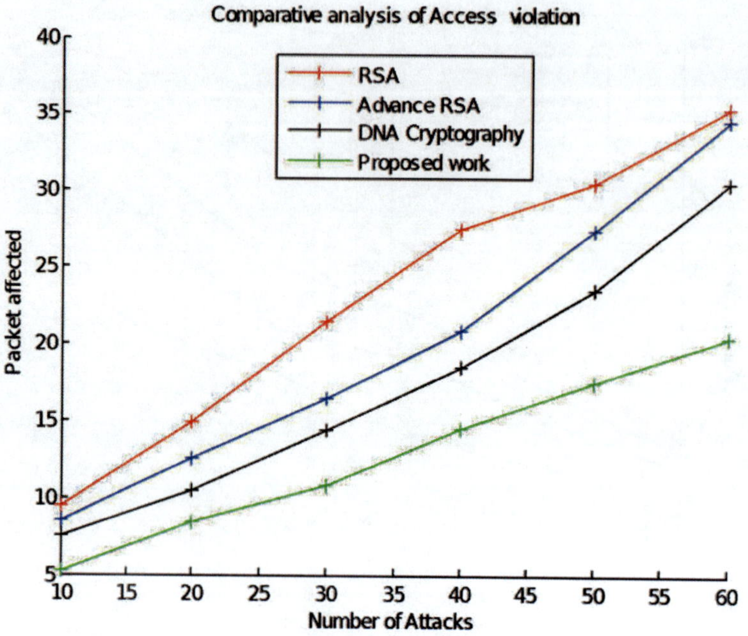


**Fig. 9** Comparative analysis in case of access violation

**Application Level Attack**

Attacks at the application level have been reduced by providing users with a special user interface. The chances of sending and receiving data without using that user interface are negligible. But previous research work has not provided a special user interface. Figure 10 presents the comparative analysis of traffic hijacking in the case of RSA, Advance RSA, DNA cryptography and proposed work.

**Attack by Malicious Insider**

The possibility of a malicious insider attack has been reduced by allowing different keys to encode and decode data. Previous researches have used same key in different sessions. As a result, the possibility of an attack by a malicious insider exists. Figure 11 presents the comparative analysis of attacks by malicious insiders in the case of RSA, Advance RSA, DNA cryptography and proposed work.

**Attack on Cloud Services**

Data compression and user-defined port numbers have reduced the likelihood of various attacks on cloud services. Figure 12 presents the comparative analysis of the attack on cloud services in the case of RSA, Advance RSA, DNA cryptography and proposed work.

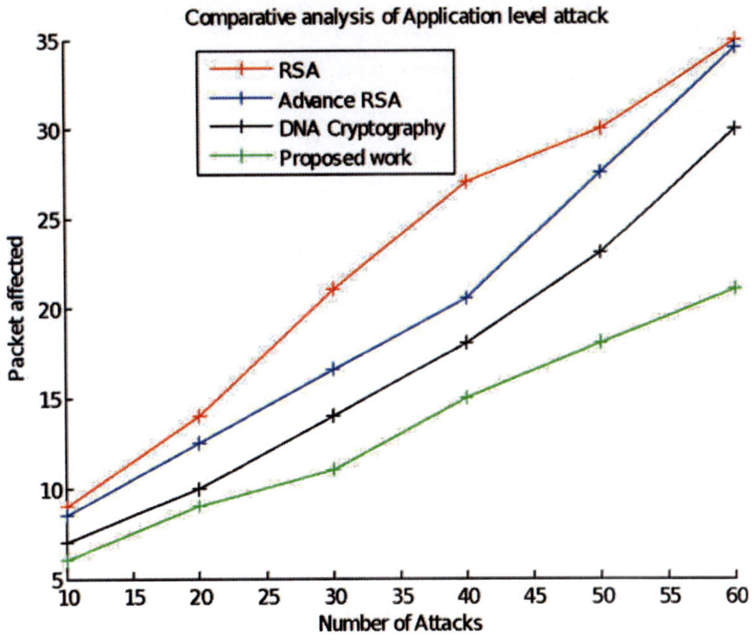


**Fig. 10** Comparison analysis of the application level attack

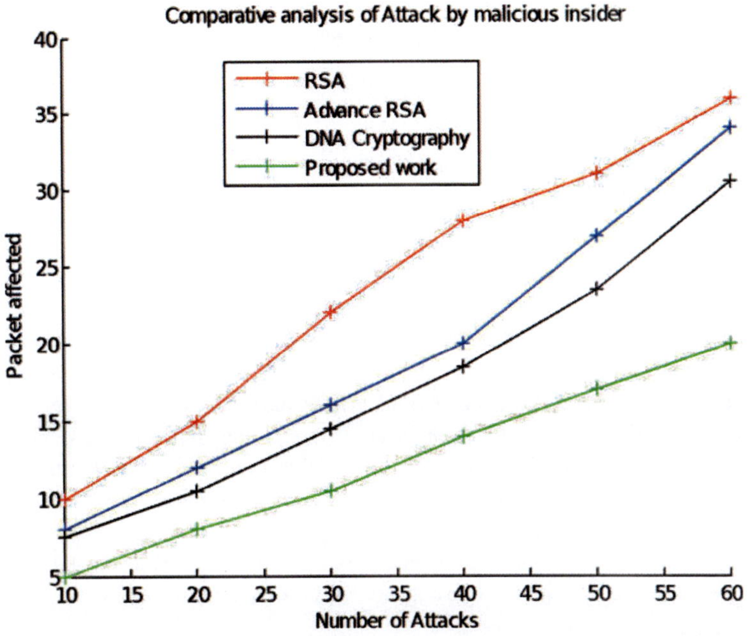


**Fig. 11** Comparative analysis for attack by malicious insider

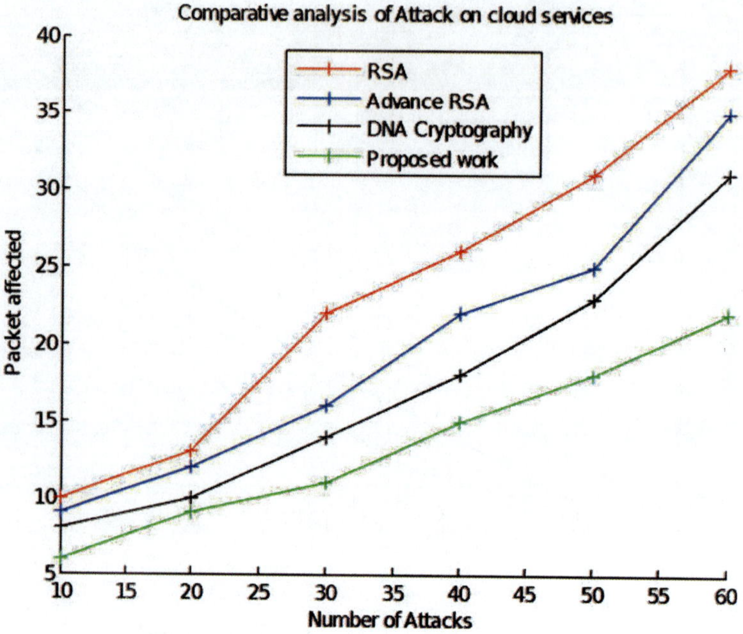


**Fig. 12** Attack on cloud services

# 6 Conclusion and Scope of Research

The major limitations of existing distance learning models are slow performance and a lack of security. It has been observed that if the security parameters are attached to it, then the performance gets degraded. The integrated method has been investigated, in which the content replacement mechanism decreased the size of the data packet while the XOR operation provided security via encryption. The suggested study has ensured the security and performance of educational cloud systems. The results of the simulations indicate that the suggested cloud-based education system outperforms conventional solutions. Because data is compressed first and then encrypted on the sender's side, it is safe and quick. The data is decrypted and decompressed on the receiving end. Because the data amount is less during transmission, the problem of error rate and delay is no longer a concern. In addition, the packet dropping ratio has decreased. When compared to conventional security systems, the proposed mechanism is more resistant to various types of attacks, such as man-in-the middle, Denial-of-Service, brute force attack, and attacks on cloud services, as well as attacks from the malicious and application layers. In comparison to RSA-based and DNA cryptography-based methods, the suggested approach is more secure.

## 7 Future Work

The future work may provide better compression mechanism. In the future, security can be improved. Future research might provide better performance along with a reduced error rate by integrating advanced cloud services and optimization mechanisms. The use of soft computing techniques could improve the reliability and quality of services. To improve the dependability of the cloud in distant learning, researchers may examine its high availability and zero downtime.

## References


1. Patil, P.: A study of E-learning in distance education using cloud computing. Int. J. Comput. Sci. Mob. Comput. **5**(8), 110–113 (2016)
2. Bouyer, A., Arasteh, B.: The necessity of using cloud computing in educational system. CY-ICER Elsevier **143**, 581–585 (2014)
3. Tugrul, A., Atun, H.: The cloud systems used in education: properties and overview. Eng. Technol. Int. J. Educ. Pedag. Sci. **10**(4) (2016)
4. Claral, A.: Implications, risks and challenges of cloud computing in academic field—a state-of-art. Int. J. Sci. Technol. Res. **8**(12) (2019)
5. Ali, A., Bajpeye, A.: E-learning in distance education using cloud computing. Int. J. Comput. Tech. **2**(3) (2015)
6. Kumar, S., Goyal, N., Singh, M.: Distance education technologies: using E-learning system and cloud computing. Int. J. Comput. Sci. Inf. Technol. **5**(2), 1451–1454 (2014)
7. Shi, Y., Hao, H.: Trends of cloud computing in education. In: Cheung, S.K.S., Fong, J., Zhang, J., Kwan, R., Kwok, L.F. (eds.) Hybrid Learning. Theory and Practice. ICHL Lecture Notes in Computer Science, vol. 8595. Springer (2014)
8. Karak, S., Adhikary, B.: Cloud computing as a model for distance learning. Int. J. Inf. Sources Serv. **2**(4), 32–38 (2015)
9. Mishra, J., Panda, S.: A novel observation on cloud computing in education. Int. J. Recent Technol. Eng. **8**(3), 5262–5274 (2019)
10. Balobaid, A., Debnath, D.: A novel proposal for a cloud-based distance education model. Int. J. e-Learn. Secur. **6**(2), 505–513 (2016)
11. Zhihong, X., Jun, Z.: Expand distance education connotation by the construction of a general education cloud. In: International Conference on Advanced Information and Communication Technology for Education (2013)
12. Pandey, G.P.: Implementation of DNA cryptography in cloud computing and using Huffman algorithm, socket programming, and new approach to secure cloud data. Socket Programming and New Approach to Secure Cloud Data (2019)
13. Suresh, P.: Secure cloud environment using RSA algorithm. Int. Res. J. Eng. Technol. **3**(2), 143–148 (2016)
14. Bandara, I., Ioras, F., Maher, K.: Cybersecurity concerns in e-learning education. In: Proceedings of ICERI 2014 Conference, pp. 728–734, Spain (2014)
15. Singh, S.K., Manjhi, P.K., Tiwari, R.K.: Data security using RSA algorithm in cloud computing. Int. J. Adv. Res. Comput. Commun. Eng. **5**(8), 11–16 (2016)
16. Kumar, G., Chelikani, A.: Analysis of security issues in cloud-based e-learning. University of Board/School of Business and IT (2011)
17. Osman, S.: Performance analysis of cloud-based web services for virtual learning environment systems integration. Int. J. Innov. Sci., Eng. Technol. **3** (2016)

18. Garrison, G., Kim, S., Wakefield, R.L.: Success factors for deploying cloud computing. Commun. ACM **55**(9), 62–68 (2012)
19. Herhalt, J., Cochrane, K.: Exploring the cloud: a global study of governments adoption of cloud. Sales Force (2012)
20. Venters, W., Whitley, E.A.: A critical review of cloud computing: researching desires and realities. J. Inf. Technol. **27**(3), 179–197 (2012)
21. Yang, H., Tate, M.: A descriptive literature review and classification of cloud computing research. Commun. Assoc. Info Syst. **31** (2012)
22. Marston, S., Li, Z., Bandyopadhyay, S., Zhang, J.: A cloud computing—the business perspective. Decis. Support Syst. **51**, 176–189 (2011)
23. Nirmala, V., Sivanandhan, R.K, Lalshmi R.S.: Data confidentiality and integrity verification using user authenticator scheme in cloud. In: 2013 International Conference on Green High-Performance Computing. IEEE, India (2013)
24. Kumar, A., Lee B.G., Lee H., Kumari, A.: Secure storage and access of data in cloud computing. In: International Conference on ICT Convergence. IEEE, India (2012)
25. Tribhuwan, M., Bhuyar, V., Prizade, S.: Ensuring data storage security in cloud computing through two-way handshake based on token management. In: International Conference on Advances in Recent Technologies in Communication and Computing, pp. 386–389. IEEE (2010)
26. Rewagad, P., Pawar, Y.: Use of digital signature with Diffie Hellman key exchange and AES encryption algorithm to enhance data security in cloud computing. In: International Conference on Communication Systems and Network Technologies, pp. 437–439. IEEE (2013)
27. Meslhy, E.: Data security model for cloud computing. J. Commun. Comput. **10**, 1047–1062 (2013)

# Hate-Speech Detection in News Articles: In the Context of West Bengal Assembly Election 2021

**Prasanta Mandal, Apurbalal Senapati, and Amitava Nag**

**Abstract** A hate speech represents an expression or phrase of offensive language. The intention behind the usage of hate speech is to abuse, dehumanize, disparage, or harass a person or a group of persons based on their race, color, gender, religion, caste, ethnicity, disability, language, belief, nationality, or other factors. Hate speech is also used to express violence, harm, or hatred against the targeted people. Nowadays, the amount of hate speech is overgrowing in social media, online newspapers, etc. It becomes very difficult to moderate the data containing hate speeches manually, as it is a tedious and time-consuming task. Therefore, an automated hate-speech detection technique is very essential. Numerous works have been done to develop the technique or tool for automatic detection of hate speeches in Twitter, Facebook, and other social media data. This paper studies hate speeches on political news articles in the context of the West Bengal Legislative Assembly Election 2021. In the computational aspect, this task is carried out in three phases. First, a political news corpus has been created, and next from that corpus a key word/phrase-based hate-speech identifier is developed. A semi-automated approach has been used to find out the set of hate-speech-related key words/phrases. Finally, the system's performance is evaluated, and the results are investigated.

**Keywords** Hate speech · West Bengal · Election · News article

P. Mandal (✉)
Department of Computer Science and Engineering, Govt. College of Engineering and Textile Technology, 12, William Carey Road, Serampore, Hooghly, West Bengal 712201, India
e-mail: prasanta.anshin@gmail.com

A. Senapati · A. Nag
Department of Computer Science and Engineering, Central Institute of Technology Kokrajhar, Kokrajhar, Assam 783370, India
e-mail: a.senapati@cit.ac.in

A. Nag
e-mail: amitava.nag@cit.ac.in



D. Gupta et al. (eds.), *Pattern Recognition and Data Analysis with Applications*, Lecture Notes in Electrical Engineering 888,
https://doi.org/10.1007/978-981-19-1520-8_19

## 1 Introduction

Since the last decade, social media has been growing at an exponential rate. It has become a great platform for communication/sharing information irrespective of their social status [1]. The revolution of communication technologies, cheap and user-friendly devices such as smartphones, tablets, and laptops has accelerated the engagement of people in various social media like Facebook, Twitter, etc. As a public communication in an open platform, there are various expressions reflecting in the critical discourse [2], and sometimes the content includes abusive languages in different forms and modes. A common form of offensive and abusive language is hate speech. The text that is intended toward a group of people for harm, violence, or social chaos is considered as hate speech [3]. Hate speech in society or social media can lead to harm, disturbance, disrespect, insult, anger, etc., that affect the harmony of conversations and disturb the social stability. According to the United Nation, hate speech is defined as any expression in spoken, written, or behavioral form which assaults or uses deprecatory or prejudicial language relating to a person or group of persons depending on their identity, origin, cast, creed, sex, religious conviction, ethnic group, national belonging, etc. Often it promotes discrimination, intolerance among people and dislodges social harmony and unity [4]. The hate-speech-related issues have also become a serious problem in India. National Crime Records Bureau (Ministry of Home Affairs) shows with statistics that the cyber-crime is rapidly increasing including hate-speech content [5]. According to the Supreme Court of India, resolving hate-speech-related issues needs deeper consideration and amendment to the Indian Penal Code (IPC), 1860 and Code of Criminal Procedure, 1973 titled with "Hate Speech" [6]. The Gazette notification, dated February 25, 2021, the Government of India, has directed to implement the Intermediary Guidelines and Digital Media Ethics Code Rules, 2021 [7]. According to that notification, the social media platforms, such as Twitter, Facebook, and others, require to conform to the rules.

## 2 Related Work

At the very beginning, several countries and agencies identified the dangers and consequences of spreading the hate speech. Agencies like United Nations, European Commission, etc., initiated various awareness, propaganda, against hate speech, racism, and intolerance [4, 8]. Different countries imposed various strategies and restrictions on hate-speech content in social media, and on the violation, there is a provision of fine/punishment against the hateful postings [9]. As a result, researchers tried to identify hate-speech content from a post automatically. The companies like Google, Facebook, etc., initiated research in that domain. Several shared tasks have been initiated since last few years related to the identification of hate speech [10–14]. In most of the approaches, machine learning and deep learning have been used.

The classical classifiers like Naive Bayes, SVM, decision tree, etc., are also being used [15, 16]. Mubarak et al. [17] have focused on detecting the vulgar and pornographic obscene with a phrases-based approach in Arabic social media, whereas Mohaouchane et al. [18] have used the deep learning-based approach. Sood et al. [19] have used the Yahoo! social news data, and various SVM classifiers have been used for classification. Nobata et al. [20] have used a feature-based n-gram model. Some of them tried to incorporate the behavioral phenomenon in their models [21, 22].

The common thing of all the existing works is that almost all are related to the domain of social media and in other domains, and it is hardly found. But, other domain like print media also contains the hate-speech content. In that concern, we have tried to address the hate-speech-related news articles in the print media.

## 3 Hate Speech in Other Domain

The main focus of hate-speech detection is concentrated in various social media. Nowadays, it has become a serious issue in visual as well as print media. Especially in the election period, so many hate speeches-related texts can be found in print media [23]. In India too, the usage of hate speech during the election is increasing gradually. But to the best of our knowledge, the work of hate-speech detection in the news articles is still not explored. In that context, this task is dedicated to the hate-speech identification in the news domain. There are several additional problems in finding hate speech in the news domain. One important issue is whether to consider the hate speech at sentence level or article level and we have tried to address this issue.

## 4 Contribution of This Work

This work is different from the traditional hate-speech detection domain. This is an attempt to address hate speech in the news domain. Because of the socioeconomic–cultural situation, the use of hate speech is increasing during the election in state or nation levels. This problem is more challenging in comparison with the social media text. One major issue is that in social media the text of the message is written by the user himself/herself, whereas in the newspaper the article is written by the third person, and as a result, there is an information gap and a biasness may be introduced. For the time being, we have ignored such difficulties and tried to find the hate-speech content in a news text with a phrase-based matching approach.

## 5 Data Preparation and Methodology

Since the work is dedicated to identify hate-speech content in a news article, it is worthwhile to collect data from news articles in the election period. We have collected all the political news articles from a daily Bengali newspaper "Anandabazar Patrika" [24]. The election date of West Bengal has been declared on February 26, 2021, by the Election Commission of India [25], and hence, we have collected the news articles from February 20, 2021, to March 30, 2021, (total 39 days' news articles) and considered this data as a political news corpus for West Bengal Assembly Election 2021. After building the political news corpus, we have prepared a list of key phrases related to hate speech in a semi-automated manner. Finally, a phrase-based matching system detects the news articles containing hate speeches.

### 5.1 *Data Preparation*

To prepare the political news corpus, all the political news article related to West Bengal Assembly Election 2021 are collected up to date March 30, 2021. According to the notification of the Election Commission of India [25], the election will be conducted in eight phases (dated March 27, 2021, April 1, 2021, April 6, 2021, April 10, 2021, April 17, 2021, April 22, 2021, April 26, 2021, and April 29, 2021). In this context, all the political news articles of a leading Bengali daily newspaper "Anandabazar Patrika" from February 20, 2021, to March 30, 2021, (i.e., of 39 days in total) have been considered. Note that, the date is considered up to March 30, 2021, i.e., it covers some news after the first phase of the election. Figure 1 shows the block diagram of the system used for building the political news corpus.

**Web scraping**

- To access the past dated political news articles, first we have prepared a list Uniform Resource Locators (URLs) from "Anandabazar Patrika" online web portal.

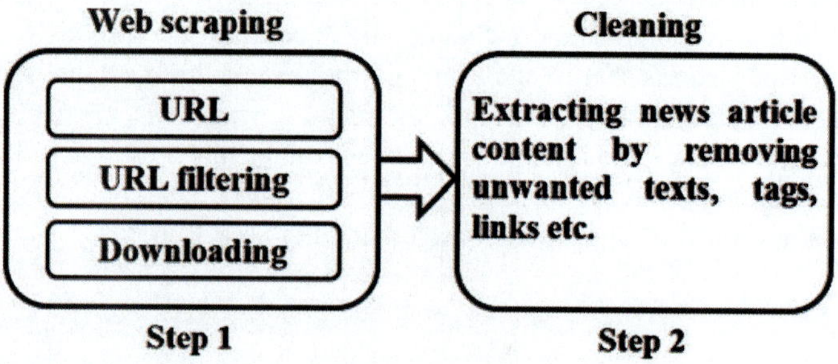


**Fig. 1** Block diagram of the system used to retrieve data for building political news corpus

**Table 1** Volume summary of political news corpus

| S. No. | Heading | Value |
|---|---|---|
| 1 | Total number of files | 1642 |
| 2 | Total number of news articles | 1642 |
| 3 | Total number of sentences | 57,755 |
| 4 | Total number of tokens | 602,141 |
| 5 | Total size of the corpus | 10 MB |

- To select only the political news articles, a filter is imposed on the URLs. To incorporate the filter, we have prepared a list of West Bengal election-related key words {"west-bengal-election", "west-bengal-election-2021", "wb-election", "wb-election-2021", "west-bengal-polls-2021", and "west-bengal-assembly-election"}. The filter will take any URL and return as it is, if the URL contains any of the enlisted keywords, otherwise return nothing. Now we pass the already prepared list of URLs through this filter, and we get the final list of West Bengal election-related news articles' URLs.
- To download the news articles, we have written a Python-based web scraper using the library Beautiful Soup. Now we have downloaded the news articles by using the web scraper and saved each article's raw text as a separate .txt file.

**Cleaning**

- In this step, we take each .txt file containing raw news data, remove the unwanted HTML tags, texts, links, advertisement texts, etc., from the raw data.
- After eliminating all the undesirable things, we get the necessary details and the actual content of the news article. Now we take only the news headline as well as the content and save it in a .txt file using UTF-8 encoding.
- Actually, each file will contain the news article's headline followed by the news article's content for a specific news article. Therefore, the total number of files in our corpus will be equal to the total number of news articles present in the corpus.

**Corpus summary**

- Table 1 shows the brief of the political news corpus in size.

## 5.2 *Methodology*

Our political news corpus has already been prepared. To detect the news articles containing hate speeches, two steps are used sequentially. First, a list of hate-speech-related key phrases is prepared from the corpus in a semi-automated manner, and then a phrase-based matching technique is used to find out the news articles which contain at least three phrases from the enlisted hate-speech-related key phrases. Following steps (shown in Fig. 2) are used for finding out the news articles containing hate speeches. As said earlier, this experiment tried to identify hate speech at a text level

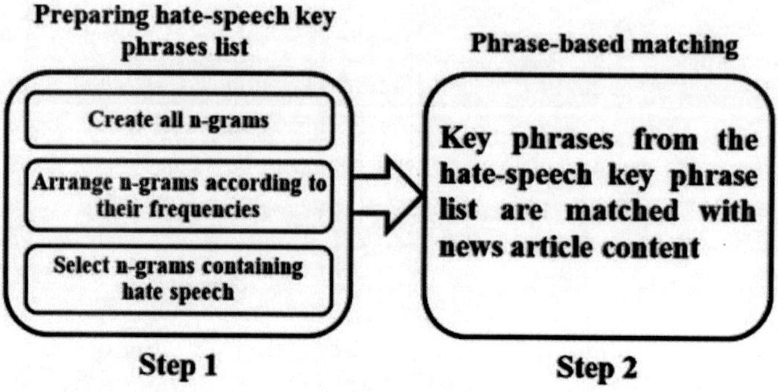


**Fig. 2** Block diagram of the system used for detecting the news article with hate speeches

rather than a sentence level. On analyzing the hate-speech content news articles, it is found that the article contains hate speech multiple times in all the cases. It is also found that sometimes non-hate-speech content news article still contains the hate-speech-related key phrases. To avoid such types of errors, a threshold value of three is considered, i.e., if the article contains three or more hate-speech-related terms/phrases, then only the article is considered as a hate-speech content article.

**Preparing hate-speech key phrases list**

- To prepare the list of hate-speech-related key phrases, a semi-automated technique is used where first n-grams (for $n = 1, 2, 3, 4, 5$) are created from the entire corpus.
- Then those n-grams are arranged in descending order based on their frequencies in the corpus.
- Finally, the n-grams containing hate speeches are selected, and a list of hate-speech-related key phrases is prepared. The list of hate-speech-related key phrases contains total 217 key phrases. The sample key phrase set is {[মিথ্যেবাদী] / mithyebadi (liar), [পিসি] / pisi (aunt), [দাঙ্গাবাজ] / dangabaz (rioter), [গদ্দার] / gaddar (traitor), [বিশ্বাসঘাতক] / biwasghatak (traitor), [ভাইপো] / bhaipo (nephew), [হুমকি] / humki (threat), [গুন্ডার] / gundar (hooligan), [মাফিয়া] / maphiya (a closed group of people involved in criminal activities), [লড়াই হবে] / larai hobe (there will be fight), [ধান্দাবাজ] / dhandabaj (grabby), [বেইমান] / beimaan (betrayer), [পাচারকারী] / pacharkaree (trafficker), [হুঁশিয়ারি] / hunshiari (warning), [বাঁদর] / bandar (monkey), [বহিরাগত] / bahiragata (external), [বিজেমূল] / bijemul (a term used to represent a political party badly), [পাগল] / pagal (crazy), [হার্মাদ] / harmad (a goon, antisocial element, or armed cadre), [দুষ্টু লোক] / dustu loke (naughty guy), [খেলা হবে] / khela hobe (there will be play), [এক ঝুড়ি লোক] / ek jhuri loke (one basket people), [ঘরে ছেলে ঢুকিয়ে দেব] / ghare chele dhukiye debo (a threat to let in boys in the house), [চোর] / chore (thief), [দুর্নীতিবাজ] / durneetibaz (corrupt), [মানসিক ভারসাম্যহীন] / manasik bharshamyaheen (mentally imbalanced), …}.

**Phrase-based matching**

- Bengali language is a rich language with respect to inflectional variations. Here we have used a heuristic-based prefix-matching approach for matching the list of hate-speech-related key phrases with each news article to detect whether it contains hate speeches or not.
- Actually in this step, we take each news article (containing news headline and content) and use phrase-based matching technique to check whether at least three of the hate-speech key phrases are found there. If it is found, then it is declared that the news article contains hate speech, i.e., the news article is a hate-speech content article.

## 6 Experiment and Result

Our political news corpus contains a total of 39 days' political news articles of February 20, 2021, to March 30, 2021, from the leading Bengali newspaper "Anandabazar Patrika". From the entire period, we have randomly selected three days' (February 26, 2021, March 9, 2021, and March 30, 2021) political news articles and evaluated the performance of our proposed system. Tables 2, 3, and 4 show the confusion matrices for the respective three days separately, and Table 5 shows the performance measure of the system in terms of precision, recall, and F1-score.

**Table 2** Performance evaluation of the system for February 26, 2021

| Total no. of political news articles = 29 | | News article contains hate speech (detected) | | Date: 26/02/2021 |
|---|---|---|---|---|
| | | Yes | No | |
| News article contains hate speech (actual) | Yes | 9 (TP) | 6 (FN) | Total no. of actual hate-speech content news articles = 15 |
| | No | 4 (FP) | 10 (TN) | Total no. of actual non-hate-speech content news articles = 14 |
| | | Total no. of detected hate-speech content news articles = 13 | Total no. of detected non-hate-speech content news articles = 16 | |

**Table 3** Performance evaluation of the system for March 9, 2021

| Total no. of political news articles = 36 | | News article contains hate speech (detected) | | Date: 09/03/2021 |
|---|---|---|---|---|
| | | Yes | No | |
| News article contains hate speech (actual) | Yes | 9 (TP) | 9 (FN) | Total no. of actual hate-speech content news articles = 18 |
| | No | 2 (FP) | 16 (TN) | Total no. of actual non-hate-speech content news articles = 18 |
| | | Total no. of detected hate-speech content news articles = 11 | Total no. of detected non-hate-speech content news articles = 25 | |

**Table 4** Performance evaluation of the system for March 30, 2021

| Total no. of political news articles = 48 | | News article contains hate speech (detected) | | Date: 30/03/2021 |
|---|---|---|---|---|
| | | Yes | No | |
| News article contains hate speech (actual) | Yes | 24 (TP) | 7 (FN) | Total no. of actual hate-speech content news articles = 31 |
| | No | 6 (FP) | 11 (TN) | Total no. of actual non-hate-speech content news articles = 17 |
| | | Total no. of detected hate-speech content news articles = 30 | Total no. of detected non-hate-speech content news articles = 18 | |

**Table 5** Performance measure of the system

| Date | Precision | Recall | F1-Score |
|---|---|---|---|
| February 26, 2021 | 0.692 | 0.600 | 0.643 |
| March 9, 2021 | 0.818 | 0.500 | 0.621 |
| March 30, 2021 | 0.800 | 0.774 | 0.787 |
| In total | 0.778 | 0.656 | 0.712 |

## 7 Error Analysis

In order to find out the weakness of the system, the source of errors is investigated. The confusion matrices show that the false negative (FN) is higher compared to the false positive (FP). The two main causes are identified for these errors as follows:

1. Phrase-based matching error: The complex morphology of the language is the root cause of the matching problem. This error could be minimized if our system considered the morphological variations with high precision.
2. List of hate-speech-related key phrases: This key phrases list is not complete, i.e., in this list, all possible hate-speech-related key phrases are not included.

Apart from these, some other problems are also identified. Some instances are found in both (FN and FP) cases which signify that the context information is needed to identify the hate speech from the text.

## 8 Conclusion

In the literature, it is seen that most of the works have been done for hate-speech detection in Facebook, Twitter, and other social media data, but it is hardly found in the news domain. Therefore, our work is one of the pioneering attempts to identify hate speeches in political news articles. The complexity to detect hate speech in the news article compared to other social media Facebook, Twitter, etc., is also explored. Because the news content may include direct or indirect speeches and moderated by editors. The error analysis gives hints to improve the accuracy. This result can be considered as a benchmark and helps for further improvement.

## References

1. Pitsilis, G.K., Ramampiaro, H., Langseth, H.: Effective hate-speech detection in Twitter data using recurrent neural networks. Appl. Intell. **48**, 4730–4742 (2018). https://doi.org/10.1007/s10489-018-1242-y
2. Habermas, J., McCarthy, T., McCarthy, T.: The Theory of Communicative Action, vol. 1. Beacon press, Boston (1984)
3. Sigurbergsson, G.I., Derczynski, L.: Offensive language and hate speech detection for Danish. arXiv preprint arXiv:1908.04531 (2019)
4. United Nations Strategy and Plan of Action on Hate Speech: Detailed Guidance (2020). https://www.un.org/en/genocideprevention/documents/UN Strategy and PoA on Hate Speech_Guidance on Addressing in field.pdf. Last accessed 2021/03/30
5. Crime in India, 2019 Statistics, Volume I, National Crime Records Bureau, (Ministry of Home Affairs), Government of India, National Highway-8, Mahipalpur, New Delhi
6. Law Commission of India, Report No. 267 Hate Speech. https://lawcommissionofindia.nic.in/reports/Report267.pdf

7. Notification dated, the 25th February, 2021 G.S.R. 139(E): The Information Technology (Intermediary Guidelines and Digital Media Ethics Code) Rules, 2021. https://www.meity.gov.in/writereaddata/files/Intermediary_Guidelines_and_Digital_Media_Ethics_Code_Rules-2021.pdf. Last accessed 25 May 2021
8. European Commission against Racism and Intolerance (ECRI) report, https://www.coe.int/en/web/european-commission-against-racism-and-intolerance/. Last accessed 2021/03/30
9. Thomasson, E.: German cabinet agrees to fine social media over hate speech. World News April 5, 2017. https://www.reuters.com/article/idUKKBN1771FK. Last accessed 2021/03/30
10. OSACT4 2020: Shared task on Arabic Offensive Language Detection (2020). http://edinburghnlp.inf.ed.ac.uk/workshops/OSACT4/
11. Wiegand, M., Siegel, M., Ruppenhofer, J.: Overview of the germeval 2018 shared task on the identification of offensive language (2018)
12. Zampieri, M., Malmasi, S., Nakov, P., Rosenthal, S., Farra, N., Kumar, R.: Semeval-2019 task 6: Identifying and categorizing offensive language in social media (offenseval). arXiv preprint arXiv:1903.08983 (2019)
13. Mandl, T., Modha, S., Majumder, P., Patel, D., Dave, M., Mandlia, C., Patel, A.: Overview of the hasoc track at fire 2019: hate speech and offensive content identification in Indo-European languages. In: Proceedings of the 11th Forum for Information Retrieval Evaluation, pp. 14–17 (2019). https://doi.org/10.1145/3368567.3368584
14. Mandl, T., Modha, S., Kumar M, A., Chakravarthi, B.R.: Overview of the HASOC Track at FIRE 2020: Hate Speech and Offensive Language Identification in Tamil, Malayalam, Hindi, English and German. In: Forum for Information Retrieval Evaluation, pp. 29–32 (2020). https://dl.acm.org/doi/10.1145/3441501.3441517
15. Abozinadah, E.A., Mbaziira, A.V., Jones, J.: Detection of abusive accounts with Arabic tweets. Int. J. Knowl. Eng. **1**(2), 113–119 (2015)
16. Abozinadah, E.: Detecting abusive Arabic language twitter accounts using a multidimensional analysis model. Doctoral dissertation (2017)
17. Mubarak, H., Darwish, K., Magdy, W.: Abusive language detection on Arabic social media. In: Proceedings of the first workshop on abusive language online, pp. 52–56. Association for Computational Linguistics, Vancouver, Canada (2017)
18. Mohaouchane, H., Mourhir, A., Nikolov, N.S.: Detecting offensive language on Arabic social media using deep learning. In: 2019 Sixth International Conference on Social Networks Analysis, Management and Security (SNAMS). IEEE, Granada, Spain (2019)
19. Sood, S.O., Churchill, E.F., Antin, J.: Automatic identification of personal insults on social news sites. J. Am. Soc. Inform. Sci. Technol. **63**(2), 270–285 (2012)
20. Nobata, C., Tetreault, J., Thomas, A., Mehdad, Y., Chang, Y.: Abusive language detection in online user content. In: Proceedings of the 25th International Conference on World Wide Web, pp. 145–153. ACM, Montréal, Québec, Canada (2016). https://doi.org/10.1145/2872427.2883062
21. Chen, Y., Zhou, Y., Zhu, S., Xu, H.: Detecting offensive language in social media to protect adolescent online safety. In: 2012 International Conference on Privacy, Security, Risk and Trust and 2012 International Confernece on Social Computing, pp. 71–80. IEEE, Amsterdam, Netherlands (2012)
22. Buckels, E.E., Trapnell, P.D., Paulhus, D.L.: Trolls just want to have fun. Personal. Individ. Differ. **67**, 97–102 (2014)
23. Hate in Elections: How Racism and Bigotry Threaten Election Integrity in the United States (2020). https://lawyerscommittee.org/wp-content/uploads/2020/09/LC2_HATE-IN-ELECTIONS_RPT_E_HIGH-1.pdf
24. Anandabazar Patrika Homepage, https://www.anandabazar.com/. Last accessed 2021/03/30
25. Chief Electoral Officer, West Bengal, Home Department, Election Branch 21, N.S Road, Kolkata-700001, Notice No. ECI/PN/16/2021 Dated: 26th Feb 2021. http://ceowestbengal.nic.in/UploadFiles/AE2021/Pressnote_AE_2021.pdf

# A Structural Equation Modeling Approach for Adoption of Big Data Analytics by SMEs in India

**Subhodeep Mukherjee, Venkataiah Chittipaka, and Manish Mohan Baral**

**Abstract** Big data means a large volume of data used and stored by different firms in their day-to-day operations. It is a field that extracts and analyzes a complex, large volume of data. This research is conducted to study the adoption of big data in Indian SMEs using the TOE framework. This research created awareness for the adoption of big data software in Indian SMEs. For this, a structured literature review was conducted. Three independent variables, technological, organizational, and environmental perspectives, are identified. Survey is carried out in the SMEs with the help of questionnaires. The target population is IT managers, plant managers, owners, and directors. For data analysis, exploratory factor analysis using SPSS 20.0 software and structural equation modeling using AMOS 20.0 software is used. The developed model using three independent variables and one dependent variable showed a good fit.

**Keywords** Big data · TOE framework · Indian SMEs · Structural equation modeling · Exploratory factor analysis

## 1 Introduction

The information has begun to produce a massive volume in various fields throughout the most recent years. It has been typical that the data will expand to a great extent. It had been portrayed that big data (BD) is the dramatic development of complex information for a vast scope as an advancing term [1]. As per [2], "major information is the data resource portrayed by its volume, velocity, variety, variability, and volatility that requires explicit innovation and logical strategies for its change into esteem."

S. Mukherjee (✉) · M. Mohan Baral
Department of Operations, GITAM SCHOOL OF BUSINESS, GITAM (Deemed to be University), Visakhapatnam, Andhra Pradesh, India
e-mail: subhodeepmukherjee92@gmail.com

V. Chittipaka
School of Management Studies, Indira Gandhi National Open University, Delhi, India



D. Gupta et al. (eds.), *Pattern Recognition and Data Analysis with Applications*, Lecture Notes in Electrical Engineering 888,
https://doi.org/10.1007/978-981-19-1520-8_20

Various government establishments just started adopting BD in their offices. Most researchers have found many advantages of using BD as it helps in handling massive information among many sectors [3]. However, many firms are not ready to adopt the latest technologies of BD due to the cost of installations and the cost of handling it [4]. The present circumstance features the requirement for another top to bottom examination to comprehend the inspirations driving the amazing cycle of huge information reception [5]. Indeed, even numerous analysts accepted that massive information appropriation could essentially upgrade firm execution [6].

In the light of digitalization, where everyone utilizes new computerized innovations, like cell phones and online media, design as far as raw data has become available principally to get a handle on, keep, examine, and use at a lowered cost [7, 8]. In this way, a pervasive and always expanding advanced record, commonly named considerable information, is getting produced by every person globally. However, notwithstanding the numerous advantages of BD, less exploration has occurred about how organizations can receive it and make business esteem from such an innovation [9]. Along these lines, there is an absence of comprehension of how organizations manage the cycle of BD, usage, and worth age [10]. Accordingly, the appropriation of creative advancements can deliver more business favorable circumstances and openings for huge companies and small and medium-sized enterprises (SMEs).

## 2 Literature Review

### *2.1 Big Data Analytics in Small and Medium Enterprises*

SMEs go about as the principal component of economic development by making open positions and being creative and profitable [11–14]. BD is a recently arisen technique for SME's development, which empowers them to settle on better choices about the market and clients' requirements by depending on analytical instruments [15]. It will also help them expand their severe status on the lookout. SMEs can get an incentive from voluminous information by accepting the help of BD specialist organizations. The selection of BD in SMEs can be productive in handling the significant difficulties of organizations. Utilizing BD and its insightful methods are not just for enormous endeavors [16–18]. These days, independent companies likewise can use the favorable circumstances and shrouded estimations of the high measures of on the web and disconnected information to settle on dependable choices following their organizations' target [19, 20].

The intense feeling of rivalry among the SMEs in the market would compel them to grasp BD's appropriation to increment operative execution [21–24]. However, advancement among SMEs ought not to be downplayed. The more significant part of the current writing underscored BD's significance in huge organizations [25]. Be that as it may, most SMEs are hesitant to use BD, strategies in their organizations, or they neglect to have useful utilization of BD ventures, which is essential because of an absence of comprehension and information about BD [26].

## 2.2 *TOE Framework and Hypothesis Development*

First proposed by Tornatzky et al. [27], the TOE system is a hypothetical structure at the association level that clarifies factors that influence the way toward receiving and rehearsing mechanical advancements from the innovative, hierarchical, and natural points of view, as opposed to factors identified with the attributes and feelings of people inside the association. Various investigations have checked the TOE structure's viability in selecting different data frameworks and innovations. Three independent variables are technological factors (TF), organizational factors (OF), and environmental factors (EF). One dependent variable is big data adoption (BDA).

### 2.2.1 Technological Factors (TF)

Alharbi et al. [28] have characterized the relative advantage (RA) "as how much a development is seen as being superior to the thought it overrides." Earlier research, for example Ahmadi et al. [29], recommends that RA is an essential component of the mechanical setting that is fit for empowering or debilitating innovation's reception. It had been characterized compatibility (COM) as "how much the development is seen as predictable with the current qualities, past encounters, and needs of the expected adopter" [30–32]. As per [33], complexity (COMP) is how much development is seen as moderately hard to comprehend and utilize. The innovative products would be less inclined to be executed if seen as the more aspiring and testing for actualizing. Trialability (TR) is the degree to which IT advancement is conceivable to attempt [34–37]. It has characterized TR "as how much a development might be explored different avenues regarding restricted premise."

H1: TF influences BDA in SMEs.

### 2.2.2 Organizational Factors (OF)

(Makena [38]) Characterizing top management support (TMS) refers to how many administrators grasp and grasp another innovation framework's innovative capacities. Likewise, Kuan and Chau [39] describe TMS as the uplifting demeanor of CEOs toward innovation reception. As indicated by Alsetoohy et al. [30], Queiroz, and Wamba [40], organizational readiness (ORN) is referred as the degree for which necessary hierarchical assets are accessible for use innovation BD [41–43]. For instance, organizations need skilled labor (SL), such as information researchers, information board specialists, and experts talented at working with huge scope data when receiving and using BD and business investigation [44–46]. Financial investment competence (FIC) refers to how firms can put resources into presenting and working BD [47]. It takes a great deal of monetary venture to receive BD in firms, including gear, programming bundle, and counseling [48–50].

H2: OF influences BDA in SMEs.

#### 2.2.3 Environmental Factors (EF)

(Abed [51], Stjepić et al. [52]) Competitive pressure (CP) is one of the transcendent precursors of IT advancements selection inside firms. Rivalry in an industry is generally seen to affect the reception of IS advancements positively [53–55]. The external support (ES) is re-evaluating had been shown as the primary drivers in the IT development achievement, which can undoubtedly impact IT advancement reception [56, 57]. Government regulations (GR) have been acknowledged as another essential component in advancement selection [58, 59].

H3: EF influences BDA in the SMEs.

## 3 Research Methodology

### *3.1 Sampling*

The data is collected through a structured questionnaire. Qualified academicians and researchers checked the questionnaire. The questionnaire was sent to the employees working in SMEs across India. The target populations were plant managers, IT managers, directors, and owners. The sample is selected simple random method from each strata because it enables population harmony from the subpopulation [60]. Four hundred fifty-nine respondents received the questionnaires; however, only 288 respondents provided usable, insightful questionnaires. For avoiding the biasness of the data, few precautions are being taken. It is mentioned in the first page of the questionnaire that the survey is for research purpose and it will not be used for anything else. We used the Harman test to calculate the single factor after the data was collected. The first factor showed a percentage of 26.445 which is below the recommended threshold level of 50% [61]. So, we can say that the data collected is not biased.

## 3.2 Demographics of the Respondents

Table 1 shows the characteristics of the respondents for the survey. The firm in which a total number of employees are in the range 51–100 respondents' percentage was 18%, which is the highest. Followed by employees in the range 151–250, respondent's percentage was 17%. The rest are in the range 26–50 employee's respondent's percentage with 16%, the range 1–9 employee's respondent's percentage was 15%, the range 101–150 employee's respondent's percentage was 14%, the range 10–25 employee's respondent's percentage was 12%, and the range 251 and above employee's respondent's percentage was 8%. The percentages of respondents who are directors are 29%, which is the highest. The plant manager is 26%, the IT manager is 25%, and the owners are 20%. The percentages of respondents from the type of medium enterprises are 39%, which is the highest. Followed by microenterprises are 31%, and small enterprises are 30%.

**Table 1** Characteristics of the respondents for the survey

| | Characteristics | Percent |
|---|---|---|
| I. *Total number of employees* | | |
| A | 1–9 employees | 15 |
| B | 10–25 employees | 12 |
| C | 26–50 employees | 16 |
| D | 51–100 employees | 18 |
| E | 101–150 employees | 14 |
| F | 151–250 employees | 17 |
| G | 251 and above | 8 |
| II. *Respondents current position* | | |
| A | Owner | 20 |
| B | Director | 29 |
| C | Plant manager | 26 |
| D | IT manager | 25 |
| III. *Type of firms* | | |
| A | Microenterprises | 31 |
| B | Small enterprises | 30 |
| C | Medium enterprises | 39 |

**Table 2** Cronbach's alpha, composite reliability, rotated component matrix, and AVE for the variables

| Latent variable | Indicators | Cronbach's alpha (α) | Composite reliability (CR) | Rotated component matrix | AVE |
|---|---|---|---|---|---|
| TF | RA | 0.849 | 0.830 | 0.840 | 0.510 |
| | COMP | | | 0.869 | |
| | TR | | | 0.882 | |
| | COM | | | 0.720 | |
| OF | TMS | 0.886 | 0.838 | 0.843 | 0.541 |
| | ORN | | | 0.886 | |
| | FIC | | | 0.892 | |
| | SL | | | 0.838 | |
| EF | CP | 0.847 | 0.794 | 0.886 | 0.514 |
| | ES | | | 0.898 | |
| | GR | | | 0.840 | |

# 4 Data Analysis

## *4.1 Reliability and Validity*

### 4.1.1 Cronbach's Alpha

A reliability test is being performed with the data for each factor. Cronbach's alpha is considered as a measure of the scale reliability. Cronbach's alpha ($\alpha$) is being calculated for all three factors. The values should be higher than 0.70 [62, 63]. Hence, all the values are within the threshold, as shown in Table 2.

### 4.1.2 Composite Reliability

For all the components, composite reliability (CR) was measured. It is measured by its ability to provide better results in terms of internal consistency [64]. Three CR constructs have > 0.7, which indicates the reliability of the composite reliability measures [60, 62], as shown in Table 2.

## *4.2 Exploratory Factor Analysis (EFA)*

Evaluation of the sample size was the first step of the EFA. For EFA, SPSS 20.0 has been used. Bartlett's sphericity test had inspected the correlations between the items [65]. For current investigations, the KMO value is 0.741 which is greater than 0.60,

**Table 3** Discriminant validity matrix

| | CR | AVE | MSV | MaXR (H) | TP | OP | EP |
|---|---|---|---|---|---|---|---|
| TF | 0.830 | 0.510 | 0.151 | 0.881 | 0.714 | | |
| OF | 0.838 | 0.541 | 0.010 | 0.894 | 0.046 | 0.735 | |
| EF | 0.794 | 0.514 | 0.151 | 0.862 | 0.388 | 0.100 | 0.717 |

i.e., the minimum acceptance level. The principal axis factoring is the extraction method used. Only values with values more significant than one have been extracted because the maximum variance is explained. For these components, components 1 (32.396%), 2 (26.310%), and 3 present the share of the total variance (14.788%). The cumulative proportion of all three components explained is 73.494%.

For interpreting the analysis results, the rotated component matrix is essential. Rotation helps to group items, and the structure is simplified by at least more than two items for each group. This is, therefore, the objective of the rotation objective. We have achieved this goal in this research. Total 11 variables are grouped into three components, as shown in Table 2.

## 4.3 *Construct Validity (CV)*

CV is the measure in which a test quantifies the idea or development that should be quantified. CV does not have a cutoff [65].

### 4.3.1 Convergent Validity

This is measured by the help of the average variance extracted (AVE). As per [66], the convergent validity AVE > 0.5. For the constructions, Table 3 shows AVE values. Every value is more than 0.5, which satisfies all the building structures' convergent validity.

### 4.3.2 Divergent or Discriminant Validity

Fornell and Larcker [66] suggested that the AVE construct must be more than one square for this validity to be calculated by the relationship between these constructs and the other constructs. Table 3 represents the values for discriminant validity matrix. Hence, in Table 3, we can see that all the constructs, i.e., TF, OF, and EF values for MSV, are lesser than AVE, which satisfies the discriminant validity of all the constructs.

## 4.4 *Structural Equation Modeling (SEM)*

For testing the proposed hypothesis taken in the study, SEM is used using the software AMOS 22.0 [67]. This shows the results of the model. The final model and latent variables and their indicators and their dependent variable are represented in Fig. 1. TF: technological factors have four indicators RA, COMP, TR, and COM; OF: organizational factors have four indicators TMS, ORN, FIC, and SL; EF: environmental factors have three indicators CP, ES, and GR. One dependent variable is BDA: big data adoption, which has four indicators: BDA1, BDA2, BDA3, and BDA4. Table 4 shows model parameters.

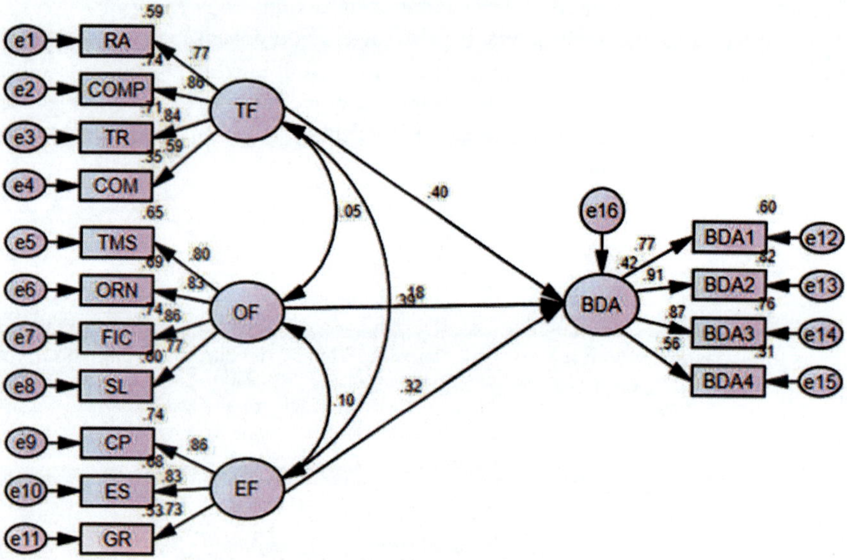


**Fig. 1** Final model for the adoption of BDA

**Table 4** Model fit measures for the confirmatory factor analysis

| Goodness-of-fit Indices | Default Model | Benchmark |
|---|---|---|
| *Absolute goodness-of-fit measure* | | |
| χ2/df (CMIN/DF) | 2.898 | Lower limit: 1.0<br>Upper limit 2.0/3.0 or 5.0 |
| GFI | 0.908 | >0.90 |
| *Incremental fit measure* | | |
| CFI | 0.934 | ⩾0.90 |
| IFI | 0.925 | ⩾0.90 |
| TLI | 0.917 | ⩾0.90 |

Table 5 Path analysis result for structural model

| | Estimate | SE | CR | P | Hypothesis |
|---|---|---|---|---|---|
| TF ← BDA | 0.405 | 0.138 | 2.94 | *** | Supported |
| OF ← BDA | 0.180 | 0.075 | 2.40 | *** | Supported |
| EF ← BDA | 0.318 | 0.086 | 3.69 | *** | Supported |

Table 5 shows the path analysis result. Three hypotheses support the *P*-value [68]. Hence, the three factors TF, OF, and EF have a positive impact on BDA. The structural model explains 41.6% of the variance of BDA.

## 5 Discussion

The current research found that the TOE perspective plays an important role for BDA in Indian SMEs. The indicators which had a significant impact are RA, COMP, TR, COM, TMS, ORN, FIC, SL, CP, ES, and G.R. From the results, it is obvious that the three components suggested by the framework help in BDA in Indian SMEs. The KMO value is 0.741, which is greater than 0.6, which is within the threshold level [60], which allows the data for factor analysis. The values below 0.4 were suppressed in the rotated component matrix table. Only the values more than 0.4 were displayed as output. The component TF relates to the technological aspects for adopting BDA. It comprises four sub-components: RA, COMP, TR, and COM, and each loading is 0.840, 0.869, 0.882, and 0.720. OF relates to organizational aspects for adopting BDA. It comprises four sub-components: TMS, ORN, FIC, and SL, and each loading is 0.843, 0.886, 0.892, and 0.838. EF relates to the environmental aspects of adopting BDA. It comprises three sub-components: CP, ES, and GR, and each loading is 0.886, 0.898, and 0.840. Hence, the loadings of sub-components are >|0.40|. In the present circumstances, BDA will play an important role in the smooth running of the Indian SMEs.

Construct validity is also an essential component of the analysis. Hence, AVE was calculated, which is >0.5 for all the three constructs TF, OF, and EF, which satisfies the convergent validity for all the constructs. Further divergent or discriminant validity was also checked for all the three constructs, which shows MSV < AVE. Hence, this criterion was also satisfied. Earlier research conducted by Lai et al. 2018 in logistics and supply chain management for BD supported this research work. Another study using the TOE framework in Korean firms for BD has supported this research [69]. Another study was conducted in the SMEs of Iran using the TOE framework for BD and supported the results and other studies [70]. Hence, the current research is based on a survey method, and a structured questionnaire was developed to collect data from the respondents from various Indian SMEs. Finally, SEM was performed to get the model fit.

## 6 Conclusion

This study's main aim is to find out the role of BD in the Indian SMEs using the TOE framework. For this, a structured literature review was conducted from the available literature. TOE framework was identified for the research as many earlier IT-related innovation adoptions studies being conducted using TOE. The target population was owners, plant managers, IT managers, and directors. Three independent variables were TF, OF, and EF. The dependent variable was BDA. For the analysis, EFA and SEM were used. The model developed showed a good fit, and the three hypotheses were accepted. This research was supported by other research work conducted in different countries.

Further this study can be extended to other sectors or some other countries.

**Annexure: Questionnaire**

1. Name of the employee (optional):
2. Designation:
3. Total number of employees:
4. Type of firms

   a. Microenterprises
   b. Small enterprises
   c. Medium enterprises

Please rate the following factors for your firm on the scale of 1–7, 1 for strongly disagree, 2 for disagree, 3 for partially disagree, 4 for neutral, 5 for partially agree, 6 for agree, and 7 for strongly agree.

| Questionnaire | Please mark | | | | | | |
|---|---|---|---|---|---|---|---|
| Big data adoption | 1 | 2 | 3 | 4 | 5 | 6 | 7 |
| Our firm is interested in adopting big data | | | | | | | |
| Our firm intends to adopt big data | | | | | | | |
| I would not hesitate to recommend to adopt big data | | | | | | | |
| I feel comfortable to recommend big data for my firm | | | | | | | |
| Technological factors | | | | | | | |
| Relative advantage | | | | | | | |
| Complexity | | | | | | | |
| Trialability | | | | | | | |
| Compatibility | | | | | | | |
| Organizational factors | | | | | | | |
| Top management support | | | | | | | |
| Organizational readiness | | | | | | | |
| Skilled labor | | | | | | | |
| Financial investment competence | | | | | | | |

(continued)

(continued)

| Questionnaire | Please mark | | | | | | |
|---|---|---|---|---|---|---|---|
| Big data adoption | 1 | 2 | 3 | 4 | 5 | 6 | 7 |
| Environmental factors | | | | | | | |
| Competitive pressure | | | | | | | |
| External support | | | | | | | |
| Government regulations | | | | | | | |

## References


1. Addo-Tenkorang, R., Helo, P.T.: Big data applications in operations/supply-chain management: a literature review. Comput. Ind. Eng. **101**, 528–543 (2016)
2. Akter, S., Wamba, S.F., Gunasekaran, A., Dubey, R., Childe, S.J.: How to improve firm performance using big data analytics capability and business strategy alignment? Int. J. Prod. Econ. **182**, 113–131 (2016)
3. Choi, T.M., Wallace, S.W., Wang, Y.: Big data analytics in operations management. Prod. Oper. Manag. **27**(10), 1868–1883 (2018)
4. Gunasekaran, A., Papadopoulos, T., Dubey, R., Wamba, S.F., Childe, S.J., Hazen, B., Akter, S.: Big data and predictive analytics for supply chain and organizational performance. J. Bus. Res. **70**, 308–317 (2017)
5. Ji-fan Ren, S., Fosso Wamba, S., Akter, S., Dubey, R., Childe, S.J.: Modelling quality dynamics, business value and firm performance in a big data analytics environment. Int. J. Prod. Res. **55**(17), 5011–5026 (2017)
6. Lugmayr, A., Stockleben, B., Scheib, C., Mailaparampil, M.A.: Cognitive big data: survey and review on big data research and its implications. What is really "new" in big data? J. Knowl. Manage. (2017)
7. Prescott, M.E.: Big data and competitive advantage at Nielsen. Manage. Decis. (2014)
8. Wang, G., Gunasekaran, A., Ngai, E.W., Papadopoulos, T.: Big data analytics in logistics and supply chain management: certain investigations for research and applications. Int. J. Prod. Econ. **176**, 98–110 (2016)
9. Zhang, Y., Ren, S., Liu, Y., Si, S.: A big data analytics architecture for cleaner manufacturing and maintenance processes of complex products. J. Clean. Prod. **142**, 626–641 (2017)
10. Chen, P.T., Lin, C.L., Wu, W.N.: Big data management in healthcare: Adoption challenges and implications. Int. J. Inf. Manage. **53**, 102078 (2020)
11. Rajabion, L.: Application and adoption of big data technologies in SMEs. In: 2018 International Conference on Computational Science and Computational Intelligence (CSCI), pp. 1133–1135. IEEE (2018)
12. Pereira, J.P., Ostritsova, V.: ICT and big data adoption in SMEs from rural areas: comparison between Portugal, Spain and Russia. In: World Conference on Information Systems and Technologies, pp. 291–301. Springer, Cham (2020)
13. Azevedo, F., Reis, J.L.: Big data analysis in supply chain management in Portuguese SMEs "leader excellence". J. Inf. Syst. Eng. Manage. **4**(3), em0096 (2019)
14. Karim, S., Al-Tawara, A., Gide, E., Sandu, R.: Is big data too big for SMEs in Jordan? In: 2017 8th International Conference on Information Technology (ICIT), pp. 914–922. IEEE (2017)
15. Tien, E.L., Ali, N.M., Miskon, S., Ahmad, N., Abdullah, N.S.: Big data analytics adoption model for Malaysian SMEs. In: International Conference of Reliable Information and Communication Technology, pp. 45–53. Springer, Cham (2019)

16. Iqbal, M., Kazmi, S.H.A., Manzoor, A., Soomrani, A.R., Butt, S.H., Shaikh, K.A.: A study of big data for business growth in SMEs: opportunities & challenges. In: 2018 International Conference on Computing, Mathematics and Engineering Technologies (iCoMET), pp. 1–7. IEEE (2018)
17. Coleman, S., Göb, R., Manco, G., Pievatolo, A., Tort-Martorell, X., Reis, M.S.: How can SMEs benefit from big data? Challenges and a path forward. Qual. Reliab. Eng. Int. **32**(6), 2151–2164 (2016)
18. Shah, S., Soriano, C.B., Coutroubis, A.D.: Is big data for everyone? The challenges of big data adoption in SMEs. In: 2017 IEEE International Conference on Industrial Engineering and Engineering Management (IEEM), pp. 803–807. IEEE (2017)
19. Saleem, H., Li, Y., Ali, Z., Mehreen, A., Mansoor, M.S.: An empirical investigation on how big data analytics influence China SMEs performance: do product and process innovation matter? Asia Pac. Bus. Rev. **26**(5), 537–562 (2020)
20. Sen, D., Ozturk, M., Vayvay, O.: An overview of big data for growth in SMEs. Procedia Soc. Behav. Sci. **235**, 159–167 (2016)
21. Wang, S., Wang, H.: Big data for small and medium-sized enterprises (SME): a knowledge management model. J. Knowl. Manage. (2020)
22. Ifinedo, P.: An empirical analysis of factors influencing Internet/e-business technologies adoption by SMEs in Canada. Int. J. Inf. Technol. Decis. Mak. **10**(04), 731–766 (2011)
23. Silva, J., Hernández-Fernández, L., Cuadrado, E.T., Mercado-Caruso, N., Espinosa, C.R., Ortega, F.A., Hugo Hernández, P., Delgado, G.J.: Factors affecting the big data adoption as a marketing tool in SMEs. In: International Conference on Data Mining and Big Data, pp. 34–43. Springer, Singapore (2019)
24. Yadegaridehkordi, E., Nilashi, M., Shuib, L., Nasir, M.H.N.B.M., Asadi, S., Samad, S., Awang, N.F.: The impact of big data on firm performance in hotel industry. Electron. Commer. Res. Appl. **40**, 100921 (2020)
25. O'Connor, C., Kelly, S.: Facilitating knowledge management through filtered big data: SME competitiveness in an agri-food sector. J. Knowl. Manage. (2017)
26. Vajjhala, N.R., Ramollari, E.: Big data using cloud computing-opportunities for small and medium-sized enterprises. Eur. J. Econ. Bus. Stud. **2**(1), 129–137 (2016)
27. Tornatzky, L.G., Fleischer, M., Chakrabarti, A.K.: Processes of Technological Innovation. Lexington Books (1990)
28. Alharbi, F., Atkins, A., Stanier, C.: Understanding the determinants of cloud computing adoption in Saudi healthcare organisations. Complex Intell. Syst. **2**(3), 155–171 (2016)
29. Ahmadi, H., Nilashi, M., Shahmoradi, L., Ibrahim, O.: Hospital information system adoption: expert perspectives on an adoption framework for Malaysian public hospitals. Comput. Hum. Behav. **67**, 161–189 (2017)
30. Mukherjee, S., Chittipaka, V.: Analysing the adoption of intelligent agent technology in food supply chain management: an empirical evidence. FIIB Bus. Rev. (2021)
31. Gupta, P., Seetharaman, A., Raj, J.R.: The usage and adoption of cloud computing by small and medium businesses. Int. J. Inf. Manage. **33**(5), 861–874 (2013)
32. Chang, I.C., Hwang, H.G., Hung, M.C., Lin, M.H., Yen, D.C.: Factors affecting the adoption of electronic signature: executives' perspective of hospital information department. Decis. Support Syst. **44**(1), 350–359 (2007)
33. Rogers, E.M.: Diffusion of Innovations: modifications of a model for telecommunications. In: Die diffusion von innovationen in der telekommunikation, pp. 25–38. Springer, Berlin (1995)
34. Gangwar, H., Date, H., Ramaswamy, R.: Understanding determinants of cloud computing adoption using an integrated TAM-TOE model. J. Enterprise Inf. Manage. (2015)
35. Gide, E., Sandu, R.: A study to explore the key factors impacting on cloud-based service adoption in Indian SMEs. In: 2015 IEEE 12th International Conference on e-Business Engineering, pp. 387–392. IEEE (2015)
36. Kouhizadeh, M., Saberi, S., Sarkis, J.: Blockchain technology and the sustainable supply chain: theoretically exploring adoption barriers. Int. J. Prod. Econ. **231**, 107831 (2021)

37. Kamble, S., Gunasekaran, A., Arha, H.: Understanding the blockchain technology adoption in supply chains-Indian context. Int. J. Prod. Res. **57**(7), 2009–2033 (2019)
38. Makena, J.N.: Factors that affect cloud computing adoption by small and medium enterprises in Kenya. Int. J. Comput. Appl. Technol. Res. **2**(5), 517–521 (2013)
39. Kuan, K.K., Chau, P.Y.: A perception-based model for EDI adoption in small businesses using a technology–organization–environment framework. Inf. Manage. **38**(8), 507–521 (2001)
40. Queiroz, M.M., Wamba, S.F.: Blockchain adoption challenges in supply chain: an empirical investigation of the main drivers in India and the USA. Int. J. Inf. Manage. **46**, 70–82 (2019)
41. Xu, W., Ou, P., Fan, W.: Antecedents of ERP assimilation and its impact on ERP value: a TOE-based model and empirical test. Inf. Syst. Front. **19**(1), 13–30 (2017)
42. Wong, L.W., Leong, L.Y., Hew, J.J., Tan, G.W.H., Ooi, K.B.: Time to seize the digital evolution: adoption of blockchain in operations and supply chain management among Malaysian SMEs. Int. J. Inf. Manage. **52**, 101997 (2020)
43. Umam, B., Darmawan, A.K., Anwari, A., Santosa, I., Walid, M., Hidayanto, A.N.: Mobile-based smart regency adoption with TOE framework: an empirical inquiry from Madura Island Districts. In: 2020 4th International Conference on Informatics and Computational Sciences (ICICoS), pp. 1–6. IEEE (2020)
44. Oliveira, T., Thomas, M., Espadanal, M.: Assessing the determinants of cloud computing adoption: an analysis of the manufacturing and services sectors. Inf. Manage. **51**(5), 497–510 (2014)
45. Pateli, A., Mylonas, N., Spyrou, A.: Organizational adoption of social media in the hospitality industry: an integrated approach based on DIT and TOE frameworks. Sustainability **12**(17), 7132 (2020)
46. Premkumar, G., Roberts, M.: Adoption of new IT in rural small business. Omega **27**, 467–484 (1999)
47. Baral, M.M., Verma, A.: Cloud computing adoption for healthcare: an empirical study using SEM approach. FIIB Bus. Rev. 23197145211012505 (2021)
48. Al Hadwera, A., Tavana, M., Gillis, D., Rezania, D.: A systematic review of organizational factors impacting cloud-based technology adoption using technology-organization-environment framework. Internet of Things 100407 (2021)
49. Badi, S., Ochieng, E., Nasaj, M., Papadaki, M.: Technological, organisational and environmental determinants of smart contracts adoption: UK construction sector viewpoint. Constr. Manag. Econ. **39**(1), 36–54 (2021)
50. Ergado, A.A., Desta, A., Mehta, H.: Determining the barriers contributing to ICT implementation by using technology-organization-environment framework in Ethiopian higher educational institutions. Educ. Inf. Technol. **26**(3), 3115–3133 (2021)
51. Abed, S.S.: Social commerce adoption using TOE framework: an empirical investigation of Saudi Arabian SMEs. Int. J. Inf. Manage. **53**, 102118 (2020)
52. Stjepić, A.M., Pejić Bach, M., Bosilj Vukšić, V.: Exploring risks in the adoption of business intelligence in SMEs using the TOE framework. J. Risk Financ. Manage. **14**(2), 58 (2021)
53. Seshadrinathan, S., Chandra, S.: Exploring factors influencing adoption of blockchain in accounting applications using technology–organization–environment framework. J. Int. Technol. Inf. Manage. **30**(1), 30–68 (2021)
54. Shahzad, F., Xiu, G., Khan, I., Shahbaz, M., Riaz, M.U., Abbas, A.: The moderating role of intrinsic motivation in cloud computing adoption in online education in a developing country: a structural equation model. Asia Pac. Educ. Rev. **21**(1), 121–141 (2020)
55. Skafi, M., Yunis, M.M., Zekri, A.: Factors influencing SMEs' adoption of cloud computing services in Lebanon: an empirical analysis using toe and contextual theory. IEEE Access **8**, 79169–79181 (2020)
56. Singeh, F.W., Abrizah, A., Kiran, K.: Bringing the digital library success factors into the realm of the technology-organization-environment framework. Electron. Libr. (2020)
57. Sharma, M., Gupta, R., Acharya, P.: Prioritizing the critical factors of cloud computing adoption using multi-criteria decision-making techniques. Glob. Bus. Rev. **21**(1), 142–161 (2020)

58. Cruz-Jesus, F., Pinheiro, A., Oliveira, T.: Understanding CRM adoption stages: empirical analysis building on the TOE framework. Comput. Ind. **109**, 1–13 (2019)
59. Pal, S.K., Mukherjee, S., Baral, M.M., Aggarwal, S.: Problems of big data adoption in the healthcare industries. Asia Pac. J. Health Manag. (2021)
60. Hair, J.F., Ringle, C.M., Sarstedt, M.: PLS-SEM: Indeed, a silver bullet. J. Market. Theory Pract. **19**(2), 139–152 (2011)
61. Podsakoff, P.M., MacKenzie, S.B., Lee, J.Y., Podsakoff, N.P.: Common method biases in behavioral research: a critical review of the literature and recommended remedies. J. Appl. Psychol. **88**(5), 879 (2003)
62. Hair Jr, J.F., Sarstedt, M., Hopkins, L., Kuppelwieser, V.G.: Partial least squares structural equation modeling (PLS-SEM): an emerging tool in business research. Eur. Bus. Rev. (2014)
63. Nunnally, J.C.: Psychometric Theory 3E. Tata McGraw-Hill Education (1994)
64. Henseler, J., Ringle, C.M., Sinkovics, R.R.: The use of partial least squares path modeling in international marketing. Emerald Group Publishing Limited, In New challenges to international marketing (2009)
65. DeVellis, R. F., Lewis, M. A., & Sterba, K. R.: Interpersonal emotional processes in adjustment to chronic illness. Social psychological foundations of health and illness, 256–287 (2003).
66. Fornell, C., Larcker, D.F.: Evaluating Structural Equation Models with Unobservable Variables and Measurement Error. J. Mark. Res. **18**(1), 39–50 (1981)
67. Byrne, B.M.: Structural equation modeling with AMOS: basic concepts, applications, and programming (multivariate applications series). Taylor & Francis Group **396**, 7384 (2010)
68. Kline, R. B.: Assumptions in structural equation modeling. - PsycNET. (2012).
69. Park, J. H., Kim, M. K., & Paik, J. H.: The factors of technology, organization and environment influencing the adoption and usage of big data in Korean firms (2015).
70. Maroufkhani, P., Ismail, W. K. W., & Ghobakhloo, M.: Big data analytics adoption model for small and medium enterprises. Journal of Science and Technology Policy Management (2020).

# Optimized Distributed Job Shop Scheduling Using Balanced Job Allocation and Modified Ant Colony Optimization

**S. Vivek, Kishan Rakesh, and Biju R. Mohan**

**Abstract** Many challenges are being faced by the manufacturing industry: ensuring profitable growth, reducing costs, increasing productivity, and giving quick responses to customers. To become more productive, reduce transportation costs, and reduce bottleneck on a single factory, industrial companies are shifting from single to distributed systems. Scheduling problems like distributed job shop, distributed flow shop, and distributed process planning are becoming a popular field to study. We try to solve the distributed job shop scheduling problem (DJSP) where the allocation of jobs to different factories needs to be done and additionally, the determination of good operation schedules for each factory. The goal of DJSP is to minimize the makespan over all the factories. To solve this problem, we first use a method of allocating jobs to factories to evenly distribute the workloads among all the factories. Later, we use a bio-inspired algorithm on each factory after the allocations, namely ant colony optimization to get a solution that is close to the most optimal solution.

**Keywords** Ant colony optimization · Job scheduling · Job shop · Pheromones · Heuristic algorithms · Makespan

## 1 Introduction

One of the major problems facing the manufacturing industry is that of scheduling. The job scheduling problem is an extremely important factor in order to maximize the productivity of a company. While there are many approaches to tackling the job scheduling problem, there are now significantly more factors that are to be accounted for. Namely, the presence of distributed workshops. This is problematic for the stan-

S. Vivek · K. Rakesh (✉) · B. R. Mohan
National Institute of Technology Karnataka, Surathkal, Mangalore 575025, India
e-mail: kishan.171it120@nitk.edu.in

S. Vivek
e-mail: vivek.171it251@nitk.edu.in

B. R. Mohan
e-mail: biju@nitk.edu.in

 
D. Gupta et al. (eds.), *Pattern Recognition and Data Analysis with Applications*, Lecture Notes in Electrical Engineering 888,
https://doi.org/10.1007/978-981-19-1520-8_21

dard job scheduling problem approaches as this distributed nature is not accounted for. Therefore, there has been an increased focus on dealing with job scheduling in a distributed workspace: distributed job shop, distributed flow shop, etc. While effective approaches to the DJSP do exist, they are subject to improvement.

In this paper, we intend to improve upon the existing approach to tackling the DJSP. The DJSP problem is comparable to the job scheduling problem. The job scheduling problem deals with a factory containing m machines, and a certain number of jobs are meant to be processed on those machines. Thus, the factors to be accounted for are; the time taken by the operations within the jobs, the precedence of those operations to each other, as well as the assignment of jobs to machines. The distributed job scheduling problem further adds the factor of the assignment of specific jobs to specific factories. Therefore, DJSP is more complicated as it increases the number of decisions to be taken. The additional decision is the allocation of jobs to different factories. The second decision to be taken is similar to that of the regular JSP, i.e., the scheduling of operations on the machines within factories. This is done with the intention of minimizing a specific performance criterion. Since a job scheduling problem is strongly NP-hard, the DJSP is ordinarily NP-hard and the job scheduling problem is attained when the number of factories, $f$ is 1.

Due to the large complexity of this problem, for our model, we assume that all of the factories have the same number of machines. We aim to reduce the maximum completion (makespan) of the factories. We do this by first using a job allocation algorithm to assign the jobs to factories such that the workloads are evenly distributed among all the factories, then making use of the modified ant colony optimization (ACO) algorithm. The ant colony algorithm is a method of finding optimal paths within a graph. It does this by mimicking the approach of ants in search of food. Ants initially wander at random. After an ant finds a food source, it returns to the ant colony leaving pheromones in its path in order to identify the route to the food. When other ants come across the pheromones, there is a certain probability that they are to follow that specific path. As subsequent ants follow the path and reach the food, they, in turn, leave their own pheromones. As more ants follow that specific path, the pheromones get stronger until there is a stream of ants following that path. Since ants drop the pheromones every time they reach the food source, the strongest pheromones are likely to be along the shorter paths. We use the same approach as using pheromones to determine the best path on the graph of jobs.

In the literature survey section, we have talked about the various approaches used previously to solve the DJSP. In the methodology, we first elaborate on the problem statement and the constraints associated with it, then describe how the DJSP is converted into a graph. The job assignment algorithm to evenly distribute jobs among factories and the MACO applied to the graph is also described in the methodology. In the results section, we compare the minimum maximum makespan obtained using the proposed model and the existing models which uses traditional job assignment followed by MACO.

### 1.1 Contribution

Vivek and Rakesh conceived the research. Vivek conceived and designed the balanced job allocation algorithm and performed experimentation with parameters to optimize the model. Rakesh integrated the balanced job allocation algorithm with modified ant colony optimization and performed data accumulation and comparison with the existing model. Vivek and Rakesh conducted a literature survey, wrote and revised the paper. Mohan performed supervision over the paper. All authors read and approved the final manuscript.

## 2 Literature Survey

The job scheduling problem has been a popular topic of study for researchers. Colorni et al. [1] propose the ant system to find optimal solutions for job scheduling problems. Guo et al. [2] propose a mathematical model which even considers energy consumption to solve the distributed flexible job shop scheduling problem (FJSP). Davis [3] proposes a new look to the initial population in the genetic algorithm to enhance the effectiveness. Zhang et al. [4] have considered the transportation time to transfer a job from one machine to another and propose an improved genetic algorithm to solve the FJSP.

The DJSP has had relatively fewer researchers dealing with it. Awerbuch et al. [5] was one of the pioneers in tackling the distributed job scheduling problem. They propose an online algorithm for scheduling jobs in a competitive manner. Jia et al. [6] propose a genetic algorithm approach in order to facilitate collaboration between distributed plants. In order to solve the same problem in a multi-factory network, in their next paper, Jia et al. [7] presented a modified genetic algorithm which utilized a two-step encoding method that encodes the factory candidates and affects operations and jobs. Subsequently, Jia et al. [8] refined their prior approach and proposed a genetic algorithm integrated with Gantt Chart in order to derive the factory schedule and combination. Naderi and Azab [9] proposed six mixed-integer linear programming models and analyzed them for performance. De Giovanni and Pezzella [10] proposed an improved genetic algorithm model that utilizes gene encoding to include information on job to flexible manufacturing units (FMU) and a greedy decoding approach to determine job routings. It then uses a local search-based operator to refine the most promising individuals of each generation to improve solutions. Chaouch et al. [11] propose a modified ant colony optimization algorithm to solve the DJSP which involves a local search procedure on the solution given by ACO algorithm. The exact problem with the constraints which we are trying to solve in this paper is the distributed complete FJCP (C-FJSP) which is mentioned in [12] by Zhang et al.

It is clear that the main objective of many approaches followed to solve the DJSP is by finding the best scheduling to minimize a specified criterion. This criterion is

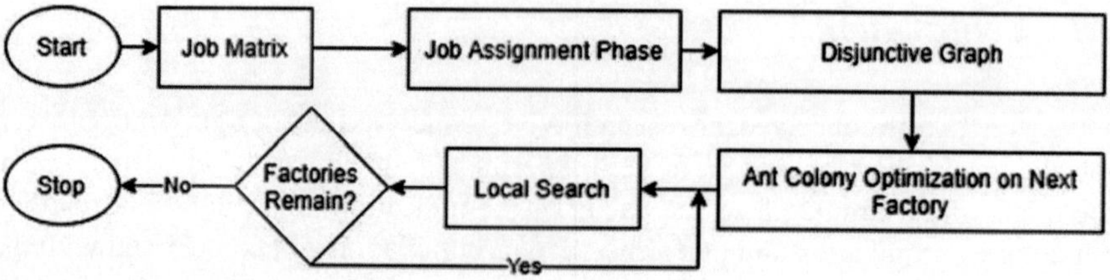


**Fig. 1** Flowchart of algorithm

generally maximum tardiness, total tardiness, or makespan. We are trying to reduce the global makespan among all the factories using a job allocation algorithm that distributes the workloads evenly among all the factories followed by a bio-inspired modified ant colony optimization algorithm (MACO) to optimally schedule the jobs in each factory.

## 3 Methodology

The following section illustrates the problem statement and research methodology that is followed for the proposed work. An overview of our model is shown in Fig. 1.

### *3.1 Problem Statement and Constraints*

The exact problem we are trying to solve is the distributed version of the C-FJSP mentioned in [12]. The distributed C-FJSP can be stated as: a set $J = j_1 \ldots j_n$ of mutually independent jobs. Each of these jobs consists of a set of operations. Operation $j + 1$ of job $i$ must be executed after operation $j$ of the same job $i$. Any operation can be executed on any machine as long as the dependency constraint between operations of the same job is met. The factories are assumed to be geographically distributed. All the factories are said to be identical; i.e., all the factories are assumed to have the same set of machines $M$. Other constraints in the distributed C-FJSP which are:

- At time 0, all jobs are autonomous and ready to be processed, and all machines are available at all times.
- A job cannot be moved to another factory after it has been allocated to one because the remaining operations must be done in the same factory.
- All factories are capable of processing any job.
- There are no precedence constraints between the operations of different jobs.
- Each operation must be completed in a consistent manner throughout the course of its processing period and on the allocated machine.

**Table 1** Processing time matrix

| Job | Machine 1 | Machine 2 | Processing route |
|---|---|---|---|
| 1 | 4 | 7 | {1, 2} |
| 2 | 3 | 5 | {2, 1} |
| 3 | 5 | 2 | {2, 1} |

- At most, one machine can process a job at a time, and one machine can process only one job at any given time.
- Machine setup times and transit times between operations are minimal.

Given these constraints, we need to schedule jobs in the factories so as to reduce the global makespan among all the factories.

## 3.2 Graphical Representation of the DJSP

The processing time matrix given as the input for each factory after the job allocation phase is transformed into an equivalent graph on which the ACO algorithm is applied to get an optimized schedule. The graph has a source and a sink node. The remaining nodes represent an operation of each job assigned to that factory. A node $N(i, j, w_j)$ represents the $j$th operation of the $i$th job, and $w_j$ represents the processing time of the operation $j$ of job $i$. Directed edges exist between operations of the same job; i.e., directed edges exist from nodes $N(i, j, w_j)$ to $N(i, j+1, w_{j+1})$ where $j$ represents the $j$th operation of the $i$th job and $j+1$ represents the $j+1$th operation of the $i$th job. Undirected edges exist between the nodes $N(i, j, w_j)$ and $N(k, m, w_m)$ where $i \neq k$. The weight of an edge from the Node $N(i, j, w_j)$ to $N(k, m, w_m)$ is equal to $w_m$ which is the processing time of the operation to which the ant wants to travel to from the operation $N(i, j, w_j)$. Figure 2 represents the equivalent graphical representation for the processing time matrix of the jobs given in Table 1.

## 3.3 Job Assignment Phase

Job assignment is an integral phase in solving the DJSP. Our job allocation algorithm aims to equalize the workloads in different factories as much as possible. In the first part of allocation, we use the first step of the job-facility assignment rule introduced in [6]. In this first step, the workload of each job $j$ is defined as follows:

$$\text{workload}(j) = \left(\sum_{i \in j} W_{i,j}\right) \tag{1}$$

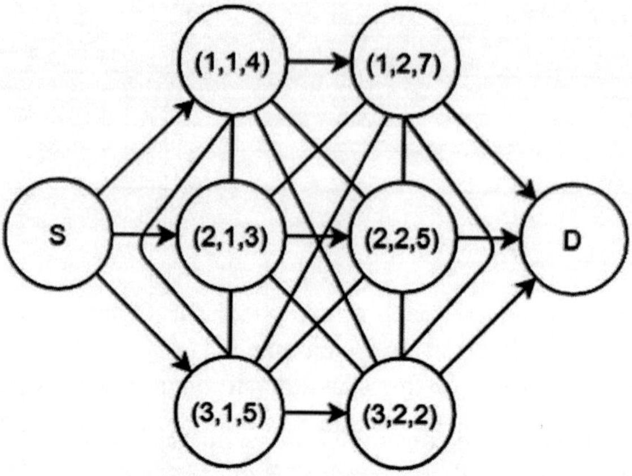


**Fig. 2** Graphical representation of the DJSP

where $W_{i,j}$ is the processing time of the $i$th operation of the $j$th job. The workload of each job is made equal to the sum of all its operation's processing time as described in Eq. (1). Then, the jobs are ranked in the descending order of their workloads. Given there are $f$ factories, the first $n$ jobs are allocated to factories $1 \ldots n$, respectively.

In the second phase of the allocation, to assign the next job, the workloads of all the machines in each factory are sorted in descending order and the operations of the job to be assigned are sorted in ascending order. These sorted workloads and operation processing times are added up to get the new workloads of the machines, and the maximum among these new workloads is calculated for each factory. The job is then assigned to the factory with a minimum new workload. This procedure repeats, and the workloads keep getting updated until all the jobs are assigned.

The pseudocode below shows the main steps involved in the second phase of job allocation:

1. Begin
2. For each job which is yet to be assigned to a factory do:
3. Sort the operations of the selected job in ascending order
4. Sort the workloads of all the machines in each factory in descending order
5. Initialize NewMinMaximumWorkload = MathMax
6. selected factory = −1
7. For each factory $f$, do:
8. Add the sorted workloads of the machines and sorted operations of the job as two vectors to get the new optimal workloads assuming the job is assigned to the factory $f$.
9. Find the maximum new workload for the factory $f$ among all its machines, let's say it is Fmax
10. Update NewMinMaximumWorkload = Minimum(NewMinMaximumWorkload, Fmax)
11. If (NewMinMaximumWorkload = Fmax) Then
12. selected factory = $f$

**Table 2** Job matrix

| Job | Operation 1 | Operation 2 | Total workload | Initial rank |
|---|---|---|---|---|
| 1 | 12 | 6 | 18 | 2 |
| 2 | 11 | 10 | 21 | 1 |
| 3 | 8 | 5 | 13 | 3 |

13. The job is assigned to the selected factory
14. Update the workloads of all the machines in the selected factory.
15. End.

Given there are two factories, two machines, and for the job matrix given in Table 2, by the allocation algorithm, Job 2 is assigned to factory 1, and Jobs 1, 3 are assigned to factory 2.

## *3.4 Ant Colony Algorithm*

The traditional ACO algorithm is applied on the graph generated for each factory after the balanced job allocation is done. A fixed number of ants are spawned at the source node of the input graph for the factory, and pheromone values of all the edges are initialized to 0. The ants make the decision of choosing the next node to move to based on the weight of the edge and the pheromone value of the edge to that node using a randomly spun roulette wheel which gives more weightage to the edge having a lesser weight and more pheromone value. An ant is permitted to move from its current Node N1 $(i, j, w_j)$ to a neighboring Node N2 $(k, m, w_m)$ if the following conditions are met:

- The ant should not have previously visited N2.
- If $i = k$, then $m = j + 1$, i.e., it can visit the next operation of the same job.
- If $i \neq k$, then the ant must have visited all the previous operations of the job $k$, i.e., all Nodes $N(k, p, w_p)$ such that $p < m$, must be visited by the ant.
- If N2 is the destination node.

The pheromone value of the edge along which the ant moves is increased. The pheromone values of edges are reduced after every movement of the ants due to evaporation.

This process is repeated until all the ants have visited all the nodes which complete one cycle giving us a possible schedule from each ant. This cycle is repeated multiple times but spawning new ants again on the existing graph with the pheromone values updated by the previous set of ants. Such multiple cycles are repeated a fixed number of times to get the final optimal schedule from the ant which recorded the minimum makespan among all the cycles. This is done for each factory to get the global

**Table 3** Table of results

| No. of jobs | No. of factories | Avg max makespan (existing) | Avg max makespan (proposed) | Difference (%) |
|---|---|---|---|---|
| 7 | 2 | 376.13 | 357.80 | 4.87 |
| 12 | 3 | 384.87 | 368.87 | 4.16 |
| 15 | 4 | 404.53 | 387.13 | 4.30 |

makespan which is the maximum makespan among all the minimum makespan of each factory.

### 3.5 *Modified Ant Colony Optimisation*

The solution given by the ant colony algorithm is fed into the MACO algorithm mentioned in [11]. The MACO uses a procedure called local search to get a better solution using the existing solution by scheduling an already scheduled operation to a machine that is free at that time. The improvement is first done by taking the job which was last scheduled within a factory by the ant colony algorithm. The constraints to schedule this operation are mentioned in [11]. The DJSP constraints are also respected by the MACO. Good results were observed as stated in [11]; hence, we incorporate MACO at the last stage of our model to give the final schedule which gives an optimal global makespan.

## 4 Results and Analysis

The job allocation algorithm used gave better results. The maximum makespan among all the factories using the innovative distribution was lesser compared to the original distribution methodology, as the workloads were evenly distributed among all the factories in the job allocation phase. In order to assess the performance of our model, we used random values as work times for the jobs, compared the maximum makespan of our algorithm to the existing algorithm, and then applied the MACO algorithm having the same parameters of ant colony optimization. We performed this comparison with different numbers of factories and jobs (Table 3).

Our model gives an improvement of approximately 4% (Figs. 3 and 4).

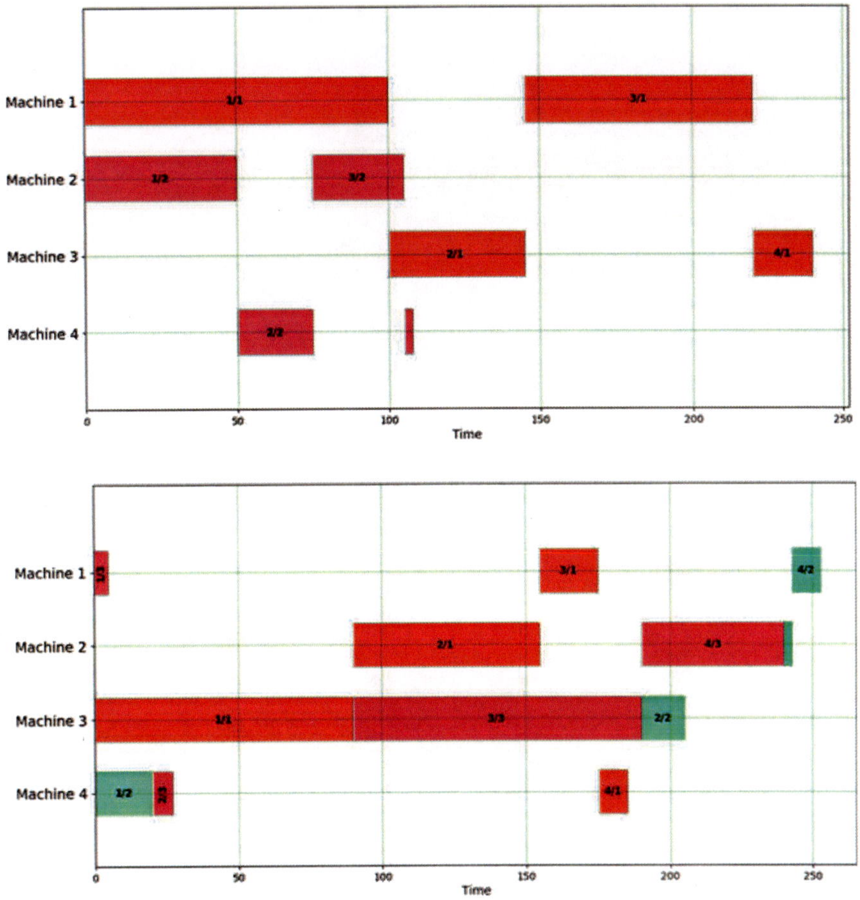


**Fig. 3** Gantt Chart of job scheduling obtained using the existing allocation algorithm followed by MACO (maxmakespan 253 units)

## 5 Conclusion

The distributed job scheduling problem is an important problem for the manufacturing sector as well as operations research. While existing models do exist, they are subject to further optimization. Our proposed model which uses an allocation algorithm that balances the workloads among different factories performs better than the existing model in all the cases that we experimented for. This improvement reduces the maximum makespan by 4% on randomly generated datasets. We believe there is potential for further improvement to the performance of the algorithm by experimenting with the different parameters, such as the initial pheromones and the rate of modification of the pheromones.

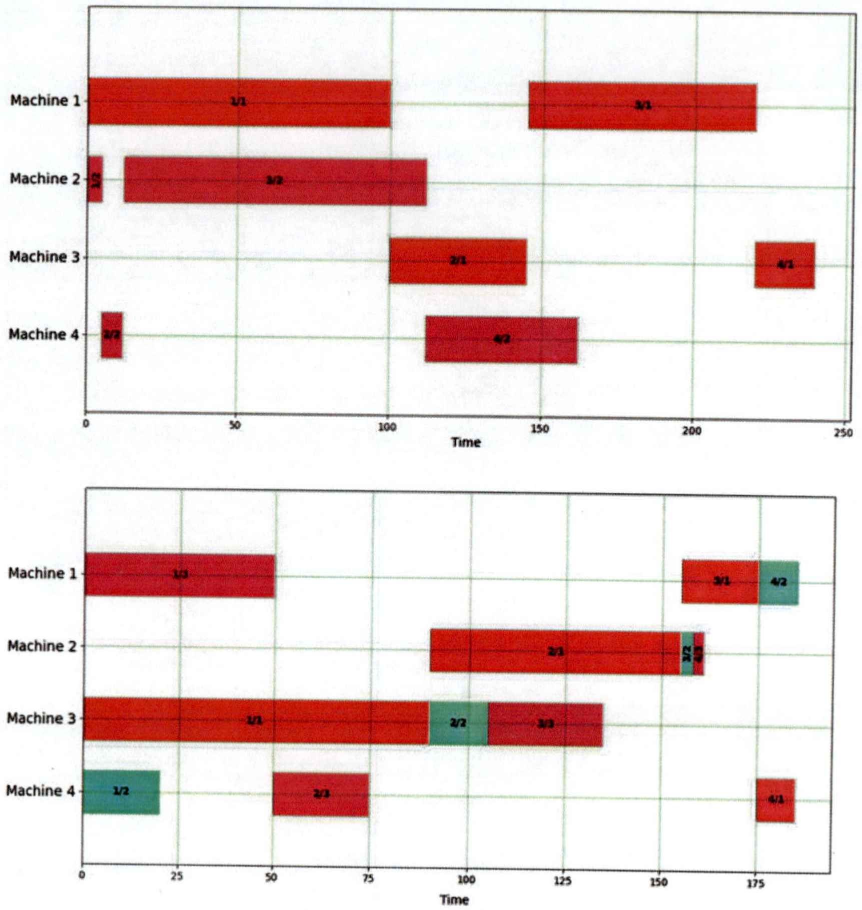


**Fig. 4** Gantt Chart of job scheduling obtained using the proposed allocation algorithm followed by MACO (maxmakespan 240 units)

## References


1. Colorni, A., Dorigo, M., Maniezzo, V., Trubian, M.: Ant system for job-shop scheduling. Belg. J. Oper. Res. Stat. Comput. Sci. **34**(1), 39–53 (1994)
2. Guo, S., Luo, W., Xu, W., Wang, L.: Research on distributed flexible job shop scheduling problem for large equipment manufacturing enterprises considering energy consumption. In: 2020 39th Chinese Control Conference (CCC) (2020), pp. 1501–1506. https://doi.org/10.23919/CCC50068.2020.9189640
3. Davis, L.: Job shop scheduling with genetic algorithms. In: Proceedings of an International Conference on Genetic Algorithms and Their Applications, vol. 140. Carnegie-Mellon University, Pittsburgh, PA (1985)
4. Zhang, G., Sun, J., Liu, X., Wang, G., Yang, Y.: Solving flexible job shop scheduling problems with transportation time based on improved genetic algorithm. Math. Biosci. Eng. **16**(3), 1334–1347 (2019). https://doi.org/10.3934/mbe.2019065

5. Awerbuch, B., Kutten, S., Peleg, D.: Competitive Distributed Job Scheduling. Association for Computing Machinery (1992)
6. Jia, H., Fuh, J.Y., Nee, A.Y., Zhang, Y.: Web-based multi-functional scheduling system for a distributed manufacturing environment. Concurr. Eng. **10**(1), 27–39 (2002)
7. Jia, H., Nee, A.Y., Fuh, J.Y., Zhang, Y.: A modified genetic algorithm for distributed scheduling problems. J. Intell. Manuf. **14**(3–4), 351–362 (2003)
8. Jia, H., Fuh, J.Y., Nee, A.Y., Zhang, Y.: Integration of genetic algorithm and Gantt chart for job shop scheduling in distributed manufacturing systems. Comput. Ind. Eng. **53**(2), 313–320 (2007)
9. Naderi, B., Azab, A.: Modeling and heuristics for scheduling of distributed job shops. Expert Syst. Appl. **41**(17), 7754–7763 (2014). ISSN 0957-4174
10. De Giovanni, L., Pezzella, F.: An improved genetic algorithm for the distributed and flexible job-shop scheduling problem. Eur. J. Oper. Res. **200**(2), 395–408 (2010). ISSN 0377-2217
11. Chaouch, I., Driss, O.B., Ghedira, K.: A modified ant colony optimization algorithm for the distributed job shop scheduling problem. Procedia Comput. Sci. **112**, 296–305 (2017). ISSN 1877-0509
12. Zhang, J., Ding, G., Zou, Y., Qin, S., Fu, J.; Review of job shop scheduling research and its new perspectives under Industry 4.0. J. Intell. Manuf. **30** (2019). https://doi.org/10.1007/s10845-017-1350-2

# Detection of Copy–Move Image Forgery Applying Robust Matching with K-D Tree Sorting

**Partha Chakraborty, Sabakun Nahar Tafhim, Mahmuda Khatun, Md. Abu Sayed, Sabab Zulfiker, Priyanka Paul, Md. Farhad Hossain, and Tanupriya Choudhury**

**Abstract** Digital images contribute significantly to the field of visualization. Using stronger technology, digital image forgery is easier. The most common method of image forgery is to re-create a portion of a person's location or to conceal a portion of an image. In our paper, we worked on detecting region duplication forgery using COMOFORD databases by utilizing the discrete cosine transform (DCT), k-dimensional tree (k-d tree) for sorting efficiently, and a robust matching method. Here, the size of the block will be $16 \times 16$, and it will be divided into four blocks. This study can detect forged portions for PNG images with better performance by

P. Chakraborty (✉) · S. N. Tafhim · M. Khatun · Md. Abu Sayed
Department of Computer Science and Engineering, Comilla University, Cumilla 3506, Bangladesh
e-mail: partha.chak@cou.ac.bd

S. N. Tafhim
e-mail: sn.tafhim@gmail.com

M. Khatun
e-mail: mahmuda@cou.ac.bd

Md. Abu Sayed
e-mail: sayeed.cse.bd@gmail.com

S. Zulfiker
Department of Computer Science and Engineering, Daffodil International University, Dhaka, Bangladesh
e-mail: sabab.rumc@gmail.com

P. Paul · Md. Farhad Hossain
Department of Statistics, Comilla University, Cumilla 3506, Bangladesh
e-mail: priyanka.paulbd@gmail.com

Md. Farhad Hossain
e-mail: farhad390ju@gmail.com

T. Choudhury
Department of Informatics, School of Computer Science, University of Petroleum and Energy Studies, Dehradun, India
e-mail: tanupriya1986@gmail.com

 
D. Gupta et al. (eds.), *Pattern Recognition and Data Analysis with Applications*, Lecture Notes in Electrical Engineering 888,
https://doi.org/10.1007/978-981-19-1520-8_22

highlighting the images with a quality factor of 0.5 and a threshold value of 10, as well as gives good results for JPEG images.

**Keywords** Image detection · Robust matching · K-D tree sorting · Image forgery · DCT

## 1 Introduction

We are living in such an era where people are able to share any kind of information with each other, located on every side of the world, with the least amount of cost by using the Internet. Widespread accessibility of technology and the negligible cost of equipment make it very convenient for daily life. The photography system provides a sophisticated medium for image modification with a top-notch appearance. For this reason, we are highly at risk of facing numerous threats related to our identity, financial security, as well as many national safety issues. We cannot completely avoid or be free of these dangers. People can share images on a variety of different Internet platforms. Digital images are assigned to various information and can be skillfully modified as a result of a limited but adequate protection system. The easy accessibility of editing software tools on all devices means that image editing has become an easy-going job at present. Edited images alter the original affection provided by the real image, which may contain threats to information security for people.

Many procedures exist to detect forgery in images, making it difficult to find more practical and perfect implementation procedures. Which algorithm has the best performance? It could have a high rate of false-positive detection. Furthermore, runtime capabilities are a critical component in determining how efficiently an algorithm works and ensuring the algorithm's usability. They are different in performance. Some provide good real-time performance, while others provide better results through modifications, and still others detect different geometrical modifications. The purpose of the research work is to inspect the existing forgery detection procedures for images for complexity reduction [1].

## 2 Literature Review

For image forgery detection, many researchers have done related work at different times. This paper established a combined procedure for copy–movement forgery supported by scale invariant transformation features as well as the Fourier-Mellin technique [2]. This paper worked on a detection method named blind copy–move forgery by deploying the KS and SVD testing methods [3]. They proposed a DCT and cellular automata-based robust copy–movement fraud descriptor in [4]. Parveen et al. [5] suggested block-based copy–move picture fraud detection using DCT. Prakash et al. [6] worked on detecting copy–move forgeries using AKAZE and SIFT key point extraction. Hegazi et al. [7] discovered a density-based clustering approach with satisfied outlier dismissal based on a copy-moving forgery approach. Tan et al. [8] conducted a review of digital image copy forgery detection for localization using passive procedures. The paper worked on a copy-moving forgery approach based on modified key point extortion and pairing [9]. Mushtaq and Mir [10] worked on copy movement forgery detection for pictures. Mahmood et al. [11] suggested a robust stationary wavelet and DCT approach in order to detect and localize copy movement forgery. Ouyang et al. [12] developed a comprehensive copy movement forgery approach by combining Zernike moments and the pyramid model. Emam et al. [13] developed a two-stage key point detection system capable of detecting region duplication forgeries in digital images. Rasse [14] examine the detection of digital picture splicing forgeries using illumination color estimation. Kaur and Sharma [15] are working on improving the prevention of duplicate fraud technique. In this work, the author has experimented with the DCT and wavelet transformations [16]. This paper has conducted extensive research into various types of picture forgeries [17]. This paper investigated features to detect forgeries using a copy-moving forgery [18]. These study investigated the identification of copy–move frauds in digital pictures [19]. The emphasis of the research was on passive forensics for copy–move forgeries utilizing a DCT-based technique [20]. They investigated the copy–move forgery approach using cellular automata in [21]. This paper has worked on copy–move detection by merging cellular automata with regional binary patterns [22]. Create a system that uses template and HOG features for object detection in [23]. Create a method for robots to compute the degree of visual focus of human attention [24, 25], which tries to find object instances in unknown image sources. Make a system that uses face detection to obtain automatic student attendance [26].

## 3 Methodology

The proposed system used in this study is described in Fig. 1.

### *3.1 Taking Input Image*

An input picture is a pixel-by-pixel modification of a source images. Detection of that input image in Fig. 2 has a higher value (less match) and a lower value (stronger match). In the input image, the threshold number of pitches appears to be copied together for it to be considered a forged region by the algorithm.

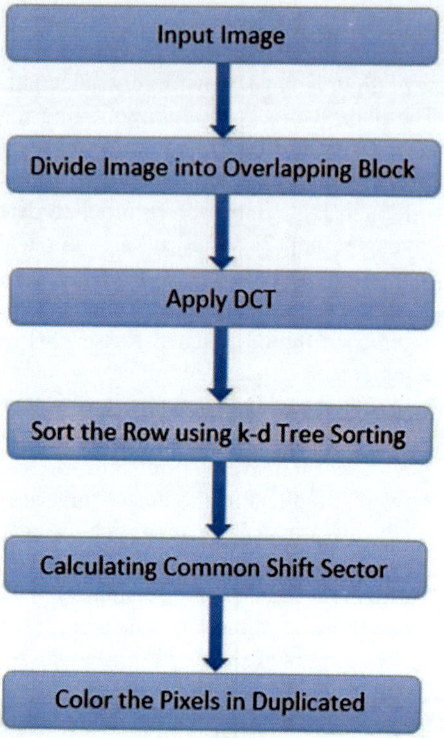


**Fig. 1** Proposed system model

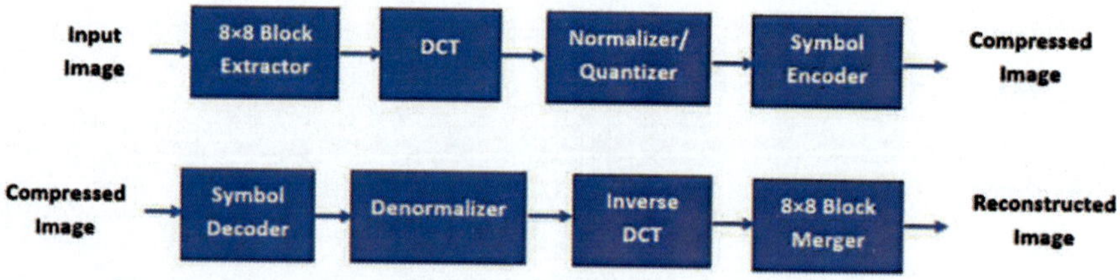


**Fig. 2** Image compression block diagram using DCT

## 3.2 *Divide Image into Overlapping Block*

In order to detect copy–move fraud in pictures, the standardized shifting matrix counter $C$ must be increased by one for each identical pairing of blocks:

$$C_{(d1,d2)} = C_{(d1,d2)} + 1$$

Displacement vectors are designed as well as in the ordered matrix $A$. Counter $C$ is augmented in every combination of progressively fitting rows. Before this process begins, the displacement vector $C$ is reset to null. Finally, in the pairing procedure, timer $C$ represents the frequency through which separate scaled displacement vectors emerge. The program then searches for any standardized displacement vectors—$d(1)$, $d(2)$ ... $d(n)$, wherein frequency surpasses a consumer threshold $K$:

$$C_{(d)}^{(r)} > K$$

For all $r = 1, n$.

The size of the smallest part that can be recognized by the method is proportional to the threshold $K$ value. Larger numbers may cause the algorithm to overlook some blocks that aren't quite so tightly matched.

### 3.3 DCT Calculation

The DCT is a transform that is connected to the Fourier series. The Fourier series constants of a periodically and symmetrically long sequence are frequently related to the DCTs. There are eight standard DCT variations, with four of them being the most prevalent. The most common discrete cosine transform variation is the kind-II DCT, which is often referred as "essentially the DCT." The type-III DCT, as its inverse, is indeed referred to as the inverse DCT or the IDCT. Because of its solid power density, the DCT, and particularly the DCTII, is widely used in signal and image processing, particularly for overfitting compression. In typical applications, a significant amount of channel estimation is contained in a small group of low DCT processes [2].

### 3.4 K-D Tree Sorting Calculation

K-D trees are a suitable data structure for a wide range of purposes, particularly searches involving connecting multidimensional search keys. It has been discovered that the other methods' weakest point is that if too many blocks are incorrectly paired, this leads to incorrect hypotheses on the dominant shift vectors, rendering the result only usable large images as well as on minor images.

This performance completely changes when using a k-d tree. Here, the moment features detect operations in all images. Only, FMT and DCT crops have better detection in large image in Fig. 3. Among the color-based methods, COLOR 3 is the most effective. The structures of COLOR 1 and COLOR 2 have very high false-positive rates and cannot be constantly used for detection. Actually, the more balanced distance computation of the k-d tree makes the nature of most feature vectors better (Fig. 4).

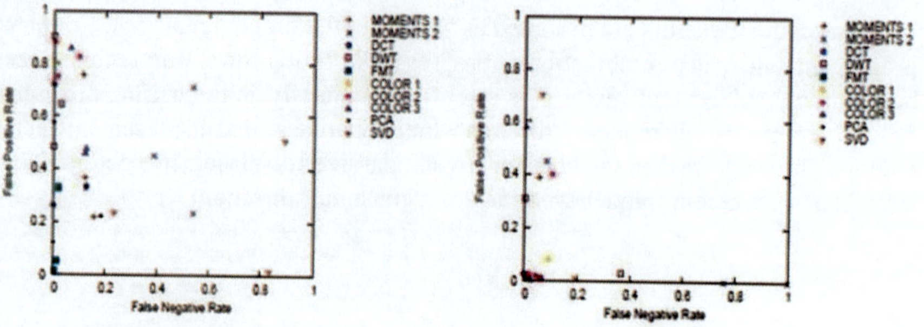


(a) Large images (b) Small images

**Fig. 3** Feature test K-D tree sorting

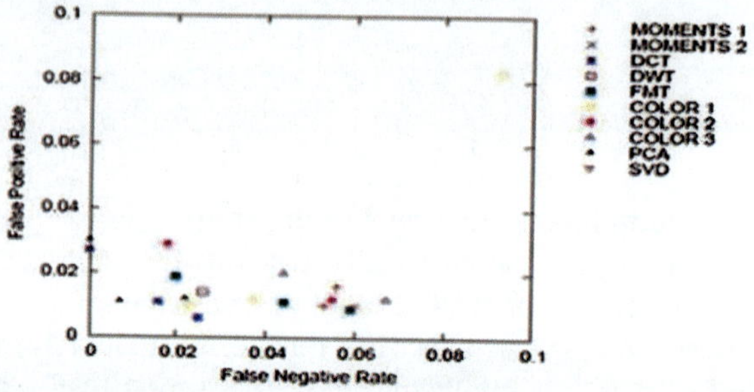


**Fig. 4** Feature test k-d tree representation with small images

In Fig. 3, every feature can be used to detect copy–move forgeries with a k-d tree. Only, MOMENTS 2 and COLOR 3 presented difficulties with a single small image. Upon closer examination, one can see that the error rates of the small images exhibit a very low false-positive rate for all methods [3].

## 3.5 Common Shift Sector Analysis

All AC frequency components for $16 \times 16$ blocks are 2.5 times greater on average than for $8 \times 8$ blocks, according to experiments, and the DC factor is two times as large. As a consequence, in the following of the $16 \times 16$ blocks, the frequency modulation matrix (for the $Q$-factor $Q$) required to compute the frequency domain has a different shape in Fig. 5.

$I$ is an $8 \times 8$ unit matrix, and $q_{ij}$ is a standard JPEG normalization matrix with a center frequency of $Q$ where all components are equivalent to one. This is understandable because it is an impromptu test, yet the matrix functioned well throughout the practical tests and subtle tweaks to the matrix had a small impact on the outcomes. We stopped looking into the quantization matrix selection [4].

$$Q_{16} = \begin{pmatrix} Q'_8 & 2.5q_{18}I \\ 2.5q_{81}I & 2.5q_{88}I \end{pmatrix}, where\ Q'_8 = \begin{pmatrix} 2q_{00} & 2.5q_{12} \cdots\cdots & 2.5q_{18} \\ 2.5q_{21} & 2.5q_{22} \cdots\cdots & 2.5q_{28} \\ \cdots & \cdots\cdots & \cdots \\ 2.5q_{81} & 2.5q_{82} \cdots\cdots & 2.5q_{88} \end{pmatrix}$$

**Fig. 5** Common shift sector analysis

## *3.6 Forgery Detection Using Pixels*

Full steps in Fig. 6 are described below:

I. Image splicing is a technique that uses a mixture of multiple or perhaps more mutual photos to generate a fake image, or a method that uses a mixture of two or perhaps more mutual photos to make a convincing photograph.

II. The images are modified less in image retouching. It only highlights a few of the image's many facets.

III. The replica movement forgery is one of the best solid forgeries. This is the most major form of picture tampering, in which information is added or removed by coating a section of an image. A copy–move operation is one in which a segment of a picture is cut and pasted into another segment of a comparable image. In the following, there is an overview of copy–move forgery: (1) Copy–move without reflection; (2) Copy–move with different scaling; (3) Copy–move with different scaling; (4) Rotate the copy–move. According to the literature review, copy–move and forgery sequence information are classified into two types. (1) A method based on blocks, (2) A method focused on key points.

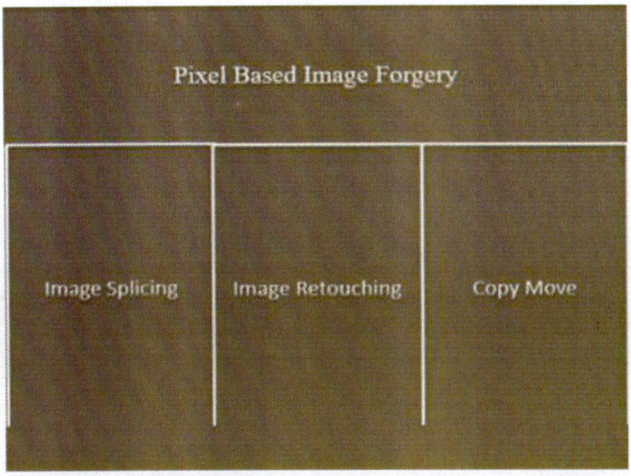


**Fig. 6** Forgery detection using pixels

### 3.7 Color the Pixels in Duplicated Region

The following requirements for the detection algorithm can be directed at:

I. An approximate comparison of image object segments must be possible with the detecting technique.
II. There must work for a fair length of time and produce a minimal false positive rate, such as locating erroneously linked regions.
III. It is also worth mentioning that instead of a mixture of tiny patches or single pixels, the fabricated portion will most likely be a linking element.

### 3.8 Exact and Robust Match

The user requires the smallest segment size that should be considered for the match at first. Assume this segment is a *BB* pixel square. A matrix has rows which are lexicographically sorted (as *BB* numeric data points) in order to classify the identical rows. In $MN\log_2(MN)$ phase, this can be done. $MN\log_2(MN)$ phases can be used to accomplish this. The matching rows in Fig. 7 can be found by searching for the sorted matrix *A* presenting *MN* rows for two comparable concurrent rows.

Robust match detection is the same as targeted search detection in that we sort and compare the blocks' robust representation, which includes quantized DCT coefficients, rather than their pixel representation. The algorithm also considers each matching block pair's shared positions and only produces an exact block pair when there are several other corresponding combinations in about the same reciprocal location (shift vector). The technique maintains the positions of matching blocks in a different section. As an example—the dimensions of a block's uppermost left pixel are used to determine its own location and increase a displacement vector counter *C* if two following rows of the ordered matrix *A* are detected. Let's denote the locations of the two identical blocks $(i1, i2)$ and $(j1, j2)$, respectively.

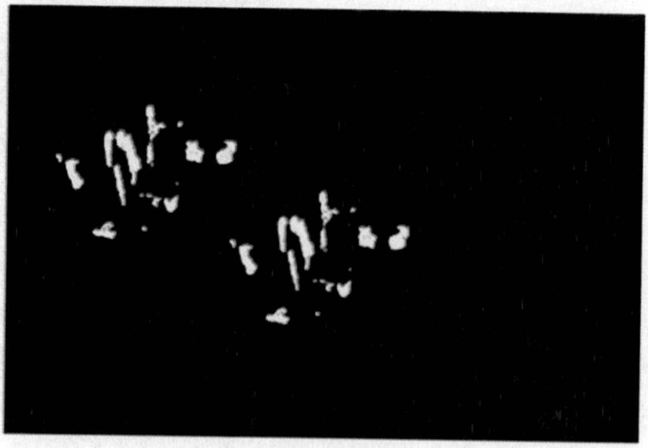

**Fig. 7** Results of the experimental algorithm

$$D = (d1, d2) = (i1 - j1, i2 - j2)$$

is the displacement vector between the two matched blocks.

The displacement vectors $d$ is regularized since the displacement vectors—$d$ as well as $d$ correlated to the same displacement.

## 4 Experimental Details

In this experiment, we used an AMD A8 processor. It is a very simple processor but works smoothly. For problem solving, we use a 64-bit operating system, 2.00 GHz graphics, and the MATLAB software; the three parts of the coding section are as follows:

I. In the first section, a color image known as a suspected image is printed. The first section is linked to the second and third sections.
II. In the second section, overlapping blocks are divided using a robust matching DCT matrix design. The size of the block will be a $16 \times 16$ matrix, and it will be divided into four blocks and a compute DCT matrix.
III. In the third part, we detect a forged region. Before computing the shift vector, the data are sorted using the k-d tree. We convert the detected part into an RGB color image.

In the test results, the sample result is displayed shown in Figs. 8 and 9. Here, three test results are found by applying the above methods. The image has been tempered with a quality factor and a threshold for locating common shift vectors. This JPEG image was divided into overlapping blocks, DCT'd, and then sorted using a k-d tree. The shift vector is then computed. In this PNG-formatted image, two common shift vectors are obtained.

The detection of the input image has a higher value (less match) and a lower value (stronger match) in each of the three input images. The threshold number of pitches appears to be copied together in the given picture of this computation to consider it a forged region divided into overlapping blocks. Here, we use MATLAB software.

The size of the block in the second portion of the input picture will be a $16 \times 16$ matrix, split into four blocks. This work is able to detect forged portions after modification. In this case, we are going to use a quality factor of 0.5 and a threshold of 10. We computed the DCT matrix here.

Here, we detect forged regions. The k-d tree sorts and then computes the shift vector. We make an RGB image out of it shown in Fig. 9. In this PNG-formatted image, two common shift vectors are obtained.

Two common shift vectors are obtained in this image. In the detected forged region of the image, a color is assigned to the forged region. The image is highlighted with a quality factor of 0.5 and a threshold value of 10, which gives better performance in detection. We used the COMOFORD database in our research, which included 60 tempered photos. In these photos, this technique correctly and efficiently detects

**Fig. 8** Tempered image and highlighting with quality factor and threshold for finding common shift vector

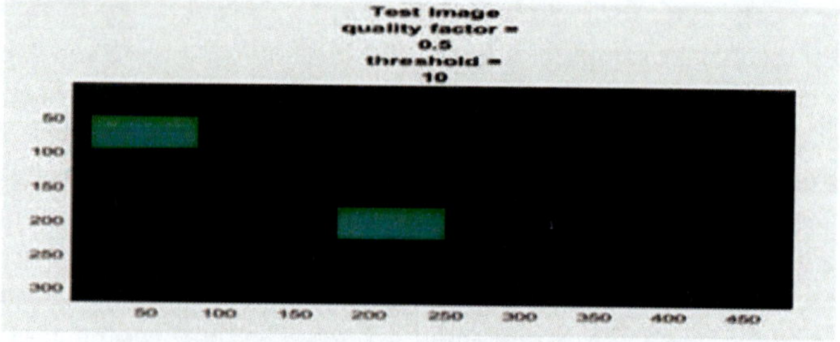


**Fig. 9** Forged region detection

most copy–move forgeries. We can see that the previous approach failed to detect fraud in the jpeg format of images. However, there is another method to detect forgery in jpeg format images. DCT may also be used to identify image forgeries, with good results for jpeg images, as seen in the figure. Detecting forged digital photographs is being accomplished in a variety of ways. In this case, we discovered a region of copy picture forging, which is a technique for duplicating and pasting a section of an image.

# 5 Discussion

Copy–move forging is one of the most frequent counterfeit techniques. Several researchers have defined a variety of methods for detecting altered photos.

However, before being pasted, the duplicated portions are sometimes rotated or flipped. In our paper work, we adopted an effective method for digital images in order to perceive the identical area in the image. To begin, the image is subdivided into adjoining rectangular blocks, which are then used to generate overlapping blocks. Second, the DCT transformation is used to restrict the search area and makes the search unit more resistant to post-processing operations like compression and rotation. Finally, the feature vectors are sorted using a k-d tree after they have been transformed. The output is shown in Figs. 10 and 11.

**Fig. 10** Tempered image and highlighting with quality factor and threshold for finding common shift vector

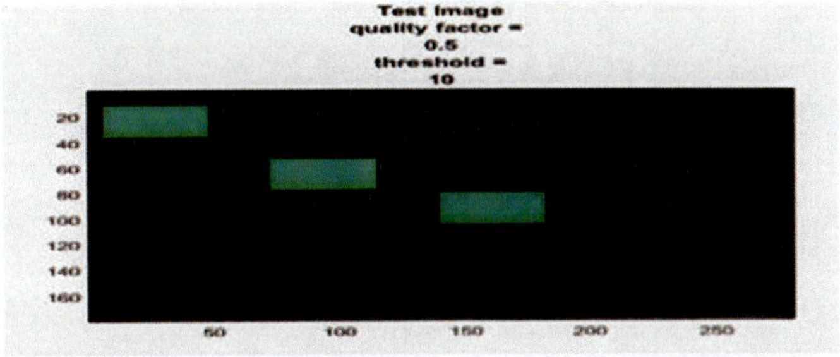


**Fig. 11** Forged region detection

We are now working on detecting copy–move image fraud utilizing robust matching and k-d tree sorting. It only works with compressed photos; original images are not supported. To improve query performance and accuracy, we want to modify the data structures even more. Even if the pasted region has been rotated or translated, this method still works.

## 6 Conclusion

On the COMOFORD database, we utilized a DCT transformation technique and a robust matching method to identify region replication forgery and obtain efficient results. The obtained result is shown in Figs. 12 and 13.

Our future target is to work on our own large dataset and modify the DCT and k-d tree sorting data structures for jpeg images to acquire noticeable results. We will also try to modify the model in such a way that it can work with the original images.

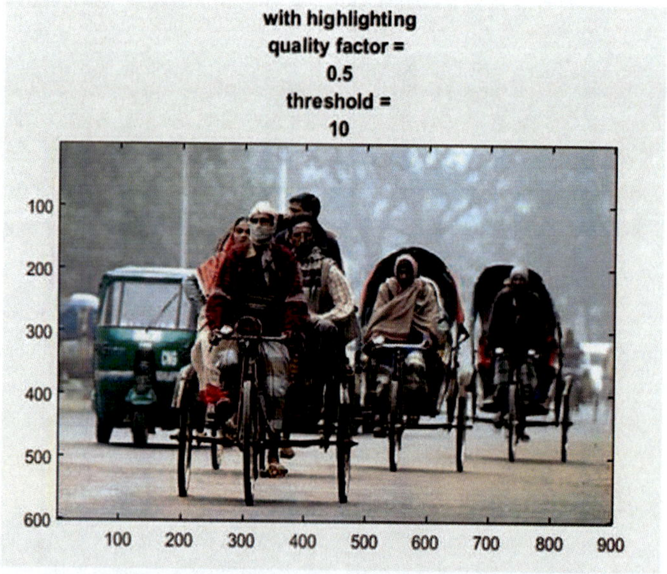


**Fig. 12** Tempered image and highlighting with quality factor and threshold for finding common shift vector

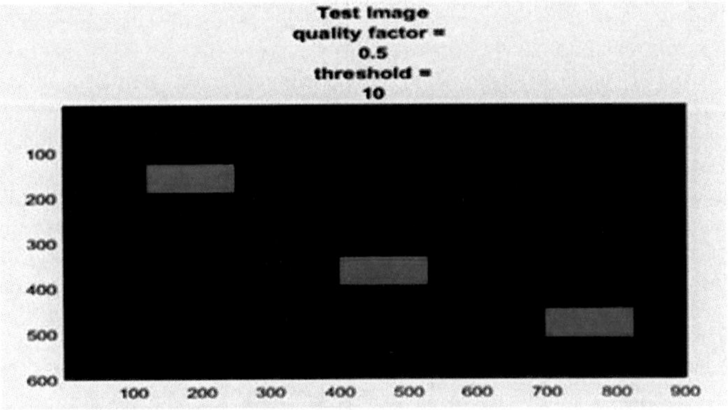


**Fig. 13** Forged region detection

## References


1. Head, J., Lai, Y.-K.: Image forgery detection (2015)
2. Meena, K.B., Tyagi, V.: A hybrid copy-move image forgery detection technique based on Fourier-Mellin and scale invariant feature transforms (2020)
3. Ahmed, B., Gulliver, T.A., Zahir, S.A.: Blind copy move forgery detection using SVD and KS test (2020)
4. Gani, G., Qadir, F.: A robust copy move forgery detection technique based on discrete cosine transform and cellular automata (2020)
5. Parveen, A., Khan, Z.H., Ahmad, S.N.: Block-based copy–move image forgery detection using DCT (2019)
6. Prakash, C.S., Panzade, P.P., Om, H.: Detection of copy move forgery using AKAZE and SIFT key point extraction (2019)
7. Hegazi, A., Taha, A., Selim, M.M.: An improved copy move forgery detection based on density based on clustering and guaranteed outlier removal (2019)
8. Tan, W., Wu, Y., Wu, P., Chen, B.: A survey on digital image copy move forgery localization using passive techniques (2019)
9. Yang, H.Y., Qi, S.R., Niu, Y., Niu, P.P., Wang, X.Y.: Copy move forgery detection based on adaptive keypoints extraction and matching (2019)
10. Mushtaq, S., Mir, A.H.: Image copy move forgery detection (2018)
11. Mahmood, T., Mehmood, Z., Shah, M., Saba, T.: A robust technique for copy move forgery detection and localization in digital images via stationary wavelet and discrete cosine transform (2018)
12. Ouyang, J., Liu, Y., Liao, M.: Robust copy-move forgery detection method using pyramid model and Zernike moments (2018)
13. Emam, M., Han, Q., Zhang, H.: Two-stage key point detection scheme for region duplication forgery detection in digital image (2017)
14. Rasse, S.G.: Review of detection of digital image splicing forgeries with illumination color estimation. Int. J. Emerg. Res. Manag. Technol. (2017)
15. Kaur, A., Sharma, R.: Optimization of copy-move forgery detection technique. Int. J. Adv. Res. Comput. Sci. Softw. Eng. (2017)
16. Telagarapu, P., Naveen, V.J., Prasanthi, A.L., Santhi, G.V.: Image compression using DCT and wavelet transformations (2017)
17. Chakraborty, P., et al.: A human-robot interaction system calculating visual focus of human's attention level. IEEE Access **9** (2021)

18. Christlein, V., Riess, C., Angelopoulou, E.: A study on features for the detection of copy move forgeries. In: Sicherheit (2017)
19. Chakraborty, P., Nawar, F., Chowdhury, H.A.: Sentiment analysis of Bengali Facebook data using classical and deep learning approaches. In: Innovation in Electrical Power Engineering, Communication, and Computing Technology, pp. 209–218. Springer, Singapore (2022)
20. Hasan, M.R., et al.: Reliable identity management system using Raspberry Pi. In: 2020 2nd International Conference on Sustainable Technologies for Industry 4.0 (STI). IEEE (2020)
21. Chakraborty, P., Sultana, S.: IoT-based smart home security and automation system. In: Micro-Electronics and Telecommunication Engineering, pp. 497–505. Springer, Singapore (2022)
22. Feroz, M., Sultana, M., Hasan, M., Sarker, A., Chakraborty, P., Choudhury, T.: Object detection and classification from a real-time video using SSD and YOLO models. In: Computational Intelligence in Pattern Recognition, pp. 37–47. Springer, Singapore (2022)
23. Sultana, M., Ahmed, T., Chakraborty, P., Khatun, M., Hasan, M.R., Uddin, M.S.: Object detection using template and HOG feature matching. Int. J. Adv. Comput. Sci. Appl. **11**(7), 233–238 (2020)
24. Chakraborty, P., Yousuf, M.A., Rahman, M.Z., Faruqui, N.: How can a robot calculate the level of visual focus of human's attention, pp. 329–342 (2020)
25. Muzammel, C.S., Chakraborty, P., Akram, M.N., Ahammad, K.M.: Zero-shot learning to detect object instances from unknown image sources. Int. J. Innov. Technol. Explor. Eng. **9**(4), 988–991 (2020)
26. Chakraborty, P., Muzammel, C.S., Khatun, M., Islam, S.F., Rahman, S.: Automatic student attendance system using face recognition. Int. J. Eng. Adv. Technol. **9**(3), 93–99 (2020)

# Recognize Meaningful Words and Idioms from the Images Based on OCR Tesseract Engine and NLTK

**Partha Chakraborty, Md. Rakib Mia, Humayun Kabir Sumon, Aditi Sarker, Al Imtiaz, Md. Mahbubur Rahman, Mohammad Abu Yousuf, and Tanupriya Choudhury**

**Abstract** OCR means optical character recognition, which is a text extraction technology that works with photos, scanned data, and PDF documents. By extracting text data, OCR systems typically convert non-editable, non-searchable documents into editable, searchable files. As a result, information finding and identification from digitized files is simplified. R bindings are provided by the Tesseract package. Tesseract is a strong optical character recognition (OCR) engine with over 100 languages supported. The engine is highly customizable, allowing you to fine-tune the detection algorithms to achieve the best possible results. With the help of Tesseract OCR technology, a method for extracting texts from photos was created. Any image can be used as input for the proposed OCR system, which converts it into a searchable

P. Chakraborty (✉) · Md. Rakib Mia · H. K. Sumon · A. Sarker
Department of CSE, Comilla University, Cumilla 3506, Bangladesh
e-mail: partha.chak@cou.ac.bd

Md. Rakib Mia
e-mail: rj.rakib2525@gmail.com

H. K. Sumon
e-mail: kabirsumon73@gmail.com

A. Sarker
e-mail: aditisarker407@gmail.com

A. Imtiaz
Department of CSE, University of Information Technology and Sciences, Dhaka, Bangladesh
e-mail: al.imtiaz@uits.edu.bd

Md. Mahbubur Rahman
Software Engineer, Crowdrealty, Tokyo, Japan
e-mail: mahbuburrahman2111@gmail.com

M. A. Yousuf
Institute of Information Technology, Jahangirnagar University, Savar, Dhaka 1342, Bangladesh
e-mail: yousuf@juniv.edu

T. Choudhury
Department of Informatics, School of Computer Science, UPES, Dehradun, India
e-mail: tanupriya1986@gmail.com



D. Gupta et al. (eds.), *Pattern Recognition and Data Analysis with Applications*,
Lecture Notes in Electrical Engineering 888,
https://doi.org/10.1007/978-981-19-1520-8_23

text document. Furthermore, this system can search for words within the generated text and display the Bengali meaning terms. It finds the words and lines first, then identifies the words, then the static character classifier classifies the character, then does analysis, and finally an adaptive classifier. It is a framework which also includes a natural language processing approach for classifying commonly used terms with Bangla meanings from the output text, in addition to OCR.

**Keywords** Tesseract · OCR · NLP · Bangla translation · NLTK

## 1 Introduction

In the field of information technology, image processing has become a hot topic. It falls within the category of digital signal processing. One of the most common applications of image processing is optical character recognition (OCR). Tesseract is an open-standard OCR engine that can adapt to a variety of scripts and languages. Aside from providing a corpus of text, very little customization is necessary for a different language [23]. When we introduce an object to ourselves, our brains or our generic identification system begins retrieving important characteristics of the object, such as color, width, form, and scale. In the memory region, certain attributes are saved. The brain is currently attempting to find the closest match for those words.

Attributes are retrieved from the whole collection of objects already included in it. This is what we should call a regular library. When it identifies a match, it returns the associated entity or signal from the standard library as the final result. Character interpretation looks to be a simple operation for humans, but teaching a system to evaluate and eventually recognize a character accurately is a difficult undertaking. OCR is such an invention that gives computers the ability to use their vision to extract data from images in order to extract useful information and makes it editable by computers. OCR technology allows scanned pictures of text or symbols (for example, a dictionary page) to be converted into text that a computer model can recognize and modify. The most well-known example is the ability to understand a paper document on a device and subsequently edit it with common word processors like Microsoft Word. Furthermore, the OCR technique can be used in a variety of ways, as well, such as part of a large-scale system integrating recognition methods. For example, the recognition system of number plates or tools for creating materials from written text for SALT development. Badla proposed [4] the optical character recognition (OCR) approach has been used to turn written/image texts into editable text in a range of applications using scanners, computers, tablets, and other devices. If programmers aim to increase the OCR system's performance so that it can work effectively on mobile devices, they will.

Tesseract is a famous OCR engine at the moment. Tesseract is called an accessible OCR engine developed by HP from 1984 to 1994 [17]. It appeared out of nowhere, blazed brilliantly with its discoveries, and then vanished behind the same veil of identity that it had been established behind. HP declared the Tesseract engine open-source software with information about its design in late 2005. Tesseract began as a Ph.D. research study [20] at HP Research labs in Bristol, and it quickly gained popularity as a possible hardware besides the software components add-on for HP's flatbed scanner series. The Tesseract OCR engine was developed in response to the fact that the professional OCR engine at the time was still in its infancy and had disastrously failed with all but high-resolution prints. This proposal's goal is to make document handling and finding simple for users by allowing them to query through pictures or non-text content. This project creates an OCR technology that allows users to search for sheet-based information very quickly instead of for hours. It cuts or removes time-consuming data input by mechanically retrieving info from paper and placing it as it is needed. It allows totally new methods to handle documents, removing the need for 'personal touch' and reducing costs and response time substantially. The suggested project is helpful for Bangla speakers and writers since it displays the meaning of a word in Bangla from OCR-extracted texts. The objectives of our approach are

- Capture words, text, or sentences from the input image and convert them into human readable, editable text, and sentences.
- It will also provide the Bangla meaning of the output text/sentence. Hopefully, in the future, we will be able to add many more languages to translate into.

In this paper, Sect. 2 described the literature review of the OCR engine; Sect. 3 showed the overall methodology of how this approach works; Sect. 4 discussed the analysis; Sect. 5 was for performance results and finally Sect. 6 conclusions and the limitations.

## 2 Related Work

Kumar et al. [12] suggested an OCR-initiated segmentation approach for handwritten Gurumukhi scripts, detailing the whole segmentation method as well as the digital operation and preprocessed techniques for contact character identification and segmentation. Kumar and Singh [11] created an OCR model and put it to the test in a variety of publications. The results were encouraging, and they were proven with excellent precision. They present a method for segmenting a scanned document image into words, lines, and characters only. To deconstruct complex letters into conspicuous form elements for Bangla OCR, Pramanik and Bag [14] devised a unique structure conformational change segmentation approach [1, 2, 8, 15, 16, 24]. To breakdown compound features into significant form components, they presented a unique shape decomposition-based segmentation technique in this study. Their strategy is to concentrate solely on the deconstruction of complex characters. Through testing, the recommended approach was found to have acceptable recognition quality. A sophisticated OCR engine has only recently been developed [10] which will give a proper guideline for which font will provide better OCR recognition. GOCR is now a freeware OCR engine which transfers photos to various types of texts. For this engine speed, correctness, and its ease, 'GOCR' is extremely useful, Dhiman and Singh [9] with a handwritten text in the background, their approach is difficult to execute. Pozo et al. [13] have created a system that uses a commercially available mobile device with an attached camera to convert and understand printed publications. Whenever applied to letters with a minimal level of 20-pixel height, it has the best potential. Cuneiform is a free and open OCR engine created by cognitive technology [6]. It supports a wide range of languages while preserving the format of the text. This generated text can be in HTML, RTF, email, or other formats. ABBYY FineReader is a famous OCR application that uses expertise to identify words in electronic materials [3]. This tool can extract content from a wide range of electronic assets, including PDFs, images, and even video frames. The FineReader employs a smart technique to ensure the picture's shape during the text extraction method. Easy OCR is also another common OCR engine. Easy OCR outperforms all these other OCR engines when it comes to number detection and recognition. The handcraft method [7] is used for detecting, while the CRNN [18] is used for identification. Extraction of features, pattern tagging (LSTM), but also decoders, are the three basic phases (CTC).

## 3 Methodology

Using the Tesseract OCR engine, [22] was utilized to recover sentences from photos throughout the proposed model. The Tesseract OCR system [22] derives textual content from its approach part. Figure 1 shows how it performed in the five following phases. The following steps are as follows.

### 3.1 *Word and Line Finding*

To begin, a picture is fed into the Tesseract OCR engine [22]. It then attempts to find every line and each word in the picture. Those stages in Fig. 2 can also be divided into lines as well as words discovery.

**Line Finding**: The line recognition method [21] is meant to identify warped paper without needing to de-skew it, preserving the quality of the image. Blob sorting and line creation are critical aspects of the procedure. First, it examines the page

**Fig. 1** System diagram

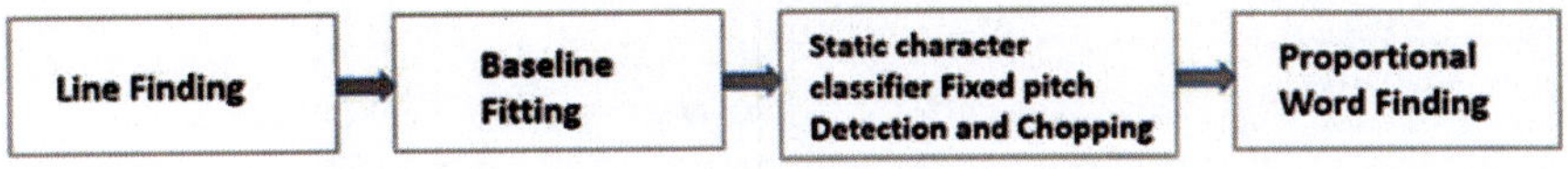


**Fig. 2** Word and line finding

**Fig. 3** Sample for baseline curved fitted

layout and calculates a text size that is fairly uniform by taking the meaning of all word heights. The OCR engine then recognizes the patterns based on their estimated height.

**Baseline Fitting**: The baselines employing a quadratic spline fitted more closely after text lines were recognized. The Tesseract engine for such an OCR machine was the first to use baseline estimation. Tesseract could be used to extract data from curved baseline pages of documentation for this purpose. The starting points/baseline are suited to categories with a fairly steady displacement of the linear baseline original by dividing the blobs. Figure 3 depicts a text line with appropriate starting points, an ascender line, a descender line, and a meaning line.

**Fixed-Pitch Detection and Chopping**: Because all letters in fixed-pitch texts are the same width, chopping occurs at predetermined intervals, as well as the chopped blobs are presented for acknowledgment. The pitch of the words causes the Tesseract to divide them into characters. Figure 4 demonstrates a common fixed-pitching word [25].

**Proportional Word Finding**: It is tough to parse words with text spacing with a fixed pitch or proportionality of textual spacing. The work isn't easy. Figure 5 denotes some tough word spacing is demonstrated by some hard word spacing samples. To address these issues, with the problematic space of words, Tesseract then tests changes in restricted vertical ranges across starting point lines [22].

**Fig. 4** Fixed pitching chopped word

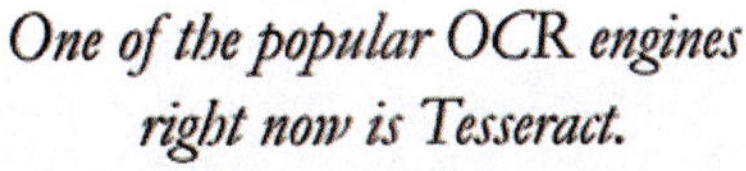


**Fig. 5** Difficult spacing word

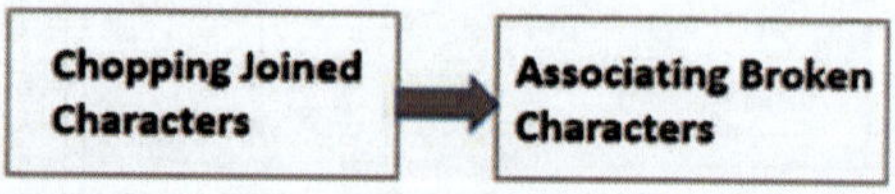


**Fig. 6** Recognition approach of word

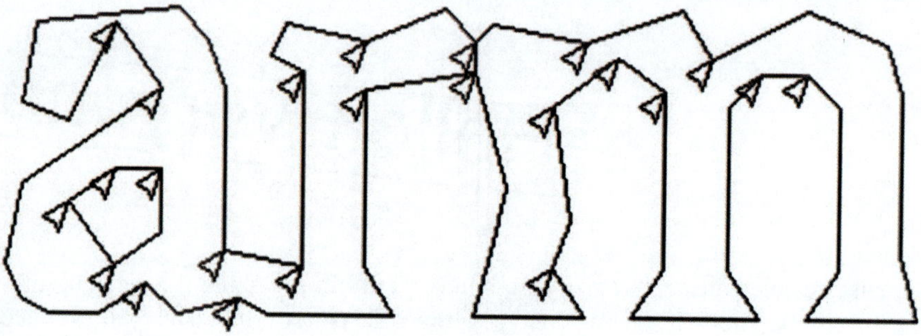

**Fig. 7** Chop figures and chop marks for the applicant [22]

## 3.2 *Recognized Words*

The approach to recognizing words is dependent on the textual format [22]. The result of the line finding is first classified. The anti-fixed pitched is the only part of the word recognition stage that is relevant. Figure 6 shows how the word recognition approach works in two phases.

**Chopping Joined Characters**: Because categorizing a word alone produces insufficient results, Tesseract improves the output by slicing the words from the character classifier with the lowest accuracy. Possible chop sites are found by looking at all the concave vertices of the polygonal representation of the portion outlines. In Fig. 7, the areas for candidates are depicted as triangles.

**Associating Broken Characters**: If the term additionally requires valuable details, it is granted to the associator once all chops have been made. Throughout this stage, disconnected blobs are grouped for the search using an A * into suitable characters technique. In that move, Tesseract's use of the A * algorithm provides a significantly higher accuracy point than several OCR techniques.

## 3.3 *Classification of Static Character*

**Functions**: Tesseract iterations [5, 19] employed a static classifier that was unaffected by font face or size. However, they were insufficiently trustworthy to address the issues encountered in real-world photos. The training step includes the usage of

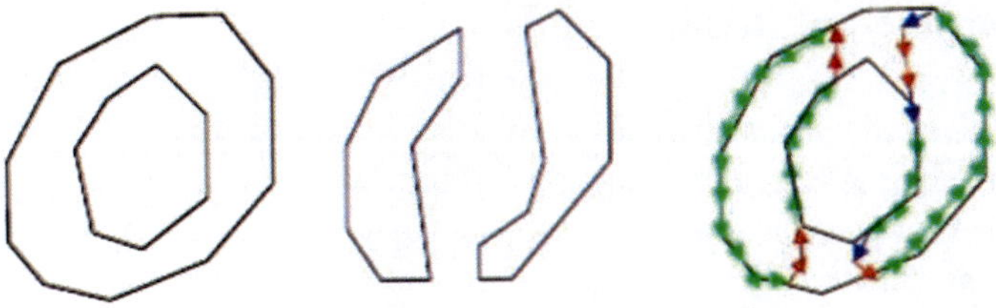

**Fig. 8** Classify static character: prototypes are matched in terms of features

polygonal approximated components as characteristics. However, in the identification process, few adjusted characteristics are retrieved from the outline, and linked multiform-to-one with linked model characteristics from the results after training is displayed in Fig. 8. The unidentified character's attributes are three-dimensional: *x*, *y*, and angle; model features are four-dimensional: *x*, *y*, angle, and length [22].

**Training Data**: Tesseract's algorithm was not trained on fragmented characters because it could not properly distinguish between cracked and detached characters. Instead, it was trained on 20 samples of 94 characters from eight typefaces. Other OCR engines require a far larger number of training samples than this [22].

## 3.4 Linguistic Analysis

Tesseract has a relatively low linguistic dependency when compared to many other OCR engines. Tesseract only considers linguistics when a new segmentation is being evaluated. The Tesseract, on the other hand, would pick one chop over the other to construct a phrase.

## 3.5 Adaptive Classifier

Tesseract's static classification must be proficient at generalizing multiple character sets, but it fails to distinguish between individual characters among the characters and the anti-characters. The constant classifier's output is used to train a more font-sensitive adaptable classifier for further granular identification. When the amount of font-specific data on a Web page is minimized, the quantity of differentiation increases.

## 4 Experimental Details

The legal basis of a mechanism means that the system operates, including what the input and output are, how the output is formed from the input, and so on. Figure 9 depicts the suggested OCR system's sequential phases with Bangla word meanings. Those are as follows:

**Input Image**—any size image can be used as an input by the system.

**OCR Engine**—optical character recognition (OCR) system. The input image is processed for text extraction using the Tesseract OCR engine.

**Text Extraction**—extract the text for next processing.

**Bangla Word Meanings**—each word is identified using NLTK, and the meaning in Bangla is displayed.

**Phrase Identification**—NLTK is used to identify phrases from extracted text and display their meaning in Bangla.

During character recognition, the suggested system uses an input image. Using the file explorer, the picture can grab the contents of any folder on the PC. The first algorithm demonstrates how the image is fed into the system. Following the selection of an image, the suggested system starts the OCR engine to capture the textual format of it. The boundary of the image is checked first by this OCR engine. The engine can also compute the text's average height by estimating the outline, allowing it to identify the lines. It splits the words and tries to identify them after recognizing the line. It outputs all of the detected words as text after satisfactory recognition. These steps are briefly shown in Algorithm 2. There are several additional processing processes for the retrieved output text, such as the Bangla meanings of each word used in the phrase discovered.

The system includes a database-stored English to Bangla dictionary. The system scans the database for phrases and words, then displays the Bangla translations of

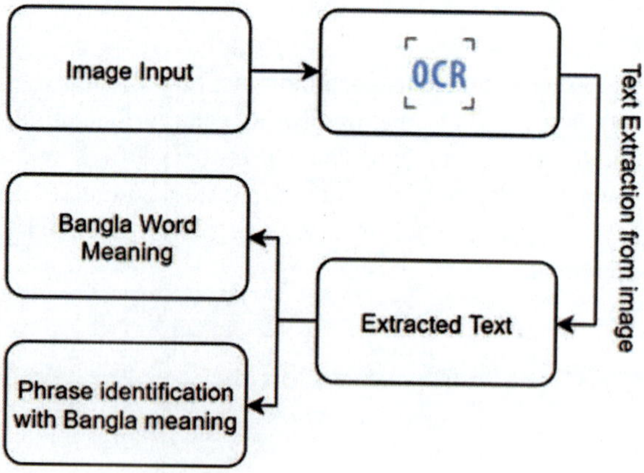


**Fig. 9** Systematic sequential steps

those phrases and words. The processes of extracted text processing are shown in Algorithm 3.

**Algorithm 1** Input image

initialization of the applications;
click the 'input image' button for getting the required image;
**while** *explorer of file is opened* **do**
    mark image $J$;
    **if** *expected image is marked out* **then**
        Optical Character Recognition ();
    **else**
        go back to main menu;
    **end process**
**Stop**

# 5 Result

To make it easier to interface with the system, the proposed system was developed as a desktop application in Python. Graphics user interface (GUI) implementation was done with Tkinter6 and Kivy7; characters recognition system was done with PyTesseract8, and textual analysis was done with NLTK9.

The proposed OCR system includes all of the primary functionalities that may be accessible from the desktop application's home window. These are the following steps:

1. Upload the image.
2. Query for words.
3. Adding a new word to the database.

The user must press the 'Upload image' icon from the home window to enter a picture into the OCR system for further processing.

**Algorithm 2** Using Tesseract OCR engine extract the text

**Input is**: Picture/image
**Output will be**: Readable text format
initialize the OCR engine $K$;
take the image $J$;
analyze the image outlines and keep them as Blobs $B$;
take $B$, find the text lines $L$;`/*` $L = \{L_1, L_2, L_3, L_4, \ldots, L_{n-1}, L_n\}$ *it* `represents each lines identified, here,` $n$ `is the number of total line */`
break the text lines $L$ into words $W$; `/*` $W = \{W_1, W_2, W_3, W_4, \ldots, W_{m-1}, W_m\}$ *it* `represents each words identified, here,` $m$ `is the amount of total word */`

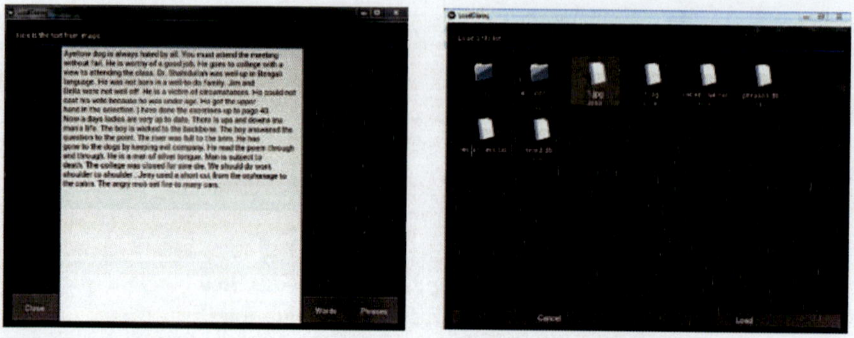


(a) Insert Image File (b) Extracted Text Format from Image File

**Fig. 10** Insertion of image file and conversion into text

**6** recognize the words *W*;
**7** classify words *W is* using the Adaptive Classifier *C*;
**8** show all the extracted texts as a result *O*;

A screen similar to the one shown here in Fig. 10 appears for the user (a). Users can enter or browse the system files for required images to load an image file format for character identification. If the user selects an image and then clicks the 'Load' icon, the photo will be loaded into the Tesseract OCR engine for text extraction. The text recovered from the picture file is shown in Fig. 10b. The user can examine the Bangla definitions of each word identified from the image presented in Fig. 12 by looking at the generated text from the image (a).

It has been accomplished by integrating a Bangla vocabulary database with the system. From any textual input or image file, including text, the program can also identify idioms and phrases. It also includes the idioms and phrases from Fig. 12 that have Bangla translations (b).

A user can also use the system's built-in English to Bangla Dictionary to look up the meaning of any English term. The system may display the searched Bangla word's definition along with English synonyms and Bangla synonyms. For system upgrades, anyone can also insert different words to that database.

**Algorithm 3** Extracted the approach of text processing

**1** firstly initialize the database *Dbase*;
**2** then initialize the NLTK;
**3** taking an output texts *OT*;
**4** showing the functions;
**5 if** *Words is marked out* **then**
**6** NLTK was used to break the text into words;
**7** searching those words through the database *Dbase*;
**8** found words shown with Bangla meaning;
**9 else if** *the phrase is selected by program* **then**
**10** then, NLTK is used to extract regularly used phrases from an output text;

searching from database as a reason Bangla word meaning *Dbase*;
after that show the meaning as Bangla of phrases those are found;
**else**
Terminated;
**Stop**

## *5.1 Performance Analysis of Tesseract OCR*

The Tesseract (OCR) engine is among the most common and trusted OCR engines for extracting text from photos in the current era. Several researchers have investigated the performance of the Tesseract engine on various tasks. Khormi et al. [10] has published a report evaluating the performance of various OCR systems in extracting source code from photos (captured and video). The analysis of performances in Tesseract OCR is provided in Fig. 11.

Font types were the first performance evaluation criterion. Fonts such as Times New Roman, Arial, Consolas, and others were used in the data captured for the test. These photos were used to test the Tesseract OCR engine. The images using the font

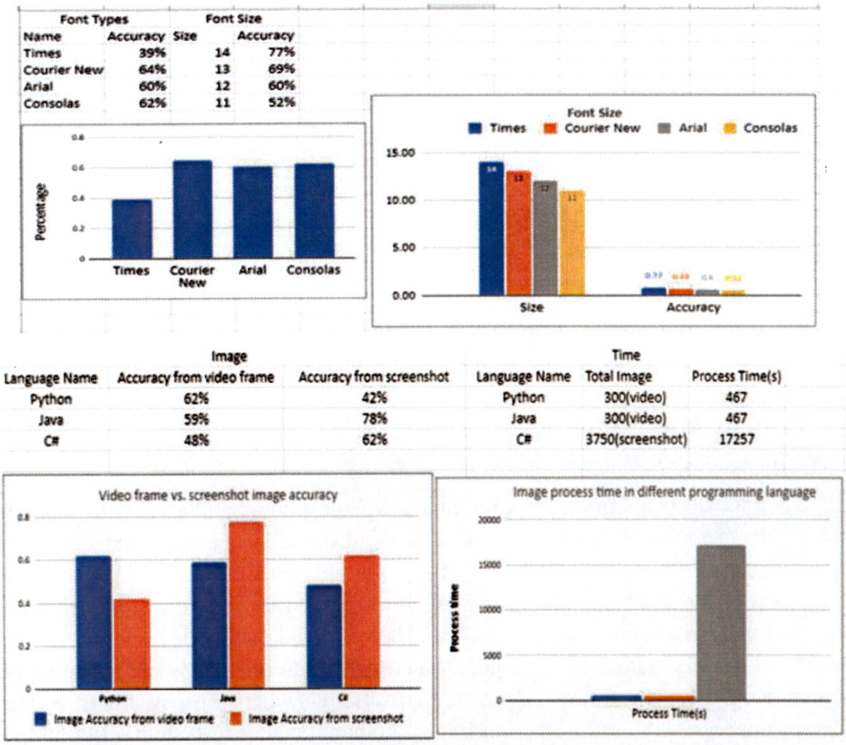


**Fig. 11** Performance analysis of Tesseract OCR with graph [10]

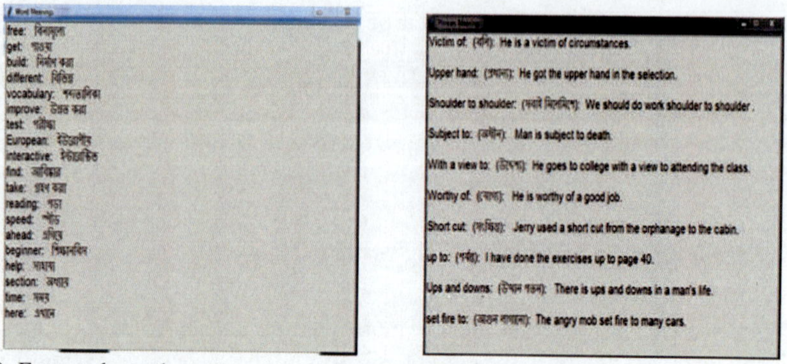


(a) Extracted text shows Bangla Meaning (b) Phrases Identification and Bangla Meaning

**Fig. 12** Word meaning into Bangla and phrases detection

'Courier New' had a maximum efficiency of 68%. The 'Times' font had the lowest accuracy, at 39%. The tesseract engine was then tested with images of various font sizes. Fonts that are larger are easier to read. With a larger font size, such as 14pt, the OCR engine achieved the highest level of correctness, whereas with a smaller font size, such as 11pt, it achieved the lowest level of accuracy. The tesseract engine was then put to the test with images from two separate sources: video frames and screenshots. The source codes for various programming languages are contained in these images (Fig. 12).

Screenshot photos for 'Java' source codes have the highest level of accuracy. When we use Java source code, we achieve the best average accuracy of sources. In screenshot photographs, the lowest accuracy was for the 'Python' programming codes. After that it calculated the time of extraction. The total number of photos from video frames was 300, and the processing took 467 s on average. The screenshot included a total of 3750 photos, which took 17,262 s to complete the process, an average of 4.60 s.

## 6 Conclusions

OCR is a useful technique for extracting data from photos and also from scanned documents that aren't editable or searchable. Mechanical data extraction technology saves time and effort, and this system is also simple to use and quick to add new characters to the database. OCR, any user can easily alter the info in the files and can use the form of edited data as needed. The proposed method is hopefully a very beneficial tool for native Bengali people because it not only finds out information from photographs, it also shows the meaningful Bangla where those words are found. Identifying regularly used terms might also be beneficial. The system's limitations are that it only works with photos. Text can be extracted from a variety of other documents. Because the Tesseract OCR is used to recognize English characters, the proposed technique works with them. For Bengali character recognition, we will develop our own Tesseract engine in the upcoming months.

## References


1. Ahammad, K., et al.: Recognizing Bengali sign language gestures for digits in real time using convolutional neural network. Int. J. Comput. Sci. Inf. Secur. (IJCSIS) **19**(1) (2021)
2. Ahmed, M., Chakraborty, P., Choudhury, T.: Bangla document categorization using deep RNN model with attention mechanism. In: Cyber Intelligence and Information Retrieval, pp. 137–147. Springer, Singapore (2022)
3. Arshad, H., Abidin, R.Z., Obeidy, W.K.: Identification of vehicle plate number using optical character recognition: a mobile application. Pertanika J. Sci. Technol. **25**, 173–180 (2017)
4. Badla, S.: Improving the efficiency of Tesseract OCR engine (2014)
5. Blesser, B., Kuklinski, T., Shillman, R.: Empirical tests for feature selection based on a psychological theory of character recognition. Pattern Recognit. **8**, 77–85 (1976)
6. Carter, R., Meggs, P.B., Day, B.: Typographic Design: Form and Communication. Wiley (2011)
7. Chakraborty, P., et al.: A human-robot interaction system calculating visual focus of human's attention level. IEEE Access **9**, 93409–93421 (2021)
8. Chakraborty, P., Nawar, F., Chowdhury, H.A.: Sentiment analysis of Bengali facebook data using classical and deep learning approaches. In: Innovation in Electrical Power Engineering, Communication, and Computing Technology, pp. 209–218. Springer, Singapore (2022)
9. Dhiman, S., Singh, A.: Tesseract vs GOCR a comparative study. Int. J. Recent Technol. Eng. **2** (2013)
10. Khormi, A., Alahmadi, M., Haiduc, S.: A study on the accuracy of OCR engines for source code transcription from programming screencasts, pp. 65–75 (2020)
11. Kumar, R., Singh, A.: Algorithm to detect and segment Gurmukhi handwritten text into lines, words and characters. Int. J. Eng. Technol. (2011)
12. Kumar, M., Jindal, M., Sharma, R.: Segmentation of isolated and touching characters in offline handwritten Gurmukhi script recognition. Int. J. Inf. Technol. Comput. Sci. **6**, 58–63 (2014)
13. Pozo, A.P., et al.: A method for translating printed documents using a hand-held device. In: 2011 IEEE International Conference on Multimedia and Expo. IEEE (2011)
14. Pramanik, R., Bag, S.: Shape decomposition-based handwritten compound character recognition for Bangla OCR. J. Vis. Commun. Image Represent. **50**, 123–134 (2018)
15. Rahman, S., Chakraborty, P.: Bangla document classification using deep recurrent neural network with BiLSTM. In: Proceedings of International Conference on Machine Intelligence and Data Science Applications. Springer, Singapore (2021)
16. Rahman, M.M., et al.: Bangla documents classification using transformer based deep learning models. In: 2020 2nd International Conference on Sustainable Technologies for Industry 4.0 (STI). IEEE (2020)
17. Rice, S., Jenkins, F., Nartker, T.: The Fourth Annual Test of OCR Accuracy. Information Science Research Institute (2012)
18. Shi, B., Bai, X., Yao, C.: An end-to-end trainable neural network for image-based sequence recognition and its application to scene text recognition. IEEE Trans. Pattern Anal. Mach. Intell. **39**, 2298–2304 (2017)
19. Shillman, R.: Character recognition based on phenomenological attributes: theory and methods (1974)
20. Smith, R.: The extraction and recognition of text from multimedia document images (1987)
21. Smith, R.: A simple and efficient skew detection algorithm via text row accumulation. In: Proceedings of the 3rd International Conference on Document Analysis and Recognition (1995)
22. Smith, R.: An overview of the Tesseract OCR engine. In: Ninth International Conference on Document Analysis and Recognition (ICDAR 2007), vol. 2, pp. 629–633 (2007)

23. Sultana, M., Ahmed, T., Chakraborty, P., Khatun, M., Hasan, M.R., Uddin, M.S.: Object detection using template and HOG feature matching. Int. J. Adv. Comput. Sci. Appl. **11**(7), 233–238 (2020)
24. Sultana, M., Chakraborty, P., Choudhury, T.: Bengali abstractive news summarization using Seq2Seq learning with attention. In: Cyber Intelligence and Information Retrieval, pp. 279–289. Springer, Singapore (2022)
25. Thakkar, A., Shah, V.: Review on Tesseract OCR engine and performance. Int. J. Innov. Emerg. Res. Eng. **4** (2017)

# Identification of Four Major Dialects of Assamese Language Using GMM with UBM

**Hem Chandra Das and Utpal Bhattacharjee**

**Abstract** The Assamese language is spoken by the people of Assam, which is located in India's north-east corner. The Indo-European language family includes the Assamese language. The pronunciation, grammar, and vocabulary of Assamese are vary in different sections of the state, resulting in different regional dialects of the language. There are four major regional dialects of the Assamese language, namely Central Assamese spoken in and around Nagaon district, Eastern Assamese dialect spoken in the Sibsagar and its neighboring districts, Kamrupi dialect spoken in Kamrup, Nalbari, Barpeta, Kokarajhar and some parts of Bongaigaon district and Goaplari dialect spoken in the Goaplara, Dhuburi and part of Bongaigaon district. Therefore, to develop a universal Assamese speech recognition system that seamlessly recognizes the words spoken in the Assamese language and its dialects, the identification of the dialect is a necessary condition. Using the Gaussian Mixture Model (GMM) and the Gaussian Mixture Model with Universal Background Model, this research proposes a novel technique for recognizing Assamese dialects (GMM-UBM). To extract spectral information from collected voice sample, the Mel-Frequency Cepstral Coefficient (MFCC) is used. Modeling is done using the GMM and GMM-UBM modeling techniques.

**Keywords** MFCC · GMM · GMM-UBM · Dialect identification

## 1 Introduction

The Assamese language, which is derived from the Indo-Aryan family of languages, is spoken by the majority of people in Assam and parts of neighboring states such as Meghalaya, Nagaland, and Arunachal Pradesh. Assamese is the state's official

H. C. Das (✉)
Bodoland University, Kokrajhar, Assam 783370, India
e-mail: hemchandradas78@gmail.com

U. Bhattacharjee
Rajiv Gandhi University, Doimukh, Arunachal Pradesh 791112, India
e-mail: utpal.bhattacharjee@rgu.ac.in



D. Gupta et al. (eds.), *Pattern Recognition and Data Analysis with Applications*, Lecture Notes in Electrical Engineering 888,
https://doi.org/10.1007/978-981-19-1520-8_24

language and designated it as a Major Indian Language in Schedule-viii of the Indian Constitution. The Assamese language evolved from Sanskrit, although ancient Assamese peoples such as the Bodos and Kacharis had a significant influence on its lexicon, phonology, and grammar [1]. According to recent study, the Assamese language is divided into four dialect groups. The Eastern group dialect is spoken by the people belongs to district of Sibsagar and its surrounding areas, while the Central group is spoken in and around the present-day Nagaon district and its surrounding areas. The Kamrupi language is talked in unincorporated Kamrup, Nalbari, Barpeta, Darrang, and a portion of Bongaigaon. Goalpara, Dhubri, and portions of Kokrajhar and Bongaigaon districts are home to the Goalparia group. However, the Central Assamese dialect is now largely regarded as the dominant or standard dialect [2]. In the field of dialect translation, there is no significant and systematic study of Assamese dialects. In linguistics, a dialect is a type of language that is socially distinct and is spoken by a specific group of native speakers who have a similar pattern of pronunciation, syntax, and vocabulary [3]. Human intelligence includes the ability to distinguish between spoken languages [4, 5]. First stage in developing a dialect-independent voice recognition system for any language is dialect identification. Furthermore, dialect identification can aid in the improvement of the quality of remote access services such as e-health, e-marketing, e-learning, and so on. Because all dialects are descended from a single language, they share information. Dialect identification is a more difficult task than language identification due to the considerable mutual information among the dialects.

## 2 Literature Survey

Wenker undertook a series of studies to determine dialect regions in 1877, which started the field of dialect identification [6]. Baily [7] was one of the pioneers in the identification and establishment of the Midland dialect as a distinct dialect. Following the findings of the study, it was concluded that dialects should not be classified only on the basis of vocabulary, because vocabulary might vary significantly amongst groups or classes within a particular geographic area [7]. Davis and Houck [8] have attempted to assess whether the Midland dialect region could be deemed distinct. The researchers effectively extracted phonological and lexical characteristics from 11 cities along a north–south line [9].

Many studies in the field of Arabic dialect recognition have been published in recent years [10–12]. Diab and Habash [13] and Watson [14] examined the Arabic dialect, enumerated its characteristics, established a link between the Standard language and regional dialects, and categorised the major regional dialects. Ibrahim et al. [15] use GMM to identify Arabic dialects. Authors consider Malaysian Quranic speakers. Spectral and prosodic characteristics were employed for this. They found a 5.5–7% improvement in accuracy when combining spectral and prosodic characteristics. The accuracy ranged from 81.7 to 89.6% for MFCC and prosodic features such as pitch, duration, etc.

Dialect recognition researches have been done in many Indian languages. Shivaprasad and Sadanandam [16] used GMM and HMM to identify regional Telugu dialects. The authors generated a dataset of Telugu dialects for this purpose. Recognition was done using MFCC and its variants such as ΔMFCC and ΔΔMFCC features. The study extracts 39 feature vectors from each spoken utterance and evaluates them with GMM and HMM models. The GMM model outperforms the HMM model. However, certain words with identical acoustic characteristics are not distinguished.

Chittaragi and Koolagudi [17] use the closest neighbor approach to identify Telugu dialects using just prosodic characteristics and a few lines from each dialect. Authors attained 75% accuracy by just considering prosodic features.

Chittaragi et al. [18] discovered 5 Kannada dialects using spectral and prosodic characteristics. To recognize dialects, the authors employed SVM and Neural Networks. With text-independent data, the Neural Network produces good results in the shortest amount of time.

Spectral and prosodic traits were used by Rao and Koolagudi [19] to differentiate five Hindi dialects: Chhattisgarhi, Bengali, Marathi, General, and Telugu. Their database comprises ten (10) individuals speaking spontaneously for 5–10 min each, totaling 1–1.5 h.

Despite the enormous potential for using an Assamese dialect identification system, no substantial attempts have been undertaken in this regard. Using GMM and GMM-UBM based recognizers, this study provides a technique for identifying Assamese dialects in real-time.

## 3 Speech Database

The review of the current literature reveals that there is no standard database for the Assamese language and its dialects. A new database has been created with speech samples from all the dialect groups. The same numbers of speakers have been recorded for each dialect region. Speech samples from ten speakers (5 male and 5 female) representing each dialect region make up the speech data. A phonetically rich script was prepared to record the speech samples. The same script was used to record all the dialects, including the standard Assamese. The recording was made at a sample rate of 16 kHz, 16-bit resolution. Subjective listening test of the recordings has been done using listeners from the respective dialect groups who were not involved in the recording process (Table 1).

## 4 Experiment Setup

Mel-frequency cepstral coefficients (MFCCs) are most commonly associated with the human peripheral auditory system. The MFCC processor's primary function is to replicate the behavior of human hearing. Instead of a linear scale, human hearing

**Table 1** Statistical representation of the speech database

| | |
|---|---|
| Number of speakers | 10 (five male and five female) for each dialect group |
| Number of sessions | 02 |
| Intersession interval | At least one week |
| Types of data | Speech signal |
| Speech type | Read speech |
| Sampling frequency | 16 kHz |
| Format for sampling | Monophonic, resolution of 16 bits |
| Speech duration | Each speaker is recording is for minimum 30 min in each session |
| Microphone | Zoom H4N portable voice recorder microphone |
| Acoustic environment | Laboratory |
| Total duration of speech data | Minimum 10 h for each dialect, including standard Assamese |

follows a Mel-spectrum scale with linear spacing below 1 kHz and logarithmic scaling above 1 kHz [20]. MFCC features are derived from the recorded speech signal in this investigation. $H(z) = 1 - 0.96z^{-1}$ was employed as a pre-emphasis filter before framing. The pre-emphasized voice stream is divided into frames with a frame frequency of 100 Hz and a length of 20 ms. Using a Hamming window, each frame is smoothed. A bank of 20 triangular filters separated on Mel-scale are used to filter the magnitude spectrum and limited 300–3400 Hz frequency range through making use of Fast Fourier Transformation (FFT) derived from windowed frame. Discrete Cosine Transformation (DCT) is accustomed to transform the log-compressed filter outputs to cepstral coefficients. Thus a 20-dimensional MFCC features have been obtained. Because the 0th cepstral coefficient correlates to the full frame's energy [21], it's never used in the cepstral feature vector. As a result, only 19 MFCC coefficients were kept. The 1st order derivatives of the cepstral coefficients were generated to keep the speech signal's time-varying characteristics. Three samples were used to approximate the 1st-order derivatives. Putting the MFCC coefficient together with the 1st-order derivative yields a 38-dimensional feature vector. To decrease the effect of channel mismatch, cepstral mean subtraction was performed each and every features.

Sum of Gaussian component densities represents by GMM (Gaussian Mixture Model), which is known as parametric probability density model. GMMs are often used to describe the probability distribution of continuously measured data as a parametric model [13]. The weighted sum of $M$ component densities is represented by a GMM as:

$$P(x|\lambda) = \sum_{i=1}^{M} w_i b_i(X) \tag{1}$$

$x$ is an $M$-dimensional random vector, $b_i$, $i = 1, 2 \ldots M$, stands for component densities, and $w_i$ $i = 1, 2, \ldots, M$, stands for mixture weights. The probability of all components is represented by:

$$b_i(X) = \frac{1}{(2\pi)^{\frac{D}{2}} \left|\sum_i\right|^{1/2}} \exp\left\{-\frac{1}{2}(x - \mu_i)' \sum_i^{-1}(x - \mu_i)\right\} \tag{2}$$

Using $\mu_i$ as the mean vector, $\sum_i$ as the covariance matrix and $D$ as the dimension of the feature vector. The weight of the mixture meets the condition $\sum_{i=1}^{M} w_i = 1$.

The mean vectors, covariance matrices, and mixture weights from all component densities are used to parameterize the entire Gaussian mixture model. The following notation may be used to express all of these parameters:

$$\theta = \{w_i, \mu_i, \Sigma_i\}, \quad \text{for} \quad i = 1, 2 \ldots, M \tag{3}$$

Each dialect in a dialect identification system a GMM model is used to represent, which is mentioned in the above model $\theta$.

When making a decision accept or reject in a identification system of dialect, a Universal Background Model (UBM) or World Model represents broad, dialect-independent, channel-independent feature characteristics that are evaluated against a model of dialect-specific feature characteristics. In this example, the UBM is a dialect-independent GMM that was trained to represent general speech characteristics using speech samples from a wide range of dialects. While training the dialect specific model, the UBM is also utilized as a prior model in Maximum a posteriori probability (MAP) parameter estimation. The block diagram of the suggested Dialect Identification system is depicted in Fig. 1.

For each test sample, the DTE (Detection Error Trade-off) curve was plotted by using the log likelihood proportion of genuine dialect with false dialect models, and the EER (Equal Error Rate) generated from the DTE curve was used as a unit of measurement of dialect identification system's performance.

## 5 Results and Discussion

Database provided in Sect. 3 was accustomed to conduct all of the experiments described in this paper. The speech database is isolated from the training and testing datasets. The system has been trained with 60% of the total samples, while the remaining 40% is being utilized for testing. Prior to feature extraction, the silence frames are detected and discarded using an energy-based silence detector. Each test

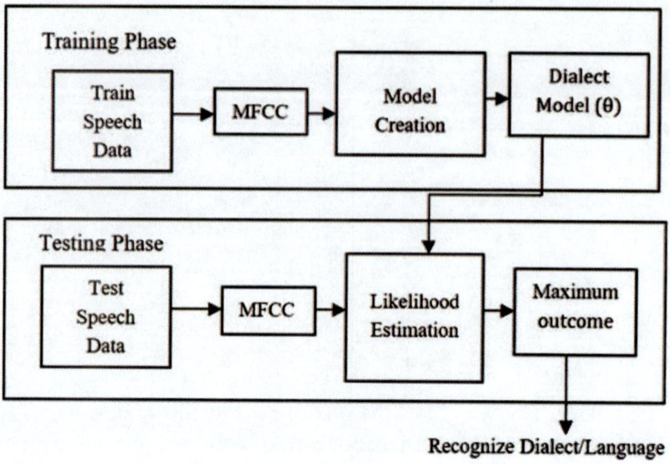


**Fig. 1** Proposed dialect identification system

segment is tested against all dialect models. There are total 930 test segments of different length. Two sets of feature vectors are constructed using MFCC features taken from the voice samples. The first set includes just 19 MFCC coefficients, whereas the second set includes 19 MFCC coefficients as well as their 1st order derivatives. The second feature set has a dimension of 38.

In the first experiment MFCC and MFCC + ΔMFCC features have been used to train separate dialect models for each dialect. The dialect models are created using 512 Gaussian components. Table 2 summarizes the outcome of the research and Fig. 2 show the DET curve for GMM-based dialect identification system by making use of MFCC and MFCC + ΔMFCC features.

In the second experiment, MFCC and MFCC + ΔMFCC features have been used to train the GMM-UBM based dialect models. All of the speech samples in the training set, regardless of dialect, were used to train the Universal background model. The dialect models were then obtained from the UBM using MAP (Maximum a Posteriori) probability. The models are put to the test on the same data set. Table 3 summarizes the results of the tests, and Fig. 3 depicts the performance of the GMM-UBM based dialect identifier for MFCC and MFCC + MFCC features.

**Table 2** Recognition accuracy for GMM-based dialect identification system

| Feature vector | Recognition accuracy |
|---|---|
| MFCC | 93.28 |
| MFCC + ΔMFCC | 95.48 |

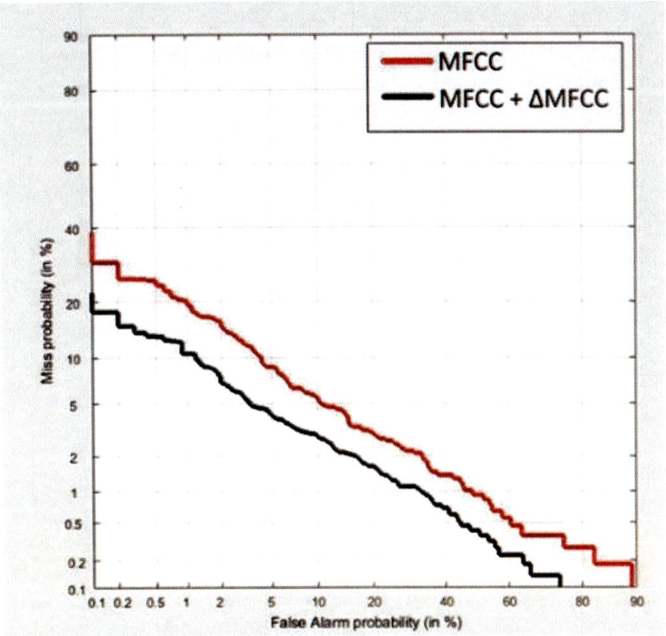


**Fig. 2** DET plot for the GMM-based dialect identification system

**Table 3** Recognition accuracy for GMM-UBM-based dialect identification system

| Feature vector | Recognition accuracy |
|---|---|
| MFCC | 95.07 |
| MFCC + ΔMFCC | 97.57 |

## 6 Conclusion

We introduced a Dialect Identification system for recognizing Assamese dialects based on MFCC and MFCC with MFCC feature vectors in this paper. The Assamese language has four major dialects. Two classification models, GMM and GMM-UBM, were used to identify the dialects. It has been discovered that employing the GMM-UBM model produces better results. When the GMM-UBM model is applied with the combined features of MFCC and ΔMFCC, it achieves an identification accuracy of 97.57%, which is higher than previously published results [16]. To increase the accuracy of the dialect identifier in the future, additional feature combinations will be employed in conjunction with deep learning technology. Furthermore, dimensionality reduction techniques will be employed in order to minimize the computational complexity of the dialect recognition system's reaction time.

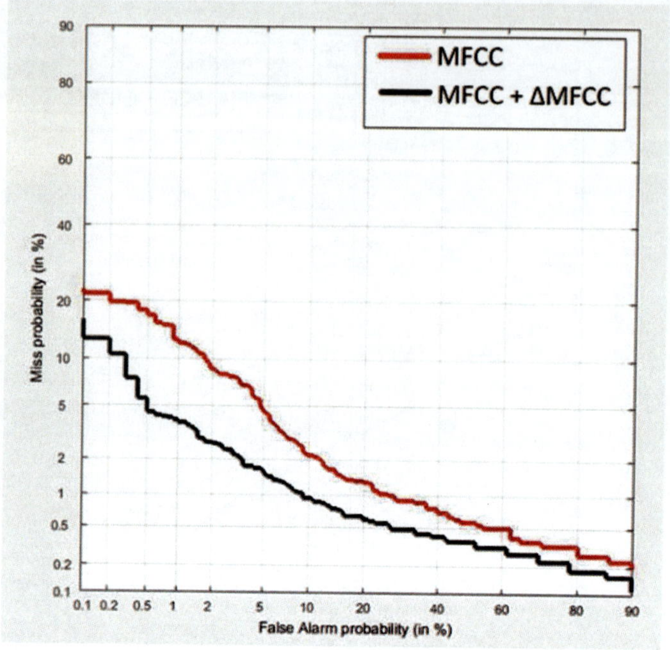


**Fig. 3** DET plot for the GMM-UBM based dialect identification system

## References


1. Assamese Dialect Translation System—A Preliminary Proposal. http://himangshu.net/docs/iconacc.pdf. Accessed 2020/11/10
2. Assamese Language. Available: https://en.wikipedia.org/wiki/Assamese_language. Accessed 2019/10/10
3. Liu, G.A., Hansen, J.H.: A systematic strategy for robust automatic dialect identification. In: 19th European Signal Processing Conference, pp. 2138–2141. IEEE (2011)
4. Li, H., Ma, B., Lee, K.A.: Spoken language recognition: from fundamentals to practice. Proc. IEEE **101**(5), 1136–1159 (2013)
5. Zhao, J., Shu, H., Zhang, L., Wang, X., Gong, Q., Li, P.: Cortical competition during language discrimination. NeuroImage **43**(3), 624–633 (2008)
6. Nti, A.A.: Studying Dialects to Understand Human Language. Massachusetts Institute of Technology (2009)
7. Bailey, C.J.N.: Is There a "Midland" Dialect of American English? ERIC Clearinghouse. Distributed by ERIC Clearinghouse (1968)
8. Davis, L.M., Houck, C.L.: Is there a Midland dialect area?—Again. Am. Speech 61–70. Duke University Press (1992)
9. Etman, A., Beex, A.L.: Language and dialect identification: a survey. In: SAI Intelligent Systems Conference (IntelliSys) 2015, pp. 220–231. IEEE (2015)
10. Shoufan, A., Alameri, S.: Natural language processing for dialectical Arabic: a survey. In: Proceedings of the Second Workshop on Arabic Natural Language Processing, pp. 36–48 (2015)
11. Guellil, I., Saâdane, H., Azouaou, F., Gueni, B., Nouvel, D.: Arabic natural language processing: an overview. J. King Saud Univ.-Comput. Inf. Sci. **33**(5), 497–507 (2021)

12. Elnagar, A., Yagi, S.M., Nassif, A.B., Shahin, I., Salloum, S.A.: Systematic literature review of dialectal Arabic: identification and detection. IEEE Access **9**, 31010–31042 (2021)
13. Diab, M., Habash, N.: Arabic dialect processing tutorial. In: Proceedings of the Human Language Technology Conference of the NAACL, Companion Volume: Tutorial Abstracts, pp. 5–6 (2007)
14. Watson, J.C.: 50. Arabic dialects (general article). In: The Semitic Languages, pp. 851–896. De Gruyter Mouton (2011)
15. Ibrahim, N.J., Idris, M.Y.I., Yakub, M., Yusoff, Z.M., Rahman, N.N.A., Dien, M.I.: Robust feature extraction based on spectral and prosodic features for classical Arabic accents recognition. Malays. J. Comput. Sci. 46–72 (2019)
16. Shivaprasad, S., Sadanandam, M.: Identification of regional dialects of Telugu language using text independent speech processing models. Int. J. Speech Technol. 1–8 (2020)
17. Chittaragi, N.B., Koolagudi, S.G.: Acoustic features based word level dialect classification using SVM and ensemble methods. In: Tenth International Conference on Contemporary Computing (IC3) 2017, pp. 1–6. IEEE (2017)
18. Chittaragi, N.B., Limaye, A., Chandana, N., Annappa, B., Koolagudi, S.G.: Automatic text-independent Kannada dialect identification system. In: Information Systems Design and Intelligent Applications, pp. 79–87. Springer (2019)
19. Rao, K., Koolagudi, S.G.: Identification of Hindi dialects and emotions using spectral and prosodic features of speech. Int. J. Syst. Cybern. **9**(4), 24–33 (2011)
20. Davis, S., Mermelstein, P.: Comparison of parametric representations for monosyllabic word recognition in continuously spoken sentences. IEEE Trans. Acoust. Speech Signal Process. **28**(4), 357–366 (1980)
21. Zhao, X., Shao, Y., Wang, D.: Robust speaker identification using a CASA front-end. In: IEEE International Conference on Acoustics, Speech and Signal Processing (ICASSP) 2011, pp. 5468–5471. IEEE (2011)

# Digitization Through SNS: Issues, Challenges, and Recommendations—A Case Study

Urmila Pilania, Rohit Tanwar, Mamta Arora, and Manoj Kumar

**Abstract** In today's digital era, everything relies on the Internet which poses so many concerns on security. Growth of the Internet results in danger for transferring information online through multimedia devices. Information transmitted through these digital devices can be hacked by hackers. The objective of this study is to throw light on problems that arise due to a lack of digital awareness for social networking sites among users and propose an automated solution to minimize these crimes. Communication sites have many areas of exploitation such as progressive promoting, online commercials and marking, etc. Many users are not much skilled in digital terms so these users are not aware of the negative side of these social network sites. The safety concerns and complications on social network sites such as character misuse, malware, phishing attacks, and unknown request risks have been discussed. Though introducing the management actions to governor these major concerns, the proposed paper also suggests some appropriate measures which could be convenient to every user as the assembly in the collaboration by private area for providing digital secure electronic world.

**Keywords** Digital multimedia · Social networking sites · Safety concerns · Security attacks · Safety measures · Cyber-crime

## 1 Introduction

In the present era of the digital world, all personal and professional information is communicated online through PCs and mobiles. People of all age groups are using the Internet today. Social networking sites (SNS) contain personal information of

U. Pilania · M. Arora · M. Kumar
Department of Computer Science and Technology, Manav Rachna University, Faridabad 121004, Haryana, India

R. Tanwar (✉)
School of Computer Science, University of Petroleum and Energy Studies, Dehradun 248007, India
e-mail: rohit.tanwar.cse@gmail.com



D. Gupta et al. (eds.), *Pattern Recognition and Data Analysis with Applications*, Lecture Notes in Electrical Engineering 888,
https://doi.org/10.1007/978-981-19-1520-8_25

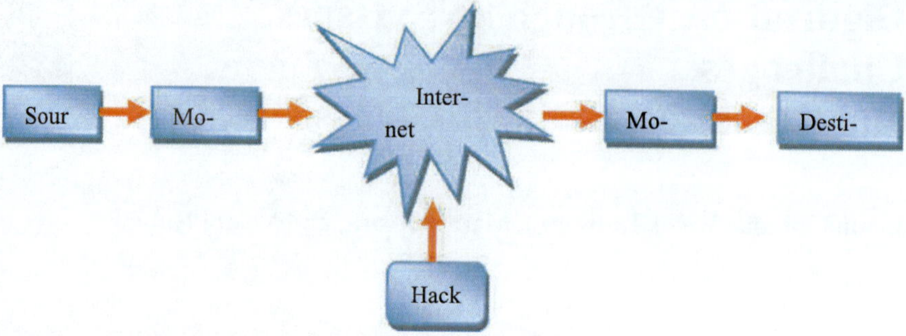


**Fig. 1** Information transmission between source and destination [3]

users like personal identity, user ID, password, etc., because of this online information safety faces many challenges [1]. Many times, the online stored information is shared with others without the knowledge of the user. In a survey in 2019, approximately 64% of SNS users have shared personal details online because of sharing of information many challenges like a scam; theft of information may take place. Many Internet users are not aware of the security policies. Digital security becomes a major concern in the world of the computer. Though security is provided to these online communicating devices utilizing encryption, digital signature, and watermarking techniques. In Fig. 1, information is being transmitted through the Internet between sender and receiver [2]. The source is the user who wants to communicate secret information in digital form through the Internet. The destination is the person who is going to receive the secret information transmitted by the source user. Hacker is an unauthorized skilled person in digital terms that can interrupt the secret information in between.

Figure 2 uses firewall for security purpose as firewall ban the unauthorized users to enter a particular site. But hackers are smart enough to break all these techniques. So, information safety became a critical task in the present time of the multimedia world [4].

As shown in Fig. 3, billion people are using SNS every day. Facebook is one of the most widely used SNS among these six top SNS. From the figure, it can be easily concluded that these sites have become part of daily life for people these days. The count of social networking site user's increasing on daily basis.

## 2 Literature Review

Every year, there is a continuous growth in the number of users using SNS. In 2010 approximately 1 billion users were there who are accessing SNS. But this number got increased by 2 billion within less than 5 years. The growth of SNS leads to an increase

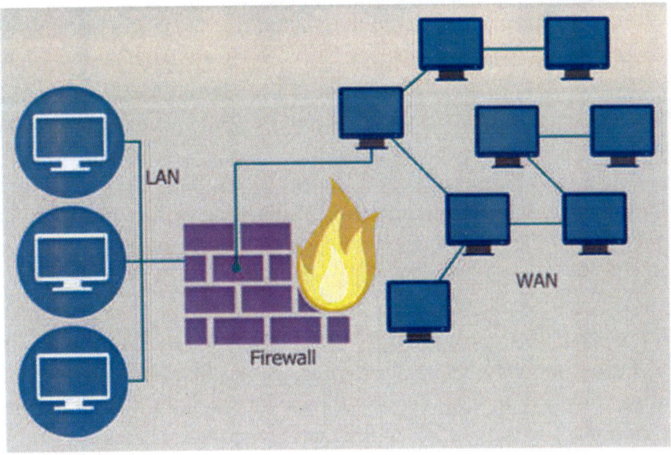


**Fig. 2** Firewall system

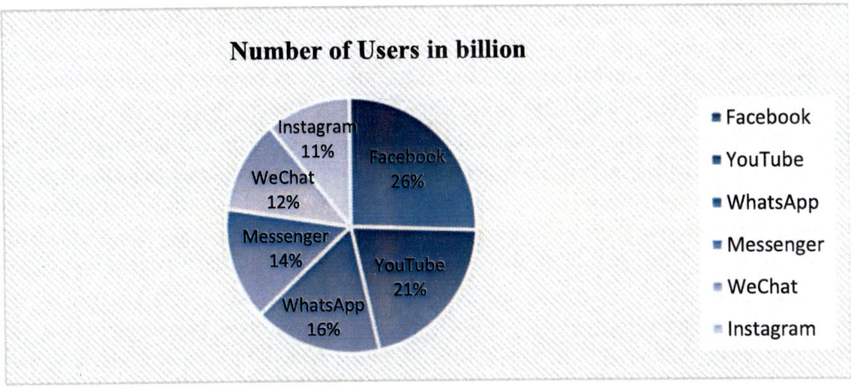


**Fig. 3** Popularity of SNS among users

in cyber-crime to a large extends. Cyber-crime consists of impairment and demolition of data, whipped money, mislaid productivity, robbery of someone's personal and financial property, misuse of rights, scams, post-attack disruption to the normal course of business, criminal inquiry, refurbishment and erasure of stolen information and systems, and reputational harm. In 2016, cybersecurity endeavors foretold that cyber-crime will cost worldwide 6 trillion dollars in a year by 2021, active from 3 trillion dollars per year in 2015 according to [5]. The current situation represents the extreme transmission of financial prosperity in history, dangers the incentives for invention and venture. Because of COVID-19, a large number of employees are working from home. At this time, everything is in digital form on the web which leads to the increase in cyber-crime continuously [6]. Digital wrongdoing has expanded continuously in the last few years. Ransomware is the most popular cyber-crime these days. According to cyber-crime magazine in every 5 s, a case of ransomware is reported. According to [7], because of COVID-19, approximately seven out of ten transactions are initiated from mobile. Computerization and identity stealing was the key attack during COVID-19 in 2020. Out of 24.6 billion transactions, 58% are declined by human-initiated attacks only.

Internet Crime Complaint Center (IC3) recorded data is shown in Fig. 5. IC3 has reported different types of frauds to steal information online. Phishing is the most popular fraud among these entire shown in figure. IC3 is continuously making efforts in collaboration with the industry to overcome these frauds.

It was noted that its damage cost will increase by 57 times more than it was in 2015. Cyber-crimes growth for 2017–2019 recorded by National Crime Record Bureau (NCRB) is shown in Fig. 4. The figure shows that frauds like data theft, fake profile, online transaction, and many more are increasing day-by-day. India comes at number three among the top 20 international victim countries.

To promote awareness in public, IC3 provides an annual report on the data reported during the year. IC3 has reported about 300,000 more cyber-crime complaints in the year 2020 as compared to complaints recorded in the year 2019. Approximately 28,500 complaints are related to COVID-19 aiming at businessmen and individuals.

The vision [10] of this strategy is to build up a protected and solid Internet for residents, organizations, and the legislature. The strategy attempts to ensure individual data, budgetary, and banking information. Service of transferring data and information technology is built a safe digital biological system in the nation. Furthermore, to protect data in the process, dealing with, capacity and travel to shield the security of resident's data and lessening monetary misfortunes due to digital wrongdoing or information robbery. The ministry has additionally characterized the techniques [11] of the strategy as follow:

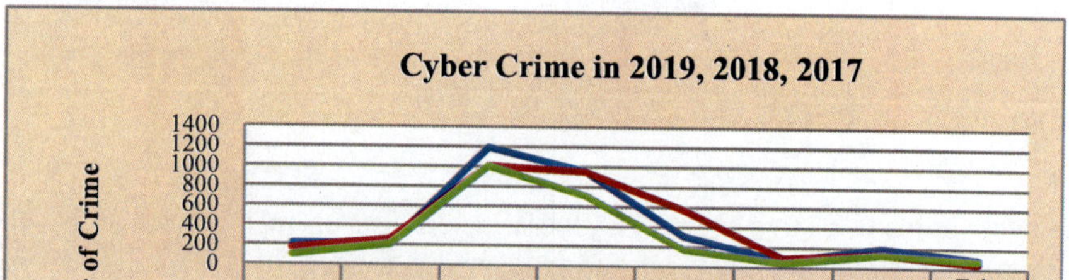


| | Data Theft | Credit/Debit Card Fraud | ATM Fraud | Online Banking Fraud | OTP Fraud | Fake Profile | Cyber Blackmailing | Fake News on Social Media |
|---|---|---|---|---|---|---|---|---|
| Year 2019 | 214 | 213 | 1194 | 953 | 309 | 65 | 203 | 87 |
| Year 2018 | 168 | 262 | 1006 | 953 | 575 | 92 | 138 | 45 |
| Year 2017 | 99 | 208 | 1006 | 711 | 174 | 46 | 133 | 75 |

**Fig. 4** Cyber-crime growth [8]

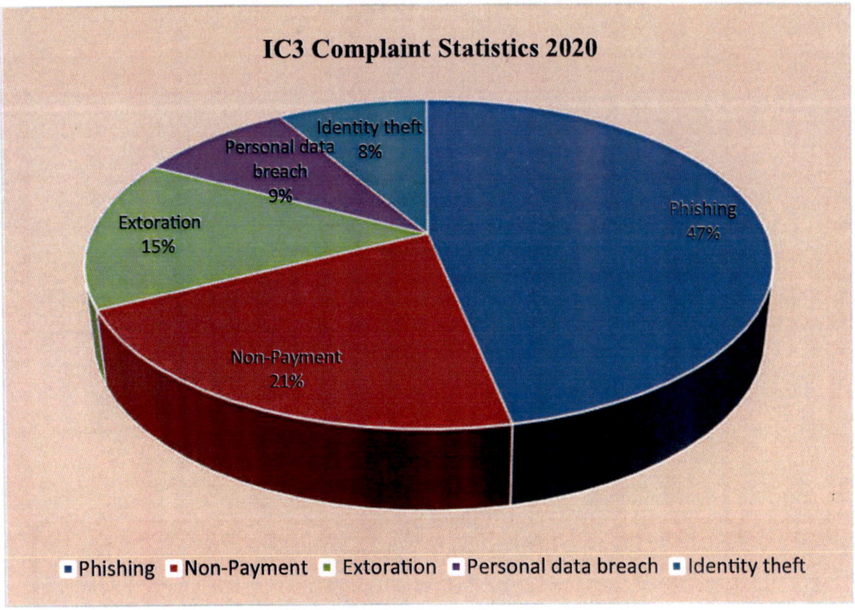


**Fig. 5** Frauds recorded by ICT in 2020 [9]

- To make a safe digital biological system.
- To make an ensured system.
- To encourage open standards.
- To make sure about e-governance administrations.

## 3 Proposed Model

The proposed model is depicted in Fig. 6. The working of the model starts with cyber-crime data collection. The data can be obtained from the publicly available data set. The data will then pass to the exploratory data analysis (EDA) module that will give the insight of data. EDA is applied to data set by the researchers as well scientists to find the features. Data pre-processing techniques then extract the best features from the set of features analyzed by EDA.

After EDA, various data pre-processing techniques then clean the data. The data obtained from the data pre-processing step will be then split into train and test data sets. The train data set is then used to develop the predictive model. The resultant model will then evaluate against the test data set.

The model will be evaluated using evaluation metrics like accuracy, root mean squared error, etc., the error rate of the model is then compared with the threshold. If the calculated error rate is high, then the model will be re-trained in the backward pass. If the error rate will be less than the threshold, then the proposed predicted model can

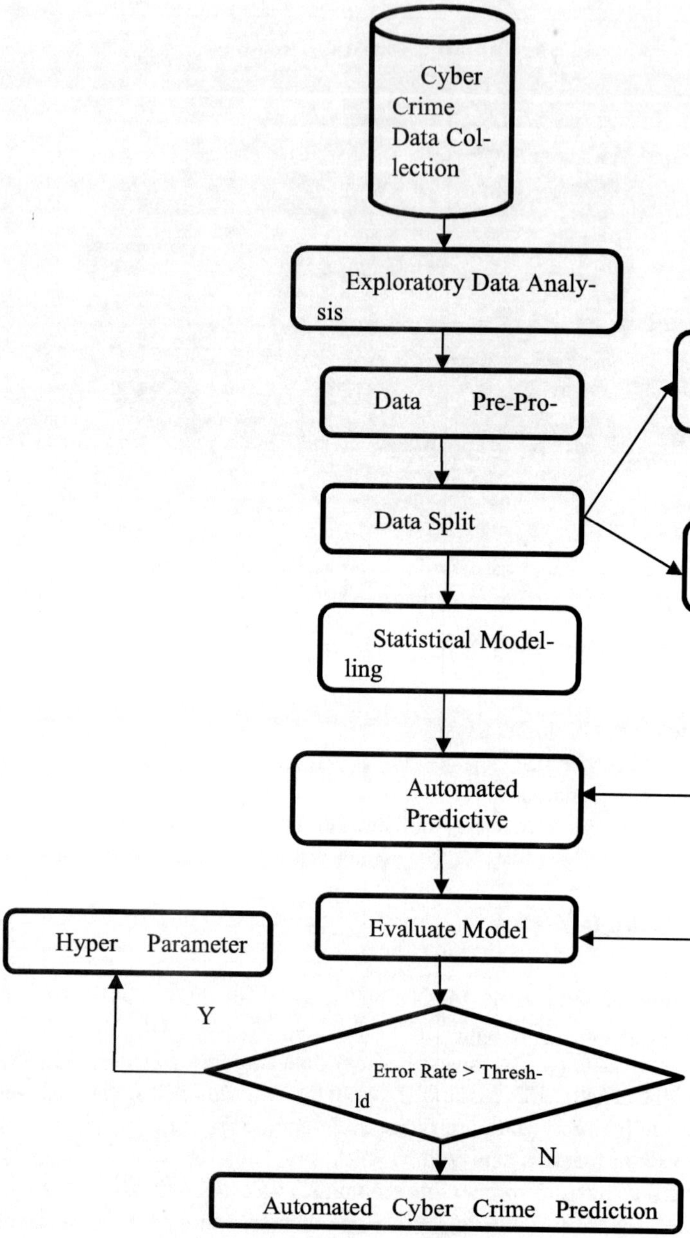


Fig. 6 Proposed model

be used in decreasing the cyber-crime rates. So, after applying this proposed model to any data set rate of cyber-crimes can be reduced.

## 4 Issues and Challenges in Existing Techniques

Security issues like protection, profiling, and credibility likely to be observable by hackers in SNS as compared to other personal web sites and blogs. This is because SNS provides a sense of intimacy created by being among online friends. The principal motivation behind person-to-person communication destinations is to associate individuals and associations. It has additionally evolved numerous business open doors for organizations and firms. Long-range informal communication destinations draw out a particular concern identified with the protection and security of the client [12]. In SNS, there are many privacy features like who can see your profile, you can tag your post with your friends only, got an alarm when someone wants to access your profile, etc. Some of the issues and challenges with these sites are given below.

### *4.1 Issues with SNS*

Establishing communication with large numbers of users results in poor safety of information being transferred. Several users have an indirect relationship with the security of information. Some of the problems are listed in Fig. 7 along with their detail as follow:

- **Misusing Personal Information**: The aggressor mimics the personality of the user brings out abusing personality [13]. The aggressor's personality tries to access information by stealing someone else authorization. At the same time, the owner of the application is not aware of his/her own real identity. The owner allows such person to access the application, and they will take all confidential information and later on that information could be misused without knowledge of the owner.
- **Using Outsider Applications**: Initially, such type of applications asks for general consent from the owner to find personal information for many dissimilar games and applications. Owner permits application to a particular extends of authorization depending on client information. Afterward, a few of these applications that are running in the front area might download firmware on the client's personal computer/mobile without the knowledge of the owner [14].
- **Trusting SNS Users**: Whenever the owner of the post shares his/her personal information on the web, the information can be simply access by system administrators.
- **Awareness:** There is no provision to make users aware of privacy policies. Training or some videos must be provided to the user to make them aware of the policies.

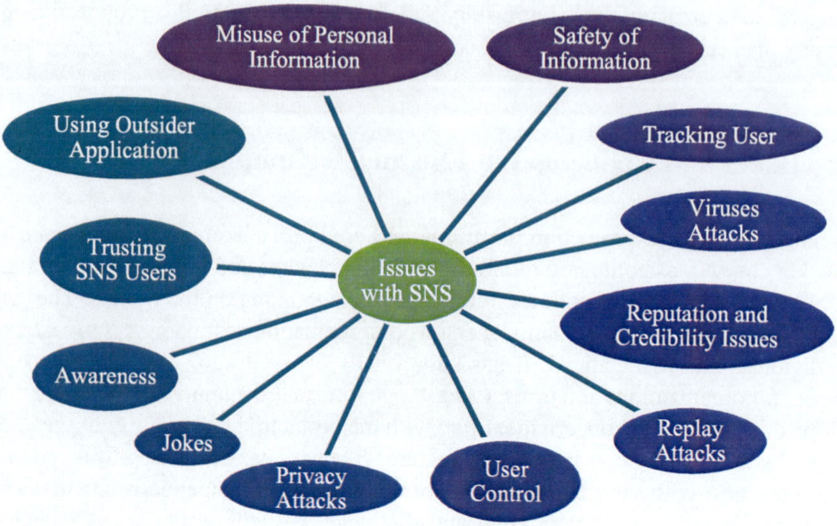


**Fig. 7** Issues with SNS

- **Jokes:** Sometimes a specific user jokes on other people present on the SNS just attracting the attention of all other users. But these jokes sometimes bored the users and can result in an adverse effect.
- **Privacy Tools**: These are the tools that are available with SNS, but the user is not aware of these tools. They do not know much about these tools. From the survey, it has been found that these tools are much complex to understand their working user must need training [15].
- **Viruses Attacks:** Viruses got entry inside anyone's personal computer through emails and promotional advertisements. When they enter someone's computer, they are capable to access all personal information of that particular computer. Internet is the source through which these unwanted threads got access right to someone's personal information.
- **Tracking Users:** By tracking someone's personal information like password, IP address, MAC address, etc. An unauthorized user can get access to his/her personal information. It may result in physical safety concerns for the owner, as an unauthorized user might get access to the information of the owner [16].
- **Safety of Information:** Users post their personal posts on social networking sites and this information may be hacked by hackers resulting in insecurity [6]. If settings are not done properly by the user, then anyone can access user information for wrongdoing.

- **User Control:** Users can only control their profile but cannot control their known one profile. Friends on SNS cannot be trusted. Friends can access our profile. Many times, they can misuse our personal information.

- **Replay Attack**: It takes place when a cybercriminal wants to interfere with online information communicated by a third party. It can edit the information transmitted and can resend it to mislead the receiver.
- **Reputation and Credibility Issues:** Reputation and credibility are very important specifically for individuals, for an organization, and also for a group of persons. It plays a major role in fields like companies, organizations, social status, etc. Lots of people depend on these SNS for keeping in relation with their known ones. These days many organizations pick information from SNS for selecting workers for them. Many people among us update their personal information on the SNS for getting in touch with organizations. But sometimes hackers can use this information for wrong means [17].

### *4.2 Challenges with SNS*

As the number of users increases on the web, security decreases. As information got transmitted among a large number of users, it may be misused by unauthorized persons. A few of the challenges are shown in Fig. 8:

- **Phishing Attacks**: These are the type of social attacks mainly applied to get user personal information, containing user names, passwords, and CC numbers. These attacks got entry when the user accesses an email, instant messages, advertisement, etc. [18].
- **XSS:** By doing some settings XSS can be produced for web sites. Hackers can find information on cookies with the help of XSS. URLs connected with XXS draw the attention of users. When the user clicks on the URL attached with XXS user information is saved on the hacker's computer.

**Fig. 8** Challenges with SNS

- **Safety Risks:** Security of personal information decreases as the number of users increase on a particular SNS. When the user accesses a particular SNS his/her detail is saved on that site. Which later on can be misused by the hackers?
- **Issue with Identity**: It is planning utilized to share the user certifications on many applications. Many applications ask the user to log in with their Google account or Facebook ID pretending to them that it is beneficial to users and users does not need to fill in their records every time [19].
- **Sale of Fake Products:** These days' hackers attract clients by providing or offering a big discount. By seeing these attractive offers clients click on the product and the personal information of that particular client is saved by the hacker. When client click on payment button, then their card detail is also saved on hacker's system. Later on, the hacker can misuse this information [20].
- **Identity Theft:** Almost all SNS ask for permission for accessing the personal information of clients. If the user does not allow accessing his/her personal information, then he is not allowed to access that specific site.
- **"LOL" Virus**: This virus got access when the user visits any SNS. This virus automatically sent to the user showing "lol" with the link. Whenever the user clicks on the link, this virus got downloaded to the user's computer. Now, this virus can access all the personal information of the host computer.
- **Zeus**: This virus is also known as a Trojan horse. It will also spread with the help of SNS. When information is accessed on the web, then sometimes it got downloading through a link. After download, it can access all information of that particular device. It mainly focuses on the bank certifications of the user.
- **Antivirus**: Using antivirus user can keep safe his/her device. As hackers are experts in creating new viruses so antivirus can keep away from these viruses.

## 5 Recommendations

How cybersecurity system works to detect, recover, and avoid the cyber-attacks. In this segment, a few suggestions are given to make sure about the data of the client:

- The company should make some policies for mails so that there is clarity for the user between mails and spam mails, attacks, or viruses.
- Verification needs to be done at every step for sites; you are accessing, and user verification is also very necessary.
- Security systems like cryptography, digital signature, hashing, etc., can be used for providing safety to digital information [21].
- Firewalls are also helpful for maintaining the security of digital information. Firewalls filter the information before it enters a particular network.
- Designing some projects to guide Internet users is also helpful for maintaining the safety of digital information. These projects help in spreading awareness among Internet users.

- Also need to focus on digital technologies, how they work and what danger can be associate with a particular technology, challenges associated with that technology, prevention method, etc.
- Users must apply some limits on the post like not allowing to access personal information: user ID, account number, address, etc.
- Always keep in mind that SNS is public anyone anywhere anytime can access the web. So, keep information that is not personal on the SNS.
- Always keep in mind that the information you post is not read-only. Before performing some action, first though carefully and take needful precautions to keep it safe [22].
- Using a strong password client can keep safe herself or himself. For making a strong password, you need to combine alphabets, numbers, and special symbols. Password must not be kept on the site.
- By applying cryptography, clients can change secret information in some other form that is not understandable by the hacker. By doing so secret information can be protected from hackers [23].
- The digital signature provides authenticity to the source person and the third party. By using a digital signature, users can avoid some of the risks associated with SNS.
- By using steganography, secret information is concealed inside a carrier file. So, the hacker must not be aware of whether some information is transmitting or not through SNS.
- Some personal interactive sites need to be designed for interaction among users. Organizing some talks to explain protection security settings for tools might be useful for securing your account [24]. Permission for your stuff: This feature is provided for Facebook users. Users can set a constraint for the other users who can or cannot check his/her posts. By setting permissions security for this social site can be improved.
- When anyone tries to access your account Facebook, you get an alert on your device that someone is trying to access your account [25].
- There is a feature in Facebook that can show who all visited your account and tried to access which information.
- The client might stay with the latest and programmed updates ought to be empowered for the program. The client must block modules, pop-ups, and advertisements for providing security, otherwise, through these advertisements viruses may enter your computer [26].

## 6 Conclusion

With the growth of SNS, security has become the most important task for the users. Hackers' main objective is to violatee and assaults digital media. Digital media violation is becoming a measure problem to the society and nation also. All government and private organizations facing issues mainly in digital applications like information communication, banking, medical, military, etc. So, some measure needs to be taken to provide security to all these digital media applications. All Internet users must coordinate with each other to improve the security level of the SNS. Training must be provided to users who are not aware of security policies. The use of antivirus also enhances the security of the existing SNS. Users can apply cryptography, digital signature, and watermarking techniques to improve the security of the SNS. The proposed model helps to reduce cyber-crime by analyzing the data set.

## References


1. Hajli, N., Lin, X.: Exploring the security of information sharing on social networking sites: the role of perceived control of information. J. Bus. Ethics **133**(1), 111–123 (2016)
2. Kumar, A., Gupta, S.K., Rai, A.K., Sinha, S.: Social networking sites and their security issues. Int. J. Sci. Res. Publ. **3**(4), 1–5 (2013)
3. Malware insights. Accessible: https://www.av-test.org/en/measurements/malware/
4. Das, R., Patel, M.: Cyber security for social networking sites: issues, challenges, and solutions. Int. J. Res. Appl. Sci. Eng. Technol. (IJRASET) **5**(4), 833–838 (2017)
5. Morgan, S.: Global cybercrime damages predicted to reach $6 trillion annually by 2021. cybersecurityventures.com (2021). Accessed 11 April 2021
6. Lawler, J.P., Molluzzo, J.C.: A study of the perceptions of students on privacy and security on social networking sites (SNS) on the internet. J. Inf. Syst. Appl. Res. **3**(12), 3–18 (2010)
7. Lexis-Nexis Risk Solution. https://risk.lexisnexis.com/. Accessed 11 April 2021
8. National Crime Record Bureau. https://ncrb.gov.in/. Accessed 10 April 2021
9. Internet Crime Complaint Center. https://www.ic3.gov/media/pdf/annualreport/. Accessed 10 April 2021
10. Facebook Privacy Basics. Accessible: https://www.facebook.com/about/nutsandbolts
11. Huber, M., Mulazzani, M., Weippl, E.: Social networking sites security: Quo Vadis. In: 2010 IEEE Second International Conference on Social Computing, pp. 1117–1122. IEEE (2010)
12. Joshi, A.P., Han, M., Wang, Y.: A survey on security and privacy issues of blockchain technology. Math. Found. Comput. **1**(2), 121 (2018)
13. Doleck, T., Lajoie, S.: Social networking and academic performance: a review. Educ. Inf. Technol. **23**(1), 435–465 (2018)
14. Zhou, W., Jia, Y., Peng, A., Zhang, Y., Liu, P.: The effect of IoT new features on security and privacy: new threats, existing solutions, and challenges yet to be solved. IEEE Internet Things J. **6**(2), 1606–1616 (2018)
15. Ismagilova, E., Hughes, L., Rana, N.P., Dwivedi, Y.K.: Security, privacy and risks within smart cities: literature review and development of a smart city interaction framework. Inf. Syst. Front. 1–22 (2020)
16. Tabrizchi, H., Rafsanjani, M.K.: A survey on security challenges in cloud computing: issues, threats, and solutions. J. Supercomput. **76**(12), 9493–9532 (2020)
17. Zeebaree, S., Ameen, S., Sadeeq, M.: Social media networks security threats, risks and recommendation: a case study in the kurdistan region. Int. J. Innov. Creativity Change **13**, 349–365 (2020)
18. Hawkins, S., Yen, D.C., Chou, D.C.: Awareness and challenges of Internet security. Inf. Manag. Comput. Secur. (2000)
19. Aldowah, H., Rehman, S.U., Umar, I.: Security in internet of things: issues, challenges, and solutions. In: International Conference of Reliable Information and Communication Technology, pp. 396–405. Springer, Cham (2018)

20. Ahn, G.J., Shehab, M., Squicciarini, A.: Security and privacy in social networks. IEEE Internet Comput. **15**(3), 10–12 (2011)
21. Zhou, J., Cao, Z., Dong, X., Vasilakos, A.V.: Security and privacy for cloud-based IoT: challenges. IEEE Commun. Mag. **55**(1), 26–33 (2017)
22. Dhami, A., Agarwal, N., Chakraborty, T.K., Singh, B.P., Minj, J.: Impact of trust, security and privacy concerns in social networking: an exploratory study to understand the pattern of information revelation in Facebook. In: 2013 3rd IEEE International Advance Computing Conference (IACC), pp. 465–469. IEEE (2013)
23. Browser Security Settings. Accessible: http://its.ucsc.edu/programming/discharge/programsecure.html
24. Ashibani, Y., Mahmoud, Q.H.: Cyber physical systems security: analysis, challenges, and solutions. Comput. Secur. **68**, 81–97 (2017)
25. Cavus, N., Sani, A.S., Haruna, Y., Lawan, A.A.: Efficacy of social networking sites for sustainable education in the era of COVID-19: a systematic review. Sustainability **13**(2), 808 (2021)
26. Newman, L., Stoner, C., Spector, A.: Social networking sites and the experience of older adult users: a systematic review. Ageing Soc. **41**(2), 377–402 (2021)

# FPGA Implementation of Optimized Code Converters with Reversible Logic Gates

**Kommalapati Rajesh, Mayuri Kundu, Argha Sarkar, M. Sreenath, and Prasenjit Deb**

**Abstract** In today's technology development, there is vast advancement present the low power VLSI; it is possible to decrease the power dissipation. The reversible concept is one of the significant technologies for attaining the more interest because of low power dissipation and high speed operation. In electronics, the code converters are widely used because it enhances the security of the information, reducing the arithmetic operations complexity and thus decreasing the required hardware, power saving, the high speed of operation, etc. In this paper, an optimized design of gray to binary, binary to gray, Ex-3 to BCD, and BCD to Ex-3 code converters with reversible logic gates is proposed. The simulation and synthesis process can be done with Xilinx ISE software, and it is dumped on FPGA Spartan-6E.

**Keywords** Aggressive VLSI · Reversible computation · Code converters · ISE · FPGA

## 1 Introduction

The reduction of power consumption is one of the main key factors in modern electronics. In 1960, R. Landauer study reveals that the conventional hardware computations and irrespective of the realization technique, the power dissipation is because

K. Rajesh
Department of Electronics and Communication Engineering, Audisankara Institute of Technology, Gudur, Andhra Pradesh, India

M. Kundu (✉)
School of Computer Science and Engineering, REVA University, Bangalore, Karnataka, India
e-mail: kundu.mayuri@gmail.com

A. Sarkar
Department of Electronics and Communication Engineering, Audisankara College of Engineering Technology, Gudur, Andhra Pradesh, India

M. Sreenath · P. Deb
School of Electronics and Communication Engineering, REVA University, Bangalore, India



D. Gupta et al. (eds.), *Pattern Recognition and Data Analysis with Applications*, Lecture Notes in Electrical Engineering 888,
https://doi.org/10.1007/978-981-19-1520-8_26

of data loss [1]. So, to overcome this problem present in the conventional methods, in 1973, C. Bennett proved that to evade the KT*ln2 of power dissipation in a circuit, it should be fabricated with reversible gates only [2, 3]. The power dissipation is ideally zero for reversible computing, because it does not lose information. The main condition for reversibility is the total numbers of inputs are equals to a total number of outputs, and there should be a one-to-one connection among all the inputs and outputs, then only we determine the inputs from the outputs also [4–6].

## 2 Overview

The reversible computing has attained more attention in few decades. It has many applications in bio-informatics, nanotechnology, thermodynamic technology, and optical computing. The construction of quantum circuits is not possible without the reversible gates. Designing of reversible logic gates (RLG) is more problematic than the typical or irreversible logic gates since there is no feedback and fan-out [4] in reversible gates.

The performance constraints in reversible computing are termed below.

1. Garbage Outputs: Existence of unused outputs available in reversible gates.
2. Delay: The total time taken by the propagation of input to the output.
3. Constant Inputs: Maintaining of inputs at either constant '0' or constant '1' to get the proper output.
4. Quantum Cost (QC): It is referred as the number of $1 \times 1$ and $2 \times 2$ reversible logic gates present in the design.

### *2.1 Reversible Logic Gates*

#### 2.1.1 Feynman Gate (FG)

FG is basically a $2 \times 2$ **RLG** with corresponding QC is '1'. Figure 1 displays the logic diagram of logic gate of Feynman gate.

**Fig. 1.** Logic diagram

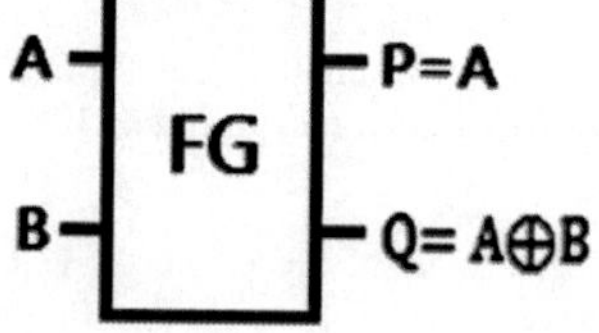

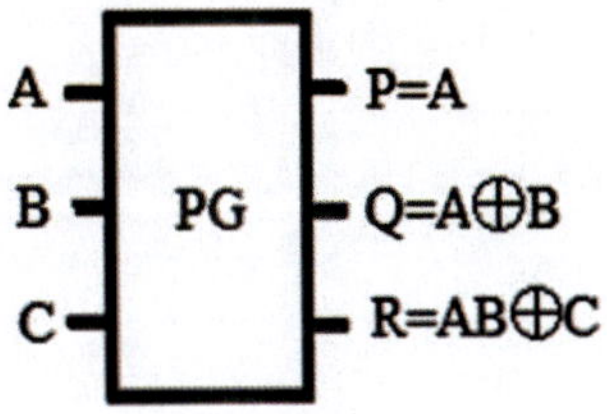


**Fig. 2** Logic diagram

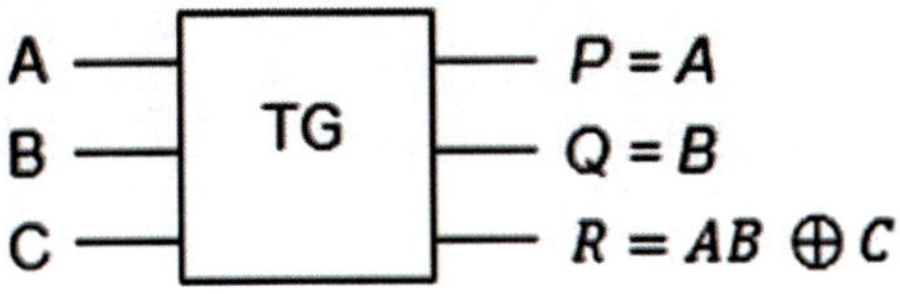


**Fig. 3** Logic diagram

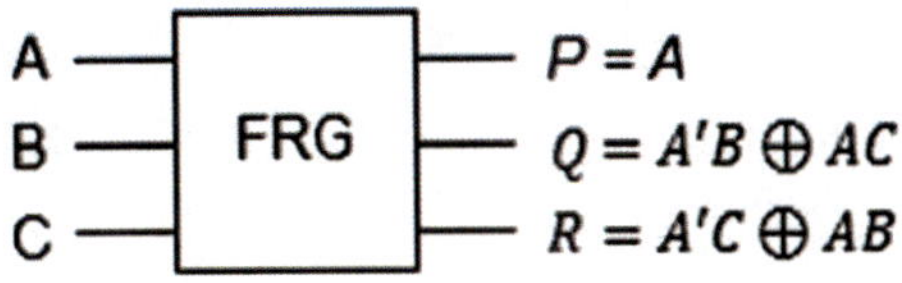


**Fig. 4** Logic diagram

#### 2.1.2 Peres Gate (PG)

PG describes a 3 × 3 **RLG** with corresponding QC is '5'. Figure 2 displays the logic diagram of logic gate of Peres gate.

#### 2.1.3 Toffoli Gate (TG)

TG describes a 3 × 3 **RLG** with corresponding QC '5'. Figure 3 displays the logic diagram of logic gate of Toffoli gate.

#### 2.1.4 Fredkin Gate (FRG)

FRG is basically a 3 × 3 **RLG** with corresponding QC is '5'. Figure 4 displays the logic diagram of logic gate of Fredkin gate.

#### 2.1.5 Universal Reversible Gate (URG)

URG is a 3*3 **RLG** with corresponding QC is '5'. Figure 5 displays the logic diagram of logic gate of URG gate.

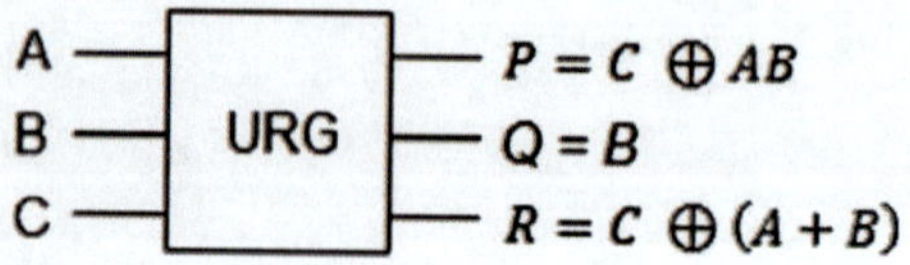


**Fig. 5** Logic diagram

**Table 1** Truth table of FG gate

| *A* | *B* | *P* | *Q* |
|---|---|---|---|
| 0 | 0 | 0 | 0 |
| 0 | 0 | 0 | 1 |
| 1 | 1 | 1 | 1 |
| 1 | 1 | 1 | 0 |

# 3 Proposed Design of Reversible Code Converters

The designs of reversible circuits are an interesting task, and the construction of reversible code converters is very crucial one because of reversible processor requires its building blocks would be reversible. In digital domain, the data represented with a group of 0s and 1s. Basically, the code is a sequence of 0s and 1s are used to denote the data. It is a part of combinational circuits which are used to alter one form to another form. Any code converter design can be implemented with the basic operations of AND, OR, and NOT gates.

## *3.1 Gray–Binary (GTB) and Binary–Gray (BTG) Reversible Code Converter*

The BTG and GTB converters are used to decrease the switching activity by attaining the one-bit change among the logical sequences. The $I(A, B, C, D)$ are the input vectors and $O(W, X, Y, Z)$ are the output vectors, respectively. It is designed with Feynman gate [7–14]. Table 1 displays the Feynman gate truth table. Figures 6 and 7 display the implementation of GTB and BTG converters, respectively.

## *3.2 Ex-3-BCD Code (E3TB) and BCD-Ex-3 (BTE3) Reversible Code Converter*

The ETB and BTE converters are used in the arithmetic circuits, and it is decreases the complexity of the hardware. It is designed with URG gate [9] and CNOT gate. Table 2 displays the URG truth table. Figures 8 and 9 display the design of E3TB and BTE3 code converters, respectively.

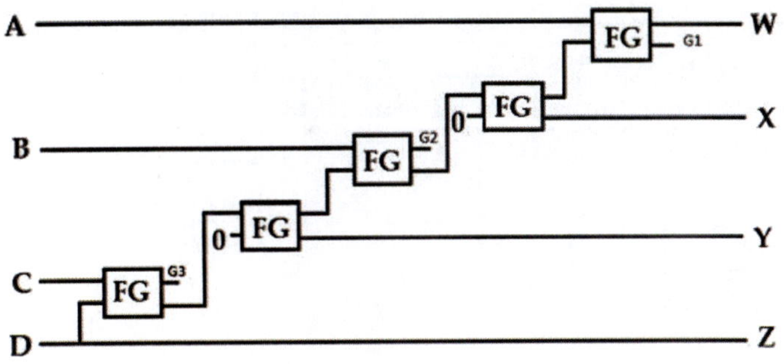


**Fig. 6** Implementation of reversible GTB converter

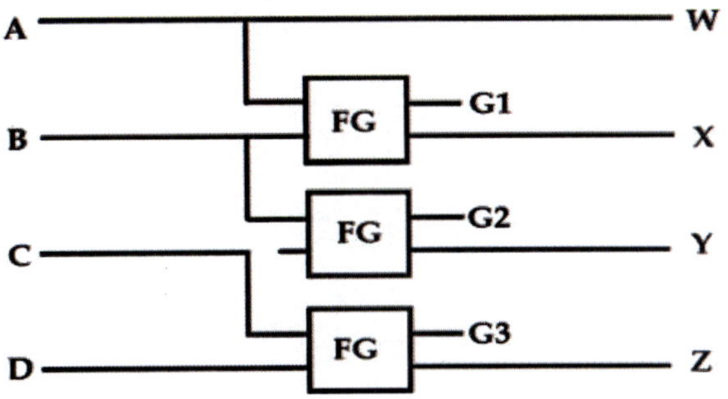


**Fig. 7** Implementation of reversible BTG converter

**Table 2** URG gate truth table

| *A* | *B* | *C* | *P* | *Q* | *R* |
|---|---|---|---|---|---|
| 0 | 0 | 0 | 0 | 0 | 0 |
| 0 | 0 | 1 | 1 | 0 | 1 |
| 0 | 1 | 0 | 0 | 1 | 1 |
| 0 | 1 | 1 | 1 | 1 | 0 |
| 1 | 0 | 1 | 1 | 0 | 0 |
| 1 | 1 | 0 | 1 | 1 | 1 |
| 1 | 1 | 1 | 0 | 1 | 0 |

## 4 Results and Discussion

The proposed code converters are optimized than existing code converters. Table 3 displays comparative analysis of reversible code converters, and Fig. 10 displays the graphical representation of reversible code converters. The logical operations are used in a proposed converters and are intended by subsequent logical assignments.

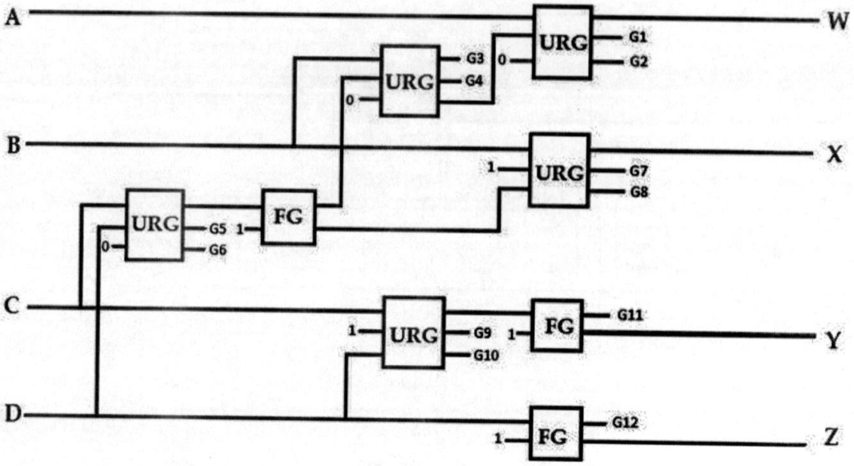


**Fig. 8** Implementation of reversible E3TB converter

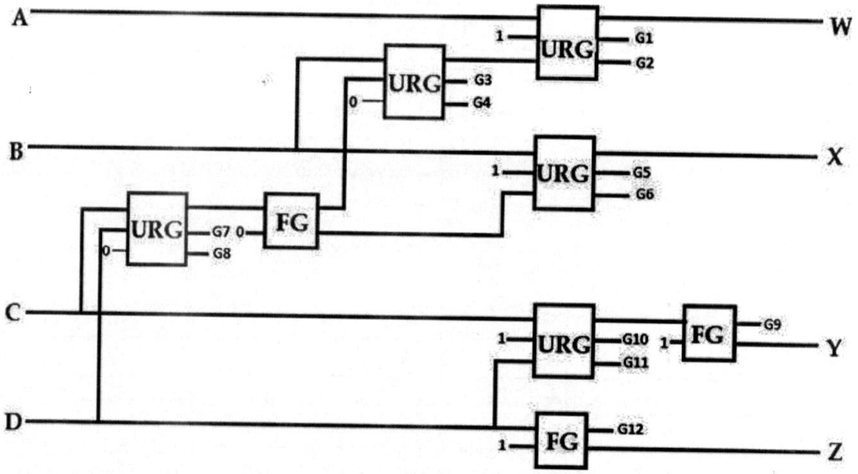


**Fig. 9** Implementation of reversible BTE3 converter

**Table 3** Comparative analysis of reversible code converters

| S. No. | Code converters with reversible gates | Number of logic gates | Number of garbage outputs | Number of constants | Total logical calculations |
|---|---|---|---|---|---|
| 1 | Excess-3 to BCD | 8 | 12 | 8 | $2p + 3r + 1\,s + 2t$ |
| 2 | BCD to Excess-3 | 8 | 12 | 8 | $3p + 1q + 2r + 1\,s + 1t$ |
| 3 | Gray to binary | 2 | 3 | 2 | $3p + 2q$ |
| 4 | Binary to gray | 3 | 3 | 0 | $3p$ |

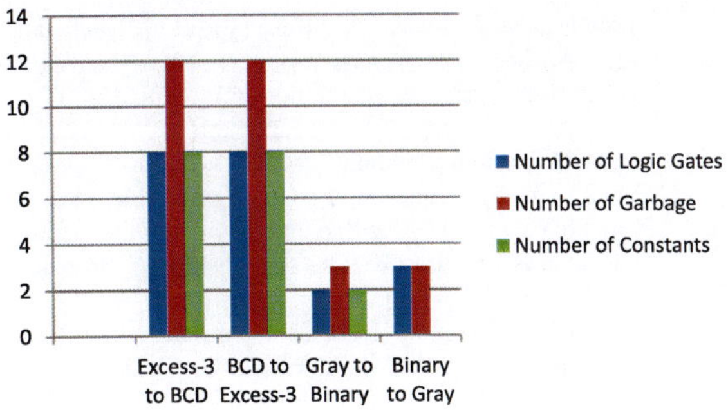


**Fig. 10** Graphical representation of reversible code converters

$p =$ AND logic
$q =$ OR logic
$r =$ NOT
$s =$ Buffer
$t =$ OR logic

For instance, if $T = 2T + 3R + 1Q + 2P$, then the circuit comprises 2-XOR operations, 3-NOT operations, 1-OR operation, and 2-AND operations (Fig. 10).

## 5 Conclusion

The proposed code converters are successfully designed with RLG, and it is optimized than the existing code converters. If the reduction of total numbers of reversible gates is ensured in a circuit, obviously it reduces the power dissipation. The proposed designs are designed with help of new FG and URG gates, might be in the future these gates are more optimized because there is no fixed number of reversible gates present in the reversible computing, then all the future projected designs are more optimized, and it affects all the performance parameters. The synthesis and simulation processes are effectively completed with Xilinx ISE software, and it is successfully dumped on FPGA Spartan-6E.

## References

1. Landauer, R.: Irreversibility and heat generation in the computing process. IBM Res. Dev. **5**(3), 183–191 (1961)
2. Bennett, C.: Logical reversibility of computation. IBM Res. Dev. **17**(6), 525–532 (1973)
3. Kerntopf, P., Perkowski, M., Khan M.H.: Universality of general reversible multiple valued logic gates. In: 34th International Symposium on Multiple Valued Logic, IEEE, pp. 68–73 (2004)

4. Perkowski, M., Jozwiak, L., Kerntopf, P., Mishchenko, A., Al-Rabadi, A., Coppola, A., Buller, A., Song, X., Yanushkevich, S., Shmerko, V.P., Chrzanowska-Jeske, M.: A general decomposition for reversible logic. In: Proc. RM'2001, pp: 119–138. Starkville (2001)
5. Perkowski, M., Kerntopf P.: Reversible logic invited tutorial. In: Proceedings URO-MICRO, pp. 31–42. Warsaw, Poland (2001)
6. Srinivas, M. B., Himanshu, T.: Novel reversible TSG gate and its application for designing reversible carry look ahead adder and other adder architectures. In: Proceedings of the 10th Asia-Pacific Computer Systems Architecture Conference, pp. 775–786. Springer, Heidelberg (2005)
7. Feynman, R.: Quantum mechanical computers. Opt. News **11**(2), 11–20 (1985)
8. Cholan, K., Saravanan, M., Abhishek, G.: Design of novel reversible multiplier using MKG gate in nanotechnology. In: Proceedings of National Conference on Automation Control and Computing, IEEE, pp. 1–6. India (2013)
9. Veezhinathan, K., Mahammad, S.N.: Constructing online testable circuits using reversible logic. IEEE J. Instrum. Meas. **59**(1), 101–109 (2010)
10. Toffoli, T.: Reversible computing. Tech Memo MIT/LCS/TM-151. MIT Lab for Computer Science (1980)
11. Peres, A.: Reversible logic and quantum computers. Phys. Rev. A **32**(6), 3266–3276 (1985)
12. Azad Khan, Md.M.H.: Design of full adder with reversible gate. In: International Conference on Computer and Information Technology, Sci. Res. pp. 515–519. Bangladesh (2002)
13. Navi, K., Haghjparsat, M.: A novel reversible full adder circuit for nanotechnology based systems. J. Appl. Sci. **7**(4), 3995–4000 (2007)
14. Navi, K., Haghparast, M.: Design of a novel fault tolerant reversible full adder for nanotechnology based systems. World Appl. Sci. J. **3**(1), 114–118 (2008)

# A Security Provocation in Cloud-Based Computing

Aritra Dutta, Rajesh Bose, Swarnendu Kumar Chakraborty, and Sandip Roy

**Abstract** Cloud computing makes a massive trend for research organizations because highly demand. There is different type of causes that many organizations depend on it and take advantage from cloud computing service. Cloud computing appears a model that requires all the primary elements of evaluation such as end-user contraption, transmission web channel, access conduction, and cloud architecture. Moreover, with the exposure of new appearance in the computing world like the 5G Internet, Internet of an object (IoB), and elegant town, the bit part of cloud evaluation will be added essential for operation and storage more data than always preparatory to. The principal advantage of cloud computing is ambidexterity, domain agreement, and superior service for the end user. So, the application of cloud evaluation in a non-identical realm is the principal stability with the ease of cloud-based resource collaboration. These incorporate many IT infrastructure, remote access servers from anywhere, we can use it through the Internet. The related safety and protection challenges in the cloud needed further investigation. Scientists from the different domains provided a possible blend to these provocations in the previously published learning area. As the expanding request of cloud computing makes a huger range toward end-user data safety. In this paper, we observe meta-inspection has been given toward cloud computing certainty. It also incorporates complete analysis based on the outcome achieved in the same. The paper gives us a logical and experiential perspective of cloud computing safety.

**Keywords** Cloud computing · Security · Threats · Vulnerabilities · Data protection · AI

---

A. Dutta (✉) · R. Bose · S. Kumar Chakraborty · S. Roy (✉)
Brainware University, West Bengal, India
e-mail: dutta.aritra09@gmail.com

S. Roy
e-mail: sandiproy86@gmail.com

S. Kumar Chakraborty
e-mail: swarnendu@nitap.ac.in



D. Gupta et al. (eds.), *Pattern Recognition and Data Analysis with Applications*,
Lecture Notes in Electrical Engineering 888,
https://doi.org/10.1007/978-981-19-1520-8_27

## 1 Introduction

Cloud computing has introduced heaps of advantages like different technology depends on it. Because it made to supply a huge number of data and different ministrations. Moreover, this principle resolved the matter of restricted supplies and lessen the value of ministration by dividing valuable resources contribution among end users. In current years, based on the majority cloud ecosystem has become one of the most important and valuable topics in the cloud data safety researches world and the analyst incorporates user- data safety in data server, web safety, and system safety. The National Institute of Standards and Technology (NIST) explains cloud ecosystem as, "a model for authorizing suited, resource contribution, omnipresent, on-request access which can be very fast delivered to the non-identical types of the service provider" [1]. The strategy of cloud-based computing follows wages as you want (WAYW), in which the client only pays for the facility they use [2]. WAYW architecture provides customize services to end user for storage, software, development platform, elasticity, stability very economically, and so on.

Virtualization technique in cloud architecture is the ground of distributing infrastructure as a service (IaaS) that distribute information and resource, network, approach, and different tools from hardware limitation. The cloud ecosystem is becoming admire among many organizations because of its acrobatic, workable, and a very low budget statement providing at software, principle, and configuration level. Software as a service (SaaS) permits end users to access requests organized by the different cloud service providers (CSP) on cloud over the Internet. Platform as a service (PaaS) allows innovators to create their platform applications, take a look at and deploy their applications on IaaS [3]. Cloud computing allows sharing maneuvers amid the individual devices or local hosts in the network. The cause and creation of the cloud based are connected with the classification model. The positioning model in the cloud ecosystem incorporate three types, namely public-based cloud, private-based cloud, and hybrid-based cloud. Security protection is a crucial challenge that grows enormously with increases in end user [4]. In this paper, we are profundity view is conducted on the recent procedure of the cloud data storage safety related to the cloud ecosystem. The survey of the cloud ecosystem and safety issues is analyzed in this paper. The main security precondition such as data integrity, availability, and confidentiality are described in this paper. Security issues in the recent strategies of cloud computing are also discussed in this paper (Fig. 1).

## 2 Literature Review

In the present-day, cloud ecosystem is an appear computing example that brings in distributed system new challenges for information safety, different type approach controls, etc., a lot of observation look over and aims at the safety provocation in the cloud-based computing ecosystem. Moreover, it is indisputable that majority of

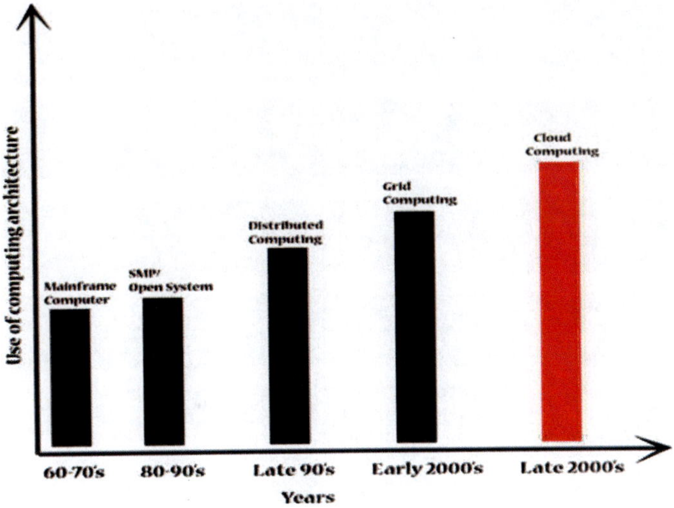


**Fig. 1** Different computing based on year

the handover assessed papers play an important part in cloud safety issues and these noticeable analysis works had been a ton of significant advantage of a research area.

Lee and Kim in their paper, they are introduced SIEM planning is that can be established to the SECaaS policy where we have been producing for examining and perceiving insightful virtual-threat based on virtualization technologies in the cloud ecosystem [5].

Wei and Xiao, they are implements new tools of load runner and AppScan, this will protect the personal private data of various education departments and teachers, and students. This will be improving teaching quality in education fields [6].

TAO and LEE, they will focus their paper that the security gaps in bridging MTCS. Healthcare IT security policy (HITSecP) in Singapore that brings forth CSPs who have been affirmed to MTCS, can realize how well and best they could meet the security prerequisites of IT frameworks for the healthcare industry using cloud ecosystem [7].

Liu et al., they are a focus in their paper private cloud network security, they are analyzing the pertinent assessment guide, and establishes a new assessment framework model and also other key applied, that elaborates the private cloud system network safety in new venture [8].

Markandey and Gajmal say their paper, that nowadays the information security assurance strategies are very big difficulty on the Internet. Availability of data in the cloud is useful for certain appeal, yet it stances dangers by introducing data to appeal which may as of now have safety condition in them [9]. Below we are discussing the cloud-based architecture in Fig. 2.

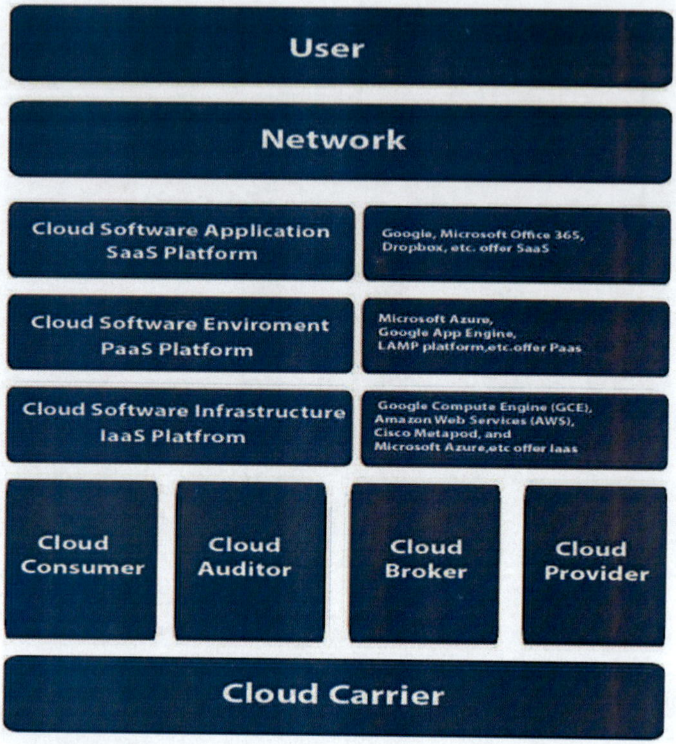


**Fig. 2** Cloud-based architecture

# 3 Cloud Organization

## *3.1 Cloud End User*

A cloud consumer or end users is a combination of human that has a traditional agreement with a cloud provider to use service principle that made available by the cloud provider. A cloud end user can option the most add-on assistance by search the services offered by the cloud contributor and closing the agreement. Cloud end users use service level agreements (SLAs) to define the technical presentation demand to be fulfilled by a cloud provider. A cloud provider may also list in the SLAs a set of limitations and commitments that cloud end users must accept. In a development cloud environment, cloud end users can openly select a cloud provider with a better rate with more favorable terms [10].

## 3.2 Cloud Service Contributor

A cloud service contributor (CSC) is an individual that provides service to the user at a different levels using a virtual cloud framework. Cloud service providers sort out cloud software by getting and dealing with the cloud framework. Many software organization or vendors sell their products on cloud-based software premises through the install, configuring, maintaining, and updating the software applications. In IaaS model, the cloud service provider provides infrastructure components including networks, storage, servers, and hosting infrastructure for end user. And the last type of service provider is PaaS, this provides a platform for software development and commonly performs the various function that are commonly used in software development.

## 3.3 Cloud Checker

A Cloud checker is responsible for monitoring the service of the service provider individually. The checker verifies the standards of the provider depend upon some affirmation. The service provider provides service access by the cloud checker concerning privacy, security control, production, etc.

## 3.4 Cloud Assent

Cloud assent compliance is a conference that has meets the client requirements or criteria in a definite type of certification or substructure. There are a variety of different types of conformity that may be needed for industry, request for proposal, client, etc. The type of cloud safety and conformity requirements will help us to determine the cloud conformity that is right for an organization or not [11].

## 3.5 Cloud Dealer

In cloud ecosystem developments, integration of cloud service for cloud consumers is, too, complex and complicated to manage the service. A cloud end user may demand cloud-based computing from a cloud dealer, instead of contacting directly to the cloud contributor. The cloud dealer is answerable for dealing with the information use, service performance, and delivery of cloud services to the end user additionally the relationships among the cloud end user and cloud providers. Cloud dealer it's divided into three category bases on service.

- Service intervention.
- Service accumulation.
- Service arbitrage.

### 3.6 Cloud Carrier

Cloud carrier is an intermediate state between the cloud end user and cloud provider. A cloud carrier is an environment that is operated by a traditional telecommunication system. Cloud carrier gives the access right to the consumer through the access network. As per service level agreement (SLA) with cloud carrier, cloud provider distributes service to the cloud consumer.

## 4 Cloud Security Service

### 4.1 Data Confidentiality

Cloud services provide different services to the data user and also have become different kinds of the solution provided. The cloud contributor offers to end-users' divergent resources like different type program-based system, hardware, and web to the clients to be able to manage and administer the available database. Data confidentiality ensures or provides trust to the user that data is not disclosed or available to an unauthorized user. Data should be sent to authorize users using a secure network. Data encryption provides data confidentiality to the end user.

### 4.2 Data Integrity

Cloud computing provides implementation software and database in cloud computing are proceed to consolidate large data warehouses, where the management of the data is store and services may not be fully reliable. We have planned a cloud computing security lifecycle to achieve safety and enable the user to take advantage of this technology as much as possible of safety. Data integrity ensures that received data not be modified by an authorized person.

### 4.3 *Availability*

Cloud computing has been operating by disparate types of a user by reason of it has several benefits as well as reducing of infrastructure money, and its main possession is adaptability, which permits administrations to be scaled up or down as per the client need. In the cloud contributors' opinion, there are innumerable provocations to be get the better service to bring cloud services that meet all needs defined in service level agreements (SLAs). High resource availability in cloud servers has been perhaps the greatest test for suppliers, and numerous administrations can be utilized to improve the accessibility of an assistance to end users, such as check pointing, load balancing, and redundancy.

### 4.4 *Data Authentication*

Data attestation or authentication of data is an important issue in cloud-based computing safety. This is the procedure to protect user valuable data in opposition to different sorts of attacks where the main aim is to confirm the specification of a user and authenticate user invocation services from cloud-based servers. Many authentication algorithms have been put forward as yet that confirm user specification before permitting to access authenticate resources. All algorithms are verified (identifier name or login name and password, multi-factor validation, mobile trusted module, open or public key infrastructure, single sign-on, and biometric authentication) is from the start depicted in here. The various strategies introduced will at that point be looked at.

### 4.5 *Non-repudiation*

Non-repudiation is a procedure that ensures the identity of user that cannot be denied by senders or receivers after sent the data. There are two types of ways follow—source abrogation and destination abrogation. In the former, the transmitter cannot prevent the transmission from getting a message, and in the last mentioned, they cannot keep the conveyance from getting a message.

# 5 Security Attacks and Threads in Cloud Computing

## 5.1 *Threat Model*

Cloud computing is a growing trend nowadays that has eliminated the burden of hardware and software infrastructure in case using virtual machines via the Internet. The infrastructure of cloud computing has accepted several safety issues that emerging from the current and new threats from hackers, anything that can base serious vandalization to a computer technology is called a threat. Cautioning can fuse everything from infections, trojans, and indirect accesses to out and out assaults from programmers.

## 5.2 *Spoofing Attack*

Spoofing is the take an action of copying a transmission of data from a dark beginning stage as being from a known, confided in source. Satirizing can engage messages, calls, and destinations, or can be more specific way, for instance, a PC mimicking an IP address, Address Resolution Protocol (ARP), or Domain Name System (DNS) worker. Parodying can be used to get to a point's customer singular information, help out malware through contaminated affiliations or affiliations, stay away from network access controls, or legitimate traffic to lead a renouncing of-organization attack (DOS-assault). A spoofing attack is a harmful way for an agitator to get access to execute a greater virtual-attack like an undeniable level tireless threat or a man-in-the-center attack [12].

## 5.3 *Tampering Attack*

Nowadays web promoting, Internet banking, online reservation, and different online administrations are acquiring prominence among clients, the specialist organizations offer support according to require. The headway in web mechanical technology is giving interface and easy to use administrations to the client. In the web system, altering assault is such a delicate assault that can be effectively used by utilizing some gadget like Tamper Data, Webscarab, Paros Proxy, and Burp Suite, and so on, web structure altering assault depends on the temper of boundaries traded among customer and information worker to control application information, for example, framework subtleties like value, the quantity of items, client qualifications, and consents, and so on, i.e., shipped off the application through a post solicitation in worker. For the most part, user data is put away in PC treats, covered up structure fields, as well as URL query strings, and is utilized to expand application execution and control.

Numerous security conventions are utilized there like SSL, TLS for giving well-being administrations like trustworthiness and validation of client information. Yet, they do not give any symbolic method to stop boundary altering assaults [13].

## 5.4 Repudiation Attack

In the systems administration model vehicle layer, organization coating safety is not get to forestall the assailant to assault the hubs in the organization. Repudiation is the virtual assault where the assailant is skirted from the vehicle and organization layer. Repudiation attacks allude to refusal of cooperation in the transmission. A disavowal assault can be viewed as a malware assault though an aggressor hub continues to utilize the PC or framework asset hub as a childish hub and renounce any leading activity which is coming from framework asset to transmission on the PC organization. Consequently, the arrangement is that taken to tackle validation or non-renouncement assaults in the organization layer or transport layer is not sufficient to forestall assaults. Illustration of renouncement assault on a business structure: an egocentric individual could deny leading a procedure on a charge card buy or deny any online exchange [14].

## 5.5 Backdoors Attack

A backdoor attack is a kind of malware assault where permits the programmer to get to the unaccredited site. The hacker embeds the malware through unstable marks of passage program, for example, obsolete modules or information fields. When they enter through the secondary passage of the organization, they approach all your significant information, including clients' very own recognizable data, introducing undesirable programming, or in any event, assuming responsibility for the whole PC.

## 5.6 Direct-Access Attack

A prohibited user who achieves physical access to a PC is undoubtedly ready to straightforwardly duplicate information from the PC or asset. They may also agree to system safety by making framework adjustments, embedding's programming worms, keyloggers, covert listening devices, or using distant mice. In any case, when the design is ensured by standard prosperity tries, these might have the decision to be by-passed by booting another working framework or device from a CD-ROM or other bootable media [15].

## 5.7 *Pretexting Attack*

Pretexting is another kind of friendly masterminding where aggressors target appearing, or a caused situation, that they use to attempt to take their difficulties' own unique data. In such assaults, the joke artist by and large says they need certain bits of information from their goal to ensure their character. They take that data and use it to submit discount mutilation or stage helper attack.

Numerous cutting-edge assaults at times endeavor to fool their points into accomplishing something that manhandles a worry's advanced and additionally actual shortcomings. For instance, an aggressor would conceivably mimic an external IT contributions evaluator so they can talk a coal organization's real security bunch into allowing them to into the structure [16].

## 5.8 *Watering Hole Attack*

A watering hole attack is a chosen assault was intended to bargain clients inside a chose industry or association of clients through method of methods for tainting sites they by and large go to and drawing them to a pernicious site. The ultimate objective of this assault is to contaminate the client's PC and obtain entrance right to the association's organization. Watering hole assaults, otherwise called key site bargain assaults, are limited in scope as they rely on detail of karma. They do yet come to be extra powerful, while blended in with email actuates to bait clients to sites.

# 6 Current Movement in Cloud Computing

Presently, new trends and survey have been shown and presented below, there appears the user information holding and cloud-based computing threats mechanism. There is also discuss the current need of today's cloud computing environment by the end user. It appears the need for data holding algorithm and resource sharing with the preservative that view from internal and external threats.

## 6.1 *Virtual Web Allocation*

We are already talking about the virtual web (VW) deployment strategy for the survey of several provocations in the cloud ecosystem. Their advanced approach is based on the target, i.e., data safety in cloud computing. The conveyed primarily based totally on the optimization for the actual period applicability.

### 6.2 Forensic Recovery of Cloud Evidence

We are mentioned beforehand that the format advantages of the debate mending of cloud-based proof. They have suggested that the structure should be explicit, reasonable, and adaptable.

### 6.3 Cloud Databases Under Multiple Keys

We are talking about the cloud database previously. We are known that the encoded data has been uploaded in cloud server because of the data confidentiality of the user data. This state of affairs occurs when a great deal of the time of network orientated quantifiable scrutiny, wherein the facts provider and professional are numerous elements. At that factor, both the statistics provider should find its encryption key and the professional should find the personal inquiry. They have proposed a secure cloud database for end users.

### 6.4 Cloud Systems Data Mobility

We are recommended that conventional techniques fail in a better stage of safety. They have proposed for enormous information that is utilize current safety algorithms are deficient in the system. So, they have proposed a steady massive data earlier than executing data potency.

### 6.5 Cryptographic for Secure Information Transmission

We are investigating that cloud-based admittance cryptography which can be useful in secure and stable cloud-based processing engineering, and it contains a gathering of inconsiderate virtualized cautiously versatile and directed effects like refining controls more noteworthy room degree and organizations. They have also suggested that the abuse biometric coding besides plan the security.

## 7 Conclusion

Cloud computing at present a vital part of the communal lifestyle, bringing an important chance to advance corporate strategy through their capacity of data to quickly Scutum, permit us to be supple with our resources for others, and give us a new opportunity for alliance. Cloud ecosystem brings many advantages for companies, organizations, and even countries. This is the reason well-being of the important information is the fundamental issue in the appointment of the cloud-based figuring. The end user and vendors are very much aware of security dangers. Particularly, the significant points of the current study are to introduce all the conceivable well-being incitement need in the cloud-based processing climate and give a potential answer for resolve these issues. In cloud-based computing stages, numerous sorts of developing security stages have been fused, if we implement artificial intelligence and machine learning procedures to mechanize undertakings and carry a more significant level of insight to recognize insider or outcast attacks for more safety of end user data. In this paper, we are mainly focusing an observation on cloud-based safety matter and provocation that emerge from the particular property of the cloud biological system. A widespread point of view on these issues has been acquainted here with redesign here for better understanding the less security of the cloud computing design and imagine conceivable counteractant for them. From the examination perspective, we propose a survey of late security design carries out in cloud-based computing as far as lessening weaknesses to turn away plausible assaults. We classify the safety demanding situations and carry out a comparative evaluation of safety difficulty, and the countermeasures advised to deal with those difficulties.

## References


1. Mell, P.M., Grance, T.: The NIST definition of cloud computing. Spec. Publ. (NIST SP) **6**(9), 145–300 (2011)
2. Tabrizchi, H., Rafsanjani, M.K.: A survey on security challenges in cloud computing: issues, threats, and solutions. In: Proceedings of the 2018 ACM Conference on Computer and Communications Security, pp. 26–31, ACM, Raleigh, NC, USA (2018)
3. Singh, M.: Virtualization in cloud computing—a study. Nethost sensor: investigating the capture of end-to-end encrypted intrusive data. Comput. Secur. **25**(6), pp. 205–251 (2020)
4. Bamiah, M.A., Brohi, S.N.: Towards intrusion detection for encrypted networks. In: Proceedings of the 2017 International Conference on Availability, Reliability and Security, IEEE, pp. 540–545, Fukuoka, Japan (2017)
5. Lee, J.H., Kim, J.K.K.: A new approach for the security of VPN. In: Proceedings of the 2nd International Conference on Information and Communication Technology for Competitive Strategies ICTCS'17, vol. 13, pp. 1–5, ACM (2017)
6. Nie, W., Xiao, X., Wu, Y., Luo, X.: A model of virus infection dynamics in mobile personal area network. J. Telecommun. Electron. Comput. Eng. **10**(2–4), 107–117 (2018)
7. Tao, Y.S., Lee, H.Y.: Why eve and mallory love android: an analysis of android SSL (in) security. In: Proceedings of the 2017 ACM Conference on Computer and Communications Security, pp. 50–61, ACM, Raleigh, NC, USA (2017)
8. Qing, L., Boyu, Z., Jinhua, W.: Virutal private network. Int. J. Adv. Res. Comput. Sci. Softw. Eng. **2**(10), 428–432 (2017)
9. Markandey A., Gajmal, Y.: Virtual private network as a service-a need for discrete cloud architecture. In: proceedings of the 7th International Conference on Reliability, Infocom Technology and Optimization (Trends and Future Directions) (ICRITO 2018), IEEE, pp. 526–532 (2018)

10. Zhang, Z., Chandel, S., Sun, J., Yan, S., Yu, Y., Zang, J.: VPN: a boon or trap? a comparative study of MPLs, IPSec, and SSL virtual private networks. In: Proceedings of the 2018 2nd International Conference on Computing Methodologies and Communication (ICCMC), IEEE, pp. 510–515, Erode, India (2018)
11. Huang, C., Smith, P., Sun, Z.: Secure network solutions for enterprise cloud services. Cloud Technology: Concepts, Methodologies, Tools and Applications. IGI Global Chapter 67, 1464–1496 (2017)
12. Kuldeep, K., Singh, V.V., Gupta. H.: A new approach for the security of VPN. In: Proceedings of the 2nd International Conference on Information and Communication Technology for Competitive Strategies ICTCS'16, vol. 13, pp 1–5. ACM (2016)
13. Cornetto, G., Mateos, J., Touhafi, A.N., et al.: Design, simulation and testing of a cloud platform for sharing digital fabrication resources for education. J. Cloud Comput. **8**(12), 117–120 (2019)
14. Yamada, A., Miyake, Y., Takemori, K., Studer, A., Perrig, A.: Intrusion detection for encrypted web accesses. In: Proceedings of the 21st International Conference on Advanced Information Networking and Applications Workshops (AINAW'17), vol. 1, pp. 569–576, Niagara Falls, Ont., Canada (2017)
15. Fahl, S., Harbach, M., Muders, T., Baumgartner, L., Freisleben, B., Smith M.: Why eve and mallory love android: an analysis of android SSL (in) security. In: Proceedings of the 2017 ACM Conference on Computer and Communications Security, pp. 52–61, ACM, Raleigh, NC, USA (2017)
16. Jyothi, K.K., Reddy, I.B.: Study on virtual private network (VPN), VPN's protocols and security. Int. J. Sci. Res. Comput. Sci. Eng. Inf. Technol. **3**(5), 99–102 (2018)

# Ocean Current Rigid Localization for Seabed WSN

Sumit Kumar, Neera Batra, and Shrawan Kumar

**Abstract** At Seabed, localization of the sensor-nodes is performed by using GPS-enabled buoyant-nodes and autonomous underwater vehicles (AUVs) usually. The deployment of AUVs intensifies the network cost and thus affects scalability terribly. Therefore, after eradicating AUVs the absolute localization will depend on buoyant-nodes exclusively. Whereas the persistent repositioning of GPS-enabled moored buoyant-nodes (anchor-nodes) due to ocean current, force it to share its displaced position in the network. The situation is furthermore worsened due to the slow propagation speed of the acoustic waves. It necessitates localizing the unknown-nodes in AUVs deprived network and by defying the effect of ocean current also. It motivates to propose an algorithm OCR, i.e., Ocean Current Rigid range-free localization. It (i.e., OCR) localizes the unknown-nodes by estimating their distances from the anchor-nodes using hop counts between them. Since hop-based distance estimation is an erroneous assessment, therefore, the problem of localization in OCR is defined by the theory of optimization in obtaining a location where distance error should be minimized using the position of anchor-nodes. OCR does not consider the current position of the anchor-nodes rather estimates their position at the time of the field observations by the seabed deployed unknown-nodes. The simulation validates OCR by localizing with a 12% localization error on average.

**Keywords** Localization · Deep-sea · WSN · AUV · Optimization

---

S. Kumar (✉) · N. Batra
Department of Computer Science and Engineering, Maharishi Markandeshwar Engineering College, Mullana, Ambala 133207, India
e-mail: sumiter@gmail.com

N. Batra
e-mail: neera.batra@mmumullana.org

S. Kumar
Department of Computer Science, Indira Gandhi National Tribal University, Regional Campus Manipur, Kangpokpi, India
e-mail: shrawankatiyar@gmail.com



D. Gupta et al. (eds.), *Pattern Recognition and Data Analysis with Applications*, Lecture Notes in Electrical Engineering 888,
https://doi.org/10.1007/978-981-19-1520-8_28

# 1 Introduction

The sensor-nodes in a group by drawing wireless sensor network (WSN) has the ability to collect and communicate their field observations for useful analysis. It helps to explore some dicey avenues like a seabed environment. The seabed WSN assists in a vast marine domain like deep-sea geomorphological data collection, ocean mining, bathymetric measurements and navigation, aquatic life and ecology, dynamics of plate tectonics, geological survey of the seabed, naval combat operations, marine habitat monitoring, etc. [1]. All these domains require scattering the low-cost disposable sensor-nodes in the field of interest to have an anisotropic WSN. In this way, the sensor-nodes communicate whenever they sense the data above a predefined threshold limit or periodically as per the requirements. The receiver of this data is able to conclude what has happened but unable to understand where it has happened precisely. The sensor-nodes at the seabed which has formed the WSN to sense the environment or happenings without any knowledge of their location are known as unknown-nodes. The unknown-nodes take the help of some GPS-enabled nodes to estimate their absolute location.

For localization, the GPS-enabled nodes include some buoyant-nodes and autonomous underwater vehicles (AUVs) [2] usually [3]. AUV is a cost-intensive hardware option that detriment the financial viability. Thus, AUVs affect network scalability adversely. It encourages drawing WSN by deploying GPS-enabled buoyant-nodes only, known as anchor-nodes, along with the unknown-nodes, as shown by Fig. 1. Here, the interaction among anchor-nodes and unknown-nodes is established through acoustic transceivers known as relay-nodes.

Further, it (Fig. 1) shows that the unknown-nodes have only the anchor-nodes to determine their absolute location. Therefore, unknown-nodes determine their distances from the anchor-nodes, i.e., intermediate distances. To draw the intermediate distances either takes the assistance of radio signal characteristics in a form of range-based techniques or opposite to it takes range-free techniques [4]. The range-free technique is a low-cost, energy-savvy, and less complex solution [5]. Therefore, the proposed localization is based upon a range-free technique using a hop matrix.

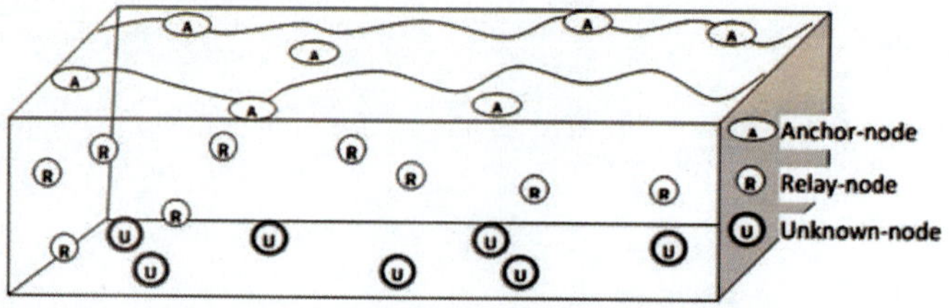


**Fig. 1** Sensor-nodes' topological random arrangement

Conclusively, the localization depends upon two terms- intermediate distances and anchor-node position. Both the terms need to consider carefully for precise localization. The hop matrix-based intermediate distance estimation is an erroneous estimation due to poor hop size approximation and the existence of non-Euler shortest. Therefore, for precise localization, OCR considers the error terms in intermediate distance values. Likewise, the position of the anchor-nodes is dubious because of the ocean current. Further, the unknown-nodes spend a significant time period to receive back the position of the anchor-node due to the slow propagation speed of acoustic waves. This time lag allows the anchor-nodes to share their displaced location with the unknown-nodes. Therefore, the localization algorithm must be rigid to counter the ocean current effect.

The proposed algorithm OCR localizes the unknown-nodes by signifying ocean current and intermediate distance error factors. Further, OCR is established with four more sections in this paper. Section 2 emphasizes some of the recent localization approaches. Section 3 presents a detailed description of the proposed OCR. Thereafter, Sect. 4 validates the proposed localization algorithm with the result and discussion. Finally, the whole discussion is concluded in Sect. 6 with its future scope also.

## 2 Literature Review

In recent times, localization in underwater WSN is getting a rich contribution. Gradually, the algorithms of localization are gaining the maturity to shift away from the need for fixed infrastructure support like baseline station deployment. The baseline station localization system requires installing adequate stationary transponders at the predefined known locations. The baseline localization is acceptable where the cost of infrastructure is justified by its long-term and repeated use. It has no or a limited scope to qualify for the low-cost disposable ad-hoc networks like WSN. Other than the baseline station localization approach, some of the proposed localization algorithms consider the WSN architecture with GPS-enabled buoyant sensor-nodes and AUVs. While beside the equipment cost of AUVs [6], Doppler velocity logs also dissent its deployment [7]. However, Wenxu et al. in [8] localize the sensor-nodes using neural networks in a predictive manner. The use of an iterative approach like a neural network requires more computations and necessitates more energy to consume. Therefore, any such approach is detrimental for the network lifetime because the network will burn out its energy harshly in the localization algorithm. Similarly, in [9] Liu et al. localizes in deep-ocean by drawing a learning function using a multi-task learning model which again insists on high energy requirements. Another, the approach of deep-ocean localization is performed by using Kalman Filter in [10]. Kalman filter is a known state-space model which filters a dynamic stochastic process. Therefore, localization based on the Kalman filter shows poor robustness intrinsically. In this way, some algorithms are able to localize the sensor-nodes with more accuracy while they compromise the productive lifetime of networks.

The energy-efficient localization is also proposed in the literature for the localization in acoustic underwater WSN. Principally, these are based upon Niculescu and Nath DV-Hop algorithm [11], in general. DV-Hop algorithm localizes in three steps. The distance between the unknown-node of interest and the anchor-nodes is estimated by using average hop size and hop count. The average hop size determination caused poor distance estimation [5, 12]. The DV-Hop algorithm is exercised by Wang et al. [13] with the improvement of the hop size values of the anchor-nodes in acoustic sensor networks. Another approach for the precise localization is proposed in IDV by Shen et al. [14] by improving the hop size of each anchor-node and identifying the most suitable three anchor-nodes with respect to the intended unknown-node. It implies that the network size is a dominating factor. Furthermore, the localization is précised by selecting a nearest set of anchor-nodes to the unknown-node of interest in ODR [5]. ODR performs localization by taking centroid of the set of nearest possible anchor-nodes.

Similarly, the intermediate distance correction for the DV-Hop algorithm is suggested by applying genetic engineering and particle swarm optimization by Han et al. [15]. Conclusively, the various variants of the DV-Hop algorithm try to estimate the intermediate distance precisely and thus achieve better localization accuracy. The imprecise distance estimation is due to the poor hop size estimation [12] and the consideration of the non-Euler shortest path [5].

Although, DV-Hop is an energy-efficient localization algorithm whereas DV-Hop also requires having some fixed positioned anchor-nodes. The anchor-nodes in the ocean are moored with bungee and drift with the ocean current. Further, the anchor-nodes share their current position whenever they get a request from the unknown-nodes only. Here, the slow propagation of the acoustic wave creates a significant time lag between the request for the location by the unknown-node and its reply from the anchor-node. During this time lag anchor-node further displaced its location and shares its current position besides sharing the position where it was when the unknown-node submits the request for the location. Therefore, the ocean current is a non-trivial constraint that must be considered for the localization, however, we find a modest contribution in this domain only. Zhao et al. [16] consider four buoyant anchor-nodes. All the four anchor-nodes are floating with the ocean current and only one anchor-node updates its location periodically using GPS. Rest three anchor-nodes determine their location with the help of the GPS-connected anchor-node and thus participate in localization of the unknown-nodes. However, the proposed algorithm requires the range-based Time of Arrival (ToA) parameter for the localization. Further, the sub-surface ocean current is modeled by Meandering Current Mobility (MCM) model in [17] which is implemented for localization by Kayalvizhi et al. in [18]. MCM modeled the mobility of the immersed sensor-node with ocean current; however, both [17, 18] have no considerations for the positional uncertainty of the anchor-nodes due to ocean current.

Therefore, the proposed algorithm OCR considers the mobility of the anchor-nodes due to ocean current and performs localization in a range-free manner using the hop matrix of the sensor-nodes.

# 3 Proposed Model: Ocean Current Rigid Localization (OCR)

The proposed OCR estimates two-dimensional coordinates $(x, y)$ of the unknown-nodes; as the $z$ coordinate can be estimated with the help of the pressure sensors which assists to keep the complexity of OCR low.

In OCR, the moored anchor-nodes get their GPS location updates periodically. In between updates they store their directional movement and estimate their location by backtracking with velocity calculations. In this way, the anchor-nodes are compass equipped and able to estimate their location at the time when the request for their location was raised by the unknown-nodes. After ascertaining the location of the anchor-nodes, the unknown-nodes estimate the intermediate distances between them and the anchor-nodes. The calculated intermediate distances are imprecise due to poor hop size estimation and by considering the non-Euler shortest path. Therefore, some hypothetical correction factors are defined and then passed to the constrained linear programming to get the location of the unknown-node of interest.

In this manner, broadly OCR performs in two steps-

1. Backtracking of Anchor-nodes
2. Localization of Unknown-nodes

The working of the two steps of OCR is explained below:

## 3.1 *Backtracking of Anchor-Nodes*

The ocean current takes due to geological and climatic factors. Therefore, in a specific region, the ocean current moves in a particular direction most often. It implies that the angle of movement of the anchor-nodes observed from between the consecutive GPS locations will not deviate much. Further, the speed of the ocean current is also a local consistent factor. Therefore, we can estimate the location of the anchor-nodes a certain time back.

Let a buoyant anchor-node positioned by GPS at $(x_a, y_a)$ displaced due to ocean current at $(x_b, y_b)$ any localization instance. Then the location $(x_b, y_b)$ is estimated with the help of the following Eq. (1)

$$\left.\begin{aligned} x_b &= x_a + s\cos(\theta) \\ y_b &= y_a + s\sin(\theta) \end{aligned}\right\} \quad (1)$$

where $\theta$ is the angular movement of the anchor-node with a distance $s$.

### 3.2 Localization of Unknown-Nodes

The unknown-node of interest bears a geometric arrangement with the anchor-nodes as shown in Fig. 2.

Figure 2 shows that the unknown-node of interest at $(x,\ y,\ z)$ requires estimating $x$ and $y$ coordinates, as $z$ is estimated with the help of the pressure sensor. The unknown-node of interest estimates its intermediate distances $d_i$ from the $n$ anchor-nodes. Now, let the unknown-node of interest is located at $(x,\ y,\ z)$, which is to be estimated; and the anchor-nodes are positioned at $(x_1,\ y_1,\ z_1), (x_2,\ y_2,\ z_2), \ldots, (x_n,\ y_n,\ z_n)$ where $z_i = 0;\ \forall i \in n$. Therefore, a set of distance equations for the unknown-node of interest is obtained by Eq. (2), as shown below.

$$(x - x_i)^2 + (y - y_i)^2 + (z - 0)^2 = (z + \varphi_i)^2;\ \ \forall i \in n \tag{2}$$

where $0 \leq \varphi_i \leq \infty$ is the distance correction factor of $i$th anchor-node.

Now, after converting Eq. (2) of quadratic equations set into linear equations set, we obtain Eq. (3) as follows.

$$\begin{aligned} &x(x_n - x_i)/z + y(y_n - y_i)/z + (-\varphi_i + \varphi_n) \\ &= \left(x_n^2 - x_i^2 + y_n^2 - y_i^2\right)/2z;\ \ \forall i \in (n-1) \end{aligned} \tag{3}$$

where $\left(\varphi_i^2 - \varphi_n^2\right)/2z$ is a trivial term by considering $\varphi^2 \ll z$, we get another set of equations Eq. (4).

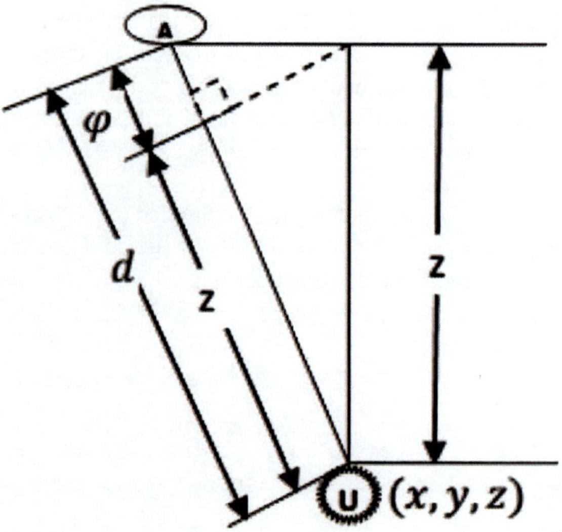


**Fig. 2** Geometric arrangement of unknown-node with anchor-node

$$A_i x + B_i y - \varphi_i + \varphi_n = C_i;\ \ \forall i \in (n-1) \tag{4}$$

where $A_i = (x_n - x_i)/z$, $B_i = (y_n - y_i)/z$,$C_i = \left(x_n^2 - x_1^2 + y_n^2 - y_i^2\right)/2z$.

Equation (4) can be written by Eq. (5), as,

$$\alpha X = \beta; \tag{5}$$

where $\alpha$, $X$, and $\beta$ are as following,

$$\alpha = \begin{bmatrix} A_1 & B_1 & -1 & 0 & \cdots & 0 & 1 \\ A_2 & B_2 & 0 & -1 & \cdots & 0 & 1 \\ \vdots & \vdots & \vdots & \vdots & \cdots & \vdots & \vdots \\ A_{n-1} & B_{n-1} & 0 & 0 & \cdots & -1 & 1 \end{bmatrix},\ X' = \begin{bmatrix} x\ y\ \varphi_1\ \varphi_2 \cdots \varphi_n \end{bmatrix},\ \text{and}\ \beta' = \begin{bmatrix} C_1\ C_2 \cdots C_{n-1} \end{bmatrix}.$$

To solve Eq. (5), we take the constrained least square linear programming technique. Therefore, to define the constraints, we estimate the intermediate distance with the assistance of the hop matrix. It implies that the product of hop count hc and the hop size hs between a node pair derives the intermediate distance. The hop size of an anchor-node is obtained using the Eq. (6) [11], as follows.

$$\text{hs}_j = \left(\sum_{i=1, i\neq j}^{n} \sqrt{\left(x_i - x_j\right)^2 + \left(y_i - y_j\right)^2}\right) \Bigg/ \left(\sum_{i=1, i\neq j}^{n} \text{hc}_{ji}\right); \tag{6}$$

where the hop size $\text{hs}_j$ of the anchor-node $j$ positioned at $\left(x_j,\ y_j\right)$ out of the total anchor-nodes $n$ is estimated as an average value by using hop count $\text{hc}_{ji}$ between the anchor-nodes $j$ and $i$.

The distance hd obtained with the assistance of the hop matrix is a representation of the non-Euler shortest path. Therefore, the intermediate distance equations can be written by Eq. (7).

$$(x - x_i)^2 + (y - y_i)^2 + (z - 0)^2 \le (\text{hd}_i)^2;\ \ \forall i \in n \tag{7}$$

where $\text{hd}_i = \text{hs}_i \times \text{hc}_i$ and $\text{hc}_i$ is the hop count between the anchor-node $i$ and the unknown-node of interest.

After applying the same operation to obtain Eq. (4) from Eq. (3), we obtain Eq. (8) from Eq. (7) as given below.

$$p_i x + q_i y \le r_i;\ \ \forall i \in (n-1) \tag{8}$$

where $p_i = x_n - x_i$, $q_i = y_n - y_i$,$r_i = \left(\text{hd}_i^2 - \text{hd}_n^2 + x_n^2 + y_n^2 - x_i^2 - y_i^2\right)/2$.

The set of equations Eq. (8) can be written in a following form by Eq. (9).

$$BX \leq b; \tag{9}$$

where

$$B = \begin{bmatrix} p_1 & q_1 & 0\,0 \cdots 0 \\ p_2 & q_2 & 0\,0 \cdots 0 \\ \vdots & \vdots & \vdots\,\vdots \cdots \vdots \\ p_{n-1} & q_{n-1} & 0\,0 \cdots 0 \end{bmatrix}_{(n-1)\times(n+2)}, X = \begin{bmatrix} x \\ y \\ \varphi_1 \\ \varphi_2 \\ \vdots \\ \varphi_n \end{bmatrix}, b = \begin{bmatrix} r_1 \\ r_2 \\ \vdots \\ r_{n-1} \end{bmatrix}$$

Therefore, Eq. (6) can be solved under the constraint of Eq. (9). Conclusively, the location of the unknown-node $(x, y)$ can be obtained by solving the following linear programming Eq. (10).

$$\left.\begin{array}{l} \alpha X = \beta \\ \text{such that} \\ BX \leq b \\ lb \leq X \\ lb\prime = \begin{bmatrix} -\infty & -\infty & 0\,0 \cdots 0 \end{bmatrix}_{1\times(n+2)} \end{array}\right\} \tag{10}$$

However, Eq. (10) is unable to deliver a unique solution for $(x, y)$. Therefore, we apply linear programming and minimize the summative magnitude of the distance correction factor as given by Eq. (11).

$$\left.\begin{array}{l} \min \sum_{i=1}^{n} |\varphi_i| \\ \text{such that} \\ \alpha X = \beta \\ BX \leq b \\ lb \leq X \end{array}\right\} \tag{11}$$

Equation (11) can be solved by rewriting in Eq. (12), as follows.

**Table 1** Complexity comparison

| Complexity | DV-Hop | ODR | OCR |
|---|---|---|---|
| Communicational complexity | $O(nT^2)$ | $O(nT^2)$ | $O(nT^2)$ |
| Computational complexity | $O(n^3)$ | $O(n^{3.5})$ | $O(n^{3.5})$ |

$$\left.\begin{array}{l} \min \sum_{i=1}^{n} t'X \\ \text{such that} \\ \alpha X = \beta \\ BX \leq b \\ lb \leq X \end{array}\right\} \quad (12)$$

where $t = \begin{bmatrix} 0 \; 0 \; 1 \; 1 \cdots 1 \end{bmatrix}'_{1\times(2+n)}$.

## 4 Complexity Comparison of OCR

The complexity of OCR is compared with DV-Hop [11], and ODR [5] in terms of communicational and computational efforts required by the localization algorithms. The comparative analysis is shown in Table 1.

Table 1 shows the complexity for a network of total $T$ sensor-nodes out of which $n$ are the anchor-nodes.

## 5 Simulation Result and Analysis

The performance of the proposed OCR validates by simulating with Matlab R2014a. The simulation setup derives the localization error [5, 12] Le as shown by Eq. (13) for the following trials.

$$\text{Le} = \left(\frac{1}{rm}\right) \times \left(\sum_{i=1}^{m} \sqrt{(x_i - x_i\prime)^2 + (y_i - y_i\prime)^2}\right) \times 100\%; \quad (13)$$

where total $m$ unknown-nodes establishes a network with communication range $r$ having actual coordinates at $(x, y)$ and estimated at $(x', y')$.

Trial 1: To observe the impact of communication range
Trial 2: To observe the impact of total number of sensor-nodes
Trial 3: To observe the impact of density of anchor-nodes
Trial 4: To observe the impact of communication noise.

**Table 2** Parameters and values for simulation

| Parameter term | Parameter value(s) |
|---|---|
| Boundaries of covered region | 50 × 50 × 600 |
| Communication range ($r$) | 15, 20, 25, 30 |
| Total number of sensor-nodes ($T = n + m$) | 200, 230, …, 500 |
| Density of anchor-nodes (% of $T$) ($n$) | 5%, 10%, …, 25% |
| Total participating unknown-nodes ($m$) | $T - n$ |
| Communication noise | 0–10%, 0–20%, 0–30% of $r$ |

Each trial is performed within a closed boundary of 50 × 50 × 600 sq. area with 100 relay-nodes. The results discussed below are drawn after taking a mean of 500 random setups under each experimental trial setup. All the parameters required for the simulation are shown in Table 2. The execution of the proposed OCR is analyzed in comparison with DV-Hop [11] and ODR [5] algorithms.

## 5.1 Trial 1: To Observe the Impact of Communication Range

The communication range of the sensor-nodes varies in 15, 20, 25, 30, 35, 40 units to analyze the effect of communication range over localization performance. All the other parameters remain constant, like, total sensor-nodes are taken 200 with only 20 anchor-nodes.

The effect of varying communication ranges is shown in Fig. 3. With every improvement in communication range, OCR shows better localization results. OCR

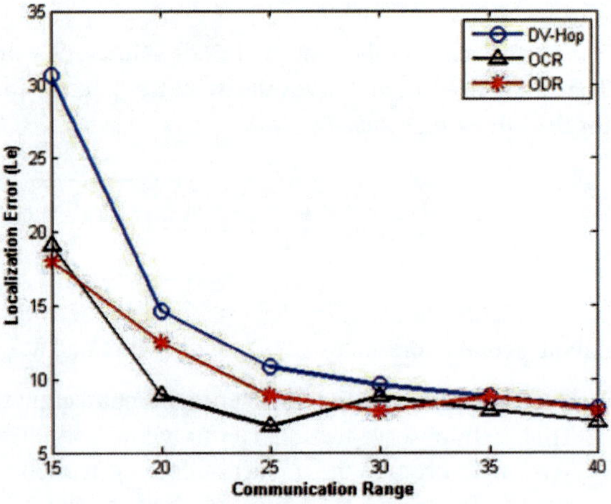


**Fig. 3** Effect of communication range

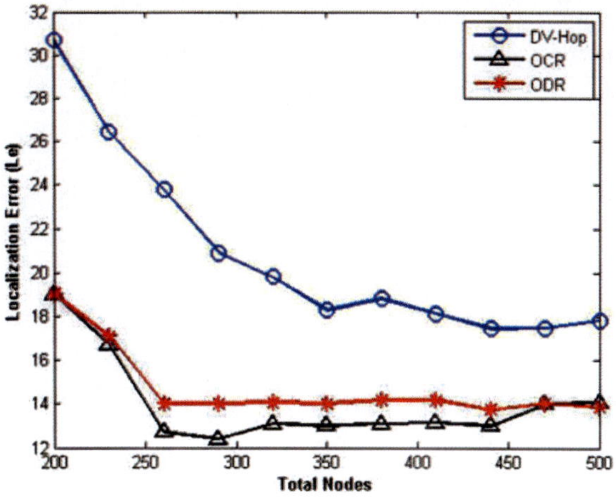


**Fig. 4** Effect of total number of sensor-nodes

improves because, with the increase in communication range, the unknown-nodes get more anchor-nodes with fewer hop counts and thus results in higher localization accuracy.

The trial shows that OCR localizes with 9.5 localization error on average whereas DV-Hop and ODR localize with 16.5 and 10.5 average localization error, respectively.

## 5.2 Trial 2: To Observe the Impact of Total Number of Sensor-Nodes

This trial is focused to understand the effect of the network size on localization performance. The total number of sensor-nodes is varied as 200, 250, …, 500 by keeping all other variables constant, similar to Trial 1.

Figure 4 shows the effect of increasing the total number of sensor-nodes in the network. Since more number of sensor-nodes implies more number of anchor-nodes also, therefore localization error downfalls.

OCR exhibits a localization error of 13.8 on average with the varying total number of sensor-nodes while DV-Hop and ODR display 20.8 and 15.2 on average, respectively.

## 5.3 Trial 3: To Observe the Impact of Density of Anchor-Nodes

The anchor-nodes are crucial for localization. Therefore, the density of anchor-nodes keeps varying in the range of 5%, 10%, …, 25% of the total sensor-nodes. The total

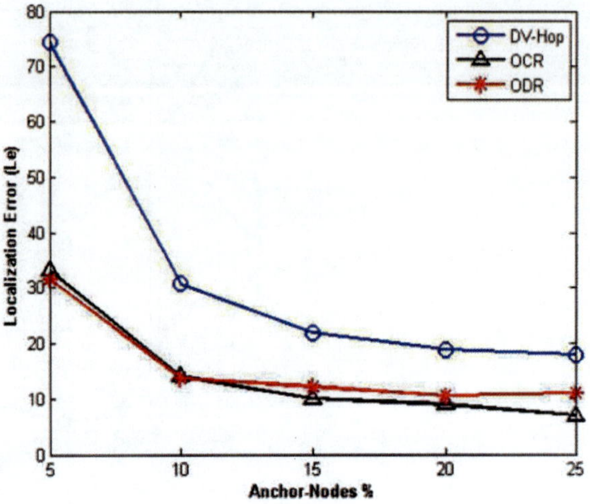


**Fig. 5** Effect of total number anchor-nodes

number of sensor-nodes is kept constant at 200 in this trial. All other variables maintain their values constant as in the previous two trials.

The results of this trial are drawn in Fig. 5 shows that the improvement in the density of the anchor-nodes has a positive effect on the localization performance. The localization error falls with the increase in the anchor-nodes because an unknown-node gets more reference points to estimate its position.

OCR outperforms the other two algorithms DV-Hop and ODR by producing the localization error with 14.4 on average in comparison to 18.2 and 16, respectively.

## 5.4 *Trial 4: To Observe the Impact of Communication Noise*

The communication noise affects the effective communication range of the sensor-nodes unfavorably by attenuating the signals substantially. Therefore, this trial focuses to examine the effect of the communication noise on localization error. The communication noise is simulated by affecting the communication range 20 with random attenuation from the range $0-10\%$, $0-20\%$, $0-30\%$ of $r$. The total network size is kept at a total of 200 sensor-nodes with 20 anchor-nodes.

The effect of communication noise by attenuating the communication range is shown in Fig. 6. The localization error for DV-Hop, ODR, and OCR rises with increasing attenuation. However, OCR is more robust by showing localization error of 38.07%, 13.17% less from DV-Hop and ODR on average, respectively.

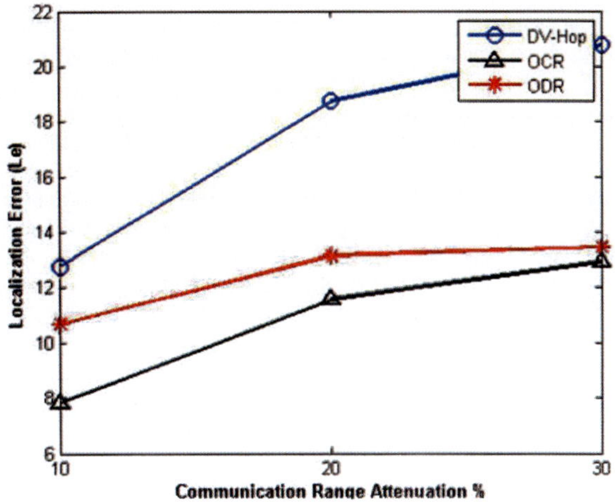


**Fig. 6** Effect of communication noise

## 6 Conclusion

The seabed localization faces some constraints due to ocean current and the slow propagation speed of acoustic waves. The uncertainty in the position of the anchor-nodes makes the seabed localization quite different from the already known localization algorithms. Therefore, the proposed localization algorithm OCR without AUVs presents a novel approach in the range-free paradigm. In this way, OCR is a cost-effective and ocean current rigid localization algorithm. The comparison of OCR with the other algorithms DV-Hop and ODR shows that OCR is more robust while both the other two algorithms require obtaining the position of the anchor-nodes through GPS at each localization instance. On average, the OCR localized with 12% localization error whereas DV-Hop and ODR localize with 20%, and 16% localization error, respectively.

In the future, we will extend it to the energy constraint network where anchor-nodes will be in a sleep mode periodically to save their energy.

## References


1. Gola, K.K., Gupta, B.: Underwater sensor networks: comparative analysis on applications, deployment and routing techniques. IET Commun. **14**(17), 2859–2870 (2020). https://doi.org/10.1049/iet-com.2019.1171
2. García, G., Gómez, E.J., Urquizo, C.A., Valdovinos, C.E., Salgado, G.L., Cabello, T., Jesús, A.E.: Autonomous under-water vehicles: localization, navigation, and communication for collaborative missions. Appl. Sci. **10**(4), 1256 (2020). https://doi.org/10.3390/app10041256
3. Fattah, S., Gani, A., Ahmedy, I., Idris M.Y.I., Hashem, T.: A survey on underwater wireless sensor networks: requirements, taxonomy, recent advances, and open research challenges. Sensors (Basel) **20**(18), 5393 (2020). PMID: 32967124; PMCID: PMC7570626. https://doi.org/10.3390/s20185393

4. Islam, T., Park, H.S.: A comprehensive survey of the recently proposed localization protocols for underwater sensor networks. IEEE Access **8**, 179224–179243 (2020). https://doi.org/10.1109/ACCESS.2020.3027820
5. Kumar, S., Kumar, S., Batra, N.: Optimized distance range free localization algorithm for WSN. Wireless Pers. Commun. **117**, 1879–1907 (2021). https://doi.org/10.1007/s11277-020-07950-7
6. Awan, M.K., Shah, A.P., Iqbal, K., Gillani, S., Ahmad, W., Nam, Y.: Underwater wireless sensor networks: a review of recent issues and challenges. Wireless Commun. Mob. Comput. **2019** (2019). https://doi.org/10.1155/2019/6470359
7. Medagoda, L., Williams, S.B., Pizarro, O.: Mid-water current aided localization for autonomous underwater vehicles. Auton. Robot. **40**, 1207–1227 (2016). https://doi.org/10.1007/s10514-016-9547-3
8. Wenxu, L., Yixin, Y., Mengqian, X., Liangang, L., Zongwei, L., Yang, S.: Source localization in the deep ocean using a convolutional neural network. J. Acoust. Soc. Am. **147**(4) (2020). https://doi.org/10.1121/10.0001020
9. Liu, Y., Niu, H., Li, Z.: A multi-task learning convolutional neural network for source localization in deep ocean. J. Acoust. Soc. Am. **148**(2), 873–883 (2020). https://doi.org/10.1121/10.0001762
10. Zhang, B., Sun, L., Fan, L., Meng, L.: Localization of moving source in the shallow layer of deep ocean based on the arriving time difference between the first and second waves. In: IEEE 20th International Conference on Communication Technology (ICCT), pp. 547–551. Nanning, China (2020). https://doi.org/10.1109/ICCT50939.2020.9295719
11. Niculescu, D., Nath, B.: Ad hoc positioning system (APS). In: Proceedings of IEEE Global Telecommunications Conference, vol. 5, pp. 2926–2931 (2001). https://doi.org/10.1109/GLOCOM.2001.965964
12. Kumar, S., Lobiyal, D.K.: An enhanced DV-Hop localization algorithm for wireless sensor network. Int. J. Wirel. Networks Broadband Technol. **2**(2), 16–35 (2012)
13. Wang, S., Lin, Y., Tao, H., Sharma, P.K., Wang, J.: Underwater acoustic sensor networks node localization based on compressive sensing in water hydrology. Sensors (2019). https://doi.org/10.3390/s19204552
14. Shen, S., Yang, B., Qian, K., She, Y., Wang, W.: On improved DV-Hop localization algorithm for accurate node localization in wireless sensor networks. Chin. J. Electron. **28**(3), 658–666 (2019). https://doi.org/10.1049/cje.2019.03.013
15. Han, D., Yu, Y., Li, K.-C., de Mello, R.F.: Enhancing the sensor node localization algorithm based on improved DV-Hop and DE algorithms. Sensors **20**(2), 343 (2020). https://doi.org/10.3390/s20020343
16. Zhao, C., Qiao, G., Zhou, F., Ahmed, N.: Underwater localisation correction method for drifting anchor nodes with an extra floating anchor node. IET Radar Sonar Navig. **14**, 1494–1501 (2020). https://doi.org/10.1049/iet-rsn.2020.0117
17. Caruso, A., Paparella, F., Vieira, L.F.M., Erol, M., Gerla, M.: The meandering current mobility model and its impact on underwater mobile sensor networks. In: IEEE INFOCOM 2008—The 27th Conference on Computer Communications, pp. 221–225 (2008). https://doi.org/10.1109/INFOCOM.2008.53
18. Kayalvizhi, C., Bhairavi, R., Sudha, G.F.: Localization of nodes with ocean current mobility model in underwater acoustic sensor networks. In: Smys, S., Bestak, R., Chen, J.Z., Ko-tuliak, I. (eds.) International Conference on Computer Networks and Communication Technologies. Lecture Notes on Data Engineering and Communications Technologies, vol 15. Springer, Singapore (2019). https://doi.org/10.1007/978-981-10-8681-6_11

# Detection of Motor Activity in Visual Cognitive Task Using Autoregressive Modelling and Deep Recurrent Network

**Shankar S. Gupta and Ramchandra R. Manthalkar**

**Abstract** The Recognition of brain activity related to motor execution is necessary for building a brain-computer interface. However, building a framework for detecting simple motor activity such as key press detection for various cognitive levels is challenging as the EEG signal quality is strongly dependent on the user's cognitive state. In this work, EEG signal from 44 subjects is recorded for the various visual cognitive load. Four levels of cognitive workload are imposed by showing geometrical shapes for recognition and counting. The Autoregressive (AR) modelling is computationally efficient and adjusts inter-subject variations in the presence of noise in the signal. The AR coefficients are utilized as features and given to deep structure utilizing Bidirectional long short-term memory (BLSTM) and LSTM for detecting motor activity. The precision, recall, and accuracy obtained using the proposed method is 95.3%, 91.2%, and 97.1%, respectively.

**Keywords** Electroencephalogram (EEG) · Cognitive workload · Brian computer interface · LSTM

## 1 Introduction

The EEG activity is one of the physiological measures to access motor, emotional, and cognitive activity [1, 2]. The electroencephalogram (EEG) activity can be acquired in a non-invasive way with millisecond resolution. The EEG signal can be recorded in an online manner without affecting the main task. The association between working memory and the number of resources required by the user to complete the task is known as cognitive workload [3]. The human experiences cognitive workload because of limited working memory capacity. The use of BCI allows the human being to communicate with the machine using brain signals [4]. The performance of BCI is affected by the cognitive state of the user [5].

S. S. Gupta (✉) · R. R. Manthalkar
Shri Guru Gobind Singhji Institute of Engineering and Technology, Nanded, India
e-mail: shankargupta99@gmail.com



D. Gupta et al. (eds.), *Pattern Recognition and Data Analysis with Applications*, Lecture Notes in Electrical Engineering 888,
https://doi.org/10.1007/978-981-19-1520-8_29

Researchers have used EEG signal to convey actual and motor imagery tasks of different body parts such as fingers, hands, feet for commanding multiple devices such as wheelchairs and various computing devices. The movement related cortical potential is extracted using discriminant canonical pattern matching to detect pre-finger movement for BCI [6]. In [7], the fusion of EEG and eye movement activity is used for motor imagery tasks. It is observed that the classification accuracy obtained by fusing both signals is larger than with EEG data alone. In literature, it is found that classification of workload and motor activity detection is rarely studied. Motor activity such as pressing a keyboard may trigger artefacts because of body movements. The classification accuracy of binary cognitive workload without considering key pressing activity is larger [8]. In [9], for cognitive workload classification, the keystroke activity does not change the statistical features but affects the morphological features such as signal length and the number of peaks. In [10], a pilot's workload is dynamically assessed by obtaining spectral EEG features for different flying conditions. In [11], the key pressing condition is predicted 100 to 230 ms before the key is pressed. The frequency content of the EEG signals changes with time due to the nonstationary nature of the EEG signal. As a result, analysing the EEG signal without considering non-stationarity can fail to capture their spectral characteristics. In [12], the Renyi entropy and energy features are extracted using quadratic time–frequency distribution to decode finger movements. In [13], the effect of theta band for finger movement detection is investigated. The results obtained are 2–4% more accurate than simply alpha and beta bands. The connections between computing units create a directed graph over a time sequence in a Recurrent Neural Network (RNN). RNN such as LSTM is utilized to analyse time-dependent data such as EEG to improve classification accuracy in motor imagery tasks [14].

The EEG signal is complex, and the signal quality is strongly associated with the subject's cognitive state; detecting motor activity in various cognitive workloads would be challenging. This study hypothesized that simple motor activity such as key pressing can be detected in multiple cognitive workloads. The paper is structured in the following manner. Section 2 gives information about materials and methods. Section 3 describes feature extraction and classification, and Sect. 4 describes the results. Finally, Sect. 5 presents the study's conclusion.

## 2 Material and Methods

EEG data of 44 subjects were recorded with the permission of the SGGS ethical committee. The participants have an age of $20.7 \pm 2.2$. EEG data is collected using the ENOBIO 32 channel wireless EEG equipment. The experimental setup is described in detail in [15]. The subjects were asked to count the number of squares, circles, and rectangles. Figures 1, 2, 3 and 4 shows a sample of every cognitive level. In level 1, any two trigonometric shapes are utilized, whereas three shapes are utilized in other levels. The complexity of levels is dependent on the number of shapes and decoding of the shapes. In level 2, the total number of shapes are six, and no shapes

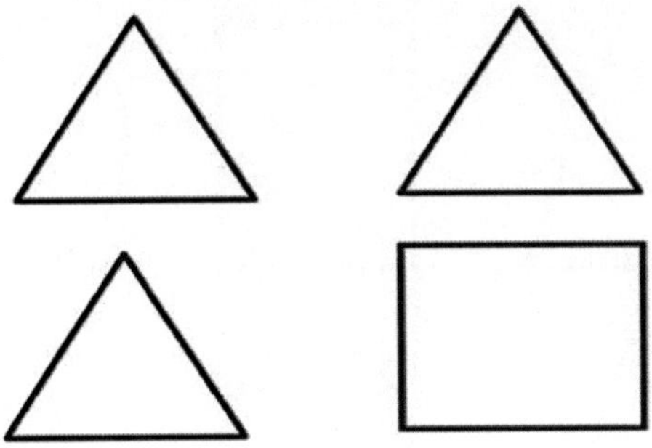

**Fig. 1** Sample of level 1

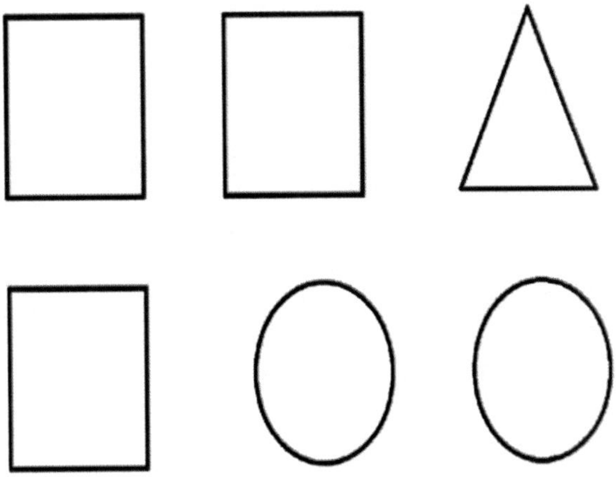

**Fig. 2** Sample of level 2

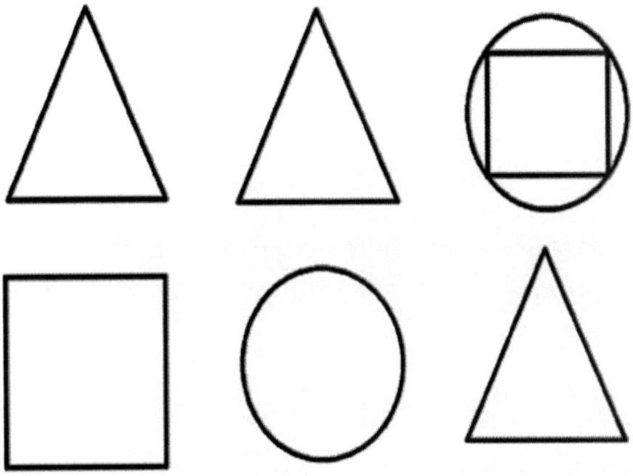

**Fig. 3** Sample of level 3

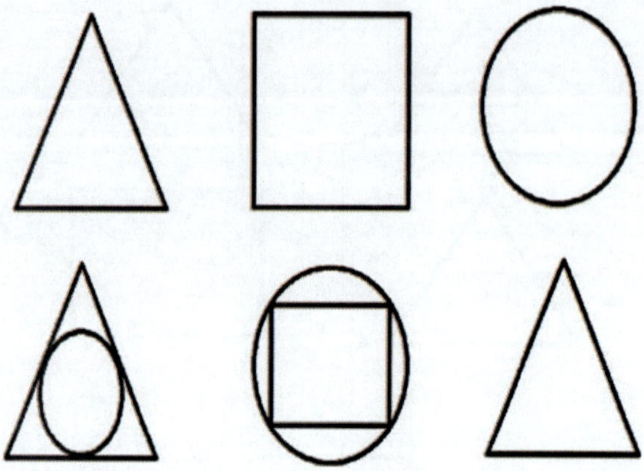

**Fig. 4** Sample of level 4

are combined. In level 3, one inscribed shape is utilized, and the total number of shapes are seven. In level 4, two inscribed shapes are utilized, and the total number of shapes are eight. At every level, ten different variations of shapes are utilized for inducing cognitive workload. The stimulation of the trigonometric shape image is seen for 7 s, followed by 2 s of a blank screen. The count of any shape from a set of three is asked, and three choices are given. The subject gives a response by pressing the key. The maximum duration from question to response is 5 s. EEG signal corrupted with eye-blinks and muscle activity are suppressed by ADJUST algorithm [16].

# 3 Methodology

## *3.1 Autoregressive Modelling*

Autoregressive (AR) modelling is the simplest way to describe EEG data since it considers the noise in the signal. The AR coefficients do not require scaling, which eliminates the issue of inter-subject differences. Small intervals of EEG data are well adapted by AR model [17]. The EEG interval can be modelled with $p$ order as

$$x_n = \sum_{i=0}^{p} s_i x_{n-1} + e(n) \tag{1}$$

where $s_1, s_2, \ldots, s_p$ are AR coefficients and $e(n)$ represents fitting error. The coefficients are calculated using the burg algorithm, which is described in [18].

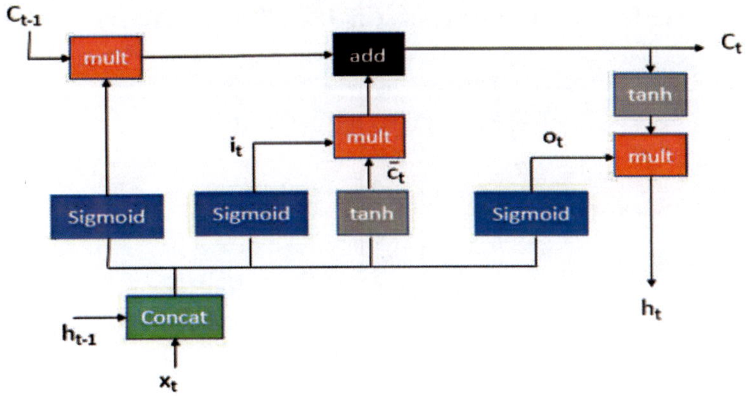


**Fig. 5** LSTM architecture

## 3.2 *Classification*

Cognitive loads vary with time, and these variations are expressed in EEG signal temporal correlations. The variations of EEG data can be correlated using LSTM that can learn long and short-term temporal associations from the sequential data. The LSTM structure is operated iteratively with five parameters ($f_t$, $i_t$, $c_t$, $o_t$, $h_t$), as shown in Fig. 5.

The LSTM framework mentioned in Fig. 5 is unidirectional, which can only learn information between current and subsequent time instants, whereas bidirectional LSTM allows to learn temporal information in forward and backward directions simultaneously. In this analysis, deep RNN with two BLSTM and one LSTM stacked with 512,128 and 64 hidden units, respectively, are utilized. The dropout layer with a dropout rate of 0.2 is utilized after each BLSTM and LSTM layer. Finally, binary classification is achieved using a densely connected layer. Figure 6 represents the block diagram of the proposed system. The training phase of classification uses a batch size of 300, 150 epochs and a learning rate of $5e^{-4}$. The EEG features are extracted for four frames. In this work, the effect of frame length and the size of window is also discussed.

# 4 Results and Discussions

The EEG signal is decomposed using Morlet wavelet into four clinical subbands as delta (0.1–4 Hz), theta (4–8 Hz), alpha (8–13 Hz), and beta (13–30 Hz). In this work, the order of AR coefficient utilized is four which is selected empirically. The AR coefficients are computed for all four bands and all 32 electrodes. Therefore, the length of the feature vector is 32(channel) $\times$ 4(bands) $\times$ 4(AR coefficients) =

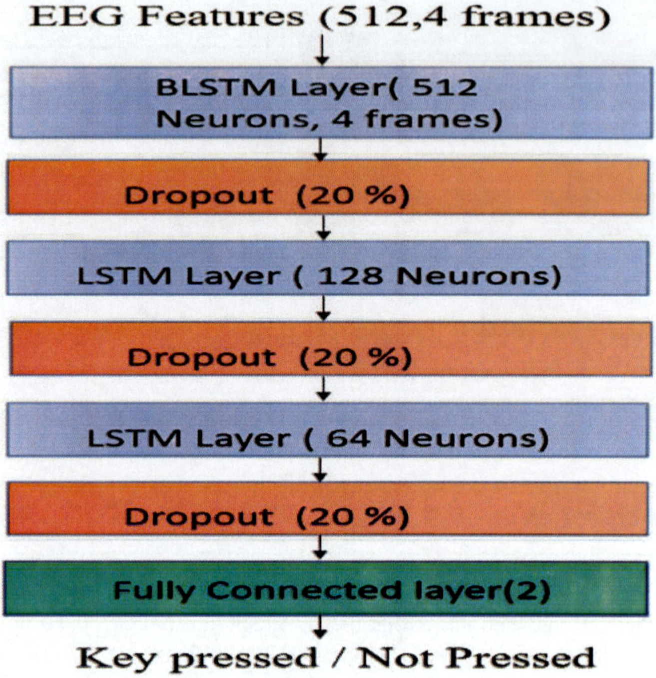


**Fig. 6** Classification using deep recurrent network

512. The EEG time frames are extracted from stimulus to response time. The non-key press activity frames are randomly obtained from stimulus to 500 ms before the keypress activity, as shown in Fig. 7. The keypress activity frames are extracted by leaving a few mill-seconds (50–170 ms) where the actual key is recognized. The total non-key activity frames extracted are 8800, and the keypress activities are 1760.

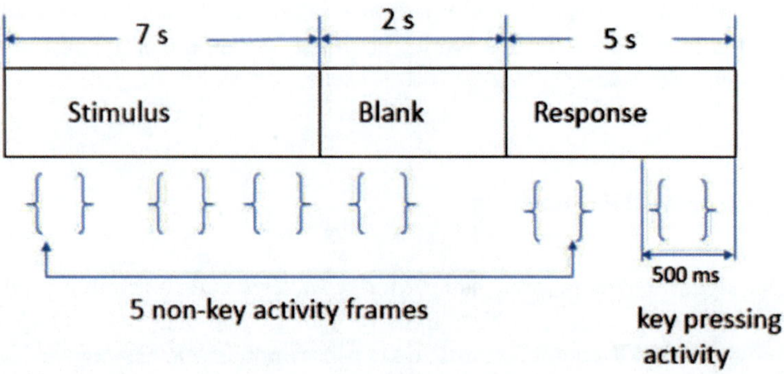


**Fig. 7** Key and non-key frame extraction

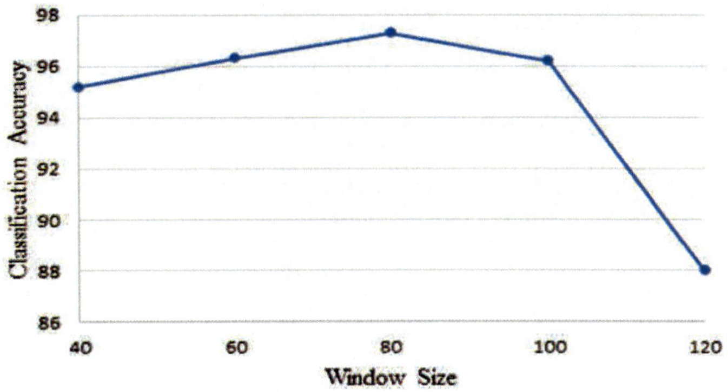


**Fig. 8** Window size versus classification accuracy

Figure 8 shows the effect of window length on classification accuracy. The highest accuracy is obtained with a window size of 80 ms. The classification accuracy is also dependent on the number of time frames. Figure 9 depicts the effect of increasing the number of frames on accuracy with an 80 ms window size. When the number of frames is four, the accuracy is maximum.

The proposed system is evaluated using precision, recall, F1 score, and accuracy. When the dataset is imbalance, accuracy should not be the only factor to consider because it may lead to an incorrect conclusion because of many true negatives. Precision is measured as the ratio of true positive to total positives predictions and reflects the classifiers credibility. Recall is the ratio of true positives to the total positives of the testing data. The classifier's overall performance can be reflected by the F1 score, which is a weighted mean of precision and recall. The classification parameters are given in Eqs. 2–5. The details for the same is given in Table 1. The highest classification metrics (bold values) are obtained when the early detection duration is 140 ms. The result of the classification accuracy of key press activity in various levels is given in Table 2. The classification accuracy of 97.8% is obtained in

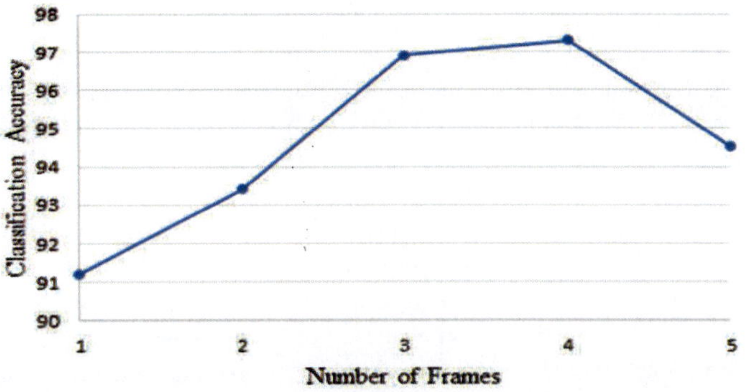


**Fig. 9** Number of frames versus classification accuracy

**Table 1** Results of classification

| Early detection (ms) | Precision | Recall | F1 score | Accuracy |
|---|---|---|---|---|
| 50 | 92.3 | 89.7 | 90.9 | 96.2 |
| 80 | 93.2 | 90.2 | 91.6 | 96.4 |
| 110 | 94.6 | 90.3 | 92.4 | 96.7 |
| 140 | **95.3** | **91.2** | **93.2** | **97.1** |
| 170 | 94.9 | 90.7 | 90.7 | 95.6 |

**Table 2** Results of classification in various levels

| Levels | Precision | Recall | F1 score | Accuracy |
|---|---|---|---|---|
| 1 | 94.1 | 92.4 | 93.2 | 96.9 |
| 2 | 95.7 | 93.2 | 94.4 | 97.8 |
| 3 | 94.5 | 92.9 | 93.7 | 97.1 |
| 4 | 92.1 | 89.4 | 90.7 | 95.6 |

level 2, whereas in level 4, the value is 95.6%, showing that motor activity detection is challenging in high cognitive tasks.

$$\text{precision} = \frac{\text{TP}}{\text{TP} + \text{FP}} \tag{2}$$

$$\text{recall} = \frac{\text{TP}}{\text{TP} + \text{FN}} \tag{3}$$

$$F_1 = \frac{2 \times (\text{precision} \times \text{recall})}{\text{precision} + \text{recall}} \tag{4}$$

$$\text{accuracy} = \frac{\text{TP} + \text{TN}}{\text{TP} + \text{FP} + \text{FN} + \text{TN}} \tag{5}$$

Table 3 gives the classification accuracy of each subject utilizing the leave-one-out approach. A paired sample t-test is used to compare the average accuracy achieved using tenfold cross-validation and the accuracy of each subject. The $p$-value obtained using the t-test is 0.34, which shows that the subject-wise and average accuracy are statistically insignificant.

Table 4 shows a comparison of research conducted on motor activities detection. It is essential to mention that the experimental protocol and the intention of doing motor activities differ amongst different methods. In [5], the effect of visual distraction on motor imagery tasks is investigated using various band powers and performance measures. The classification accuracy obtained is below 75%. In [13], the classification accuracy obtained in detecting finger movements is investigated using PSD features and support vector machine (SVM) classifiers. The average detection

**Table 3** Classification accuracy of individual subject

| Subject | Accuracy | Subject | Accuracy | Subject | Accuracy |
|---|---|---|---|---|---|
| 1 | 97.9 | 16 | 97.1 | 31 | 94.6 |
| 2 | 98.3 | 17 | 98.8 | 32 | 98.3 |
| 3 | 95 | 18 | 95.8 | 33 | 97.9 |
| 4 | 95.4 | 19 | 96.7 | 34 | 97.1 |
| 5 | 97.9 | 20 | 97.5 | 35 | 96.7 |
| 6 | 98.3 | 21 | 94.6 | 36 | 98.3 |
| 7 | 98.8 | 22 | 96.2 | 37 | 96.2 |
| 8 | 95.8 | 23 | 96.7 | 38 | 98.8 |
| 9 | 97.5 | 24 | 97.5 | 39 | 98.3 |
| 10 | 97.9 | 25 | 97.9 | 40 | 97.9 |
| 11 | 96.7 | 26 | 98.3 | 41 | 96.2 |
| 12 | 98.3 | 27 | 95.8 | 42 | 98.3 |
| 13 | 96.2 | 28 | 97.1 | 43 | 94.6 |
| 14 | 97.5 | 29 | 96.2 | 44 | 98.3 |
| 15 | 97.9 | 30 | 96.7 | | |

**Table 4** Comparison of the current work with previous studies

| Authors | Features | Classifiers | Accuracy (%) |
|---|---|---|---|
| Emami and Chau [5] | Band power | – | 75 |
| Ketenci and Kayikcioglu [13] | PSD | SVM | 80.7 |
| Wang et al. [6] | ERD and MRCP | FDA | 91.5 |
| Proposed | AR coefficients | Deep RNN | 97.1 |

accuracy is 80.7%. In [6], 91.5% accuracy is obtained using event related synchronization (ERD) and movement related cortical potential (MRCP) features using fisher discriminant analysis (FDA) classifier. In [5, 6, 13], multiple finger movements are recognized. The key is pressed with the index finger only in the current work, but key pressing activity is recognized under various cognitive workload levels.

The proposed system can detect key pressing activity in various cognitive workloads. The non-key pressing activity is extracted from visual stimulus, blank display and whilst reading a question. Thus, the proposed method can identify motor activity in various cognitive activities. The number of features required by the classifier can be reduced by employing a feature selection algorithm.

## 5 Conclusion

This study shows the detection of motor activity in various levels of visual cognitive tasks. The proposed framework decomposes EEG signal into four clinical bands, and AR coefficients of order four are utilized as a feature vector. The windowing technique is utilized for obtaining features, and the classification is achieved using a deep recurrent network. The highest classification metrics in key pressing activity are achieved 140 ms before the system recognizes the key. The highest classification accuracy is obtained in level 2, and the lowest is obtained in level 4. Precision, recall, and accuracy obtained for four frames are 95.3%, 91.2%, and 97.1%, respectively.

## References

1. Berka, C., Levendowski, D.J., Lumicao, M.N., Yau, A., Davis, G., Zivkovic, V.T., Olmstead, R.E., Tremoulet, P.D., Craven, P.L.: EEG correlates of task engagement and mental workload in vigilance, learning, and memory tasks. Aviat. Space Environ. Med. **78**(5), B231–B244 (2007)
2. Gupta, V., Chopda, M.D., Pachori, R.B.: Cross-subject emotion recognition using flexible analytic wavelet transform from EEG signals. IEEE Sens. J. **19**(6), 2266–2274 (2018)
3. Hart, S.G., Staveland, L.E.: Development of NASA-TLX (task load index): results of empirical and theoretical research. Adv. Psychol. **52**, 139–183 (1988)
4. Wang, X.-Y., Jin, J., Zhang, Y., Wang, B.: Brain control: human-computer integration control based on brain-computer interface approach. Acta Automatica Sinica **39**(3), 208–221 (2013)
5. Emami, Z., Chau, T.: The effects of visual distractors on cognitive load in a motor imagery brain-computer interface. Behav. Brain Res. **378**, 112240 (2020)
6. Wang, K., Xu, M., Wang, Y., Zhang, S., Chen, L., Ming, D.: Enhance decoding of pre-movement EEG patterns for brain–computer interfaces. J. Neural Eng. **17**(1), 016033 (2020)
7. Cheng, S., Wang, J., Zhang, L., Wei, Q.: Motion imagery-BCI based on EEG and eye movement data fusion. IEEE Trans. Neural Syst. Rehabil. Eng. **28**(12), 2783–2793 (2020)
8. Zhang, P., Wang, X., Zhang, W., Chen, J.: Learning spatial–spectral–temporal EEG features with recurrent 3D convolutional neural networks for cross-task mental workload assessment. IEEE Trans. Neural Syst. Rehabil. Eng. **27**(1), 31–42 (2018)
9. Wang, S., Gwizdka, J., Chaovalitwongse, W.A.: Using wireless EEG signals to assess memory workload in the n-back task. IEEE Trans. Human-Mach. Syst. **46**(3), 424–435 (2015)
10. Mohanavelu, K., Poonguzhali, S., Adalarasu, K., Ravi, D., Chinnadurai, V., Vinutha, S., Ramachandran, K., Jayaraman, S.: Dynamic cognitive workload assessment for fighter pilots in simulated fighter aircraft environment using EEG. Biomed. Signal Process. Control **61**, 102018 (2020)
11. Blankertz, B., Curio, G., Müller, K.R.: Classifying single trial EEG: towards brain computer interfacing. In: Advances in Neural Information Processing Systems, pp. 157–164 (2002)
12. Alazrai, R., Alwanni, H., Daoud, M.I.: EEG-based BCI system for decoding finger movements within the same hand. Neurosci. Lett. **698**, 113–120 (2019)
13. Ketenci, S., Kayikcioglu, T.: Investigation of theta rhythm effect in detection of finger movement. J. Exp. Neurosci. **13**, 1179069519828737 (2019)
14. Wang, P., Jiang, A., Liu, X., Shang, J., Zhang, L.: LSTM-based EEG classification in motor imagery tasks. IEEE Trans. Neural Syst. Rehabil. Eng. **26**(11), 2086–2095 (2018)
15. Gupta, S.S., Manthalkar, R.R.: Classification of visual cognitive workload using analytic wavelet transform. Biomed. Signal Process. Control **61**, 101961 (2020)
16. Mognon, A., Jovicich, J., Bruzzone, L., Buiatti, M.: ADJUST: an automatic EEG artifact detector based on the joint use of spatial and temporal features. Psychophysiology **48**(2), 229–240 (2011)

17. Wang, Q., Sourina, O.: Real-time mental arithmetic task recognition from EEG signals. IEEE Trans. Neural Syst. Rehabil. Eng. **21**(2), 225–232 (2013)
18. Jansen, B.H., Bourne, J.R., Ward, J.W.: Autoregressive estimation of short segment spectra for computerized EEG analysis. IEEE Trans. Biomed. Eng. **9**, 630–638 (1981)

# An Approach for Secure Data Sharing in Cloud and Fog-Based IoT Network

**Deeksha Arya and Mayank Dave**

**Abstract** The glorification of the Internet of Things (IoT) and its application with Cloud computing and Fog computing has revolutionized and simplified every sphere of human life. However, the implementation of IoT, Cloud, and Fog devices introduces several new challenges. Privacy of sensitive data shared on the Internet is one such issue. The data is vulnerable to attackers, hackers as well as Cloud/Fog owners. This article proposes a new secure data sharing scheme that prevents unauthorized access to users' data. The proposed scheme keeps the data secure from malicious users and Cloud/Fog owners by providing complete access control to owner of the data. Additionally, it prevents the leakage of data sharer's lists to Cloud/Fog owners and handles user revocation. Further, it uses lazy re-encryption, which makes it more efficient in terms of resource utilization.

**Keywords** Internet of Things · Fog computing · Cloud computing · Security · Proxy re-encryption · Secure data sharing

## 1 Introduction

In the era of smart technology, Internet of Things (IoT) [1, 2] no longer needs an introduction. IoT has brought a revolution in the field of automation and has become the base for several applications. However, IoT suffers from a plethora of underlying security and privacy risks along with the benefits [3–6], giving rise to the need for a mechanism that can prevent unauthorized access to sensitive data. In most cases, the IoT applications work by storing the data on the public Cloud, either directly or via the Fog-cum-edge devices [7–9].

D. Arya (✉) · M. Dave
Department of Computer Engineering, National Institute of Technology, Kurukshetra, Kurukshetra, Haryana, India
e-mail: deekshasheokand@gmail.com

M. Dave
e-mail: mdave@nitkkr.ac.in



D. Gupta et al. (eds.), *Pattern Recognition and Data Analysis with Applications*, Lecture Notes in Electrical Engineering 888,
https://doi.org/10.1007/978-981-19-1520-8_30

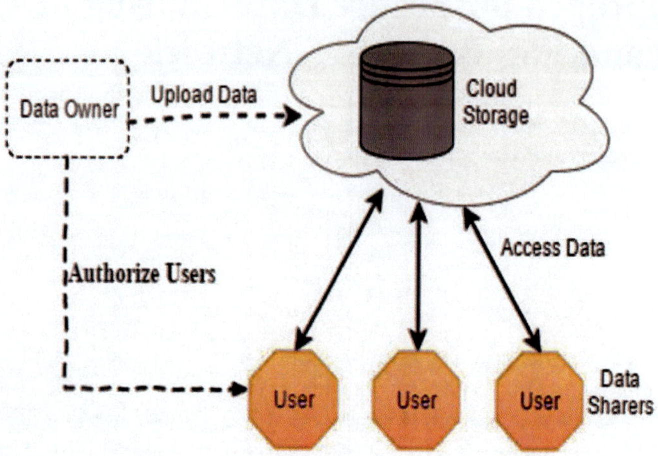


**Fig. 1** Set-up for data exchange between data owner and data sharers using Cloud

Figure 1 shows the set-up for data exchange between data owner and data sharers directly through Cloud. In Cloud Computing, centralized processing is involved and in most cases, the server/Cloud is located far away from the users. To cope with this, the Fog computing was introduced, which performs data processing and storage closer to data sources/users than Cloud computing, making it more efficient and location-aware. The system model where the users of IoT share the data on the Internet by using Cloud and Fog devices, for faster access, is shown in Fig. 2.

To keep the data stored in public Cloud storage confidential, the data owners should encrypt their data before uploading them to the Cloud or any Fog device. However, the encryption introduces a new challenge of sharing the data effectively. The traditional approach for sharing the data stored in encrypted form on Cloud involves the following sequence of steps to be performed by data owner: download his/her data from the storage server on Cloud, decrypt the data, re-encrypt the data using public key of person with whom the data is to be shared (data sharer) and then send the re-encrypted data to that person or re-upload the re-encrypted data to the Cloud. This approach imposes high overhead on data owner and is thus highly inefficient. Additionally, it also loses the merit of the public Cloud storage.

In 2009, Satyanarayanan et al. [10] introduced the concept of Cloudlet, a Cloud data centre, to handle the computation-intensive tasks offloaded from the user machine, with lower latency. And thus the Cloudlet provided a solution to overcome the limitations of traditional approach for secure data sharing with the assumption that the Cloudlets are trusted [11]. Blaze et al. [12] proposed proxy re-encryption (PRE) scheme for solving the problem of secure delegation for the decryption rights. Similar to Cloudlet architecture, PRE system also uses a three-tier architecture in which the proxy plays the role of middle layer between User and Cloud. But, unlike Cloudlet, the PRE system assumes the proxy to be semi-trusted. In a PRE system, the data owner employs the proxy to convert a ciphertext encrypted under the data owner's public key into a new ciphertext of the same message encrypted under the data sharer's public key without letting the proxy learn the underlying message. Many researchers (see [13] and the references therein) extended the work of Blaze et al. [2] for providing secure data sharing in Cloud along with some extra features. We propose a new scheme for sharing the data securely in Fog-computing-based IoT. The main contributions of our scheme are:

1. It prevents the data from unauthorized access from malicious users as well as from the owners of Cloud/Fog devices used to store and share the data on the Internet.
2. It keeps the data sharer's list confidential to data owner.
3. It handles user revocation and provides the complete access control to data owner by giving it the right to apply timing constraints on data sharers' rights.

The remaining paper is organized as follows. Section 3 discusses the existing literature. Section 4 provides the details about the main factors which can be considered for comparing various secure data sharing schemes. Section 4 covers detailed working of the proposed approach followed by its security analysis in Sect. 5. Section 6 concludes the current work and provides directions for future.

## 2 Related Work

This section briefly discusses the existing work related to data sharing schemes in IoT. The authors in [1, 2] address the Internet of Things in detail and describe the various aspects of life, the IoT has high impact on. Fog computing was introduced by Bonomi et al. in 2011, and its role in IoT is described in [7].

Vurukonda and Rao present a study on data storage security issues in Cloud computing [14]. Satyanarayanan et al. [10] discuss the virtual machine-based Cloudlet in mobile computing which is another important aspect of IoT. The authors in [11] proposed a protocol for data security in Cloudlet-based architecture. The protocol provides perfect forward secrecy but has the limitation that the Cloudlets are assumed to be trusted. Another solution for securely sharing the data on Cloud involves the use of proxy re-encryption [12]. In re-encryption, a ciphertext $M_X$ (message $M$ encrypted using user '$X$' public key) can be converted into message $M_Y$ which has the same plaintext as $M_X$ but is encrypted using user '$Y$' public key. The main advantage of proxy re-encryption lies in the fact that the conversion of ciphertexts from one form to another is done without letting the converter/proxy learn the underlying plaintext.

The proxy re-encryption technique was widely adopted and extended by adding some extra features along with providing confidentiality to users' data. A survey of these techniques for secure data sharing in Cloud is given in [13]. Jun et al. used bidirectional aspect and proposed a secure proxy re-encryption method for cryptographic Cloud storage [15]. A bidirectional scheme has the property that the

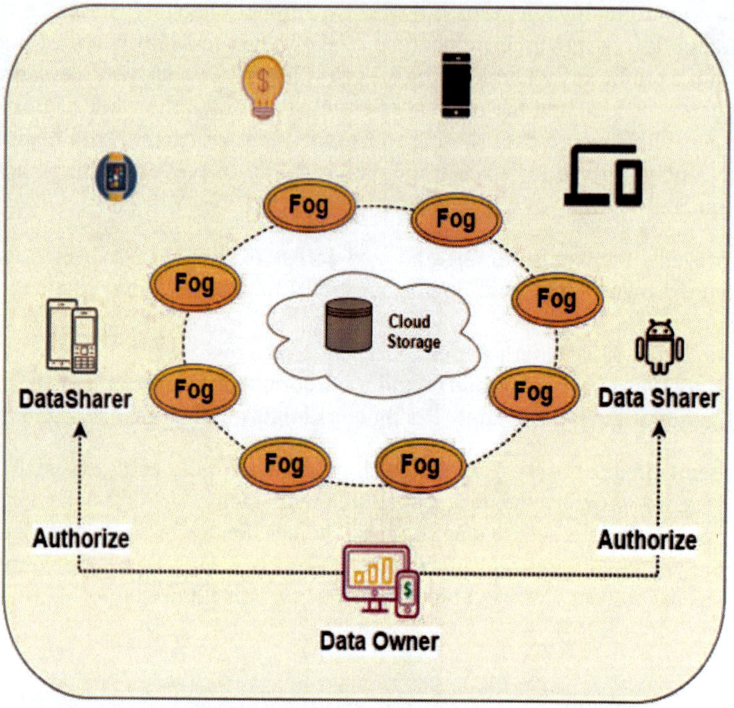


**Fig. 2** Data sharing in Cloud and Fog-based IoT network

same re-encryption key can be employed to re-encrypt the data in both the directions, from ciphertext for user '$X$' to cipher-text for user '$Y$' and vice versa. In addition to being bidirectional, their scheme has the property of keeping the ciphertext size same regardless of the number of transformations applied to it.

Li et al. put forward a pairing-free scheme for securely sharing the data on public Cloud [16]. This scheme is computation efficient since it does not use the time-consuming bilinear paring operations.

Some authors have focused on providing fine-grained access control. For instance, Yang et al. proposed a Cloud-based data sharing scheme with fine-grained proxy re-encryption which provides the data owner the option to share only a subset of its ciphertexts without re-encrypting the data as a whole [17]. Similarly, Yu et al. offered a secure and scalable data sharing scheme considering fine-grained access control in Cloud [18]. Lin and Tzeng proposed a secure data forwarding scheme for a distributed system by integrating threshold proxy re-encryption with decentralized erasure code [19].

Fan et al. proposed a secure timed-release proxy conditional re-encryption scheme which allows the data owner to apply time constraints on data sharer's access to data [20]. Likewise, Liu et al. address the problem of user revocation by using a clock-based re-encryption [21]. This technique allows the Cloud owners to re-encrypt the data in synchronization with their internal clock even when no command is received from the data owner. Our scheme, discussed in Sect. 4 of this paper, is computationally more efficient since it handles user revocation without requiring multiple re-encryption unlike the aforementioned scheme.

Furthermore, some authors explored mobile Cloud Computing and identity-based re-encryption. For instance, Wei et al. proposed an anonymous data sharing protocol for Cloud [22]. The authors in [23] discuss an identity-based proxy re-encryption for making mobile access easy in Cloud. Mollah et al. present a scheme for secure data sharing and searching at the edge of Cloud-assisted Internet of Things [24]. A lightweight scheme for securely sharing the data in Mobile Cloud computing is presented in [25]. A recent survey on challenges related to security and privacy concerning Mobile Cloud computing is presented in [26]. Further, the authors in [27] address the security for smart wearable systems, one of the most trendy application of IoT.

The following section discusses the factors which can be used for comparing various data sharing schemes present in the literature.

## 3 Comparison Factors

The main factors that can be considered for comparing various secure data sharing schemes are:

1. ***Unidirectional Versus Bidirectional***: A proxy re-encryption scheme is termed as unidirectional, if for a given re-encryption key, the proxy is capable to convert user A's ciphertext to the ciphertext intended for user B, but not the vice versa. That is, for a unidirectional scheme, the re-encryption keys used to translate A's ciphertext to that of B and B's ciphertext to that of A are different. For a bidirectional scheme, same re-encryption key can be used to perform the translation in both the directions [13].
2. ***Interactive Versus Non-interactive***: It relates to the involvement of data owner at the time of data sharing. For an interactive data sharing scheme, data owner's secret key is required to generate the re-encryption key.
3. ***Transparency***: It relates to proxy invisibility in case of re-encryption schemes. A scheme is termed as transparent if both the data owner and data sharer are kept unaware of the presence of any proxy between them. For a transparent scheme, the data sharer must not be able to differentiate between the ciphertext generated by encryption using his public key and the ciphertext generated by re-encryption key at proxy.
4. ***Key Optimality***: It relates to the storage overhead of keys. In a key optimal scheme, a user is required to store only a small constant number of keys irrespective of number of decryption delegations it has accepted or delegated.
5. ***Collusion Safety***: In a collusion safe scheme, even a proxy colluding with data sharer cannot learn the data owner's private key.

6. ***Transitivity***: It relates to re-delegation of decryption rights by proxy. Consider the three Cloud users A, B and C sharing the data using proxy re-encryption. If a proxy has the re-encryption key $rk_{AB}$ to translate A's ciphertext to that of B and $rk_{BC}$, to translate B's ciphertext to that of C, then for a non-transitive scheme, the proxy should not be able to generate the re-encryption key $rk_{AC}$, the key to translate A's ciphertext to that of B.
7. ***Conditional Sharing***: It relates to the data owners' ability to control the access rights of data sharer based on some conditions. For instance, the data owner may wish to share different subsets of the data with different sharers. A good data sharing scheme should provide a mechanism to implement conditional sharing.
8. ***Static Versus Dynamic***: A dynamic scheme let the data owner decide the data sharer any time.
9. ***Time Constraints***: It involves providing the data owners' with an option to apply timing constraints on data sharers' decryption rights.

The following section describes the approach proposed in this paper.

# 4 Proposed Approach

The following sections describe the system architecture and the step-by-step procedure of the proposed approach for secure data sharing in IoT.

## 4.1 System Architecture

The presented work considers the system comprising following five components:

1. ***Cloud***, the unlimited storage device which provides the backbone for efficient working of IoT. It is assumed that the Cloud Owner (CO) too could be malicious and thus the users' data must not be revealed to Cloud Owner.
2. ***Data Owner*** (DO), the source of data to be shared and stored in IoT.
3. ***Data Sharer*** (DS), the user who wants to access the data shared by data owner. Any authorized or unauthorized user may wish to access the data; however, the approach needs to ensure that only authorized users are granted the access.
4. ***Fog***, the edge devices present in the proximity of users. Similar to Cloud Owner, the Fog Owners (FO) could be malicious too and are thus exempted from the right to access users' data.
5. ***Trusted Third Party*** (TTP), which generates different keys (public key, private key, re-encryption key) required for encryption–decryption of the data shared on the Internet.

Figure 3 shows the system architecture and the communication between different components involved in data sharing in IoT. Users including data owners and sharers

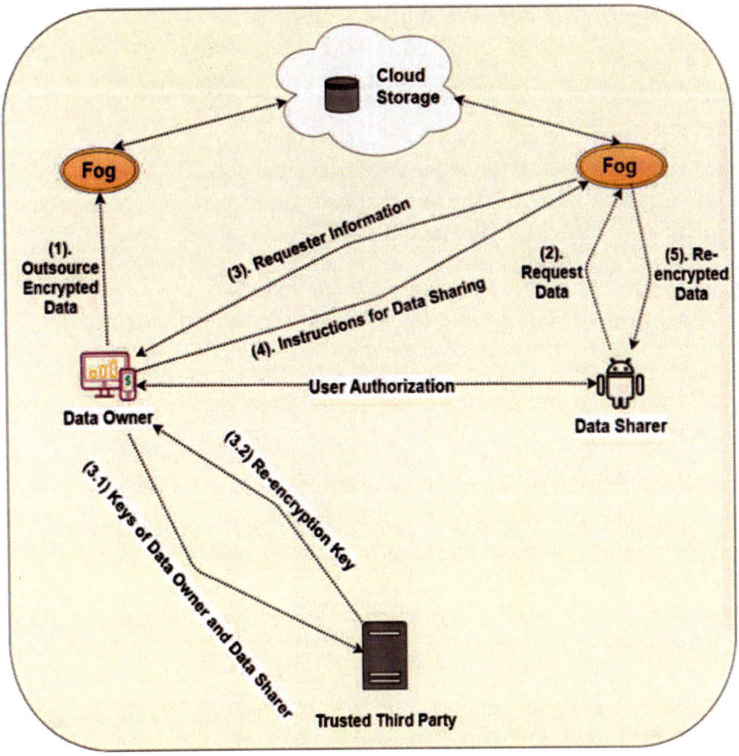


**Fig. 3** Proposed approach for securely sharing the data in Cloud and Fog-based IoT network using trusted third party

are connected via Fog and Cloud. To keep the sensitive data confidential, the Data Owner outsource the data after encryption. This is the only time when the Data Owner performs the encryption. All the other encryptions (termed as re-encryptions) required in the later stages, are performed by the Fog devices, thus reducing the burden on Data Owners. The encryption is considered to be homomorphic so that any operation which was intended to be carried out on user's data can be executed on this encrypted data. Further, to prevent unauthorized access to the data stored on Cloud/Fog, our approach works by informing Data Owner about every request made to access that data. To provide complete access control of the data, the Data Owner is given an option to share the data partially by instructing the Fog device to apply the re-encryption only on a subset of the data. The following section describes the working of our approach.

## 4.2 *Step-by-Step Procedure*

The proposed approach comprises five steps as described below.

1. The Data Owner outsources the encrypted data.
2. The Data Sharer sends the request to access the data.
3. The Fog device receiving the request sends the information of requester to Data Owner along with the following query:
   - *Do you want to share your data with the requester?*
   - *If yes, do you want to share the whole data or only a subset of it?*
   - *Please share the required information.*
4. The Data Owner, on receiving the above message, decides whether the requester is authorized or not.
   - *If not*, it sends a *decline* response to the Fog device which sent the query. The *decline* message indicates that the Data Owner does not want to share the data with the requester, and therefore, the request should be declined by the corresponding Fog device.
   - *In case of authorized requester*, the Data Owner contacts the trusted third party (TTP).
     - The TTP generates re-encryption key for converting the ciphertext intended for Data Owner into the ciphertext for the requester as per the Data Owner's guidelines.
     - The Data Owner, then, passes this re-encryption key to the Fog device which sent the query.
5. The Fog device downloads the data from Cloud, re-encrypt it (as a whole or only a subset of it, as instructed by Data Owner) using the key provided by the Data Owner. This re-encrypted data is then shared with the data requester.

# 5 Security Analysis

The proposed data sharing protocol has the following security characteristics:

- ***Provides Confidentiality***
  - Only the encrypted data is uploaded to Fog and Cloud in the proposed scheme. This ensures that only the user having access to the key can access the data, and hence keeps the data confidential.

- ***Prevents Unauthorized Access from Malicious Users***
  - The presented scheme grants data access to only the users who have been authorized by the Data Owner. Whenever a request is received to retrieve the stored data, the Fog device receiving the request sends a query to Data Owner. The data is transferred to the requester only after Data Owner's approval. Moreover, the data is shared only in encrypted form which further enhances the security.
- ***Prevents Unauthorized Access from Cloud and Fog Owners***
  - The Data Owner uploads the data in encrypted form, thus preventing the Cloud and Fog owners from learning the plaintext.
- ***Provides Collusion Safety***
  - The presented scheme is collusion safe, i.e. even if the Fog device having the re-encryption key collides with the corresponding data sharer, it would not be able to learn the data owner's private key.
- ***Handles User Revocation***
  - Each time someone request the access to data, the Fog device first take the permission from Data Owner before sharing the data with the requester. Thus, the Data Owner can decline the request of revoked users easily.
- ***Provides User Anonymity by Preventing the Leakage of Data Sharer's List to Cloud/Fog Owners***
  - The proposed approach does not require the sharing of list of data sharer with Cloud or Fog. Instead the data owner itself maintains the list and is allowed to change it any time.

It can be noted that the scope of the current work is limited to establish the theoretical framework. The future work may consider providing experimental support for the same.

## 6 Conclusion and Future Work

Among the several challenges associated with Cloud and Fog-computing-based IoT, this paper addresses the security of the data shared on Internet. Firstly, the factors that can be used to compare different data sharing approaches are discussed. Secondly, a new scheme to share the data securely using Cloud and Fog devices is proposed. Thirdly, the security analysis for the proposed approach is carried out in the presented manuscript.

The proposed scheme provides complete access control of the data to its owner and prevents the unauthorized access to the same when stored on public Cloud storage in

IoT. Further, the problem of user revocation is addressed and information valuable to the users of Cloud and Fog Computing, along with IoT, including the data owners and algorithm designers is presented.

In future, other aspects of data sharing on IoT, or data storage on Cloud and Fog devices can be considered to extend the current work. One such aspect is ensuring the integrity of the data. Furthermore, the current work assumes the third party used in the system model as trusted. However, practically, it needs to be considered that TTP is also connected through the Internet to provide services and may also have malicious intentions. Hence, the future work may explore the protocols for the trusted third party used in the presented approach.

## References


1. Atzori, L., Iera, A., Morabito, G.: The internet of things: a survey. Comput. Netw. **54**(15), 2787–2805 (2010)
2. Li, S., Da Xu, L., Zhao, S.: The internet of things: a survey. Inf. Syst. Front. **17**(2), 243–259 (2015)
3. Tawalbeh, L.A., Muheidat, F., Tawalbeh, M., Quwaider, M.: IoT privacy and security: challenges and solutions. Appl. Sci. **10**(12), 4102 (2020)
4. Hameed, S., Khan, F.I., Hameed, B.: Understanding security requirements and challenges in internet of things (IoT): a review. J. Comput. Netw. Commun. (2019)
5. Sadique, K.M., Rahmani, R., Johannesson, P.: Towards security on internet of things: applications and challenges in technology. Procedia Comput. Sci. **141**, 199–206 (2018)
6. Maple, C.: Security and privacy in the internet of things. J. Cyber Policy **2**(2), 155–184 (2017)
7. Bonomi, F., Milito, R., Zhu, J., Addepalli, S.: Fog computing and its role in the internet of things. In: Proceedings of the First Edition of the MCC Workshop on Mobile Cloud Computing, pp. 13–16. ACM (2012)
8. Arya, D., Dave, M.: Priority based service broker policy for fog computing environment. In: International Conference on Advanced Informatics for Computing Research, pp. 84–93. Springer, Singapore (2017). https://doi.org/10.1007/978-981-10-5780-9-8
9. Arya, D., Dave, M.: Security-based service broker policy for FOG computing environment. In: 8th International Conference on Computing, Communication and Networking Technologies (ICCCNT), pp. 1–6. IEEE (2017). https://doi.org/10.1109/ICCCNT.2017.8204036
10. Satyanarayanan, M., Bahl, P., Caceres, R., Davies, N.: The case for VM-based cloudlets in mobile computing. IEEE Pervasive Comput. **8**(4), 14–23 (2009)
11. Jindal, M., Dave M.: Data security protocol for cloudlet based architecture. In: Recent Advances and Innovations in Engineering (ICRAIE), pp. 1–5. IEEE (2014)
12. Blaze, M., Bleumer, G., Strauss, M.: Divertible protocols and atomic proxy cryptography. In: Advances in Cryptology—EUROCRYPT'98, pp. 127–144. Springer (1998)
13. Zhiguang, Q., Hu, X., Shikun, W., Jennifer, B.: A survey of proxy re-encryption for secure data sharing in cloud computing. IEEE Trans. Serv. Comput. (2016)
14. Vurukonda, N., Rao, B.T.: A study on data storage security issues in cloud computing. Procedia Comput. Sci. **92**, 128–135 (2016)
15. Jun, S., Rongxing, L., Xiaodong, L., Kaitai, L.: Secure bidirectional proxy re-encryption for cryptographic cloud storage. Pervasive Mob. Comput. 113–121 (2016)
16. Lu, Y., Li, J.: A pairing-free certificate-based proxy re-encryption scheme for secure data sharing in public clouds. Futur. Gener. Comput. Syst. (2016)
17. Yang, Y., Zhu, H., Lu, H., Weng, J., Zhang, Y., Choo, K.K.R.: Cloud based data sharing with fine-grained proxy re-encryption. Pervasive Mob. Comput. **28**, 122–134 (2016)

18. Yu, S., Wang, C., Ren, K., Lou, W.: Achieving secure, scalable, and fine-grained data access control in cloud computing. In: 2010 Proceedings IEEE INFOCOM, pp. 1–9. IEEE (2010)
19. Lin, H.Y., Tzeng, W.G.: A secure erasure code-based cloud storage system with secure data forwarding. IEEE Trans. Parallel Distrib. Syst. **23**(6), 995–1003 (2012)
20. Fan, C.I., Chen, J.C., Huang, S.Y., Huang, J.J., Chen, W.T.: Provably secure timed-release proxy conditional re-encryption. IEEE Syst. J. (2015)
21. Liu, Q., Tan, C.C., Wu, J., Wang, G.: Reliable re-encryption in unreliable clouds. In: Global Telecommunications Conference (GLOBECOM 2011), pp. 1–5. IEEE (2011)
22. Wei, G., Lu, R., Shao, J.: EFADS: efficient, flexible and anonymous data sharing protocol for cloud computing with proxy re-encryption. J. Comput. Syst. Sci. **80**(8), 1549–1562 (2014)
23. Zhou, Y., Deng, H., Wu, Q., Qin, B., Liu, J., Ding, Y.: Identity-based proxy re-encryption version 2: making mobile access easy in cloud. Futur. Gener. Comput. Syst. **62**, 128–139 (2016)
24. Mollah, M.B., Azad, M.A.K., Vasilakos, A.: Secure data sharing and searching at the edge of cloud-assisted internet of things. IEEE Cloud Comput. 34–42 (2017)
25. Li, R., Shen, C., He, H., Gu, X., Xu, Z., Xu, C.Z.: A lightweight secure data sharing scheme for mobile cloud computing. IEEE Trans. Cloud Comput. **6**(2), 344–357 (2017)
26. Mollah, M.B., Azad, M.A.K., Vasilakos, A.: Security and privacy challenges in mobile cloud computing: survey and way ahead. J. Netw. Comput. Appl. **84**, 38–54 (2017)
27. Nikooghadam, M., Amintoosi, H., Kumari, S.: On the security of "secure and lightweight authentication with key agreement for smart wearable systems". Wireless Pers. Commun. 1–8 (2021)

# Privacy Protection of Edge Computing Using Homomorphic Encryption

**Ganesh Kumar Mahato and Swarnendu Kumar Chakraborty**

**Abstract** The emergence in the field of Internet of Things and the success of the cloud providers have extended the horizon of the new technological model that is Edge computing. This technology has got the ability to handle problems such as improving the response time, increasing battery life, cost saving of the network, along with ensuring protection and privacy of the data. To manifest the idea of Edge Computing, here a detailed case study is presented starting from cloud architecture to the security issues, further extending to its security countermeasures using Homomorphic encryption. This article explores the security challenges and the smart way to handle them and ensure privacy to the data with nicely explained algorithms. The algorithms perform computation on the encrypted form without decryption of the ciphertext. This does not require the sharing of a secret key. The opportunities in the field of edge computing is also emphasize with the expectation that this paper will pique interest and encourage further study in this area.

**Keywords** Cloud security · Edge computing · Homomorphic encryption · Internet of Things

## 1 Introduction

With the emergence of cloud computing, our way of living and working has transformed tremendously. Having different services of the cloud computing, it has made the life smoother in different fields. For instance, Software as a Services (SaaS) such as Facebook, Twitter, Google Apps, etc. has brought the things closer to us [1]. Platform as a Service (PaaS) allows the users to develop, run and manage the software applications on a single platform. Another service model is Infrastructure as a Service

G. K. Mahato (✉) · S. K. Chakraborty
National Institute of Technology Arunachal Pradesh, Papum Pare 791112, India
e-mail: ganesh.phd20@nitap.ac.in

S. K. Chakraborty
e-mail: swarnendu@nitap.ac.in

 
D. Gupta et al. (eds.), *Pattern Recognition and Data Analysis with Applications*,
Lecture Notes in Electrical Engineering 888,
https://doi.org/10.1007/978-981-19-1520-8_31

(IaaS) which provides us with a wide range of services such as servers, storage of information, networking, hardware, and many more [2]. Later on, Internet of Things (IoT) connected the various devices that used to process and exchange the data over the Internet [3]. IoT is basically introduced to make a computer sense like a human and gather information from the edge community. Some of the IoT applications need an environment that can communicate very quickly with short response time. In order to facilitate these applications, cloud computing is not effective enough. Due to the advent of the IoT and mass penetration of wireless networks, the devices of edge computing and data gathered from the edge has been exponentially rising in recent years. By 2025, the International Data Corporation (IDC) predicts that globally the data is supposed to exceed 180 zettabytes (ZB), with 70% of it provided by IoT being utilized at the network's edge. IDC also estimates that by 2025, more than 150 billion computers will be connected globally. In this situation, cloud computing's clustered processing mode is insufficient to manage the data provided by the edge. The edge computing paradigm transfers all data across the network to the cloud data center, which then uses its fast-computing capacity to solve computation and storage issues, allowing cloud providers to generate economic benefits [4]. Traditional cloud storage, on the other hand, has many drawbacks in the sense of security, latency, bandwidth, availability, resources, stability, and privacy [5].

This paper covers the most recent developments in edge computing, as well as the emerging problems and potential prospects in this field. Our primary focus lies on ensuring security to the edge computing devices. The article has been organized in the following way. In the first part we have tried to describe the works already done in the field of security of edge computing, then we have given a light on what is edge computing, its function and architectural model. Next, we described the homomorphic encryption and its principles followed by its implementation in the security of edge computing. Then experimental analysis is explained. Finally, we conclude this paper describing the scope of the mentioned algorithm in the future that can ensure privacy to the edge computing devices.

## 2 Related Works

The idea of connecting the devices can be traced back to 1970's. Then it was called as embedded Internet. The word IoT was used in 1999 for the first time by Kevin Asthon. IoT started gaining popularity after 2010. Being a new concept in connecting the devices via Internet, it is expected that 35 billion devices will be connected by the end of 2021 [6]. And so, the vulnerability will multiply. Being a new technology there has been very less work done till now in the field of ensuring security to the edge devices, but the researchers are striving continuously to secure this emerging technology in a better way. Security implementation is a basic and critical issue in edge computing, and it has been addressed in a number of previous studies [7, 8]. In this section the literature survey of edge computing and its security is presented. Since 2010 many industries and organizations started implementing this emerging technology but did not emphasize on the security issues. Many researchers have presented their papers, mentioning the threats and vulnerability but that remained theoretical [3].

Wu et al. in his paper [9] discussed the use of homomorphic encryption for ensuring security to the edge devices. He further describes the use of proxy re-encryption and ciphertext policy attribute-based encryption. Abawajy [10] in his article proposed a model that determines the android malware, provided a mobile malware detection model and punishes them automatically. Murugesan et al. in his paper [11] used hybrid RSA Elliptic Curve Cryptography, this scheme is confined to small key size and limited storage. RSA being a slow algorithm consumes much time in encryption and decryption. Later on, Yan et al. in his article [12] presented a model to secure the edge computing using RSA algorithm and proxy re-encryption of the ciphertext. The new edge computing data privacy security scheme cannot alone fulfill the requirements for safe data storage and retrieval details. The primary aim of this paper is on the protection of edge devices from various threats. Further we focus on the application of homomorphic cryptography to edge devices, including implementation of encrypted data and performing computation on that under various attacks.

## 3 Proposed Model

In this paper we have proposed a novel design of Homomorphic Encryption for ensuring security to the edge computing-based IoT devices from various vulnerability. We have used hybrid model comprises of RSA and proxy re-encryption schemes that ensures double fold protection to the sensitive data in the edge. As data is processed in the network edge then it is sent to the data center or cloud for further storage, proxy re-encryption plays a vital role in encapsulating the data before being stored for further use.

## 4 Edge Computing

Since 2015, edge computing has been on the rise, drawing a lot of interest from both academia and business. In the year 2016, the National Science Foundation (NSF) based in United States identified edge computing as a highlighted field of computer systems study. This technology has been transforming the way of processing, handling and delivering the data over the millions of devices around the globe. The exponential growth of IoT in the past one decade, needs the real time computing resources that drives the edge computing technology. Being able to respond to data almost instantaneously, it eliminates the lagging time, reduces the Internet bandwidth usage and lowers the cost. Since the data is processed in the real time in the network edge without transferring it to the public cloud, the sensitive data remains out of threat. Prior to data processing the sensitive data is sent to the cloud for further storage. Edge computing systems are accelerating the development and support of real time applications, such as self-driving cars, video conferencing, AI virtual assistance, augmented reality, and many more.

### 4.1 Defining Edge Computing

Edge computing is a novel technique allowing the users to process, store and manage the data at the edge of the network and provides intelligent services by the collaboration with cloud computing [13]. In a broader sense, all the computations taking place outside the cloud where real time processing of data is necessary that respond to the request generated by sensors or users at low latency is the edge [14]. According to Shi et al. [15] Edge computing can be referred as the technology that permits computation to be carried out at the network edge, on downstream data on behalf of cloud computing and upstream data on behalf of IoT services.

### 4.2 Functions of Edge Computing

There are two processing streams of edge computing: one is upstream where computation is from applications or devices to the cloud and another is downstream where computation is from cloud to the devices or application [15]. The devices of the edge computing serve as the data producers as well as data consumers. At the edge level, services and contents are requested from the cloud as well as computations are performed from the cloud. Edge has the capability to process the data, store it safely, request and deliver services from the cloud to the end users. Moreover, it is reliable, ensures security and delivers privacy to the sensitive data and meets the end user's requirement efficiently [16].

### 4.3 Edge Computing Three Tier Model and Architecture

**Three Tier Model**. Edge computing is a three tier model that comprises IoT, edge and cloud. IoT is the first layer, which includes drones, connected health cameras, smart home systems and appliances, industrial Internet equipment, etc. IoT and the second layer that is edge, are connected using a variety of communication protocols. For instance, drones can be connected to a network tower via 4G/LTE, and smart home devices can communicate through Wi-Fi. The cloud's massive processing and storage capabilities are used to complete complex tasks at the edge, which includes self-driving vehicles, network towers, gateways, and edge servers. Low power consumption and short distance are common characteristics of the protocols between IoT and

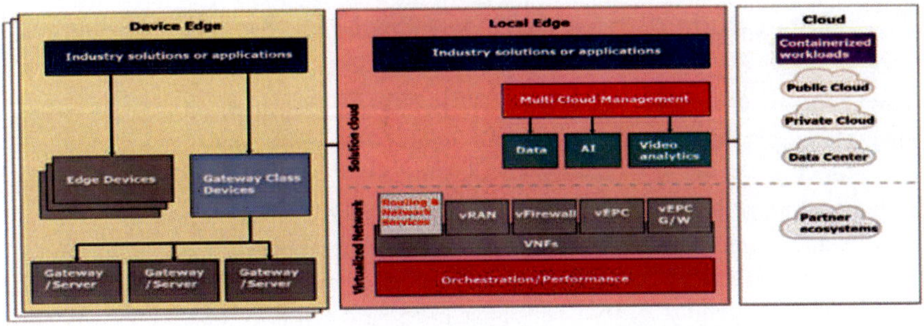


**Fig. 1** Architecture of edge computing

Edge. Larger throughput and high speed are common characteristics of protocols between edge and cloud [3]. The networking protocols between the edge and the cloud are basically Ethernet, optical fiber, and the oncoming 5G network.

**Architecture**. There are basically three primary nodes in the architecture of edge computing: Device Edge, Local Edge, and the Cloud [4]. Each of these nodes are explained here using Fig. 1 along with its overall design of edge computing.

*Device Edge*. Edge devices are the hardware components that collect data or communicate with the edge data such as security cameras, drones, RFID readers, digital signage, medical implants, and other connected items [9]. The devices serve us with numerous benefits like: transmitting, routing, processing, monitoring, filtering, translating and storing data passing between the networks. The working principle of the edge is very simple, it acts a medium to connect two different networks and translates one protocol to another. Hence serves as a network entry or exit. Being the connection at the network edge, the devices eliminate the latency issues and serves the data processing at the real time. Cost saving is one of the advantages as bandwidth usage is optimized since data is manipulated within the edge.

*Local Edge*. The applications that operate on-premises or at the network's edge comes under the local edge. The edge server and its network can be a separate or an integrated entity existing on a different or same location. This architectural layer mainly operates on Application layer and Network layer where the device edge applications are placed. In the Application layer the applications that cannot run at the system edge due to the device's footprint being too big can run here. Complex video analytics and IoT analysis are two examples of such applications. In the Network layer the physical devices are mainly not used in order to avoid complication in controlling or monitoring them. Hence the complete Network layer is on a virtual mode. Routers and switches are the main devices used in this layer.

*Cloud*. The Cloud is the most important segment of this entire architecture. Bearing a huge data load and managing computations of the various applications at a time makes it ideal when we compare to edge [17]. It is responsible in deploying the workloads to the various edge nodes with the help of systematic management.

### 4.4 Edge Computing Security and Privacy

With the increase in edge computing devices, the vulnerability is also growing parallelly. Having various layers of the edge computing architecture, we are threatened by different levels of security issues. Since the edge devices are in direct communication with the edge data center, this may bypass the central monitoring system. It may lead to system hacking, primary leakage, data tampering, injection of information, and many more [18]. Though edge computing having distributed characteristic, the above-mentioned security issues have become a matter of major concern. Since the IoT devices and various sensors are in interconnection and communicate via the network channels such as wireless network and mobile network, this may lead to network attacks like Denial of Services (DoS). So, ensuring security to the devices has become our key responsibility [7].

## 5 Homomorphic Encryption

Homomorphic Encryption (HE) is a cryptographic method focused on the computational complexity theory of mathematical problems [1]. We can get an output that is identical to the original data when dealing with homomorphic encryption data. HE is an encryption technique that allows you to execute operations on protected (encrypted) data without having to decode it [17]. It helps us to execute operations on encrypted data without the use of a secret key. On encrypted files, any mathematical procedure of any complexity can be performed without hampering the protection. "Homo" is a Greek word that means "same," and "Morphic" means "structure". When performing related mathematical operations on encrypted files, the HE method produces the same result. The performance is the same after decrypting the data, implying that the operations were done on unencrypted data [19]. On the encrypted files, algebraic operations are used to perform a number of computations. HE will be the encryption scheme of the future of cloud and edge computing and allowing multiple companies to store encrypted data in a decentralized cloud without risk of compromising its security and provide benefit to the users in availing protected services [20].

### 5.1 HE Definition

Here we consider $P$ as the plain text that is $P = \{0, 1\}$, it contains the message tuple $(M_1, M_2, \ldots M_n)$ as input. The Boolean circuit is represented by $C$, whereas the ordinary function as $C(M_1, M_2, \ldots M_n)$ to show the computation on the message tuple [21]. The HE is defined as follows:

**Key Generation**: $\text{Gen}(1^{\lambda}, \alpha)$ is the algorithm that produces key triplets as secret keys (sk and pk) and the evaluation key (evk), where $\lambda$ is security parameter and $\alpha$ is auxiliary input. It is given as:

$$(\text{sk}, \text{pk}, \text{evk}) \leftarrow \text{KeyGen}(\$) \tag{1}$$

**Encryption**: $\text{Enc}(\text{pk}, M)$ this is to encrypt the message $(M)$ using the public key (pk) and generates a ciphertext $c$. The relation is denoted as:

$$(c \in C), c \leftarrow \text{Enc}(\text{pk}, M) \tag{2}$$

**Decryption**: $\text{Dec}(\text{sk}, c)$ this is to decrypt the ciphertexts (encrypted plaintext) using the secret key (sk) and gets back the original message $(M)$ as the desired output.

$$M \leftarrow \text{Des}(\text{sk}, c) \tag{3}$$

**Evaluation**: $\text{Eval}(\text{evk}, C, c_1, c_2, \ldots c_n)$ generates computational value while considering $evk$ key as the input, where the circuit $c \in C$ and input tuple ciphertexts that is, $c_1 \ldots c_n$ as the already computed results. Evaluation is performed as:

$$c^* \leftarrow \text{Eval}(\text{evk}, C, c_1, c_2, \ldots c_n) \tag{4}$$

## 5.2 *HE Properties*

The operation on the plain text using homomorphic encryption can be represented as below [22].

$$E(M_1) = M_1^e \text{ and } E(M_2) = M_2^e \tag{5}$$

So, Additive Homomorphic property is expressed as

$$E(M_1) + E(M_2) = M_1^e + M_2^e = (M_1 + M_2)^e = E(M_1 + M_2) \tag{6}$$

And Multiplicative Homomorphic property is expressed as:

$$E(M_1) * E(M_2) = M_1^e * M_2^e = (M_1 * M_2)^e = E(M_1 * M_2) \tag{7}$$

## 5.3 Homomorphic Encryption in Edge Computing

It is critical to incorporate different forms of protection and privacy mechanisms, as well as avoid any attraction from hostile adversaries [23], that one may build a robust edge computing environment of security and accessible platform. The current protection and privacy protocols that can be used in the edge computing framework are presented in this subsection. In addition, an edge computing data protection analysis using Homomorphic Encryption is presented here. This algorithm follows the basic concept of RSA Homomorphism over proxy re-encryption that ensures data privacy in edge computing.

**Step 1: Key Generation**: Let us take two prime numbers $p$ and $q$ randomly to achieve the given condition.

$$gcd(pq, (p-1)(q-1)) = 1 \tag{8}$$

Find the modulus

$$n = \text{pq.} \;=\; \text{lcm}(p-1, q-1) \tag{9}$$

where lcm = least common multiple of $p-1$ and $q-1$.

Now select any arbitrary number $g \in \left(g \in Z^*_{n2}\right)$ and we get

$$\mu = \left(L\left(g^\lambda \bmod n^2\right)\right)^{-1} \bmod n \tag{10}$$

Find greatest common divisor of $L\left(g^\lambda \bmod n^2\right)$ and $n$.
$Z^*_{n2}$ represents a set of integers coprime to $n^2$ in $Z_{n2}$.
For the function $L(x) = x - 1/n$.
Public key is $(n, g)$ and private key is $(\lambda, \omega)$.
While encrypting and decrypting, take the plain text as $m(m \in Z_{n2})$.
where $m < n$.
And integer as $r\left(r \in Z^*_{n2}\right)$.

**Step 2: Encryption Process**: Let us take $m \in Z_{n2}$ as the message and encrypt it. Compute the ciphertext $c = m \bmod n$.

The encryption is performed as follows:

$$c = E(m) = g^m \cdot r^n \bmod n^2 \tag{11}$$

where $c$ is the ciphertext for the plain text $m$ and $c \in Z^*_{n2}$.

It is to be noted that for the same ciphertext $m$, the integer $r$ is selected randomly in the process of encryption, so $r$ may be different. Hence the corresponding ciphertext may vary in order to provide security to the ciphertext.

**Step 3: Proxy Re-encryption Process**: Ciphertext is generated after computing public key ($R_{sk}$) and private key $\left(R_{pk}\right)$ using RSA algorithm. The public key $\left(R_{pk}\right)$ is kept safely in the server after re-encryption [24] for further use.

**Step 4: Decryption Process**: Decryption of the ciphertext $c$ is performed as follows:

$$m = D(c) = L\left(c^{\lambda} \bmod n^2\right) * (\omega \bmod n) \tag{12}$$

After getting $E(d_i)(i \in p_\tau)$, it is decrypted to obtain $(d_i)(i \in p_\tau)$, sign it and send the $\text{Sign}_q(d_i)(i \in p_\tau)$ to $EN_q$.

**Step 5**: After $EN_q$ gets $(d_i)(i \in p_\tau)$, divide the dispersion degree into two types. Create a set of users of the same type with elements as $Q$ and a set of users with degree of dispersion as $G$. When any threatening user takes a portion, $Q$ that includes a normal user, the value of target increases at the end of the task whereas $G$ decreases. Hence, two parameters $\mu$ and $v$ are introduced to control the updated values of target when increases or decreases. The value of the target varies according to the below equation:

$$r_i^{\text{new}} \begin{cases} r_i + (1 - r_i) \cdot \mu \text{ if } \in \ Q \\ r_i \cdot (1 - \upsilon) \qquad \text{if } i \ \in G \end{cases} \tag{13}$$

where $\mu$ and $v$ are both positive and $\upsilon < 1$.

The RSA algorithm [11] is used when the execution side transmits the address of the storage and the secret key of data. The decryption of the ciphertext can be done by one who knows the RSA public key, this secret public key is shared to the intended person. Hence, this scheme provides the security while transferring the data storage address and the secret decryption key. Distributed file system helps in fetching the data using the given address of the file where it is saved. Both the parties never collude each other, hence privacy is maintained. While delivering the data, other than the two parties, do not hold any information of the data file. They can get only ciphertext, which is not possible to alter. Therefore, in the process of data delivery, the data is secured and the privacy is also maintained.

## 6 Test and Analysis

In order to carry out the experiment and get the details of data analysis, Ubuntu 64-bit OS is used. Here 10 D vector data is used. Then after the data is encrypted and send the ciphertext to the edge node. When accurate data is supplied, $\mu$ gives the increasing rate of target value, whereas $v$ is for denoting the decreasing rate when malicious data is supplied. In Fig. 2, we can see the changes occurring in the target value against task (n) when correct data is supplied. Here we take 10 D vector data as 0.02, 0.04, 0.06, 0.08, 0.10, 0.13, 0.15, 0.17, 0.19, and 0.2. The value of $\mu$ converges

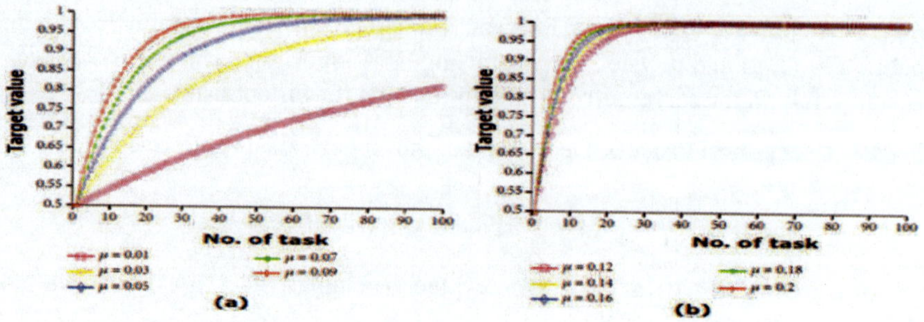


**Fig. 2** Variation of $\mu$ on different target values

to 1 when high value of $\mu$ is supplied, it attains the target value easily and converges to 1. Moreover, if the value of $\mu$ is too high, then only few accurate data is needed to attain the value of target. However, the usual value of $\mu$ ranges from 0.1 to 0.2.

When continuously wrong data (malicious) is provided the $v$ converges to 0. Let us supply wrong value as 0.1, 0.2, 0.3, 0.4, 0.5, 0.6, 0.7, 0.8, 0.9, and 1. If high value of $v$ is taken then it converges to 0 more easily as shown in Fig. 3. In general, $v$ ranges from 0.2 to 0.5.

Figure 4 shows the relation between the number of tasks ($n$) and its time required for encryption. The graph grows linearly when the task increases. It is obvious that the higher the no. of tasks, more time is consumed for converting the plain text to ciphertext. Furthermore, the time taken in encryption and decryption has been compared with the similar schemes in Table 1. It was found that our scheme consumes less time thereby enhances the performance.

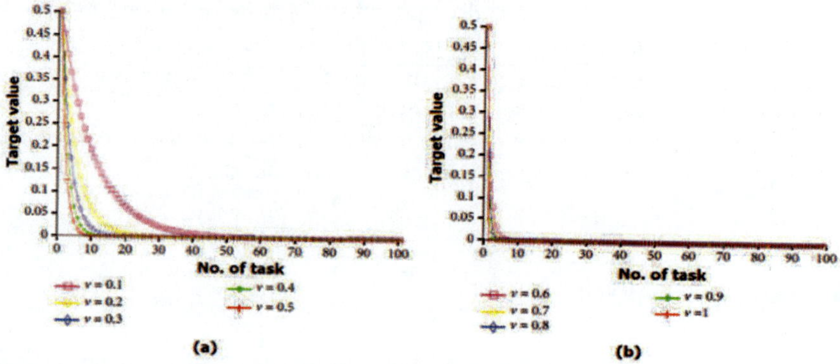


**Fig. 3** Variation of $\nu$ on different target values

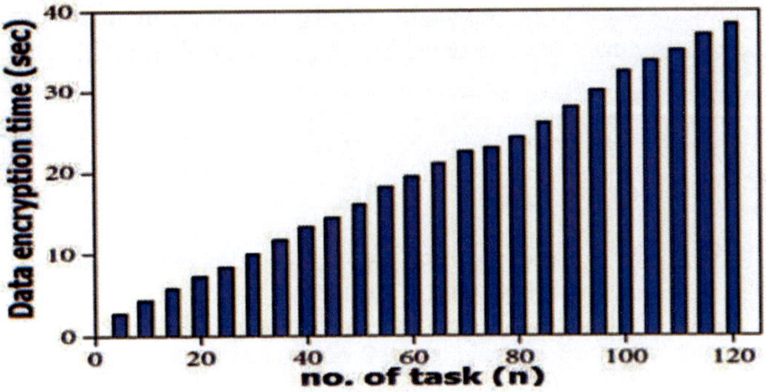


**Fig. 4** Encryption time over given task

**Table 1** Comparison of existing schemes with the proposed model

| Ref. No. | Security parameter (bits)/schemes | Encryption time (s) | | | Decryption time (s) | | |
|---|---|---|---|---|---|---|---|
| | | 80 | 128 | 256 | 80 | 128 | 256 |
| [11] | RSA_ECC_PRE | 18.7 | 33.9 | 892 | 200 | 908 | 1097 |
| [11] | FH_ECC_PRE | 6.5 | 30.4 | 573 | 3.7 | 18.3 | 231.7 |
| [12] | RSA_PRE | 5.4 | 26.2 | 437 | 10.5 | 15.7 | 157.2 |
| | Our scheme | 3.2 | 7.5 | 237 | 4.2 | 6.2 | 97.3 |

*Note RSA* Riverst Shamir Aldamen, *ECC* elliptic curve cryptography, *PRE* proxy re-encryption, *FH* fully homomorphic encryption

## 7 Conclusion

Most of the networks which were earlier under cloud network are now moving to edge network because of having more advantages as compared to the cloud. The users are experiencing faster response and higher efficiency in edge as all the computation and the processing are taking place at the network edge. There is a huge saving of the bandwidth as maximum part of the data manipulation is being performed at the edge network instead of uploading and performing the data computation in the cloud server. With the advancement in the IoT devices and mobile users, the edge computing has become data producer, earlier it was data collector. We established our definition of edge computing in this article, based on the idea that the computation can take place close to the data sources. Then we explored few scenarios where edge computing can be implemented like IoT devices, mobile computing, smart homes, etc. We focused on the collaborative edge which can connect the users and the cloud that can minimize the communication gap between the distant networks for storing and computing the data. Security being one of the major issues drawn our focus on the homomorphic encryption that can be implemented in the edge computing to ensure privacy in data sharing through the communication channel. The algorithm mentioned here needs a proper implementation to come up with a secured communication in the edge devices. Hoping that this paper will create an awareness to the security of edge devices and its real application.

## References


1. Alaya, B., Laouamer, L., Msilini, N.: Homomorphic encryption systems statement. Trends and challenges. Comput. Sci. Rev. **36**, 100235 (2020)
2. Li, J., Song, D., Chen, S., Lu, X.: A simple fully homomorphic encryption scheme available in cloud computing. In: IEEE 2nd International Conference on Cloud Computing and Intelligence Systems, pp. 214–217 (2012)
3. Khan, M.A., Salah, K.: IoT security: review, blockchain solutions, and open challenges. Future Gener. Comput. Syst. **82**, 395–411 (2018)
4. Ai, Y., Peng, M., Zhang, K.: Edge computing technologies for Internet of Things: a primer. Digital Commun. Networks **4**(2), 77–86 (2018)
5. Magesh, S., Indumathi, J., Radha RamMohan, S., Niveditha, V.R., Shanmuga Prabha, P.: Concepts and contributions of edge computing in Internet of Things (IoT): a survey. Int. J. Comput. Networks Appl. **7**(5), 146–156 (2020)
6. Faisal, M., Ali, I., Khan, M.S., Kim, S.M., Kim, J.: Establishment of trust in Internet of Things by integrating trusted platform module: to counter cybersecurity challenges. Complexity (2020)
7. Caprolu, M., Pietro, R.D., Lombardi, F., Raponi, S.: Edge computing perspectives: architectures, technologies, and open security issues. In: IEEE International Conference on Edge Computing (EDGE) 2019, pp. 116–123 (2019)
8. Alwarafy, A., Al-Thelaya, K.A., Abdallah, M., Schneider, J., Hamdi, M.: A survey on security and privacy issues in edge-computing-assisted Internet of Things. IEEE Internet Things J. **8**(6), 4004–4022 (2021)
9. Wu, H., Zhang, Z., Guan, C., Wolter, K., Xu, M.: Collaborate edge and cloud computing with distributed deep learning for smart city internet of things. IEEE Internet Things J. **7**(9), 8099–8110 (2020)
10. Abawajy, J.: User preference of cyber security awareness delivery methods. Behav. Inf. Technol. **33**(3), 237–248 (2014)
11. Murugesan, A., Saminathan, B., Al-Turjman, F., Kumar, R.L.: Analysis on homomorphic technique for data security in fog computing. Trans. Emerg. Telecommun. Technol. 1–16 (2020)
12. Yan, X., Wu, Q., Sun, Y.: A homomorphic encryption and privacy protection method based on blockchain and edge computing. Wirel. Commun. Mobile Comput. **2020** (2020)
13. Mannanuddin, K., Kumar, M.R., Aluvala, S., Nagender, Y.: Fundamental perception of EDGE computing. In: International Conference on Recent Advancements in Engineering and Management, vol. 981. Warangal, India (2020)
14. Parikh, S., Dave, D., Patel, R., Doshi, N.: Security and privacy issues in cloud, fog and edge computing. Proc. Comput. Sci. **160**, 734–739 (2019)
15. Shi, W., Cao, J., Zhang, Q., Li, Y., Xu, L.: Edge computing: vision and challenges. IEEE Internet Things J. **3**(5), 637–646 (2016)
16. Faruque, M.A., Vatanparvar, K.: Energy management-as-a-service over fog computing platform. IEEE Internet Things J. **3**(2), 161–169 (2016)
17. Alabdulatif, A., Khalil, I., Yi, X.: Towards secure big data analytic for cloud-enabled applications with fully homomorphic encryption. J. Parallel Distrib. Comput. **137**, 192–204 (2019)
18. Lin, H., Bergmann, N.W.: IoT privacy and security challenges for smart home environments. Information **7**(3), 44 (2016)
19. Elhassani, M., Boulbot, A., Chillali, A., Mouhib, A.: Fully homomorphic encryption scheme on a non-commutative ring R. In: International Conference on Intelligent Systems and Advanced Computing Sciences (ISACS) 2019, pp. 1–4 (2019)

20. Wang, B., Zhan, Y., Zhang, Z.: Cryptanalysis of a symmetric fully homomorphic encryption scheme. IEEE Trans. Inf. Forensics Secur. **13**(6), 1460–1467 (2018)
21. Zhang, Y.J., Shang, T., Liu, J.W., Wu, W.: Quantum homomorphic encryption based on quantum obfuscation. In: International Wireless Communications and Mobile Computing (IWCMC) 2020, pp. 2010–2015 (2020)
22. Morampudi, M.K., Prasad, M.V.N.K., Raju, U.S.N.: Privacy-preserving iris authentication using fully homomorphic encryption. Multimed. Tools Appl. **79**, 19215–19237 (2020)
23. Varghese, B., Wang, N., Barbhuiya, S., Kilpatrick, P., Nikolopoulos, D.S.: Challenges and opportunities in edge computing. In: IEEE International Conference on Smart Cloud (SmartCloud) 2016, pp. 20–26 (2016)
24. Manzoor, A., Braeken, A., Kanhere, S.S., Ylianttila, M., Liyanage, M.: Proxy re-encryption enabled secure and anonymous IoT data sharing platform based on blockchain. J. Network Comput. Appl. **176** (2021)

# Cybercrime Detection Using Live Sentiment Analysis

**Balvinder Singh Gambhir, Jatin Habibkar, Anjesh Sohrot, and Rashmi Dhumal**

**Abstract** Cyberbullying is a continual act that harasses, humiliates, threatens, or hassles people through electronic devices and online social networking Websites. Cyberbullying through the Internet is considered additionally more dangerous than any form of bullying done in the past, because it can probably amplify the humiliation to a vast online audience. Current models have a variety of issues which our proposed system try to address through our proposed work. Many of the other models use small, heterogeneous datasets, without a thorough evaluation of applicability. At the same time, many models yield small datasets that fail to capture the required complex social dynamics and impede direct comparison of progress. Our model uses real-time data from Twitter API which is preprocessed using regex. It is then fed to the LSTM neural network which will filter out negative tweets and also passed through a sentiment analyzer. After passing through these components, if the sentence is found to be negative in nature and contains an abusive word, then we classify it as an act of cyberbullying. In the coming years, we may see people trying to find different ways to harass each other on social media, so by an over proposed method, we have the model to detect sentiment analysis of sentences as well as we have the reference as to why the sentence is passing negative impressions to people using sentiment analyzer.

**Keywords** Cyberbullying · LSTM neural network · Twitter API · Sentiment analyzer · Stemmer · Machine learning · Web app

B. S. Gambhir (✉) · J. Habibkar · A. Sohrot · R. Dhumal
Ramrao Adik Institute of Technology, Mumbai, India
e-mail: balvindersi2@gmail.com

R. Dhumal
e-mail: rashmi.dhumal@rait.ac.in



D. Gupta et al. (eds.), *Pattern Recognition and Data Analysis with Applications*, Lecture Notes in Electrical Engineering 888,
https://doi.org/10.1007/978-981-19-1520-8_32

## 1 Introduction

Cybercrime is a crime which uses digital technologies to do crime. There are many types of cyberbullying happening online today. We are focusing on detecting hate or bullying speech on social media platforms like Twitter. The proposed system focuses on creating a model that will be able to differentiate between hate/bullying text or normal text, and then, if the text is detected as a bullying text, we will show why that text is considered as a bullying text. The proposed system focuses on developing a Web application through which users can track topics in which they want to detect cyberbullying. The system monitors these topics for cyberbullying and then informs the user if a cyberbullying text is found. Then, the user can take the required actions. The proposed system will be used by people who monitor social media Websites like Twitter and Facebook. Through this application, they will be able to filter out bullying texts and will be able to take appropriate actions.

## 2 Literature Review

This section describes the methodology adopted for the literature review. This paper represents an exploration of the contributions that have already been made in the academic field.

Nurrahmi and Nurjanah discussed the cyberbullying detection for Indonesian tweets to identify cyberbullying text and actors in Twitter [1]. They proposed a system based on texts and credibility analysis of users and notify them about the harm of cyberbullying. They have applied SVM and KNN to learn and detect cyberbullying text on the data collected from Twitter. The data collected from Twitter are unlabeled, so the author designed a Web-based tool to classify it into two classes: cyberbullying and non-cyberbullying. The SVM shows better results than KNN and also categorized users into four categories based on credibility analysis. Foong and Oussalah used sentiment140 training data using Twitter database and proposed a method to improve classification using Naive Bayes [2]. Their accuracy is around 58.40%. Their main focus was on doing sentiment analysis on tweets related to movies. They provided negativity, positivity, and objectivity of a tweet using Naive Bayes and SVM. They implemented their system using NLTK and Python Twitter API. Sarlan et al. focused on discovering public opinion by performing sentiment analysis on real-time Twitter data [3]. They used Hadoop, Hive warehouse, and Apache Flume for storing and analyzing tweets data. After doing sentiment analysis on the tweets, the tweets were classified as positive, negative, and neutral, and these were used for decision making. The analyzed tweets were then plotted on histogram and bar chart. Bahrainian and Dengel used a deep learning model for detecting cyberbullying on various platforms like Twitter and Wikipedia [4]. They used many machine learning techniques and found out that CNN and BLSTM were best for detecting cyberbullying. One limitation of their model was that it takes too much time to analyze one tweet so it could not be used on a real-time system. Yazğılı and Baykara mention how cybercrime is affecting physical/mental health of a person and which lead to suicidal tendencies [5]. They also tried to create a model using SVM, KNN to detect cybercrime. Hang and Dahlan created a dataset of words that are used in cyberbullying [6]. Their strategy is made up of several steps, namely understanding of cyberbullying exclusion principles, word list selection, recognition of a keyword, classes and subclasses identification of ontology and lexicon and lastly, cyberbullying detection. Bertot et al. have written about the effects of cyberbullying on various topics like political, technical, and crowdsourcing and how to overcome such a problem [7]. They also explain how privacy and social data of users are important in detecting cyberbullying. Banerjee et al. used word vectors with CNN to detect cyberbullying tweets [8]. Their accuracy is 81.6% on Twitter tweets. They also suggested making parents track their kids' social media activity. They focused on finding cyberbullying tweets in Arabic language [9]. They used a dataset of abused words that were used to detect cyberbullying tweets. Their system allowed users to add their own abuse words that should be considered while marking a tweet as cyberbullying or not.

## 3 Limitation of Existing System

- The proposed system needs human labeled data to train a model to detect various types of harassment.
- Less advanced algorithms like SVM, decision tree, etc., are not providing good results, and advanced algorithms take more time to predict the results.
- As we have to keep our model up to date with the changing the way people harass each other, so we have to keep training the model on new labeled datasets.
- Not able to detect spammers from different accounts harassing the same person.
- Unsupervised learning can be useful but how can we be so sure about the result as it may lead to biased prediction.

## 4 Methodology

### *4.1 Proposed Work*

This proposal is aimed at development of an application system through which the user enters keywords which is then passed to Twitter API which will then fetch all tweets related to those keywords, and these tweets are then passed to our model Fig. 1. The main objective of the project is the development of an application system through which the users can monitor cyberbullying on a particular topic. Real-time analysis of tweets is done, and then, the sentiment of each tweet is calculated. The sentiment can be positive, neutral, or negative. The main functions include

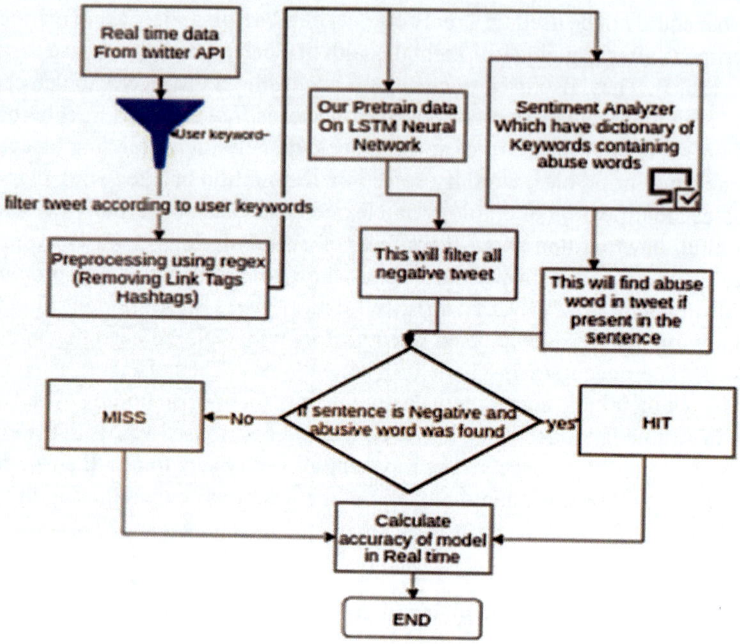


**Fig. 1** Proposed system workflow

- Detecting negative sentence tweets
- Detecting harassing tweets
- Give feedback on various tweets to improve the ML model.
- Real-time model accuracy on testing data.
- Creating word-cloud in our Web application to understand filtered tweets in real time.
- Sentiment analysis as well as extracting harassing words from tweet help us to understand people's thoughts in real time.

**Proposed System Workflow Explanation**

As shown in Fig. 1, the user will supply keywords that it wants to filter the tweets on. These topics will be sent to our backend which will then fetch the related tweets from Twitter. The fetched tweets are preprocessed and then is passed to our classifier which will classify them as positive or negative. After the classifier gets the result, we need to analyze if the tweet is harassing or not, so we used sentiment analyzer to predict the accuracy of the model in real time. This analyzer has dictionary which contains harassing words. This software will detect/underline words in sentence if the tweet contains any kind of harassing word. As analyzer is our reference to calculate accuracy of model in real time so we come to a conditional statement where we check if our model predicts a negative statement as well as it contains any abuse word, then we will consider it as HIT. If the text is negative, and it does not contain any abusive

Layer (type) Output Shape Param #
==========================================================

embedding_1 (Embedding) (None,300,300) 87171600
________________________________________________ dropout_4
(Dropout) (None,300,300) 0
________________________________________________ lstm_4 (LSTM)
(None,100) 160400 ________________________________________
dense_4 (Dense) (None,1) 101
====================================================== Total
params: 87,332,101 Trainable params: 160,501 Non-trainable params: 87,171,600

**Fig. 2** Neural network architecture

words, it will be a MISS. After this process, we will calculate to total number of HIT and MISS and calculate the accuracy of model in real time using this formula, e.g., 1.

$$\text{HIT}/(\text{HIT} + \text{MISS}) \tag{1}$$

### 4.2 *Classifier Implementation*

**Preprocessing the data**: As data are fetched with the help of Twitter API, it contains hyper link, numbers, stop words. So we remove all these useless features.

**Word to vector**: As computers do not understand words, so we converted words to vectors using NLTK software. It contain size of vocab dictionary ie30520 W2VSIZE = 300 weights = initially embedding matrix by random float number SEQUENCELENGTH = 300 same as W2VSIZE.

**Create tokenizer**: After converting all the words to vectors, we tokenize whole words so that all words are understood by a computer.

**Creating neural network**: After creating tokenized words, the word is passed to the (LSTM) neural network. Fig. 2 refers to model architecture.

### 4.3 *LSTM Networks*

**LSTM Networks**: Long short-term memory networks (LSTM) are a special kind of RNN that are capable of learning from long-term dependencies. They work well with many types of problems and are widely used. LSTMs are designed to avoid the long-term dependency problem. They are able to remember information for a long duration of time. The key to LSTMs is the cell state. LSTM has the ability to remove or add information to the cell state, through structured gates. Gates are made through sigmoid neural network and pointwise multiplication. The sigmoid layer has output between zero and one which can be used to know how much information should be transferred. A value of 0 means that no information should be passed through, while value of 1 means every information should be passed through. We are passing preprocessed text to our LSTM model which then classifies the text as negative or positive.

- Sentence is preprocessed and pass to lemmatization
- Lemmatizing convert all words to lemma words
- Convert all unique words to tokens
- Convert words to vector–score every word and averaging the score.

**Sentiment analyzer** is a rule-based model for sentiment analysis.

There are over 9000 features that are rated from extremely negative to extremely positive. −4 is used for extremely negative, and 4 is used for extremely positive. 0 is considered as neutral. Our model kept features that had a mean rating not equal to 0, and whose standard deviation was less than 2.5. So after removing these features, 7500 features were left. For example, okay has a positive score 0.9, good has 1.9, and great has 3.1, whereas the score for the word horrible is −2.5. We are passing the preprocessed text to the sentiment analyzer for getting abuse words. The sentiment analyzer score is calculated by adding all of the scores for each term in the lexicon and then normalizing the result between −1 and 1.

## 5 Data Description and Data Cleaning

We have scraped 1.6 million tweets by using Twitter. The tweets have been classified as negative and positive. These tweets are then used to detect sentiment of new tweets. We have used the following fields:

1. ids: tweet ids that are randomly created
2. Target: polarity of the tweet. It has the following values Neg, Neu, and positive.
3. Users: username of the user (balvinderzuser).
4. Dates: date of the tweet (Sat May 21 10:13:44 UTC 2022).
5. Texts: content of tweets like hello there.

## 6 Experimental Design

As shown in Fig. 3, the proposed system has a Web application (developed in vue) and a backend (developed in flask). Initially, the user will select three 3 topics that he wants to detect cybercrime on. These topics will be sent to the backend which will

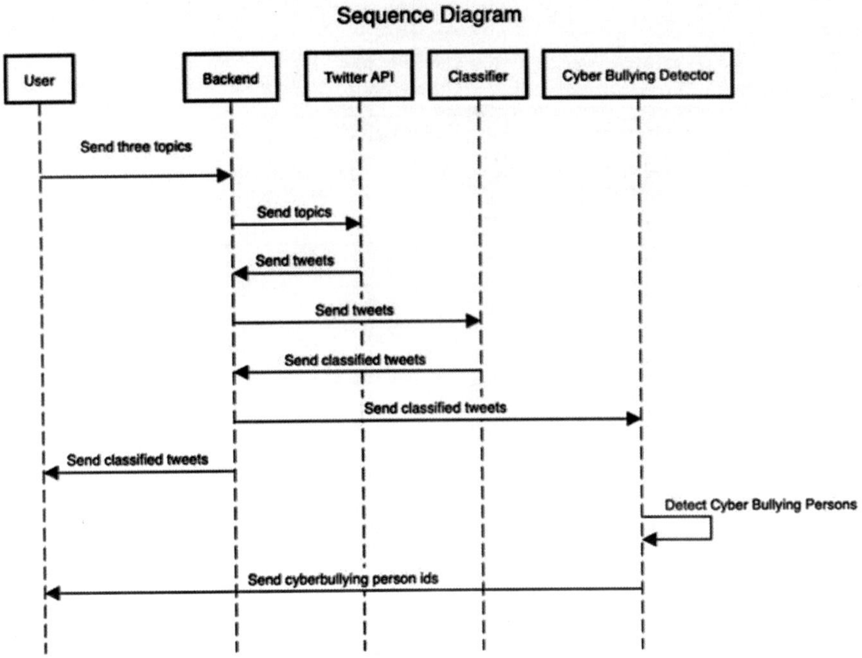


**Fig. 3** Sequence diagram of application

then fetch the related tweets from Twitter. The fetched tweets will then be passed to our classifier which will classify them as positive or negative. After the classifier gets the result, we need to analyze if the tweet is harassing or not, so we use sentiment analyzer to predict if the tweet contains harassing words or not. This analyzer has a dictionary which contains harassing words. This software will detect/underline words in sentences if the tweet contains any kind of harassing word.

In Fig. 4, we have considered an example from the live tweets. "@RahulGandhi is a shameful idiot @NationalistCol". The tweet is first preprocessed (RahulGandhi is a shameful idiot nationalistcol). After preprocessing, the tweet is tokenized. Then, it is converted into a word to vector model which is then passed to our trained model. The trained model classifies the tweet as positive or negative. The tweet is also passed through NLTK sentiment analyzer. In this case, our model classifies the tweet as negative, and the NLTK sentiment analyzer also found harassing words ("shameful", "idiot"). So it is a hit.

## 6.1 *Result and Analysis*

The proposed system focuses on designing a Web-based app that will monitor the Twitter Website for detecting cyberbullying tweets and also justifying as in why the

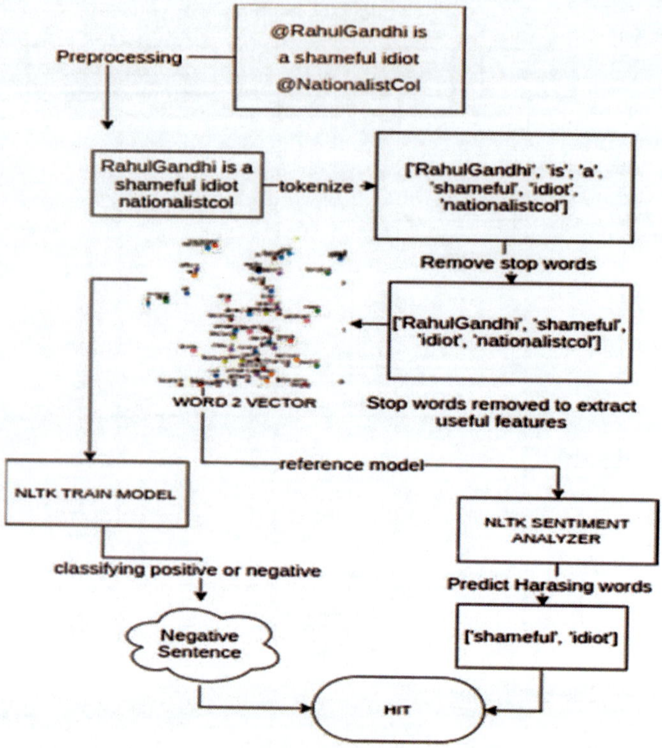


**Fig. 4** Text example

tweet is harassing. The model is able to filter out negative tweets in real time using the LSTM model with an accuracy of around 70–80%. We are using sentiment analyzer to detect harassing words that will act as a reference to get the accuracy of our model in real time (Fig. 5).

Reference to Fig. 6, as we are working on real-time data, we are not considering if our model predicted positive and it contain a harassing word as most of the people on Twitter is talking about same topic, and this may lead to biased dataset containing only positive sentences. Reference to Fig. 5, negative sentences which do not contain harassing words according to sentiment analyzer.

**Example 1: "Why oh why did I read the comments. The level of idiocy is just … I cannot even".**

So our model may have identified idiocy as harassing, but the sentiment analyzer may not have identified it as harassing because of two reason: (1) We have currently set sentiment analyzer to greater than 0.0, and we can increase it to get more predictions which contain less harassing words. (2) It is dictionary may not contain that word.

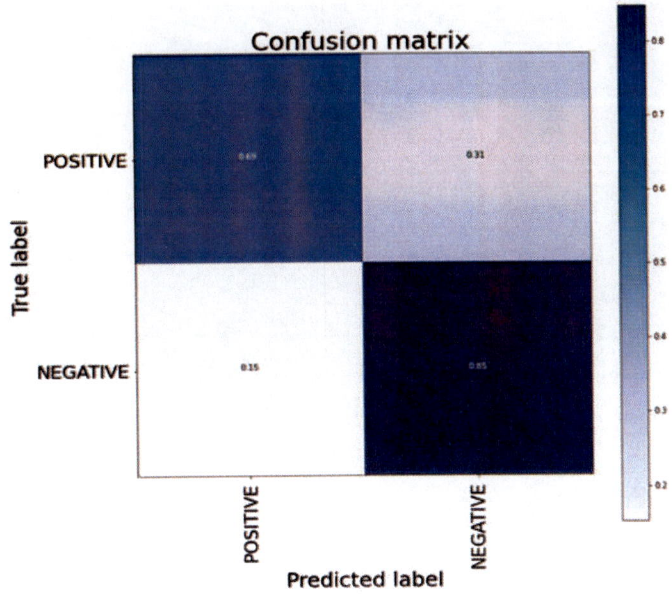


**Fig. 5** Confusion matrix of testing data (0.69 | 0.31 | 0.15 | 0.85)

<u>Training Phase</u>
Training data : 1280000 Sentences
Testing data : 320000 Sentences
Total words TOKENS 290572

ACCURACY On training phase: 81.3 %

---

<u>Testing Phase on real Data</u>
Topics considered ["disha ravi","farmer protest","trump"]

```
count of our model
negative  200
actual count of harasing words in sentences
sentence count (TP):  163
```

```
In [8]: accuracy=len(sentNeg)/len(countneg)
        print("Accuracy is : ",accuracy*100,"%")

Accuracy is :  81.5 %
```

**Fig. 6** Accuracy on 200 new real-time data

| | precision | recall | f1-score |
|---|---|---|---|
| NEGATIVE | 0.82 | 0.69 | 0.75 |
| POSITIVE | 0.73 | 0.85 | 0.79 |
| accuracy | | | 0.77 |
| macro avg | 0.77 | 0.77 | 0.77 |
| weighted avg | 0.77 | 0.77 | 0.77 |

**Fig.7** F-score of training and testing data

### *6.2 Precision and Recall*

Reference to Fig. 7, calculating the f-score of our deep learning model which was trained on 12.80 k training and 320 k testing dataset has and (if you ask what proportion of positive identifications was actually correct?) precision of 82% on positive dataset and 73% on negative dataset. (If you ask what proportion of actual positives was identified correctly?) Recall, also called as sensitivity, our model has classified the tweets with a recall of 69% on negative tweets and 85% on positive tweets. F1 score passes on the harmony between the precision and the recall the formula to calculate f1 score is 2*((precision*recall)/(precision + recall)).

## 7 Conclusion and Future Work

By our proposed work, we have detected harassing tweets on real-time system also shown real-time accuracy of our system. We have also justified, if the tweet is negative or not using custom sentiment analyzer. As in real life, we cannot just rely on neural networks or we cannot just rely only on dictionaries, so we have to create such methodology and techniques to solve this problem with minimum resources. In future work, we can also detect users who are harassing other users and give those users details to concerned authorities.

## References

1. Nurrahmi, H., Nurjanah, D.: Indonesian twitter cyberbullying detection using text classification and user credibility. In: 2018 International Conference on Information and Communications Technology (ICOIACT), pp. 543–548. IEEE, Indonesia (2018)

2. Foong, Y.J., Oussalah, M.: Cyberbullying system detection and analysis. In: European Intelligence and Security Informatics Conference (EISIC), pp. 40–46. IEEE, Athens (2017). https://doi.org/10.1109/EISIC.2017.43
3. Sarlan, A., Nadam, C., Basri, S.: Twitter sentiment analysis. In: Proceedings of the 6th International Conference on Information Technology and Multimedia, pp. 212–216. Putrajaya (2014). https://doi.org/10.1109/ICIMU.2014.7066632
4. Bahrainian, S., Dengel, A.: Sentiment analysis and summarization of twitter data. In: 16th International Conference on Computational Science and Engineering, pp. 227–234. IEEE, Sydney (2013). https://doi.org/10.1109/CSE.2013.44
5. Yazğılı, E., Baykara, M.: Cyberbullying and detection methods. In: 1st International Informatics and Software Engineering Conference (UBMYK), pp. 1–5. IEEE, Turkey (2019). https://doi.org/10.1109/UBMYK48245.2019.8965514
6. Hang, O.C., Dahlan, H.M.: Cyberbullying lexicon for social media. In: 6th International Conference on Research and Innovation in Information Systems (ICRIIS), pp. 1–6. IEEE, Malaysia (2019). https://doi.org/10.1109/ICRIIS48246.2019.9073679
7. Bertot, J.C., Jaeger, P.Y., Hansen, D.: The impact of polices on government social media usage: issues, challenges, and recommendations. Gov. Inf. Q. **29**(1), 30–40 (2012)
8. Banerjee, V., Telavane, J., Gaikwad, P., Vartak, P.: Detection of cyberbullying using deep neural network. In: 5th International Conference on Advanced Computing and Communication Systems (ICACCS), pp. 604–607. IEEE, India (2019). https://doi.org/10.1109/ICACCS.2019.8728378
9. Mouheb, D., Abushamleh, M.H., Abushamleh, M.H., Aghbari, Z.A., Kamel, I.: Real-time detection of cyberbullying in Arabic twitter streams. In: 2019 10th IFIP International Conference on New Technologies, Mobility and Security (NTMS). IEEE, Spain (2019). https://doi.org/10.1109/NTMS.2019.8763808

# Analysis of Power Quality for TOSA-PID Controller-Based Hybrid Power Generation System

**Subhadip Goswami, Tapas Kumar Benia, and Abhik Banerjee**

**Abstract** The Hybrid Renewable Energy Systems (HRESs) are recommended as a suitable means to deliver electricity onto the remote and also off-grid regions existent in the nations developing. Effectual and reliable energy is not offered by the Stand-alone technique. However, energy efficacy enhancement is required in the HRES. Prevalent research protocol shave taken measures aimed at enhancing energy efficacy; however, apt charging, as well as discharging maintenance has not been done. Therefore, this study protocol utilizes the Taxicab Owl Search Algorithm (TOSA) centred Proportional Integral Derivative's (TOSA-PID's) controller-centred hybrid Power Generation (PG) system and examines the system's power quality. This technique comprises the solar-wind-diesel generator's hybridization that possesses the prime ability to offer power. Utilizing this hybrid PG system, the battery, together with the loads is integrated. Herein, utilizing Maximal Power Point Tracking (MPPT), the DC–DC and DC–AC buck-boost converter are employed to attain the regulated DC voltage as well as the require AC output. Next, aimed at controlling the system's charging and discharging, the Proportional Integral Derivative (PID) controller is utilized to tune the parameter for maximal energy utilizing TOSA. Lastly, the TOSA-PID controller's performance is analogized with the existent controller-centred hybrid PG system. The controller-centred hybrid PG system proposed yields superior power analogized to the other existent controllers-centred PG system.

**Keywords** Hybridization of the solar-wind-diesel generator · Taxicab owl search algorithm-based proportional integral derivative (TOSA-PID) controller · DC–DC buck-boost converter · DC–AC buck-boost converter · Maximum power point tracking (MPPT)

S. Goswami · T. K. Benia · A. Banerjee (✉)
Department of Electrical Engineering, NIT Arunachal Pradesh, Arunachal Pradesh 791112, India
e-mail: abhik@nitap.ac.in

S. Goswami
e-mail: subhadip.phd@nitap.ac.in



D. Gupta et al. (eds.), *Pattern Recognition and Data Analysis with Applications*,
Lecture Notes in Electrical Engineering 888,
https://doi.org/10.1007/978-981-19-1520-8_33

## 1 Introduction

The world's recent prime research topic is electricity generation utilizing renewable energy resources. Aimed at incrementing the renewable and also sustainable energy's share, vital efforts are being implemented universally [1]. Individual renewable energy's resource's applications analogized with the HRESs propose that the HRESs are helpful to overcome energy shortages [2]. The HRES encompasses diverse resources that incorporate diesel source and also renewable resources, namely, the solar Photovoltaic (PV) system, battery systems with wind energy turbine generator [3]. These resources provide diverse loads linked to the storage elements aimed at compensating for the renewable energy source intermittency and also obtain maximal overall energy efficacy [4]. Popular renewable resources, like wind and also solar energy are employed on account of their ecological benefits. Wind and PV sources' distinctive feature is their complementary nature as wind is adequately obtainable in the night period and also cloudy circumstances but it is extremely inadequate during sunny times [5]. Moreover, charitable subsidized policies are executed by diverse nations aimed at accelerating and stimulating investment in the wind as well as solar energy's development [6]. Diesel generators are the main supplying power resource in the remotely situated regions; nevertheless, they are costly to fabricate and also maintain. Renewable energy's resources, namely, solar, as well as wind, have been incorporated with the diesel PG systems aimed at boosting the power supply's ability and dependability [7]. Currently, numerous scholars have learned the techno-economic performance of the stand-alone WT-centred, PV-centred, and also wind/PV-centred hybrid power systems [8]. Such systems' key drawbacks are the regions comprising elevated wind speed. The modern technology termed power electronic converter is utilized for resolving this issue. The converters aimed at diverse MPPT stratagems, for instance, buck, boost and also a bi-directional converter, have been employed in wind and also solar PV's applications [9]. In general, aimed at the voltage together with the current regulation in inner as well as outer loops, Proportional Integral is utilized to resolve the power's quality problems [10]; conversely, it comprises a saturation issue. This work employs the TOSA-PID controller aimed at resolving that issue and incrementing the power quality.

The presented work's structure is arranged as: Sect. 2 details the top-notch mechanisms linked to the hybrid PG system; Sect. 3 explicates the TOSA-PID centred power quality; Sect. 4 inspects the presented research's outcome; Sect. 5 winds up the work with the future development.

## 2 Related Work

Ding et al. [11] proffered a wind turbine together with a solar thermal power system for building a wind-solar hybrid PG system. Aimed at maximizing the economic performance, the technique utilized a capacity configurations optimization design centred on a Particle Swarm Optimization (PSO) technique. As the study case, an 80 MWe hybrid system existent in Zhangbei China was picked. The outcome exhibited that the capacity ratio of the wind PG to the solar thermal PG, the thermal energy storage system's capacity, solar multiple, and also electric heater's (EH's) capacity were 1.91, 13 h, 2.9, and then 6 MW, correspondingly. The hybrid PG system comprised $27.67 M the greatest net present value. The outcomes exhibited that the EH efficiently decremented the wind curtailment and also enhanced the system's overall stability. However, as per the perspective of the system's economic return, EH's integration was not beneficial always.

Mosobi et al. [12] presented the hybrid renewable energy resources comprising solar PV, wind energy's system, together with a micro-hydro system. Utilizing '2' power converters, the solar PV system was designed, the 1st one is the DC to DC converter together with a maximal power point tracking aimed at attaining a regulated DC outputted voltage, and then the 2nd one is the DC to AC converter for attaining AC output. Aimed at operating a Self-Excited Induction Generator (SEIG), the wind energy system had been build utilizing the wind turbine's prime mover with changing wind speed but fixing pitch angle. Aimed at driving a SEIG, the micro-hydro system was built utilizing a stable inputted power. Aimed at boosting the load voltage and also current profiles, a statistical compensator was utilized; it also mitigated the voltage's and the current's harmonic contents. The attained simulation outcomes exemplified the system's feasibility and were satisfying. However, controlling the charging and discharging was not maintained correctly.

Moghaddam et al. [13] established a hybrid renewable energy's PV/wind/battery system to boost the load supply's reliability over the study prospect pondering the Net Present Cost (NPC) to be the objective function aimed at decrementing. The NPC incorporated the costs regarding the hybrid system's investment, operation, maintenance, and also replacement. The pondered reliability index had been the load demand's scarce power-hourly interruption possibility. The decision variables incorporated the number of wind turbines, PV panels, together with batteries, the inverter transmitted power's capacity, PV panel's angle, and also wind tower height. A technique termed the Improved Crow Search Algorithm (ICSA) was employed aimed at resolving the optimization issue. The ICSA's performance was analogized with the Crow Search Algorithm (CSA) and PSO techniques in the diverse amalgamation of systems. Therefore, the analogy exhibited that the ICSA yielded efficient performance analogized to the other existent techniques. However, the ICSA comprised a premature convergence issue.

## 3 Power Quality Analysis of Hybrid Renewable Energy System

Recently, HRESs are amidst the key research areas prevalent in the sustainable energy fields. Renewable energy sources have been more needed with the constantly incrementing demand aimed at energy's necessity since they have been as well eco-friendly whilst analogized with the non-renewable energy systems. Here, aimed at the solar-wind-diesel-centred hybrid PG system, the power quality is examined. Herein, the Buck-Boost converter is utilized; the PID controller is implemented aimed at controlling the voltage. The PID's parameters are adjusted utilizing the TOSA to yield superior energy. Figure 1 exhibits the proposed research protocol's block diagram.

### 3.1 PV Array Modelling

Single-diode or else '2'-diode designs are extensively utilized to fundamentally construct the PV design. On the whole, the single-diode design is much apt for utilization. Aimed at incrementing the power output, numerous solar cells are linked in series, or else parallel manner, and then mounted upon the surface creating a solar cell unit or else a PV unit. The PV unit designing is executed utilizing the Eq. (1),

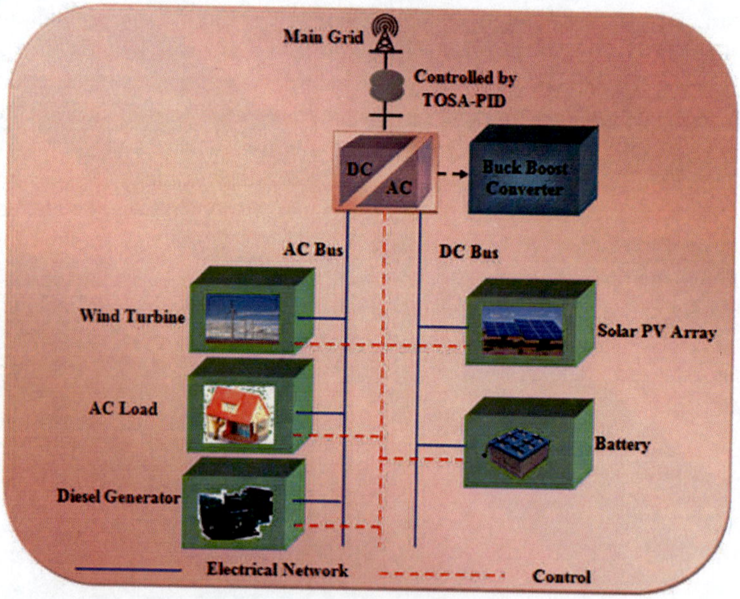


**Fig. 1** Block diagram for the presented research methodology

$$C = C_{\mathrm{pv}} - C_0\left[\exp\left(\frac{G + E_s C}{G_t}\right) - 1\right] - \frac{G + E_s C}{E_p} \quad (1)$$

Here, $C_{\mathrm{pv}}$ signifies the light created current; $E_s$ implies the cell sequence's resistance; $E_p$ signifies the cell shunt's resistance; $C$ implies the current; $C_o$ symbolizes the dark saturation current's value; $G$ signifies the solar cell's output voltage; $\zeta$ implies the linearity factor; $G_t$ symbolizes the thermal voltage of the array comprising $T_s$ linked in the series that is equated as,

$$G_t = \frac{T_s \cdot \eta \cdot \mathrm{Temp}}{e_c} \quad (2)$$

Herein, $\eta$ implies the Boltzmann gas constant; Temp symbolizes the absolute temperature (Kelvin); $e_c$ signifies the electron charge; $T_s$ implies the number of PV units prevalent in the series.

### 3.2 WECS Modelling

Wind power engages wind energy's conversion to electricity utilizing the wind turbines. It changes the kinetic energy as mechanical energy. A wind turbine comprises several propellers similar to blades termed as the rotor. As of the variation in the air movement around the earth's surface, the wind emerges. A turbine's power output is stated as the function of the cube of the wind's speed. Whilst the wind's speed increments, the outputted power increments. The wind generator's output eqn. is,

$$H = \frac{1}{2} \times a_d \times w_a \times o_o \times \lambda^3 \quad (3)$$

Here, $a_d$ signifies the air's density; $O_c$ symbolizes the power coefficient; $w_a$ implies the swept area; $\lambda$ signifies the wind's velocity.

### 3.3 Diesel Generator Modelling

Whilst the energy requirement goes beyond the total energy created by the hybrid PG system, always the diesel generator is prepared to supply power onto the load. A diesel generator's energy generation ($R_{\mathrm{DG}}$) is equated as,

$$R_{\mathrm{DG}} = \O_{\mathrm{DG}} \cdot \mathrm{Op}_t \int J_{\mathrm{DG}} \cdot dt \quad (4)$$

Here, $J_{\mathrm{DG}}$ signifies the DG's read power output, $\varphi_{\mathrm{DG}}$ implies its efficacy; $\mathrm{OP}_t$ symbolizes the operating time. The generator's capacity should be selected concerning the maximal expected load demand aimed at decrementing the fuel's consumption.

## 3.4 Battery Modelling

Batteries have been amidst the energy storage system's fundamental components. It is stated as one or more electrochemical cells' electrical connection where the ions are generated as a consequence of the oxidation reaction in the cell operation. The battery's self-discharging occurs whilst the electrolyte is compiled with the electrons since the electrons comprise poor conductivity aimed at ions. Therefore, the internal circuit is built in betwixt the electrodes. Herein, the solar, wind, and also diesel have been interlinked that is equated as,

$$\Psi = C + H + R_{\mathrm{DG}} - \mathrm{ac} \tag{5}$$

Here, $\Psi$ signifies the battery; ac implies the hybrid DC together with the AC dynamic loads.

## 3.5 Converter

Next, the solar output and also the wind system's outputs are fed into the Buck-boost converter as input. The DC–DC buck-boost converter utilizes the MPPT to attain the regulated DC voltage and the DC–AC buck-boost converter is employed to yield the AC output. These converters are employed to yield the DC as well as AC outputs. In the converter, whilst $Z_x = Z_{\mathrm{in}}$, the transistor is ON, or else whilst $Z_x = Z_o$, the transistor is OFF. The average voltage alongside the inductor is '0' aimed at zero net current variation over a period.

$$Z_{\mathrm{in}} K_{\mathrm{ON}} + Z_0 K_{\mathrm{OFF}} = 0 \tag{6}$$

The converter's voltage ratio is,

$$\frac{Z_o}{Z_{\mathrm{in}}} = -\frac{D_r}{(1 - D_r)} \tag{7}$$

Next, the converter's current ratio is,

$$\frac{W_o}{W_{\mathrm{in}}} = -\frac{(1 - D_r)}{D_r} \tag{8}$$

Here, $Z_{\text{in}}$ and $Z_o$ signifies the voltage-in and voltage-out; $W_{\text{in}}$ and $W_o$ implies the current-in and current-out; $D_r$ symbolizes the duty ratio; $K_{\text{ON}}$ and $K_{\text{OFF}}$ signifies the transistor's ON as well as OFF phases; $Z_x$ implies the voltage-level.

## 3.6 Controller

Next, the system's charging as well as discharging is administered by the PID controller. The PID controller is engaged as the specific regulator in the loop feedback, which is extensively utilized in the industrial regulation system. PID comprises '3' constant parameters: $L_{\text{pro}}$ aimed at proportional, $L_{\text{int}}$ aimed at integral, and then $L_{\text{der}}$ aimed at derivative control. The '3' parameters' summation is equated as,

$$A(t) = L_{\text{pro}} \cdot r(t) + L_{\text{int}} \int r(t)\mathrm{d}t + L_{\text{der}} \frac{\mathrm{d}r}{\mathrm{d}t} \tag{9}$$

Here, $A(t)$ signifies the PID's control variable; $r(t)$ implies the error value; d$r$ symbolizes the variation in the error value; d$t$ signifies the variation in time. Utilizing the TOSA, the PID's parameters are adjusted aimed at attaining superior energy. Owl Search Algorithm (OSA) is stated as a nature-enthused population-centred technique where an owl group functions collectively iteratively to discover the global optimal solutions. However, in the prey searching procedure, the distance information is computed utilizing the Euclidean distance calculation; nevertheless, it does not offer the correct outcome aimed at large-scale data. Consequently, the taxicab distance's calculation is pondered here. If there prevail total $M$ owls that begin with arbitrary positions, then the $i$th owl's initial position is equated as,

$$Q_i = Q_1 + \alpha(0, 1) \times (Q_u - Q_1) \tag{10}$$

Here, $Q_l$ and $Q_u$ signifies the owl's lower, as well as upper bounds; $\alpha(0, 1)$ implies the random number created in the (0, 1) range; $Q_i$ symbolizes the owls' initial positions. Next, the fitness value is enumerated via the Integration of the Time-weighted Absolute of the Error (ITAE) that is equated as,

$$\mathrm{OB}_f = \int_0^{\infty} (t \cdot |r(t)|) \cdot \mathrm{d}t \tag{11}$$

Here, $\mathrm{OB}_f$ implies the objective function that is computed aimed at every initial parameter (i.e. population).Next, the owl's fitness is enumerated; the evaluated outcome is articulated as,

$$I_i = \mathrm{fit}([Q_1, Q_2, \ldots, Q_n]) \tag{12}$$

Here, $I_i$ signifies the fitness function; fit($[Q_1, Q_2, \ldots, Q_n]$) implies all the populations' fitness function. After that, the $i$th owl's normalized intensity is enumerated utilizing the Eq. (13),

$$F_i = \frac{I_i - Q_p}{Q_e - Q_p} \tag{13}$$

Here, $F_i$ implies the normalized intensity level; $Q_p$ signifies the worst owl; $Q_e$ signifies the best owl (i.e. the owl comprising the minimal fitness value is signified as the worst owl; the owl comprising the maximal fitness value is signified as the finest owl). The owls vary their position continuously centred on their prey's movements. The technique simulates the prey's movement centred on the probability that creates their new positions' updation centred on the Eq. (14):

$$Q_i^{g+1} = \begin{cases} Q_i^g + \delta \times V_i \times \left|\tau D - Q_i^g\right|, & \rho < 0.5 \\ Q_i^g - \delta \times V_i \times \left|\tau D - Q_i^g\right|, & \rho \geq 0.5 \end{cases} \tag{14}$$

Here, $Q_i^{g+1}$ implies the novel position; $Q_i^g$ signifies the present position; $\rho$ signifies the prey's movement probability; $\tau$ implies a uniformly distributed random number; $\delta$ symbolizes a decrementing linear constant; $D$ implies the prey's position that is attained by the fittest owl; $V_i$ signifies the owl's intensity variation; this intensity variation is enumerated centred on taxicab calculation that is equated as,

$$\text{Dist} = |Q_i - Q_{i+1}| + |V_i - V_{i+1}| \tag{15}$$

Herein, $Q_i$ signifies the $i$th prey's location that is attained by the best owl.

Figure 2 exhibits the TOSA's pseudo code. The population's initialization, the prey's movement and also the position updation aimed at the hunting procedure are detailed here.

# 4 Result and Discussion

Herein, the proffered hybrid PG system's performance is examined. In MATLAB/SIMULINK's working platform, the hybrid PG system presented is imposed.

## 4.1 *Performance Analysis*

Table 1 exhibits the presented TOSA-PID controller's parameters and their corresponding values,

**Input:** Parameters of PID, $L_{pro}$, $L_{int}$, and $L_{der}$
**Output:** Tuned parameters

**Begin**
  **Initialize** population, initial position of the owl $Q_i$ and maximum iteration $m_t$.
  **Evaluate** fitness function $OB_f$ by ITAE, $I_i = fit([Q_1, Q_2, \ldots\ldots Q_n])$
  **Set** iteration $i_t = 1$
  **While** $(i_t \leq m_t)$ **do**
    *// prey movement*
    **if** $(\rho < 0.5)$ {
      **Update** the new position for hunting by,
      $Q_i^{g+1} = Q_i^g + \delta \times V_i \times \left|\tau D - Q_i^g\right|$
    } **else** {
      **Update** the new position for hunting by,
      $Q_i^{g+1} = Q_i^g - \delta \times V_i \times \left|\tau D - Q_i^g\right|$
    } **end if**
    **Calculate fitness**
    **Set** $i_t = i_t + 1$
    **Return** tuning parameters
  **End while**
**End**

**Fig. 2** Pseudo code for TOSA

**Table 1** Analyse the parameters of the TOSA-PID controller

| S. No. | TOSA-PID parameters | Values |
|---|---|---|
| 1 | $L_{pro}$ | 0 |
| 2 | $L_{int}$ | 11 |
| 3 | $L_{der}$ | 0.2980 |
| 4 | Population size | 50 |
| 5 | Number of iterations | 50 |
| 6 | Objective function value | 517.48 |

Here, the engaged population's count is 50; the iteration level is 50. The proportional, integral, and also derivative parameters' values are 0, 11, and then 0.2980, correspondingly; the fixed objective function's value is 517.48.

Figure 3 exhibits the PG via the TOSA-PID controller-centred hybrid PG system. Herein, the presented system's PG is examined centred on the time intervals. Aimed at 0.1 ms, the system presented yields power above 2250 W; likewise, for 0.4–0.5 ms, the power yielded is 200 W.

Figure 4 exhibits the analogy of the proposed TOSA-PID controller-centred hybrid PG system namely the amalgamation of solar, wind, and also diesel generator with the prevalent techniques like PID, Genetic Algorithm (GA) centred PID (GA-PID), PSO-centred PID (PSO-PID), and then Grey Wolf Optimization (GWO) centred PID (GWO-PID). Centred on the time interval variations, the PG is examined. Herein, the proposed PG system attains superior power. Aimed at 2 s time interval, the

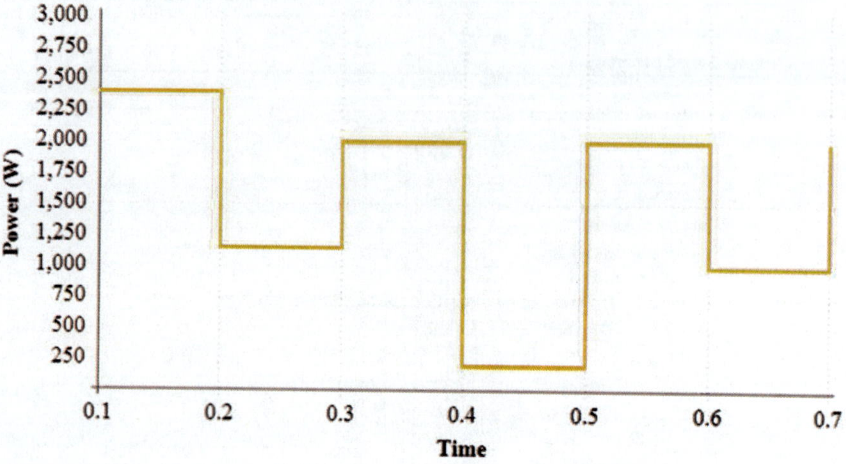


**Fig. 3** Analysis of the power generation for proposed TOSA-PID controller-based hybrid power generation system

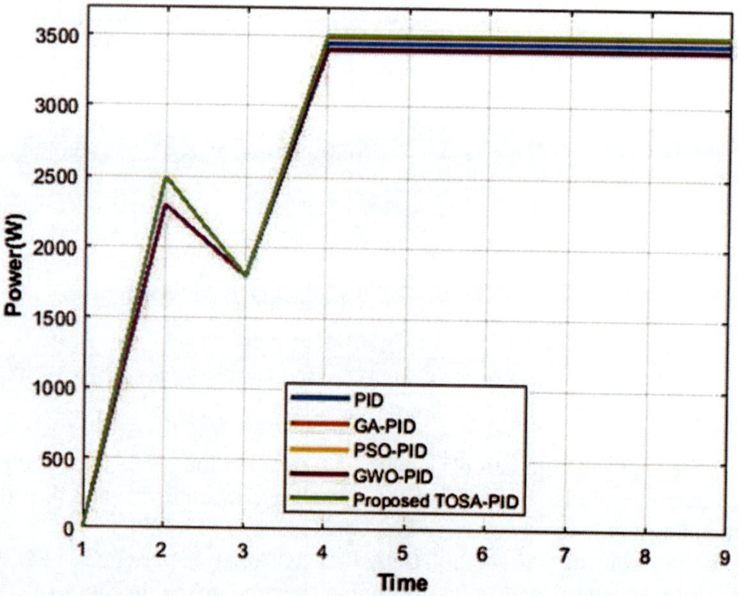


**Fig. 4** Power generation analysis for TOSA-PID-based hybrid power generation system with the existing controller-based power generation system

**Table 2** Fitness versus iteration analysis

| Iteration | Proposed TOSA | OSA | PSO | GA | GWO |
|---|---|---|---|---|---|
| 10 | 0.797 | 0.734 | 0.699 | 0.634 | 0.708 |
| 20 | 0.831 | 0.774 | 0.721 | 0.684 | 0.736 |
| 30 | 0.873 | 0.801 | 0.769 | 0.706 | 0.752 |
| 40 | 0.914 | 0.841 | 0.815 | 0.731 | 0.821 |
| 50 | 0.972 | 0.878 | 0.852 | 0.789 | 0.891 |

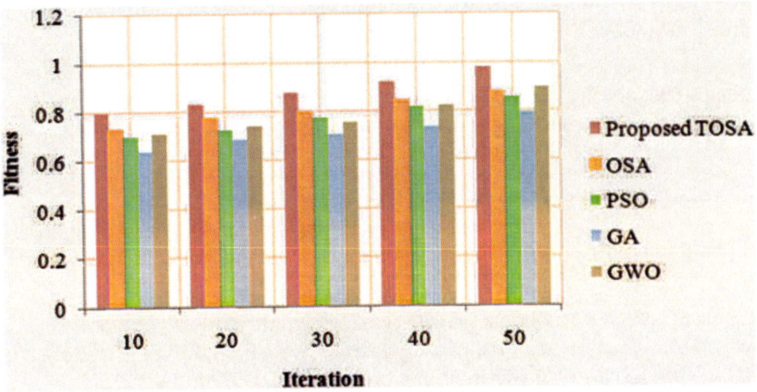


**Fig. 5** Fitness versus iteration analysis

TOSA-PID controller proposed yields 2500 W power that is greater analogized to the other existent PID controller-centred hybrid PG system. Therefore, it finalized that the controller-centred hybrid PG system proposed yields efficient performance analogized to the other existent controller-centred hybrid PG system.

Table 2 exhibits the fitness versus iteration examination aimed at the proffered TOSA along with the existent OSA, PSO, GA, and GWO techniques. Centred on the number of iterations, the performance is examined. Herein, the iteration level is altered as of the iterations 10–50. Aimed at the iteration count '50', the TOSA proposed yields 0.972 fitness function; however, the existent techniques yield lesser fitness function values. Therefore, this validates that the TOSA yields efficient performance analogized to the other existent techniques. Figure 5 exhibits Table 2's graphical depiction.

## 5 Conclusion

Utilizing HRESs is a smart choice to decrement the carbon emitted by power plants. These systems' efficiencies rely on picking the right renewable source combination, their sizes, and the generating units' suitable scheduling. The work hybrids the wind, solar, and also diesel generator along with the controlling of the charging together with discharging via the TOSA-PID controller. Herein, utilizing MPPT, the DC–DC, as well as DC–AC buck-boost converters, are employed aimed at controlling the electrical network. Next, to control the charging together with discharging function, the PID controller is utilized that adjusts the parameters to attain superior energy utilizing TOSA. In the experiential examination, the proposed TOSA-PID-centred hybrid PG system's performance is examined and is analogized with the existent PID, GWO-PID, PSO-PID, and also GA-PID centred on the PG metric. Herein, the TOSA-PID proposed yields efficient outcomes analogized to the other techniques. The proposed TOSA's fitness is analogized with the prevalent optimization techniques like GWO, PSO, GA, and also OSA. The proposed TOSA comprises a greater fitness value aimed at every iteration. Therefore, the system proposed offers a greater power quality. The protocol proposed can be lengthened in the upcoming future by implementing added renewable energy utilizing advanced controllers.

## References

1. Ramesh, M., Saini, R.P.: Dispatch strategies based performance analysis of a hybrid renewable energy system for a remote rural area in India. J. Clean. Prod. **259**, 1–19 (2020). https://doi.org/10.1016/j.jclepro.2020.120697
2. Saheli, M.A., Fazelpour, F., Soltani, N., Rosen, M.A.: Performance analysis of a photovoltaic/wind/diesel hybrid power generation system for domestic utilization in winnipeg, Manitoba, Canada. Environ. Prog. Sustain. Energy **38**(2), 548–562 (2019). https://doi.org/10.1002/ep.12939
3. Anh, H.P.H., Kien, C.V.: Advanced intelligent fuzzy control of standalone PV-wind-diesel hybrid system. In: International Conference on System Science and Engineering (ICSSE), pp. 129–135. IEEE (2019)
4. Benlahbib, B., Bouarroudj, N., Mekhilef, S., Abdeldjalil, D., Abdelkrim, T., Bouchafaa, F.: Experimental investigation of power management and control of a PV/wind/fuel cell/battery hybrid energy system microgrid. Int. J. Hydrogen Energy **45**(53), 29110–29122 (2020). https://doi.org/10.1016/j.ijhydene.2020.07.251
5. Mahesh, A., Sandhu, S.K.: A genetic algorithm based improved optimal sizing strategy for solar-wind-battery hybrid system using energy filter algorithm. Front. Energy **14**(1), 139–151 (2020). https://doi.org/10.1007/s11708-017-0484-4
6. Ramli, A.M.M., Bouchekara, H.R.E.H., Alghamdi, S.A.: Optimal sizing of PV/wind/diesel hybrid microgrid system using multi-objective self-adaptive differential evolution algorithm. Renew. Energy **121**, 400–411 (2018). https://doi.org/10.1016/j.renene.2018.01.058
7. Shiraliyan, M., Sharma, P., Sharma, C.: Automatic reactive power control of isolated wind–diesel hybrid power system using artificial bee colony and gray wolf optimization. Int. J. Green Energy **15**(14–15), 889–904 (2018). https://doi.org/10.1080/15435075.2018.1529584
8. Li, C., Zhou, D., Wang, H., Lu, Y., Li, D.: Techno-economic performance study of stand-alone wind/diesel/battery hybrid system with different battery technologies in the cold region of China. Energy **192**, 116702 (2020). https://doi.org/10.1016/j.energy.2019.116702
9. Rufus, A.A., Kalaivani, L.: A GOA–RNN controller for a stand-alone photovoltaic/wind energy hybrid-fed pumping system. Soft. Comput. **23**(23), 12255–12276 (2019). https://doi.org/10.1007/s00500-019-04224-8
10. Rezkallah, M., Chandra, A., Saad, M., Tremblay, M., Singh, B., Singh, S., Ibrahim, H.: Composite control strategy for a PV-wind-diesel based off-grid power generation system

supplying unbalanced non-linear loads. In: IEEE Industry Applications Society Annual Meeting (IAS), pp. 1–6. IEEE (2018)
11. Ding, Z., Hou, H., Yu, G., Hu, E., Liqiang, Zhao, J.: Performance analysis of a wind-solar hybrid power generation system. Energy Convers. Manage. **181**, 223–234 (2019). https://doi.org/10.1016/j.enconman.2018.11.080
12. Mosobi, W.R., Chichi, T., Gao, S.: Power quality analysis of hybrid renewable energy system. Cogent Eng. **2**(1), 1005000 (2015). https://doi.org/10.1080/23311916.2015.1005000
13. Moghaddam, S., Bigdeli, M., Moradlou, M., Siano, P.: Designing of stand-alone hybrid PV/wind/battery system using improved crow search algorithm considering reliability index. Int. J. Energy Environ. Eng. **10**(4), 429–449 (2019). https://doi.org/10.1007/s40095-019-00319-y

# Investigation of Sensing Ability of Double-Slot Hybrid Plasmonic Waveguide for Liquid Analyte

**Lokendra Singh, Prakash Pareek, Bahija Siddiqui, and Eswara Prasad Konakalla**

**Abstract** This paper focuses on studying the potential of plasmonic technology for sensing liquid analyte. In this work, a double-slot hybrid plasmonic waveguide on $SiO_2$ substrate layer is considered. The waveguide consists of two narrow slots between metal (silver) and dielectric (silicon) blocks. The width of slots is chosen to facilitate quasi-transverse electric mode. The mode field distribution of proposed waveguide structure with $SiO_2$ filled slots revealed that this hybrid plasmonic waveguide can serve as a reliable candidate for sensing liquid analyte effectively. Effective mode area for propagating optical signal is obtained as 0.025 per square at operating wavelength of 1550 nm. Moreover, peak sensitivity of 910 nm/RIU was achieved for 150 nm thick liquid and 300 nm thick silicon filled slots.

**Keywords** Plasmonic waveguide · Effective mode area · Optical sensors · Optical energy · Sensitivity

## 1 Introduction

Nowadays, an increase in the requirement as well as feasibility of compact devices with low power consumption and broad bandwidth is the reason for the origin of photonic integrated circuits (PICs) [1]. In this perspective, to allow the device integration on nanoscales, for the confinement and better results, beyond the diffraction limit is one of the prime issues in the current scenario [2, 3].

L. Singh · B. Siddiqui
Department of Electronics and Communication Engineering, Koneru Lakshmaiah Educational Foundation, Vaddeswaram, Andhra Pradesh, 522302, India

P. Pareek (✉)
Department of Electronics and Communication Engineering, Vishnu Institute of Technology, Vishnupur, Bhimavaram, Andhra Pradesh, 534202, India
e-mail: prakash.p@vishnu.edu.in

E. Prasad Konakalla
Department of Physics and Electronics, B.V. Raju College, Vishnupur, Bhimvavaram, Andhra Pradesh, 534202, India



D. Gupta et al. (eds.), *Pattern Recognition and Data Analysis with Applications*, Lecture Notes in Electrical Engineering 888,
https://doi.org/10.1007/978-981-19-1520-8_34

Although few geometries of waveguides are proposed to realize the confinement of surface plasmons (SP) beyond the diffraction limit [4]. The diffused filled properties of surface plasmons at the interface of metal and dielectric are capable of reducing the propagation loss, which further enhances the possibility of fabrication of nanoscale all optical devices [5]. SP-based pure metallic waveguides are capable of confining the optical field to subwavelength scale, but their susceptibility to ohmic loss again limit their further applications [6, 7].

Hence, in order to alleviate these issues, a new kind of waveguide geometry has been proposed named as hybrid plasmonic waveguide (HPWG). A HPWG is a combination of dielectric and plasmonic waveguide, which has been designed to attain subwavelength confinement of SPs with longer propagation length. Some waveguides were also proposed for better optical confinement or infiltrating materials as to increase the sensitivity such as dielectric waveguide, hollow core waveguide, and plasmonic slot waveguide [8].

Recently, optical sensing is of huge interest and hence, various geometries were analyzed to implement it such as dielectric and subwavelength-based grating sensors, hollow core waveguide sensors, and slot waveguide-based sensors [9]. Nanoslots waveguides-based sensors are capable of providing the larger optical sensitivity to the infiltrating materials of the waveguide. The scheme of nanoslot plasmonic waveguides is somewhat different than those of normal index guiding waveguide geometries.

In a dielectric slot waveguide, high–low–high index structure provides the optical confinement where as in plasmonic, the optical confinement occurs in low index medium and confinement gained by plasmonic optical increment. Moreover, in plasmonic waveguides, propagation, and excitation of modes take place at metal surface.

Plasmonic waveguide has better optical confinement properties than its dielectric counterpart but at a cost of high propagation loss. Hence, a double-slot hybrid plasmonic waveguide (DSHPWG) has been proposed to utilize the benefits of both plasmonic as well as dielectric waveguide. DSHPWG waveguide structure shows the benefits of less propagation loss, high optical confinement over the conventional silicon on insulator (SOI) technologies.

Hence, in this paper, DSHPWG studied for its possible application as liquid analyte sensor. The rest of the paper is organized as follows. Section 2 briefly describes proposed waveguide structure along with its design considerations. It also provides theoretical formulation to obtain the key parameter for assessing the potential of considered waveguide as an optical sensor. Section 3 highlights obtained salient results supported by discussions. Finally, Sect. 4 provides conclusions and future scope of this work.

## 2 Device Structure and Theoretical Formulation

The cross section view of waveguide is shown in Fig. 1. The dielectric slots between the Au and Si ring are narrow enough that the quasi-transverse electric (TE) mode can be supported. It consists of dielectric material in between the metal and silicon to guide and confine the hybrid plasmonic mode. Initially, air is treated as dielectric material between the slots of metal and dielectric, silicon dioxide ($SiO_2$) is used as the substrate layer.

A block of silicon (Si) is sandwiched between the two nanoslots. The width and height of Si block are denoted by $w_{\text{si}} = 300$ nm and $h_{\text{Si}} = 250$ nm, respectively. Then, to create the nanoslots, the blocks of gold are placed on both sides of silicon block.

The mode field distribution through the waveguide is shown in Fig. 2, which is captured when silicon dioxide (width and height of slot are 150 nm and 300 nm) is taken as dielectric material in the slots sandwiched between the gold and silicon.

In order to verify the nature of localized field, the effective mode area of guided fundamental mode is plotted as a function of wavelength in Fig. 3 [10]. The mode area can be calculated by using Eq. (1), where $x$ and $y$ are representing the longitudinal and direction of propagation of the field, respectively. The power distribution of power is integrated over the region of length of waveguide.

$$A_m = \frac{\int_{-\infty}^{+\infty} p(x, y)\mathrm{d}x\mathrm{d}y}{\max[p(x, y)]}\left(\frac{1}{\mu m^2}\right) \tag{1}$$

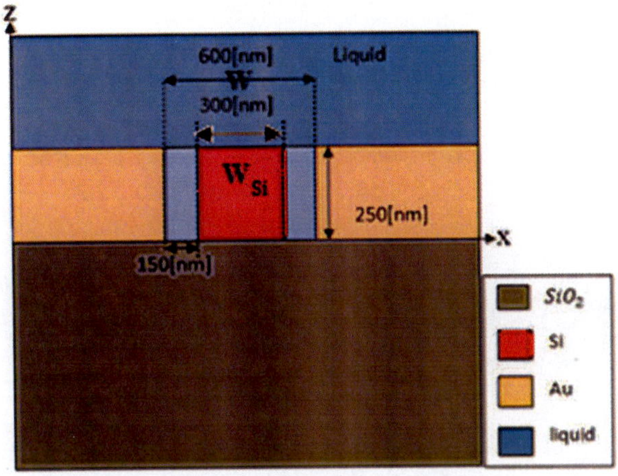


**Fig. 1** Cross section view of double-slot hybrid plasmonic waveguide

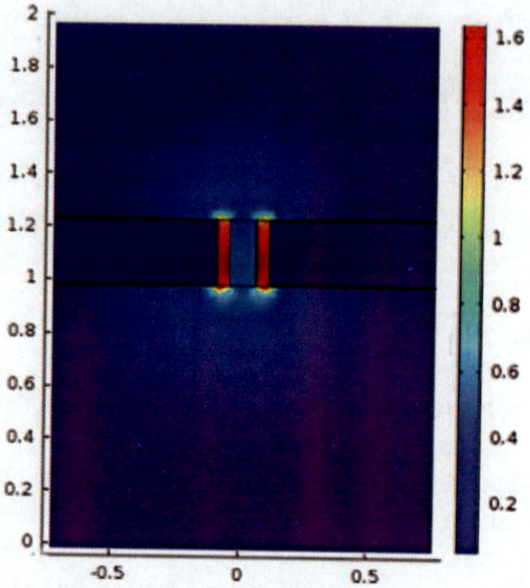


**Fig. 2** Propagation of optical fields through the double slots of Au

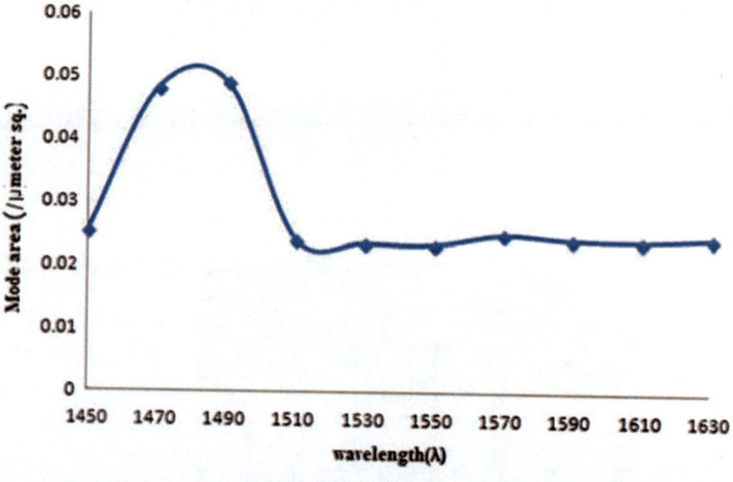


**Fig. 3** Effective mode area with respect to wavelength when $SiO_2$ is used as dielectric material

## 3 Results and Discussion

The confinement of optical signal for the designed double-slot waveguide-based sensor structure as shown in Fig. 2, satisfied the mathematical formulation given in Eq. (1). Therefore, the analysis of proposed sensor structure was done in terms of evaluation of effective mode area and sensitivity with respect to the width of slots.

The trend of effective mode area with respect to operating wavelength under the presence of $SiO_2$ as dielectric material is presented in Fig. 3. It represents that for the

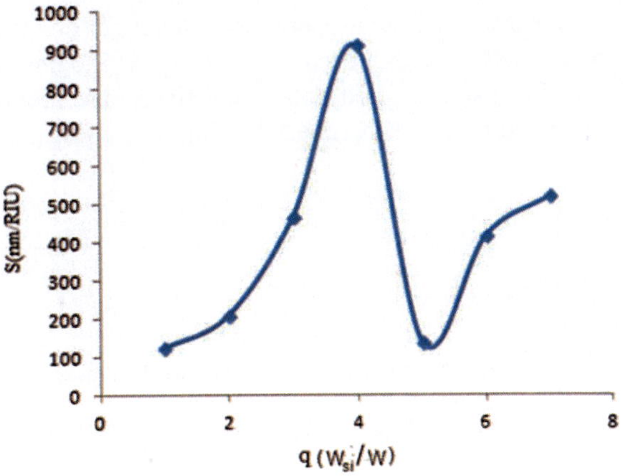


**Fig. 4** Variation sensitivity with respect to *q*-parameter when liquid is used in the nanoslots

operating wavelength of 1550 nm, the effective mode area for propagating optical signal is 0.025 per square meter. Thereafter, the analysis of sensitivity was carried out with respect to slot widths. The slot width was considered in the ratio of width of silicon to the width of liquid filled slots.

The attained results are presented in the form of plot as shown in Fig. 4. The maximum sensitivity of 910 nm/RIU was attained at the slot width of liquid and silicon filled slots were considered 150 nm and 300 nm, respectively. The calculation of sensitivity of proposed sensor structure was evaluated in terms of effective mode area, and its mathematical formulation is given in Eq. (2).

The sensitivity evaluation was done in terms of figure of merit (FOM) of the developed sensor model with respect to working wavelength. The working wavelength for the proposed sensor structure was set to equal to third window of telecommunication because of its low loss characteristics. The quality factor denotes the working capabilities of the sensor model.

$$S = \frac{\text{FOM}\,\lambda}{Q} \tag{2}$$

## 4 Conclusion

This work investigates the viability of the double-slot hybrid plasmonic waveguide for sensing liquid analyte. In the waveguide, dielectric slots between the Au and Si ring are narrow enough that the quasi-transverse electric (TE) mode can be supported. It consists of dielectric material in between the metal and silicon to guide and confine the hybrid plasmonic mode. Simulation-based analysis of proposed sensor structure was done in terms of evaluation of effective mode area and sensitivity with respect to the width of slots. Effective mode area for propagating optical signal is obtained as 0.025 per square at operating wavelength of 1550 nm. Maximum sensitivity of 910 nm/RIU was achieved at dimension of 150 nm and 300 nm, which are widths of liquid and silicon filled slots, respectively. The proposed structure can be fabricated in the future and may prove to be vital for sensing quality of liquid analyte like water, blood, etc.

## References


1. Soref, R.A.: Silicon-based optoelectronics. In: Proceedings of the IEEE, vol. 81, pp. 1687–1706. IEEE, USA (1993)
2. Gramotnev, D.K., Bozhevolnyi, S.I.: Plasmonics beyond the diffraction limit. Nat. Photon. **4**, 83–91 (2010)
3. Schörner, C., Lippitz, M.: Single molecule nonlinearity in a plasmonic waveguide. Nano Lett. **20**, 2152–2156 (2020)
4. Kim, S., Yan, R.: Recent developments in photonic, plasmonic and hybrid nanowire waveguides. J. Mater. Chem. C **6**, 11795–11816 (2018)
5. Kim, H.-M., Park, J.-H., Lee, S.-K.: Fabrication and measurement of optical waveguide sensor based on localized surface plasmon resonance. Micro Nano Syst. Lett. **7**, 7 (2019)
6. Zhang, Y., Zhang, Z.: Ultra-subwavelength and low loss in V-shaped hybrid plasmonic waveguide. Plasmonics **12**, 59–63 (2017)
7. Desiatov, B., Goykhman, I., Levy, U.: Experimental demonstration of locally oxidized hybrid silicon-plasmonic waveguide. In: CLEO 2011-Laser Applications to Photonic Applications, p. JTuI53, OSA, Baltimore, Maryland (2011)
8. Melikyan, A., Alloatti, L., Muslija, A., Hillerkuss, D., Schindler, P.C., Li, J., et al.: High-speed plasmonic phase modulators. Nat. Photon. **8**, 229–233 (2014)
9. Barrios, C.A., Gylfason, K.B., Sánchez, B., Griol, A., Sohlström, H., Holgado, M., et al.: Slot-waveguide biochemical sensor. Opt. Lett. **32**, 3080–3082 (2007)
10. Mere, V., Muthuganesan, H., Kar, Y., Kruijsdijk, C.V., Selvaraja, S.K.: On-chip chemical sensing using slot-waveguide-based ring resonator. IEEE Sens. J. **20**, 5970–5975 (2020)

# Prediction of the Final Rank of the Players in PUBG with the Optimal Number of Features

Diptakshi Sen, Rupam Kumar Roy, Ritajit Majumdar, Kingshuk Chatterjee, and Debayan Ganguly

**Abstract** PUBG is an online video game that has become very popular among the youths in recent years. Final rank, which indicates the performance of a player, is one of the most important feature for this game. This paper focuses on predicting the final rank of the players based on their skills and abilities. In this paper, we have used different machine learning algorithms to predict the final rank of the players on a dataset obtained from Kaggle which has 29 features. Using the correlation heatmap, we have varied the number of features used for the model. Out of these models, GBR and LGBM have given the best result with the accuracy of 91.63% and 91.26%, respectively for 14 features and the accuracy of 90.54% and 90.01% for eight features. Although the accuracy of the models with 14 features is slightly better than eight features, the empirical time taken by eight features is 1.4× lesser than 14 features for LGBM and 1.5× lesser for GBR. Furthermore, reducing the number of features any more significantly hampers the performance of all the ML models. Therefore, we conclude that eight is the optimal number of features that can be used to predict the final rank of a player in PUBG with high accuracy and low run-time.

**Keywords** PUBG · Machine learning · Light gradient boosting method (LGBM) · Gradient boosting regressor (GBR)

D. Sen · R. K. Roy
Department of Computer Science an Engineering, Universityof Calcutta, Kolkata, India

R. Majumdar
Advanced Computing and Microelectronics Unit, Indian Statistical Institute, Kolkata, India

K. Chatterjee · D. Ganguly (✉)
Department of Computer Science and Engineering, GCECT, Kolkata, India
e-mail: debayan@gcelt.org



D. Gupta et al. (eds.), *Pattern Recognition and Data Analysis with Applications*, Lecture Notes in Electrical Engineering 888,
https://doi.org/10.1007/978-981-19-1520-8_35

## 1 Introduction

PUBG is an online Battle Royale video game which is a multiplayer shooter game where the players have to fight to remain alive till the end of the game. In this game, a maximum of 100 players are allowed. Players can choose to enter the match solo, duo, or with a team of up to four (squad). Players are dropped empty handed from a plane on one of the four maps at the beginning of the match. Once they land, the players start searching for weapons and armors which are periodically distributed throughout the game. The players then fight one on one and the last player, or team, alive wins the match.

The rank of a player or a team is an important aspect of the game because the rank of a player is the position at which the player or the team gets eliminated, and this rank is required to calculate the tier of the player. Machine learning (ML)-based approach to predict the final rank of the players have been studied in some papers [1–3]. These papers use different ML-based techniques with more than 15 features for this task.

In this paper, we have predicted the rank of the players or teams in PUBG using both previously used algorithms, such as multiple linear regression, LGBM, random forest, and some other algorithms as well such as gradient boosting regression (GBR) [4], lasso and ridge regression [5], decision tree [6], K-nearest neighbours (KNN) [7] on a dataset from Kaggle [8]. We find that light gradient boosting method (LGBM) [9, 10] and GBR [11] provide the highest accuracy of ~91.63% and an MAE of 0.06, which is at par with the earlier studies on this. However, we further show that the number of features can be reduced to eight (the top eight features of the correlation heatmap) without hampering the performance of the model. Nevertheless, this reduction in the number of features provides an approximate 1.5× speedup to the empirical run-time of this algorithm. We, further, numerically have shown that reducing the number of features any more have a significant role on the performance of the ML algorithms.

Remaining paper is arranged as follows:

We did *data cleaning in Sect. 2, feature engineering and feature selection in Sect.* 3. Further, we have discussed our *findings in result and discussion in Sect. 4*, and finally, *concluded our paper in Sect. 5*. We have uploaded our code to GitHub and have provided its link with this paper in *code availability section.*

## 2 Data Cleaning

As discussed earlier, the maximum number of players allowed in solo, duo, and squad match types are 1, 2, and 4, respectively. But the dataset has categorized the match types into 16 different types which are variations of these three primary types. We have mapped the number of players into its proper match type (Fig. 1) as follows:

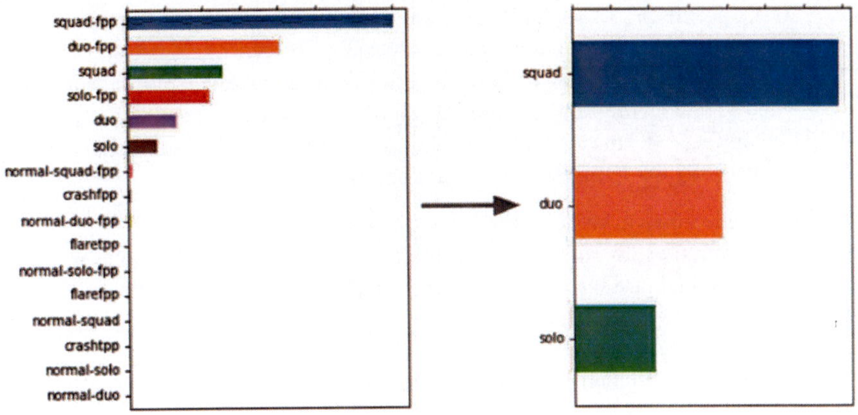


**Fig. 1** Mapping of different match types into core match types

$$\text{Players in game type } j = \sum_i \text{players in game type } j_i \tag{1}$$

where $j_i$ are the different sub-formats of the primary game format $j$.

We have further removed anomalous data from the dataset. The criteria for removal of a data are one or more of the following: (i) Number of players in a team for a particular match type is greater than the allowed number of players and (ii) (possibly offline) players who have either 0 kill and have not covered any distance and have not picked up any weapon.

## 3 Features Engineering and Selection

PUBG allows a maximum of 100 players in a match, but it is not always necessary to have 100 players. If there are 100 players in a match, then it might be easier to find and kill enemies as compared to 90 players. Using this notion, we have normalized the features

$$\text{FEATURES} * \frac{(100 - \text{number of players})}{100} + 1 \tag{2}$$

This provides a higher score to a player for their achievement when the total number of players is less.

Furthermore, we have created some new compound features by combining existing features. These features lead to a higher accuracy of the ML algorithms. The new features that we have created are:

1. Assist_Revive = Assist and revive are the part of teamwork so we have
2. Taken both as a single parameter, i.e. Assist + Revive
3. Total_Distance = Total distance is the distance travelled by the players by walking and swimming, i.e. walk distance + swim distance
4. Players_in_a_team = Number of players in team (based on groupID)
5. Headshot/kill = Number of headshot per kill

All these features have been normalized for the prediction purpose.

All the attributes do not have equal impact on the final rank of a player. So we have performed feature selection to select those features which will affect the result. In Fig. 2, we show the correlation heatmap [12] which is used to select the top 14 features which have high correlation with the target variable.

We have further studied how reducing the number of features affects the accuracy and MAE of the models. In order to do so, we have varied the number of features from 5 to 8 which are the subsets of previously taken 14 features. We see that up to eight features, the performance remains more or less steady, but drops significantly for a lower number of features.

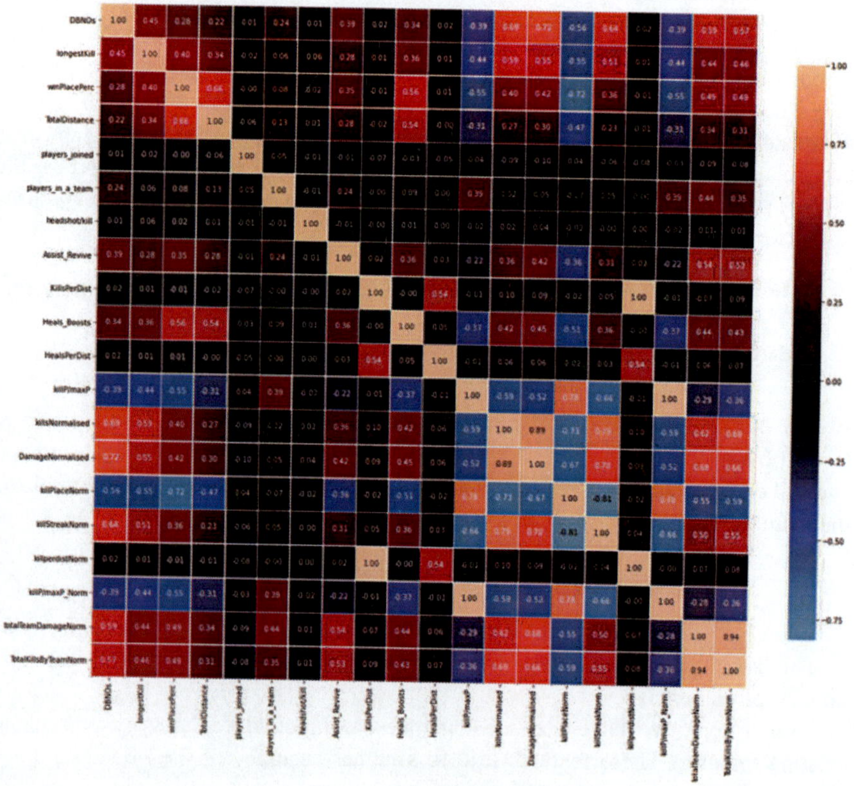

**Fig. 2** Correlation matrix representing the correlation of the features

## 4 Result and Discussion

We have varied the number of features from 14 to 5 while applying several ML models. We have closely observed the accuracy achieved, empirical time taken and the reduction of MAE while varying the number of features. We have shown our results by comparing the MAE, accuracy, and empirical time taken by various methods for a particular set of features. In Tables 1, 2 and 3, we explicitly show the actual features considered and the accuracy, time, and MAE for different ML algorithms with 14, 8, and 7 features, respectively. Note that all these are standard PUBG features, and we, therefore, do not explain them further. However, as their names suggest, each of these features have been normalized as discussed in Sect. 3. Henceforth, in Figs. 3 and 4, we, respectively, show the accuracy and the empirical time required for different ML models as the number of features is varied from top 14 to top 5.

**Table 1** Comparison of MAE, accuracy, time of different models with 14 features

| Models | LGBM | Random forest | GBR | Decision tree | KNN | Ridge | LASSO | Linear regression |
|---|---|---|---|---|---|---|---|---|
| MAE | 0.063 | 0.065 | 0.062 | 0.088 | 0.100 | 0.110 | 0.115 | 0.110 |
| Accuracy (%) | 91.26 | 90.45 | 91.63 | 82.59 | 79 | 75 | 73 | 75 |
| Time (in s) | 12.73 | 9.20 | 23.37 | 1.38 | 0.79 | 0.03 | 0.30 | 0.05 |

**Table 2** Comparison of MAE, accuracy, time of different models with eight features

| Models | LGBM | Random forest | GBR | Decision tree | KNN | Ridge | LASSO | Linear regression |
|---|---|---|---|---|---|---|---|---|
| MAE | 0.067 | 0.069 | 0.065 | 0.091 | 0.093 | 0.126 | 0.128 | 0.13 |
| Accuracy (%) | 90.01 | 89.49 | 90.54 | 81.20 | 81 | 69 | 69 | 69 |
| Time (in s) | 8.89 | 6.86 | 15.38 | 0.91 | 0.54 | 0.02 | 0.06 | 0.03 |

**Table 3** Comparison of MAE, accuracy, time of different models with seven features

| Models | LGBM | Random forest | GBR | Decision tree | KNN | Ridge | LASSO | Linear regression |
|---|---|---|---|---|---|---|---|---|
| MAE | 0.089 | 0.078 | 0.081 | 0.103 | 0.116 | 0.144 | 0.14S6 | 0.14 |
| Accuracy (%) | 84.30 | 86.26 | 86.23 | 75.65 | 73 | 63 | 63 | 63 |
| Time (in s) | 6.76 | 6.71 | 14.07 | 0.86 | 0.55 | 0.45 | 0.18 | 0.10 |

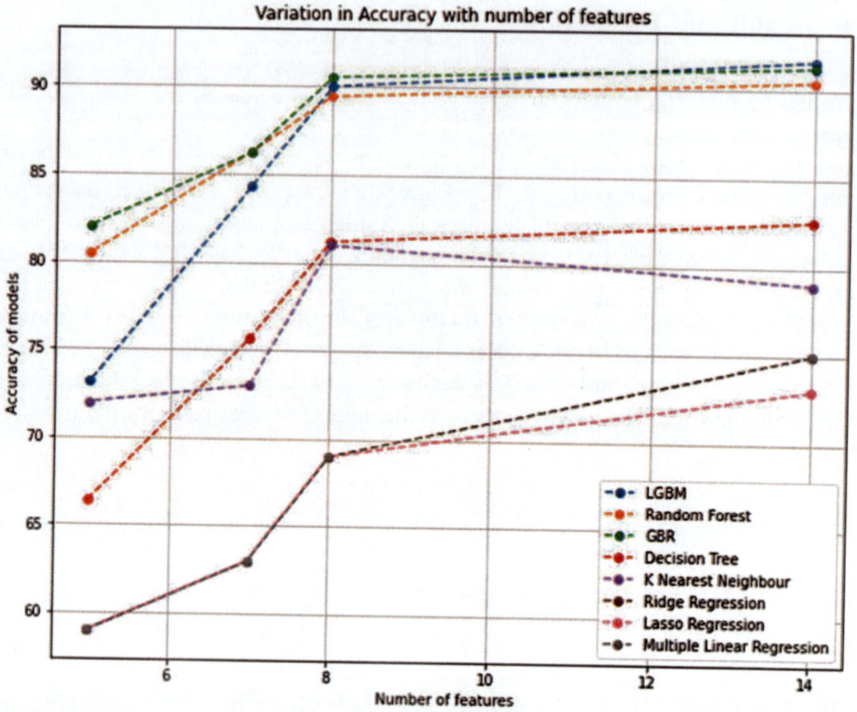


**Fig. 3** Accuracy of the models versus the number of features

14 FEATURES: The used features are:

DBNOs, killPlaceNorm, killStreakNorm, longestKill, TotalDistance, killperdistNorm, HealsPerDist, Assist_Revive, killP/maxP_Norm, totalTeamDamageNorm, TotalKillsByTeamNorm, killsNormalised, DamageNormalised.

8 FEATURES: The used features are:

TotalDistance, TotalKillsByTeamNorm, killsNormalised, DamageNormalised, Heals_Boosts, killPlaceNorm, killP/maxP_Norm, longestKill.

7 FEATURES: The used features are:

TotalDistance, TotalKillsByTeamNorm, killsNormalised, DamageNormalised, Heals_Boosts, killP/maxP_Norm, longestKill.

From Tables 1, 2, and 3, we have observed that LGBM and GBR are giving the best result with respect to the MAE and accuracy. From Tables 1 and 2, we can say that there is a nominal change in MAE and accuracy for eight features as that of 14 features although empirical time for the former is much lesser than 14 features. We have further tried to reduce the empirical time by reducing the number of features, but there is a significant degradation of accuracy and the MAE (observed from Table 3) and the trend persists (as illustrated in Figs. 3 and 4).

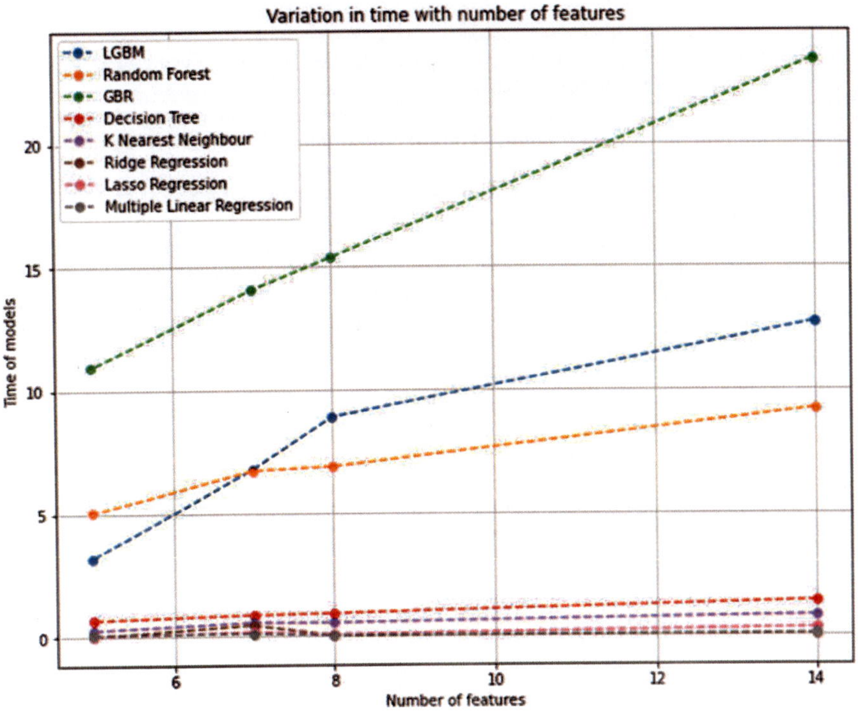


**Fig. 4** Empirical time taken by different features

## 5 Conclusion

In this paper, we have used several ML models to predict the final placement or rank of the players in the PUBG. Out of these models LGBM and GBR have given the best result. It can be concluded that eight features are more preferable than 14 features because it reduces the empirical run-time of the ML algorithm. Furthermore, eight is the lower threshold, since further reduction in the number of features significantly degraded the performance. Eventually, although this study is based on PUBG, the technique can be extended to any multiplayer game by using suitable features.

**Code Availability**

Code is available at this link:
https://github.com/Diptakshi/PUBG

## References

1. Rokad, B., Karumudi, T., Acharya, O., Jagtap, A.: Survival of the fittest in player unknown's

battlegrounds. arxiv:1905.06052, (2019)
2. Chatterjee, S.: PUBG data analysis using Python [online] (2020). Available: https://www.mygreatlearning.com/blog/pubg-data-analysis-using-python/
3. Wei, W., Lu, X., Li, Y.: PUBG: A Guide to Free Chicken Dinner. Stanford University (2018)
4. Huang, W., Liu, Y., Li, C.: GBRTVis: online analysis of gradient boosting regression tree. J. Visuali. **22**, 125–140 (2019)
5. Pereira, J.M., Basto, M., Silva, A.F.: The logistic lasso and ridge regression in predicting corporate failure. Proc. Econ. Finan. **39**, 634–641 (2016)
6. Friedman, J.H.: Stochastic gradient boosting. Comput. Stat. Data Anal. **38**(4), 367–378 (2002)
7. Zhang, Z.: Introduction to machine learning: K-nearest neighbours. Big-data Clin. Trial Column. Ann. Transl. Med. **4**(11) (2016)
8. PUBG finish placement prediction. https://www.kaggle.com/c/pubg-finish-placement-prediction. Last accessed 29 Apr 2021
9. Omar, K.B.A.: XGBoost and LGBM for PortoSeguro's Kaggle challenge: a comparison. Preprint Semester Proj. ETH Zurich (2018)
10. Ke, G., Meng, Q., Finley, T., Wang, T., Chen, W., Ma, W., Ye, Q., Liu, T.: LightGBM: a highly efficient gradient boosting decision tree. In: 31st conference on Neural Information Processing System (NIPS), Long Beach (2017)
11. Friedman, J.H.: Greedy Function Approximation: A Gradient Boosting Machine. IMS (1999)
12. Szabo, B.: How to create a seaborn correlation heatmap in python? Medium (2020)

# Cryptanalysis of a Security Scheme for Smart Traffic Lighting System Based on Fog Computing

**Uddalak Chatterjee, Maddirala Venkat, and Sangram Ray**

**Abstract** Security is our primary goal for any intelligent systems that are connected to the wireless networks. Researchers are constantly working to improve the security of various applications of these systems like smart home, smart city, and smart transportation. In 2018, Khalid et al. have proposed a security framework to enable secure communication and authentication between smart vehicles and road-side units (RSUs) of intelligent traffic control system such as vehicle to vehicle (V2V), vehicle to infrastructure (V2I), and infrastructure to vehicle (I2V) in the context of intelligent transportation systems. In Khalid et al. scheme, the authors have claimed that their proposed scheme is providing secure communication between smart vehicles and RSUs of intelligent traffic control system and is capable to withstand against various security attacks such as replay attack, DoS attack, Sybil attack, and impersonation attack. In this paper, we have done a thorough security analysis of Khalid et al. authentication scheme and found that their scheme is insecure against some relevant security attacks. The scheme addressed by Khalid et al. is prone to possible security attacks such as side-channel/smart device attack and code injection attacks.

**Keywords** Intelligent traffic light control systems (ITLCSs) · RSUs (road-side units) · Smart cities · Intelligent transportation systems (ITSs) · VANETs (vehicular ad hoc networks) · OBU (on-board unit)

## 1 Introduction

**Significance of smart traffic lighting system in a smart city**: Transportation systems are very important and play a key role in our daily life. As the number of vehicles are increasing rapidly, but transportation systems are not improved in such a way to provide an efficient and an effective transportation facility to the vehicles. Because of this, traffic congestion and traffic-related problems such as pollution

U. Chatterjee (✉) · M. Venkat · S. Ray
Department of Computer Science and Engineering, National Institute of Technology Sikkim, Ravangla, Sikkim 737139, India
e-mail: uddalak.udi@gmail.com



D. Gupta et al. (eds.), *Pattern Recognition and Data Analysis with Applications*, Lecture Notes in Electrical Engineering 888,
https://doi.org/10.1007/978-981-19-1520-8_36

has been increasing with further affects the various markets both environmentally and financially all-around world [1–3]. From the last few years, the advancement in wireless communication-related technologies and VANETs has given a way to implement intelligent traffic control systems [4, 5]. An intelligent traffic control system is a system which mainly aimed at managing transportation efficiently during emergency situations with the help of cutting-edge technologies and intelligent systems. Intelligent traffic control systems consist smart traffic lights that are basically adapt depending on traffic condition for smooth and effective traffic flow by running an efficient scheduling algorithm on traffic lights to change signals which reduce waiting period of vehicles based on their speed, position, and direction [3, 6]. Many countries are also adopting ITSs because of its wide advantages as compared with traditional transportation systems; at the same time, there are few disadvantages of ITSs compared with traditional transportation systems. Security is one of the biggest issues associated with intelligent traffic control systems [7]. The intelligent traffic control systems use wireless network technologies in order to establish communication between smart vehicles and RSUs, and the use of wireless communication will increase the more security threats [6]. So, it is mandatory to any wireless communication technology-based solution to handle security threats efficiently. Therefore, it is very much essential to preserve data confidentiality, data integrity, and authentication effectively.

**Our contribution**: In this paper, we have thoroughly analyzed, and we found that claim made by authors in [8] is not fully acceptable as the scheme is prone to possible various attacks that an attacker can make. We have found and proved using some mathematical assumptions that Khalid et al. authentication scheme is insecure against side-channel attacks, code injection attacks which makes scheme insecure against all the attacks claimed by authors.

**Organization of our paper**: We have arranged the remaining paper as: Section 2 provides literature review of various security attacks and research surveys of different schemes of ITLCSs addressed by various authors. Section 3 provides detailed review of Khalid et al. authentication scheme [8]. Security attacks and vulnerabilities of Khalid et al. authentication scheme are presented in Sect. 7. Section 10 provides conclusion.

## 2 Literature Review

We provided an overview and a review of various security attacks and research surveys of many schemes that are related to intelligent traffic control systems addressed by various authors such as [1–4, 7, 9–19]. In [9], the authors highlighted the side-channel attacks and presented that an efficient authentication scheme is very much needed to provide a strong secure communication and authentication between smart vehicles and their infrastructure that fulfills the security goals in vehicular ad hoc networks (VANETs) and also proposed a scheme to prevent side-channel threat by continuously updating sensitive information inside a tamper proof devices such as on-board units (OBUs) of the vehicles. In [10]. Fushan wei et al. presented the risks faced by the Internet of connected vehicles that it is very important to protect vehicle intelligent terminals such as OBUs by providing secure authentication such as biometric characteristics such as fingerprints. In [11], Muhammad Awais Javed et al. are mentioned about data integrity threats in ITSs. Main components in ITSs are road-side units (RSUs), OBUs that are smart devices capable of sending and receiving information. If an adversary inserts a false information in data that are being transmitted over the network it results in making wrong decisions by ITSs and also presented a scheme to mitigate the data integrity attacks; injection attacks are also one of biggest security threats to the Web-based applications; the attacker can retrieve sensitive data by inserting malicious sql code into the database through input parameters [13]. There are various authors so far [12–16] presented input-based analysis for sql injection threats to the Web-based applications and for smart devices. In [14], the authors mentioned about code injection attacks such as sql injection and xss injection attacks. In [1–4, 7, 17], various authors presented various research surveys and schemes to provide secure communication and authentication between vehicles and their infrastructure in the context of intelligent transportation systems. In [18], the authors highlighted a point that attackers can insert or drop some malicious packets by monitoring data that are being transmitted over the network; ultimately, it is easy for the attacker to compromise the RUSs and vehicles, so it is important to protect private data when transmitted over unreliable networks. In [19], the authors also highlighted smart device attacks in the context of smart grid communication that is if a smart device is stolen or lost, it is easy for attackers to retrieve secret information inside the device. In 2018, Khalid et al. [8] proposed security scheme using symmetric and asymmetric cryptography for intelligent traffic control systems based on fog computing for providing secure communication and authentication between vehicles and their infrastructure. In this scheme, the authors made a claim that their scheme is providing strong and secure authentication against different security attacks, namely replay attack, DoS attack, Sybil attack, impersonation attack, and is having less communication and computational overheads.

## 3 Review of Khalid et al. Authentication Scheme

This section provides a brief explanation of Khalid et al. authentication scheme [8]. The preliminaries of symmetric and asymmetric cryptography, VANETs and ITSs can be found in [3–7, 17–20]. The Khalid et al. authentication scheme proposed architecture model of the system mainly consists three entities: (a) DMV, (b) RSU, and (c) vehicle.

1. **DMV**: It is a trusted government entity, and its responsibilities include installation of RSUs having storage, communicational, and computational capacities.

**Table 1** Notations used in Khalid et al. scheme

| Notation | Description |
|---|---|
| DMV | Dept. of motor vehicles |
| OBUs | On-board units |
| RSUs | Road-side units |
| $T_{\mathrm{p}}$ | Travel plan |
| $C_1, C_2$ | Cipher texts |
| Sk | Symmetric key |
| ID | Unique vehicle's ID |
| $E_{\mathrm{sk}}$ | Encryption using symmetric key |
| $K_{\mathrm{sess}}$ | Session key |
| $_{\mathrm{sign}}$(token) | Token signed by DMV |
| $K_{\text{pub-DMV}}$ | Public key of DMV |
| $K_{\text{pub-RSU}}$ | Road-side unit public key |
| $K_{\text{priv-DMV}}$ | Private key of DMV |
| $K_{\text{priv-RSU}}$ | Road-side unit private key |
| $E_{\text{Kpub-RSU}}$ | Encryption using public key of road-side units |
| $D_{\text{Kpriv-RSU}}$ | Decryption using private key of road-side units |
| $D_{\mathrm{sk}}$ | Decryption using symmetric key |
| $M_1, M_2$ | Messages |
| Rs | Random secret |
| UTM | Universal transverse Mercator |

And it also provides registration to all the vehicles of the system.

2. **RSU**: It is also a trusted government entity, and its responsibilities include broadcasting of secret messages, certificates to every vehicle in monitored area, and verification of messages received from vehicles in monitored area.
3. **Vehicle**: All the vehicles of the system are equipped with inbuilt OBU having storage, communicational, and computational capacities. Each vehicle needs to be registered first with DMV in order to establish communication with RSU to get travel assistance.

The proposed scheme [8] mainly consists of three phases: (i) installation (ii) registration (iii) communication. All symbols used in this scheme are provided in Table 1, and corresponding proposed architecture model of the system in [8] is shown in Fig. 1.

## 3.1 Installation

This is the initial phase, where the trusted third-party entity called DMV is responsible for installing RSUs in a particular position in the map having certain longitude and

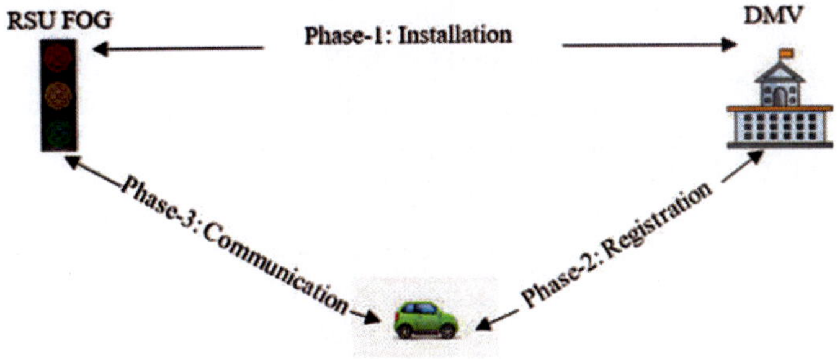


**Fig. 1** Khalid et al. authentication scheme system architecture

latitude which are capable of having computational and storage capabilities. DMV also stores Sk, $K_{\text{pub-RSU}}$, $K_{\text{priv-RSU}}$, a signed certificate with $K_{\text{priv-DMV}}$, and $K_{\text{pub-DMV}}$ on every installed RSU.

## *3.2 Registration*

In this phase, all the vehicles need to be registered with the DMV. On successful completion of registration, the trusted third-party DMV allots an individual unique ID to every vehicle and stores a unique ID, $K_{\text{pub-DMV}}$, signed token [with DMVs secret key ($K_{\text{priv-DMV}}$)] on every vehicle.

## *3.3 Communication*

This phase has been further divided into four sub-phases which are broadcasting, key generation, verification, and decision phase.

**Broadcasting**

In this phase, RSUs are responsible to broadcast ciphertexts $C_1$, $C_2$, and its public-key certificate which is signed by trusted third-party DMV to each vehicle in a monitored area. RSU masks Geolock and symmetric key (Sk) to get $C_1$, and RSU performs X-OR operation for Geolock and symmetric (Sk) masking as described below. RSU first calculates the Geolock based on its position in the map with longitude and latitude; after calculating the position of RSU, it gets converted into the UTM. The UTM value is further divided by 100 which is the range of the RSU-monitored region. Further, it only considers the integer value by discarding decimal value. Then, values are concatenated or multiplexed. At last, SHA-1 is calculated on concatenated values. SHA-1 is considered as Geolock value for generating $C_1$.

Position

| Latitude (N) | Longitude (E) |
|---|---|
| 34.164532 | 73.222982 |

↓

Then it is converted into UTM as follows

UTM

| Easting (E) | Northing(N) |
|---|---|
| 336203.53/100 | 3781825.90/100 |
| 3362 E | 37818 N |

Afterward, UTM values are concatenated as shown below

$$3362 + 3781 = 33623781$$

Finally, SHA-1 is calculated

$$\text{SHA} - 1(33623781) = 262969\text{BE}943207\text{DA}35\text{AA}781\text{B}8\text{A}0\text{E}967\text{A}6787\text{BA}5\text{B}$$

and it is a Geolock value which is further used to generate $C_1$.

$$C_1 = \text{GeoLock} \oplus \text{Sk}$$

and RSU generates a random secret key called rs, and it encrypts rs using Sk for generating ciphertext $C_2$.

$$C_2 = \text{Esk(rs)}$$

On receiving $C_1$,$C_2$, and a personal certificate of RSU, each vehicle in the monitored region performs the following operations to get Sk, rs is described below

$$\text{Sk} = \text{GeoLock} \oplus C_1$$

$$\text{rs} = D_{\text{sk}}(C_2)$$

After calculating, the Sk and rs vehicles verify the public-key certificate of RSU with the help of public of key of the trusted third-party DMV which was installed on

OBUs of smart vehicles at time of registration. On successful verification, vehicles use the verified RSUs public key in the next sub-phase for generating $M_1$.

**Key generation**

In the key generation phase, each vehicle in the monitored area generates $M_1$ by concatenating signed token with rs and then encrypts $M_1$ with the public key of RSU. After encryption, every vehicle sends an encrypted message $M_1$ to the RSU.

$$M_1 = E_{K\text{pub - RSU}}\left(_{\text{sign}}(\text{token})||\text{rs}\right)$$

After receiving the message $M_1$, RSU performs decryption of $M_1$ using its private key as follows

$$_{\text{sign}}(\text{token})||\text{rs} = D_{K\text{priv}-\text{RSU}}(M_1)$$

and then verifies the token which is signed by DMV, before using it to generate session key. Once verification is done successfully at each end, both vehicle and RSU perform masking operation to generate session key is as follows

$$K_{\text{sess}} = {}_{\text{sign}}(\text{token})||\text{rs}$$

Same session is generated at both ends. After generating session key the generated session key is used in the next phase.

**Verification**

In verification phase, the vehicle sends $M_2$ which is encrypted using session key generated in previous phase to RSU, and $M_2$ is treated as a reply to challenge that has been sent by RSU in initial phase. On receiving $M_2$, RSU performs decryption and then verifies message integrity. On completion of verification successfully, the reply is treated as genuine;, each vehicle generates $M_2$ as follows:

$$M_2 = E_{K\text{sess}}\left(\text{rs}\middle|\middle|T_p\middle|\middle|\text{ID}||\text{hash}(\text{rs, Tp, ID})\right).$$

And decisions are made in the next phase.

**Decision**

On successful verification of message in previous phase, RSU enters in the decision phase, as the vehicles are not cheated RSU, so RSU starts running traffic scheduling algorithm based on vehicles travel plan (Tp) to provide desired signals to vehicles, and RSU provides equal assistance to every vehicle in a monitored area.

# 4 Cryptanalysis of Khalid et al. Authentication Scheme

Khalid et al. [8] authentication scheme, which is discussed above, provides an authentication scheme for establishing secure communication between RSUs and vehicles using symmetric and asymmetric cryptography. Although there are various authentication schemes, proposed by different researchers using symmetric and asymmetric cryptography [1, 9–11, 21]. However, there are many flaws still present in this scheme are discussed below.

## 4.1 *Side-Channel Attack*

In intelligent transportation systems, the main target for attackers is its infrastructure such as OBUs and RSUs. In cryptography, a side-channel attack is used to retrieve sensitive information from secure smart devices such as tamper proof devices (TPDs) (such as OBUs, RSUs) rather than finding weaknesses in the scheme [9]. In this scheme, the DMV which is government-trusted third-party stores sensitive information such as public and private key pair of DMV ($K_{\text{pub-DMV}}$, $K_{\text{priv-DMV}}$) along with unique ID of vehicle in OBUs and public private key pairs of RSU ($K_{\text{pub-RSU}}$, $K_{\text{priv-RSU}}$) and DMV and Sk in RSUs during its installation for communication and verification purposes. If the smart device/TPD (such as either OBUs or RSUs in this scheme) is somehow stolen by an attacker, then it is easy for attackers to retrieve sensitive information using power analysis attack [19].

Due to possibility of power analysis attack, it is easy for attacker to create multiple identities using public and private key pair of DMV ($K_{\text{pub-DMV}}$, $K_{\text{priv-DMV}}$) by registering vehicle with a fake unique ID, which makes the way easy for attacker to perform Sybil attack by creating Sybil nodes, DoS attack by overloading RSU using malicious nodes and similarly attacker performs replay attack and impersonation attack by pretending it is a legitimate user. Hence, Khalid et al. authentication scheme is more vulnerable to this attack which makes the scheme insecure against all the attacks addressed by authors.

## 4.2 *Code Injection Attacks*

In this Internet world, information is very crucial for any attacker to perform cryptographic attacks on any smart device or Web-based applications that are connected wirelessly to the Internet [16]. In this scheme, the smart devices such as OBUs and RSUs are connected to the Internet wirelessly for communicating and sharing information between them. Due to growing concern about different cyberattacks, it is essential to maintain information confidential. This scheme proposed by Khalid et al.[8] fails to provide confidentiality because of poor authentication, mainly symmetric key (Sk) used for generating ciphertext. It is easy for anyone to calculate Geolock by simply using position of RSU which makes the way easy for attacker to get Sk as shown below

$$\text{Sk} = \text{GeoLock} \oplus C_1$$

Which further leads to the possibility of code injection attacks such as SQL injection and XSS injection attacks by using snortIdS or by inserting malicious code in input parameters [13, 14]. As Sk, Geolock is available to the attacker, so attacker can perform code injection attacks on smart devices by inserting malicious code in input parameters. If attacker somehow compromised the smart device using code injection attacks, then it is easy for attacker to gain all sensitive data. Later, attacker uses this information to perform various cryptographic attacks such as Sybil attack, replay attack, DoS attack, and impersonation attack, which makes this scheme totally insecure against many cyber-threats. Hence, Khalid et al. authentication scheme is not secure and vulnerable to code injection attacks.

## 5 Conclusion

The Khalid et al. authentication scheme uses both symmetric and asymmetric cryptography to provide secure communication between vehicles and its infrastructure with the help of smart traffic lighting systems in smart cities. The authors claimed that their scheme is providing secure communication between vehicles and its infrastructure and is capable of withstanding against different security attacks such as replay attack, DoS attack, Sybil attack, and impersonation attack mainly. However, in our paper, we have shown that their scheme completely fails to provide security against side-channel attacks and code injection attack which further makes the scheme insecure against many other cryptographic attacks. In future, we are motivated to propose a strong, lightweight, and secure authentication scheme for communication between vehicles and its infrastructure of smart traffic lighting systems in smart cities as extension of this work.

## References


1. Liu, J., Li, J., Zhang, L., Dai, F., Zhang, Y., Meng, X., Shen, J.: Secure intelligent traffic light control using fog computing. Futur. Gener. Comput. Syst. **78**, 817–824 (2018)
2. Cunha, J., Batista, N., Cardeira, C., Melicio, R.: Wireless networks for traffic light control on urban and aerotropolis roads. J. Sens. Actuator Netw. **9**(2), 26 (2020)
3. Mandhare, P.A., Kharat, V., Patil, C.Y.: Intelligent road traffic control system for traffic congestion a perspective. Int. J. Comput. Sci. Eng. **6**(07) (2018)

4. Kumar, P., Verma, A., Singhal, P.: VANET protocols with challenges—a review. In: 2019 6th International Conference on Computing for Sustainable Global Development, pp. 598–602. IEEE (2020)
5. Nellore, K., Hancke, G.P.: A survey on urban traffic management system using wireless sensor networks. Sensors **16**(2), 157 (2016)
6. Kafi, M.A., Challal, Y., Djenouri, D., Doudou, M., Bouabdallah, A., Badache, N.: A study of wireless sensor networks for urban traffic monitoring: applications and architectures. Procedia Comput. Sci. **19**, 617–626 (2013)
7. van der Heijden, R.W., Dietzel, S., Leinmüller, T., Kargl, F.: Survey on misbehavior detection in cooperative intelligent transportation systems. IEEE Commun. Surv. Tutor. **21**(1), 779–811 (2018)
8. Khalid, T., Khan, A.N., Ali, M., Adeel, A., Shuja, J.: A fog-based security framework for intelligent traffic light control system. Multimedia Tools Appl. **78**(17), 24595–24615.2 (2019)
9. Ali, I., Li, F.: An efficient conditional privacy-preserving authentication scheme for vehicle-to-infrastructure communication in VANETs. Veh. Commun. **22** (2020)
10. Wei, F., Zeadally, S., Vijayakumar, P., Kumar, N., He, D.: An intelligent terminal based privacy-preserving multi-modal implicit authentication protocol for internet of connected vehicles. IEEE Trans. Intell. Transport. Syst. (2020)
11. Javed, M.A., Khan, M.Z., Zafar, U., Siddiqui, M.F., Badar, R., Lee, B.M., Ahmad, F.: An efficient protocol to mitigate data integrity attacks in intelligent transport systems. IEEE Access **8**, 114733–114740 (2020)
12. Rankothge, W.H., Randeniya, M., Samaranayaka, V.: Identification and mitigation tool for Sql injection attacks (SQLIA). In: 2020 IEEE 15th International Conference on Industrial and Information Systems (ICIIS), pp. 591–595. IEEE (2020)
13. Jana, A., Bordoloi, P., Maity, D.: Input-based analysis approach to prevent SQL injection attacks. In: 2020 IEEE Region 10 Symposium (TENSYMP), pp. 1290–1293. IEEE (2020)
14. Alnabulsi, H., Islam, R.: Protecting code injection attacks in intelligent transportation system. In: 2019 18th IEEE International Conference on Trust, Security and Privacy in Computing and Communications/13th IEEE International Conference on Big Data Science and Engineering (TrustCom/BigDataSE), pp. 799–806. IEEE (2019)
15. Su, G., Wang, F., Li, Q.: Research on SQL injection vulnerability attack model. In: 2018 5th IEEE International Conference on Cloud Computing and Intelligence Systems (CCIS), pp. 217–221. IEEE (2018)
16. Joshi, P.N., Ravishankar, N., Raju, M.B., Ravi, N.C.: Encountering sql injection in web applications. In: 2018 Second International Conference on Computing Methodologies and Communication (ICCMC), pp. 257–261. IEEE (2018)
17. Gauher, A., Umrani, A., Javed, Y.: Communication security in VANETs. In: 2020 IEEE 17th International Conference on Smart Communities: Improving Quality of Life Using ICT, IoT and AI (HONET), pp. 63–67. IEEE (2020)
18. Basudan, S., Lin, X., Sankaranarayanan, K.: A privacy-preserving vehicular crowdsensing-based road surface condition monitoring system using fog computing. IEEE Internet Things J. **4**(3), 772–782 (2017)
19. Sadhukhan, D., Ray, S.: Cryptanalysis of an elliptic curve cryptography based lightweight authentication scheme for smart grid communication. In: 2018 4th International Conference on Recent Advances in Information Technology (RAIT), pp. 1–6. IEEE (2018)
20. Stallings, W.: Cryptography and Network Security, 4/E. Pearson Education India (2006)
21. Andreica, G.R., Bozga, L., Zinca, D., Dobrota, V.: Denial of service and man-in-the-middle attacks against IoT devices in a GPS-based monitoring software for intelligent transportation systems. In: 2020 19th RoEduNet Conference Networking in Education and Research (RoEduNet), pp. 1–4. IEEE (2020)

# Gaining Actionable Insights in COVID-19 Dataset Using Word Embeddings

**Rajat Aayush Jha and V. S. Ananthanarayana**

**Abstract** The field of unsupervised natural language processing (NLP) is gradually growing in prominence and popularity due to the overwhelming amount of scientific and medical data available as text, such as published journals and papers. To make use of this data, several techniques are used to extract information from these texts. Here, in this paper, we have made use of COVID-19 corpus (https://www.kaggle.com/allen-institute-for-ai/CORD-19-research-challenge) related to the deadly corona virus, SARS-CoV-2, to extract useful information which can be invaluable in finding the cure of the disease. We make use of two word-embeddings model, Word2Vec and global vector for word representation (GloVe), to efficiently encode all the information available in the corpus. We then follow some simple steps to find the possible cures of the disease. We got useful results using these word-embeddings models, and also, we observed that Word2Vec model performed better than GloVe model on the used dataset. Another point highlighted by this work is that latent information about potential future discoveries are significantly contained in past papers and publications.

**Keywords** NLP · COVID-19 · Embeddings · Word2Vec · GloVe

## 1 Introduction

The vast majority of scientific and medical information is written as text, which makes conventional statistical analysis and modern machine learning approaches difficult to analyze. In comparison, the primary source of machine-interpretable data in various medical and science fields is the structured property databases. Although these databases contain just a small portion of the total knowledge found out in the research literature. Supervised natural language processing, which necessitates large amount of hand-labeled data, was primarily used to retrieve information from these

R. A. Jha (✉) · V. S. Ananthanarayana
Department of Information Technology, National Institute of Technology Karnataka, Srinivas Nagar, Mangalore, Karnataka 575025, India
e-mail: rjgenuis123@gmail.com



D. Gupta et al. (eds.), *Pattern Recognition and Data Analysis with Applications*, Lecture Notes in Electrical Engineering 888,
https://doi.org/10.1007/978-981-19-1520-8_37

literatures. But, the advent of word embeddings has been a huge relief in this domain as it captures the most complex relations hidden in the dataset and also doesn't require in human labeling or supervision.

One of the most important methods in natural language processing is the assignment of high-dimensional vectors (embeddings) to words in a text corpus in a manner that maintains their syntactic and semantic relationships. Typically, word embeddings are created using machine learning algorithms, like, GloVe or Word2vec. They make use of knowledge about word co-occurrences in a text corpus. The main concept is that since words with common meanings are often used in comparable circumstances, hence, similar embeddings will be used as well.

In this paper, we will make use of these word embeddings to gain valuable insights into finding containment and cure of the coronavirus called SARS-CoV2. In recent years, no other epidemic has caused such extensive medical and economic damage. This is a moment when everyone from around the world, including medical experts, healthcare personnel, and even computer scientists, must band together to combat a mutual enemy. We have made use of GloVe and Word2Vec word embeddings and compared their performances on the COVID-19 dataset [1] available from Kaggle. We get useful information from our findings, and it also suggests that latent information about potential future discoveries, is significantly contained in past papers and publications. Our observations point to the potential of extracting information and relationships from a large body of scientific literature in a systematic way.

## 2 Related Work

Many different approaches are being taken by machine learning researchers and computer scientists all over the world for the same purpose of finding important information about the virus. One of the major challenges in this research is the lack of already existing database regarding the virus. So, all the machine learning researchers and computer scientists are collecting and compiling their own datasets.

A collection of hundreds of CT scans and chest X-ray images has been compiled and distributed by a researcher from the University of Montreal [2]. The photographs come from reports on the disease that are freely accessible. Similarly, Johns Hopkins University has created a remarkable dashboard [3] of well-sourced data that are constantly updated, including a global view of the disease's distribution and mortality. Also, the dataset which has been used in this paper, COVID-19 [1], is available on Google's data science competition platform Kaggle and is updated with new cases daily. Other datasets have come straight from clinics and hospitals that are treating patients and have attempted to rapidly turn around machine learning algorithms to help physicians searching for symptoms of illness.

Deep learning and computer vision are also used extensively in some of the researches on this subject. CT images were analyzed by researchers in Shanghai [4], making use of deep learning and computer vision, which reduced the time of analysis from hours down to four minutes. Some work has also been done on visualizing the virus's effect on the lungs of a patient [5]. Their objective was to track the progress of the virus and illness over time.

Another research [6], which was inconclusive, was done to screen COVID-19 in an auditory manner by analyzing the rate of breathing of a person. Though it was not conclusive, it presented novel approach for checking the presence of virus that is less invasive. Moreover, an algorithm [7] was built by some researchers, predicting the mortality rate for patients, using the electronic health records of almost 3000 patients in Wuhan, China. Their algorithm performed incredibly well, giving an accuracy of more than 90%.

Most of the existing works related to word embeddings are done in the cases of unstructured and large texts. Large text data were processed using Word2Vec, and word similarity was evaluated in. In addition, the similar words were further clustered to fit into a new dimension. The performance of GloVe for network embedding and node representations has been done in.

## 3 Data Specifications

The dataset used in this experiment, COVID-19, has been taken from Kaggle. The dataset used in the experiment is a collection of 47,000 scholarly articles, including 36,000 full texts related to corona virus pandemic all over the world. The full-text scholarly articles in the dataset can be categorized in four types, namely Biorxiv, PubMed Central (PMC), commercial, and non-commercial. All the articles from these categories were combined together to form a large single dataset. The size of the dataset is 6 GB.

## 4 Methodology

We analyze the unlabeled language corpus using unsupervised natural language processing. Certain techniques and methods can be used to do the same. Word2Vec and GloVe are two such techniques. They represent each word in the corpus as a vector, where these vectors are intended to represent word's context (Fig. 1).

Since each word is associated with a vector, these vectors can also be used to represent word relationships and analogies. They can be used to find the similarity

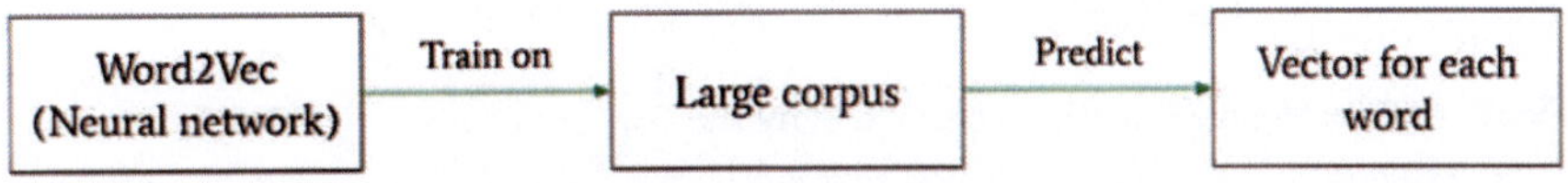


**Fig. 1** Working of word embeddings

between different words. We will use make use of this property of word embeddings to reach our objective.

We will follow the following steps to arrive at our result:

1. We will use WordCloud API available in Python to find the most commonly related words to the study of COVID-19. For example, "virus," "infection," "viral," etc. We will call these words keywords.
2. We will then find the most similar words to these keywords. Euclidean distance will be used to find the vectors closest to the vectors of these keywords.
3. We will again find words similar to the words obtained in the above step (second-order similarity). These words will be our result.

To visualize our result, we will make use of principal component analysis (PCA). Using PCA, vectors of several dimensions are compressed into a smaller vector (with two or three dimensions), while the majority of the information in the original vector is preserved (using some linear algebra). When working with high-dimensional data, PCA makes visualization simpler.

## 4.1 *Word2Vec Architecture*

It is a neural network model that has been trained on billions of tokens. The neural network seeks the best vector representation of each term in the corpus after it has been trained. Word2Vec word embeddings can be obtained using two methods (both involving neural networks): skip-gram and common bag of words (CBOW) (Fig. 2).

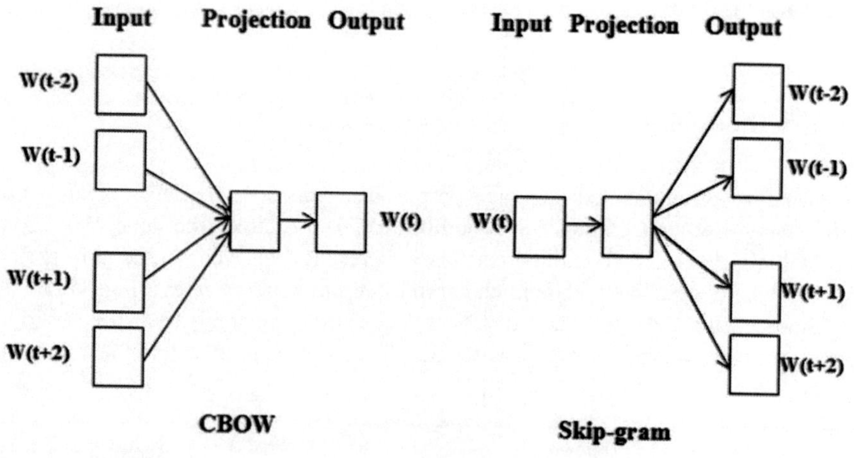


**Fig. 2** Architecture of Word2Vec model—CBOW and skip-gram models

| Probability and Ratio | $k = solid$ | $k = gas$ | $k = water$ | $k = fashion$ |
|---|---|---|---|---|
| $P(k\|ice)$ | $1.9 \times 10^{-4}$ | $6.6 \times 10^{-5}$ | $3.0 \times 10^{-3}$ | $1.7 \times 10^{-5}$ |
| $P(k\|steam)$ | $2.2 \times 10^{-5}$ | $7.8 \times 10^{-4}$ | $2.2 \times 10^{-3}$ | $1.8 \times 10^{-5}$ |
| $P(k\|ice)/P(k\|steam)$ | 8.9 | $8.5 \times 10^{-2}$ | 1.36 | 0.96 |

**Fig. 3** Table of co-occurrence probability of tokens in GloVe

### *4.2 GloVe Architecture*

GloVe is essentially a log-bilinear model with a weighted least-squares objective. The training objective of GloVe is to learn word vectors such that their dot product equals the logarithm of the words' probability of co-occurrence (Fig. 3).

## 5 Results

We used the pretrained models of Word2Vec and GloVe and trained them again on the COVID-19 dataset. We trained them just on abstract and body text of all the papers available in corpus. It was done to find latent meaning in word embeddings to find relevant drugs for COVID-19.

We then looked at the second-order word similarities to the word "antiviral" to find potential measures and cures for COVID-19. We have chosen the word "antiviral" because it is obvious that any cure to the virus will have similar word embeddings to "antiviral." We will look at the findings of both Word2Vec and GloVe. The outcomes have been presented using 2D PCA (Fig. 4).

The above diagram is the result from the Word2Vec model. In the above figure, "antiviral" is a keyword which came from using the WordCloud API in Python. This means that "antiviral" was one of the most common words in the corpus related to the study of COVID-19. We then find the six most similar words related to the keyword, i.e., "antiviral." Those six words come out to be: "daas," "gemcitabine," "repurposed," "antiviral," "anti-hbv," and "anti-parasitic." The vectors of these six words are at closest Euclidean distance to the vector of the word "antiviral." Now, in the next step, we again find the most common words related to the six words (second-order similarity) we found in the previous step. The results of this step are shown in the above figure using 2D PCA, which reduces the number of dimension of a vector and makes it easier for visualization.

"Hydroxychloroquine," "anti-malarial," "doxorubicin," are the most commonly repeating words in the above figure. Also, these three drugs have been widely used in the most of the countries on COVID-19 patients, which prove that the outcomes of these models are correct to some extent. So, we can conclude that Word2Vec gives us useful results (Fig. 5).

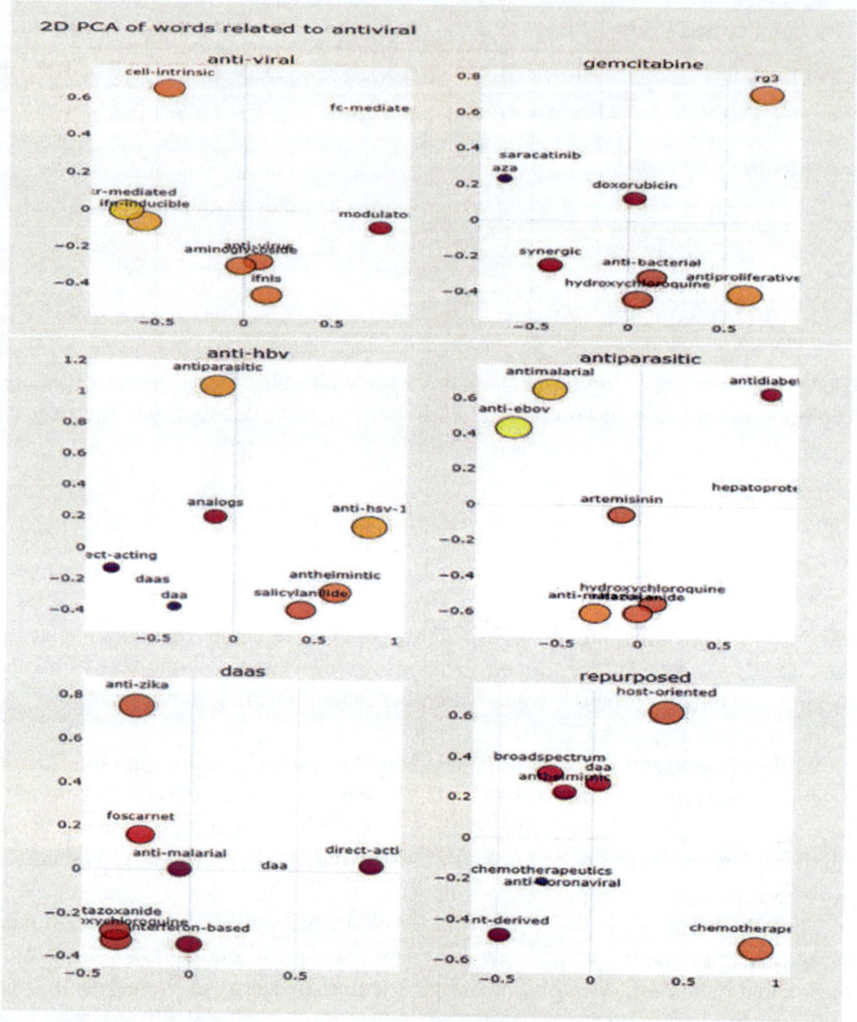


Fig. 4 Outcome from Word2Vec technique which represents the second-order similar words to "antiviral." It can be observed that drugs "anti-malarial" and "hydroxychloroquine" are repeated several times in the results

The above diagram is the result from the GloVe model. The first step here is same as in the previous model. We find the six most similar words related to the keyword, i.e., "antiviral" Those six words come out to be: "anti-virus," "interferon," "evades," "antiviral," "anti-hsv," and "anti-pathogenic." The vectors of these six words are at closest Euclidean distance to the vector of the word "antiviral." Now, in the next step, we again find the most common words related to the six words (second-order similarity) we found in the previous step. The results of this step are shown in the

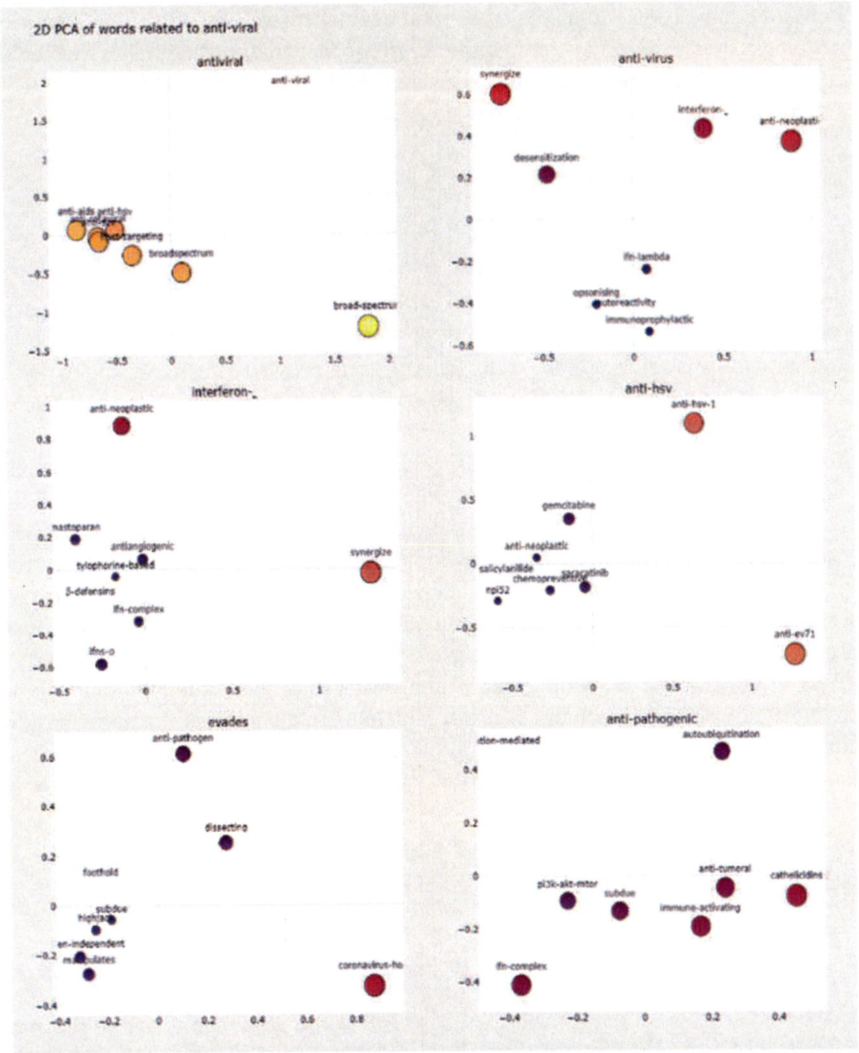


**Fig. 5** Outcome from GloVe technique which represents the second-order similar words to "antiviral." It can be observed that all the drugs' names here are not useful in the treatment of coronavirus. So, this result given by GloVe is not much useful

above figure using 2D PCA, which reduces the number of dimension of a vector and makes it easier for visualization.

All the names of the drugs and medications which came out as the outcome were not related to coronavirus treatment. So, it is correct to assume that the results from the GloVe technique were not as useful and correct.

From the above results, we can see that results from Word2Vec are more useful then results from GloVe. This is because drugs name that are present in the result of

Word2Vec have been actually used to treat against corona virus in some countries before vaccines were discovered, while the drugs that are the outcome of the GloVe are not useful in curing the patients. So, here, Word2Vec performs better than GloVe and gives more useful results and valuable insights.

## 6 Conclusion and Future Work

The objective of this work was to use word embeddings on text corpus so as to get valuable insights regarding the cure of coronavirus. We have seen that knowledge and information available in the literature can be used effectively and expertly encoded as information-dense word embeddings. We used Word2Vec and GloVe word embeddings to carry out this experiment. The outcome we got was several drugs and medications names, such as "anti-malarial," "hydroxychloroquine", and "doxorubicin," which have been used widely on COVID-19 patients all around the world, which proves that the outcomes of these models are correct. Also, we observed that Word2Vec gave more useful results compared to GloVe technique. Another point highlighted by this work is that latent information about potential future discoveries is significantly contained in past papers and publications, as we made use of published articles and texts to predict the potential medications and drugs for COVID-19.

As a future work, we would like to find ways to evaluate the results given by any unsupervised NLP technique. It would help us in quantifying the above results and make the comparison easier. Also, other word-embedding techniques such as FastText, Bert, and Bio-Bert could be used to perform the same analysis.

## References

1. Dataset COVID-19, https://www.kaggle.com/allen-institute-for-ai/CORD-19-research-challenge. Last accessed 2021/04/20
2. Database of CT scans and X-ray images, https://github.com/ieee8023/covid-chestxraydatset/blob/master/README.md?fbclid=IwAR30yTGBr55WXdCngCoICDENHycmdL2bGwlvl1ckdZM-ucjGH10Uakz7khk. Last accessed 2021/04/20
3. Johns Hopkins Dashboard, https://www.arcgis.com/apps/opsdashboard/index.html#/bda7594740fd40299423467b48e9ecf6. Last accessed 2021/04/20
4. Lung Infection Quantification of COVID-19 in CT Images with Deep Learning, arXiv e-prints.2020, https://arxiv.org/abs/2003.04655v2
5. Rapid AI Development Cycle for the Coronavirus (COVID-19) Pandemic: Initial Results for Automated Detection & Patient Monitoring using Deep Learning CT Image Analysis, arXiv e-prints. 2020, https://arxiv.org/abs/2003.05037. Last accessed 2021/04/20
6. Abnormal respiratory patterns classifier may contribute to large-scale screening of people infected with COVID-19 in an accurate and unobtrusive manner. arXiv e-prints. 2020. https://arxiv.org/abs/2002.05534v1. Last accessed 2021/04/20
7. Prediction of criticality in patients with severe Covid-19 infection using three clinical features: a machine learning-based prognostic model with clinical data in Wuhan. https://doi.org/10.1101/2020.02.27.20028027. Last accessed 2021/04/20

# Phrase-Based English–Nyishi Machine Translation

Nabam Kakum and Koj Sambyo

**Abstract** Machine translation (MT) is the sub-domain of natural language processing (NLP). It process and analyze natural language data which eliminates the language inconceivable matters and help to interact among people of different linguistic backgrounds. In our work, we used the statistical machine translation (SMT) approaches to comprehensively explore the translation accuracy, fluency, and adequacy with the context of low resources Indian dialect (language) Nyishi. SMT needs less training time and works well with highly complex long sentences than other methods, but it requires adequate parallel corpus, which is troublesome in the context of low-resource language like Nyishi. In this paper, we train 30,000 newly collected pairs of corpora and measured the translation accuracy in both directions forward and backward. Finally, results of individual n-gram of BLEU and NIST with and without tuning are calculated, and by using these results, we find out the effectiveness of translation accuracy in respect of fluency and adequacy.

**Keywords** SMT · BLEU score · Human evaluation NIST · Moses

## 1 Introduction

MT with its automatic translation method is highly helpful in a multilingual country like India which has 22 scheduled languages. Several researchers tend to work on MT using different methods to eliminate the language barrier issues with the functionality of NLP language translation method and analyze the translated text to obtain better translation accuracy. This work is an experiment to contribute an intelligible translation for the low-source Indian language which is used by the Nyishi people of Arunachal Pradesh. Arunachal Pradesh is inhabited by one of the most culturally and linguistically diverse communities. The ethnic composition of the tribes predominantly belongs to the mongoloid stock. Arunachal Pradesh is the home of 26 major tribes, and more than 100 sub-tribes, Nyishi is considered to be the largest tribe of the

N. Kakum · K. Sambyo (✉)
Department of Computer Science and Engineering, NIT Arunachal Pradesh, Jote, India
e-mail: nabam.phd19@nitap.ac.in



D. Gupta et al. (eds.), *Pattern Recognition and Data Analysis with Applications*, Lecture Notes in Electrical Engineering 888,
https://doi.org/10.1007/978-981-19-1520-8_38

state with a population of 300,000 approximately from the total population of 13.82 lakhs. Generally, the languages in Arunachal Pradesh belong to the Sino-Tibetan language family and more specifically under the Tibeto-Burman group of languages. This paper attempts to expose low-resource Nyishi Language into machine translation environment by using the properties of NLP with SMT. Many highly resourced languages are already benefitted in almost perfect translation in MT, but the language of the low resource like Nyishi might take many years to advance into MT communities. Nyishi doesn't have a written script which is the main cause of late development and evolvement into MT communities. According to the Worlds languages in Danger (2009) by UNESCO atlas, more than 26 languages of Arunachal Pradesh have been identified as endangered languages as well Nyishi language comes under definite danger language.

At present, several MT approaches such as rule-based machine translation (RBMT), statistical machine translation (SMT), NMT, and transformer model are available. Many researchers consider transformer model as a state-of-the-art model in the automatic translation world as it provides a high accuracy rate compared to all other approaches. In this paper, by the advantages of less training time and better translation accuracy in complex data, we attempt to use SMT using MOSES. MOSES is a machine translation package (toolkit) that is user-friendly and easily understandable by any user and finally evaluates the translation accuracy and adequacy of the corpus. We have used the BLEU score to evaluate the quality of translation [1, 2].

## 2 Related Works

The Nyishi language lacks research work in MT since this language doesn't have record of any previous documentation. We choose to use phrase-based MT method for the low-resourced Nyishi Language as this approach helps in meaning conservation rather than the structural conservation. The resource we used is manually translated as it is low resource, and we tend to analyze the data with the model whose priority is to recommend the meaning of the language. Several research works have been done on different low-resource languages like English–Hindi, Hindi–Nepali, English–Tamil, English–Mizo (Amarnat et al. 2018), English–Punjabi, Hindi–Nepali, and Punjabi–Hindi by using SMT; related works provide the better translation accuracy. SMT overcomes the problem of the rule-based system by supporting the n-gram model, where basic steps are finding the maximum probabilities of phrase pairs in parallel source–target sentences. Finally, BLEU is used to calculate the automatic metrics of the predicted translation [3–6].

### 2.1 SMT for Low-Resource Language

Regardless of the word length, phrase-based MT reduces the restriction of word-based translation by translating the whole sequence of words. The sequence of the words is called blocks or phrases. Among several machine translation approaches, we chose to use SMT because it doesn't depend on language attributes. This model supports training data on the small and large dataset as well as helps in preserving meaning of the language better; increasing of dataset will eventually provide better translation results. The SMT uses the Bayes theorem for probability distribution [7].

**Bayes rule**:

$$\begin{aligned}\tilde{e} &= \arg\max_{e} \frac{p_{TM}(n|e)\,p_{LM}(e)}{p(n)} \\ &= \arg\max_{e} p_{TM}(n|e)\,p_{LM}(e)\end{aligned}$$

where

- Translation model $p_{TM}(n|e)$ is the probability where the source string is the translation of the target string.
- Language model $p_{LM(e)}$ is the probability used for distinguishing that target language string.
- Where $\tilde{e}$ is the best translation concluded by choosing the one that provide highest probability?

### 2.2 Moses in PBSMT

Phrase based and tree based are the two translation models that come with Moses toolkit; in this paper, we use phrase-based approaches to obtain the translation accuracy on the parallel corpus. Moses supports several languages modeling toolkits, such as SRILM, KenLM, IRSTLM, and RandLM, and also allowed to introduce a new toolkit. GIZA supports word aligning for given parallel corpus and language model estimation. The decoder of Moses has several procedures such as chart parsing and cube pruning. In our paper, we use MGIZA for alignment, cube pruning for identifying the number of hypotheses to be covered; IRSTLM and RandLM are used for language modeling.

## 3 System Descriptions

This paper analyzes the translation accuracy by using the SMT with phrase-based approaches, and other parts of the paper are as follows: corpus preparations, language model, translation model, decoding, experiment part, and results, which are evaluated by BLEU, NIST; and finally, human evaluation is done to calculate the overall translation accuracy and adequacy.

### 3.1 Corpus Preparations

The parallel aligned sentence pair of 30,000 with both backward and forward direction has been developed, where English is the source and Nyishi is the target language and vice versa. Due to low resources and lack of script in this language, the preparation of corpus is complex and uncertain. The collected parallel sentences of Nyishi have been verified thoroughly by a native of Nyishi language experts. In PBSMT, corpus preparation includes tokenization, true casing, and cleaning. True casing helps to handle the difference between uppercase and lowercase words and finally convert all to lowercase forms. The cleaning step drops down the long and unwanted sentence along with the threshold exceeding chunk.

### 3.2 Language Model

Language models in SMT reorder the words/phrases that are suggested by the translation model to generate the target language. Sentence probability is calculated by the probability method in the SMT language model, using the n-gram model. Computation of the probability in the language model is based on single word that provides all the words that lead to make a whole sentence [8]. IRSTLM tool is used to develop the language model. Language modeling in PBSMT helps to break the probability of a sentence $P(S)$ with the probability of individual words $P(W)$ shown in the below equation:

$$
\begin{aligned}
P(S) &= P(W1, W2, W3, \ldots, Wn \\
&= P(W1)P(W2|W1)P(W3, |W1W2) \\
&\quad P(W4|W1W2W3) \ldots P(Wn|W1W2 \ldots Wn - 1)
\end{aligned}
$$

Initially, the probability of a word in a sentence is calculated following which the probability of sentences is calculated, which gives the orders of word proceeding to it and n-gram model is used to generate the probability approximation of all the previous words in the sentences [5].

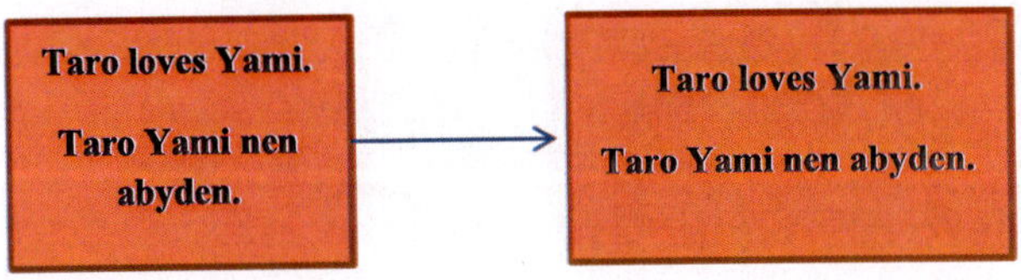


**Fig. 1** Word alignment with MGIZA

## 3.3 *Translation Model*

The main objective of the translation model is to advise a set of possible words/phrases for a given source sentence and compare the expression of meaning between languages. This model is known as the translation model because of its expression comparing properties between different languages. It shows that a large quantity of the corpus will produce a high translation score than the low resources dataset as MT supports the better result of translation accuracy with a large dataset. Alignment model building between the input and the output language is the first step of translation models [9]. In every sentence of the training parallel corpus, the job of providing the best set of alignment links is handled by the alignment models. The MGIZA in Moses toolkit has been used to align our Nyishi–English and vice versa training corpus. The word-align text in alignment model in SMT is necessary, as it helps to initialize the translation model in machine translation. Such as, finding words that correspond to each other performed automatically with the probabilistic method as elaborated below [10] (Fig. 1):

$$P(\text{Yami}|\text{Yami}) = 0.99, \quad P(\text{Taro}|\text{Taro}) = 0.97.$$
$$P(\text{loves}|\text{abyden}) = 0.46, \quad P(\text{loves}|\text{nen}) = 0.04.$$

## 3.4 *Decoding*

The main function of the decoder is to identify the better candidate translation and to search for the hypothesis which contains the best model score. The output of the LM, TM, and the input for the decoder are the source sentence. In the decoding process, computational complexity is high. We used the beam search approach for searching strategy in the decoder which supports heuristic-based algorithm strategies. Decoder generally uses mathematical approach of most probable translation, for which probability of an individual-targeted word/phrase is maximum. Figure 2 shows the PBSMT system conceptual architecture [5].

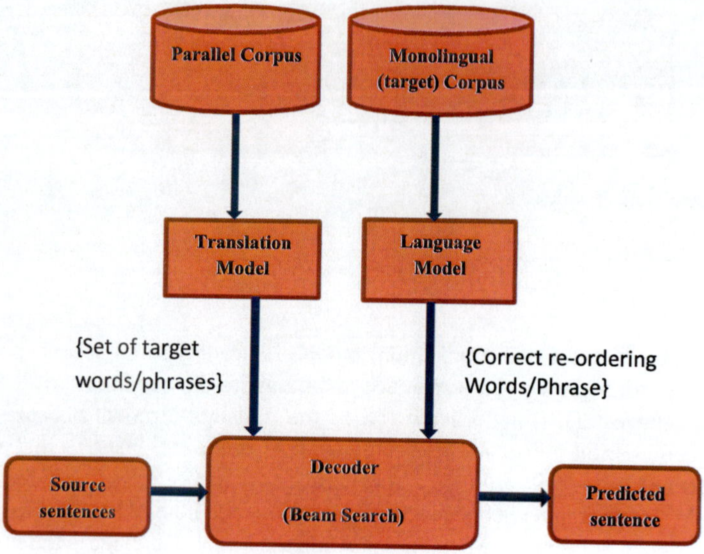


**Fig. 2** Abstract architecture of PBSMT system [5]

### *3.5 PBSMT Experimental Design for English–Nyishi*

The newly collected parallel aligned corpus of 30,000 of the English–Nyishi is divided into three parts as training, tuning, and testing data, and English sentence of the monolingual file has been used for building the language model. The language model is built with the IRSTLM toolkit in the Moses that is used to assist data structures and algorithm, where language models are suited for accessing and storing large n-gram data. Data used are shown as in Table 1.

## 4 Results and Analysis

Results that are generated from the SMT systems have been calculated by the BLEU metric, and human evaluation has also been performed for the same.

**Table 1** Corpus statistics

| Corpus type | English–Nyishi | Nyishi–English |
|---|---|---|
| Training data | 28,033 | 28,033 |
| Tuning data | 967 | 967 |
| Testing data | 1001 | 1001 |

### 4.1 About BLEU and NIST Score

BLEU used in MT is an algorithm that attempts to provide the accuracy of the text translation by a machine-translated output. The output of the BLEU is always between 0 and 1, whereas NIST ranges from 1 to 10. BLEU/ NIST metrics only depend on precision. The score of BLEU/NIST is calculated from n-grams precision [25]. Test set is required for BLEU/NIST metrics machine translation system. The equation of the BLEU score for n-gram is calculated by using the following formula:

$$\min\left(1, \frac{\text{candidate} - \text{length}}{\text{reference} - \text{length}}\right)\left(\prod_{i=1}^{n} \text{precision}_i\right)^{\frac{1}{n}}$$

Here, candidate-length stands for candidate translation length, and the reference-length stands for reference translation length, and also precision I denotes precision score [25] for *i*th gram match. NIST and BLEU identify the translation accuracy with tuning and without tuning which clearly shows that with tuning in both the translation accuracy is better than without tuning that is because tuning maximizes the performance of translation on a small set of parallel sentences by finding the optimal weights for this linear model (Tables 2 and 3).

The individual n-gram score generated from the model shows that results are comparatively better in unigram which means translation accuracy is high in short sentences compare to medium and long sentences (Table 4).

The score generated by English–Nyishi and vice versa is low compared to other high-resources language due to the high ambiguity of Nyishi words.

**Table 2** Individual n-gram by BLEU score

| No. | Corpus type | BLEU score without tuning | | | BLEU score with tuning | | |
|---|---|---|---|---|---|---|---|
| | Individual n-gram | 1-g | 2-g | 3-g | 1-g | 2-g | 3-g |
| 1 | English–Nyishi | 0.3624 | 0.1357 | 0.0744 | 0.3798 | 0.1792 | 0.1184 |
| 2 | Nyishi–English | 0.3497 | 0.1479 | 0.0988 | 0.3699 | 0.1937 | 0.1402 |

**Table 3** Individual n-gram by NIST score

| No. | Corpus type | NIST score without tuning | | | NIST score with tuning | | |
|---|---|---|---|---|---|---|---|
| | Individual n-gram | 1-g | 2-g | 3-g | 1-g | 2-g | 3-g |
| 1 | English–Nyishi | 1.6744 | 0.2717 | 0.0213 | 2.3398 | 0.4510 | 0.0534 |
| 2 | Nyishi–English | 2.2148 | 0.3617 | 0.0571 | 2.3660 | 0.4861 | 0.0820 |

**Table 4** N-gram scoring English–Nyishi

| No. | Corpus type | NIST score without tuning | BLEU score without tuning | NIST score with tuning | BLEU score with tuning |
|---|---|---|---|---|---|
| 1 | English–Nyishi | 1.9700 | 0.0802 | 2.8512 | 0.1411 |
| 2 | Nyishi–English | 2.6408 | 0.1419 | 2.9444 | 0.1849 |

### 4.2 *Human Evaluation*

Human evaluation evaluates the adequacy and fluency as well as the overall rating of translated results. Human evaluation is necessary for MT as BLEU scores fail to meet the quality of translated score in terms of adequacy and fluency; these two factors help to identify the translation quality of MT. Adequacy helps to identify the total number of similar meanings between reference translation and candidate translation. Fluency is measured by the translated sentence of reference sentence irrespective of candidate sentence, where fluency is considered when reference sentence is translated fluently. As the BLEU fail to enclose all the features of evaluating the candidate translation. The output of quality translation is measured with the reference translation; the basic criteria of measurements are totally on fluency and adequacy. In this experiment, Nyishi language expert is from (NBCC) Nyishi Baptist church council and a native of Nyishi, who have a great knowledge of the Nyishi language and also a core in charge of Nyishi language development in all Nyishi elite society and other who contributed in translation. The core of the Nyishi language and where the translation is to be used is well conscious of the experts; however, the human evaluation process is very expensive, time-consuming.

## 5 Analysis of Translation

To analyze the translation quality from the output of the SMT system in both directions, we have used few sample sentences from translated output where the quality is judged from various issues like under-translation, over-translation, and wrong translation of name entities. Therefore, the selected translated sentence has been judged on two factors, i.e., adequacy and fluency where system translations are examined in case of best or worst. Although corpus is analyzed in both the direction but most common error that we encounter are mixed translation of source word and target word, **English Gold (EG)** in English is the reference sentence, **Nyishi Test (NT)** as a test sentence, **and English Predicted (EP)** in English the predicted sentence.

1. Best performance of translation which is adequate and fluent.
   **NT: "Svka"**
   **EG: "Assist"**
   **EP: "Help"**

2. The system has predicted the unigram and tri-gram correctly "I, him," and therefore, it is not *adequate* but is *fluent.*

   **NT: "Ngo mwvn kapapa"**
   **EG: "I saw him"**
   **EP: "I met him"**

3. Prediction result with *partially adequate* and *perfectly fluent* translation.

   **NT: "Nyem ko"**
   **EG: "daughter"**
   **EP: "Girl"**

4. Prediction result predicted as neither *fluent* nor *adequate.*

   **NT: "Tom danypa"**
   **EG: "Tom is simple"**
   **EP: "Tom is ordinary"**

5. Prediction result that is *adequate* as well as *fluent* and automatically adds "7" which is numerically predicted from sentences.

   **NT: "Hv sija kvn paku"**
   **EG: "It is seven now"**
   **EP: "Now it is 7"**

6. The SMT system has made a prediction that mixed word of target and source to the predicted sentence.

   **NT: "No Tom nen svka numyv?"**
   **EG: "Did you help Tom?"**
   **EP: "You numyv let tom help?"**

From the above analysis of the translations predicted by the PBSMT system, human evaluation is done to evaluate the translated results in factors of adequacy and fluency. Further, it shows that score degrades as the sentence length increases, and scores are comparatively better in short and medium sentences than of long sentences.

## 6 Conclusion and Future Work

MT is a communication tool to bridge among the people belonging to the different linguistic backgrounds around the world. This work attempts to introduce the low-resource language Nyishi into a machine translation environment with few corpora, and we successfully analyzed by using SMT; furthermore, we intend to increase our corpus to get better prediction result as it requires large corpus to provide better translation using MT methods. To minimize the error, we aim to identify more statistical approaches as well as the neural method to improve adequacy and fluency.

## References

1. Dey, M.: Negation in Nyishi. NEHU J. **15**(2), 79–100 (2017)
2. John, S.S., Lomdak, L.: Language Endangerment in Arunachal Pradesh: Current Issues and Future Prospects. Centre for Endangered Languages (CFEL) Rajiv Gandhi University (2017)
3. Laskar, S.R., Dutta, A., Pakray, P., Bandyopadhyay, S.: Neural machine translation: English to Hindi. In: IEEE Conference on Information and Communication Technology, pp. 1–6 (2019)
4. Laskar, S.R., Dutta, A., Pakray, P., Bandyopadhyay, S.: Neural Machine Translation: Hindi⇔Nepali. In: Proceedings of the Fourth Conference on Machine Translation (WMT), Vol. 3 Shared Task Papers, pp. 202–207. Association for Computational Linguistics, Italy (2019)
5. Pathak, A., Pakray, P., Bentham, J.: English-Mizo Machine Translation using neural and statistical approaches. Neural Comput. Appl. **31**(11), 7615–7631 (2019)
6. Rushanti, K., Sambyo, K.: Phrase-based machine translation of Digaru-English. In: Electronic Systems and Intelligent Computing, pp. 983–992. Springer (2020)
7. Graham Neubig Homepage, http://www.phontron.com/slides/building-smt-en-20120510.pdf. Last accessed 2021/03/11
8. Baruah, K.K., Das, P., Hannan, A., Sarma, S.K.: Assamese-English Bilingual Machine Translation. Int. J. Nat. Lang. Comput. (IJNLC) **3**(30), 73–82 (2014)
9. Zin, T.T., Soe, K.M., Thein, N.L.: Translation model of Myanmar phrases for statistical machine translation. In: International Conference on Intelligent Computing, pp. 235–242. Springer, Heidelberg (2011)
10. Hu, Y.: Statistical machine translation based on translation rules. J. Chem. Pharm. Res. **6**(7), 1628–1635 (2014)
11. Romdhane, A.B., Jamoussi, S., Hamadou, A.B., Smaïli, K.: Phrase-based language model in statistical machine translation. Int. J. Comput. Linguist. Appl. (2016)
12. Raghavendra, U.U., Tanveer, A.F.: An English-Hindi statistical machine translation system. In: International Conference on Natural Language Processing 2004, IJCNLP, pp. 254–262. Natural Language Processing (2004)
13. Cho, K., Merriënboer, B.V., Bahdanau, D., Bengio, Y.: On the properties of neural machine translation: encoder-decoder approaches. In: Proceedings of SSST-8, Eighth Workshop on Syntax, Semantics and Structure in Statistical Translation, pp. 103–111. Association for Computational Linguistics, Doha (2014)
14. Kishore, P., Roukos, S., Ward, T., Zhu, W.J.: Bleu: a method for automatic evaluation of machine translation. In: Proceedings of the 40th Annual Meeting of the Association for Computational Linguistics, pp. 311–318. Association for Computational Linguistics (2002)
15. Denkowski, M., Lavie, A.: Meteor universal: language specific translation evaluation for any target language. In: Proceedings of the Ninth Workshop on Statistical Machine Translation, pp. 376–380. Association for Computational Linguistics, Baltimore (2014)
16. Koehn, P., Hoang, H., Birch, A., Burch, C.C., Federico, M., Bertoldi, N., Cowan, B., Shen, W., Moran, C.: Moses: open source toolkit for statistical machine translation. In: Proceedings of the ACL 2007 Demo and Poster Sessions, pp. 177–180. Association for Computational Linguistics, Prague (2007)
17. Dugonik, J., Boskovic, B., Maucec, M.S., Brest, J.: The usage of differential evolution in a statistical machine translation. In: IEEE Symposium on Differential Evolution (SDE), pp. 1–8. IEEE, Orlando (2014)
18. Nabhan, A.R., Rafea, A.: Tuning statistical machine translation parameters using Perplexity. In: IRI—IEEE International Conference on Information Reuse and Integration, IEEE, Las Vegas (2005)
19. Vedantam, R., Zitnick, C.L., Parikh, D.: Cider: consensus-based image description evaluation. In: Proceedings of the IEEE Conference on Computer Vision and Pattern Recognition, pp. 4566–4575. IEEE, Boston (2015)
20. Ganguly, D., Leveling, J., Jones, G.J.F.: Bengali (Bangla) information retrieval. Technical Challenges and Design Issues in Bangla Language Processing, IGI Global (2013)

21. Faili, H.: An experiment of word sense disambiguation in a machine translation system. In: International Conference on Natural Language Processing and Knowledge Engineering, pp. 1–7. IEEE, Beijing (2008)
22. Sudhahar, S., Cristianini, N.: Detecting shifts in public opinion: a big data study of global news content. In: International Symposium on Intelligent Data Analysis, pp. 316–327. Springer Cham (2018)

# Capacitated Vehicle Routing Problem Using Genetic Algorithm and Particle Swarm Optimization

Vrinda Sharma and Niharika Varshney

**Abstract** Vehicle routing problems (VRPs) have several applications in logistics such as transportation and supply chain where finding an optimal route would not only help in saving the resources but will also provide profit as reduced journey time and distances, causing an overall reduction in overhead cost. In this paper, a CVRP is solved using GA and PSO. In GA, tournament selection, one-point crossover and swap mutation operators are used. To convert the real-valued vector into discrete solution for PSO, the random key method is employed. The proposed algorithms were tested on 74 benchmark instances. For all 74 benchmark instances, GA achieves better outcomes as compared to the PSO regarding total travelled distance.

**Keywords** Vehicle routing problem · Genetic algorithm · Particle Swarm Optimization · Heuristics · Optimization

## 1 Introduction

The vehicle routing problem (VRP) deals with routing of different vehicles in which we have set of paths assigned to them to serve customers as per their respective demands. The main aim of VRP is to achieve minimized distance travelled by vehicles to serve "n" customers present at different coordinates that directly results in lowering of cost to serve all the customers.

Capacitated Vehicle Routing Problem (CVRP) is originally explained by Dantzig and Ramser in 1959 [1]. The definition of CVRP can be formally stated as [2, 3]. The problem is concerned with the vehicles fixed in numbers with uniform capacity constraints imposed on them. All the vehicles start their journey from a common depot and they should serve according to the known demands of the customer (that are

V. Sharma (✉)
Department of Computer Science, AMU, Aligarh, India
e-mail: vrinda.sharma.amu@gmail.com

N. Varshney
Department of Computer Applications, GLBITM, Greater Noida, India



D. Gupta et al. (eds.), *Pattern Recognition and Data Analysis with Applications*, Lecture Notes in Electrical Engineering 888,
https://doi.org/10.1007/978-981-19-1520-8_39

non-negative [4]) at minimum capital possible. The distance between two locations can be calculated easily as the customers and depot locations are known in advance. CVRP is considered as a Non-Deterministic Polynomial hard problem in which long operational time is required to achieve optimized solutions. Henceforth, more evolved computing approaches have been applied to CVRP so that we can achieve an ideal solution in a short span of time [5].

Heuristic algorithms are those algorithms which are used in place of classic methods to solve complex problems quickly and efficiently. They are also reliable when classical methods fail in achieving exact solution by providing approximating solutions. Heuristics are mostly used to solve NP-hard problems. They are used in various fields like computer science, machine learning and mathematical optimization. Moreover, they do not guarantee about achieving exact solution [6]. For example, genetic algorithm is a type of heuristic algorithm which is applied in this research to solve Capacitated Vehicle Routing Problem to achieve near-optimal routes. Also, we used another optimization algorithm that can identify an optimal solution in an acceptable time frame which is Particle Swarm Optimization (PSO), it is Population-based Swarm Optimization which is an iterative algorithm that engages a number of simple entities—particles (solutions)—iteratively over the search space of objective functions. In PSO, particle represents a solution, and a decoding method is used to change it into problem specified solution [5].

## 2 Related Work

There had been several studies and research in this area and many researchers have contributed to optimize and improve this transportation problem by using and applying different algorithms with many improved variants and techniques. Several related works and the research in this area are as follows—Nora Niazy (2020) works on solving Capacitated Vehicle Routing Problem using Chicken Swarm Optimization with genetic algorithm, their approach is to use the hieratical order of Chicken Swarm Algorithm to find paths after using the moving equations. Then they will rearrange the hieratical order according to the paths cost. In an attempt to improve results for some chickens, they then used the genetic algorithm because it has the advantage that it searches in the neighbourhood to find the best solution [7]. Thibaut Vidal (2020) introduced a HGS specialized to CVRP [8]. Farida Ramadhani (2021) solves Vehicle Routing Problem using Particle Swarm Optimization with Fuel Consumption Minimization [9]. Yousefikhoshbakht (2021) works on a hybrid-modified version of PSO algorithm which was presented to solve the CVRP problem, their results were compared with other meta-heuristic algorithms. Wisittipanich W (2021) works on Particle Swarm Optimization (PSO) and differential evolution (DE), and delivery routings with minimum travel distances were find [10]. Stanley Jefferson (2020) shows an analysis of the influence of different encoding schemes on the population behaviour when genetic algorithm is applied to the CVRP [11]. Sajid (2020) works on solving Capacitated Vehicle Routing Problem using hybrid genetic and

**Table 1** Summary of recent literature survey

| Year | Authors | Problem type | Algorithm used |
|---|---|---|---|
| 2020 | Niazy et al. [7] | CVRP | Chicken Swarm Optimization with genetic algorithm |
| 2020 | Vidal [8] | CVRP | Hybrid genetic search algorithm |
| 2021 | Ramadhani et al. [9] | VRP with fuel consumption minimization | Particle Swarm Optimization |
| 2021 | Yousefikhoshbakht [16] | VRP | Hybrid improved particle swam optimization |
| 2021 | Wisittipanich et al. [10] | Postman delivery routing problem | Particle Swarm Optimization and differential evolution algorithms |
| 2020 | Lima et al. [11] | CVRP | Genetic algorithms with different encoding schemes |
| 2020 | Sajid et al. [12] | CVRP | Hybrid genetic and simulated annealing algorithm |
| 2020 | Sbai et al. [13] | CVRP with 2D loading constraint | Genetic algorithm |
| 2019 | Lin et al. [14] | CVRP | Order-aware hybrid genetic algorithm |
| 2021 | Daham et al. [15] | VRP with backhauls | Genetic algorithm |

simulated annealing algorithm, the proposed hybrid genetic and simulated annealing combined (HGSA) algorithm employs novel nearest-neighbour crossover operator which generates solutions based on nearest-neighbours to obtain minimum possible distance [12]. Ines Sabai (2020) proposed a new heuristic based on an adaptive genetic algorithm for solving the CVRP with 2D loading constraint [13]. Na Lin, Yanjun Shi, et al. worked on the CVRP in the IoT by using an effective order-aware hybrid genetic algorithm [14]. Hajem Ati Daham, Husam Jasim Mohammed worked on the vehicle routing problems with backhauls [15] (Table 1).

## 3 Problem Formulation

In VRP, group of uniform vehicles are represented by "$v$". A digraph "$g$" = $(C, a)$, consisting a group of consumers, $C$. The exiting store is shown by the vertex 0, and the returning store is shown by the vertex $m + 1$ [17]. The set $N$ having $m$ vertices denotes customers. The edge set "$a$" represents all the edges through the vertices (which includes the store vertex also). At vertex 0, no edge will end and at vertex $m + 1$, no edge will begin. All paths will begin from 0 and finish at $m + 1$ [17]. A cost variable "$C_{i,j}$" is assigned with each edge $(i, j) \in a$. All the vehicles have alike capacity restraint "$Q$". Each consumer, $i$, has a demand $d_i$, $i \in C$.

This mathematical model has only one conclusion variable "X". The conclusion variable $X_{ijk}$ equates to 1 for edge $(i, j)$ if each vehicle "k" travels from a vertex "i" to "j" where $i$ is not equal to $j$, $i$ is not equal to $m + 1$ and $j$ is not equal to 0 (if not then $X_{ijk} = 0$). Main objective of VRP is to deliver required commodities using $V$ vehicles to all the customers $C$ by covering as minimum distance as possible with some constraints imposed [17], such as vehicle capacity constraint is considered, each consumer is only served at one time, vehicle initiates its journey from vertex 0 and returns to vertex $m + 1$. This CVRP model is numerically formulated below

$$\text{Minimum} \sum_{k \in v} \sum_{i \in N} \sum_{j \in N} C_{ij} X_{ijk} \tag{1}$$

$$\sum_{k \in v} \sum_{j \in N} X_{ijk} = 1 \text{ for all } i \in C \tag{2}$$

$$\sum_{i \in C} d_i \sum_{j \in N} X_{ijk} \leq Q \text{ for all } k \in v \tag{3}$$

$$\sum_{j \in N} X_{ojk} = 1 \text{ for all } k \in v \tag{4}$$

$$\sum_{i \in N} X_{ihk} - \sum_{j \in N} X_{hjk} = 0 \text{ for all } h \in C, \text{ for all } k \in v \tag{5}$$

$$\sum_{i \in N} X_i, m + 1, k = 1 \text{ for all } k \in v \tag{6}$$

$$X_{ijk} \in \{0, 1\} \text{ for all } i \in N, j \in N, \text{ for all } k \in v \tag{7}$$

where vehicle set "v" = $\{1, 2, \ldots, k\}$ and customer set "C" = $\{1, 2, \ldots, m\}$.

$0, m + 1$ vertex, $d_i$ is client $i$ demand, and $q_k$ is vehicle $k$ capacity.

# 4 Proposed Algorithms

## 4.1 Genetic Algorithm

The main purpose of GA is to produce new populace or generation from the parent or preceding generation. This latter generation is better than the former one in many aspects as it comprises high-quality solutions. All the generations have solutions which are known as "chromosomes". All the chromosomes are unique within each generation. These represent the travelling network of vehicles involved and comprised of "genes". Genes represent the total number of customers that are being served in our problem.

## 4.2 VRP Using GA

Initially, we generate a population of "$n$" customers randomly in the form of a matrix of size $n \times n$. In that matrix, each row has customers starting from 1 to $n$, none of them would repeat itself. Each customer represents a node and have some distance "$d$" related to it from the depot. Along with the distance, each customer has associated demand. Now, this demand would be fulfilled by number of "$k$" vehicles. Each vehicle has some identical capacity "$q$". Each and every customer would be served once. No two vehicles can visit the same customer. From the matrix, we will select each row that represents one "chromosome" and we will calculate the distance travelled by vehicles after visiting customers and returning back to depot. There are "$n$" customers, so initially, we have "$n$" chromosomes and "$n$" distances related with them. One chromosome represents one solution of VRP (Fig. 1).

The selected chromosomes are chosen so that their genes are passed to the new generation and hence new population becomes more fit with the help of tournament selection. Every chromosome has an associated fitness. It is evaluated by the summation of distances between adjacent genes and the chromosome having the minimum distance will be treated as the fittest amongst all the other chromosomes as there is a constraint in our vehicle routing problem which states that "We are searching for those routes which serve all the customers or cities with the least distance travelled or

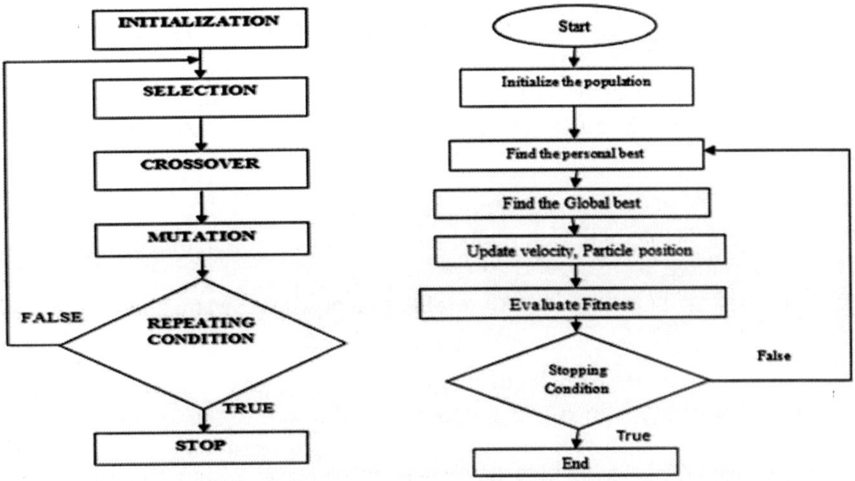


Fig. 1 GA and PSO flow chart

we can say the optimized ones." Two chromosomes will be selected from the population pool (random population), and we apply operators of GA. There are two GA operators: crossover and mutation. In our population, we use one-point crossover. So, two chromosomes are split at one point each and then combined from start of one to the end of other. From two chromosomes, we get two new chromosomes. We will make one new crossover population. Similarly, there are several ways of applying mutation. After crossover, we apply swap mutation. In swap mutation, two genes in the same chromosomes are swapped to make a new population. After mutation, we discard the previous crossover population and apply the same algorithm again and again up till three or more times. Every time we combine new population and previous (random) population and sort it in ascending order and consider the first "*n*" solutions which have minimum distance.

The solution with minimum distance having low cost of transportation will be considered as the best solution out of the "*n*" solutions. This solution can be one of the solutions from random population or from the population generated by genetic algorithm.

### *4.3 Particle Swarm Optimization*

PSO was designed in 1995 as a stochastic optimization algorithm by Kennedy and Eberhart, a computational method that relies on duplicating human social behaviour in different situations and habitat, i.e. the social simulation model [18]. The Algorithm has search points moving in any random motion throughout the whole population, retaining in their memory the best experience or position that they attained. Their experience then interacts with part or whole of the population, on basis of which the movement is shifted to the most favourable and reassuring regions, so far, searched. As the name suggests, here the motivational factor is group of birds, i.e. "Swarm." The particles in PSO moves around the search space in a motion similar to that of Swarm of birds, these particles progress in a guided direction which is determined by

- The earlier velocity of particle termed as inertia.
- Distance of the particle from its best position found so far which is known as Cognitive force.
- Distance of particle from best-known location of whole Swarm known as Social force [19].

It is important for the particles to interact with each other so that they can explore the search area faster which will supposedly find a much superior solution [19]. Swarm's direction can be changed at any point during the searching, if the particle's own best position comes out to be better than the whole Swarm's best position.

## 4.4 VRP Using PSO

In PSO, a particle is treated as a solution. These particles when combined forms the Swarm. Initially, random solutions are created over search space. Mathematically, a particle has two vectors, position vector and velocity vector [20].

*Position Vector.* It shows position matrix of particles and is represented by, say, $P$ [particle, $N$] where $N$ represent no. of customers. It depicts the current position of particle (solution). Initially, the number of particles for given $N$ customers can be obtained using the given Eq.

$$P(i, j) = \text{rand1} * \text{rand2} \tag{8}$$

where rand1 is random number in uniform form lying in between zero and 1 and rand2 is uniform random integer number between −30 and 30.

*Velocity Vector.* Velocity is required to update the position on each iteration. It represents the velocity matrix and is represented by say, $V$ [particle, $N$]. The velocities of each particle are initialized using the equation,

$$V(i, j) = \text{rand1} * \text{rand2} \tag{9}$$

where rand1 is random number in uniform form lying in between zero and 1 and rand2 is uniform random integer number between −30 and 30.

*Particle's personal best matrix.* The algorithm generates the personal best (Pb) matrix of order particle $X$ $N$. This matrix consists of the best position values for each particle searched so far. The Pb for each particle is initialized with the initial value of position of the particle. In successive iterations, the value of Pb changes according to the best position of the particles.

*Global best matrix.* This matrix shows the particle having the best position in the whole Swarm so far, i.e. the best particles searched amongst Pb. Initially, the best position amongst all the personal best is the global best, i.e. Gb. In successive iterations, the Gb changes its value as per the position of the best particle found in the swarm. After initializing, algorithm executes all the steps in while loop until stopping condition does not meet. Particle's new velocity is determined by taking in account, the particle's personal best and the Gb by employing the following Eq.

$$\begin{aligned} V(i, j) = {} & C1 * r * (Pb(i, j) - p(i, j)) \\ & + C2 * r * (Gb(i) - P(i, j)) + w * V(i, j) \end{aligned} \tag{10}$$

where $W$ = inertia constant and $W$ belongs to [0.4, 0.9] with the increment of (0.9–0.5), $C1$ = Cognitive force and $C2$ = Social force.

The widely accepted values of $C1$ and $C2$, i.e. ($C1 = C2 = 2.05$) has been used in this study. The "$r$" belongs in the range of 0 and 1. Next, updated particle's position is given by,

$$P(i, j) = V(i, j) + P(i, j) \tag{11}$$

*Particle-Customer Sorting.* Since the particle in PSO has continuous values and VRP needs a path representation, a method is developed to change the continuous values to values which are discrete in nature so that we can determine the task to particle-vehicle mapping [21]. A sorting method is used to get solution on every iteration where customers have their respective position vectors. Particle's positions are arranged in an increasing manner and their correlated customers, i.e. current solution set is obtained through it.

*Evaluating Fitness Value.* After we obtain customer set from sorting their particles' position, we assign the vehicles as per the demand of each customer and the total capacity of each vehicle, keeping in note that, total customer's demand cannot exceed the capacity of the vehicle, for that, a new vehicle will be assigned. Assignment of vehicles to their respective customers forms a cluster set which is actually the number of customers assigned to each vehicle. After finding required vehicles and cluster set, we calculate the particles' fitness value which is the distance travelled during whole route network. Similarly, the remaining particles' fitness value can be evaluated on every iteration, and the stopping criterions are checked against their conditions. If they are satisfied, the loop is terminated; otherwise, whole loop is repeated until the stopping criterions do not fulfil.

## 5 Experimental Study

The GA and PSO algorithms are applied, respectively, to the three CVRP's data set named *A*, *B* and *P*having 74 instances [22]. The instances have different location coordinates with vehicles having alike capacities. Data set A comprising 27 instances with capacity of vehicle being 100, no. of customers being different in number in different instances with given (*x*, *y*) position coordinates. The data set *B* consists of 23 instances with capacity of vehicle being 100, no. of customers being different in number in different instances with given (*x*, *y*) position coordinates. The data set *P* having 24 instances have variable capacities as well as no. of customers are different for different instances. The GA and PSO methods are executed over all the instance problem 200 times, i.e. 200 generations of every instance problem were taken into account, and final fitness value obtained was recorded in the tabular form. This whole process is implemented in Windows 10 OS environment, an Intel core i3 processor having speed of 2.10 GHz with 8 Gb RAM. The analysis of both algorithms was successfully done in Python 3.7 programming language. The code written for the proposed algorithms to solve VRP is provided [23]. The graphical results of

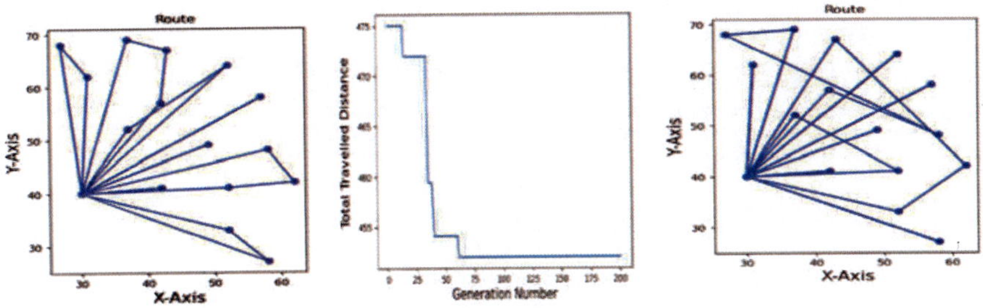


**Fig. 2** Network path created by vehicles using GA on instance P-n101-k4, the relationship between distance and generation number when GA is applied on instance P-n101-k4 and network path created by vehicles using PSO on instance P-n101-k4

benchmark instance from set *P* are shown in Fig. 2, they represent vehicle's whole network journey to different customers indicated by *x*, *y* coordinates. The outcomes for data sets A (27 instances), B (23 instances) and P (24 instances) are given in Tables 3 and 4, respectively. Table constitutes of the fitness values (near-optimal distance) obtained, i.e. solutions by genetic algorithm as well as PSO method along with the instance data set over which they are applied. When examining the fitness values (distance) calculated, it is observed that genetic algorithm works well on combinatorial optimization CVRP and gave better results, when compared with meta-heuristic PSO which is originally designed to solve continuous optimization problems where path representation is not always required as in case of discrete vehicle routing problem.

When considering the running time of both the proposed algorithms, PSO have faster running time than GA. The new and better variants of these algorithms can be developed for many more real-life problems that have applications in areas like machine learning, dynamical systems [24] bio-informatics [25], etc.

## 5.1 *Simulation Study*

The algorithms GA and PSO separately focus on two objectives, i.e. obtaining the ideal distance travelled by vehicles as well as length of the longest route. Sole purpose of our work is to find how GA contrasts with PSO in terms of performance applied on same set of instances. These two algorithms were developed in Python 3.7 on Dell-Intel core i3 7th generation (Table 2).

The results obtained in Tables 3 and 4 are the computed distance using GA and PSO applied separately on VRP instances, respectively, for the comparative analysis of both algorithms.

**Table 2** System parameter

| Parameter | Values |
|---|---|
| CVRP instances | 74 |
| Population size | 200 |
| Generations | 200 |
| Crossover probability | 0.9 |
| Mutation probability | 0.1 |
| rand1 | (0, 1) |
| rand2 | (−30, 30) |
| Inertia constant ($w$) | [0.4, 0.9] |
| Cognitive force ($C1$) | 2.05 |
| Social force ($C2$) | 2.05 |

**Table 3** Results obtained as total travelled distance when GA and PSO are applied over instance set A and B, respectively

| Instances | GA | PSO | Instances | GA | PSO |
|---|---|---|---|---|---|
| A-n32-k5 | 911 | 1677 | A-n65-k9 | 1332 | 3325 |
| A-n33-k5 | 702 | 1427 | A-n69-k9 | 1358 | 3568 |
| A-n33-k6 | 833 | 1452 | A-n80-k10 | 2035 | 4370 |
| A-n34-k5 | 830 | 1654 | B-n31-k5 | 696 | 1196 |
| A-n36-k5 | 892 | 1734 | B-n34-k5 | 845 | 1432 |
| A-n37-k5 | 805 | 1660 | B-n35-k5 | 998 | 1784 |
| A-n37-k6 | 1039 | 1774 | B-n38-k6 | 843 | 1494 |
| A-n38-k5 | 789 | 1732 | B-n39-k5 | 582 | 1440 |
| A-n39-k5 | 924 | 1873 | B-n41-k6 | 872 | 1910 |
| A-n39-k6 | 889 | 1810 | B-n43-k6 | 826 | 1649 |
| A-n44-k7 | 1014 | 2129 | B-n44-k7 | 946 | 2132 |
| A-n45-k6 | 1018 | 2369 | B-n45-k5 | 824 | 2057 |
| A-n45-k7 | 1262 | 2290 | B-n45-k6 | 791 | 1608 |
| A-n46-k7 | 1041 | 2212 | B-n50-k7 | 837 | 2321 |
| A-n48-k7 | 1251 | 2380 | B-n52-k7 | 823 | 2373 |
| A-n53-k7 | 1199 | 2635 | B-n56-k7 | 802 | 2336 |
| A-n54-k7 | 1294 | 2805 | B-n57-k7 | 1256 | 3311 |
| A-n55-k9 | 1215 | 2712 | B-n57-k9 | 1696 | 3007 |
| A-n60-k9 | 1514 | 3115 | B-n63-k10 | 1701 | 3586 |
| A-n60-k9 | 1527 | 3199 | B-n64-k9 | 1009 | 2690 |
| A-n62-k8 | 1413 | 3014 | B-n66-k9 | 1430 | 3100 |
| A-n63-k9 | 1822 | 3582 | B-n67-k10 | 1170 | 2963 |
| A-n63-k10 | 1492 | 3067 | B-n68-k9 | 1360 | 3296 |
| A-n64-k9 | 1580 | 3295 | B-n78-k10 | 1390 | 3885 |

**Table 4** Results obtained as total travelled distance when GA and PSO are applied over instance set P

| Instances | GA | PSO | Instances | GA | PSO |
|---|---|---|---|---|---|
| P-n16-k8 | 452 | 512 | P-n51-k10 | 845 | 1560 |
| P-n19-k2 | 651 | 1277 | P-n55-k7 | 690 | 1604 |
| P-n20-k2 | 221 | 397 | P-n55-k8 | 662 | 1545 |
| P-n21-k2 | 221 | 399 | P-n55-k10 | 779 | 1668 |
| P-n22-k2 | 246 | 379 | P-n55-k15 | 1056 | 1694 |
| P-n22-k8 | 612 | 832 | P-n60-k10 | 876 | 1814 |
| P-n23-k8 | 548 | 644 | P-n65-k10 | 918 | 1965 |
| P-n40-k5 | 552 | 1089 | P-n70-k10 | 1018 | 2119 |
| P-n45-k5 | 614 | 1307 | P-n76-k4 | 710 | 2190 |
| P-n50-k7 | 688 | 1412 | P-n76-k5 | 771 | 2291 |
| P-n50-k8 | 731 | 1369 | P-n101-k4 | 879 | 3004 |
| P-n50-k10 | 816 | 1469 | | | |

## 6 Conclusion

The main aim was to minimize or obtain a near-optimal travelled distance across the whole network. The proposed algorithms were tested with 74 benchmark instances. It was concluded that over problems like vehicle routing which is a discrete combinatorial optimization problem, GA proves to be more effective in obtaining a near-optimal distance than PSO method which is originally designed for solving continuous optimization problems. Not only the algorithms were applied to the instances, they were also precisely elaborated and explained as to how these proposed algorithms works on the classical CVRP, what are the methods by which the algorithm is working. Random key method was required in case of PSO algorithm since VRP needs path representation. Different variants of these algorithms could also give more effective results on the same problem as well as could be applied on more such problems of the same domain. Our proposed algorithms could further be applied to different variants of CVRP such as OVRP, VRPB VRPTW, etc.

## References


1. (Online) Available. Vehicle routing problem—Wikipedia
2. Cordeau, J.F., Gendreau, M., Laporte, G., Potvin, J.Y., Semet, F.: A guide to vehicle routing heuristics, J. Oper. Res. Soc. **53**(5), 512–522 (2002)
3. Prins, C.: A simple and effective evolutionary algorithm for the vehicle routing problem. Comput. Oper. Res. **31**(12), 1985–2002 (2004)
4. (Online). Available: AlvimTaillard2013.pdf (heig-vd.ch)
5. The, J.A., Voratas, K.: Particle swarm optimization and two solution representations for solving the capacitated vehicle routing problem. Comput. Ind. Eng. **56**(1), 380–387(2009)

6. (Online). Available: https://en.wikipedia.org/wiki/Heuristic_(computer_science)
7. Nora, N., Ahmed, E.S., Gadallah, M.: Solving capacitated vehicle routing problem using chicken swarm optimization with genetic algorithm. Int. J. Intell. Eng. Syst. **13**(5) (2020)
8. Vidal, T.: Hybrid Genetic Search for the CVRP: Open-Source Implementation and SWAP* Neighborhood (2020). https://arxiv.org/abs/2012.10384
9. Ramadhani, B.N.I.F., Garside, A.K.: Particle swarm optimization algorithm to solve vehicle routing problem with fuel consumption minimization. Jurnal Optimasi Sistem Industri **20**(1), 1–10 (2021)
10. Wisittipanich, W., Phoungthong, K., Srisuwannapa, C., Baisukhan, A., Wisittipanit, N.: Performance comparison between particle swarm optimization and differential evolution algorithms for postman delivery routing problem. Appl. Sci. **11**(6) (2021)
11. Lima, S.J.D.A., Araújo, S.A.D.: Genetic algorithm applied to the capacitated vehicle routing problem: an analysis of the influence of different encoding schemes on the population behavior. Am. Sci. Res. J. Eng., Technol., Sci. **73**(1) (2020)
12. Sajid, M., Jafar, A., Sharma, S.: Hybrid genetic and simulated annealing algorithm for capacitated vehicle routing problem. In: 2020 Sixth International Conference on Parallel, Distributed and Grid Computing (PDGC), pp. 131–136 (2020)
13. Sbai, I., Limam, O., Krichen, S.: An effective genetic algorithm for solving the capacitated vehicle routing problem with two dimensional loading constraint. Int. J. Comput. Intell. Stud. **9**(1/2) (2020)
14. Lin, N., Shi, Y., Zhang, T., Wang, X.: An effective order-aware hybrid genetic algorithm for capacitated vehicle routing problems in Internet of Things. IEEE Access **7**, 86102–86114 (2019)
15. Dahama, H.A., Mohammed, H.J.: An evolutionary algorithm approach for vehicle routing problems with backhauls. Mater. Today: Proc. (2021)
16. Yousefikhoshbakht, M.: Solving the vehicle routing problem by a hybrid improved particle swarm optimization. Int. J. Optim. Civ. Eng. **11**(1), 75–99 (2021)
17. Ombuki, B., Ross, B.J., Hanshar, F.: Multi-objective genetic algorithms for vehicle routing problem with time windows. Appl. Intell. **24**, 17–30 (2006)
18. Parsopoulos, K., Vrahatis, M.: Particle Swarm Optimization and Intelligence: Advances and Applications (2010). https://doi.org/10.13140/2.1.3681.1206
19. (Online). Available: https://towardsdatascience.com/particle-swarm-optimisation-in-machine-learning-b01b1d2ad8a8
20. Marinakis, Y., Marinaki, M., Migdalas, A.: Particle swarm optimization for the vehicle routing problem: a survey and a comparative analysis. In: Martí, R., Panos, P., Resende, M. (eds.) Handbook of Heuristics. Springer, Cham (2017). https://doi.org/10.1007/978-3-319-07153-4_42-1
21. Tasgetiren, M.F, Liang, Y.C., Sevkli, M., Gencyilmaz, G.: Particle swarm optimization algorithm for makespan and total flowtime minimization in the permutation flowshop sequencing problem. Eur. J. Oper. Res. **177**(3), 1930–1947 (2007)
22. (Online). Available: https://neo.lcc.uma.es/vrp/vrp-instances/capacitated-vrp-instances/
23. Available: https://drive.google.com/file/d/1glubhEx1PzMET-sRAE6zKiOQJ0sd-DVM/view?usp=sharing
24. Cui, X., Charles, J.S., Potok, T.E.: A simple distributed particle swarm optimization for dynamic and noisy environments. In: Nature Inspired Cooperative Strategies for Optimization (NICSO 2008), pp. 89–102 (2018)
25. Agrawal, S., Silakari, S.: A review on application of particle swarm optimization in bioinformatics. Curr. Bioinform. **10**(4) (2015)

# A Secure Health Management Framework with Anti-fraud Healthcare Insurance Using Blockchain

**Sourav Mahapatra, Noorjahan, Ditipriya Sinha, and Ayan Kumar Das**

**Abstract** Telemedicine and electronic health record management are playing an inevitable role in today's era. The smart healthcare system suffers from security issues due to no control of patients over their personal data. On the other hand, insurance companies are suffering from falsifying information where the claimant produces false or tampered diagnosis report during their claiming. In e-health system, secure communication among patient, healthcare provider, and insurance companies are essential part. Blockchain can overcome the aforesaid challenges due to its numerous advantages such as distributed database, transparency, trusted networks, decentralization, and autonomous connectivity. Authentication and the privacy of health data are ensured through cryptographic key exchange using blockchain. In this paper, a healthcare management framework is proposed where patient, healthcare provider, and insurance companies are sharing their transactions in blockchain. In our proposed approach, e-health management system is protected from different types of security attacks such as replay and man-in-the-middle attacks.

**Keywords** Health care · Telemedicine · EHR · Anti-fraud · Blockchain · Key analysis

---

S. Mahapatra (✉)
Techno International Newtown, Kolkata, West Bengal 700156, India
e-mail: souravm.phd20.cs@nitp.ac.in

S. Mahapatra · Noorjahan · D. Sinha
National Institute of Technology Patna, Patna, Bihar 800005, India
e-mail: noorjahanjuli05@gmail.com

D. Sinha
e-mail: ditipriya.cse@nitp.ac.in

A. Kumar Das
Birla Institute of Technology, Mesra, Patna Campus, Patna, Bihar 800014, India
e-mail: das.ayan@bitmesra.ac.in

 
D. Gupta et al. (eds.), *Pattern Recognition and Data Analysis with Applications*, Lecture Notes in Electrical Engineering 888,
https://doi.org/10.1007/978-981-19-1520-8_40

## 1 Introduction

Advancement of technology in the healthcare sector creates a new era in telemedicine and electronic health records (EHR) management. Smart healthcare systems not only improve the efficiency and accuracy of diagnosis but also break the geographical barrier. Sharing of EHR can provide benefit to the patients, researchers, and all the stakeholders of healthcare system. Due to storage and energy limitation of smart devices, common practice is that the digital health information (EHR) produced by smart health monitoring devices is usually sent to a centralized unit of a health institute for further processing. These are stored either in their local storage or share a cloud-based service to provide a shareable environment. Due to security, ownership, and integrity of medical data, cross-institute operation of records is not allowed. Though patients are the owner of the "EMR," they have very less control over their data. Though the different encryption techniques are followed to provide security and authenticity to the EMR by the central authority, they suffer from single-point failure, compromises data security, transparency, integrity, and key distribution challenges.

Nowadays, healthcare services are essential for all people in our society, and cost is increasing day-by-day. Healthcare insurance can play an inevitable role to provide a quality service during crisis time. The major challenge for an insurance company is to identify fraud during claim settlement. According to Gill et al. [1], fraud in the insurance domain can be defined as "knowingly making a fictitious claim, inflating a claim, or adding extra items to a claim or being in any way dishonest with the intention of gaining more than legitimate entitlement." It is observed by the allied institutions and researchers that fraud detection is difficult and not cost effective. National Health Care Anti-Fraud Association (NHCAA) has reported that more than 3% of US annual health care expenditure was lost due to false claim identification and management.

To provide the best service to the community or to perform any historical data analysis for the benefits of the healthcare system, data need to be shared with various entities among the system with a tamper-proof, transparent, and distributed nature. Cloud storage is the most efficient data sharing technique where security is maintained through cryptographical encryption technique. All the cloud-based services are centralized in nature, which leads them to single-point failure. Centralized key distribution is a target point to the attacker for tampering the records. A transparent, temper-proof, and distributed system is needed to handle all these issues. Fraud management can also be controlled efficiently when the agencies are working together on a tamper-proof distributed ledger. Blockchain has the potential to overcome the challenges like interoperability, non-standardization of information which leads the system more stable. Every transaction inside the blockchain is accountable in distributed ledger form, which eliminates the single point of failure or bottleneck of cloud services. EMR, diagnosis report, and insurance claim as a transaction can be used to create a new block in the blockchain. A new block is consisting of the hash value of the previous block, new transaction, timestamp, and a nonce. A set of nodes (miner) participate in a consensus mechanism to validate and add a new block

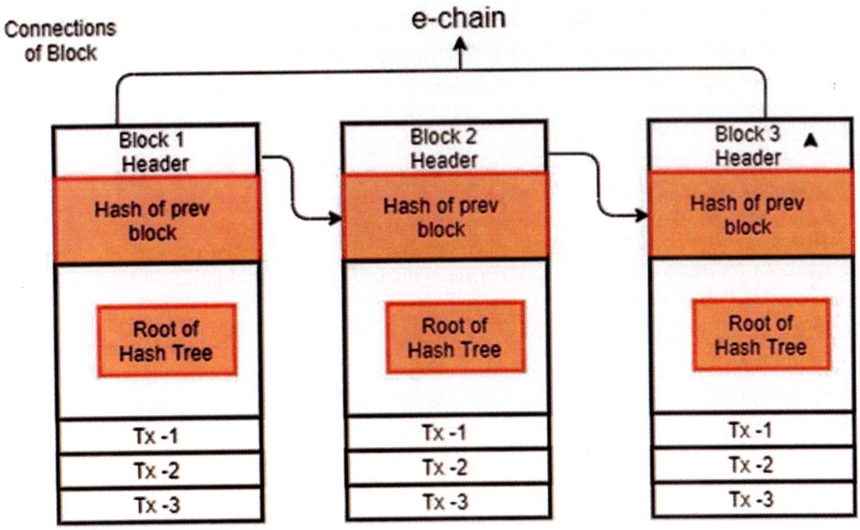


**Fig. 1** Structure of blockchain

into the chain. Figure 1 shows a basic structure of a block in the blockchain. Hash of all the hashes of all transactions for a block forms a Merkel root. Merkel root is a summary of all the transactions in a block. It can be used to verify a large volume of transactions in a single instance. Smart contracts are used in blockchain to provide fine control on the records.

Blockchain can improve the performance of the whole healthcare management system by automating claim processes and settlements, streamlining the business, and also cut off the operational and transaction cost of the system.

In this paper, we propose a healthcare management scheme where patients, healthcare providers, and insurance institutions take part to share their transactions with three blockchain. It is a trustless environment where every transaction is immutable. Users of these chains, namely patients, can upload only the hash value of the encrypted IoT or other EMR to e-chain, caregivers can read those transactions and decipher the original data from an IPFS for further diagnosis, the diagnosis reports are again stored in IPFS, and its hashed values are stored in another blockchain called d-chain. The insurance agency can access the immutable records from the e-chain and d-chain to verify the genuineness of the claim and perform a required transaction in the i-chain for its claim settlements. To make our system robust and secure, keys are also shared through blockchain. IPFS stores encrypted EMR, diagnosis report, and claim report, and their respective hash values are recorded into three blockchains. In this framework, blockchain, IPFS, and cryptographic hash function combinedly achieve security and access control among the stakeholders.

The flow of the paper is as follows. In Sect. 2, we discuss the related works on the blockchain-based healthcare system. The preliminary concept of blockchain, EMR, and health insurance is defined in Sect. 3. Detail description of our proposed framework is described in Sect. 4. Security analysis of our framework with probable attacks is explained in Sect. 5. A performance analysis is mentioned in Sect. 6. At last but not least, the conclusion is drawn in Sect. 7.

## 2 Related Works

This section discusses the related works on the security of telehealthcare management systems using blockchain.

Xu et al. describe a blockchain-based privacy-preserving scheme [2] (Healthchain) for fine-grained access control of large-scale health data. Flexible key management is designed in this paper. This paper does not provide security between the insurance company, patient, and hospital in a blockchain-based environment. On the other hand, Liu et al. proposed a blockchain-based anti-fraud scheme [3] for healthcare insurance. This application addresses the issues related to healthcare insurance anti-fraud and detects the fraud. This paper does not concern about cryptography algorithms applying on blockchain technology. Wan [4] uses searchable symmetric and attribute-based encryption techniques to achieve privacy with fine-grained access control. They are not concerned about the consensus process. The granular access rule is imposed by Shahnaz et al. [5], to provide safe electronic record storage. This paper does not detect how to defend against various forms of attacks [5, 6]. In "EdgeMediChain," Akkaoui [7] introduces edge computing and blockchain to provide security for a healthy ecosystem. Bhattacharya et al. [8] design "BinDaaS" which operates in two phases and combines blockchain and deep learning techniques to maintain security and also predict potential risks for sharing EHR records among multiple healthcare users. Cao [9] designs a cloud-based e-health system where the primary objectives are to protect EHR and patient privacy. Kumar [6] proposes a smart healthcare system, based on the convergence and interoperability of Blockchain 3.0 and Health care 4.0. Fan [10] constructs "MedBlock" which is an efficient privacy-preserving and sharing scheme based on blockchain. It combines customized access control protocols and symmetric cryptography. The term "Smart Contract" is not stated in this article. To strengthen the hospital's electronic health infrastructure, Liu [11] proposes a medical data exchange and security scheme based on the hospital's private blockchain, the PBC, and OpenSSL libraries. Melih Kirlidog [12] proposes blockchain-based data mining tools on large sets of insurance claim data to detect fraud. For detecting health insurance fraud, Saldamli [13] also proposes a blockchain-based solution. It does not concentrate on encryption or attack prevention. Xia [14] presents MeDShare, a blockchain-based framework that addresses the issue of medical data sharing in a trustless environment among medical big data caretakers.

The aforesaid studies describe the numerous issues surrounding blockchain-based healthcare data sharing. Table 1 represents a comprehensive study of recent research trends on this domain with their characteristics. From the state of the art, it is concluded that most of the blockchain-based healthcare systems are not concerned

**Table 1** Comparative analysis of state-of-the-art blockchain-based healthcare systems that incorporate blockchain technology

| Author | Year | Consensus algorithm | Smart contract design | Dapp development | Cryptography algorithm | Security issues resolved | Fine-grained access control | Attack resistance | Insurance fraud management |
|---|---|---|---|---|---|---|---|---|---|
| Xu and Xue [2] | 2019 | ✔ | ✘ | ✘ | ✔ | ✔ | ✔ | ✔ | ✘ |
| Liu [3] | 2019 | ✔ | ✔ | ✘ | ✔ | ✔ | ✘ | ✘ | ✔ |
| Wan [4] | 2019 | ✘ | ✔ | ✔ | ✔ | ✔ | ✔ | ✘ | ✘ |
| Shahnaz [5] | 2019 | ✘ | ✔ | ✔ | ✔ | ✔ | ✘ | ✘ | ✘ |
| Akkaoui [7] | 2020 | ✔ | ✔ | ✔ | ✔ | ✔ | ✘ | ✔ | ✘ |
| Bhattacharya [8] | 2019 | ✘ | ✔ | ✔ | ✔ | ✔ | ✘ | ✔ | ✘ |
| Cao [9] | 2020 | ✘ | ✔ | ✘ | ✔ | ✔ | ✘ | ✔ | ✘ |
| Kumar [6] | 2020 | ✔ | ✔ | ✔ | ✔ | ✔ | ✘ | ✘ | ✘ |
| Fan [10] | 2018 | ✔ | ✘ | ✘ | ✔ | ✔ | ✘ | ✔ | ✘ |
| Liu [11] | 2019 | ✔ | ✘ | ✘ | ✔ | ✔ | ✘ | ✘ | ✘ |
| Saldamli [13] | 2020 | ✔ | ✔ | ✘ | ✘ | ✔ | ✘ | ✘ | ✔ |
| Xia [14] | 2017 | ✘ | ✔ | ✘ | ✔ | ✔ | ✘ | ✔ | ✘ |
| Proposed approach | 2021 | ✔ | ✔ | ✔ | ✔ | ✔ | ✔ | ✔ | ✔ |

combinedly privacy and security among patients, doctors, and health insurance companies.

## 3 Preliminaries

This section deals with some basic-related technology used to design our scheme.

**Blockchain Technology**: It was first developed to maintain an immutable public distributed ledger for a cryptocurrency exchange like bitcoin and get popularity. Blockchain is constructed following the principle of a link list (Fig. 1). A new block contains information about the current transaction and the hash value of the previous block which provides a temper-proof immutable environment. In any kind of transaction, following feature of blockchain should retain.

1. *Decentralization*: Every node in this network can store the updated transaction. So, single-point failure or network bottleneck should be eliminated. All nodes are involved in the verification process to include a new node.
2. *Immutable*: The transaction cannot be rewind. Hash of the previous block stored in the current block. So, any update in the previous block needs to modify all the linked blocks.
3. *Anonymity*: All transactions carried out in blockchain are verified by a well-constructed consensus algorithm.
4. *Autonomy*: Every blockchain network has a smart contract that provides free and secure communication among users. It is an automated process where human intervention is not required.

**Inter planetary File System (IPFS)**: It provides a distributed approach to store our data in a peer–peer network. IPFS uses content-based addressing to identify the uploaded file in global space. Each IPFS object has followed a data structure with two fields—data and link. A distributed has table (DHT) is maintained among the participating node to store the cryptographic hash. When a node searches for content, DHT helps to locate that.

**Smart Contract**: It is an executable code inserted into the blockchain, which runs automatically to validate and verify a transaction with some predetermined agreement. No human or centralized intervention is needed.

**Electronic Health Record**: A structured collection of health information of an individual or population through a digital platform to share with the practitioner for better service. In a shareable environment, EHR suffers to provide security, authentication, and integrity to our sensitive information. We aim to establish a scheme, where all these challenges should be overcome.

**Health Insurance**: All human beings need to be assisted by a health insurance company to get a better medical facility. Insurer supports their customer when they face a medical crisis. Policyholders' claims for reimbursement will be verified by the insurer. This verification process is a costly and long-term process. Broadly, we can categorize fraud in health insurance in five categories as follows.

1. Application fraud: Insured provides wrong information related to his/her current health status, DOB, claims, etc.
2. Eligibility fraud: It is related to employment status, age, and preexisting diseases.
3. External and internal frauds: False claim raised by policy holder, agents, or by their internal employee.
4. Deliberate and opportunity fraud: A group of people trying to withdraw the benefit in favor of the policyholder by providing a false claim request.

Majority of the insurance company bears the huge loss due to fraud management. Their research groups also struggling to find an effective solution. We highlighted the immutable characteristic of blockchain to provide a temper-proof transaction within the system.

# 4 Proposed Model

In this section, we introduce the architecture of our model and the workflow of the model in detail.

## *4.1 System Model*

As we can see in Fig. 2, our model includes mainly five entities: patient, healthcare organization, health insurance company, IPFS, and finally, blockchain. All these entities are described as follows.

**Patients**: The owner of the EMR collects all personal health records either through medical smart devices or from the diagnostic center. He is responsible for the generation and distribution of symmetric keys and encrypting the collected EMR. The patient can perform a transaction on both e-chain and i-chain.

**Healthcare Organization**: They can provide services to the patients by sharing the EMR of a patient and generate a diagnosis report. They can record their transaction only in d-chain.

**Health Insurance Company**: They can issue a policy and provide support to the patients during his/her medical crisis. All their transactions are stored in i-chain.

**IPFS**: It is a p2p file storing process that stores all transactions in encrypted form in a distributed way and provide an index to the user for reference. It is more faster and cost-effective compared to other distributed data storage. Due to its high download speed, cheaper storage space, and independent to the backbone network, it increases the acceptability with blockchain.

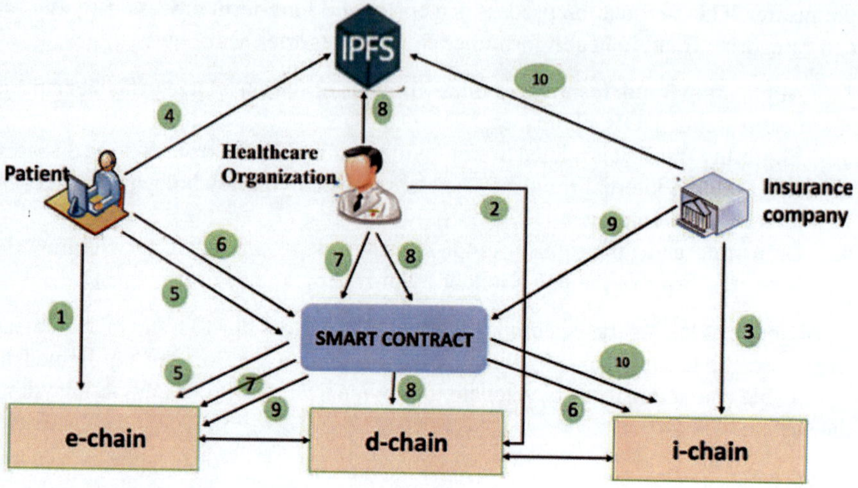


**Fig. 2** Workflow diagram

**Blockchain**: We proposed three different chains, namely e-chain, d-chain, and i-chain to maintain transparency on the transaction by the user of this scheme. The purpose of every chain is given below.

*e-chain*: Patients can add a new block to the chain only when he/she has either a new EMR or want to change the access permission to the caregivers. As it is a public chain, so any time anyone can take part or leave the chain.

*d-chain*: It is a consortium blockchain, where a registered healthcare organization is the only participant to generate a new block. The purpose of this chain is to store all kinds of activity performed by a health organization to serve a patient including diagnosis, bills, etc.

*i-chain*: A private chain between the insurance company and the registered patient is formed to provide a smooth and authenticated claim settlement process. The participated insurance company can easily classify the fraud users of their policies.

All these three chains follow the basic principle of blockchain, i.e., each new block stores the hash of the previous block with the current transaction to maintain the temper-proof in nature. Each new block stores the hash value of the encrypted original transaction, whereas the actual transaction stores at IPFS in encrypted form, and an index is provided to the user of the system for future access.

## *4.2 Workflow of the Model*

All the steps involved in this process are subdivided into four levels—initialization, EHR transaction, EHR diagnosis, claim management.

1. **Initialization**: At first, participants of three chains (patients, health organization, insurance company) initialize their chains, initiate their contracts, and store the contract address, and application binary interface (ABI) is involved in steps 1, 2, and 3 in Fig. 2.
2. **EHR transaction**: It includes all the activities that take place on the patient's side. These are described in steps 4–6 in Fig. 2.

   Step 4 Patients generate a symmetric key *(*$EMRk_i$ *)* to encrypt the EMR and also store the encrypted EMR in IPFS. It receives one hash value for that transaction to IPFS.

   Step 5 Patients create another symmetric key *(*$dk_{ij}$*)* which is used by healthcare organizations to encrypt their transactions. Both these keys $EMRk_i$, *dkij*, and a hash of encrypted *EMRi* are stored in the *e-chain* in two different transaction forms, namely—EMR transaction and key transaction.

   Step 6 Whenever a patient wants to buy an insurance policy, he/she can register themselves to take part in the *i-chain.* To initiate a claim, the patient performs a transaction ($t_{clrq}$) in the i-chain.

3. **EHR diagnosis:** In this level steps 7–8 are included in Fig. 2

   Step 7 Healthcare providers can access e-chain and retrieve the EMR transaction for which he is responsible.

   Step 8 After investigating EMR, healthcare organization generates a report on action taken, encrypts it by $dk_{ij}$, and uploads it in IPFS. The hash value of the encrypted reports is store in the d-chain. Patients and insurance companies can be able to view that from the d-chain.
4. **Claim Management:**
5. Step 9 Insurance company searches on *i-chain* to find a claim request transaction generated by the patient. If found, verify all the related information from the e-chain and d-chain.

Step 10 If the claim request is valid, it generates a response transaction encrypted by a symmetric key and shares this key with both the parties through a digital envelop in its claim response transaction.

After verification, if any fraud transaction is identified, then the patient will be blacklisted.

Except for initialization, all the steps are passed through the smart contract to verify the integrity and access control.

## 5 Key Analysis and Distribution

*e-chain* includes EMR and key transactions to upload the patients' information.

*EMR transaction*: The hash value of the encrypted EMR of a patient is added the *p-chain.*

$$t_{EMR} = \{ID_{pi}, ts1, HEEMR, S_i, htEMR_i\} \tag{1}$$

$S_i = Sign(Sk_{pi}, H(ID_{pi}, ts1, HEEMR))$   $htEMR_i = H(ID_{pi}, ts1, HEEMR_i, S_i)$.

$t_{EMR}$—Transaction of EMR   $ts1$ = time stamp.

$S_i$ = Signature of patient by its private key $Sk_{pi}$   $ID_{pi}$—*ID* of patient *i*.

*HEEMR*—hash of encrypted EMR   $htEMR_i$—hash of whole EMR transaction.

*Key transaction:* Hash of encrypted EMR key and diagnosis key are added by patient to *e-chain.*

$$t_{key} = \{ID_{pi}, ts2, Env_{ij}, Envp_i, Sig_i, htk_i\} \tag{2}$$

$$Env_{ij} = (ID_{Dj}, htEMR_i, Enc(PK_{Dj}, (EMRK_i, dk_{ij}))$$

$$Sig_i = Sign(Sk_{pi}, H(ID_{pi}, ts2, Env_{ij}, Envp_i))$$

$$Envp_i = Enc(PK_{pi}(EMRK_i, dk_{ij}))$$

$$htk_i = H(ID_{pi}, ts2, Env_{ij}, EnvIn_k, Envp_i, Sig_i)$$

$t_{key}$—Transaction of key   $ID_{Dj}$ = ID of doctor *j*

$Env_{ij}$—Envelop for doctor contains encrypted $EMRK_i$, $dk_{ij}$

$EMRK_i$—*EMR* key generated by $\text{patient}_i$   $dk_{ij}$—diagnosis key

$PK_{Dj}$—Public key of $\text{doctor}_{Dj}$   $PK_{pi}$—public key of $\text{Patient}_{pi}$

$Envp_i$—Envelop for patient authentication, encrypted by $PK_{pi}$

$htk_i$—hash of whole $t_{key}$

Healthcare organizations add a new block to *the d-chain.*

*Diagnosis transaction*: Hash of an encrypted diagnosis report is added by doctor in *d-chain.*

$$t_{dia} = \{ID_{Dj}, ts3, htEMR_i, HEdm, S_{Dj}, htD_j\} \tag{3}$$

$$S_{Dj} = Sign(Sk_{Dj}, H(ID_{Dj}, ts3, htEMR_i, HEdm)$$

$$ht_{Dj} = H(ID_j, ts3, htEMR_i, HEdm, S_{Dj})$$

$S_{Dj}$—Signature of $\text{doctor}_{Dj}$   $Sk_{Dj}$—private key of $\text{doctor}_{Dj}$

$t_{dia}$—diagnosis transaction $ID_{Dj}$—ID of doctor$_{Dj}$
$HEdm$—hash of encrypted diagnosis $htD_j$—hash of whole $t_{dia}$

*Claim transaction*: Patient and insurance company perform claim request ($t_{clrq}$) and claim settlement ($t_{cl}$) transaction, respectively, on *i-chain*.

$$t_{clrq} = \{ID_{pi}, ts4, htxEMR_i, HEclrq, Env_{In}, S_{pi}, htclrq\} \quad (4)$$

$$S_{pi} = Sign(SK_{pi}, H(ID_{pi}, ts4, htxEMR_i, HEclrq, Env_{In})$$

$$Env_{In} = (ID_{In}, htxEMR_i, Enc(Pk_{In}, (EMRK_i, dk_{ij}))$$

$$htclrq = H(ID_{pi}, ts4, htxEMR_i, HEclrq, S_{pi})$$

$HEclrq$ = hash of encrypted ($Pk_{In}$) claim requested generated by patient.
$ID_{In}$ = ID of insurance company $Pk_{In}$ = public key of insurance company.

After the verification of information in Eq. (4), claim settlement is recorded in i-chain.

$$t_{cl} = \{ID_{In}, ts5, htEMR_i, Env_{cm}, HECm, S_k, htcl\} \quad (5)$$

$$S_k = Sign(Sk_{In}, H(ID_{cl}, ts4, htEMR_i, HECm)$$

$$htcl = H(ID_{cl}, ts5, htEMR_i, HECm, S_k)$$

$$Env_{Cm} = (ID_{In}, Enc(PK_{pi}, K)||Enc(PK_{Dj}, K)||Enc(K, Cm))$$

$t_{cl}$—transaction for claimsettlement $ID_{In}$—*ID* of insurance company.

$HECm$—hash of encrypted claim report cm.

$Sk_{In}$—private key of insurance company.

$Env_{Cm}$—encrypted *Cm* with key sharing.

In our model, patient can generate two transactions, one for EMR transactions and the other for the key transaction. In *p-chain*, EMR transaction contains ID of that patient, timestamp, a hash of encrypted EMR, signature (*Si*) of the sender, and a hash of whole EMR transaction. The signature of the patient is encrypted with its private key to represent who is the owner of that transaction. Hash of whole EMR transaction ($htEMR_i$) is used to control the integrity.

We grant permission to the doctor and the insurance provider to access the data generated by key transactions ($t_{key}$). $Env_{ij}$ is a digital envelope for a doctor which is encrypted by the public key of the doctor contains the EMR key and diagnosis key.

$Env_{In}$ is a digital envelope for an insurance company that also contains the EMR key and diagnosis key, encrypted by the public key of the insurance company. $Envp_i$ is the patient's digital envelope that is encrypted with the patient's public key to provide authentication.

An authorized doctor can decrypt their envelope and access the EMR key and diagnosis key from the e-chain. The diagnosis transaction is used to determine which EMR data doctors have finished their diagnoses. Through this EMR, key doctor can decrypt the EMR data stored in IPFS. After diagnosis, the hash value of the encrypted diagnosis report (HEdm) signed by the doctor is stored in the *d-chain*.

The hash value of the encrypted claim request *(HEclrq)* is stored in the *i-chain*. Hash of encrypted claim settlement *(HECm)*, encrypted claim report (Cm), and a hash of whole EMR transaction is participating in claim settlement transaction ($t_{cl}$). Any miss match of information collected in Eq. (4) leads the $t_{clrq}$ as fraud transaction. The insurance company shares a secrete key, which is encrypted by the patient's and doctor's public keys separately in an envelope $Env_{Cm}$. Doctor and patient can access that $t_{cl}$ to know the status of a claim.

# 6 Performance Analysis

In this section, we are trying to measure the efficiency of our framework. We use remix as a development tool and solidity to write our smart contracts and deploy them in the Ethereum network. Computational and operational performance can determine the efficiency of a system. Various network attacks can degrade the computational performance of the system. In Sect. 6.1, various attacks and their solutions are justified.

## 6.1 Different Attacks

1. Replay and impersonation attacks:

   Attackers capture the traffic and resend the packet, acting as an original sender without any modification on the message. This type of attack can be prevented by using some extra information like a session key, timestamp. In our model, to authenticate the transaction, every transaction in Eqs. (1)–(5) includes digital signature *Si* which maintains a session key *Sk* and timestamp *ts* .
2. Man-in-the-Middle Attack:

   Personal information or identity theft is one aim of the attacker. It is just like a mailman open and stole the bank details from your bank statement, resealing the envelope, and delivered it to you. MITM can be overcome through multichannel authentication without exchange secrets. All our transactions in *p-chain, d-chain, and i-chain* are encrypted by symmetric key $EMRK_i$, $dk_{ij}$, and K, respectively. These keys are also shared with the receiver through a digital envelope encrypted by receiver's public key.

3. User Anonymity:

   It protects user identity from the network. Medical data are more sensitive and attackers are interested in that. In our model, original information collected from patients at different times is encrypted and stored in IPFS. Blockchain holds only the hash value of encrypted medical records, diagnosis reports, or claim settlements. Retrieval of information through investigating blocks is not possible by attackers.

4. Parallel Session and Reflection Attacks:

   Attackers send a duplicate challenge to the sender as their challenge and get the response. After that, it returns that responds to the original sender as a response to the challenge and wins the game.

5. Mutual Authentication: Before original communication occurs, two parties mutually authenticate each other by verifying the certificate. It is required to establish a secure channel. Blockchain is a trustless system where the transaction is immutable. In the key transaction (Turkey) phase, digital signature is used to authenticate the patient and healthcare organization.
6. Ephemeral Secret Leakage (ESL) Attack or Known Key Secrecy/Forward Secrecy:

Forward secrecy is achieved by frequently changing the key used for encryption and decryption purposes. Sometimes EMR data may be exposed to fraud doctors or insurance companies. We can revoke the new key transaction with another hospital or insurance company to overcome this situation.

## 6.2 Computational Performance

The performance of these three chains proportionally depends on the number of transactions. Before we add a new block to the chain, original records are encrypted through asymmetric key cryptography. The computation overhead of our system is reflected in Fig. 3. Whereas it is remarkable, when compared to traditional systems mentioned in Table 1. It is clear from Fig. 3 that computation cost is increasing as time increases. But the inclination of the curve has a significant acceptance.

In our proposed model, there are four types of transaction (mentioned in Eqs. 1–4) which are considered. We implement them in Ganache local blockchain, and performance of these transactions is represented in Fig. 4. Key transaction ($t_{\text{key}}$) takes 33%, claim transaction ($t_{\text{cl}}$) takes 26%, diagnosis transaction ($t_{\text{dia}}$) takes 24%, and EMR transaction ($t_{\text{EMR}}$) takes 17% of total time.

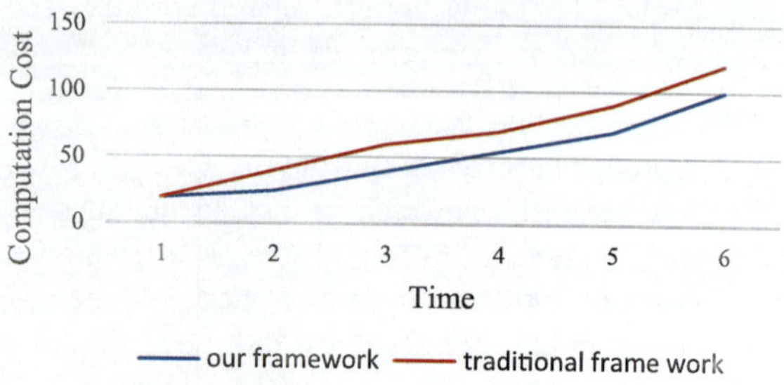


Fig. 3 Performance analysis

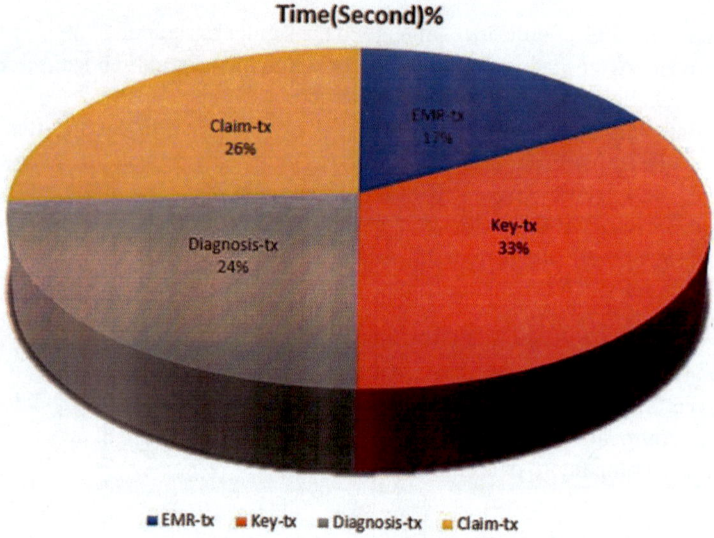


**Fig. 4** Computation time

## 7 Conclusion

In this paper, a schematic approach to protect our digital health records in a sharable environment is proposed. An approach for quick settlement of claim and identification of frauds are introduced through blockchain. This paper proposes secure transactions among the stakeholders of the telemedicine healthcare system applying inherent features of blockchain technology. Our proposed model prevents online healthcare systems to defend different types of attacks such as man-in-the-middle, replay attack, mutual authentication attacks. It also provides fine access control to the patients. In this model, any healthcare management system can share their information in a secure way and patients can have full control over their personal records.

Within less time and minimum involvement, a claim for reimbursement can finalize. To find out a suitable consensus algorithm, further research is needed.

## References


1. Gill, K.M., Woolley, K.A., Gill, M.: How the detection of insurance fraud succeeds and fails. Psychol., Crime Law **12**(2), 163–180 (2006). https://doi.org/10.1080/10683160512331316325
2. Xu, J., Xue, K., Li, S., Tian, H., Hong, J., Hong, P., Yu, N.: Healthchain: a blockchain-based privacy preserving scheme for large-scale health data. IEEE Internet Things J. **6**(5), 8770–8781 (2019)
3. Liu, W., Yu, Q., Li, Z., Li, Z., Su, Y., Zhou, J.: A blockchain-based system for anti-fraud of healthcare insurance. In: 2019 IEEE 5th International Conference on Computer and Communications (ICCC), Chengdu, Dec 2019, pp. 1264–1268. IEEE, China (2019)
4. Wang, S., Zhang, D., Zhang, Y.: Blockchain-based personal health records sharing scheme with data integrity verifiable. IEEE Access **7**, 102887–102901 (2019). https://doi.org/10.1109/ACCESS.2019.2931531
5. Shahnaz, A., Qamar, U., Khalid, A.: Using blockchain for electronic health records. IEEE Access **7**, 147782–147795 (2019). https://doi.org/10.1109/ACCESS.2019.2946373
6. Kumar, A., Krishnamurthi, R., Nayyar, A., Sharma, K., Grover, V., Hossain, E.: A novel smart healthcare design, simulation, and implementation using Healthcare 4.0 processes. IEEE Access **8**, 118433–118471 (2020). https://doi.org/10.1109/ACCESS.2020.3004790
7. Akkaoui, R., Hei, X., Cheng, W.: EdgeMediChain: a hybrid edge blockchain-based framework for health data exchange. IEEE Access **8**, 113467–113486 (2020). https://doi.org/10.1109/ACCESS.2020.3003575
8. Bhattacharya, P., Tanwar, S., Bodke, U., Tyagi, S., Kumar, N.: Bindaas: blockchain-based deep-learning as-a-service in healthcare 4.0 applications. IEEE Trans. Network Sci. Eng. **8**(2), 1242–1255 (2021). https://doi.org/10.1109/TNSE.2019.2961932
9. Cao, S., Zhang, X., Xu, R.: Toward secure storage in cloud-based eHealth systems: a blockchain-assisted approach. IEEE Network **34**(2), 64–70 (2020). https://doi.org/10.1109/MNET.001.1900173
10. Fan, K., Wang, S., Ren, Y., Li, H., Yang, Y.: Medblock: efficient and secure medical data sharing via blockchain. J. Med. Syst. **42**(8), 1–11 (2018). https://doi.org/10.1007/s10916-018-0993-7
11. Liu, X., Wang, Z., Jin, C., Li, F., Li, G.: A blockchain-based medical data sharing and protection scheme. IEEE Access. **7**, 118943–118953 (2019). https://doi.org/10.1109/ACCESS.2019.2937685
12. Kirlidog, M., Asuk, C.: A fraud detection approach with data mining in health insurance. Procedia Soc. Behav. Sci. **62**, 989–994 (2012). https://doi.org/10.1016/j.sbspro.2012.09.168
13. Saldamli, G., Reddy, V., Bojja, K.S., Gururaja, M.K., Doddaveerappa, Y., Tawalbeh, L.: Health care insurance fraud detection using blockchain. In: 2020 Seventh International Conference on Software Defined Systems (SDS), Paris, Apr 2020. IEEE Computer Society, pp. 145–152. IEEE, France (2020)
14. Xia, Q.I., Sifah, E.B., Asamoah, K.O., Gao, J., Du, X., Guizani, M.: MeDShare: trust-less medical data sharing among cloud service providers via blockchain. IEEE Access **5**, 14757–14767 (2017). https://doi.org/10.1109/ACCESS.2017.2730843

# An Energy-Aware Fog-Enabled Optimized VM Consolidation Scheme for Real-Time Applications

K. Hemant K. Reddy, Rajat S. Goswami, and Shubham

**Abstract** As the fifth generation computing paradigm is innovative, maturing, and envisioned toward real-time services, conventional centralized network infrastructure architecture based upon cloud is unable to meet the demands for real-time responses. Whereas,low latency-based fog architecture showing a promising alternative but it cannot meet the huge number of service demands in real-time. Neither fog computing alone becomes a solution nor neither centralized cloud. In this paper, we envision employing two-layered service architecture for handling upcoming real-time IoT applications. However,centralized cloud service layer is for handling huge service demand of delay tolerable applications and decentralized fog service layer is to cater the real-time delay sensitive applications. With the purpose to cater the fog nodes energy consumption and service completion of delay sensitive applications, an optimized (GA) virtual machine (VM) consolidated scheme was employed. VM consolidation includes various parts such as the host underload/overload detection, VM placements, and VM collections. The simulation results presents the efficacy of the proposed methodology.

**Keywords** Real-time application · Fog computing · VM consolation · Genetic algorithm

---

K. H. K. Reddy (✉) · R. S. Goswami
National Institute of Technology Arunachal Pradesh, Jote, India
e-mail: khemant.phd20@nitap.ac.in

R. S. Goswami
e-mail: rajat@nitap.ac.in

Shubham
National Institute of Science and Technology Berhampur, Berhampur, Odisha, India
e-mail: shubham.it.2017@nist.edu



D. Gupta et al. (eds.), *Pattern Recognition and Data Analysis with Applications*,
Lecture Notes in Electrical Engineering 888,
https://doi.org/10.1007/978-981-19-1520-8_41

# 1 Introduction

Fog computing is described as a computing system which is distributed in nature; it expands the cloud computing services to the network's edge. Fog computing, which is viewed as a supplement to cloud infrastructure, enables storage of data, task management and networking among the mobile users and cloud data centers [1]. Several computing services and applications that do not ensemble with the cloud, such as soft wares that needs predictable and low latency, for example, video conferencing, globally dispersed applications such as wireless sensor networks, and the transportation applications which are intelligent in nature, may all be performed by the fog such as smart connected vehicles and smart traffic lights [2].

Components such as routers, smart gateways, and data centers along with the network infrastructure edge components such as fog nodes to support a fog computing customer could be implemented by cloud architectures. Fog nodes are the devices which are typically constrained on resources including the routers, base stations, access points, and set-top boxes which aids the storage and computational resources, mobility, and variety of interface types and transmission protocol. In order to meet the needs of computational application which demands for low communication overhead, large-scale delivery, low latency, and density, the fog nodes are used. Fog devices, such as Cisco's IOx products, assist developers in developing the IoT applications such as cybersecurity, data aggregation, and control systems. On the guest operating system, these IoT applications run which in turn allows running of compiled code on the network edge.

New operating and management roles, like work scheduling, are added to support the fog services. A fog server's collection located at the network's edge in a chosen locations, such as shopping malls, local government offices, or service stations, may provide these new fog functions. Every fog server is fitted with a computing computer, data storage cards, and a wireless communication unit, making it a highly virtualized computing resource. A mobile consumer can interact with fog servers directly via a single-hop wireless link, such as 4G LTE devices, WiFi, Bluetooth, and other wireless interfaces. As a result, fog servers offer applications which are predefined and information which is precached to mobile users which are independent of cloud resources. Additionally, to get hold of more application tools and computing resources, a direct link between the cloud infrastructure and the fog servers is frequently built through wireless or wired connections such as via cellular networks. Demand from mobile clients can be supported by several VMs built in many fog servers using the fog computing infrastructure. As a result, the service which is requested could be broken down into a collection of primitive services primitives such as job tasks that would then be performed on newly generated VMs after calculation of optimal scheduling among the machines. Mobile devices are paid per CPU hour per virtual machine, which is offset by a subscription fee. The job scheduling issue necessitates the use of an approach which is optimal, also known as a broker approach that ensures the division of tasks across multiple fog devices in order to satisfy tens thousands of mobile user queries per second. Consolidating VMs is a useful strategy for improving resource efficiency and lowering energy consumption in fog data centers. It can be applied either centralized or decentralized. Consolidation of VMs is one of the most important mechanisms for developing an energy-efficient dynamic fog system for resource management. It is based on the assumption that consolidating VMs into less physical machines (PMs) will achieve both optimization goals: increasing fog server utilization while lowering fog data center energy consumption. However, since VMs share the PM's underlying physical resources, packing more VMs into a single server can result in poor quality of service (QoS). To address this, VM consolidation (VMC) algorithms are designed to dynamically pick VMs for migration while taking into account the effect on QoS as well as the optimization objectives listed above. Since VMC is an NP-Hard problem, researchers have proposed a variety of heuristic and metaheuristic VMC algorithms with the goal of achieving near-optimality.

The residue of the paper is as follows. Section 2 represents the relevant works shown. The optimization model and scheme for VM consolidation are shown in Sect. 3. Simulation setup and obtained results are discussed in Sect. 4. In Sect. 5, extensive experiment findings are presented, accompanied by winding up remarks and future work in Sect. 6.

## 2 Related Works

A VM consolidation scheme should decide where VMs migrate from and to, as well as which PMs can be disabled; in other words, it should solve the problem of defining the source and destination fog node for live VM migration. VM placement [3], host overload detection, and VM migration selection [4] are among the many works [5] that address this problem from various perspectives. We concentrate on optimization-based VM consolidation methods in this paper, so we primarily address VM placement and VM consolidation methods.

Recently, few researchers focused on computation at the network edge to address the real-time applications and modeled real-time applications like smart city applications, smart medical, smart agricultural and smart home applications using Fog nodes for addressing latency issues like centralized cloud. In [6], Malik et al. presented a fog-based energy-efficient model for massive IoT applications. In our earlier papers [7, 8], we introduced context-aware computing and incorporated fog nodes located at the network edge to address the issues related to latency for IoT applications. We presented a heuristic approach to schedule the service requests to VM of fog nodes along with an efficient service migration approach. Yousefipour et al. [9] proposed a statistical model for reducing power usage and costs in the cloud data center by implementing successful VM consolidation. They then proposed an energy and cost-aware VM consolidation genetic algorithm-based metaheuristic algorithm to solve the problem. Abdelsamea et al. [10], to improve VM consolidation, proposed the use of hybrid variables. For host overload detection, they originated an algorithm based on multiple regression which utilizes the CPU, memory and bandwidth. Multiple Regression Host Overload Detection (MRHOD), the introduced algorithm, greatly reduces energy utilization. Masoumzadeh and Hlavacs [11] proposed a model to give attention to the VM selection mission, as well as a Fuzzy Q-learning (FQL) technique for making optimal virtual machine selection decisions. Using the real-world PlanetLab workload and CloudSim toolkit, they validated their approach. The effect of the cloud simulator option on the implementation of the algorithms and the evaluation results was investigated by Mann [12]. Portaluri et al. [13] compared a bunch of allocators of Virtual Machine for Cloud Data Centers (DCs) that do network resource allocation and joint computing.

# 3 Optimized VM Consolidation Model

## *3.1 Fog Layer Architectural Design*

The device model has three major roles: consumer, cloud broker, and the cloud provider (fog provider). The broker of the cloud is the most important component of the model associated with the framework, as it serves on the account of a middleman among the providers and users to manage the usage as well as the distribution of services of the cloud while keeping efficiency in mind. The cloud broker can also provide device transparency by making cloud providers deceptive to client while allowing users to communicate with the broker rather than directly with the providers.

A user sends a request in the form of cloudlets to the broker and the broker assigns it to the cloud provider. The component of the virtual infrastructure manager checks with the component of the registry on a regular basis for any changes in resource availability. Each cloud provider provides the registry with the necessary details. A new provider that wants to enter the environment fills out its own registration form in the registry, which is explained in Sect. 4.

This segment outlines our VM consolidation strategy. It consists of two sections: (1) Energy-aware GA-based VM placement, (2) overload/under loaded host detection.

These two algorithms work together to provide live VM migration with guaranteed QoS, better resource usage, and lower energy consumption. Figure 1 depicts the

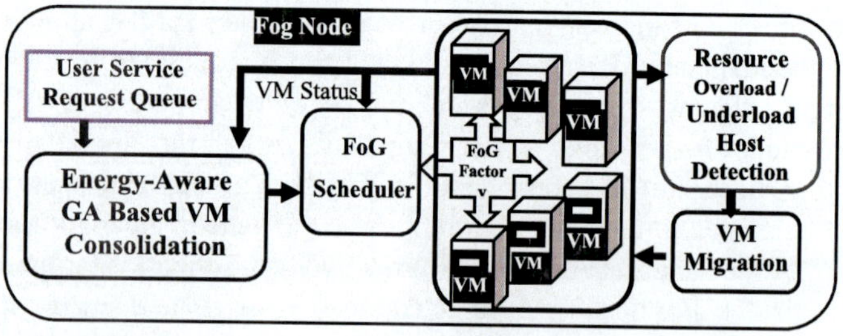


**Fig. 1** Fog node architecture

system model of VM consolidation approach. GA VM placement algorithm is used to schedule the incoming service request to different virtual machines in effective way to minimize the energy consumption. In order to improve the QoS further, resource over load and under load detector is incorporated to detect such conditions and migrates virtual machines to other hosts depending on the situation.

In most data centers, it is important to move some VMs from overloaded PMs to under loaded PMs in order to reduce the possibility of overloading. As a result, this paper proposes two requirements for VM selection. (i) After VM migration, the workload on source PMs becomes more stable; (ii) the period for live VM migration should be as short as possible to minimize its negative impact on QoS. After VM migration, the first criterion can be calculated in terms of changes in overloading probability on source and destination PMs. The precopy mechanism of [14] is used for the second criterion; that is, the smaller the memory consumption, shorter the migration time, effectively reducing the negative impact of VM migration on QoS.

After the VM migration, the source PM must continue to evaluate the overloading risk to ensure that the new source PM has relieved the overloading risk. If the current VM migration does not eliminate the risk of host overloading, the above steps must be repeated before the risk of host overloading is eliminated.

### *3.2 Power Model*

The cumulative power usage of physical servers is referred to as data center power consumption. Each server's power consumption is determined by taking into account two server states: static and running. The static state denotes a server in which no VMs are running but the server is still operational. The VM allocation process is in progress in the running state. Therefore, $P_i$ represents consumption of server's power located in the data center which can be designed by a function of power as given below

$$\text{Power} = P_{\text{idle}} + P_{\text{placement}} + P_{\text{vm}} + P_{\text{switch}} \tag{1}$$

Power denotes the total computation of power of the data center's server $S_i$; $P_{\text{idle}}$ represents the consumption of server's power $S_i$ which is in inactive mode. The consumption of power by the VM instances which are running on the server $S_i$ is denoted by $P_{\text{placement}}$.

The performance of a recent VM which is running on a server without execution of the job is denoted by $P_{\text{vm}}$. The power consumption on a processor when the server switches between on and off states is represented by $P_{\text{switch}}$.

### 3.3 GA-Based VM Consolidation

- **Encoding Schema**

The chromosome is one of the most significant components of the genetic algorithm.

The performance is influenced by the chromosome's representation or encoding. Each chromosome in our proposed methodology has two segments and is $2N + L + M - 2$ in length. Integer encoding was used to describe this situation since every virtual machine must be allocated with a specific instance type. Apart from this, each VM must be assigned to PM at the end of the issue. The $|N + L - 1|$ gene, which represents integer numbers permutation ranging from 1 to $N + L - 1$, is located on the first segment of the chromosome. Delimiters are used for numbers greater than $N$. As a result, the first segment's decoding schema, from the start of the list to the initial delimiter, shows the list of VMs who chose the primary instance type for its own execution. This method of selecting the instance type continues with respect to the delimiter. Here, $N$ denotes the total sum of VMs and $L$ refers to the total count of instance types. If two of the delimiters appear consecutively, it represents that none of the VMs choose the instance form listed.

STAGE 1: VM CHOOSE THE INSTANCE TYPE.

STAGE 2: VM CHOOSE THE PHYSICAL MACHINE [PM].

- **Initialization**

The initial population is created at random by considering the count of servers, virtual machine instance types, and VM itself. Integer encoding schema is used to predetermine the set of values for each gene in order to reduce the time of computation of GA execution.

- **Evaluation Function**

The created chromosome will be compared to a power and cost function; the lower the objective value, the better the chromosome's output. The consumption of power and the cost of placement plan depended upon the information received from the chromosomes is determined by the fitness function.

- **Selection Strategy**

The roulette-wheel selection strategy is used by us in our proposed model, which allocates the selection of each chromosome's probability depending upon the fitness value.

- **Crossover Operator**

In GA, the new people are created from two parents using crossover operators. The crossover percentage in each generation is the parameter used to determine the count of offspring appended to the population by this operator. A segment-oriented crossover operator dependent upon a single-point crossover is utilized. In order to produce offspring, from each segment of the chromosome, a gene is selected on the random basis, subsequently mixing of parents of each produced fragment takes place.

- **Mutation Operator**

Mutation operator, similar to that of crossover operator, is utilized to avoid the betimes convergence and find a new fix. Mutation operator normally shifts a gene's value unlike the crossover operator. The mutation percentage in each generation is the parameter used to determine the count of offspring appended to the population by this operator. According to the proposed algorithm, a parent selects two genes from each section at random from the population. Finally, in order to create a new human, we exchange gene values.

- **Data Center Model**

Assume that a data center consists of a collection of fog nodes denoted by $F$, $F = \{f_1, f_2, \ldots, f_j, \ldots, f_n\}$, and the VMs are denoted by $V$, $V = \{v_1, v_2, \ldots, v_i, \ldots, v_m\}$. In VM deployment relationship, each vector $m_i$ represents a mapping relation between the virtual machine $v_i$and all FNs in data centers.

For the virtual machine $v_i$, the vector $P_i$ is referred to as the deployment vector. The $j$th variable in the vector $P_{i,j} = 1$ if the virtual machine $v_i$ is deployed on the physical machine $f_j$; otherwise, $P_{i,j} = 0$. Since each VM can only be assigned to a single FN, the deployment vector $P_i$ satisfies the restriction condition for VM placement.

$$\sum_{j=1}^{n} p_{i,j} = 1 \tag{2}$$

The configured resource capacity of virtual machine $V_i$ is expressed by $V_i^{\text{res}}$ for each resource, where resource {cpu, mem, and band}. $F_j^{\text{res}}$ stands for each form of resource power on the fog node $f_j$. The utilization of each resource on a FN is calculated using the following formula: (1),

$$v_j^{\text{res}} = \sum_{v_i = V_j} \frac{r_i^{\text{res}}}{f_j^{\text{res}}} \tag{3}$$

where $V_j$ denotes the collection of virtual machines (VMs) installed on the physical machine $f_j$, and $V_i^{\text{res}}$ denotes the actual resource consumption on virtual machine $v_i$ in the FN.

In the case of a fog node $f_j$, the cumulative consumption of all deployed VMs for a particular form of resource cannot exceed the VMs' resource power, i.e.

$$r_i^{\text{res}} < v_i^{\text{res}} \tag{4}$$

As a result, the total requested resources from the deployed VMs must satisfy the constraint condition [3, 5, 6, 9] for the destination PM.

$$\sum_{v_i \in V_j} v_i^{\text{res}} < f_j^{\text{res}} \tag{5}$$

- **Energy Consumption Model**

The fog node's consumption of energy is primarily calculated on four factors:

(1) Consumption of energy at the time of data transmission, (2) consumption of energy at the time of data reception, (3) consumption of energy at the time of computation, and (4) consumption of energy at the time of idle time. Fog node energy consumption $\text{FN}_i$:

$$\text{En}(\text{FN}_i) = \text{En}_{\text{trs}} + \text{En}_{\text{rec}} + \text{En}_{\text{comp}} + \text{En}_{\text{idle}} + \text{En}_{\lambda} + \text{En}_{\text{sleep}} \tag{6}$$

where $\text{En}_{\text{trs}}$ is the total sum of energy consumption in the course of a sole cycle of transmission, $\text{En}_{\text{rec}}$ represents the total quantity of energy consumption during a reception of sole data, $\text{En}_{\text{comp}}$ represents the quantity of energy consumption in the course of computation period, and $E_{\lambda}$ represents the sum of energy consumption during a single on/off cycling period of the fog node, $\text{En}_{\text{idle}}$ is the total sum of consumption of energy in the course of a fog node's idle mode in a set of given time period, and $\text{En}_{\text{sleep}}$ is the sum of the consumption of energy while a fog node's sleep mode for a set period of time.

$$\text{En}_{\text{trs}} = N_{\text{trs}} * E_{\text{amnt}} + S_{\text{level}} \tag{7}$$

During a single information transmission, $N_{\text{trs}}$ is the count of transmissions of a packet with the specified size, $\text{En}_{\text{amnt}}$ represents the quantity of energy absorbed during a sole transmission of information, and $S_{\text{level}}$ represents the level of protection maintained while the transmission of information.

$$\text{En}_{\text{rec}} = N_{\text{trs}} * \text{En}_{\text{amnt}} * S_{\text{level}} \tag{8}$$

The count of times a packet of fixed size is transmitted in the course of a single transmission of data is denoted by $N_{\text{trs}}$.

$$\text{En}_{\text{comp}} = \text{Nc} * E_{\text{Apptype}} * T_{\text{amnt}} \tag{9}$$

Nc is the count of the processed context instances to satisfy the upcoming request during transmission, the sum of energy expended for a specific request from the processed context is denoted by $\text{En}_{\text{amnt}}$ while the time it takes to complete the end user request processing is represented by $T_{\text{amnt}}$.

$$E_{n\lambda} = N_{\text{cycles}} * E_{\text{broadcast}} + N_{\text{cycles}} * E_{\xi} \tag{10}$$

$N$ cycles is the count of on/off patterns performed by a given fog node in a given span of time, $E_{\text{broadcast}}$ represents the energy sum used to send a wake up note to its neighbors, and $E$ represents the energy overhead used by every on/off pattern.

$F_{i,j}$ is a variable in form of binary that indicates, whether the $j$th virtual machine's $i$th fog node is busy or not.

$$E_{i,j}\left(R_r^N\right) = \begin{cases} 1, & \text{if } j\text{th vm of } i\text{th Fog node is busy with a request } R \\ 0, & \text{other wise} \end{cases} \tag{11}$$

where '$R$' denotes the resource type, '$r$' represents the count of necessary slots, and '$N$' denotes the count of necessary slots to complete.

The total amount of energy consumed by an $i$th fog node along with $\text{VM}_j$ virtual machine can then be determined as follows:

$$\text{Total } E(\text{FN}_i) = \text{En}_{\text{trs}} + \text{En}^i_{\text{res}} \sum_{\text{vm}_j}^{n} E_{\text{vm}_j} * E_{i,j}\left(R_r^N\right) \tag{12}$$

It can be observed from Eq. (6) that:

$$\text{if } E_{i,j}\left(R_r^N\right) \text{ is zero, than } \sum_{\text{vm}_{j=0}}^{n} E_{i,j}\left(R_r^N\right) = 0 \tag{13}$$

If all the VMs are not doing anything, then the fog node can go into sleep mode.

The algorithm introduced by us effectively deals with such situations by redirecting arriving requests to the further fog nodes through an effective on/off pattern mechanism, allowing for a longer sleep mode stay. If all the VMs are not doing anything, then the fog node can go into sleep mode.

**GA-Based VM Consolidation Algorithm**

**INPUT:** *VMList, PMList, InstanceTypeList*

**OUTPUT:** *allocation of VMs*

*1. Initially Pop Size, CP, MP /* Num of population, crossover and mutation percentage */*

*2. t ← 0, Termination Condition (t > 1000 OR Nochange in step-6)*

*3. ncß Pop size * CP;   flag = 0 // number of offspring*

*4. nmß Pop size * MP;   repValue = 10 // number of mutants*

*5. population ß InitializePopulation(Pop Size); /* Generation of initial random population */*

*6. curVal = min(EvaluatePopulation(population));*

*7. While (t < 1000 OR flag < repValue) do*

*8.   t = t + 1;*

*9.   For i = 1...nc/2   // Apply selection and crossover*

*10.     parents ß selection(population,2);*

```
        offspringsß Crossover(parents);
        curVal1 = min(EvaluatePopulation(offsprings));
        if(curval1 = = curVal)than
            flag = flag + 1;
        else
            flag = 0;
        endif
        Population.add(offsprings);
    Endfor
    For i = 1...nm        // Apply mutation
        parents ß selection(population,1);
        offspringsß mutation(parent);
        EvaluatePopulation(offsprings);
        Population.add(offsprings);
    Endfor
End While
Population.sort();        /*-- sort the individuals according to their fitness*/
Allocationßpopulation.get(first);    /*-selection of best individual--*/
Return allocation;
End
```

# 4 Simulation Setup and Result Discussion

## 4.1 Simulation Details

The above presented fog framework simulated using CloudSim. Cloudsim is a simulated toolkit which is used for simulation and designing the environments of cloud and fog computing as well as it is used for resource provisioning algorithms evaluations. CloudSim consists of classes which cater as a simulator for different cloud computing components like broker, CIS which stands for cloud information service, VM, cloudlet which represents task/job in cloud computing, data center, and PMs. Scheduling the cloudlet on VMs and in the data center is dependent on two policies—time shared and space shared.

The time-shared scheduling policy assigns one or more processing elements to a VM and requires multiple VMs to share those processing elements. This is demonstrated by the sharing of the compute resources such as the logical processor, central processing unit, and so on. It handles several requests at once and shares the machine's computing resources, so they affect each other's processing time, resulting in degradation of performance.

The space-shared policy assigns one or more processing elements to each VM and does not allow them to be shared. The allocation fails if there are no free processing components. Space sharing, in other words, refers to the sharing of memory space

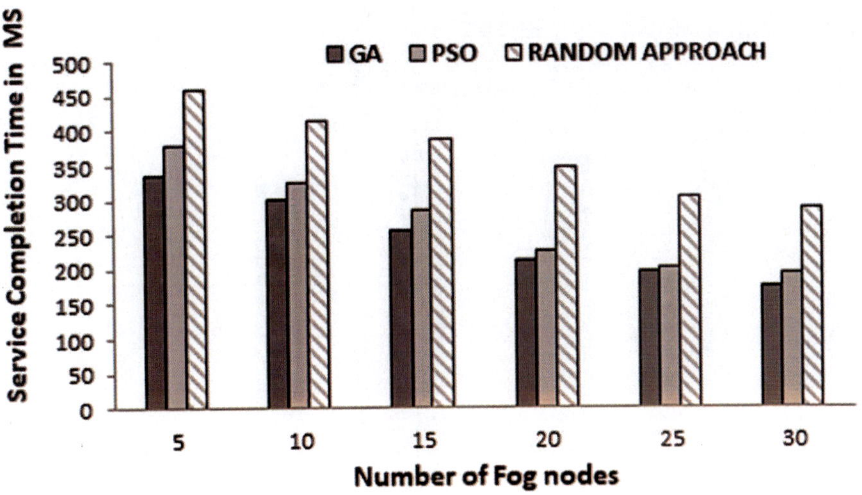


**Fig. 2** Analysis of service completion time with respect to varying fog nodes

such as hard disks, RAM, and so on. For the experiments, we used workload traces data of PlanetLab.

### *4.2 Result Discussion*

The above presented fog environment is simulated in CloudSim, and the following figures depict the efficacy of the proposed model. Figure 2 depicts the service completion time with respect to varying fog nodes for fixed number of service requests, whereas Fig. 3 depicts the fitness value of two optimization techniques. Figure 4 depicts the energy consumption of GA and GA with load balancer along with a randomized approach. Load balancer drastically reduces the energy usage by reducing the number of active fog nodes without violating QoS constraints.

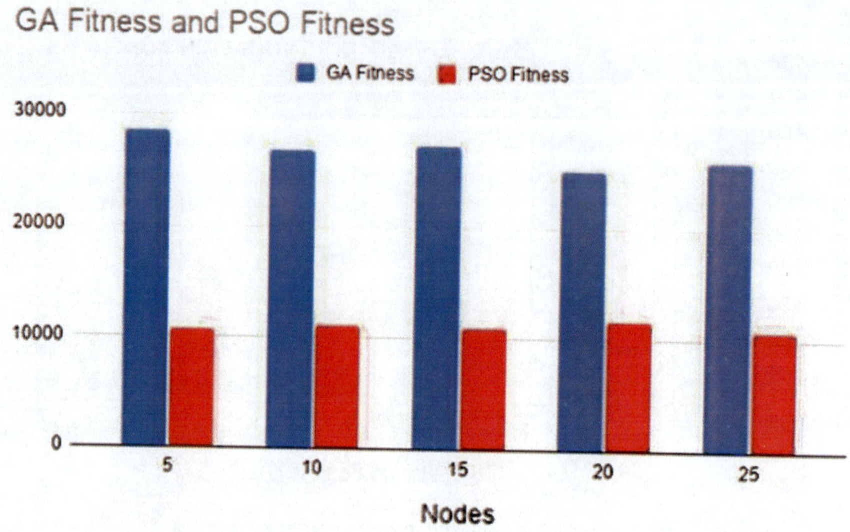


Fig. 3 Analysis of fitness value with respect to varying fog nodes

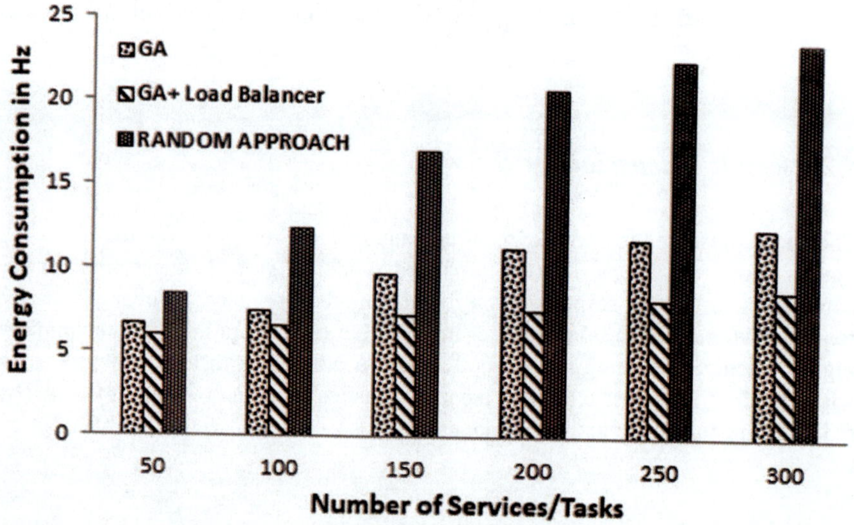


Fig. 4 Analysis of energy consumption with respect to varying tasks

## 5 Conclusion

The conservation of power was extensively considered in fog computing along with an aim to reduce the overall fog center power consumption. Researchers have suggested a variety of methods to achieve this goal which includes both the hardware-based solutions such as DVFS and software-based strategies like server consolidation. We focused on software-based optimizations using genetic algorithm-based techniques in this paper, with a specific emphasis on successful VM placement techniques for VM consolidation. The proposed algorithm uses an objective function based on fitness value to evaluate power and cost, which was implemented using the help of the CloudSim simulation platform.

## References

1. Dastjerdi, A.V., Gupta, H., Calheiros, R.N., Ghosh, S.K., Buyya, R.: Fog computing: principles, architectures, and applications. In: Internet of Things, pp. 61–75. Morgan Kaufmann (2016)
2. Bonomi, F., Milito, R., Natarajan, P., Zhu, J.: Fog computing: a platform for internet of things and analytics. In: Big Data and Internet of Things: A Roadmap for Smart Environments, pp. 169–186. Springer, Cham (2014)
3. Masoumzadeh, S.S., Hlavacs, H.: A cooperative multi agent learning approach to manage physical host nodes for dynamic consolidation of virtual machines. In: 2015 IEEE Fourth Symposium on Network Cloud Computing and Applications NCCA, pp. 43–50. IEEE (2015)
4. Shaw, S.B., Singh, A.K.: Use of proactive and reactive hotspot detection technique to reduce the number of virtual machine migration and energy consumption in cloud data center. Comput. Electr. Eng. **47**, 241–254 (2015)
5. Li, M., Bi, J., Li, Z.: Improving consolidation of virtual machine based on virtual switching overhead estimation. J. Netw. Comput. Appl. **59**(C), 158–167 (2015)
6. Malik, U.M., Javed, M.A., Zeadally, S., ul Islam, S.: Energy efficient fog computing for 6G enabled massive IoT: recent trends and future opportunities. IEEE Internet Things J. (2021)
7. Roy, D.S., Behera, R.K., Reddy, K.H.K., Buyya, R.: A context-aware fog enabled scheme for real-time cross-vertical IoT applications. IEEE Internet Things J. **6**(2), 2400–2412 (2018)
8. Reddy, K.H.K., Behera, R.K., Chakrabarty, A., Roy, D.S.: A service delay minimization scheme for QoS-constrained, context-aware unified IoT applications. IEEE Internet Things J. **7**(10), 10527–10534 (2020)
9. Yousefipour, A., Rahmani, A.M., Jahanshahi, M.: Energy and cost-aware virtual machine consolidation in cloud computing. Softw. Pract. Exp. **48**(10), 1758–1774 (2018)
10. Abdelsamea, A., El-Moursy, A.A., Hemayed, E.E., Eldeeb, H.: Virtual machine consolidation enhancement using hybrid regression algorithms. Egypt. Inform. J. **18**(3), 161–170 (2017)
11. Masoumzadeh, S.S., Hlavacs, H.: Integrating VM selection criteria in distributed dynamic VM consolidation using fuzzy Q-learning. In: Proceedings of the 9th International Conference on Network and Service Management (CNSM 2013), Dec 2013, pp. 332–338
12. Mann, Z.Á.: Cloud simulators in the implementation and evaluation of virtual machine placement algorithms. Softw. Pract. Exp. **48**(7), 1368–1389 (2018)

13. Portaluri, G., Adami, D., Gabbrielli, A., Giordano, S., Pagano, M.: Power consumption-aware virtual machine placement in the cloud data center. IEEE Trans. Green Commun. Netw. **1**(4), 541–550 (2017)
14. Ibrahim, K.Z.: Optimized pre-copy live migration for memory intensive applications, in High Performance Computing, Networking, Storage and Analysis, vol. 26, pp. 1–11. IEEE (2011)

# Spatio-temporal Analysis of Flood Hazard Zonation in Assam

**Sanjiban Roy, Sanjiv Kumar Ojah, Nilay Nishant, Pankaj Pratap Singh, and Dibyajyoti Chutia**

**Abstract** Flood, the most catastrophic natural disaster of the globe is affecting the life of the people each year in both aspects; economically and socially. It has been causing enormous damage to livelihood, infrastructure, property, etc. Assam is the most flood-affected state of Northeast India which is inundated by floods are caused due to the Brahmaputra and its tributaries. The traditional approach of flood monitoring necessitates ground truth information, surveys, manpower, and other resources; this procedure is time-consuming, labor-intensive, and requires continuous monitoring. Satellite remote sensing data, serve as beneficial input for monitoring the flood. It is known that the continuous monitoring of vast flood-affected areas is only possible using remote sensing (RS) technology. This work's framework is exhibited for Assam's several districts. The result of this framework shows the spatial hotspots zone of different flood hazards and vulnerability assessment of the districts in a statistical manner with Moderate Resolution Imaging Spectroradiometer (MODIS). Assam has a flood hazard zone covering 25.44% of its territory. The very flood sensitive area was predicted to cover 8% of the total land, with districts like Barpeta and Morigaon having the most acreage under flood zone.

**Keywords** MODIS · Near real-time flood data · Flood percentage · Assam · Flood hazard

---

S. Roy (✉) · S. Kumar Ojah (✉) · P. Pratap Singh
Central Institute of Technology, Kokrajhar, India
e-mail: p20cse1003@cit.ac.in

S. Kumar Ojah
e-mail: ojahsanjiv@gmail.com

P. Pratap Singh
e-mail: pankajp.singh@cit.ac.in

N. Nishant · D. Chutia
North Eastern Space Application Centre, Umiam, India
e-mail: d.chutia@nesac.gov.in



D. Gupta et al. (eds.), *Pattern Recognition and Data Analysis with Applications*, Lecture Notes in Electrical Engineering 888,
https://doi.org/10.1007/978-981-19-1520-8_42

## 1 Introduction

Floods which are considered as significant natural hazards [1] encountered perennially almost each year in several parts of the North-Eastern states of India most often from April to September which results in significant loss of livelihood. The most flood-affected state in North-East India is Assam, particularly in Brahmaputra valley.

Earlier reported data shows that Assam has positioned itself as the worst flood-affected state in India. Every year in the monsoon season the Brahmaputra and its tributaries have been causing floods in the valley, which leads to a huge loss of lakhs of hectares in agricultural lands [2]. According to RBA (viz., Rashtriya Barh Ayog), the flood-affected area is 31.05 lakh hectares (approx., 39.58% total land).

The state of Assam is located in a heavy rainfall region and the mighty River Brahmaputra along with its tributaries flowing through the state is the main reason of occurring yearly floods which affect the agriculture economics and livelihood of the valley. Thus it can be observed that the current situation as well as days to come is worsened due to the effect of climate change and manmade hazard.

The floods in Assam occurred every year. The river all over the valley encounters the highest level of water and strongest flows during the monsoon season. Due to heavy rainfall along the foothill of the bordering valley the water of the Brahmaputra rose; flooding its tributaries resulting in a flood all over the valley. Thus, the severe cases of flooding impacted by various reasons call for monitoring of flood and control measures in the state [3].

The requirement of frequent monitoring and mapping of flood calls for a satellite-based remote sensing approach. Due to recent development in RS technology, it can capture high-resolution with reasonable accuracy flood data [4]. Earlier studies conducted all over the globe show that satellite images are a rich source of information for capturing any disaster events and controlling them [5].

A flood hazard map is considered required by several departments in order to mitigate and plan for possible flood-related damages [6]. Administrators and planners in developing countries where a considerable proportion of the population lives in flood-prone areas can utilize flood hazard maps to identify areas of flood hazard and priorities mitigation operations [7]. The traditional approach of creating flood hazard zone maps necessitates extensive field surveys and the incorporation of near real-time information regarding flood plains, flood duration, river configuration, and other factors, which is a time-consuming, complex, and costly task. In the realm of flood disaster management, remote sensing has emerged because of its cost-effective and data in close to real-time [8].

In this study, our main focus was on the Assam MODIS near real-time flood data. The threshold to flood data such that we consider only above 50% flood-affected area and also consider only the significant districts. Our research study's major goal is to identify the flood's impact on various resources and to act to avoid it in the future.

## 2 Study Area and Data Used

Assam is the North-Eastern state of the country India surrounded by Manipur and Nagaland to the east, west by Meghalaya and Bangladesh, north by the neighboring country Bhutan and the NE state Arunachal Pradesh, and south by the state Tripura and state Mizoram. It has three physical regions, the first one is the Brahmaputra valley, the second one is the Barak valley and the third one is an elevated region surrounded by Nagaland and Meghalaya. The monsoon rain starts from April to September. Assam receives an average of 230 cm of rainfall during this time [9]. Due to the unrelenting monsoon, Assam experiences floods every year affecting the wildlife, livelihood, and infrastructure of the valley (Fig. 1).

MODIS is a sensor onboard TERRA (aka EOS AM-1) and AQUA (aka EOS PM-1) satellite of NASA [1]. In the morning, the satellite TERA revolves around the earth north to the south equator whereas the satellite AQUA revolves from south to north in the afternoon. Terra and AQUA sensors can capture data in 36 different bands, capturing data of the earth every day. These data help with the study of land, ocean, and lower atmosphere.

MODIS NRT global flood mapping product essentially consists of global daily surface water and flood water maps at 250 m spatial resolution [10, 11]. Fifteen days composite flood water percentage data from the year 2013 to 2018 was used in this research study. For a year, we used 24 MODIS flood data sets. We have total of

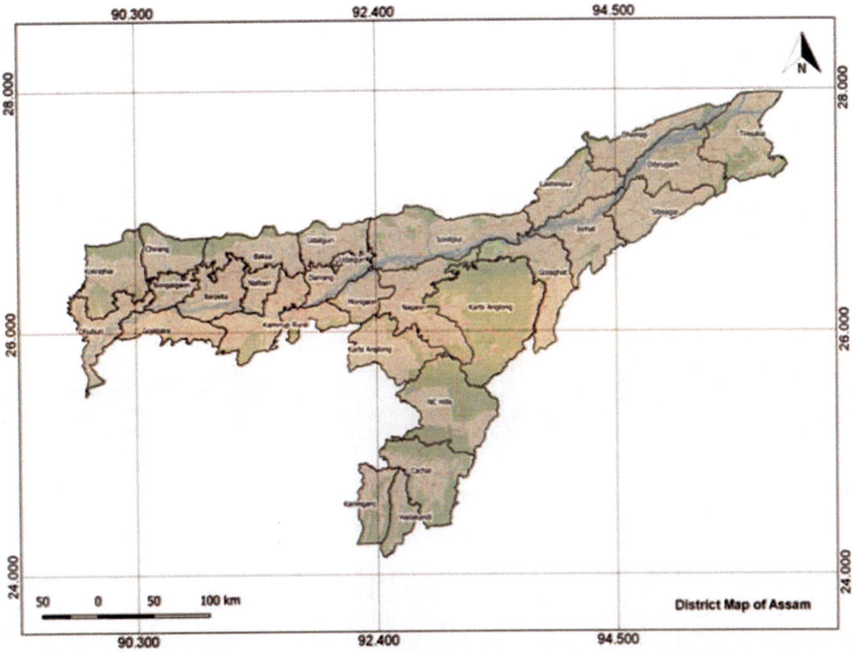


**Fig. 1** Study area map

120 5-year dates ranging from 2013 to 2018. It is a single band composite flood percentage data product.

# 3 Methodology

Figure 2 shows that data preprocessing, research activity, and output of the research to identify flood hazard vulnerability in Assam districts. The flow charts describe different disciplines in this research.

Flood is a dynamic event which changes every year so NRT data is helpful for identification of flood hazard zones.

## 3.1 *Flood Hazard Zonation Schema*

The flood layer was created from 120 satellite dataset obtained from 2013 to 2018. The hazard layer shows different areas based on how many times they have been flooded in the last five years. It is classified into 5 different categories (Table 1), with very high indicating that they possess been flooded 4 to 5 times in past five years. High implies 3–4 times flooded, moderate suggests 2–3 times flooded, low represents 1–2 times flooded, and very low indicates 0–1 time flooded.

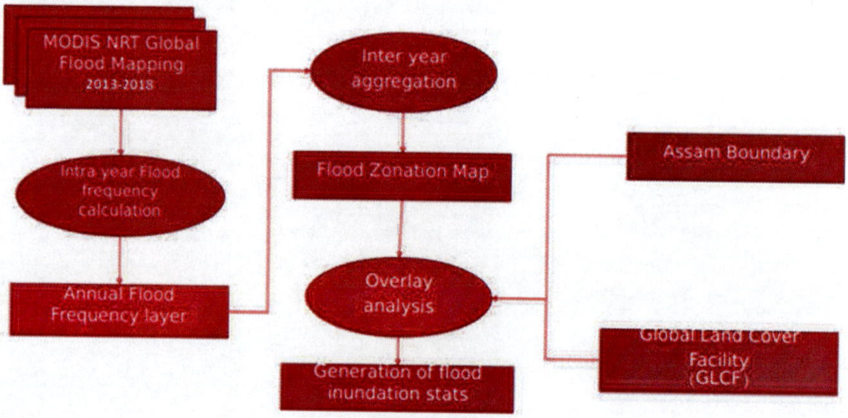


**Fig. 2** Schematic workflow of the methodology

**Table 1** Flood hazard classification

| Flood hazard classification | Number of times/years the area was subjected to flood inundation during 2013–2018 |
|---|---|
| Very high | 4–5 (almost every year) |
| High | 3–4 |
| Moderate | 2–3 |
| Low | 1–2 |
| Very low | 0–1 |

## 3.2 *Flood Hazard Index*

In addition to designating flood zones, the following flood hazard index attempts to estimate the severity of flood in various districts.

$$\text{Flood hazard Index} = \sum \text{Hazard Category(H)} \times \text{Hazard Area(A)}$$

1. Each type of flood hazard (H) was assigned a weighting, as shown in Table 2.
2. As stated in Table 3, weightages were also assigned depending on the percentage of flood hazard area (A) in the district.

**Table 2** Weightage for flood hazard category

| Hazard zones | Weightage for hazard zones (H) |
|---|---|
| Very high | 5 |
| High | 4 |
| Moderate | 3 |
| Low | 2 |
| Very low | 1 |

**Table 3** Weightage for % submerged area

| Percentage of district hazard area (%) | Weightage (A) | Percentage of district hazard area (%) | Weightage (A) |
|---|---|---|---|
| 0–5 | 1 | 46–50 | 10 |
| 6–10 | 2 | 51–55 | 11 |
| 11–15 | 3 | 56–60 | 12 |
| 16–20 | 4 | 61–65 | 13 |
| 21–25 | 5 | 66–70 | 14 |
| 26–30 | 6 | 71–75 | 15 |
| 31–35 | 7 | 76–80 | 16 |
| 36–40 | 8 | 81–90 | 17 |
| 41–45 | 9 | 91–100 | 18 |

### 3.3 Computation of Intra-Year Flood Frequency

Per-pixel cumulative analysis was computed on the NRT flood data to identify the areas survey affected by the flood. The operation was performed using python with the help of the Geospatial Data Abstraction Library (GDAL). GDAL is an open-source software library package used for read and write vector and raster image data. Masking of the River Brahmaputra from each flood frequency raster layer has been processed because it was observed that the river itself was identified as high flooded zones. So, for the removal of the river, a pixel-based approach has been performed.

### 3.4 Generation of Flood Hazard Zonation

Intra-year flood frequency layers are reclassified and assimilated to generate the different flood hazard maps for the state. Reclassification is performed to quantify the intra-annual information of flood statistics. In reclassification, we have created some rules which will identify the entire pixel in between the range 0–100%. The flood area which has a pixel value below 30 was categorized as low, 30–60 as moderate, and above 60 as high. Post reclassification, the flood hazard layer is prepared by aggregating the raster-based on categories of the raster map. It was done using Quantum GIS (QGIS), an open-source cross-platform. The flood hazard zone map is prepared by aggregating all the intra-year flood frequency datasets in an empirical equation. District level statistics are computed based on the regions falling in different categories of flood hazard, the zonal statistics are computed using an open-source software tool viz., Quantum GIS by overlaying the district boundary of Assam with the Assam flood hazard map.

## 4 Results and Discussion

Assam has experienced floods in past due to high slopes with the dense drainage system. The present research study shows the satellite data observations from MODIS TERRA and AQUA platform obtained during the flood season of 2013–2018 can be utilized as a quantitative proxy for estimating the flooded acreage of flood-affected areas.

The study shows that the flood situation was occurred because of heavy rainfall from April to September over the valley with high-intensity rainfall. As a result, the water level of the Brahmaputra and its tributaries was raised causing full or partial submerge of various land use land cover (LULC) all over the region. Flood analysis is done for the years from 2013 to 2018 and hazard areas were estimated for the state. The flood severity level is shown in five classes: no flood zone, 10–20% indicates less flood, 20–30% indicates moderate flood, 30–40% indicates severe flood, and above 40% indicates a very severe flood. Much of the flood happening in the Cachar,

Karimganj, Hailakandi, and KarbiAnglong shows high flood percent in 2013, 2015, 2016, and 2017, among them the year 2015 shows the highest flood percent. The year 2014 shows the least flood it is because 2014 was declared as a drought year.

About 25.44% of land in Assam is under flood during 2013–2018 (Table 1). Out of total flood-affected area (19.96 lakh hectares), around 2% of land comes under very high flood inundated area (more than 40% times), high flood inundated area (30–40% times) in flood hazard categories. Within flood-affected zones, the percentage area of each flood hazard category varies from 8 to 21%. Fig. 1 shows the graphical distribution of area under different hazard categories (Figs. 3 and 4).

A flood hazard zonation map is created based on a composite of all the years, i.e., 2013–2018.The map depicts that majority of the region along the Brahmaputra

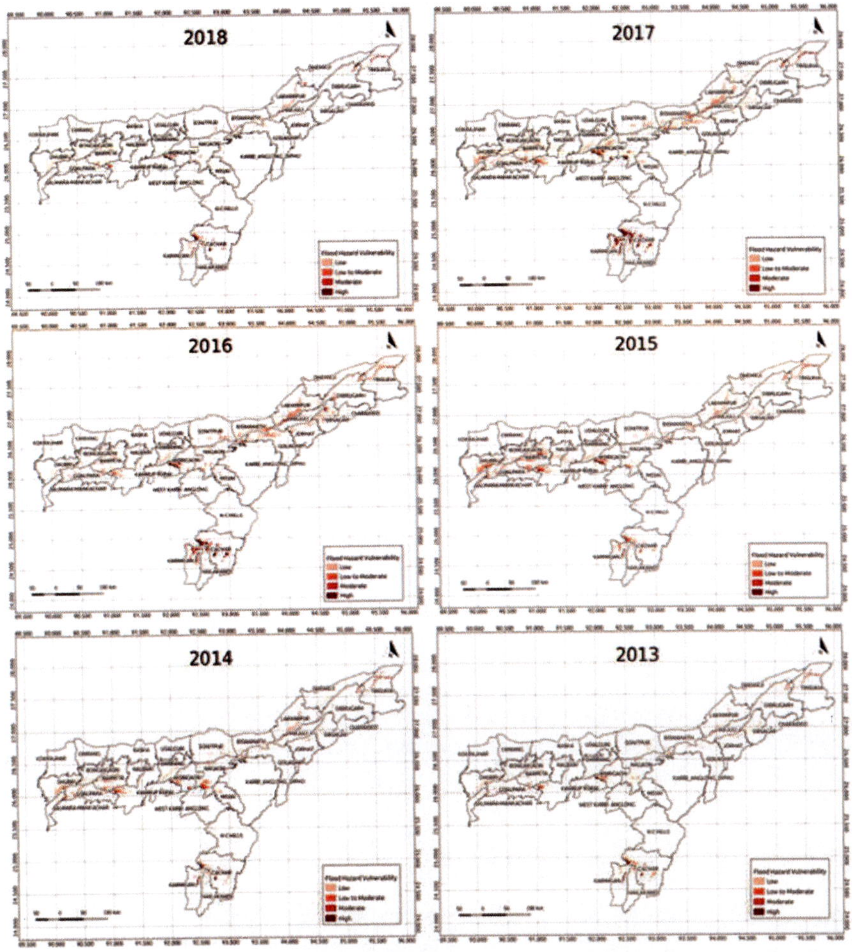


**Fig. 3** Intra-year flood frequency map generated from MODIS NRT data

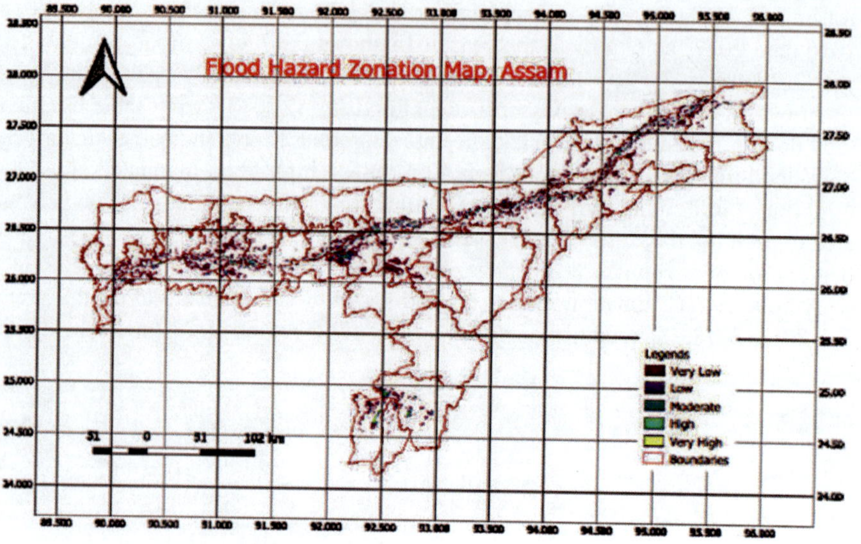


**Fig. 4** Flood hazard zone map for Assam

valley and some portions of Barak valley fall under the flood hazard zonation. Upon further investigation, it has been observed that districts such as Barpeta, Morigaon, and Dibrugarh are majorly affected by a flood. With reference to Table 4, it is seen that 25% of the flood-affected area falls under very low-risk zones, these low-risk zones comprise 6.32% of the total LULC of the state.

Due to the varied time frame, number of datasets, and resolution of the data, the discrepancy between MODIS and Bhuvan data is very considerable in very low and very high flood hazard zones, as shown in Figs. 5 and 6.

Except for the high-risk vulnerable area, it is observed that all zones are equally distributed. With all the remaining zones comprising 20–25% of the flood-affected region and a very high flood zone comprising 8% of the flooded zone. District level distribution can be observed. District level acreage distribution of hazard zones can be estimated from Fig. 7.

**Table 4** Statistics on flood hazard area under various categories

| Hazard severity | Flood hazard area (ha) | % Flood hazard (w.r.t. state geographic area) | % Flood hazard (w.r.t. total flood hazard area) |
|---|---|---|---|
| Very low | 495,765 | 6.32 | 25 |
| Low | 479,822 | 6.11 | 24 |
| Moderate | 449,496 | 5.73 | 22 |
| High | 414,672 | 5.28 | 21 |
| Very High | 157,241 | 2 | 8 |
| Total | 1,996,996 | 25.44 | 100 |

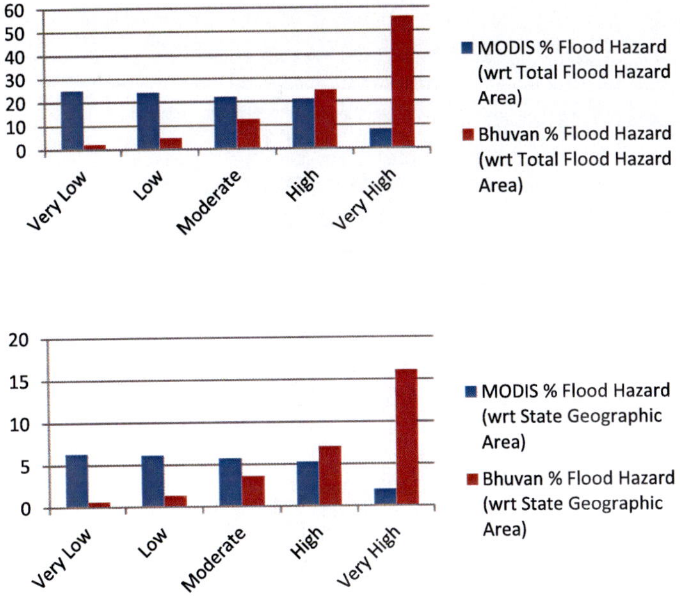


**Fig. 5** Comparison of MODIS and Bhuvan % flood hazard w.r.t. to total flood hazard area and state geographic area

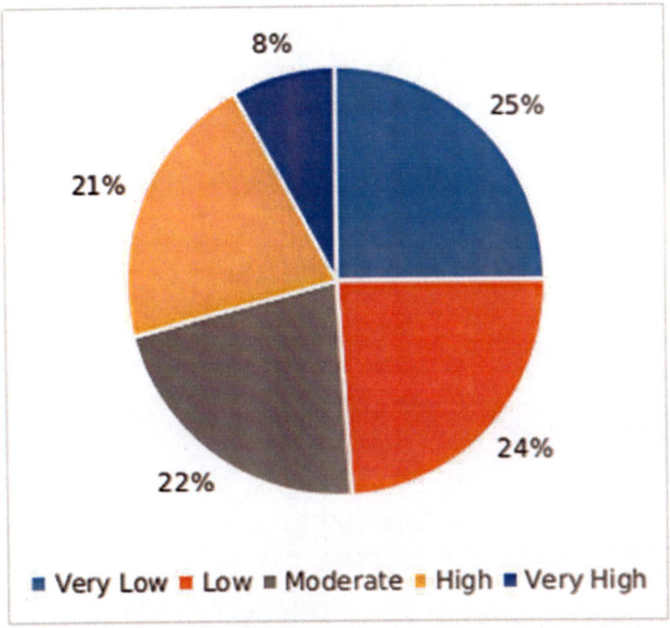


**Fig. 6** Distribution of different flood hazard zones among the state

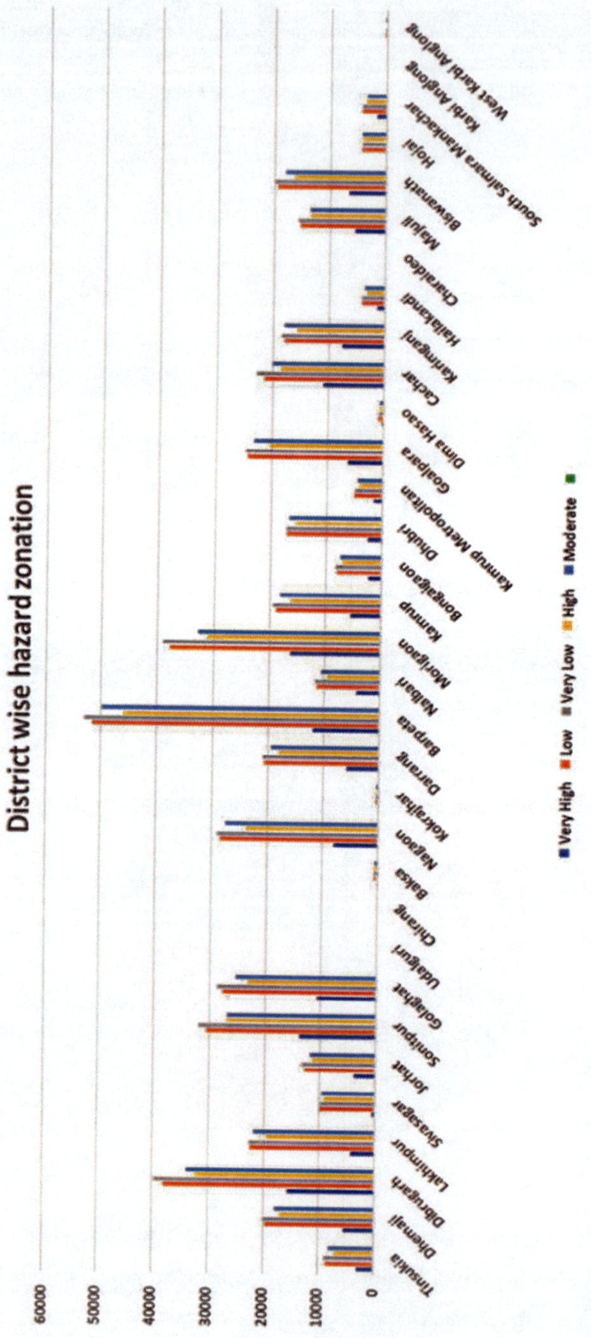


**Fig. 7** District-wise area distribution of flood hazard zonation

## 5 Conclusions

The study illustrates that space-based observations helped in the precise evaluation of flood severity and risk in Assam as well as to understand the source and factors of the flood. The accurate extent of the flood-affected area can prove a vital input for the assessment of crop damage and other geophysical analysis. Heavy rainfall under the factor of climate change may lead to similar flood conditions in the future. Therefore, the proper management to runoff the water during heavy rainfall should be done. The built-up development in the state should be limited keeping in view of the river flows. Likewise, the water flowing capacity of the river and lakes may be increased to absorb the excess rainfall from the hilly areas surrounding the valley. In the future, we can explore ML/DL techniques to extract water bodies and satellite images of flood-affected areas with high-resolution like SENTINEL/LANDSAT-8.

## References

1. Khan, S.I., Hong, Y., Wang, J., Yilmaz, K.K., Gourley, J.J., Adler, R.F., Brakenridge, G.R., Policelli, F., Habib, S., Irwin, D.: Satellite remote sensing and hydrologic modeling for flood inundation mapping in Lake Victoria basin: Implications for hydrologic prediction in ungauged basins. IEEE Trans. Geosci. Remote Sens. **49**(1), 85–95 (2010)
2. Ganguly, K., De, S.K.: Spatio-temporal analysis of flood and identification of flood hazard zone of west Tripura district, Tripura, India using integrated geospatial technique, hill Geographer. **XXXI**(1), 1–22 (2015)
3. Baky, M.A.A., Islam, M., Paul, S.: Flood hazard, vulnerability and risk assessment for different land use classes using a flow model. Earth Syst. Environ. **4**, 225–244 (2020)
4. Afifi, Z., Chu, H.J., Kuo, Y.L., Hsu, Y.C., Wong, H.K., Ali, M.Z.: Residential flood loss assessment and risk mapping from high-resolution simulation. Water **11**, 751 (2019)
5. Islam, A.S., Bala, S.K., Haque, M.A.: Flood inundation map of Bangladesh using MODIS time-series images. J. Flood Risk Manage. **3**(3), 210–222 (2010)
6. Baldassarre, G.D.: Floods in a Changing Climate: Inundation Modelling. Cambridge Univ. Press, Cambridge (2012)
7. Azizat, N., Omar, W.M.S.W.: Spatial and temporal flood risk assessment for decision making approach. IOP Conf. Ser.: Mater. Sci. Eng. **318**, 012022 (2018). https://doi.org/10.1088/1757-899X/318/1/012022
8. Zhang, F., Zhu, X., Liu, D.: Blending MODIS and Landsat images for urban flood mapping. Int. J. Remote Sens. **35**(9), 3237–3253 (2014)
9. Sharma, S.V.S.P., Rao, S.G., Bhatt, C.M., Sree, M., Veerubhotla, B.: Development of flood hazard maps for Assam state, India using historical multitemporal satellite images (36) (2012)
10. Zhan, X., Sohlberg, R.A., Townshend, J.R.G., Dimiceli, C., Carroll, M.L., Eastman, J.C., Hansen, M.C., DeFries, R.S.: Detection of land cover changes using MODIS 250 m data. Remote Sens. Environ. **83**(1–2), 336–350 (2002)
11. Brakenridge, R., Anderson, E.: MODIS-Based Flood Detection, Mapping and Measurement: The Potential for Operational Hydrological Applications, Transboundary Floods: Reducing Risks Through Flood Management, pp. 1–12. Springer, Dordrecht (2006)

# A Pilot Study on Human Pose Estimation for Sports Analysis

Pranshu Sharma, Bishesh Bikram Shah, and Chandra Prakash

**Abstract** Human pose estimation is the identification and detection of different poses of a human through the information collected from body part movements where the body parts refer to the joints and the bones. By referencing a video, it can calculate accurate poses and body movements for athletes so that they can accomplish optimum results in their performance. Pose Estimation can also be further used to identify the health condition of a particular player. We have developed a model which identifies various anatomical key points of a person in a given image or a video (frames) for Pose Estimation. We further attempt to extract insights on the body movement of an athlete to carry out analysis of their running behavior. The model accurately extracts 18 anatomical key points (like Hip Joint, Knee joint, Ankle Joint, etc.) without the need of any laboratory settings and special sensors, which makes it easy for anyone to use and implement the model. The model used is based on the MobileNet CNN architecture. For analysis we use various gait parameters such as Cadence, Knee angle, and Velocity. We further attempt to compare the results of the striding patterns with the running patterns shown by an athlete. The model is able to track the body movements of an athlete and then output the various gait parameters associated with these body movements. The implementation of the model has been made easy to assist the athletes in achieving optimal performance without the need of personal trainers and equipment, which can be quite costly.

**Keywords** Pose estimation · Gait parameters · MobileNet · Cadence

P. Sharma (✉) · B. B. Shah · C. Prakash
National Institute of Technology, Delhi, India
e-mail: 171210043@nitdelhi.ac.in

C. Prakash
e-mail: cprakash@nitdelhi.ac.in



D. Gupta et al. (eds.), *Pattern Recognition and Data Analysis with Applications*,
Lecture Notes in Electrical Engineering 888,
https://doi.org/10.1007/978-981-19-1520-8_43

## 1 Introduction

Human pose estimation is the identification and detection of the different poses of a human through the information collected from body part movements where the body parts refer to the joints and the bones. Human Pose Estimation has various useful applications like- physical therapy. It can also be used to detect postural issues (like scoliosis) by analyzing the abnormalities in a patient's posture and others, hence the domain of Human pose estimation has vast applications. In sports, Human pose estimation (combine with gait analysis) can also be used to identify the health condition of a particular player, predict optimal posture and recommend body movement behaviors to help athletes achieve better performance results. Human pose estimation is based on detecting the key body parts of a person and extracting the various coordinates associated.

Most of studies done are focused in 3-D analysis, in which multiple cameras are placed at a fixed distance from one another to capture every movement in the 3-D space. However, this method is quite expensive and hence requires a proper laboratory setting. This leaves us to use 2-D analysis to predict the behaviors which can be easily accessible to athletes. In the work presented by Zhe Cao et al., confidence maps and the concept of part affinity field have been deployed to detect the body parts of an individual and connect them to form a whole body [1]. Based on this work the OpenCV library for Python which has been open-sourced contains the state-of-the-art work with extraordinary accuracy on various public benchmarks.

In Human Pose estimation models, the objective is to identify pre-defined points of interest on a person's body (body joints and organs) and then subsequently link them to draw a computer generated "skeleton" for every person in the image [2]. The "Skeletal motion" can then be traced across various video frames and translated to study body kinematics, so as to use it directly to analyze running performances of athletes [3].

Sport is an activity where heavy physical exertion and skills are involved. For better result with minimum injuries many sports individual and teams have to hire expensive coaches. Instead of hiring them we can create a model using machine learning, more precisely, using pose estimation we can determine the perfect posture for a sprinter to achieve maximum acceleration [4], a better angle of shoulder for a javelin thrower to throw javelin at maximum distance, a better tactics for the football to achieve highest win rate in a league, etc [5].

## 2 Recent Work

Human pose estimation has been done previously by Cao [1], where it uses confidence maps along with Part Affinity Fields, a nonparametric representation that allows identifying different joints on a person's body. Then finally a greedy bottom-up parsing step outputs the 2-D key points for all people in the image. This work has been identified on various public benchmarks and is the backbone of the work carried out by us.

Another impressive approach is given by Alexander Toshev, where the pose estimation is formulated as a regression problem toward body joints based on a DNN architecture [6]. A full image and a 7-layered generic convolution DNN are taken as an input and the location of each joint of the body is regressed. This work provides a good approach but is primitive and is unable to provide high accuracy in case of a moving person, where self-occlusion is high. Similarly, Adrian Bulat, uses convolutional part heatmap regression to estimate human pose but suffers the same problem in case of occlusion [7].

The recent work by Koen van der Meijden presents his research done for Sport Analysis in which the sprinting patterns using Open Pose are studied [8]. The work identifies four behavioral sprinting features and analyzes a series of videos of athletes sprinting (100 m) by extracting coordinates of body parts identified by using pose estimation method. Thus, the results to increase performance of a person in Sprinting and various postural obstacles are discussed. This work required proper laboratory setup like using high-definition equipment for the implementation. In contrast to this our work, provides a model which does not required any special laboratory setup and can take input from a mobile phone camera. Also we provide analysis of some general gait parameters, which can easily be further used to gain detailed insights and aid an athlete to analyze and improve their performances.

## 3 Methodology

### *3.1 Dataset*

The dataset used for training the OpenPose algorithm developed by Zhe Cao et al.(2016) is the COCO dataset. The OpenPose algorithm has been used to compile the OpenCV repository in Python, which allows us to carry out the Pose estimation from a given image/frame.

For Sprint analysis, the model used is based on the MobileNet architecture which allows the model to run, even on a video recorded from a mobile phone camera, with appropriate results. Since a dedicated dataset of images or videos of athletes sprinting was not available to us, hence we have carried out the analysis of the model on videos taken from our mobile phone camera.

### 3.2 MobileNet Model

MobileNet is a streamlined architecture which uses depthwise separable convolutions to construct lightweight deep convolutional neural networks, thus providing an efficient model for mobile and embedded vision applications [9].

Depthwise convolution filters and point convolution filters constitute to form Depthwise separable convolution filters. A single convolution on each input channel is performed by the depthwise convolution filter and the output of depthwise convolution is linearly combined with 1 ∗ 1 convolutions by the point convolution filter, shown above in Fig. 1.

A standard convolution, in a single step, both filters the inputs as well as combines them into a new set of outputs. The depthwise separable convolution divides this into

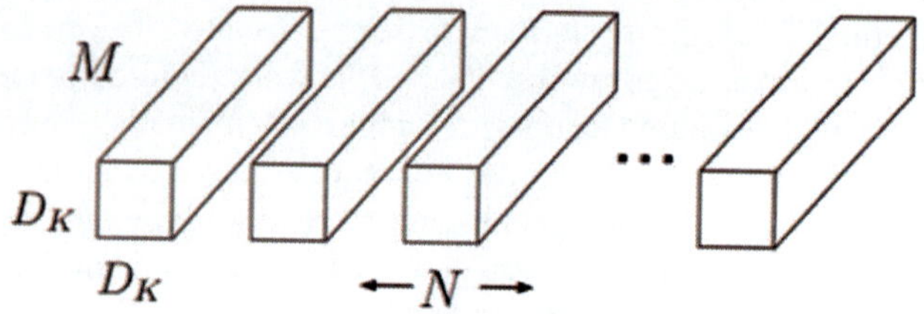


**(a) Standard Convolution Filters**

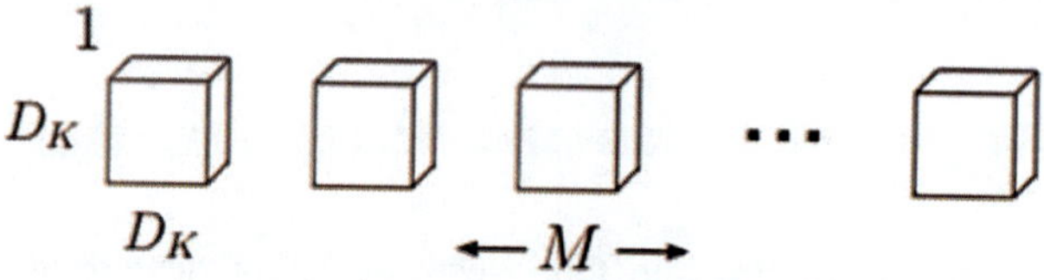


**(b) Depthwise Convolutional Filters**

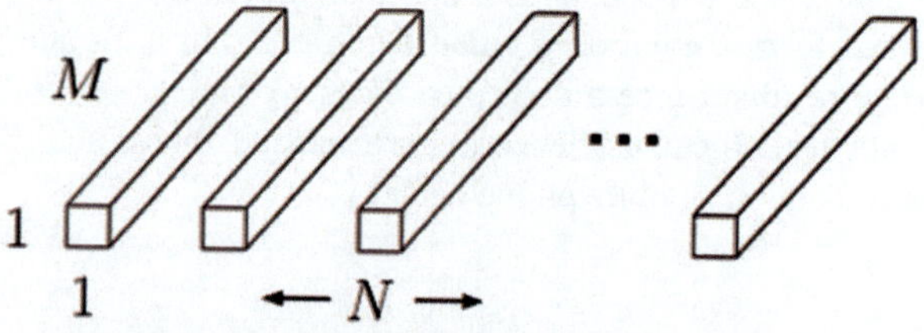


**(c) 1 × 1 Convolutional Filters called Pointwise Convolution in the context of Depthwise Separable Convolution**

**Fig. 1** Standard convolutional filters and depthwise separable filters [9]

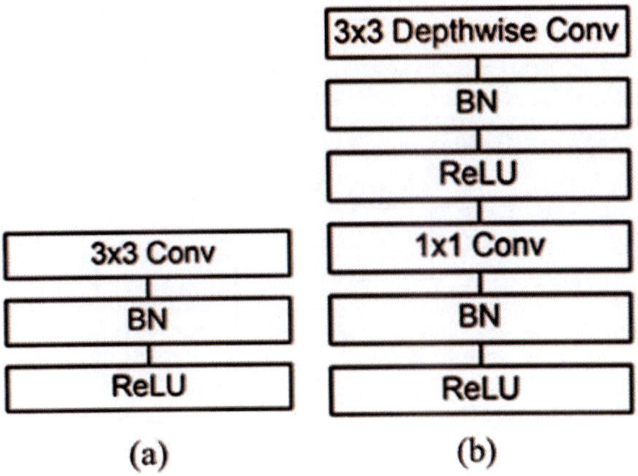


**Fig. 2** **a** Standard 3 × 3 convolutional with batch-norm and ReLU. **b** Depthwise separable convolution with pointwise and depthwise layers followed by batch-norm and ReLU [9]

2 separate layers, a discrete layer for combining and a discrete layer for filtering. This factorization results in tremendously reducing computation as well as the model size.

In contrast to a traditional CNN, where a single 3 × 3 convolution layer is followed by the batch normalization and ReLU, the MobileNet architecture is splits the convolution into a 3 × 3 depthwise convolution and a 1 × 1 pointwise convolution followed by batch normalization and ReLU non-linearity after every convolution layer, as shown in the figure below (Fig. 2).

### *3.3 Human Pose Estimation*

This work makes use of the open-source repository OpenPose available, which represents an approach to identify the 2-D human pose in a given image. A nonparametric representation (i.e., Part affinity field) is able to associate different body key points with individuals in an image (Fig. 3).

The image is taken as the input. The input image is first analyzed by a convolutional neural network (first 12 layers of MobileNet model) and produces a set of feature maps. These feature maps are the input of confidence map and part affinity fields. The confidence map shows the joints whereas part affinity fields associate the orientation of the joints. For each joint of each person we have single confidence map and part affinity field. The set of 2-D confidence maps for the various body part locations is known as part confidence maps. A confidence map is associated to every joint location is derived. This set of 2-D vector fields (Part Affinity Fields) helps in encoding the degree of association between all such body key points. Finally, these confidence maps as well as the Part Affinity Fields are processed together through greedy parsing to obtain and estimate the pose(s) for each person in the image.

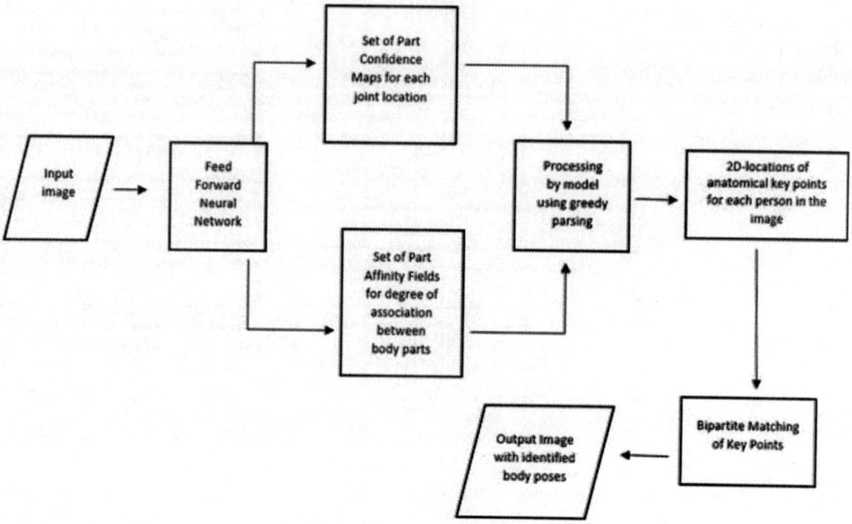


**Fig. 3** Workflow of pose estimation in the model

## 3.4 Calculations

**Knee angle**

By obtaining the Hip, Knee, and Ankle coordinates of the athlete for each leg, we can easily calculate the knee angle for both of the legs in a gait cycle. We construct 3 vectors, namely, $P12$, $P23$, and $P13$ by taking any 2 points at a time (eg. $P1(x, y)$ and $P2(x, y)$) and calculate the length of each such vector by distance formula:

$$P_{12} = \sqrt{\left((P1_x - P2_x)^2 + (P1_y - P2_y)^2\right)} \quad (1)$$

By obtaining the lengths of all 3 vectors ($P_{12}$, $P_{23}$, and $P_{13}$) we can easily find the knee angle by assuming the knee coordinate to be the central point ($P_2$) and applying law of cosines:

$$\text{Angle} = \cos^{-1}\left(\left(P_{12}^2 + P_{13}^2 - P_{23}^2\right)/(2 * P_{12} * P_{13})\right) \quad (2)$$

**Determining Velocity**

Velocity of the athletes is known to be the best comparative factor for studying the performance of the athletes, which can in turn be useful as the dependent variable in various regression models. The average velocity of a particular body part of an athlete in the video can easily be calculated with the equation:

$$v = \Delta x \Delta t \tag{3}$$

Here $\Delta x$ represents the total distance covered by the athlete and $\Delta t$ is the time taken for completing the task. Since the $\Delta x$ can be measured in pixels only (not meters) from the video, we can calculate the distance traveled of the hip joint along the $x$-axis in the video. Thus by dividing this distance(in pixels) with the time ($\Delta t$) in frames, gives us velocity ($v$) in pixels per frame. This can then be used to obtain valuable insights regarding the movement of the athletes.

# 4 Results

With the help of Transfer Learning we implement our model which makes use of the MobileNet model based on the OpenCV library, provided by ildoonet/tf-pose-estimation [10]. The MobileNet model requires only about 7 MB and can easily be implemented even on videos taken from a mobile phone camera without the need of special cameras or sensors. We estimate the body pose of a human by using the model and thus extract the coordinates of the various key points on the human identified by the model.

This implementation is carried out in Google Colab and the programming language used is Python. The model estimates the poses of a running athlete. After extracting the coordinates of all the body points we attempt to plot them on a graph so as to obtain a graphical representation of the movement of various joints in the human. In this way we can analyze the movement of an athlete and thus by analyzing his movements, the athlete can achieve better results and optimum performance, hence achieving our goal of Sports Analysis.

## *4.1 Video Input*

The videos of an athlete running as well as a video of an athlete striding are taken as input. Both the videos were taken from a mobile phone camera in the sagittal view of a particular athlete. The model processes the video frame-wise and the coordinates of various body points identified are extracted.

We extract the coordinates ($x$, $y$) of the Hip, Knee, and Ankle joints of both Right and Left legs of the athlete, for each frame of the video of athlete. Once these coordinated are extracted, we can calculate the knee angle for the left and right leg. The calculation of the angle is explained in calculations section (above).

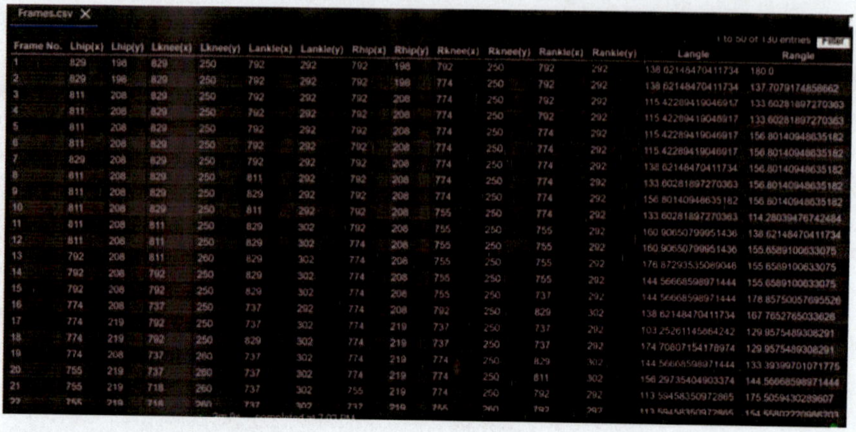

Frames.csv

1 to 50 of 130 entries

| Frame No. | Lhip(x) | Lhip(y) | Lknee(x) | Lknee(y) | Lankle(x) | Lankle(y) | Rhip(x) | Rhip(y) | Rknee(x) | Rknee(y) | Rankle(x) | Rankle(y) | Langle | Rangle |
|---|---|---|---|---|---|---|---|---|---|---|---|---|---|---|
| 1 | 829 | 198 | 829 | 250 | 792 | 292 | 792 | 198 | 792 | 250 | 792 | 292 | 138.62148470411734 | 180.0 |
| 2 | 829 | 198 | 829 | 250 | 792 | 292 | 792 | 198 | 774 | 250 | 792 | 292 | 138.62148470411734 | 137.7079174858662 |
| 3 | 811 | 208 | 829 | 250 | 792 | 292 | 792 | 208 | 774 | 250 | 792 | 292 | 115.42289419046917 | 133.60281897270363 |
| 4 | 811 | 208 | 829 | 250 | 792 | 292 | 792 | 208 | 774 | 250 | 792 | 292 | 115.42289419046917 | 133.60281897270363 |
| 5 | 811 | 208 | 829 | 250 | 792 | 292 | 792 | 208 | 774 | 250 | 774 | 292 | 115.42289419046917 | 156.80140948635182 |
| 6 | 811 | 208 | 829 | 250 | 792 | 292 | 792 | 208 | 774 | 250 | 774 | 292 | 115.42289419046917 | 156.80140948635182 |
| 7 | 829 | 208 | 829 | 250 | 792 | 292 | 792 | 208 | 774 | 250 | 774 | 292 | 138.62148470411734 | 156.80140948635182 |
| 8 | 811 | 208 | 829 | 250 | 811 | 292 | 792 | 208 | 774 | 250 | 774 | 292 | 133.60281897270363 | 156.80140948635182 |
| 9 | 811 | 208 | 829 | 250 | 829 | 292 | 792 | 208 | 774 | 250 | 774 | 292 | 156.80140948635182 | 156.80140948635182 |
| 10 | 811 | 208 | 829 | 250 | 811 | 292 | 792 | 208 | 755 | 250 | 774 | 292 | 133.60281897270363 | 114.28039476742484 |
| 11 | 811 | 208 | 811 | 250 | 829 | 302 | 792 | 208 | 755 | 250 | 755 | 292 | 160.90650799951436 | 138.62148470411734 |
| 12 | 811 | 208 | 811 | 250 | 829 | 302 | 774 | 208 | 755 | 250 | 755 | 292 | 160.90650799951436 | 155.6589100633075 |
| 13 | 792 | 208 | 811 | 260 | 829 | 302 | 774 | 208 | 755 | 250 | 755 | 292 | 176.87293535089046 | 155.6589100633075 |
| 14 | 792 | 208 | 792 | 250 | 829 | 302 | 774 | 208 | 755 | 250 | 755 | 292 | 144.56668598971444 | 155.6589100633075 |
| 15 | 792 | 208 | 792 | 250 | 829 | 302 | 774 | 208 | 755 | 250 | 737 | 292 | 144.56668598971444 | 178.85750057695526 |
| 16 | 774 | 208 | 737 | 250 | 737 | 292 | 774 | 208 | 792 | 250 | 829 | 302 | 138.62148470411734 | 167.7652765033628 |
| 17 | 774 | 219 | 792 | 250 | 737 | 302 | 774 | 219 | 737 | 250 | 737 | 292 | 103.25261145664242 | 129.9575489308291 |
| 18 | 774 | 219 | 792 | 250 | 829 | 302 | 774 | 219 | 737 | 250 | 737 | 292 | 174.70807154178974 | 129.9575489308291 |
| 19 | 774 | 208 | 737 | 260 | 737 | 302 | 774 | 219 | 774 | 250 | 829 | 302 | 144.56668598971444 | 133.39399701071775 |
| 20 | 755 | 219 | 737 | 260 | 737 | 302 | 774 | 219 | 774 | 250 | 811 | 302 | 156.29735404903374 | 144.56668598971444 |
| 21 | 755 | 219 | 718 | 260 | 737 | 302 | 755 | 219 | 774 | 250 | 792 | 292 | 113.59458350972865 | 175.5059430289607 |
| 22 | 755 | 219 | 718 | 260 | 737 | 302 | 737 | 219 | 755 | 260 | 792 | 292 | [illegible] | [illegible] |

Fig. 4 Data stored in CSV file format

## 4.2 Calculation of Data from the Frames

As the video is passed in our model, the coordinates for hip, knee, and ankle as well as the total knee angles for both the legs are stored in a CSV. Furthermore, this data is used for the calculation of cadence and velocity (pixels/sec) of the person. Following is the snapshot of the CSV file data (Fig. 4):

This file contains coordinates of knee, hip, and ankle joints for both legs and also the knee angles for both the legs.

This CSV file helps to remove the irregular coordinates, i.e., the coordinates which are outliers or which shows unexpected behavior. We use interpolation method of the pandas. Data frame class of python to fill in these irregularities. The data used for analysis consists of about 110 frames.

## 4.3 Graphical Representation

The graphs below are the plots of the coordinates (*x*, *y*) (in pixels) of the hip, knee, and the ankle joints of the left leg and right leg, respectively, for the leg-movement of a running athlete traversing 4 gait cycles (Fig. 5).

Below are the plots of the total knee angles of the right leg and left leg, respectively, for the leg-movement of the running athlete (Fig. 6).

Similarly, below are the plots of the coordinates (*x*, *y*)(in pixels) of the hip, knee, and the ankle joints of the left leg and right leg, respectively, for the leg-movement of an athlete striding, traversing through 2 gait cycles (Fig. 7).

The graphs below are the plots of the total knee angles of the right leg and left leg, respectively, for the leg-movement of the striding athlete (Fig. 8).

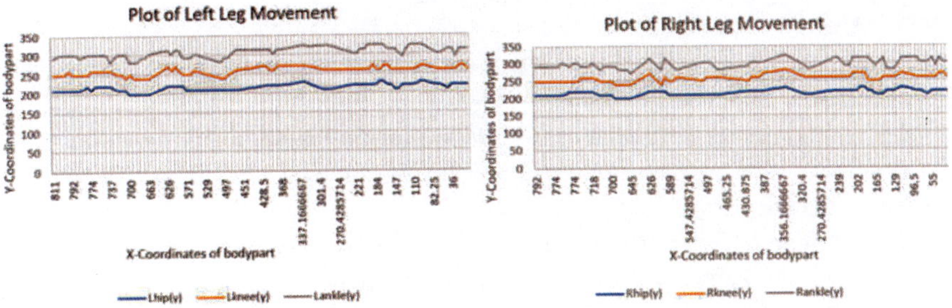


**Fig. 5** Plots of coordinates of hip, knee, and ankle joints (coordinates in pixels)

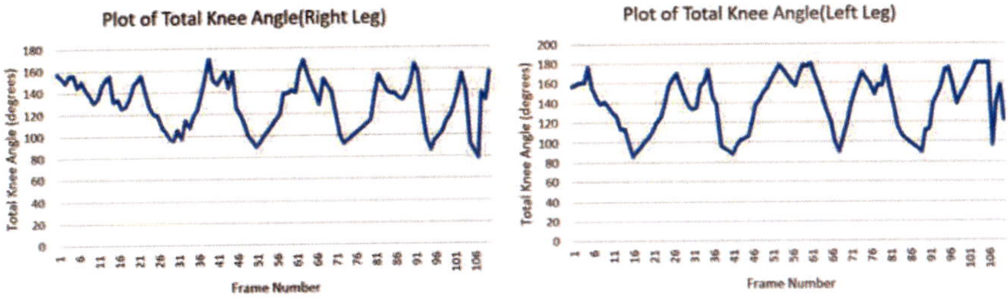


**Fig. 6** Plots of total knee angle for both legs for a running athlete (Angle in degrees vs. Frame No)

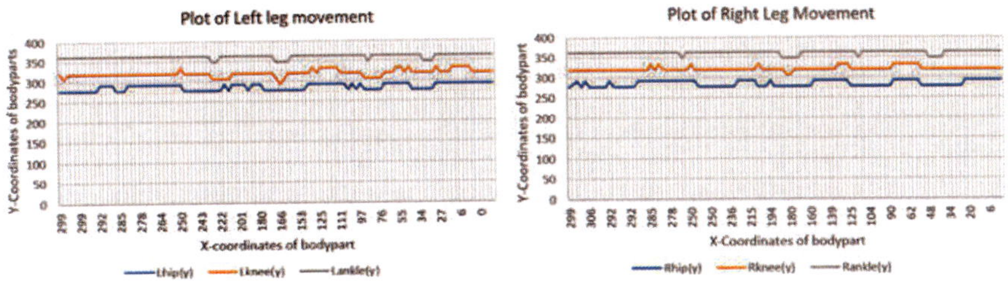


**Fig. 7** Plots of coordinates of hip, knee, and ankle joints (coordinates in pixels)

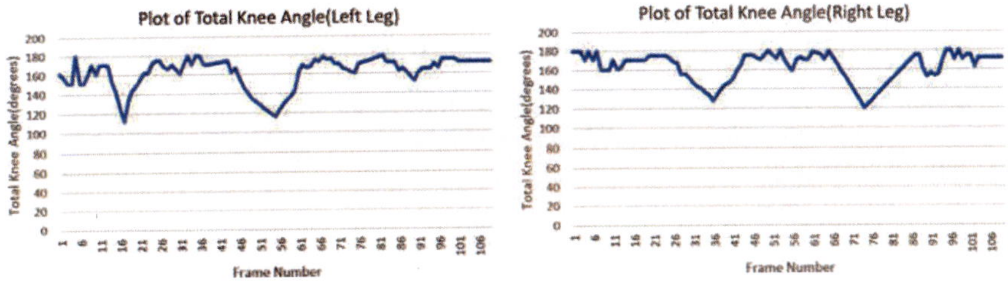


**Fig. 8** Plots of total knee angle for both legs for a striding athlete (Angle in degrees vs. Frame No)

## 4.4 Analysis

By comparing the data for running (video input-1) with that obtained for striding (video input-2) of an athlete, it was noticed that the total knee angle varies in the range of 180°–90° for the running athlete whereas in case of striding it only varies between 180 to 115°.

In running involved sports, Cadence is defined as a measure of the athletic performance of a person and is calculated as the total number of steps taken within a given period of time, often expressed in cycles or steps per minute and acts as. Cadence in gait is calculated by following formula:

$$\text{Cadence} = \text{steps taken/min} \tag{4}$$

**Input video-1**

In the input video-1 it is seen that the person is taking total of 8 steps in 4 s. The number of gait cycles is 4, for which the above data has been processed. It can be easily verified from the Knee angle plots above.

From our model we find out that body parts were traveling from 0 to 829 pixels. From this we assume that the total distance traveled by the person is 829 pixels. From the data we calculated the value of cadence,

$$\text{Cadence} = (8/4) \times 60 = 120\,\text{steps/min}$$

After cadence we find out the value of velocity from the data as:

$$\text{Velocity} = 829/4 = 207.5\,\text{pixel per s}$$

**Input video-2**

In the input video-3 it is seen that the person is taking total of 4 steps in 4 s. The number of gait cycles is 2, for which the above data has been processed. It can be easily verified from the Knee angle plots above.

From our model we find out that body parts were traveling from 6 to 361 pixels. From this we assume that the total distance traveled by the person is 356 pixels. From the data we calculated the value of cadence,

Cadence = $(4/4) \times 60 = 60$ steps/ min

After cadence we find out the value of velocity as:

$$\text{Velocity} = 356/4 = 89\,\text{pixel per s}$$

## 5 Conclusion

The above work was carried out to firstly, accurately identify the various body points of a human from a given video which was achieved successfully. The model identifies 18 key body points and our work essentially makes use of the hip, knee, and the ankle joint to track the movement of the athlete. The coordinates of these key points were successfully extracted first and then stored in a CSV file to carry out analysis of the movement on a human, so as to extend our work to Sprint Analysis for athletes.

The data obtained from the video consisted of outliers which were appropriately removed using the mathematical technique of interpolation. For analyzing sprinting behaviors, we further calculated parameters such as Cadence, Velocity (pixels/sec), and the Knee angles of the athlete, by taking a video of a sprinting athlete as input in a sagittal view to determine the vertical movement of various body points of the athletes during sprinting. Finally the results were compared with the striding patterns of the athletes in contrast to the running patterns. The range of the knee angle was found to have significantly changed. The results were consistent with the actual movement.

## 6 Limitations

The model has high runtime and when the occlusion between body parts is high then the model at some instances has difficulty detecting some body points. So practically in some cases the model misses out some of the body parts and is unable to identify then accurately. But to study and analyze the movement of the particular sports person, one such frame would not prohibit us from carrying out the analysis accurately. So in order to overcome this problem we can omit such fames to obtain accurate results.

## 7 Future Work

Since Sports Analysis using artificial intelligence is a relatively new domain and not much work has been done in it, thus a more thorough research can be carried out to determine more insights on the running features, running-related injuries and their prevention.

There are several possibilities for future directions with this work, the most imminent being an extension into further sports areas such as Golf swing dynamics, Kinematic analysis in American football, Tennis swing form, and many more.

## References

1. Cao, Z., Simon, T., Wei S., Sheikh, Y.: OpenPose: Realtime multi-person 2D pose estimation using part affinity fields. In: Proceedings of the IEEE Conference on Computer Vision and Pattern Recognition. ArXiv:1611.08050 (2017)
2. Sun, J., Wang, Y., Li, J., Wan W., Cheng D., Zhang H.: View-invariant gait recognition based on kinect skeleton feature. Multimed Tools Appl. **77**, 24909–24935 (2018)
3. Xiao, B., Wu, H., Wei, Y.: Simple baselines for human pose estimation and tracking. In: Proceedings of the European Conference on Computer Vision (ECCV), pp. 466–481 (2018)
4. McCabe, A., Trevathan, J.: Artificial intelligence in sports prediction. In: Fifth International Conference on Information Technology: New Generations (ITNG), pp. 1194–1197 (2008). https://doi.org/10.1109/ITNG.2008.203
5. Herold, M., Goes, F., Nopp S., Bauer P., Thompson C., Meyer T.: Machine learning in men's professional football: Current applications and future directions for improving attacking play. Int. J. Sports Sci. Coaching **14**(6), 798–817 (2019)
6. Toshev, A., Szegedy, C.: DeepPose: Human pose estimation via Deep neural networks. In: Proceedings of the IEEE Conference on Computer Vision and Pattern Recognition (CVPR), pp. 1653–1660 (2014)
7. Bulat, A., Tzimiropoulos, G.: Human pose estimation via convolutional part heatmap regression. In: Leibe, B., Matas, J., Sebe, N., Welling, M. (eds.) Computer vision—ECCV 2016. ECCV 2016. Lecture Notes in Computer Science, vol. 9911. Springer, Cham (2016)
8. Meijden, K.: Analyzing sprint features with 2D Human Pose Estimation, Anr. u504531—Snr. 2017494. Tilburg University (2019)
9. Howard, A., Zhu, M., Chen, B., Kalenichenko, D., Wang, W., Weyand, T., Andreetto, M., Adam, H.: MobileNets: Efficient Convolutional Neural Networks for Mobile Vision Applications. ArXiv:1704.04861 (2017)
10. Mobilenet Model.: https://www.github.com/ildoonet/tf-pose-estimation/tree/master/models/graph/mobilenet_thin, Last accessed 16 Nov 2020

# Optimal Sizing of a Hybrid System for Litan, Manipur

**Wairokpam Dhanraj, Ingudam Chitrasen Meitei, and Moirangthem Twinkle Devi**

**Abstract** This paper suggests a method for optimizing a system consisting of battery, solar and wind energy. With ever-increasing demand of electric power due to civilization of man-kind and increase in human population, we need to find an alternate means to meet the increase in load demand and to reduce the overutilization of fossil fuels. For this, we propose a WSB-HPS working in both grid-connected and stand-alone modes. This WSB-HPS combines wind and solar energy power generation and also reduces the charge and discharge time of the battery. Therefore, this system improves the reliability of the power supply and hence reduces the whole cost as the investment in battery is reduced. Compared with the standard methods, this WSB-HPS system achieves a greater power reliability with reduced battery requirement in self-supporting power system. Since grid is a two-way power system, we can either give extra power generated or take power when required. In doing so, this method proposes to achieve much lesser power fluctuation when injected. This method also optimizes the battery capacity and hence results in higher efficiency.

**Keywords** Hybrid power system · Optimization · Power generation

## 1 Introduction

Ever-increasing use of fossil fuels and the depletion of these resources have led to energy crisis and environmental pollution. The concern for environmental pollution has led to the development of systems comprising wind and solar for power generation [1, 2]. But the output from wind power generation (WPG) and photovoltaic (PV) is fluctuating in nature due the intermittence and uncertainty of the energy. Hence, the system will need battery storage unit for backup in off-grid mode. Such challenges can be minimized by using of solar–wind hybrid system [3, 4]. By using a hybrid system comprising of solar PV, wind, and battery, we will not only reduce the system cost, but it also improves the reliability of the power supply.

W. Dhanraj (✉) · I. C. Meitei · M. T. Devi
National Institute of Technology Manipur, Imphal, India
e-mail: wairokpamdhanraj@gmail.com



D. Gupta et al. (eds.), *Pattern Recognition and Data Analysis with Applications*,
Lecture Notes in Electrical Engineering 888,
https://doi.org/10.1007/978-981-19-1520-8_44

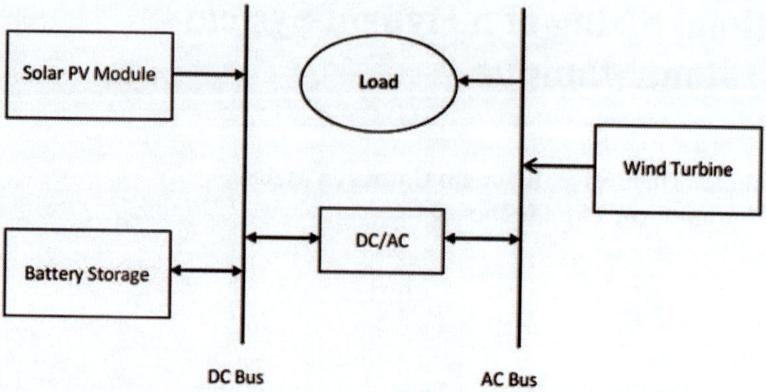


**Fig. 1** Layout of hybrid system

But there are still some major challenges faced by the generating stations when transmitting the generated electricity [5, 6]. New technologies like HVDC and facts have already been introduced in the power market for more efficient transmission of electricity. But nevertheless, the effort in transmitting power from generating stations to end users in hilly regions is always difficult because there is increase in cost due to difficult terrain. Hence, this hybrid power system consisting of wind, battery, and solar is introduced so that it can be installed near the end users so that the transmission cost is reduced as well as the dependency on conventional energy sources is also decreased [7, 8] (Fig. 1).

Renewable energy resources are abundantly available in nature although it depends on the weather conditions and locations whether these energies are accessible or not. Hybrid system employing renewable resources that combines one or more resources along with battery is more promising and has higher reliability than the conventional energy source [8, 9]. In remote and isolated places this hybrid power system is more preferable [10]. At the same time, depletion of exhaustible non-renewable energy resources is kept in check [9]. The output of photo voltaic (PV) and power generated from wind turbines usually fluctuates. This issue can be overcome by the mutual combination of solar and wind characteristics by taking into account the complementary characteristics of solar and wind energy [11, 12]. An appropriate solar, battery and wind energy system can maximize the reliability and also reduces the system cost [13]. In grid-connected hybrid system, grid is kept as a backup power system for fulfilling the required load demand [14–17]. An optimal sizing method for wind–solar battery hybrid power system located in Hohhot, China, has been carried out by using BSA algorithm, and a satisfactory result is obtained in comparison to traditional methods [18]. The analysis and optimization for a medical institute (RIMS) is obtained by using HOMER, and an optimized result is obtained [19].

## 2 Methodology and Numerical Formulation

The process of optimization is implemented using an algorithm known as back-tracking search algorithm (BSA). BSA is newly developed progressive algorithm [15, 17, 18, 20]. It has a particular mechanism to generate trial individual enabling it to perform calculation of numerical optimization problems very fast.

### 2.1 *Mathematical Modeling of the Required Components*

The power output for the photovoltaic arrays is given by

$$P_{\mathrm{pv}} = f_{\mathrm{pv}} P_{\mathrm{pv\_r}} \frac{G}{G_{\mathrm{STC}}} [1 + \alpha_{\mathrm{T}}(T - T_{\mathrm{STC}})] \tag{1}$$

where

$P_{\mathrm{pv\_r}}$ Rated power output of the PV module.
$f_{\mathrm{pv}}$ Derating factor (loss and shading considered).
$G_{\mathrm{STC}}$ Standard solar radiation on PV.
$T_{\mathrm{STC}}$ Standard temperature on PV.
$T$ and $G$ Real-time temperature and solar radiation.
$\alpha_{\mathrm{T}}$ Temperature coefficient.

The curve for the generated wind power from the turbine can be represented by

$$P_{\mathrm{wt}} = \begin{cases} 0 & v_{\mathrm{w}} \langle V_{\mathrm{ci}} \mathrm{or} v_{\mathrm{w}} \rangle V_{\mathrm{co}} \\ P_{\mathrm{wt\_r}} \frac{v_{\mathrm{w}} - V_{\mathrm{ci}}}{V_{\mathrm{r}} - V_{\mathrm{ci}}} & V_{\mathrm{ci}} \leq v_{\mathrm{w}} \leq V \\ P_{\mathrm{wt\_r}} & V_{\mathrm{r}} \leq v_{\mathrm{w}} \leq V_{\mathrm{co}} \end{cases} \tag{2}$$

where

$P_{\mathrm{wt\_r}}$ Rated power output of wind turbine.
$v_{\mathrm{w}}$ Wind speed.
$V_{\mathrm{r}}$ Rated wind speed.
$V_{\mathrm{ci}}$ Cut-in speed.
$V_{\mathrm{co}}$ Cut-out speed.

The terminal voltage of the battery is given by

$$V_{\mathrm{bs}} = E_{\mathrm{bs}} - I_{\mathrm{dch}} R_0 \tag{3}$$

where

$E_{bs}$ Effective internal voltage.
$I_{dch}$ Discharge current.
$R_0$ Internal resistance.

The effective internal voltage is given by

$$E_{bs} = E_o + AX + CX/(D - X) \tag{4}$$

where

$E_o$ Internal battery voltage at fully charged/discharged state.
$A$ Variation in initial linear internal battery voltage with charging state.
$D, C$ Increase/decrease in battery voltage during progressive charging/discharging.
$X$ Maximum normalized capacity at specified current.

## 2.2 Strategy of Energy Management

The power flow equation is given by

$$\text{(i) } P_L(t) = P_{wt}(t) + P_{pv}(t) + P_{bs\_dch}(t) \quad \text{(if total power generated > load demand)} \tag{5}$$

$$\text{(ii) } P_L(t) = P_{wt}(t) + P_{pv}(t) - P_{bs\_ch}(t) \quad \text{(if total power generated < load demand)} \tag{6}$$

## 2.3 Optimal Sizing Methodology

The reliability of the power supply is given by

$$\text{LPSP} = \frac{\sum_{i=1}^{N}\left[P_L(t_i) - \left(P_{wt}(t_i) + P_{pv}(t_i) + P_{bs\_dch}(t_i)\right)\right]}{\sum_{i=1}^{N} P_L(t_i)} \tag{7}$$

where $t_i$ $t_N$ = operating time of the system.

If the LPSP is 0, then it indicates that the load demand is always met by the system.

The rate of relative fluctuation is given by

$$D_L = \frac{1}{\overline{P_L}}\sqrt{\frac{1}{N}\sum_{i=1}^{N}\left(P_{wt}(t_i) + P_{pv}(t_i) - P_L(t_i)\right)^2} \quad (8)$$

where $\overline{P_L}$ = average power of load.

A lower value of $D_L$ implies that complementary characteristics of solar and wind are utilized efficiently.

## 2.4 Required Constraints

The maximum number of wind generator turbines, solar panels and battery, respectively, are given by

$$N_{wt} \leq \left[\frac{L}{(6-10)d} + 1\right].\left[\frac{W}{(3-5)d} + 1\right] \quad (9)$$

where

| | |
|---|---|
| $L$ and W | Length and width for the region. |
| $d$ | Rotor diameter |

$$N_{pv} \leq \left[S_2/S_{pv}\right].\alpha_{pv} \quad (10)$$

where

| | |
|---|---|
| $S_2$ | Given installation area for solar PV panels. |
| $S_{pv}$ | Area of one PV unit. |
| $\alpha_{pv}$ | Coefficient for possible shadow area |

$$N_{bs} \leq [S_3/S_{bs}] \quad (11)$$

where

| | |
|---|---|
| $S_2$ | Given installation area for battery. |
| $S_{bs}$ | Area of single battery. |

The minimum number of wind turbines, solar panels, and battery, respectively, are given by

$$N_{\text{wt}} \geq \int_{tm2}^{tm3} P_{\text{L}}(t)\text{d}t \Big/ \int_{tm2}^{tm3} P_{\text{wt}}(t)\text{d}t \tag{12}$$

where $t_{m2} - t_{m3}$ = effective operating time of wind turbine during night.

$$N_{\text{pv}} \geq \int_{tm0}^{tm1} P_{\text{L}}(t)\text{d}t \Big/ \int_{tm0}^{tm1} P_{\text{pv}}(t)\text{d}t \tag{13}$$

where $t_{m0} - t_{m1}$ = effective operating time of PV during day.

$$N_{\text{bs}} \geq \frac{\lambda . W_{\text{Ld}}}{\eta . C_{\text{bs}} . V_{\text{bs.}} DOD_{\max}} \tag{14}$$

where

$W_{\text{LD}}$ Energy consumed everyday by load.
$V_{\text{bs}}, V_{\text{bs}}$ Voltage and capacity of single battery.
$\eta$ Battery discharging efficiency.

The reserved operating capacity given by

$$\sum P_{\text{DG}} \geq (1 + \mu\%) P_{\text{L}} \tag{15}$$

where

$P_{\text{DG}}$ Total power output of the distributed generation.
$\mu$ Operating reserve ratio (10%).

The charging and discharging constraints of the battery are given by

$$\text{SOC}_{\min} \leq \text{SOC} \leq \text{SOC}_{\max} \tag{16}$$

$$r_{\text{ch}} \leq r_{\text{ch\_R}}, \quad r_{\text{dch}} \leq r_{\text{dch\_R}} \tag{17}$$

where

$r_{\text{ch}}, r_{\text{dch}}$ Charging and discharging rate.
$r_{\text{ch\_R}}, r_{\text{dch\_R}}$ Limited charging and discharging rate

$$I_{\text{ch}} \leq I_{\text{chmax}} \quad I_{\text{dch}} \leq I_{\text{dchmax}} \tag{18}$$

where

$I_{ch}$, $I_{dch}$ Charging and discharging current.
$I_{chmax}$, $I_{dchmax}$ Maximum charging and discharging current.

$$0 \leq P_{bs\_ch} \leq P_{bs\_chmax} \tag{19}$$

$$0 \leq P_{bs\_dch} \leq P_{bs\_dchmax} \tag{20}$$

where

$P_{bs\_ch}$, $P_{bs\_dch}$ Charging and discharging power.
$P_{bs\_chmax}$, $P_{bs\_dchmax}$, Maximum charge and discharge power [21, 22]

$$N_C \leq N_{Cmax} \tag{21}$$

where $N_C$,$N_{Cmax}$ = charging/discharging cycle of battery and its limited value.

## 2.5 *Total Cost*

$$\text{Initialcost of the system } C_i = (N_{pv}C_{pv} + N_{wt}C_{wt} + N_{bs}C_{bs})f_{cr} \tag{22}$$

where

$C_{pv}$, $C_{wt}$, $C_{bs}$ Cost of PV panels, wind turbine, and battery.
$f_{cr}$ Capital recovery factor

$$\text{Operating and Maintenance cost } C_{OM} = C_{pv\_OM}t_{PV} + C_{wt\_OM}t_{wt} + C_{bs\_OM}t_{bs} \tag{23}$$

where

$C_{pv\_OM}$, $C_{wt\_OM}$, $C_{bs\_OM}$ Operating and maintenance cost of PV panels, wind turbine, and battery.
$t_{PV}$, $t_{wt}$, $t_{bs}$, Operating time of PV panels, wind turbine, and battery

$$\text{Replacement Cost } C_R = C_{pv\_R} + C_{wt\_R} + C_{bs\_R} \tag{24}$$

where $C_{pv\_R}, C_{wt\_R}, C_{bs\_R}$ = replacement cost of PV panels, wind turbine, and battery.

## 2.6 Objective Function

Reducing the total cost of the hybrid power system is regarded as the objective function. It is given by

$$\min f = \min(C_i + C_{OM} + C_R - C_{gs} + C_{gp} + C_{pc}) \tag{25}$$

where

$C_{gp}, C_{gs}$ Cost of power purchased from grid and selling power to the grid.
$C_{pc}$ Penalty cost.

The flowchart of the suggested method is shown in Fig. 2. The flowchart explains the step-by-step procedure of the optimization process. The various data collected are used, and the final net total cost is found out.

# 3 Site Selection and Load Estimation

The suggested model is presented for powering the district Litan, Manipur in India. The selected site is situated at 24°56.5′ N latitude and 94°12.8′ S longitude [23] (Table 1).

In our examination, the required data of electrical load for Litan are collected from the Manipur State Power Distribution Company Limited (MSPDCL). The Manipur government is allocating large amounts of money about 800 crores every year in an attempt to meet the increasing load demand. But despite the efforts by the state government, the state is experiencing power shortage and ultimately ends up purchasing the deficit power from the neighboring states.

Here, Fig. 3 presents the load 1 profile, and the highest peak demands can be seen during 1200–2300 h.

And Fig. 4 presents the load 2 profile, and the highest peak demands can be seen during 1200–1600 h [11].

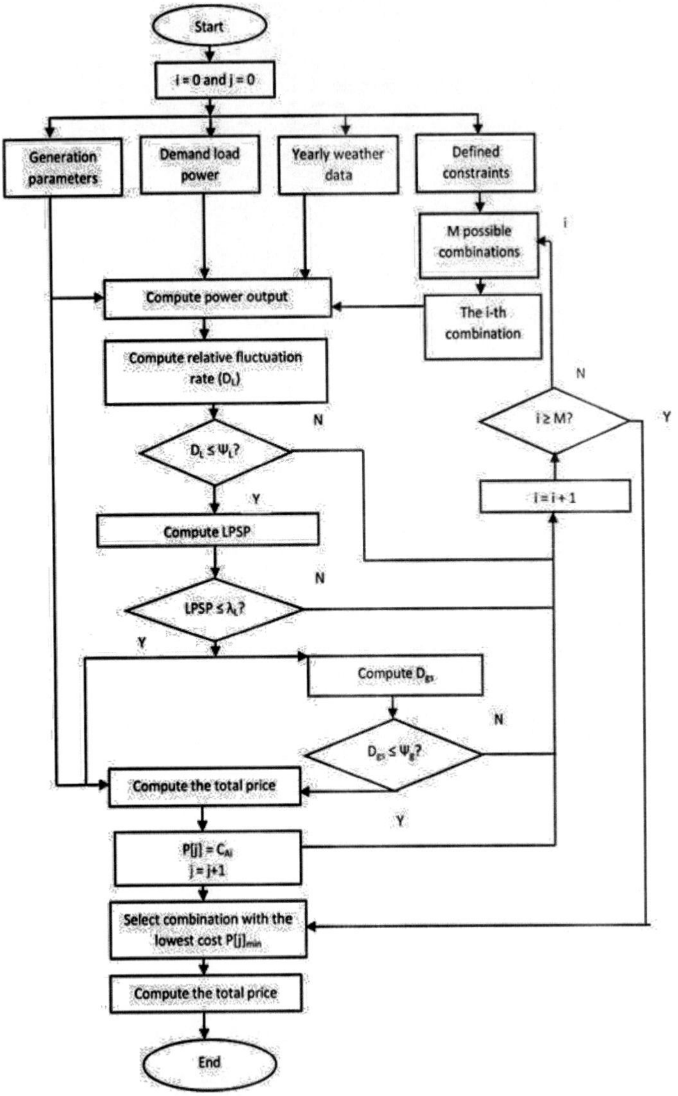


**Fig. 2** Flowchart of the proposed method

**Table 1** Details of load profile

| Sl. No | Load considered | Daily load demand | Peak load demand |
|---|---|---|---|
| Load 1 | Residential and community places | 414.75 kW/day | 77 kW |
| Load 2 | Commercial such as schools, shops, health centers, offices | 187.7 kW/day | 26.95 kW |
| Total average load demand | | 602.45 kW/day | |

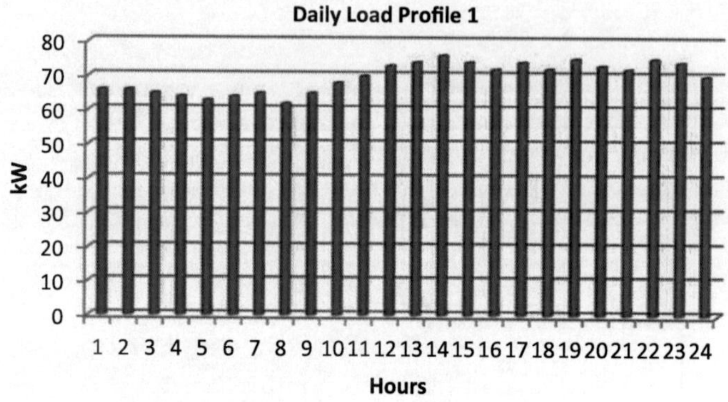


**Fig. 3** Average hourly electrical load 1 of Litan

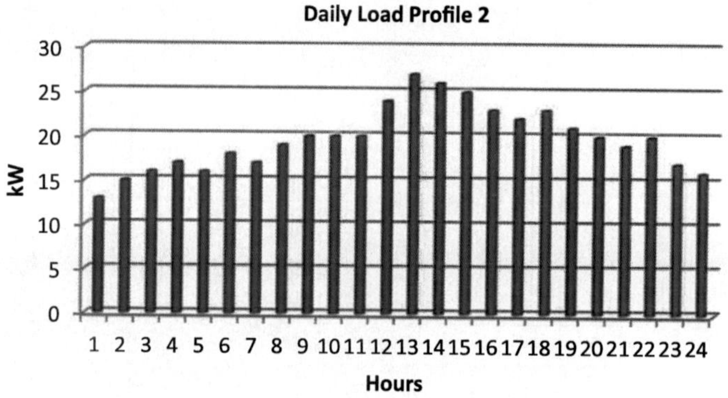


**Fig. 4** Average hourly electrical load 2 of Litan

# 4 Resources and Elements

## *4.1 The Resources and Elements that Are Viewed for the Optimization Procedure Are Listed in the Table Below*

A wind turbine which generates power less than 100 kW at rated speed is known as small wind turbines. A model wind turbine of rated power 1 kW is taken into consideration (Table 2).

Photovoltaic panels of 340 W Monocrystalline Panel of Loomsolar is taken into consideration [12].

The battery taken into consideration is a lead acid battery of 12 V/80Ah Amaron HCV620D31R [13].

**Table 2** Details of the components used

| Source | Company | Lifetime | Type | Rating | Capital cost | Replacement cost | O& M cost |
|---|---|---|---|---|---|---|---|
| Wind turbine | J.D. Engineering works | 20 years | 3-phase permanent magnet | 48 V and rated power 1 kW | Rs. 50,000/kW | Rs. 90,000/kW | Rs. 10,000 |
| Solar PV module | Loom solar | 25 years | Monocrystalline | 340 W, Derating factor of 80% | Rs. 12,400/unit | Rs. 12,400/unit | Rs. 4,000/unit |
| Battery | Amaron HCV620D31R | 8 years | Lead acid battery | 12 V, 80 Ah | Rs. 6,300/unit | Rs. 6,300/unit | Rs. 1000/battery |

## 4.2 Solar Energy and Wind Energy Resources

The solar energy radiation data of the selected location which is at 24°56.5′ N latitude and 94°12.8′ S longitude is obtained from the National Aeronautics and Space Administration (NASA) surface meteorology and database of solar energy.

Figure 5 shows the solar radiation data. The clearness index value ranges from 0 to 1 and is shown in Fig. 5. The average solar insolation of the selected site is 5.53 kWh/m$^2$/day [11].

The data of wind resource for the selected site is also obtained from the National Aeronautics and Space Administration database. Figure 6 indicates the average wind speed data. The yearly average wind speed data is 4.2 m/s.

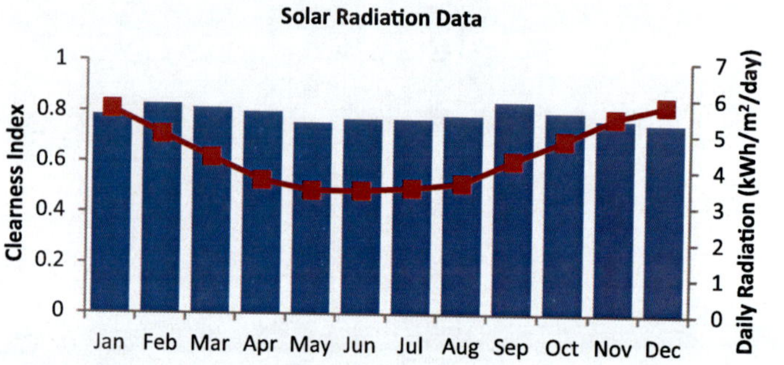


**Fig. 5** Solar insolation data

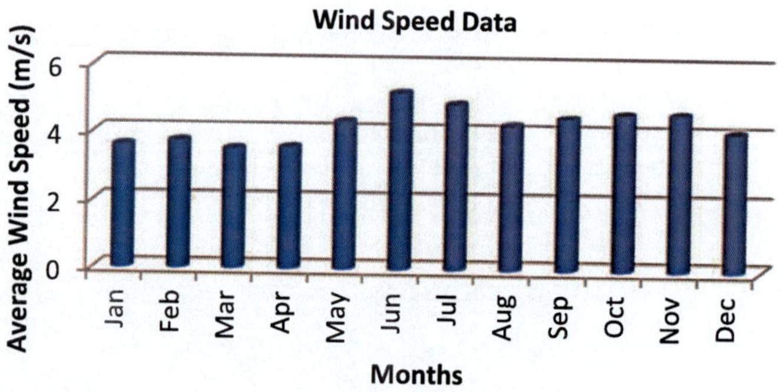


**Fig. 6** Wind speed data

**Table 3** Optimization results

| Method | No. of PV module | No. of wind turbine | No. of battery | Total cost | $D_L$ | LPSP |
|---|---|---|---|---|---|---|
| BSA | 2276 | 9 | 2889 | 92,727,709 | 0.45 | 0 |

## 5 Results and Discussion

The BSA algorithm performs several repeated iterations, and the optimum results can be seen from the final iteration. As can be seen from the iteration process, the most optimal and efficient results are the design composed of solar photovoltaic cells, wind turbine, and battery. The last stages of the iteration process are shown in the figure below (Table 3).

## 6 Conclusion

A practically feasible and satisfactory result is obtained by using backtrack search algorithm. So, we are able to satisfy the load demands of load profile 1 and 2. The results are concluded in the following:

1. In this paper, a simulation attempt is proposed to analyze the optimum size and total cost for Litan, Manipur, using backtrack search algorithm (BSA). The total net present cost (NPC) is presented at Rs. 92,727,709.
2. In order to utilize the complementary characteristics of wind and solar energy, wind turbines are included in the model. So, we can conclude that the most efficient and economic system for supplying electrical power to the above location is by using BSA algorithm comprising of the components battery, wind, and solar PV panels.

## References

1. Swarnkar, N.M., Sharma, R., Gidwani, L.: An application of HOMER Pro in optimization of hybrid energy system for electrification of technical institute. In: International Conference on Energy Efficient Technologies for Sustainability (ICEETS) (2016)
2. Chompoo-Inwai, C., Lee, W.J., Fuangfoo, P.: System impact study for the interconnection of wind generation and utility system. IEEE Trans. Ind. Appl **41**(1), 1452–1458 (2005)
3. Phurailatpam, C., Rajpurohit, B.S., Wang, L.: Planning and optimization of autonomous DC microgrid for rural and urban application in India. Renew. Sustain. Energy Rev. **82**(1), 198–204 (2018)
4. Estefania, P., Asier, G.D.M., Jon, A., Inigo, K., De, A.I.M..: General aspects, hierarchical controls and droop methods in microgrids: a review. Renew. Sustain Energy Rev. **17**, 147–159 (2013)

5. Khan, K.S., Ullah, Z., Khan, B., Sami, I., Ali, S.M., Mehmood, C.A.: Assessment of hybrid off-grid wind photovoltaic system: A case study of university campus. In: International Conference on Energy Conservation and Efficiency (ICECE) (2017)
6. Woyte, A., Van, V., Belmans, R., Nijs, J.: Voltage fluctuations on distribution level introduced by photovoltaic systems. IEEE Trans. Energy Convers. **21**(1), 202–209 (2006)
7. Aghenta, L.O, Iqbal, M.T.: Design and dynamic modelling of a hybrid power system for a house in Nigeria. Int. J. Photoenergy 1–13 (2019)
8. Fulzele, J.B., Dutt, S.: Optimium planning of hybrid renewable energy system using HOMER. Int. J. Electr. Comput. Eng. (IJECE) **2**, 68–74 (2012)
9. Rout, K., Sahu, J.K.: Various optimization techniques of hybrid renewable energy systems for power generation: a review. Int. Res. J. Eng. Technol. (IRJET) **05**(07), (2018)
10. Akher, M.A., Ali, A.A., Eid, A.M., Kishky. H.E.: Optimal size and location of distributed generation unit for voltage stability enhancement. In: Proceeding of the IEEE ECCE, pp. 104–108 (2011)
11. Inwai, C.C., Lee, W.J., Fuangfoo, P.: System impact study for the interconnection of wind generation and utility system. IEEE Trans. Ind. Appl. **41**(1), 1452–1458 (2005)
12. Woyte, A., Van, V., Belmans, R., Nijs, J.: Voltage fluctuations on distribution level introduced by photovoltaic systems. IEEE Trans. Energy Convers. **21**(1), 202–209 (2006)
13. Srivastava, R., Giri, V.K.: Optimization of hybrid renewable sources using HOMER. Int. J. Renew. Energy Res. **6**(1) (2016)
14. Chedid, R., Rahman, S.: Unit sizing and control of hybrid windsolar power systems. IEEE Trans. Energy Convers. **12**(1), 79–85 (1997)
15. Chedid, R., Akiki, H., Rahman, S.: A decision support technique for the design of hybrid solar-wind power system. IEEE Trans. Energy Convers. **13**(1), 76–83 (1998)
16. Ardakani, F., Riahy, G., Abedi, M.: Optimal sizing of a grid-connected hybrid system for north-west of Iran-case study. In: Proceeding of the IEEE EEEIC, pp. 29–32 (2010)
17. Menniti, D., Pinnarelli, A., Sorrentino, N.: A method to improve microgrid reliability by optimal sizing PV/wind plants and storage systems. In: Proceeding of the IEEE CIRED, pp. 8–11 (2009)
18. Xu, L., Ruan, X., Mao, C., Zhang, B., Luo, Y.: An improved optimal sizing method for wind solar battery hybrid power system. IEEE Trans. Sustain. Energy **4**(3), (2013)
19. Meitei, I.C., Singh, T.B., Denish, K., Meitei, H.H., Singh, NA.: Optimum design of photovoltaic system for a medical institute using HOMER. In: International Conference on Intelligent Computing and Smart Communication, pp. 1337–1346 (2020)
20. Civicioglu, P.: Backtracking search optimization algorithm for numerical ptimization problems. Appl. Math. Comput. **219**, 8121–8144 (2013)
21. Manwell, J., McGowan, J.: Lead acid battery storage model for hybrid energy systems. Sol. Energy **50**(5), 399–405 (1993)
22. Manwell, J.F.: HYBRID2-A hybrid system simulation model theory manual. Dep. Mechan. Eng. Univ. Mass., Renew. Energy Res. Lab. (2005)
23. Meitei, I.C., Pudur, R.: Analysis and optimization of hybrid renewable energy sources: A case study for Litan, Manipur. J. Adv. Res. Dyn. Control Syst. **12**(03), 200–208 (2020)

# An Intrusion Detection Approach Based on Decision Tree-Principal Component Analysis Over CICIDS2017

**Gulab Sah and Subhasish Banerjee**

**Abstract** In today's environment, an Intrusion Detection System (IDS) is becoming increasingly crucial in a network's Defense System to protect our network from any external threats or attack. The primary function of an IDS is to offer a shield for a specific host or network, as well as to examine and forecast client network access activities. The entire traffic is classified as normal or an assault based on these patterns. In order to classify traffic as valid or malicious, IDS must process all incoming communication over networks. As a result, IDS must cope with significant or enormous amounts of data. However, in IDS, not all features may be required to be processed among this data. As a result, extracting or locating the only relevant features among all features is always challenging. To address this issue, we have proposed a feature selection technique using Principal component analysis with Decision tree algorithm (DT-PCA) over real-time datasets, i.e., (CICIDS2017). The proposed classifier (Decision Tree) employing Principal component analysis performed well over CICIDS2017 datasets, according to the results presented in this research. To measure the performance or efficiency of the method, the most important metrics namely, recall, F-measure, precision, and accuracy have been used in this paper. In addition, this research examines the differences between DT-PCA and DT with all features. According to the CICIDS2017 datasets, the DT-PCA approaches can improve the IDS's performance and the accuracy rate can be achieved more than 99%.

**Keywords** Decision tree · IDS · NSL KDD dataset · Recursive feature elimination · CICIDS2017 dataset · And principal component analysis

G. Sah (✉) · S. Banerjee
Department of Computer Science and Engineering, National Institute of Technology Arunachal Pradesh, Jote, Arunachal Pradesh 791113, India
e-mail: gulab.phd@nitap.ac.in



D. Gupta et al. (eds.), *Pattern Recognition and Data Analysis with Applications*, Lecture Notes in Electrical Engineering 888,
https://doi.org/10.1007/978-981-19-1520-8_45

## 1 Introduction

The innovation and applications of the Internet are rapidly developing in today's world. These modern technologies generate a significant amount of data with volume, diversity, and velocity characteristics, resulting in big data. Side by side, as a result of technological breakthroughs, the different forms of attacks have also evolved in the modern world/society.

An IDS is one of the network security techniques that protects the network from assaults or intruders. Various machine learning algorithms have recently been used in IDS to construct/develop data-driven models. These models detect attacks by employing a variety of ways to identify aberrant behavior in a network or computer system. IDS employ the datasets for training purposes in order to cope with anomalous behavior. These datasets have a lot of attributes, however not all of them are necessary to categories data as normal or abnormal. Hence, feature selection strategies are used to select the most significant features from an initial set of all features. These features are then utilized to construct a model utilizing a variety of machine learning (ML) algorithms. In IDs, ML is used for either unsupervised or supervised learning, such as classification or regression techniques, with the goal of improving prediction ability [1].

The major contributions of this paper are defined as follow:

(i) Developed intelligent IDS that accurately detect aberrant network behavior using real-time datasets.
(ii) By selecting only the most important features rather of all features, the time complexity has been reduced.
(iii) Developed a decision-making mechanism to improve computer network security.

The following is a breakdown of the paper's structure: The next section describes the literature review for IDS research. The proposed framework and approach have been described in Sect. 3. In Sect. 4, the experiments and its outcomes, have discussed. Finally, in Sect. 5, the conclusion and future scope of this research have been addressed.

## 2 Literature Review

The IDS includes both software and hardware which can passively or actively controls networks or specific host to detect any intrusion [2]. An IDS used as defense strategies in industries or organization's security systems [3]. Many technologies, for detection system has been developed in last decades, but a still significant problems like high false-alarm rate and false-positive are exist.

Rani and Xavier [4] present a new hybrid approach by combining different classifiers for classification. Decision tree C5.0 was used to build the model and NSL KDD dataset was utilized to perform the whole experiment. The result shows the overall performance of the proposed model is good in terms of low false-alarm rate and high detection rate compare to traditional methods. Aslahi-Shahri et al. [5] proposed a hybrid technique of GA and support vector machine [6] for IDS. The proposed method utilized the feature reduction technique and minimizes the 41 features to 10 features. Later these 10 features are used to build a model. Singh et al. [7] proposed a feature selection approach called intelligent water drop (IWD). This method was used to find an optimal subset of features. Later these subset features are used by SVM to build a model. The proposed models give better detection rate, precision and accuracy compare to prevailing models. Alternatively, Nilesh et al. [8] analyzed the different classifiers (SVM, Decision Tree, k-nearest neighbor (KNN), Random Forest, Naive Bayes) performance over normal or DoS attacks based on confusion matrix and accuracy. To get better accuracy in IDS, Elhag et al. [9] proposed a fuzzy system in which an experiment was conducted on the KDD99 dataset and results are analyzed. Sah et al. [10] proposed the features selection and classification techniques for IDS in which recursive features elimination (RFE) was used as feature selection and the random forest was used as a classifier to classify the normal, DoS, probe, U2R, and R2L over NSL KDD datasets. Solani et al. [11] used supervised learning algorithm with feature selection technique (FST) onto UNSW-NB15 dataset and made a comparative study. The results shows FST reduce the false-alarm rate and improved the performance of IDS. Thakkar et al. [12] research and study the effect of different FST with machine learning (ML) technique on the performance of IDS using NSL KDD dataset and presented the comparative study. Elmasry et al. [13] performed empirical study and present the masquerade detection technique (MDT) using four datasets and studied 6 of built-in ML models in AML to examined the their effectiveness in MDT. The result shows decision forest and Decision jungle model performed well compare to other models.

In conclusion, after studying the previous works deeply, it exposed that many research carried out works using feature reduction and selection techniques but still there is gaps while we considering high dimensionality datasets, it requires higher computation cost. Therefore this problem should address properly.

## 3 Proposed Framework

The proposed framework consist of 5 parts namely, datasets (CICIDS 2017), data-initializations and pre-processing, principal component analysis, decision tree algorithms, and prediction and evaluation as shown in Fig. 1.

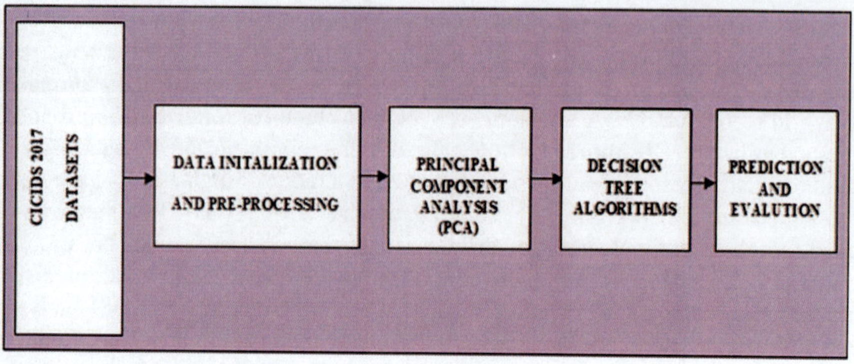


**Fig. 1** Proposed framework

## *3.1 CICIDS 2017 Datasets*

The dataset is an important part of any IDS for evaluating or measuring the performance of decision engines approaches, because of the struggle of classifying between the valid and attacks activities/abnormalities in real-time network traffics. Network traffic has been the capture and collected from real-time traffic which consists of various malicious and normal records.

With vast sizes and high bandwidth of a modern networks environment, network traffic has a feature of big data (BD). The technologies such as MySQL CGE, Hadoop tools are used to handle the BD (represented in terms of variety, velocity, and volume) because generally, old database structures cannot process or deal with BD that enclosed the problems of real-world. To process the network traffic in real-time, the data is capture or gather to examine and detect suspicious activities. Therefore, the tools such as Bro-IDS, tcpdump, are used to capture network traffic features. Finally, a Decision Engine approach in IDS is utilized for discovering Zero-day and existing attacks from the attributes. CICIDS 2017 is one of the real-time network traffic datasets which is developed by the Canadian Institute of Cyber-security. It includes different type's attacks like Web Attack, Port Scan, Infiltration, a botnet, Distributed Denial of service (DDoS), Denial of Service (DoS), Brute Force. The traffic was capture for a total of 5 working days, from Monday-Friday that is available in 8 different files in which each file has consist of 78 features. In this study, CICIDS 2017 dataset has been used which has a total of 2,830,743 numbers objects as shown in Table 1.

## *3.2 Data-Initialization and Pre-processing*

According to certain rules, processing the data in a dataset into the data warehouse is known as data-preprocessing. Basically in this phase, data has to go through a

**Table 1** Number of objects present in CICIDS2017 datasets

| S. No | Labels | No of objects |
|---|---|---|
| 1 | Normal | 2,273,097 |
| 2 | DoS/DDoS | 380,699 |
| 3 | Port scan | 158,930 |
| 4 | Web attack | 2180 |
| 5 | Botnet | 1966 |
| 6 | Infiltration | 36 |
| 7 | Brute force | 13,835 |

cleaning process to eliminate the identical records. After that conversion of features has been done in which all non-numerical features are converted into numerical features. Lastly in this phase, feature normalization was performed to avoid/escape the features (has a large value) that give weight too much in the result.

### 3.3 *Principle Component Analysis*

In the third phase, Feature selection technique is utilized to eliminate the unimportant features that are not participating in identifying the attacks instead of it, lads to increase the computation cost. Therefore to reduce the computation cost, principal component analysis as feature selection and reduction techniques is used. PCA is an unsupervised learning method that is similar to clustering that reduces the complexity of datasets by preserving the patterns and tendency. The limitation of the PCA algorithm is obtaining a set may not be ideal in the situation of the non-Gaussian method. PCA is a decent optimization technique for discovering the best performing subset of features from the original set of features. The idea is to utilize only relevant features instead of irrelevant features that improve the classifier performance in terms of detection rate.

### 3.4 *Decision Tree Algorithms*

After selecting only relevant features using PCA, the next phase is to build the model using decision tree algorithms to separate the data as normal or abnormal using information gain until the object in every leaf node has uniform class labels. In previous phase important feature are identified and selected. These important features are used to take decision between normal or abnormal traffic. Also it shows why particular feature have higher importance. Therefore, to build a model a decision tree classifier is used in this phase. Basically, the decision tree (DT) algorithm uses two measures/parameters one is information gain and the other is entropy to do a good split. DT is consists of internal decision nodes and terminals leaves. Each decision node provides a test function that leads to results in discrete by branch labeling. At every node, an input is provided to construct a test, and based on the conclusion; one branch from a set of branches will be measured. The learning process of the algorithm begins from the root and it will continue the process in a recursive manner until a node (leaf) is stretched in which each leaf node represents a class label or target class in case of classification.

### 3.5 *Prediction and Evaluation*

After building the model using decision tree classifier, the next phase is prediction and evaluation, in which prediction of model was performed using test data and various evaluation matrices such as recall, precision, F-measure, and accuracy was estimated to measure the performance proposed method. The following evaluation metrics are utilized to measure the performance and effectiveness of classification models [14].

(1) Accuracy: the accuracy ($A$) is calculated as

$$A = \frac{\text{True\_positive\_objects} + \text{False\_negative\_objects}}{\text{Total\_number\_of\_objects}}$$

(2) Precision: the precision ($P$) is calculated as

$$P = \frac{\text{True\_positive\_objects}}{\text{True\_positive\_objects} + \text{False\_positive\_objects}}$$

(3) Recall: the recall ($R$) is calculated as

$$R = \frac{\text{True\_positive\_objects}}{\text{True\_positive\_objects} + \text{False\_negative\_objects}}$$

(4) $F$- measure: the $F$-measure ($F$-m) is calculated as

$$F - m = 2 * (P * R)/(P + R)$$

## 4 Experiment and Results

This study uses the CICIDS 2017 dataset to validate the superiority of the decision tree model with the PCA method in the experiment. As our main aim to examine whether or not model with selected features, will increase the performance and detect the different types of attacks more accurately. The PCA method was used to acquire the least number of relevant features that reduce the computational cost and improve the model detection rate compare to the model with all features or get accuracy close to the model with all features. In order to increase detection rate of IDS, the important features identified and selected during the model building which contribute in improvement and reduces the computation cost. Tables 2 and 3 demonstrate the performance of the model with selected and all features, respectively. The performance and effectiveness of model are examined using matrices such as precision, recall, F-measure, and accuracy.

In Fig. 2, the x-axis represents different types of attacks, and the y-axis presents the percentage of accuracy, precision-recall, F-measure for decision tree model using selected features. In Fig. 3, the y-axis presents the percentage of accuracy, precision-recall, F-measure for the decision tree model using all features and the x-axis represents different types of attacks.

**Table 2** Decision tree classifier with selected features

| Classes | Accuracy (%) | Precision (%) | Recall (%) | *F*-measure (%) | No. of features |
|---|---|---|---|---|---|
| DoS/DDoS | 99.92 | 99.83 | 99.85 | 99.84 | 8 |
| Port Scan | 99.89 | 99.58 | 99.51 | 99.57 | 8 |
| Web Attack | 99.81 | 94.20 | 95.85 | 94.80 | 8 |
| Botnet | 99.94 | 71.98 | 69.00 | 69.81 | 8 |
| Infiltration | 99.90 | 84.90 | 85.00 | 84.90 | 8 |
| Brute force | 99.90 | 99.60 | 99.00 | 99.00 | 8 |

**Table 3** Decision tree classifier with all features

| Classes | Accuracy (%) | Precision (%) | Recall (%) | *F*-measure (%) |
|---|---|---|---|---|
| DoS/DDoS | 99.91 | 96.87 | 94.86 | 95.61 |
| Port Scan | 99.89 | 99.48 | 99.41 | 99.47 |
| Web Attack | 99.981 | 94.20 | 95.85 | 94.80 |
| Botnet | 99.93 | 68.57 | 69.00 | 67.64 |
| Infiltration | 99.90 | 84.90 | 85.00 | 84.90 |
| Brute force | 99.90 | 99.97 | 99.92 | 99.84 |

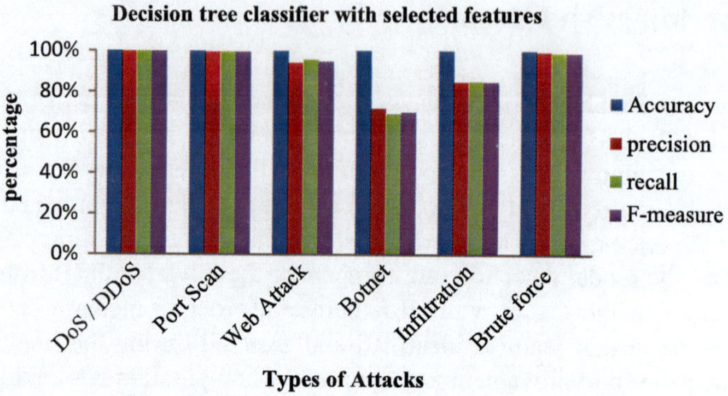


**Fig. 2** Precision, accuracy, recall, and *F*-measure for Decision tree classifier using selected features

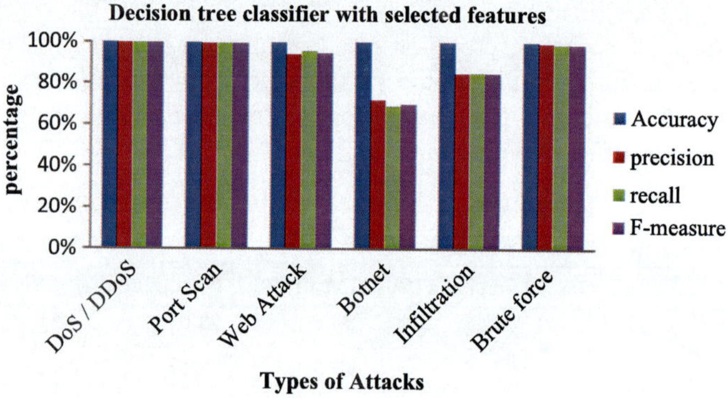


**Fig. 3** Precision, accuracy, recall, and *F*-measure for Decision tree classifier using all features

## 5 Conclusion

The results suggested that the proposed model with selected features performs better than a model with all features in this research. Furthermore, as demonstrated in Tables 2 and 3, the decision tree model with selected characteristics generated using the PCA method improves detection accuracy. Moreover, in the proposed model for IDS used only important/relevant features instead of using all features in order to detect any abnormal behavior in networks, therefore, it may reduce the computation cost also. As an extension of this research, we propose that detecting anomalous network behavior utilizing feature mining properties combined with deep learning (self-learning ability) may be effective in improving attack detection performance.

## References

1. Gupta, U., Gupta, D.: Least squares large margin distribution machine for regression. Appl. Intell. Springer, 1–36 (2021)
2. Tidjon, L.N., Frappier, M., Mammar, A.: Intrusion detection systems: A cross-domain overview. IEEE Commun. Surv. Tutorials. **21**(4), 3639 (2019)
3. Liang, W., Li, K.C., Long, J., Kui, X., Zomaya, A.Y.: An industrial network intrusion detection algorithm based on multifeature data clustering optimization model. IEEE Trans. Industr. Inf. **16**(3), 2063 (2020)
4. Rani, M.S., Xavier, S.B.: A hybrid intrusion detection system based on c5.0 decision tree and one-class svm. Int. J. Curr. Eng. Technol. **5**, 2001 (2015)
5. Aslahi-Shahri, B.M., Rahmani, R., Chizari, M., Maralani, A., Eslami, M., Golkar, M.J., Ebrahimi, A.: A hybrid method consisting of ga and svm for intrusion detection system. Neural Comput. Appl. **27**, 1669 (2016)
6. Gupta, U., Gupta, D.: Regularized based implicit Lagrangian twin extreme learning machine in primal for pattern classification. Int. J. Mach. Learn. Cyber. Springer, 1311–1334 (2021
7. Acharya, N., Singh, S.: An iwd-based feature selection method for intrusion detection system. Soft Comput. **22**, 1–10 (2017)
8. Nanda, N.B., Parikh, A.: Network intrusion detection system based experimental study of combined classifiers using random forest classifiers for feature selection. Int. J. (IJRECE). **6**(4), 341 (2018)
9. Elhag, S., Fernández, A., Altalhi, A., Alshomrani, S., Herrera, F.: A multi-objective evolutionary fuzzy system to obtain a broad and accurate set of solutions in intrusion detection systems. Soft Comput. **23**, 132 (2019)
10. Sah, G., Banerjee, S.: Feature reduction and classifications techniques for intrusion detection system. In: International Conference on Communication and Signal Processing. IEEE, 1547–1551 (2020)
11. Solani, S., Jadav, N.K.: A novel approach to reduce false-negative alarm rate in network-based intrusion detection system using linear discriminant analysis. Inventive Commun. Computat. Springer, 911–921 (2021)
12. Thakkar, A., Lohiya, R.: Attack classification using feature selection techniques: A comparative study. J. Ambient Intell. Humanized Comput. **12**, 1249–1266 (2020)
13. Elmasry, W., Akbulut, A., Zaim, A.H.: Comparative evaluation of different classification techniques for masquerade attack detection. Int. J. Inf. Comput. Secur. **13**(2), 187 (2020)
14. Sah, G., Goswami, R.S., Nandi, S.K.: Machine learning methods for predicting the popularity of forth coming objects. Int. J. Innovative Technol. Exploring Eng. (IJITEE). **9**(2S), 645 (2019)

# Detection of Epilepsy Using Graph Signal Processing of EEG Signals with Three Features

**Hemant Kumar Meena, Ramnivas Sharma, Abhinav Tailor, Harshil Verma, and Rajveer Saini**

**Abstract** Epilepsy may occur with a genetic disorder or an acquired brain injury, such as a trauma or stroke. It is a type of disorder in which activity of nerve cell in the brain is disturbed, causing seizures. Electroencephalogram (EEG) is used to analyze the Epileptic seizure which is a very serious nervous system disorder. In this work detection of epilepsy disease is approached by a Graph Signal Processing (GSP) technique (with computing the Graph Discrete Fourier Transform (GDFT)). GDFT coefficients are produced on the Eigen space of Laplacian matrix with the help of EEG data points. The Laplacian matrix is calculated from the weighted graph designed for EEG signal. The proposed GDFT based feature vectors are used to detect the epilepsy seizure class from the given EEG signal and classify by using Stationarity ratio and TIK-norm. By observing the simulated results one can analyze that the proposed GDFT based total features can discover epileptic seizure with 97% accuracy which is obtained from Gaussian Weighted Graph. To provide a nice compact format to encode the structure within the data, new tools are being developed in GSP.

**Keywords** Epilepsy · Electroencephalogram · Graph discrete fourier transform (GDFT) · Graph signal processing (GSP)

## 1 Introduction

The basic idea of graphs were first introduced by the Swiss mathematician Leonhard Euler, one of the most eminent mathematicians. His work on the famous "Seven Bridges of Konigsberg problem", are commonly quoted as origin of graph theory, *Graph Theory* is ultimately the study of relationships. Given a set of nodes and connections, which can abstract anything from city layouts to computer data, graph theory provides a helpful tool to quantify and simplify the many moving parts of dynamic systems. Studying graphs through a framework provides answers to many arrangement, networking, optimization, matching and operational problems. Graphs

H. K. Meena (✉) · R. Sharma · A. Tailor · H. Verma · R. Saini
Department of Electrical Engineering MNIT Jaipur, Jaipur, India
e-mail: Hmeena.ee@mnit.ac.in



D. Gupta et al. (eds.), *Pattern Recognition and Data Analysis with Applications*, Lecture Notes in Electrical Engineering 888,
https://doi.org/10.1007/978-981-19-1520-8_46

can be used to model many types of relations and processes in physical, biological, social and information systems, and has a wide range of useful applications [1].

Graph Signal Processing (GSP) provides the solution of irregular domain living on the nodes of a graph in place of normal periods or domain such as grids. New tools are being evolved in GSP [1] to offer a nice compact format to encode the shape in the data among diverse fields together with social community, gesture popularity, street network and many others. Human Brain offers shape which can be understanding easily and analyzed in graph signal area. It gives the most inclusive facts of mental state of someone. The psychological and physical functions is such type of human behavior analysis that influences by the Epilepsy seizure. It is a neurological disorder in which unexpected unusual reactions arise in the mind producing fluctuations that are captured in EEG signal.

Earlier detection of the epilepsy was done by manually through inspection of EEG signals [2]. To automate the technique of detection of Epilepsy, different approaches, processes and numerous strategies were developed [3–7]. In the present day technology, graph signal processing are emerging fields to research and study about the brain signals. One such graph signal based weighted graph matrix technique provides methods to capture the turbulent nature of EEG signals [8]. This weighted matrix method has endorsed us to offer a brand new epilepsy detection technique. By doing this work, Graph Discrete Fourier Transform (GDFT) is proposed primarily based feature of EEG alerts described on weight matrix. This work consist of the steps, which are given as below.

1. EEG time series data is used to obtain the weight matrix by performing a Gaussian kernel based method for defining a unique weight to the edges.
2. Detection of the epilepsy disease is done by the performing Graph Discrete Fourier Transform based approach used for obtaining features.
3. Proposed new features and libraries based on Graph Signal Processing (GSP).

The rest of this paper is prepared as follows: A quick evaluation of the related work within the vicinity of detection of epilepsy is given in Sect. 2. Section 3 gives system overview and our methodology in detail. Section 4 deals with simulation consequences, results and publicly had EEG database and ultimately Section 5 concludes the paper and offers the future avenues.

## 2 Review of Related Work

The time series signal is converted into a complicated graph for detection of epileptic seizure using graph based technique. One such mapping called weight matrix approach is proposed through Lacasa et al. [3]. To offer exclusive energy to the edges of the graph, numerous techniques have been created to develop Weighted Graph matrix [3, 4, 8]. Supriya et al. [4] proposed an edge weight given in radian feature that's the perspective among connected nodes measured by using arc tangent. To improve the detection price of epilepsy, numerous features that can be extracted from the graph had been proposed like entropy as given through Mohammadpoory et al. [6]. Further graph signal processing is implemented to mind signals in fMRI records for characteristic extraction as given through Huang et al. [9]. Rui et al. in [10] explored graph signal processing for dimensionality reduction. In addition, neural networks for graph alerts had been considered for analysis of MEG signals as in Guo et al. [11] and fMRI signals by using Ktena et al. [12]. However, GDFT based total method for detection and category of epilepsy in EEG signals is not explored prominently. Gaussian kernel function is used to obtain the new edge weights for construction of the weight matrix of EEG signal. This weight feature presents particular value to each edge thereby capturing sudden fluctuations going on in EEG in the course of seizure activity. Further Graph Discrete Fourier Transform (GDFT) is carried out on EEG signals by calculating the Laplacian matrix and their Eigen vector from weighted graph. Thus, a unique set of GDFT coefficients is obtained from the Eigen area of the Laplacian matrix for every EEG graph signal. Now Stationarity ratio and TIK-norm play an important role to serve as a feature vector for classification of EEG signals that can be calculated with the help of GDFT coefficients. Thus, exactly classification of EEG facts for epilepsy for various EEG signals are based on GDFT features. Simulation outcomes carried out at the EEG database were able to detect the epilepsy magnificence from EEG alerts. Our technique is pretty powerful with minimum complexity as Stationarity ratio and TIK-norm is used with an outcome inside the detection of epileptic seizure with 97% accuracy.

## 3 Methodology

Steps are involved in algorithm of detection of epilepsy disease as following.

- Initially EEG signal is converted into the weight matrix with assigned edges by using Gaussian kernel function.
- Calculate the Eigen vector of the Laplacian matrix by using weight matrix to obtain the graph discrete Fourier coefficient (GDFT).
- Obtain the Stationarity ratio and Tik- norm feature vector.
- Stationarity ratio and Tik-norm feature vector are used for detection of epilepsy by classifying the EEG signal (Fig. 1).

A. **Assigning Edge weight**: We are approaching the signals defined on an undirected graph, connected, weighted graph $G = \{V, E, \mathbf{W}\}$, which consists of a finite set of vertices $V$ with $|V| = N$, a set of edges $E$, and a weighted adjacency matrix $\mathbf{W}$. If there is an edge $e = (i, j)$ connecting vertices $i$ and $j$, the entry $Wi, j$ represents the weight of the edge; otherwise, $Wi, j = 0$ [2]. Weight matrix is defined as a set of nodes which is connected through edges. Distance between two nodes is known as edges and calculated by Euclidian distance formula. When the edge weights are not naturally defined by an application, one common way to define the weight of an edge connecting vertices $i$ and $j$ is via a threshold Gaussian kernel weighting function [13].

**Fig. 1** Flow chart of our proposed method of detection of epilepsy from EEG signal

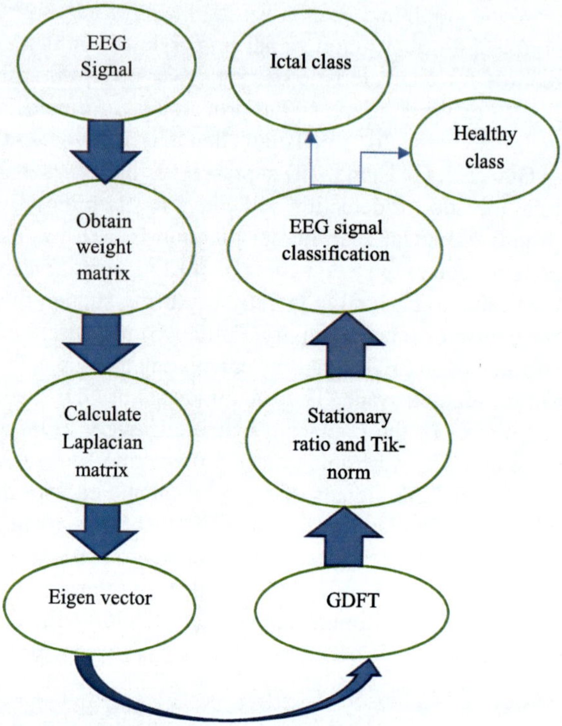


$$W_{i,j} = \begin{cases} exp\left(\frac{[dist(i,j)^2]}{2\theta^2}\right) & if\ dist.(i, j \leq k) \\ 0 & otherwise \end{cases} \quad (1)$$

For some parameters and *k*. In (1), dist (*i*, *j*) may represent a physical distance between vertices *i* and *j*, or the Euclidean distance between two feature vectors describing *i* and *j*, the latter of which is especially common in graph-based semi supervised learning methods. A second common method is to connect each vertex to its *k*-nearest neighbors based on the physical or feature space distances.

B. **Graph Laplacian matrix**: In our work Laplacian play an in important role for the evaluation of the GDFT coefficients. Laplacian matrix is defined as the equation given below [2].

$$L = D - W \quad (2)$$

where,

$D$ is the diagonal matrix
$W$ is the weight matrix.

Laplacian has an orthonormal Eigenvectors, because it is real symmetric matrix. By using the Laplacian matrix we calculate the Eigen value and Eigen vector of the matrix and it is used for calculating the GDFT coefficient. GDFT coefficient is the product of the Eigen vector of the Laplacian matrix and EEG signal and it is expressed by the equation given as below [2].

$$X_{\mathbf{GDFT}} = U^H X \tag{3}$$

where,

$U^{\mathrm{H}}$ is the Eigen vector of the Laplacian matrix.
$X$ is the EEG graph signal.

C. **Feature extraction**: Two basic features related with the GSP have been used in our work

1. Stationary ratio
2. Tik-norm

**Stationary ratio**: This can be classified into two category.

I. Strong Stationary (Healthy class): It is defined with the finite dimensional distribution of a stochastic process in which finite sub sequence of random variable of the stochastic process remains same with respect to the time. This means that it is shift invariance process with time [14].
II. Weak Stationary (Ictal class): it only requires the shift-invariance (in time) of the first moment and the cross moment (the auto-covariance). This means the process has the same mean at all-time points, and that the covariance between the values at any two time points, $t$ and $t - k$, depend only on $k$, the difference between the two times, and not on the location of the points along the time axis [14].

Stationarity ratio examines the percentage of the data variance that is not in the diagonal of the variance of the GFT of $X$. Data matrix $X$ is defined by the index of a graph i.e. how well it fits for distribution[15] (Fig. 2).

Stationary ratio calculates the ratio of energy contained into diagonal of the Fourier covariance matrix [14]:

$$T = U' * C * U \tag{4}$$

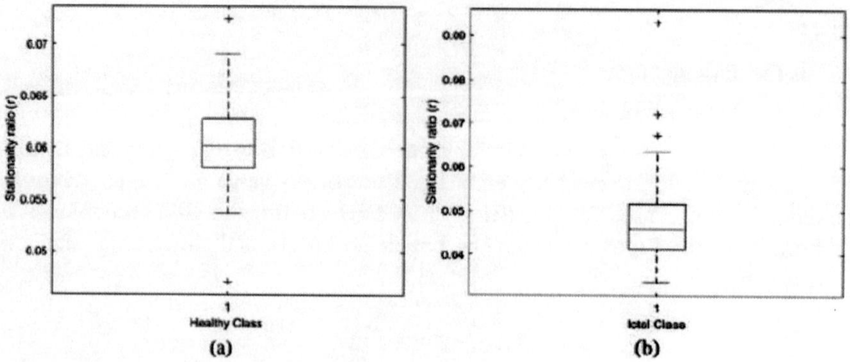


**Fig. 2** Box plot of stationarity ratio of EEG signal. **a** Healthy **b** Ictal

where,

**C** is the covariance matrix, ***U*** is the Eigen vector matrix, ***U′*** is the transpose conjugate of ***U*** and ***T*** is the Fourier covariance matrix. The stationary ratio is given by the formula:

$$r = norm(diag(T))/norm(T,' fro') \tag{5}$$

where norm (***T***, 'fro') returns the Frobenius norm of ***T***.

**Tik-norm**: Norm is defined as the real or complex non-negative values that treats as the distance from the origin. It follows the scaling, triangle inequality law. From this we can say the Euclidean distance from the origin is known as Euclidean norm. Square product of a vector with itself is also known as Euclidean distance or 2-norm [16].

Tik-norm compute the squared L2 norm of gradient on graph. If **X** is a matrix, a vector of norm is returned. It can also be used for general symmetric positive matrices.

$$P = L * X \tag{6}$$

$$\mathbf{Y} = \sum_i X_i * P_i \tag{7}$$

where,

***L*** is Laplacian matrix, X is data vector and ***Y*** is TIK- norm (Fig. 3).

The performance of the model is done by confusion matrix which is given in Table 1

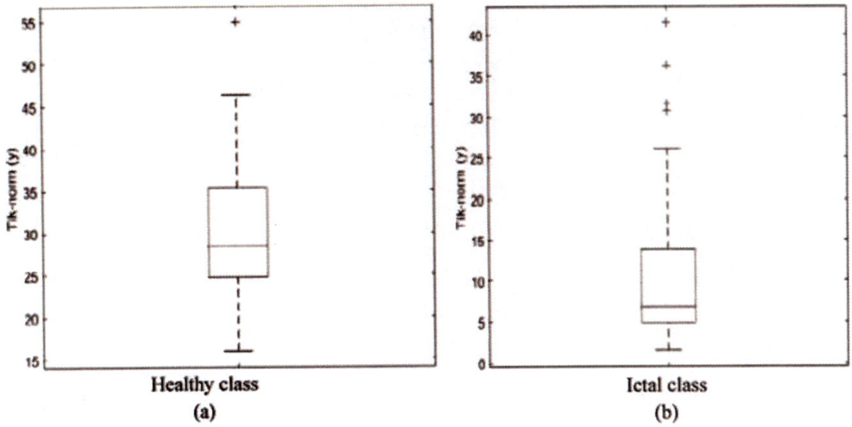


**Fig. 3** Box plot of tik-norm values of EEG signal. **a** Healthy **b** Ictal

**Table 1** Confusion matrix

| $N =$ Total prediction | Actual: No | Actual: Yes |
|---|---|---|
| Predicted: No | True negative | False positive |
| Predicted: Yes | False negative | True positive |

*True Negative* Model give prediction No, and the real or actual value was also No.
*True Positive* The model has predicted yes, and the actual value was also true.
*False Negative* The model has predicted No, but the actual value was yes.
*False Positive*:The model has predicted Yes, but the actual value was No.

# 4 Results and Discussion

## *4.1 Database*

Our proposed epilepsy detection technique usage of graph signal is examined by the online available EEG database which is furnished by center for epilepsy in University of Bonn, Germany [17]. Each group of EEG signal consists database of one hundred samples in the form of text file. There are three types of set i.e. healthy (Set A), and interictal (set C, D) and epileptic (ictal set E). Here, we have used two sets including healthy (Set A), and epileptic (ictal Set E) to elaborate the performance and techniques of our proposed methodology. We have used 4096 sample factors from every signal.

The proposed technique has been carried out in MATLAB. Figure 2 shows the ictal and healthy class of EEG signal obtained from the Stationarity ratio. GDFT is evaluated by the Eigen vector of Laplacian matrix. By observing the graph of

GDFT coefficient we can say that magnitude of GDFT coefficients of ictal class of epilepsy is higher than those of healthy class. For separation of feature vector into different classes of ictal and healthy by detecting the epileptic seizure from EEG signal, Stationarity ratio and Tik-norm special features have been used.

### 4.2 Performance Assessment

To check the performance of proposed method, we've used the following parameters:

1. Sensitivity: is defined as possibility of positive outcome in case of correct class of sample.

$$\text{Sensitivity} = \frac{\text{True Positive}}{\text{True Positive } + \text{False Ngative}}$$

2. Specificity offers the chance of negative result in case of incorrect magnificence of pattern

$$\text{Specificity} = \frac{\text{True Negative}}{\text{True Negative} + \text{False Positive}}$$

3. Accuracy is the ratio of the suitable class of samples to the total wide variety of samples

$$\text{Accuracy} = \frac{\text{True Positive} + \text{False Negative}}{\begin{array}{c}\text{True Positive} + \text{False Negative}\\ +\text{True Negative} + \text{False Positive}\end{array}}$$

where in, True Positive shows efficiently labeled healthy class, True Negative is the efficiently classified epileptic class, False Positive measures the fake detection of healthy class, and False Negative gives falsely detected epileptic class in EEG data. In this paper, 97% accuracy is completed the use of threshold based totally classification with proposed approach for detection of epilepsy in EEG database. Further in our method GDFT based features is compared with the entropy features [6] utilized by the researchers on visibility graph of EEG signals by using the same database with usage of K Nearest Neighbor (KNN) classifier. Performance of our proposed capabilities with above measures is examined using ten-fold cross-validation technique. With KNN classifier additionally, the proposed method achieves increased accuracy. Now by concluding all the analysis done by different techniques we can say that our proposed method to calculate the GDFT coefficient based feature vector with the help Gaussian weighted graph method in graph signal processing is more accurate as comparative to existing entropy based methods (Table 2).

**Table 2** Comparison between proposed method and entropy based method

| Method | Sensitivity % | Specificity % | Accuracy % |
|---|---|---|---|
| Entropy based approach | 95.23 | 97.34 | 96.7 |
| Proposed method (Stationarity ratio) | 92.3 | 92.3 | 92.3 |
| Proposed method (Tik-norm) | 98.9 | 95.6 | 97.3 |

By observing the above table we can say that the proposed method **(Tik–norm)** having more sensitivity, specificity and accuracy as compare to proposed method **(Stationarity ratio),** are significantly better to detect epilepsy in the patients.

## 5 Conclusion and Future Work

This paper offered a novel approach for detecting epilepsy in EEG signals with the help of Graph Discrete Fourier Transform (GDFT) based features, Stationarity ratio and Tik-norm used a feature extraction of EEG signal. In the first stage of analysis, the Laplacian matrix were used for obtaining the GDFT coefficient second stage of optimization in which Stationarity ratio and Tik-norm is used. Therefore it is highly accurate and smooth to feature extraction and pre-described threshold to early stumble on ictal class of epilepsy of EEG signal. The experimental consequences display that the proposed functions can detect epilepsy with 97% accuracy. Future scope of the proposed approach is to compare the performance of the proposed method with other non-visibility graph based techniques. Further the idea of Graph Signal Processing could be prolonged for detecting different brain issues and in diffusion modeling of brain signals.

## References

1. Shuman, G., Narang, S., Frossard, P., Ortega, A.,Vandergheynst, P.: The emerging field of signal processing on graphs: Extending high dimensional data analysis to networks and other irregular domains. Signal Process Mag. IEEE 30(3), 83–98 (2013)
2. Mathur, P., Chakka V.: Graph signal processing of EEG signals for detection of epilepsy. In: 7th IEEE International Conference on Signal Processing and Integrated Network (SPIN). Noida, India, pp. 839–843 (2020)
3. Zhu, G., Li, Y., Wen, P.: Epileptic seizure detection in EEGs signals using a fast weighted horizontal visibility algorithm. Comput. Methods Programs Biomed. **115**(2), 64–75 (2014)
4. Supriya, S., Siuly, S., Zhang, Y.: Automatic epilepsy detection from EEG introducing a new edge weight method in the complex network. IET Electron. Lett. **52**(17), 1430–2143 (2016)
5. Supriya, S., Siuly, S., Hua, W.: Weighted visibility graph with complex network features in the detection of epilepsy. IEEE Access **4**, 6554–6566 (2016)
6. Mohammadpoory, Z., Nasrolahzadeh, M., Haddadnia, J.: Epileptic Seizure detection in EEGs signals based on the weighted visibility Graph entropy, Seizure. Eur. J. Epilepsy Elsevier, **50**, 202–208 (Aug 2017)

7. Siuly, S., Wang, H., Zhang, Y: Analyzing EEG Signal Data for Detection of Epileptic Seizure: Introducing Weight on Visibility Graph with Complex Network Feature, Australasian Database Conference (ADC), vol. 9877 (2016)
8. Lacasa, L., Luque, B., Ballesteros, F., Luque, J., Nuno, J.C.: From time series to complex networks: the visibility graph. Proc. National Acad. Sci. **105**(13), 4972–4975 (2008)
9. Huang, W., Goldsberry, L., Wymbs, N.F., Grafton, Bassett, D., Ribeiro, A.: Graph frequency analysis of brain signals. IEEE J. Sel. Top. Signal Process. **10**(7), 1189–1203 (2016)
10. Rui, L., Nejati, H., Cheung, N.: Dimensionality reduction of brain imaging data using graph signal processing. In: Proceedings of International Conference on Image Processing, pp. 1329–1333 (2016)
11. Guo, Y., Nejati, H., Cheung, N.: Deep neural networks on graph signals for brain imaging analysis. In: IEEE International Conference on Image Processing (ICIP), pp. 3295-3299 (2017)
12. Ktena, S., Parisot, E., Ferrante, Rajchl, M., Lee, M., Glocker, B., Rueckert, D.: Distance metric learning using graph convolutional networks: Application to functional brain networks. In: Medical Image Computing and Computer Assisted Intervention ( MICCAI), Lecture Notes in Computer Science, Springer, Berlin, vol. 10433 (2017)
13. Grady, L., Polimeni, J.: Discrete Calculus. Springer, Berlin (2010)
14. Towards data science Homepage: https://towardsdatascience.com/stationarity-in-time-series-analysis-90c94f27322
15. Perraudin, N., Vandergheynst, P.: Stationary signal processing on graphs".arXiv preprint ar Xiv.1601.02522 (2016)
16. Perraudin, N., Paratte, J., Shuman, D., Kalofolias, V., Vanderghenyst, P., Hammond, D.: GSPBOX: A toolbox for signal processing on grap, ArXiv e-prints (Aug 2014)
17. Andrzejak, R., Lehnertz, K., Mormann, F., Rieke, C., David, P., Elger, C.: Indications of nonlinear deterministic and finite-dimensional structures in time series of brain electrical activity: Dependence on recording region and brain state. Phys. Rev. E **64**, 61907 (2001)

# Hybrid Approach for Fake Profile Identification on Social Media

**Shruti Shinde and Sunil B. Mane**

**Abstract** Millions of people use social media platforms like Twitter, Facebook, and Instagram all around the world. People are drawn to these social media sites, and as the prevalence of social media grow, so do the security and privacy concerns that come with it. Nowadays, it is critical to ensure that we are following the correct social media account or purchasing a product from the actual consumer, as malicious users can be extremely harmful. This paper proposes a hybrid method for detecting fake social media user accounts. For detecting fake accounts, it makes use of the Instagram social media platform's dataset. There are two steps to the hybrid approach. The first stage is to use Principal Component Analysis (PCA) which turn original variables into new uncorrelated variables, and the second stage is to use various classification algorithms, in the second stage five algorithms are used to obtain accurate results. Fake profiles are detected using naive Bayes, artificial neural networks (ANN), support vector machine (SVM), logistic regression, and K-nearest neighbors (KNN) algorithms. When the classification performances of these approaches are compared, the artificial neural network outperforms the others.

**Keywords** Hybrid · PCA · Machine learning · Instagram · ANN · Confusion matrix

## 1 Introduction

Various associative areas have given rise to the term Online Social Network. Over the past two decades, social networking has grown tremendously. One of the most pressing issues of today's online world is the steady rise in the number of false user accounts on social media platforms such as Facebook, Twitter, and Instagram.

S. Shinde (✉) · S. B. Mane
College of Engineering, Pune(COEP), Pune, India
e-mail: shrutiss19.comp@coep.ac.in

S. B. Mane
e-mail: sunilbmane.comp@coep.ac.in



D. Gupta et al. (eds.), *Pattern Recognition and Data Analysis with Applications*, Lecture Notes in Electrical Engineering 888,
https://doi.org/10.1007/978-981-19-1520-8_47

There are some types of spam profiles that can be found on social media sites.

1. Fake Profile—To carry out fraudulent acts, a fake profile is created in the name of an individual or a business that does not exist in real life on social media.
2. Duplicate Profile—Duplicate profiles are created by stealing the information of an existing user and using them to generate a new profile, a concept known as cloning.
3. Bots accounts—Social media bots are automated social media accounts developed with the aim of increasing the number of followers on a particular account.

This paper provides a solution to the issue by considering it. This study looks at Instagram as a social media platform for detecting fake user profiles. The Instagram dataset contains information for both private and public user accounts. This paper provides a hybrid solution for detecting fake profiles on Instagram. This hybrid method is divided into two stages. The original variables are transformed into new uncorrelated variables in the first step. This will be achieved using Principal Component Analysis (PCA), [1] and the fake profiles will be computed using different classification algorithms on the principal component values. The aim of this study is to come at a precise solution by implementing five classification algorithms. The confusion matrix is used to calculate the accuracy of these models.

Selecting most important features is important step for building the accurate and precise model. The respective dataset has 15 feature, but every feature is equally important is not necessary, hence using the principal component analysis we get the important feature set list. The dataset may differ for building different types of models but, selecting appropriate feature set list can become important step for every dataset in the process of model building. Hence the use of applying PCA before classification algorithms is important.

Structuring the rest of the paper: The current research work on the thesis is shown in Sect. 2. The features used in datasets and datasets will be explained in Sect. 3. The proposed model is outlined in Sect. 4. Section 5 provides an insight into model implementation. Section 6 gives the performance values outcomes and final discussion. The work is concluded in Sect. 7.

## 2 Related Work

Different approaches to detecting fake user accounts have been provided by different publishers. The hybrid method will be used in this study. For predicting fake user profiles on Instagram, the hybrid approach will use a PCA + Classification algorithm model. Following will show some of the works that have been presented in this field of segment.

Reference [2] Consider tweeting texts on twitter to identify fake users; Logistic Regression, ADA Boost, XGBoost, Random Forest algorithms that are precise to 89%. In [3–5] different machine learning, supervised algorithms and unsupervised ones are included. The bar graph shows the exact comparison of these algorithms. In [6] fake and Twitter cloning accounts are detected with algorithms for classifying and measuring distances, where the detection of clone profiles gives more precision than the detection of fraudulent profile. In [7] machine learning algorithms and the Waikato Environment for Information (WEKA) tool are used to identify duplicate accounts on social media [8]. Uses spam comment, artificial behavior, and interaction speed attributes for fake profiles recognition. They also used the Decision Tree gradient boost algorithm and offer greater accuracy when the Random Forest algorithm is modified with the Gradient boost algorithm. In [9] for classification purposes where real positive rate is 85%, they used Random Forest Algorithm.

In reference [10] supervised dataset preprocessing Entropy Minimization Decartelization Technology Discretization (EMD) is used for classification and the Naive Bayes algorithm. For false profile identification, this uses a twitter data collection [11]. Several approaches have been examined to identify fake accounts in real social networks. Effect and impact on the identification of bot by machine learning methods were also addressed. Reference [12] the methodology is focused on the versatility of the DFA (regular expression) approach to the identification of profiles. Mechanism for the proposed detection of social graph profile FPR uses regular expression notations to create a Friend Pattern (FP). In [13] by using the WalkPool pooling layer to optimize CNN computation, they set up dynamic CNN architecture. It also deals with data over fitting and under fitting issues. The reference [1] gives hybrid classification model information using cancer related dataset.

In previous researches, single classic algorithms were mostly used for malicious profiles identification; however, by presenting a hybrid approach with the use of PCA, this paper provides a new method for identifying fake users on social media sites.

## 3 Dataset

In this study, the social media site Instagram is used to recognize fake accounts. This dataset is a combination of public and private Instagram user accounts. The features set of this dataset is classified as user-based features, content-based features and time-based features. Instagram Web Scrapper is used to scrape the feature of each user account. Table 1 gives more basic information about the attributes of the dataset.

## 4 Proposed Method

The proposed framework begins by collecting information about fake and genuine accounts available on the Instagram social media platform, as seen in the system flow diagram. The features in the dataset were collected from users present on Instagram.

**Table 1** Dataset description

| No | Feature name | Type of data | Description |
|---|---|---|---|
| 1. | Profile picture | Boolean | 0 if the user does not have the profile picture, 1 if the user have the profile picture |
| 2. | Numbers/Length username | Double value | How many special characters of numeric characters the username has on its full length |
| 3. | Full name words | Numeric value | How many words are present in the full name? |
| 4. | Bio length | Numeric value | How many characters present in users description |
| 5. | External URL | Boolean value | 0 if the user does not have the external URL in the description and 1 if there is URL |
| 6. | Private | Boolean value | 0 if the profile private and 1 if profile is Public |
| 7. | Is verified | Boolean value | 0 if the users do not have the verified account, 1 otherwise |
| 8. | Is Business | Boolean value | 1 if the users do not have a business account, 1 otherwise |
| 9. | Post | Numeric value | The number of the posts presents on user Profile |
| 10. | Followers | Numeric value | The number of the followers of users |
| 11. | Following | Numeric value | The number of the following of the user |
| 12. | Last post | Boolean value | Recent 0 if the user has not published a post within 6 months, 1 otherwise |
| 13. | Post single day | Double value | How many posts have been published in the same day on the total number of the posts |
| 14. | Index of activity | Double value | In average how much post the user posts every month |
| 15. | Average of Likes | Double value | Average number of likes on the post |

The next step is to upload the dataset; however, before doing so, preprocessing is needed. The following steps are included in the preprocessing stage: handling missing values, label encoding, data standardization, and so on. This stage's performance is fed into the classification model.

There will be two stages to the hybrid model. The first step would be to turn the original variables into new ones that are uncorrelated. Principal Component Analysis (PCA) [14] is used to accomplish this. The Fake users are computed in stage two using various classification algorithms on the principal component values (Fig. 1).

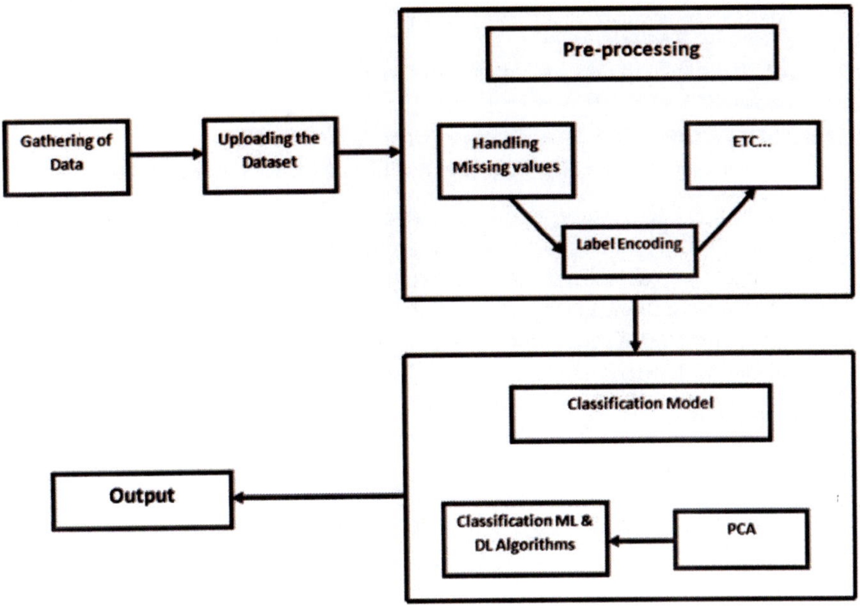


**Fig. 1** System flow diagram

# 5 Implementation

## *5.1 Principal Component Analysis*

PCA is an unsupervised machine learning algorithm for feature selection and data analysis. It is necessary to use PCA statistics and matrix algebra to analyze the results [14]. In PCA, Eigenvectors and Eigen values are numbers and vectors that are associated with matrices and provide the Eigen decomposition of a matrix, which analyses the structure of a matrix. PCA in particular is acquired through the decoding of a covariance or a matrix for correlation.

The steps for PCA [14] and constructing the respective classification hybrid model are as follows.

1. Gathering and analyzing the dataset-

   The dataset is gathered from Instagram social media platform. The data is preprocessed before proceeding to the next stage.
2. Subtract the Mean-

   The average value of all data is derived from and data calculation in order to ensure PCA works correctly. [14] The average value of each calculation is the subtracted mean. All x values are then subtracted from x and all y values from y are subtracted. This gives null average records.

3. Calculate the Covariance Matrix-

   The covariance matrix indicates if the transition takes place in the same or the opposite direction. It is a square matrix that lets each pair of attributes of a random vector relate.

4. Calculate the Eigenvectors and Eigenvalues of the Covariance Matrix-

   Calculating Eigenvectors and Eigenvalues for the matrix improves in obtaining useful information about the results. First, we have *N*-dimensions in the original data; we can choose the first *P* Eigenvectors based on their Eigenvalues by measuring the Eigenvalues and Eigenvectors. The most popular method of calculating Eigenvectors and Eigenvalues is to define an Eigenvector of the matrix *A* as a vector *u*.

   By rewriting,

$$\mathrm{Au} = \lambda \mathrm{u} \tag{1}$$

$$(\mathrm{A} - \lambda)\mathrm{Iu} = 0 \tag{2}$$

   *I* is the Identity Matrix, and lambda is a scalar that is an Eigenvalue associated with the respective Eigenvector [14]. Similarly, we can assume that the vector *u* is an Eigenvector of the matrix *A* if its length (but not its direction) changes when it is multiplied by *A*.

5. Choosing Components and Forming a Feature Vector-

   When we calculate the Eigenvalues, we get first *P* attributes from the respective data that are more descriptive, and the new dataset will only have *P*-dimensions data. The dataset's main components are the Eigenvalues with the highest values. As Eigenvectors are found in the covariance matrix, they are sorted in ascending order by Eigenvalues. The next step is to build the function vector, which is nothing more than the metrics of the vectors. To accomplish this, extract Eigenvectors to be saved from the Eigenvectors list and use these function vectors in the columns to form a matrix [14].

6. Acquiring the New Dataset-

   In this stage, we obtain our final *P*-dimensional data from the entire *N*-dimensional original data. The following PCA result, Fig. 2 provides a clearer picture of importance of attributes in dataset. *X* axis gives the list of attributes present in the dataset and *Y* axis gives its degree of importance, by considering this degree of importance principal component values are selected and given as an input to the next level.

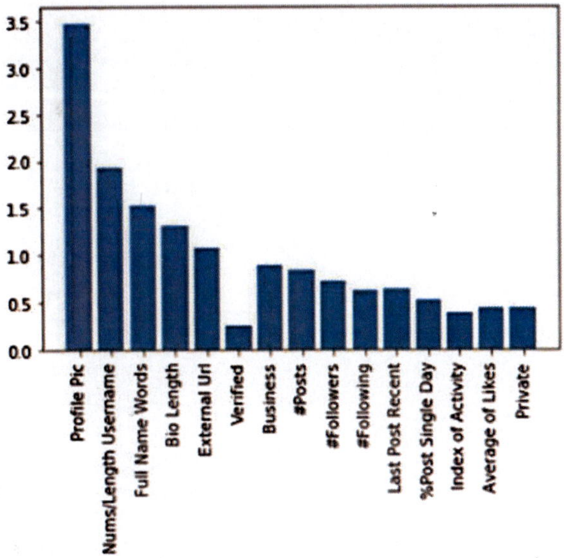


**Fig. 2** Feature selection bar graph

## 5.2 *Classification Model*

Five classification algorithms are used to build the second step of this hybrid model. Support vector machines, logistic regression, *k*-nearest neighbor, artificial neural networks, and naive Bayes are the algorithms which are used. The classification model that produces the best results will be chosen as the final model. The aim of this stage is to select the model with the highest accuracy among the five of them in order to improve the hybrid model's performance and then finalizing one model which gives highest accuracy. According to the results and discussion in Sect. 6, ANN outranks the other four algorithms in terms of accuracy. The artificial neural networks section describes why it outperforms other classification models and detailed information of algorithm implementation. Section 6 contains a detailed discussion of the results of each algorithm. Among them, SVM has the second highest accuracy.

**Artificial Neural Networks (ANN)**

A learning classification algorithm is called as an artificial neuron network. As shown in Fig. 3, a single neuron is referred to as a perceptron. It has one input and output layer, with weights assigned to each input that govern the magnitude of that respective input.

$$u_k = \sum_{j=0}^{m} W_{Kj} x_i \tag{3}$$

where

**Fig. 3** ANN

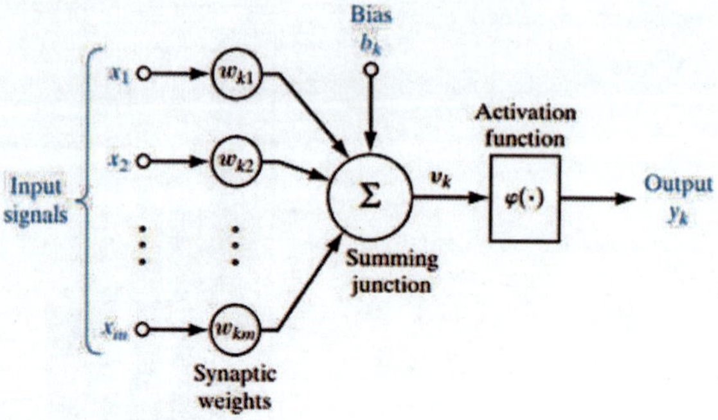


$$x_0 = 1, b_k = W_{k0} \text{ and } y_k = f(u_k) \quad (4)$$

| | |
|---|---|
| $b_k$ | Bias parameter |
| $u_k$ | Linear combination of output |
| $y_k$ | Final output of the neuron |
| $x_1, x_2, x_3\ldots.. x_m$ | Input signals |
| $W_{k1}, W_{k2}, W_{k3}\ldots..W_{km}$ | Respective weights. |

The activation function is then given the sum of the products of these input values and weights. Activation functions are important in this context because they are in possession of learning and mappings of nonlinear complex functional between the input and the corresponding response variable. Table 2 compares the activation functions, along with their benefits and drawbacks.

By considering the comparison chart (Table 2) Rectified linear unit (ReLu), a nonlinear activation is used to build the corresponding model. $f(x) = \max(0, x)$ is the ReLU function. This is typically applied element-by-element to the output of another function, including a matrix–vector product. The Sequential Classifier of Keras is used for classification task. The data is divided into 9:1 ratio, hence 10% data is used for testing purpose and 90% data is used for training purpose. For gradient

**Table 2** Comparison chart of activation function

| Activation function name | Advantages | Disadvantages |
|---|---|---|
| Binary step function | It is threshold-based activation function. Based on value of the certain threshold neuron gets fired | It do not allow multiple valued output |
| Linear activation function | Better than step function. It take weights with input for each neuron | Do not support back propagation All layers collapse into one |
| Nonlinear activation function | Allow to create complex mappings between inputs and outputs. Better than above activation function | Types of this function have some disadvantages which are overcome by one another |

descent, Adam optimizer is selected. The batch size is taken as 100 and number of epoch is 150. Trial and error were used to determine the batch size and number of epochs. The amount of samples that will be propagated across the network is defined by the batch size and a complete pass through all of the training data is referred to as an Epoch. Using all these characteristic of artificial neural network the classification model is build, which results that ANN classification algorithm gives better results compared to other four algorithms.

# 6 Results and Discussion

## *6.1 Evaluation Criteria*

A confusion matrix is an $N \times N$ matrix used to evaluate a classification model's output, where $N$ is the number of target groups. It has parameters like precision, recall, and F1-score (Fig. 4).

$$\text{Accuracy} = \text{TP} + \text{TN} / \text{TP} + \text{TN} + \text{FP} + \text{FN} \tag{5}$$

where

TP = True Positive, TN = True Negative

FP = False Positive, FN = False Negative.

The true positive value is when the predicted value corresponds to the actual value. If the predicted value was mispronounced then the predicted value is incorrect. Precision tells us how many of the correctly expected cases have been positive and Recall tells us how many of the real positive cases with our model we will properly estimate.

**Fig. 4** Confusion matrix

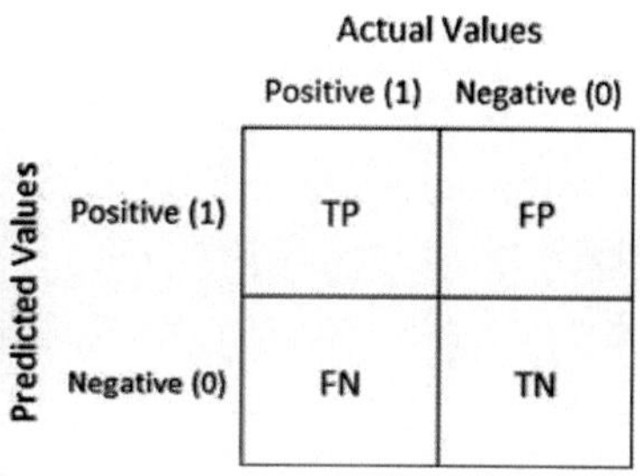

## 6.2 Results

Principal component analysis is used in the first stage of the model to consider the most relevant attributes from the feature set. By considering the ten most significant features from total 15 the new data is given to the classification algorithms. Support vector machine, artificial neural network, logistic regression, *K*-nearest neighbor, and naive Bayes are some of the five algorithms considered for the classification stage. When the results are computed as a (PCA + classification algorithm), it is observed that ANN outperforms all other algorithms. Accuracy is calculated with the confusion matrix as seen in evaluation criteria. Following table provides comprehensive information on each model's performance.

According to the Table 3 representation, the model (PCA + ANN) provides greater accuracy than the other models. For confirming the accuracy standards of the selected hybrid model that is (PCA + ANN) another dataset was used which is twitter's dataset. The same hybrid method was applied on the twitter dataset only the difference is in the data and attributes of the dataset. By considering the same methodology the accuracy achieved is 96.5%. Another dataset is used only for confirming the standards of proposed model. The accuracy comparison of Instagram and Twitter dataset is shown in the Fig. 5 in the form of bar graph by considering precision and recall parameters.

**Table 3** Accuracy comparison of the model

| Computation model | Accuracy (%) | Precision | Recall | F1-score |
|---|---|---|---|---|
| PCA + ANN | 92.83 | 0.9231 | 0.9056 | 0.9145 |
| PCA + Logistic Regression | 89.72 | 0.8845 | 0.8578 | 0.8911 |
| PCA + Naive Bayes | 85.20 | 0.8756 | 0.8525 | 0.8490 |
| PCA + KNN | 88.19 | 0.8834 | 0.8712 | 0.8831 |
| PCA + SVM | 90.20 | 0.9027 | 0.8974 | 0.8945 |

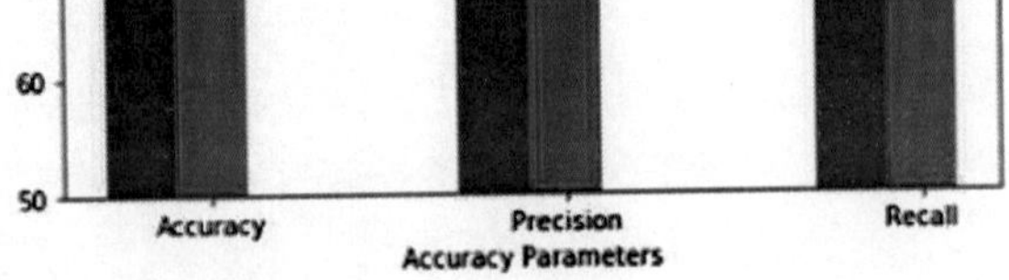


**Fig. 5** Accuracy comparison between Instagram and Twitter dataset

## 7 Conclusion and Future Work

Social media platform attract millions of Internet users because they are the most common and widely used website. This raises a number of security issues, including the possibility of a fake profile and the spread of malicious material. In this paper, we suggested a hybrid method for detecting fake profiles using PCA + classification algorithm model. In the five classification models ANN gives greater accuracy than other algorithms by considering the user and content-based feature set.

Future work may include developing a browser plug-in that can detect fake user accounts. There are several applications available that assist users in keeping track of their followers. In future by considering the account detected fake or real, we can keep the dataset populating and building the model more efficiently. Future research could focus on creating an application that alerts users to fake profiles while they are using a particular social media application. This study uses Instagram as a social media platform; however, other social media sites, such as LinkedIn and Facebook, can be used to detect fake accounts using different methodologies and feature sets in the future.

## References


1. Sahu, B., Mohanty, S., Rout, S.: A Hybrid approach for breast cancer classification and diagnosis. In: SIS, EAI (2019). https://doi.org/10.4108/eai.19-12-2018.156086
2. Pakaya, F., Ibrohim, M., Budi, I.: Malicious account detection on twitter based on Tweet account features using machine learning. In: Fourth International Conference on Informatics and Computing (ICIC), Semarang, Indonesia, pp. 1–5 (2019). https://doi.org/10.1109/ICIC47613.2019.8985840
3. Patel, K., Agrahari, S., Srivastava S.: Survey on fake profile detection on social sites by using machine learning algorithm. In: 8th International Conference on Reliability, Infocom Technologies and Optimization (Trends and Future Directions) (ICRITO), pp. 1236–1240. Noida, India (2020). https://doi.org/10.1109/ICRITO48877.2020.9197935
4. Gayathri, A., Jayalakshmi, S.: Detection of fake profile in online social networks using machine learning approach. In: Pecial Issue Published in International Journal of Advanced Networking Applications (IJANA)
5. Gupta, A., Kaushal, R.: Towards detecting fake user accounts in facebook. In: ISEA Asia Security and Privacy (ISEASP), Surat, 2017, pp. 1–6 (2017). https://doi.org/10.1109/ISEASP.2017.7976996
6. Sowmya, P., Chatterjee, M.: Detection of fake and clone accounts in Twitter using classification and distance measure algorithms. In: International Conference on Communication and Signal Processing (ICCSP), pp. 0067–0070. Chennai, India (2020). https://doi.org/10.1109/ICCSP48568.2020.9182353
7. Devmane, M., Rana N.: Detection and prevention of profile cloning in online social networks. In: International Conference on Recent Advances and Innovations in Engineering (ICRAIE-2014). pp. 1–5, Jaipur, India (2014). https://doi.org/10.1109/ICRAIE.2014.6909237
8. Maniraj, S., Harie Krishnan, G., Surya, T., Pranav, R.: Fake account detection using machine learning and data science. In: International Journal of Innovative Technology and Exploring Engineering (IJITEE), vol. 9(1) Nov (2019). ISSN: 2278-3075

9. Durga, S., Reddy, P.: Fake profile identification using machine learning. In: International Research Journal of Engineering and Technology (IRJET), vol.06(12) (2019). e-ISSN: 2395-0056
10. Erşahin, B., Aktaş, O., Kılın D., Akyol, C.: Twitter fake account detection. In: International Conference on Computer Science and Engineering (UBMK), pp. 388–392. Antalya (2017). https://doi.org/10.1109/UBMK.2017.8093420
11. Tiwari, V.: Analysis and detection of fake profile over social network. In: International Conference on Computing, Communication and Automation (ICCCA), pp. 175–179. Greater Noida, India (2017). https://doi.org/10.1109/CCAA.2017.8229795
12. Torky, M., Meligy A., Ibrahim, H.: Recognizing fake identities in online social networks based on a finite automaton approach. In: 12th International Computer Engineering Conference (ICENCO), pp. 1–7. Cairo. https://doi.org/10.1109/ICENCO.2016.7856436
13. Wanda, P., Jie, H.J.: Deep profile: Finding fake profile in online social network using dynamic CNN. J. Inf. Secur. Appl. **52**, 102465 (2020). https://doi.org/10.1016/j.jisa.2020.102465
14. Mishra, S., Sarkar, U., Taraphder, S., Datta, S., Swain, D., Saikhom, R.: Multivariate statistical data analysis- principal component analysis (PCA). Int. J. Livestock Res. **7**(5), 60–78 (2017). https://doi.org/10.5455/ijlr.20170415115235

# Electroencephalogram-Based Emotion Recognition Using Random Forest

**Nalini Pusarla, Anurag Singh, and Shrivishal Tripathi**

**Abstract** In recent years, emotion recognition based on electroencephalogram (EEG) has gained prominence due to its wide applications in the area of healthcare, affective computing, brain-computer interface, etc. Capturing the emotion quotient effectively and thus improving the recognition performance has been a major challenge in the conventional emotion recognition problem based on EEG. This work presents a new automatic emotion recognition algorithm using hybrid multi-channel EEG features and a Grid Search Random Forest (GSRF). The proposed algorithm extracts multi-domain features from different channels of the EEG signal and fused them into a hybrid feature matrix. A GSRF has been fed with a labeled feature matrix to classify the emotions into different classes. The algorithm has been validated on two widely used open-source databases DEAP and SEED. The proposed algorithm obtained an average classification accuracy of 86.3% and 97.9% using DEAP and SEED, respectively, with tenfold cross-validation. As compared to the Random Forest classifier, the proposed approach exhibited superior emotion recognition performance.

**Keywords** Emotion recognition · Electroencephalogram (EEG) · Random Forest (RF) · Grid search random forest (GSRF) · And multi-domain features

N. Pusarla (✉) · S. Tripathi
DSPM IIIT, Naya Raipur, Chhattisgarh, India
e-mail: nalini@iiitnr.edu.in

S. Tripathi
e-mail: shrivishal@iiitnr.edu.in

N. Pusarla · A. Singh
Vignan's Institute of Information Technology, Visakhapatnam, Andhra Pradesh, India
e-mail: anurag@iiitnr.edu.in



D. Gupta et al. (eds.), *Pattern Recognition and Data Analysis with Applications*,
Lecture Notes in Electrical Engineering 888,
https://doi.org/10.1007/978-981-19-1520-8_48

## 1 Introduction

Emotion plays a ubiquitous role in the everyday life and work of humans. Recognition of emotions has become a multidisciplinary research topic in neuroscience, interfaces between human computers, cognitive science, artificial intelligence, and psychology [1]. Experts in various fields have recently proposed different approaches for identifying emotions. The first approach is focused on interpreting non-physiological signs such as facial movements, gestures of the mouth, and body. This technique is also used to interpret emotions since it is very simple and does not require any special hardware. But the key problem with this approach is that people will fake their true emotional statements by masking their voice tone and facial expressions [2]. The emotion detection approach is also not trustworthy. This approach cannot yet be extended to people with disabilities or with diseases. Another way of achieving so is to examine physiological signals such as electroencephalogram (EEG) [3], electromyography (EMG) [4], electrocardiogram (ECG) [5], galvanic skin resistance (GSR) [6], heart rate, pulse rhythm [6], etc. These physiological signals are essential signatures that are out of an individual's control. Therefore they are better suited for defining human emotions and are more effective.

Owing to the growing development of non-invasive and low-cost EEG recording systems, EEG-based emotion identification gained further interest in research and diverse applications over the past few years. Several emotion models were proposed in the literature for the classification of emotions. Moreover, there are two basic models for representing the emotional behaviors known as discrete and dimensional models. The first method classifies emotions into discrete entities, such as anger, disgust, fear, happiness, sadness, and surprise in Ekman's theory [7]. The second method quantifies emotions using multidimensional scales such as valence, arousal [8]. Valence measures emotions from negative to positive, arousal reflects emotional intensity from passive to aggressive. Among other models, Russell's two-dimensional models, taking valence on the horizontal axis and arousal on the vertical axis, are mostly used. Out of the two-dimensions valence is the most vital dimension for the emotion which can discriminate the emotion between positive and negative. In therapeutic applications of specific depression types, it is important to define negative valence levels [9]. Higher-order crossing (HOC)-based features have been employed in [10] for EEG-based emotion recognition. The authors have analyzed the performance of four classifiers including quadratic discriminant analysis, KNN, Mahalanobis distance, and support vector machine (SVM). Power spectral density (PSD) and Pearson correlation coefficient (PCC) features were also explored for EEG-based emotion classification using Stacked Auto Encoder (SAE) and Long Short-Term Memory (LSTM) [11]. However, this work reported lower classification accuracy. Statistical and frequency domain features such as PSD, Discrete Cosine Transform (DCT) and STFT features are extracted from EEG and were classified using sparse discriminative ensemble algorithm [12]. Few works [13–15] have employed wavelet transform as a time–frequency analysis tool, with excellent time–frequency localization property, proved to be quite helpful in effectively capturing physiological events in EEG. Adrian et al. [13] have used DWT-based features along with artificial neural network (ANN) as a classifier for emotion recognition. Recently, Vipin et al [14] have investigated cross-subject emotion recognition by extracting channel-specific features using flexible analytical wavelet transform (FAWT). With Random Forest as a classifier, FAWT achieved 90.48% accuracy using SEED while underperformed for DEAP with 79.99%. A variant of DWT called tunable-Q wavelet transform (TQWT) is used for feature extraction which was followed by emotion classification using extreme learning machine (ELM) [15]. Moreover a combination of time domain features, power spectral density and wavelet have been extracted for lateralization and emotion recognition with an accuracy of 75.6% [16]. This method also achieved better accuracy for the four-class classification problem. A combination of time–frequency and non-linear dynamical features have helped Li et al [17] to achieve relatively higher classification accuracy. Chunmei et al. [18] extracted correlation and entropy features using ensemble classifier and reported 63.5% and 75% with DEAP and SEED, respectively. A multi-method fusion approach by authors in [19] yielded accuracy up to 72% and 89% with DEAP and SEED, respectively. Further, Li et al. [20] employed two variants of multi-source selective transfer machine (MS-STM): supervised and semi-supervised for transfer learning of Differential entropy (DE) features with SEED and a maximum of 91.3% accuracy could be achieved. In the latest developments in this domain, researchers began exploring deep neural networks specifically, convolutional neural network (CNN) and its variants to enable automatic feature extraction and classification of EEG-based emotions. Muhammad et al. [21] build a two-dimensional (2-D) time–frequency (TF) plot from EEG as AlexNet can learn features better from 2-D data. The authors employed a bag of deep feature model (BoDF) for feature selection and SVM for classification of emotions and achieved an accuracy of 77.4% and 93.8% with DEAP and SEED, respectively. Hong Zeng et al. [22] employed the SincNet-R model and achieved accuracy of 94.5% using the raw EEG from SEED. On the other hand, Yucel et al. [23] worked on both datasets and exhibited a mean cross-subject accuracy of 86.56% and 72.81% for SEED and DEAP, using raw EEG. These works based on pre-trained CNN models for emotion classification were computationally intensive and could achieve average performance.

Most of the above-discussed works either use time/frequency or time–frequency domain features to capture emotion-related information from EEG. However analyzing EEG in any one domain may not be sufficient due to its time-varying complex nature. This ignores the complementary information of the other two domains. Beyond the usage of dominant features, the correct choice of classifier boosts the prediction results during classification. Motivated from this, our work presented two strategies to enhance the classification rate of emotion recognition. First, multi-domain analysis of EEG signals, where the selected features are extracted from multiple domains and combined to form a hybrid feature tensor. Second, we have used Random Forest (RF) and Grid search Random Forest (GRF) to automatically select an optimized set of features from the hybrid feature matrix and classify the associated emotions in different classes.

## 2 Dataset Preparation

The experiments in this paper have been validated on the Database for Emotion Analysis of Physiological Signals (DEAP) [24] dataset and the SJTU Emotion EEG Dataset (SEED) [25] dataset for emotion recognition. DEAP is a multi-channel EEG database in which 32 effective channels and 8 peripheral channels were collected from thirty-two healthy participants. 32 participants are watching "40" one-minute videos to stimulate different types of emotions. Forty trails of EEG with respective to 40 videos have four emotional dimensions like arousal, valence, dominance, and liking with ratings of 0–9. High and low valence classes are known as positive and negative emotion classes that have been used in this experimentation. SEED is also a multi-channel EEG database granted by Shanghai Jiao Tong University in the year 2015. It contains 15 subjects with 15 trials. Each subject experimented with 15 emotional videos and the respective 15 EEG signals collected by placing electrodes on the head of the subjects. Each EEG is a 62 multi-channel signal recorded at a sampling rate of 1000 Hz. The EEG data is downsampled to 200 Hz. The stimuli videos duration is varied: each video is about 4 min or 240 secs, thus results in the data length of $200 \times 240$ s $= 48{,}000$ samples.

## 3 Methodology

A block diagram depicting the major steps involved in the proposed framework is shown in Fig. 1. Broadly, there are four major steps: preprocessing of the EEG signal, multi-domain feature extraction from multi-channel EEG data, the fusion of the multi-domain features, and emotion classification. A detailed description of these steps has been given in the following subsections.

### 3.1 Preprocessing

Recorded EEG has to be pre-processed to remove the EMG, EOG artifacts, and downsampled to decrease the computational burden of the experiment. In the case of DEAP, the default preprocessing approach is as follows: (1) EEG data is downsampled to 128 Hz; (2) A bandpass filter with a passband frequency range of 4.0–45.0 Hz

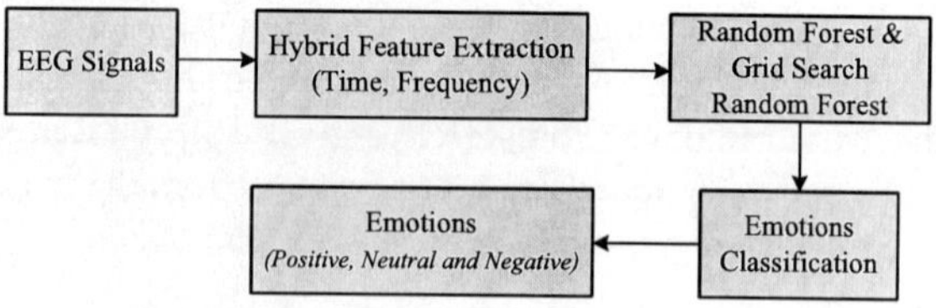


**Fig. 1** Block diagram of the proposed method

is used to remove the noise and EOG artifacts; (3) EEG signals are segmented into 60-s trails and 3-s pre-trial baseline. For SEED, the default preprocessing approach is as follows: (1) The data is downsampled to 200 Hz; (2) A bandpass filter with a cut-off frequency range of 0–75 Hz is used to filter out noise and artifacts; (3) Segments of EEG corresponding to each video are extracted.

### 3.2 *Feature Extraction*

This section describes a specific set of features, which are responsible for capturing different emotional states of the human brain through the EEG signals. It is difficult to capture the discriminative set of features in any particular domain due to the non-defined morphology and non-stationary nature of multi-channel EEG signals. So, we have extracted prominent multi-domain features in the time, frequency, and time–frequency domains and combined them to create the hybrid feature tensor.

a. Time domain features determine the characteristics of the time series EEG that oscillate between distinct emotional states. These features are captured through statistical parameters. Major statistical parameters include mean, variance, zero crossing rate, and sampling entropy. These four features are extracted from all channels of each EEG signal available in the two datasets as shown in the Table 1. Sample entropy quantifies the complexity of the EEG time series data. It helps provide hidden dynamics associated with the signal. It calculates the Chebyshev distance between two template vectors x and y formed from the EEG data points. The ZCR of EEG indicates the number of sign changes of the amplitude along with the whole duration of the signal. They are defined as follows:

**Table 1** Details of features extracted from two databases

| Feature type | DEAP database | SEED database |
|---|---|---|
| Mean | 32 (channels) × 1 = 32 | 62 (channels) × 1 = 62 |
| Variance | 32 (channels) × 1 = 32 | 62 (channels) × 1 = 62 |
| Zero crossing | 32 (channels) × 1 = 32 | 62 (channels) × 1 = 62 |
| Sampling entropy | 32 (channels) × 1 = 32 | 62 (channels) × 1 = 62 |
| Band power | 32(channels) × 5(bands) = 160 | 62(channels) × 5(bands) = 310 |
| Power difference | 14 (pairs) × 4 (bands) = 56 | 27 (pairs) × 4 (bands) = 108 |
| Total features (columns) | 344 | 666 |

$$\text{Mean}\ \mu = \frac{1}{T}\sum_{-\infty}^{+\infty}|x(t)|^2 \tag{1}$$

$$\text{Variance}\ \sigma^2 = \sum \frac{[x(t)-\mu]^2}{T} \tag{2}$$

$$\text{SampEn}(m, r, N) = -\log\frac{x}{y} \tag{3}$$

$$\text{Zero Crossings}\ Z(i) = \frac{1}{2W_L}\sum_{n-1}^{W_L}|\text{sgn}[x_i(n)] - \text{sgn}[x_i(n-1)]| \tag{4}$$

where sgn(.) is sign function, i.e.,

$$\text{sgn}[x_i(n)] = \begin{cases} 1, x_i(n) \geq 0 \\ -1, x_i(n) < 0 \end{cases}$$

b. Frequency domain features such as spectral power is a very crucial feature for emotion recognition as it captures the differences between activities of different brain regions. To calculate spectral power, Discrete Wavelet transform (DWT) is used to decompose each EEG signal into low-frequency approximation and high-frequency detail subbands. As a resultant, we have got five frequency bands, namely, delta (0–4 Hz), theta (4–8 Hz), alpha (8–13 Hz), beta (13–30 Hz), and gamma (30–70 Hz) [26]. The EEG signals corresponding to the aforementioned frequency bands represent different emotional states of a human mind such as consciousness/alertness, calmness, thought process, etc. Hence, extracting features within these frequency bands may help capture the specific emotion of a person. Power spectral density (PSD) of EEG rhythms, delta, theta, alpha, beta, and gamma may capture the insight of dominating brain emotion. We employed the Welch method to compute PSD of different EEG rhythms. A total of 5 * 62 channels = 310 PSD values (features) are extracted as shown in the Table 1. To extract correlation information between any two channels, we are computing power differences between channels by pairing the electrodes (channels) symmetrically. A total of 108 features from 27 electrode pairs and four frequency bands excluding delta band are extracted and in the same manner for DEAP 14 electrode pairs, 56 features are extracted as shown in the Table 1.

### 3.3 *Random Forest*

The most prominent classification ensemble methodologies are bagging, boosting, and stacking. These are used for both classification and regression. Random forest is a modified version of bagging, based on an ensemble of decision trees. RF grows these decision trees using random samples from the training dataset known as bootstrapping

[13]. Here the samples are drawn with replacement in each tree, in other words, the same sample is used repeatedly. Assume $N$ features are with training data and $M$ features are associated with each decision tree classifier. An $m$ features ($m \ll M$) are chosen at each node to make the decision. Secondly, RF selects a training set by selection. Each node randomly selects a subset of $f$ ($f < N$) features when raising the tree. Out of these f features, A single feature is chosen for node splitting. Trees are iteratively added in the Random Forest in the proposed algorithm known as construction pass. The process begins with an initial number of trees. We then update each step of the building through four phases to update the list of useful and optional features. First, we measure the weights of different features and test them by their weight. If the weight of the feature crosses the threshold are excluded afterward. Then, we determine the unimportant features based on a new criterion from the remaining features. Note that if a feature is listed as relevant at the building pass, it is still applicable at the end and not omitted in the next passes. Now that we have major negligible elements, at this construction direction, we are constructing a reasonable cap for adding large numbers of trees into the forest. We demonstrate that the precision of forest classification naturally improves when trees that exceed the cap are added.

### 3.4 *Grid Search Random Forest Algorithm (GSRF)*

In machine learning, model's behavior is governed by specific parameters known as hyperparameters that are to be tuned. Tuning is one of the most complicated parts of model development, but it is a task necessary to boost predictions. Fortunately, two extensively used tuning methods, Grid Search and Random Search, for picking the optimized parameters which result in the maximal performance of the model. In this study, the Grid Search tuning algorithm is used along with the RF model for increasing its prediction capability. This GSRF has proved its efficiency by achieving an improved 10% classification accuracy over the RF model. In this experiment, RF and GSRF models are validated with the cross-validation technique. The features extracted from the two datasets are partitioned into 80% train and 20% test datasets. Using the training dataset, two models were trained and tested with the unseen 20% test data. In machine learning, the model's behavior is accessed by the metrics such as confusion matrix and ROC curves. The results section emphasizes both metrics obtained by the two models.

## 4 Results

The classification performance of the Random Forest (RF) and proposed Grid Search Random Forest (GSRF) models with multi-domain features of DEAP and SEED databases have been evaluated using confusion matrix.

## 4.1 Confusion Matrix

The confusion matrix (CM) presents information about correctly classified and misclassified instances. Figure 2a depicts positive and negative accuracies of 70.09% and 79.86%, respectively, and an average classification accuracy of 75.4% achieved by RF using DEAP features. Figure 2b describes the confusion matrix for negative and positive emotions with an accuracy of 82.42% and 93.41% and an average classification accuracy of 86.3% achieved by GRF. The GRF model has very high classification capability, i.e., its overall accuracy is almost greater than 10% compared to RF. Figure 3a gives the confusion matrix for negative, neutral, and positive emotions with an accuracy of 76.9%, 78.6%, and 94.4%, and average classification accuracy of 84.4% achieved by RF. Figure 3b demonstrates the confusion matrix for negative, neutral, and positive emotion classes with accuracies of 100%, 93.3%, and 100% and average classification accuracy of 97.8% achieved by GSRF. In the overall average sense, the classification accuracy achieved by the GSRF is almost 11% higher with both datasets. Both the classification models performed better with SEED features

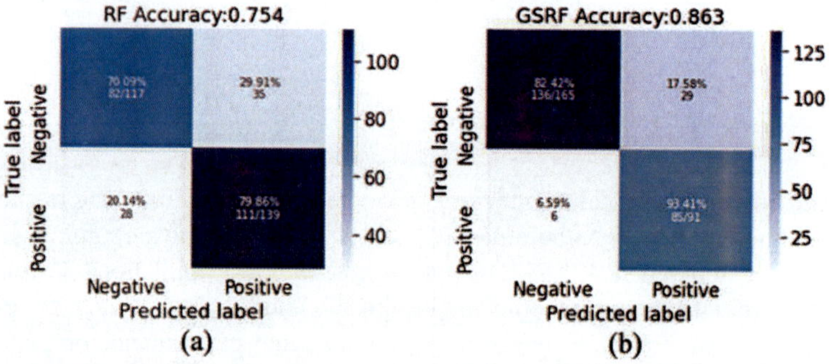


**Fig. 2** Confusion matrix obtained for DEAP features **a** RF model **b** GSRF model

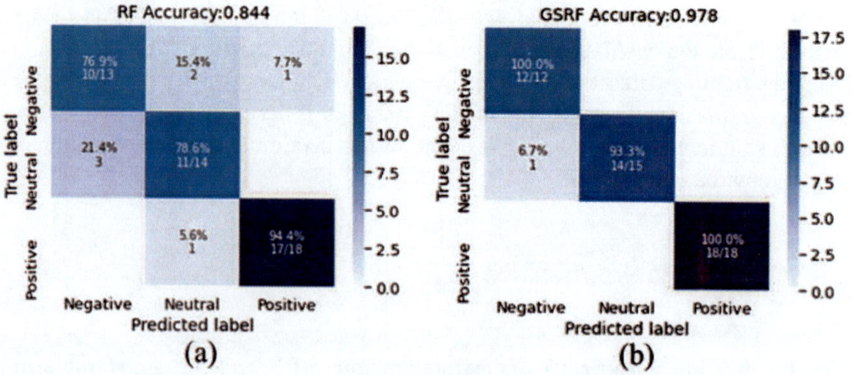


**Fig. 3** Confusion matrix obtained for SEED features **a** RF model **b** GSRF model

than DEAP. This is due to imbalances present in the DEAP dataset. Due to this, we have employed another important metric known as the ROC curve to judge the classifier's efficacy.

## 4.2 *Region of Convergence (ROC)*

In this work, the performance analysis of the RF and proposed GSRF has been carried out in terms of classification accuracy. Furthermore, accuracy alone is not adequate and robust to evaluate the classifier, as it does not take into account class imbalances and the basic feature distribution. Therefore, the two models, RF and proposed GSRF are further evaluated over a more consistent and effective metric, i.e., Region of Convergence (ROC), which relates true positive rate (TPR) and false positive rate (FPR) that the classifier achieves depending on all feasible thresholds. Besides, another important measure "Area under the Curve (AUC)" is used for model assessment that quantifies GSRF functioning across all possible thresholds. In this section, both models discriminating strength across two/three emotion classes were determined by ROC curves. Here two ROC curves were plotted for the two considered datasets on cross-validation. The ROC curve plotted in Fig. 4a depicts the reliability of RF for cross-validation of the DEAP dataset. Its effective prediction for classes 1, 2 represents negative and positive emotions with AUC = 0.75 and 0.75, respectively. Its minimum value was achieved with DEAP due to the presence of more imbalances in the no. of EEG samples for each class. The ROC curve, shown in Fig. 4b, is constructed for analyzing the performance of GRF for cross-validation on the DEAP dataset. As mentioned above DEAP has two classes, class 0, 1 corresponding to positive and negative emotion categories. The model obtained an impressive AUR of 0.88 and 0.88 for the two classes. The ROC curve in Fig. 5a is built for RF while tested on the SEED dataset, where classes 0, 1, and 2 correspond to negative, neutral, and positive emotions, respectively. The ROC curve, shown in Fig. 5b, was obtained for GSRF cross-validation on SEED. However, network performance is

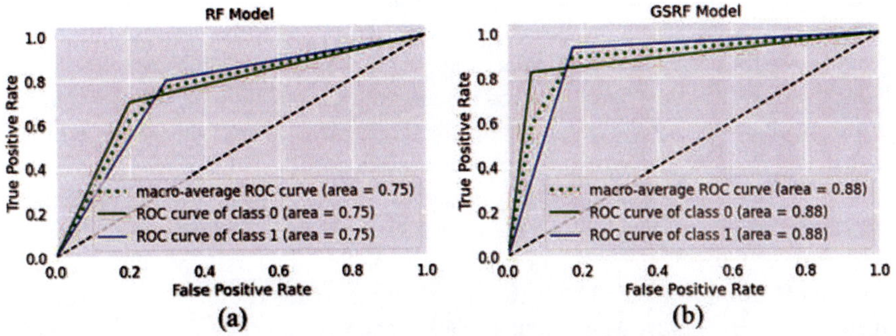


**Fig. 4** ROC curves obtained for DEAP features **a** RF model **b** GSRF model

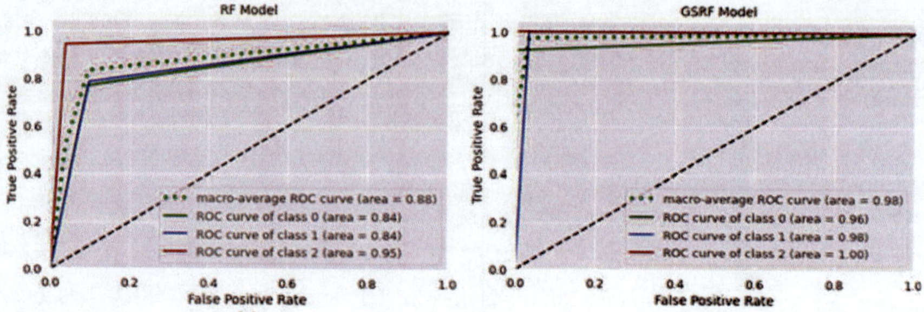


(a) (b)

**Fig. 5** ROC curves obtained for SEED features **a** RF model **b** GSRF model

higher for AUC 0.98, 0.96, and 0.98. The features extracted from SEED's EEG are more informative than DEAP. The best output on SEED is due to the presence of stronger emotion-related information in the EEG signals of around 4 min compared to the DEAP dataset.

## 5 Comparative Analysis

The proposed Grid Search Random Forest is relatively simple and consistent against the existing models compared in Table 2 such as LSTM, ST-SBSSVM, MS-STMSVM, AlexNet, SincNet-R and Inception Resnet-V2 . The emotion classification accuracy achieved with the proposed method is compared with the recent emotion recognition works and comparative results on both the datasets have been shown in Table 2. The comparative analysis is performed in terms of the accuracy claimed by current works. It can be observed from the results that the proposed algorithm performs significantly better than the existing emotion recognition works with both the datasets. The proposed multidomain feature extraction approach from multi-channel EEG and Random forest for classification of those features play a vital role in enhancing the classification accuracy and this fact has been already verified in the Table 2. It can be observed from the results that the proposed algorithm performs significantly better than the existing ML-based emotion recognition works [12, 14, 16, 19]. Some works [12, 16, 19] employed multi-domain features and few of them used feature selection techniques [12, 19] too. Moreover, GSRF achieved higher accuracy compared to complex models-based works [11, 18, 20–23]. With the best of our knowledge, the maximum accuracy reported by Hong Zeng et.al [22] is 94.5% for SEED and Vipin et al. [14] is 79.99% for DEAP. The proposed algorithm outperforms the above works and enhances the emotion classification accuracy by 7% and 6% using SEED and DEAP, respectively. Another key observation from this analysis is [22] employed a computationally expensive Inception Resnet-V2 model with 164 layer architecture and obtained slightly low accuracy compared to state-of-the-art. Therefore our GSRF model is less complex (29 layers) and efficient when

**Table 2** Comparative analysis with recent works

| Author | Classifiers | DEAP Database (%accuracy) | SEED Database (%accuracy) |
|---|---|---|---|
| Xiaofen Xing[11] | SAE and LSTM | 75.23 | – |
| Habib Ullah[12] | Sparse channel Ensemble | 78.2 | – |
| Vipin Gupta[14] | Random Forest | 79.99 | 90.48 |
| Evi Septiana Pane[16] | Random forest | 75.6 | – |
| Chunmei[18] | SAE and DT | 63.09 | 75 |
| Fu. Yang[19] | ST-SBSSVM | 72 | 89 |
| Li Jingpeng[20] | MS-STM with SVM | – | 91.3 |
| Muhammad et.al [21] | AlexNet with SVM | 77.4 | 93.8 |
| Hong Zeng et.al [22] | SincNet-R | - | 94.5 |
| Yucel Cimtay et.al [23] | Inception Resnet-V2 | 72.81 | 86.56 |
| Proposed Method | Grid Search Random Forest | 86.3 | 97.8 |

compared with Resnet-V2. Based on the experimental results and comparative analysis, we concluded that the proposed method has two advantages. First, the Grid search random forest classifies and recognizes emotions with comparable emotion recognition accuracy to the deep CNN models [21–23] as shown in Table 2. Second, GRF reduced the program runtime as it is fed with the best features which represent the emotional characteristics of EEG. Therefore it is verified that the proposed algorithm's overall performance on both the databases is better in terms of accuracy obtained.

## 6 Conclusion

A Random Forest-based emotional classifier was proposed in this work. Our proposed algorithm classified the emotions using multi-domain features extracted from multichannel EEG. Random Forest and Grid Search Random Forest is fed with features extracted from every channel of EEG, to extract a good amount of spatial information. This information helped the Grid Search Random Forest to efficiently discriminate the emotions into negative and positive classes for DEAP and negative, neutral, and positive for SEED datasets. The suggested method attained higher emotion classification accuracy when compared with the existing works on the same databases. The proposed work also revealed that tuning parameters have a greater impact on the performance of the Random Forest. The proposed emotion recognition system can be used in a real-world scenario for many applications like stress or depression detection, brain-computer interface (BCI) applications, etc.

## References

1. Nijholt, G.C.: A survey of affective brain-computer interfaces: principles, state-of-the-art, and challenges. Brain-Comput. Interfaces **1**(2), 66–84 (2014)
2. Picard, R.W.: Affective computing for HCI. HCI (1), 829–833 (1999)
3. Rajeev, S., Ram Bilas, P., Abhay, U.: Automatic sleep stages classification based on iterative filtering of electroencephalogram signals. Neural Comput. Appl. **28**(10), 2959–2978 (2017)
4. Yoo, J., Jaerock, K., Yoonsuck C.: Predictable internal brain dynamics in EEG and its relation to conscious states. Front. Neurorobotics **8**, (2014)
5. Erguzel, T,T., Gokben, H. S., Nevzat, T.: Artificial intelligence approach to classify unipolar and bipolar depressive disorders. Neural Comput. Appl. **27**(6), 1607–1616 (2016)
6. Mourlas, C.E., Nikos, E.T., Panagiotis, E.G.: Cognitive and emotional processes in Web-based education: Integrating human factors and personalization. Inf. Sci. Ref. /IGI Global (2009)
7. Jerritta, S., Murugappan, M., Nagarajan, R., Khairunizam W.: Physiological signals based human emotion recognition: a review. In: 2011 IEEE 7th International Colloquium on Signal Processing and its Applications, pp. 410–415 (2011)
8. Russell, J.A.: A circumplex model of affect. J. Pers. Soc. Psychol. **39**(6), 1161(1980)
9. Lin, C.F., Zhu, J.-D.: Hilbert–Huang transformation-based time-frequency analysis methods in biomedical signal applications. In: Proceedings of the Institution of Mechanical Engineers, Part H: Journal of Engineering in Medicine, vol. 226(3), pp. 208–216 (2012)
10. Zhuang, N., Ying, Z., Li T., Chi, Z., Hanming, Z., Bin, Y.: Emotion recognition from EEG signals using multidimensional information in EMD domain. BioMed Res. Int. (2017)
11. Xing, X., Zhenqi, L., Tianyuan, X., Lin, S., Bin, H., Xiangmin, X. SAE+ LSTM: A new framework for emotion recognition from multi-channel EEG. Frontiers Neurorobotics 13 (2019)
12. Ullah, H., Muhammad, U., Arif, M., Mohib, U., Sultan Daud, K., Faouzi Alaya, C.: Internal emotion classification using eeg signal with sparse discriminative ensemble. IEEE Access **7**, 40144–40153 (2019)
13. Ang, A.Q., Yi, Q., Wee, W.: Emotion classification from EEG signals using time-frequency-DWT features and ANN. J. Comput. Commun. **5**(3), 75–79 (2017)
14. Gupta, V., Mayur Dahyabhai, C., Ram Bilas, P.: Cross-subject emotion recognition using flexible analytic wavelet transform from EEG signals. IEEE Sens. J. **19**(6), 2266–2274 (2018)
15. Krishna, A.H., Sri, A.B., Priyanka, K.Y.V.S, Sachin, T., Varun B.: Emotion classification using EEG signals based on tunable-Q wavelet transform. IET Sci. Measur. Technol. **13**(3), 375–380 (2019)
16. Pane, E.S., Adhi Dharma, W., Mauridhi Hery, P.: Improving the accuracy of EEG emotion recognition by combining valence lateralization and ensemble learning with tuning parameters. Cogn. Process. 1–13 (2019)
17. Li, X., Dawei, S., Peng, Z., Yazhou, Z., Yuexian, H., Bin, H.: Exploring EEG features in cross-subject emotion recognition. Front. Neurosci. **12**, 162 (2018)
18. Qing, C., Rui, Q., Xiangmin, X., Yongqiang, C.: Interpretable emotion recognition using EEG signals. IEEE Access 7, 94160–94170 (2019)

19. Yang, F., Xingcong, Z., Wenge, J., Pengfei, G., Guangyuan, L.: Multi-method fusion of cross-subject emotion recognition based on high-dimensional EEG features. Front. Comput. Neurosci. **13**, 53 (2019)
20. Li, J., Shuang, Q., Yuan-Yuan, S., Cheng-Lin, L., Huiguang, H.: Multisource transfer learning for cross-subject EEG emotion recognition. IEEE Trans. Cybern. **50**(7), 3281–3293 (2019)
21. Asghar, M.A., Khan, M.J., Amin, Y., Rizwan, M., Rahman, M., Badnava, S., Mirjavadi, S.S.: EEG-based multi-modal emotion recognition using bag of deep features: An optimal feature selection approach. Sensors **19**(23), 5218 (2019)
22. Zeng, H., Zhenhua, W., Jiaming, Z., Chen, Y., Hua, Z., Guojun, D., Wanzeng, K.: EEG emotion classification using an improved SincNet-based deep learning model. Brain Sci. **9**(11), 326 (2019)
23. Cimtay, Y., Erhan, E.: Investigating the use of pretrained convolutional neural network on cross-subject and cross-dataset EEG emotion recognition. Sensors **20**(7), 2034 (2020)
24. Zheng, W.-L., Bao-Liang, L.: Investigating critical frequency bands and channels for EEG-based emotion recognition with deep neural networks. IEEE Trans. Auton. Mental Dev. **7**(3), 162–175 (2015)
25. Koelstra, S., Christian, M., Mohammad, S., Jong-Seok, L., Ashkan, Y., Touradj, E., Thierry, P., Anton, N., Ioannis, P.: Deap: A database for emotion analysis; using physiological signals. IEEE Trans. Affect. Comput. **3**(1), 18–31 (2011)
26. Singh, P., Shiv, D.J., Rakesh K.P., Kaushik, S.: Fourier-based feature extraction for classification of EEG signals using EEG rhythms. Circ. Syst. Signal Process. **35**(10), 3700–3715 (2016)

# Controlled Active Rectifier Circuit-Based Extreme Fast Charging System for Electric Vehicles

**Amit Kumar and D. Saxena**

**Abstract** This paper proposes the model of an Extreme Fast Charging System (XFC) with a Controlled Active Rectifier Circuit (CARC). In this paper controlled active rectifier circuit has been introduced in place of any kind of uncontrolled or semi-controlled rectifier. With the increasing awareness of electric vehicles (EVs), it is required to advance extreme fast charging system topology for high power charging of EV for a power greater than 350 kW. The research focused on charging the electrical vehicles (EVs) battery at unity power factor (UPF) at the input side of the system using a controlled active rectifier circuit. It will transfer maximum active power to the battery of the electric vehicle with no reactive power so that the power factor of the circuit will be unity and also have a very less harmonic distortion in current waveforms at the input side of the rectifier, with the proposed converter scheme of CARC total harmonic distortion (THD) in the system has been achieved to 4.37% which is desired result. By using the CARC, we can control the output voltage of the rectifier for a wide range of voltage from 400 to 800 V without affecting the UPF at the input side of the XFC system. Multiple partial power charging unit (PPCU) is placed at the output side so that maximum EVs battery can be charged at a time. The performance evaluation of the XFC system is done using MATLAB/Simulink and simulated results are for various steady states.

**Keywords** Controlled active rectifier circuit (CARC) · Solid state transformer (SST) · State of charge (SoC) · Partial power charging unit (PPCU) · Extreme fast charging (XFC) · Total harmonic distortion (THD) · Unity power factor (UPF) · Electric vehicles (EVs)

A. Kumar (✉) · D. Saxena
Electrical Engineering, Jaipur, M.N.I.T, India
e-mail: 2019PPD5081@mnit.ac.in

D. Saxena
e-mail: dsaxena.ee@mnit.ac.in



D. Gupta et al. (eds.), *Pattern Recognition and Data Analysis with Applications*, Lecture Notes in Electrical Engineering 888,
https://doi.org/10.1007/978-981-19-1520-8_49

# 1 Introduction

Electric Vehicles (EVs) are growing at a very faster rate as conventional modes are very polluting by transport electrification has superior performance, decrement in greenhouse emissions, and decarbonizing the transport sector. Enhancement in the battery technology and driving range has significantly given the interest in adoption toward EVs. Environment-related concern gives rises the interest in electrical vehicles. In 21 century the use of automobiles is growing day by day and which is very important in a country like India because more and more people will buy an automobile. By using the automobiles with conventional mode they have some disadvantages like more carbon emission, decreasing foreign reserve, increasing fuel price. A feature like a silent ride, immediate torque, premium performance, regenerative braking, and low maintenance cost has given more interest toward the EVs. When compared to the internal combustion engine (ICE) vehicles, they are very efficient and cost-saving in terms of fuel. As in ICE vehicles, lots of mechanical parts are there so high maintenance is required and are very less efficient [1]. As renewable technologies, smart microgrids are growing at a very faster rate because to reduce global warming, greenhouse gases emission, and the shortage of fossil fuel.

Due to Li-ion batteries has their limitation and also the EVs charging technology and infrastructure, charging time is still a big concern when compared with the existing gasoline station especially when a long-distance trip is required. Designing the charging infrastructure is complex, to serve the demand for EVs new adequate charging infrastructure is required. The most simple charging method is vehicles onboard charger that connects single-phase ac supply, these are Level-1 and Level-2 chargers 120 V for Level-1 and 240 V for Level-2, and these chargers can deliver power up to 1.92 kW for Level-1 and 19.2 kW for Level-2. The level-1 charger charges the EV battery at the rate of 3–5 miles/h, Level-2 at the rate of 60 miles/h, and Level-3 at the rate of 200 miles/h approximately. Due to limited power capability, high charging time, and less range dc fast charger comes into scope most recently up to 350 kW, so the power is transferred to converters via isolated power converter located outside the EVs [2, 3].

The extreme fast charger of 350 kW takes only 10 min to charge for 200 miles of range [4]. Extreme fast charging (XFC) system is for a power range greater than 350 kW and is required for long-range. It reduces the charging time and can be compared with gasoline refueling time. XFC can be operated up to dc voltage of 800 V. When multiple XFC systems are arranged together it forms an XFC station which can reduce the overall reduction in the capital cost so that it is economically viable [5]. This XFC charger is installed as a single cell unit or multiple cell units and each cell is rated for 50 kW so according to the required power rating they can arrange. XFC station composed single-phase or three-phase ac-dc rectification stage for experiment purpose downscaled model is useful and therefore single-phase ac is used.

The first main problem associated with the XFC is the input power factor (IPF) when the front-end converter or any other converter is used, which power factor is not high, then higher current is required by the equipment so that cost of equipment is increased. At high current copper, losses are also high, therefore the efficiency of the overall system is reduced. A higher current in the system produces a higher voltage drop across each element in the system so the voltage regulation becomes poor. As all electrical machinery are rated in kVA and power factor has the inverse relationship with the kVA so when the power factor reduced kVA rating increases which larger the cost of the converters and other equipment's. Power Supply Company enacts the penalty at the low power factor below 0.95 lagging in electric power bill so power factor must be above 0.95. Both the running and capital costs are associated with the power factor, so it also increases the cost at the low power factor [6].

The second main problem associated with the XFC system is the Total Harmonic Distortion (THD), one of the major effects of the THD in the XFC system the high current. It increases the total RMS current in the system which is mostly due to third harmonics, it introduces the zero sequence current and therefore overall increment of current in the system. These harmonics degrades the overall performance of the XFC system. Harmonics flowing into the system downgrades the quality of the input current, so there may have numerous negative effects on the operation of the power system. Overload, premature aging of different components connected to the system, losses, and noise in the system increase due increase in effective RMS current [7].

The third main problem associated to charge the EV battery is the different voltage specifications of the different batteries. Many manufactures have their standard battery specifications in terms of voltage, power, and current rating according to electric vehicles because of different EV Company, so there is need of chargers which can adjust their voltage according to the battery.

In this article, to overcome above mentioned all issues, a controlled active rectifier circuit (CARC) is introduced and studied in Sect. 2.1, which has the function to improve the input power factor of the circuit to unity and to decrease the Total Harmonic Distortion (THD). By the use of CARC power factor has been achieved to 0.99 and THD to 4.37% which is the desired result explained in Sect. 3. A battery current control scheme is used to control the battery current [8], CARC to control the output voltage of the rectifier has been also proposed in this article. The proposed CARC is connected between the MV grid and solid state transformer (SST) [9]. A partial power charging unit (PPCU) is placed after the SST [10], SST and PPCU are retained from [5].

## 2 Model of XFC System

In the proposed scheme extreme fast charging system has been employed in MATLAB/Simulink to attain unity input power factor at the input side of the rectifier, to reduce the system total harmonic distortion (THD), and to control the output voltage of the rectifier. To attain the desired result active rectifier circuit unit has been integrated into the XFC system. The active rectifier circuit unit provides the unity power factor (UPF) and controls the output voltage of the rectifier. Figure 1 shows

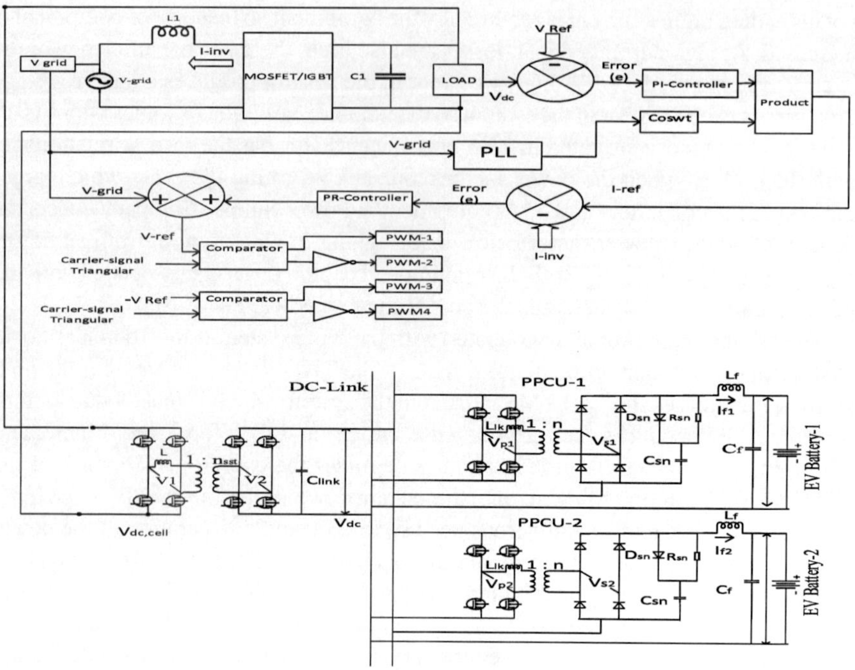


**Fig. 1** Model of extreme fast charging system with CARC

the model of an extreme fast charging system with an active rectifier circuit unit and this XFC system model consists of:

1. Controlled active rectifier circuit.
2. Solid state transformer.
3. Partial power charging unit.
4. Battery current controller.

## 2.1 Controlled Active Rectifier Circuit

In the proposed scheme in place of a front-end converter or normal active rectifier circuit, a controlled active rectifier circuit has been used. A normal active rectifier circuit or any front-end converter causes the low power factor at the input side of the converter and high harmonic due to the switching of power electronic device which introduces the harmonic and distorts the input waveforms. The proposed scheme eliminates the problem related to low power factor and harmonics. A controlled active rectifier circuit is the first component of the XFC system and its purpose is to provide the UPF and to smoothen the input distorted current and voltage waveforms. This unit is also able to control the output voltage of the rectifier to the desired range of 400–800 V without affecting the unity power factor of the circuit. With UPF at

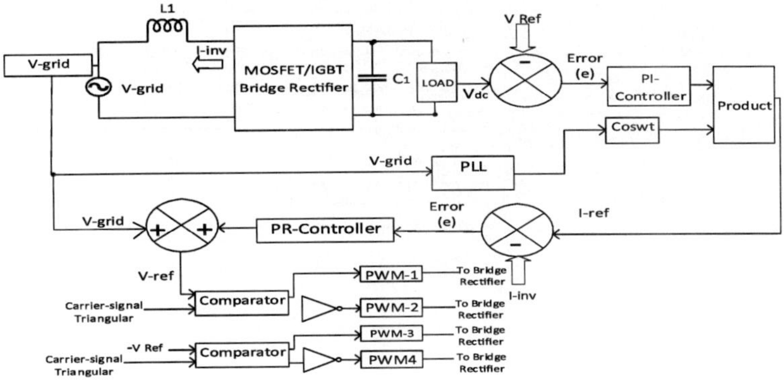


**Fig. 2** Model of controlled active rectifier circuit

the input side of the converter increases the efficiency of each component and to the overall system, provide low voltage drop at each component, reduction of the size of conductor or cable used to charge the EVs battery, reduces the copper loss in the cable, increases the amount of available power, eliminate the penalty of low power factor and most importantly saving electricity bill.

The system as MOSFET/IGBT has been used in place of normal semiconductor diode and SCRs. The normal rectifier has a significant effect on power quality due to the passive component, a passive filter having higher harmonics. These devices have a constant voltage drop of 0.5–1 V. MOSFET is a variable voltage drop device with a low voltage drop. Active rectifier circuit improves the quality of the current waveform at the input and improves the power factor.

In Fig. 2, model of the controlled active rectifier circuit (CARC) is shown, inductor $L_1$ is placed at the input side of the converter is used to boost the output voltage of the rectifier and to smoothen the current waveforms at the input side of the XFC system. Bridge rectifier converts AC into DC, capacitor $C_1$ is used to smoothen the voltage waveform. Output from the load $V_{dc}$ the signal is fed to the PI controller. A phase-locked loop (PLL) is provided in an active rectifier circuit, coswt is the unit vector generated by PLL and this component is aligned with the phase voltage. PLL circuit is a closed-loop system that synchronizes the output signal with the input signal, when phase difference becomes zero system gets locked. PLL circuit helps in attaining the UPF at the input side of the converter. Output from the PLL and PI controller is multiplied and given to the adder block, where $I_{ref}$ and $I_{inv}$ have been compared and given to the Proportional-Resonant (PR) controller. PR controller is a closed-loop controller in place of a PI controller such controller has a high gain around the resonant frequency and thus can eliminate the steady state error (Table 1).

The inductor $L_1$ value on the converter side is calculated by the following equations.

**Table 1** Controlled active rectifier circuit parameters

| System parameters | Value |
|---|---|
| Grid voltage, $V_{grid}$ | 240 V |
| Inductance, $L_1$ | 4.06 mH |
| Capacitance, $C_1$ | 550 μF |
| Reference voltage, $V_{ref}$ | 400 V |
| Switching frequency | 20 kHz |

$$L_1 = \frac{V_{ref}}{16.f_{sw}.\Delta I_{L-max}} \quad (1)$$

$$\Delta I_{L-max} = a.\frac{P_n\sqrt{2}}{V_{grid}} \quad (2)$$

Here "a" is the current ripple ratio and $\Delta I_{L-max}$ is the maximum ripple current flowing throw the inductor.

## 2.2 *Solid State Transformer (SST)*

SST is flexible can be employed in AC and DC grids to enable bi-directional power flow. For large power SST is connected to the medium voltage (MV) grid, line frequency transformer has a problem of heavyweight, large size, and volume but SST has small size as it operates at a higher frequency, high efficiency [9].

The purpose of SST in our XFC system is to provide the isolation between input and output (i.e., EV battery) provides galvanic isolation between grid and battery so that battery remain protected and provide the voltage conversion. SST has features of current limiting capabilities, better controllability, and higher efficiency at light load [2] (Table 2).

**Table 2** Solid state transformer parameters

| System parameters | Value |
|---|---|
| DC link voltage, $V_{dc}$ | 300 V |
| Inductance, $L$ | 17 μH |
| Transformer turn-ratio, $n_{sst}$ | 1 |
| Switching frequency, $f_{sw-sst}$ | 50 kHz |
| DC link capacitance, $C_{link}$ | 450 μF |

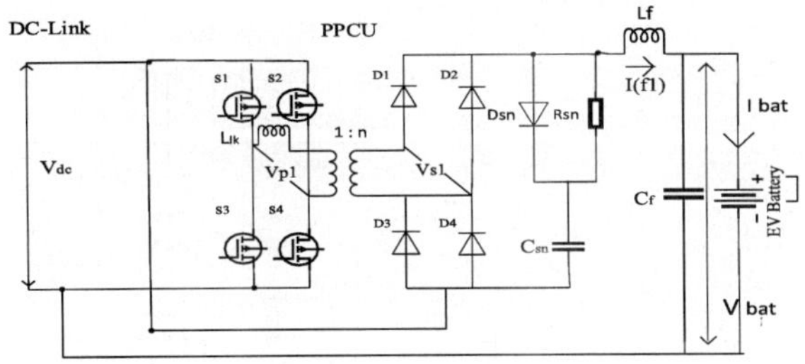


**Fig. 3** Circuit diagram of PPCU

**Table 3** Partial power charging unit (PPCU) parameters

| System parameters | Value |
|---|---|
| Leakage Inductance, $L_{lk}$ | 3.5 μH |
| Transformer turns ratio, $n$ | 0.35 |
| Filter inductance, $L_f$ | 520 μH |
| DC link voltage, $V_{dc}$ | 300 V |
| Filter capacitance, $C_f$ | 20 μF |
| Snubber resistance, $R_{sn}$ | 500 Ω |
| Switching frequency, $f_{sw}$ | 50 kHz |
| EV battery voltage, $V_{bat}$ | 360–400 V |
| Battery peak power, $P_{bat}$ | 3.2 kW |

## 2.3 *Partial Power Charging Unit (PPCU)*

It is partial power DC-DC converter interfaced between the DC link and to electric vehicles battery process, the only fraction of total power performing the regulation of the system and the rest of power is being bypassed to the load which reduces the conduction, switching, and magnetic loss, and increases the system efficiency. PPCU offers the only fraction of power to the electric vehicle battery's total charging power. Enables independently charging control to multiple electric vehicles battery with efficiency improvement. PPCU allows the gain in terms of weight volume and cost as it can be designed for reduced power [10]. The circuit diagram of the Partial Power Charging unit is shown in Fig. 3. PPCU improves overall efficiency as it eliminates redundant power conversion (Table 3).

## 2.4 *Battery Current Controller*

In Fig. 4, battery current controller circuit is interfaced between the PPCU and EV battery and the function of this controller circuit is to control the battery current flowing to the battery by adjusting the desired current value. EV battery current and desired current is being compared, this error is feed to the PI Controller and then

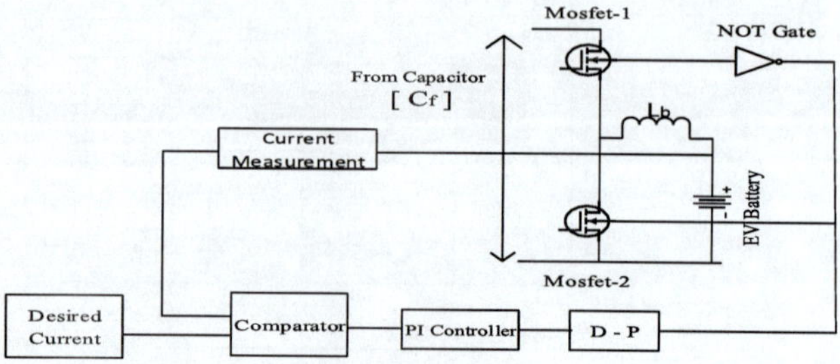


**Fig. 4** The current controller circuit for an EV battery

**Table 4** The current controller circuit for EV battery parameters

| System parameters | Value |
|---|---|
| Capacitance, $C_f$ | 20 μF |
| Inductance, $L_b$ | 1 mH |
| Proportional gain, $K_p$ | 50 |
| Integral gain, $K_I$ | 10 |
| PWM generator (D-P) | 50 kHz |

given to the PWM generator (DC-DC). The purpose of the PWM generator block is to generate the desired pulse according to the reference current and battery current then these pulses are feed to the MOSFET-1 and MOSFET-2. The battery controller provides the independent current control for charging the battery of each electric vehicle at constant current and variable current mode (Table 4).

## 3 Simulation Results

To attain the desired result MATLAB/Simulink model has been used, all simulated results are for battery rating of 3.2 kW and total system power rating of 6.4 kW. In Fig. 5, voltage and current waveform of the XFC system have been presented. In

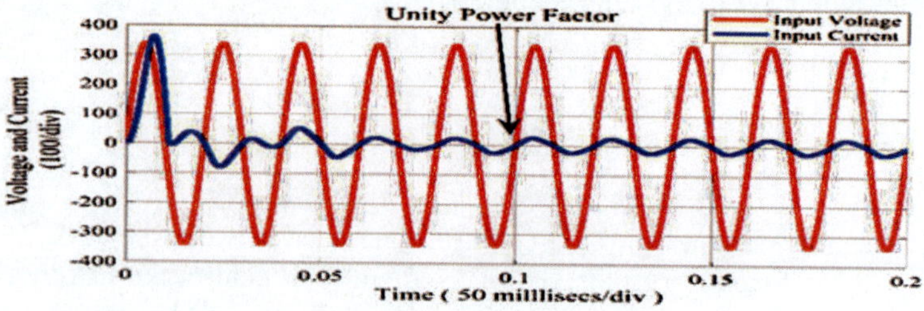


**Fig. 5** Unity power factor (UPF) at the input side of the converter

these waveforms, input voltage and current waveforms are in the same phase so that unity power factor (UPF) has been achieved successfully by the implementation of a controlled active rectifier circuit (CARC) unit in the XFC system. In Fig. 5, at initial current is high due to the switching of the MOSFET device, which introduces the harmonics in the system but after half cycle, CARC can eliminate those spikes and current waveforms come to a steady state. By the use of CARC, the power factor has been achieved to 0.99 which is very close to the unity. Improvement in the power factor, decreases the total RMS current decreases, decreases in voltage drop, lowers the copper losses, increases in available power, decrease the rating of the equipment, and therefore decreases the overall cost to the system.

The purpose of controlling the output voltage of the rectifier is based on the EV battery's different voltage specifications. Different company has their different voltage specifications and thus required a charger which is flexible to meet the required demand. To solve this purpose of a controlled active rectifier circuit (CARC) is used, which can regulate the output voltage of the rectifier to the desired range of voltage from 400 to 800 V without loss of unity power factor (UPF) of the XFC system. So that the voltage and current are in the same phase and maximum active power will flow to the circuit. In Fig. 6, at upper section controlled rectifier output voltage of the rectifier is shown which has been regulating from 400 to 800 V with the least ripple by the use of a controlled active rectifier circuit. With the change in voltage, the power factor of the system should not be changed so in Fig. 6 the input voltage and current waveform still show at unity power factor (UPF) when the output voltage of the rectifier is regulated from 400 to 800 V.

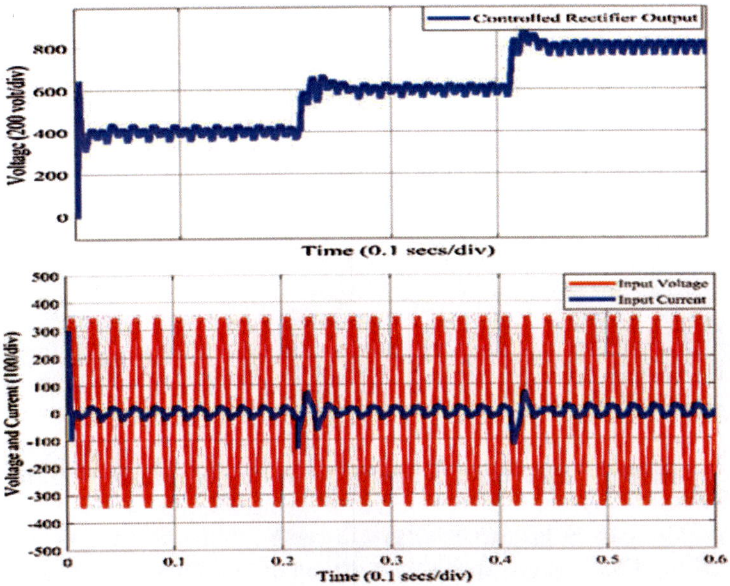


**Fig. 6** UPF at a controlled output voltage

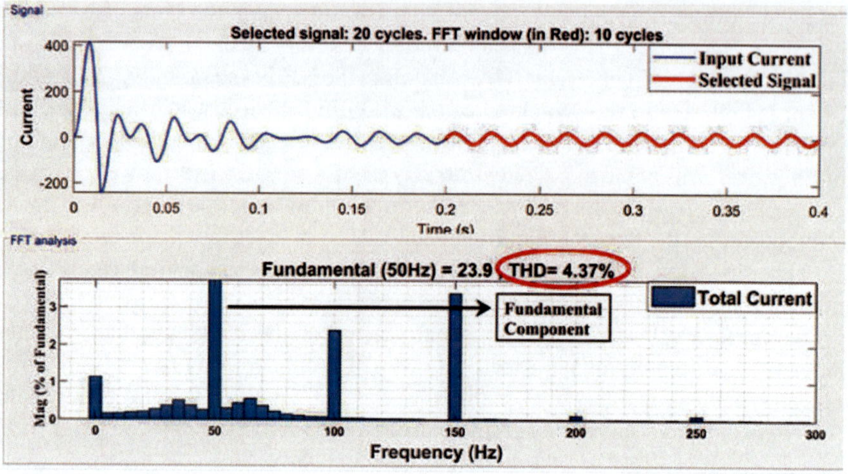


**Fig. 7** THD in XFC system

The lesser the THD, the lesser will be the effective RMS current, decreases the 3rd order harmonics, noise, and premature aging of each component. In the Fig. 7 THD in the XFC system has been shown at the input side of the converter in the current waveform, with the proposed converter scheme THD has been achieved to 4.37% which is desired THD less than 5%. From Eq. (4), for THD of 4.37% power factor has been calculated which comes to 0.99 very close to unity. Here g is the distortion factor, value of g should be close to unity, for THD to be zero distortion factor should be unity. Equation (4) also shows the relationship between THD and power factor, they are inversely related to each other. When the power factor increases THD automatically decreases and vice-versa.

$$\mathrm{THD} = \sqrt{\frac{1}{g^2} - 1} \tag{3}$$

$$\mathrm{Power-factor} = \frac{1}{\sqrt{1 + (\mathrm{THD}^2)}} \tag{4}$$

To charge the EV battery, constant current or variable current mode is required. By using the battery current controller circuit shown in Fig. 4, constant current and variable current control are achievable in the XFC system. In Fig. 8 charging of EVs battery for constant current and variable current control is shown for the battery SoC. In the left half of Fig. 8, battery SoC and battery current are shown for constant rate charging of EV battery. In this battery, the current is maintained at 4 Amp and battery SoC is changed accordingly. While the right half of Fig. 8, battery SoC and battery current are shown to charge the battery at a variable rate. The charging current rate is increased by the use of a battery current controller circuit. In this current is varied

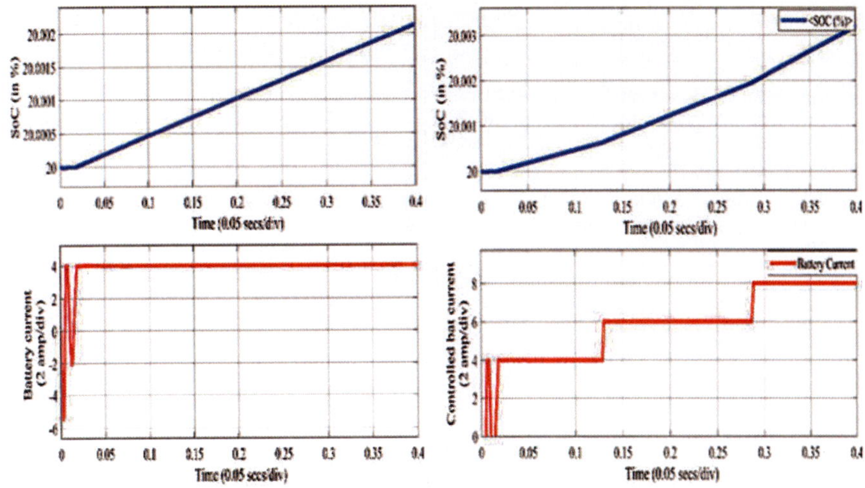


**Fig. 8** Battery SoC and current for constant charging current and variable charging current

from 4 to 6 Amp and then 6 Amp to 8 Amp, with change is battery current battery SoC is also increasing at a faster rate when compared to the constant charging current rate.

## 4 Conclusion

This paper has presented the detailed study of extreme fast charging (XFC) system with controlled active rectifier circuit (CARC) scheme in MATLAB/Simulink at the downgraded model of system power rating of 6.4 kW and battery power rating of 3.2 kW to obtain the unity power factor (UPF) at the input side of the XFC system, controlling the output voltage of rectifier and to reduce the total harmonic distortion (THD) in current waveforms at the input side to the XFC system. THD in the XFC system has been achieved to 4.37% which is less than desired 5% and the power factor is achieved to 0.99 which is very close to unity. With the proposed scheme of CARC and battery current controller circuit battery current has been also varied successfully according to the desired specification of the battery.

With the proposed scheme of the controlled active rectifier circuit, unity power factor has been achieved nearly 0.99, and the output voltage of this rectifier unit is varied from 400 to 800 V without affecting the unity power factor (UPF) of the circuit and reduction in current harmonics in the XFC system. Results show great improvement in input power factor of circuit, wide range of output voltage control at rectifier side, and excellent EVs battery current control.

## References


1. Ahmadi, M., Mithulananthan, N., Sharma, R.:A review on topologies for fast-charging stations for electric vehicles. In: 2016 IEEE International Conference on Power System Technology (POWERCON) pp. 1–6. Wollongong, NSW, Australia, (2016). https://doi.org/10.1109/POWERCON.2016.7753886.
2. Tu, H., Feng, H., Srdic, S., Lukic, S.: Extreme fast charging of electric vehicles: A technology overview. IEEE Trans. Transp. Electrification **5**(4), 861–878, (2019). https://doi.org/10.1109/TTE.2019.2958709
3. Ronanki, D., Kelkar, A., Williamson, S.S.: Extreme fast charging technology—Prospects to enhance sustainable electric transportation. Energies **12**(19), 3721 (2019). https://doi.org/10.3390/en12193721
4. Srdic, S., Lukic, S.: Toward extreme fast charging: Challenges and opportunities in directly connecting to medium-voltage line. IEEE Electrification Mag. **7**(1), 22–31 (2019). https://doi.org/10.1109/MELE.2018.2889547
5. Iyer, V.M., Gulur, S., Gohil, G., Bhattacharya, S.: An approach towards extreme fast charging station power delivery for electric vehicles with partial power processing. IEEE Trans. Industr. Electron. **67**(10), 8076–8087 (2020). https://doi.org/10.1109/TIE.2019.2945264
6. Heger, C.A., Sen, P.K., Morroni, A.: Power factor correction—A fresh look into today's electrical systems. In: 2012 IEEE-IAS/PCA 54th Cement Industry Technical Conference, pp. 1–13. San Antonio, TX, USA (2012). https://doi.org/10.1109/CITCON.2012.6215705
7. Megha, A., Mahendran, N., Elizabeth, R.: Analysis of harmonic contamination in electrical grid due to electric vehicle charging. In: 2020 Third International Conference on Smart Systems and Inventive Technology (ICSSIT), pp. 608–614. Tirunelveli, India (2020). https://doi.org/10.1109/ICSSIT48917.2020.9214096
8. Hussain, M.N.M.: Development of PI controller for battery charger using PFC rectifier. In: SPEEDAM 2010, pp. 1099–1101. Pisa, Italy (2010), https://doi.org/10.1109/SPEEDAM.2010.5542029
9. Abu-Siada, A., Budiri, J., Abdou, A.F.: Solid State transformers topologies, controllers, and applications: State-of-the-Art literature review, Electronics **7**, 298 (2018). https://doi.org/10.3390/electronics7110298
10. Rojas, J., Renaudineau, H., Kouro, S., Rivera, S.: Partial power DC-DC converter for electric vehicle fast-charging stations. In: IECON 2017—43rd Annual Conference of the IEEE Industrial Electronics Society, pp.5274–5279, Beijing, China (2017). https://doi.org/10.1109/IECON.2017.8216913

# Ambient Fine Particulate Matter and COVID-19 in India

**Amit Singhal, Arman Qamar, Shekhar Kunal, M. P . Girish, Muthiah Vaduganathan, Sameer Arora, Rakesh Yadav, Vishal Batra, Pushpendra Singh, Binish Fatimah, Anubha Gupta, and Mohit D. Gupta**

**Abstract** We examine the correlation between COVID-19 case activity and air pollution in two cities of Delhi and Mumbai in India. Data regarding air quality index (AQI) of PM2.5 and PM10 from Delhi and Mumbai were collected between July and November 2020. Within the same time period, confirmed cases and daily

A. Singhal
Department of ECE, Netaji Subhas University of Technology, Delhi, India

A. Qamar
Cardiovascular Institute, NorthShore University Health System, University of Chicago Pritzker School of Medicine, Evanston, IL, USA

S. Kunal · M. P. . Girish · V. Batra · M. D. Gupta (✉)
Division of Cardiology, GB Pant Institute of Post Graduate Medical Education and Research, Delhi, India
e-mail: drmohitgupta@yahoo.com

M. Vaduganathan
Brigham and Women's Hospital Heart and Vascular Center, Harvard Medical School, Boston, MA, USA

S. Arora
Division of Cardiology, University of North Carolina School of Medicine, Chapel Hill, NC, USA

R. Yadav
Division of Cardiology, All India Institute of Medical Sciences, Delhi, India

P. Singh
Department of Electronics and Communications Engineering, National Institute of Technology Hamirpur, Hamirpur, HP, India
e-mail: spushp@nith.ac.in

B. Fatimah
Department of Electronics and Communications Engineering, CMR Institute of Technology Bengaluru, Bengaluru, India
e-mail: binish.f@cmrit.ac.in

A. Gupta
Department of Electronics and Communications Engineering, Indraprastha Institute of Information Technology Delhi, Delhi, India
e-mail: anubha@iiitd.ac.in

 
D. Gupta et al. (eds.), *Pattern Recognition and Data Analysis with Applications*, Lecture Notes in Electrical Engineering 888,
https://doi.org/10.1007/978-981-19-1520-8_50

deaths due to COVID-19 in these two cities were also recorded. AQI levels in Delhi were worst in November (PM2.5: 446 ± 144.6 $\mu g/m^3$; PM10: 318 ± 131.7 $\mu g/m^3$) and were significantly higher as compared to Mumbai (PM2.5: 130 ± 41.2 $\mu g/m^3$; PM10: 86 ± 21.2 $\mu g/m^3$). This correlated with greater number of cases and higher mortality in Delhi (cases: 6243; deaths: 85) relative to Mumbai (cases: 1526; deaths: 35) during the same time period. This observational study shows that air pollution is associated with poor outcomes in patients with COVID-19. There is an urgent unmet need for appropriate public health measures to decrease air pollution along with strict policy change.

**Keywords** Air pollution · COVID-19 · Pneumonia

## 1 First Section

Coronavirus disease 2019 (COVID-19) is caused by a novel coronavirus SARS-CoV-2 and has led to a global pandemic affecting millions worldwide [1]. India currently has one of the highest burdens of COVID-19 cases globally, and many studies [2–4] have been done for predictive monitoring. In absence of any definitive therapy, social distancing and masking remain the key in preventing disease transmission. Multi-organ involvement is often the feature of COVID-19 infection with lungs, and cardiovascular system being most commonly affected. The spectrum of clinical presentation in COVID-19 varies with asymptomatic ones to severe life-threatening forms leading to acute respiratory distress syndrome and death [1]. Comorbidities such as diabetes, hypertension and chronic respiratory conditions like asthma and chronic obstructive pulmonary disease often lead to a higher mortality in COVID-19 [5]. Most of the severe forms of COVID-19 infection are characterized by the presence of an exaggerated immune response and raised pro-inflammatory cytokines [1]. Air pollution comprises of a mixture of gaseous as well as particulate matter which often vary in time and space. Particulate matter such as PM10 and PM2.5, and gaseous pollutants like nitrogen and sulphur oxides are generated from the combustion of fossil fuels and have an adverse impact on the cardiovascular and respiratory system [6]. Air quality of a region is reflected by the air quality index (AQI) which is a numerical index and ranges in value from 0 to 500 with higher AQI value suggesting deteriorated air quality [6]. Data from the recent studies have documented air pollution to have an adverse impact on COVID-19-related outcomes [7–12]. Higher rates of COVID-19 infection and deaths have been reported with both short as well as long-term exposures to air pollution. It has been estimated that around 15% of worldwide COVID-19-related deaths could possibly be linked to air pollution [13]. Air pollution still remains one of the major challenges in developing countries such as India. Delhi, the capital city of India, has infamously earned the title of being the most polluted capital city of the world two years in a row [14]. Air pollution trends in Delhi suggest that the winter season is often the worst with

November and December being mostly affected [15]. A rapid surge in the COVID-19 infections in Delhi was reported in the month of November 2020. We sought to examine the relation between the recent surge in COVID-19 cases in Delhi in relation to the air pollution at the same time.

## 2 Materials and Methods

In this cross-sectional analysis, data regarding the AQI of particulate matters PM2.5 and PM10, confirmed COVID-19 cases and daily deaths between 1 July2020 and 22 November2020 in Delhi were collected. Data regarding the concentrations of various particulate pollutants (PM2.5 and PM10) were obtained from the Central Pollution Control Board (CPCB) [16], Ministry of Environment, Forest and Climate Change, Government of India. The AQI values were obtained using the national ambient air quality standards prescribed by CPCB. Data regarding the daily and cumulative cases as well as deaths due to COVID-19 were collected from a publicly available, crowd-sourced database [17] which collected statistics from the reports released by local health commissions and Indian Council of Medical Research (ICMR). Air pollution and COVID-19 infections and mortality statistics of Delhi were compared with that of Mumbai (both of them being densely populated with comparable infrastructure and reasonable healthcare facilities). All the collected data were observed in the form of time series as a function of the days elapsed since the first day, i.e. 1 July2020. In order to capture the role of air pollution on the spread and mortality of COVID-19, careful inspection was required to segregate the pollution levels that can enhance the sustenance of COVID-19 virus in the air, as the virus majorly spreads through the air.

## 3 Materials and Methods

The examination of AQI levels of PM2.5 and PM10 in Delhi showed a significant increase in the months of October and November 2020. The AQI showed that people in Delhi had been breathing very poor-quality air in recent months as compared to Mumbai, where AQI remained below 200 always and even below 100 for most of the time (Fig. 1 and Table 1). Data regarding air pollution levels revealed that the month of November had been worst for Delhi with average AQI levels for PM2.5 and PM10 as $446 + 144.6$ gm$^3$ and $318 + 131.7$ gm$^3$, respectively. The same period also witnessed a significant surge in COVID-19-related cases and deaths in Delhi (Fig. 1 and Table 2). The COVID-19 deaths and confirmed cases had a strong correlation with the AQI in Delhi (Fig. 2). The AQI levels for PM2.5 and PM10 in Mumbai in the same month were $130 + 41.2$ gm$^3$ and $86 + 21.2$ gm$^3$, respectively. There was a weak correlation in the AQI levels and COVID-19 cases and mortality in Mumbai as compared to Delhi (Fig. 2).

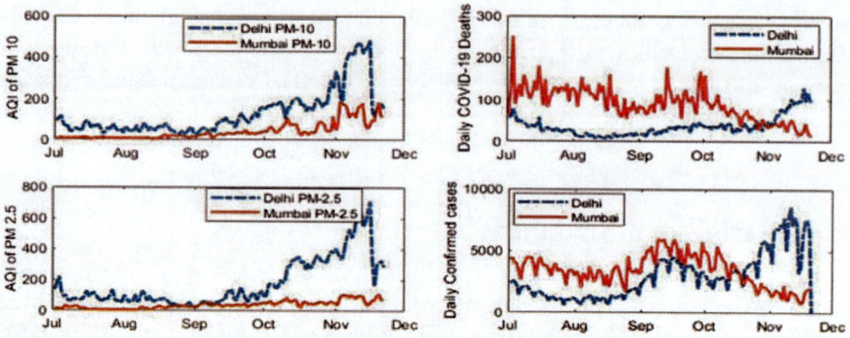


**Fig. 1** (Starting from top left in anti-clockwise direction) AQI values for PM10 and PM2.5 pollutants in Delhi and Mumbai; daily confirmed cases and deaths due to COVID-19 in Delhi and Mumbai

**Table 1** Monthly average AQI values for PM10 and PM2.5 in Delhi and Mumbai (PM 2.5 and PM 10: all values in $\mu g/m^3$)

| Month | Delhi | | Mumbai | |
|---|---|---|---|---|
| | PM10 | PM2.5 | PM10 | PM2.5 |
| July | 63 ± 20 | 95 ± 35.5 | 17 ± 8 | 8 ± 2.5 |
| August | 51 ± 14.7 | 62 ± 25.4 | 24 ± 8.2 | 12 ± 5.9 |
| September | 91 ± 34.7 | 83 ± 34.1 | 41 ± 7.9 | 31 ± 8.6 |
| October | 185 ± 44.1 | 283 ± 80.8 | 57 ± 12.2 | 61 ± 23.4 |
| November | 318 ± 131.7 | 446 ± 144.6 | 86 ± 21.2 | 130 ± 41.2 |

**Table 2** Monthly average of COVID-19 daily confirmed cases and deaths in Delhi and Mumbai

| Month | Delhi | | Mumbai | |
|---|---|---|---|---|
| | Confirmed cases | Deaths | Confirmed cases | Deaths |
| July | 1513 | 38 | 3667 | 125 |
| August | 1301 | 15 | 3056 | 102 |
| September | 3523 | 31 | 4952 | 100 |
| October | 3536 | 38 | 3191 | 67 |
| November | 6243 | 85 | 1526 | 35 |

## 4 Discussion

Air pollution has significantly increased the burden of lower respiratory tract infections, especially respiratory viral infections. Historical data from the previous Spanish flu pandemic of 1918 had suggested that cities with greater proportion of coal usage had higher deaths after considering the confounders [18]. Similarly, data from the previous SARS-CoV-1 pandemic of 2003 too revealed that in highly polluted

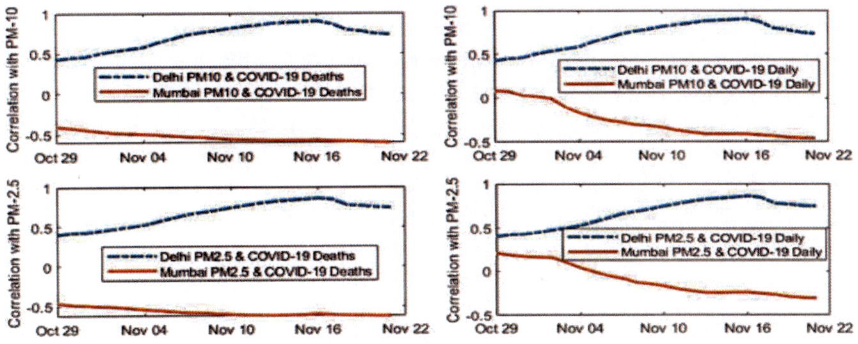


**Fig. 2** (left) Correlation of daily COVID-19 deaths with PM10 and PM2.5 AQI values; (right) Correlation of daily confirmed cases with PM10 and PM2.5 AQI values for Delhi and Mumbai.

areas there was twofold higher risk of death as compared to less polluted areas [19]. Studies concerning air pollution and COVID-19 pandemic have documented the adverse impact of both short term (less than 2 months) and long-term exposures (greater than 2 months) to air pollutants [7–12]. Data from retrospective series from various countries have suggested a significant positive correlation between particulate matter levels, AQI and COVID-19 cases [7–12]. Long-term data from developing countries regarding the impact of air pollution and COVID-19 are highly limited. In a machine learning-based model from 25 cities in India, a unidirectional causality was established between PM2.5 levels and COVID-19 mortality [20]. In a study from Lima, Peru, higher levels of PM2.5 were linked to a greater incidence of cases and deaths from COVID-19 [21].

Studies have also shown air pollution to be strongly linked to disease severity suggestive of a possible additive or synergistic effect [22]. Such an association becomes more pertinent in highly polluted cities like Delhi as was seen in our study, wherein increased COVID-19 cases and deaths strongly correlated with air pollution.

SARS-CoV-2 virus and air pollutants, primarily particulate matter, enter the lungs via respiratory tract and have potential systemic effects following its entry into the circulation [1]. Outcomes following infection with SARS-CoV-2 often depend on the presence of comorbidities which increases a patient's risk profile [5]. According to several epidemiological studies, air pollution can often predispose individuals to serious forms of COVID-19 infection since it directly damages the respiratory and cardiovascular systems [7–12]. Multiple hypothesis including endothelial damage and disruption of the immune response has been postulated to explain the potential interaction of air pollution and SARS-CoV-2 virus [23–25]. The primary response to inhalation of particulate matter is local inflammation, oxidative stress and endothelial injury. In addition, PM2.5 exposures have been associated with overexpression of various pro-inflammatory cytokines such as TNF-α, IL-1 and IL-6 [23–25]. Air pollution induced inflammation increases the susceptibility to COVID-19, and it might potentiate the underlying lung injury. The persistent exposure to pollutants may weaken the respiratory system, which may contribute to the severity of infection with

COVID-19 [26]. Long-term exposure to air pollution is also linked to chronic rhinosinusitis, which increases airway mucosal permeability and facilitating an easy entry for SARS-CoV-2 [27]. Another potential impact of exposure to particulate matter is its adverse effects on the cardiovascular system. Increased PM2.5 levels have strong correlation with cardiovascular mortality owing to systemic inflammation, oxidative stress leading to endothelial dysfunction, platelet activation and atherothrombosis [26]. Additionally, SARS-CoV-2 also exhibits a procoagulant and pro-inflammatory phenotype, resulting in synergistic effects on the cardiovascular system in patients with long-term exposure to air pollutants. In cities such as Delhi, long-term exposure to pollutants leads to subclinical inflammation and greater host vulnerability for COVID-19 infection. A second possible link between air pollution and COVID-19 may be the particulate matter, such as PM10. These particulate matters remain suspended in the air and can act as a carrier for droplets, increasing the spread of the virus. This hypothesis was supported by a study which showed that polluted air greatly increases the transmission of SARS-CoV-2 to humans apart from the human-to-human transmission [28]. The findings in our study highlighted a possible positive correlation between AQI levels and COVID-19-related cases and deaths. This was reflected in the massive surge of cases in Delhi coinciding with the rapid increase in particulate pollutants in the months of October and November 2020 as compared to the city of Mumbai. Though the cause and effect relation of air pollution in this analysis cannot be confirmed, however, it does suggest air pollution as a possible factor potentiating the systemic effect of SARS-CoV-19 virus, thereby increasing morbidity and mortality.

### *4.1 Limitations*

In this cross-sectional survey, it is difficult to establish a cause and effect relation because of the presence of multiple confounders which might be playing a significant role in the same. Further, other unaccounted sources of air pollution might also be contributing to this increase in cases. It is still difficult to interpret whether these results are due to long-term exposure to pollution or a sudden rise in pollutant levels can propagate the virus spread.

## 5 Conclusion

The findings of our study highlighted the adverse impact of air pollution in terms of COVID-19 cases and deaths in two cities in India. These findings were corroborative with few of the epidemiological studies assessing impact of air pollution and COVID-19 in developed countries. In a resource limited country like India, individual and government efforts to reduce air pollution need to be enforced. There is an urgent need to take environmental precautions and build facilities to minimize the variables resulting in PM2.5 emissions. Furthermore, dedicated, epidemiological and experimental studies are needed to further assess and clarify the role of air pollution in such situations.

## References


1. Hu, B., Guo, H., Zhou, P., Shi, Z.L.: Characteristics of SARS-CoV-2 and COVID-19. Nat. Rev. Microbiol. **19**(3), 141–154 (2021)
2. Singhal, A., Singh, P., Lall, B., Joshi, S.D.: Modeling and prediction of COVID-19 pandemic using Gaussian mixture model. Chaos, Solitons Fractals **138**, 110023 (2020)
3. Singh, P., Gupta, A.: Generalized SIR (GSIR) epidemic model: An improved frame- work for the predictive monitoring of COVID-19 pandemic. ISA Trans. (2021). https://doi.org/10.1016/j.isatra.2021.02.016
4. Singh, P., Singhal, A., Fatimah, B., Gupta, A.: An improved data driven dynamic SIRD model for predictive monitoring of COVID-19. In: ICASSP 2021—2021 IEEE International Conference on Acoustics, Speech and Signal Processing (ICASSP), pp. 8158–8162 (2021). https://doi.org/10.1109/ICASSP39728.2021.9414762
5. Sanyaolu, A., Okorie, C., Marinkovic, A., Patidar, R., Younis, K., Desai, P., Hosein, Z., Padda, I., Mangat, J., Altaf, M.: Comorbidity and its impact on patients with COVID-19. SN Compr. Clin. Med. 1–8 (2020)
6. Manisalidis, I., Stavropoulou, E., Stavropoulos, A., Bezirtzoglou, E.: Environmental and health impacts of air pollution: A review. Front Publ. Health **8**, 14 (2020)
7. Zhu, Y., Xie, J., Huang, F., Cao, L.: Association between short-term exposure to air pollution and COVID-19 infection: Evidence from China. Sci. Total Environ. **727**, 138704 (2020)
8. Jiang, Y., Xu, J.: The association between COVID-19 deaths and short-term ambient air pollution/meteorological condition exposure: a retrospective study from Wuhan, China. Air Qual. Atmos Health. 1–5 (2020)
9. Frontera, A., Cianfanelli, L., Vlachos, K., Landoni, G., Cremona, G.: Severe air pollution links to higher mortality in COVID-19 patients: The “double-hit” hypothesis. J. Infect. **81**(2), 255–259 (2020)
10. Zhang, Z., Xue, T., Jin, X.: Effects of meteorological conditions and air pollution on COVID-19 transmission: Evidence from 219 Chinese cities. Sci. Total Environ. **741**, 140244 (2020)
11. Zoran, M.A., Savastru, R.S., Savastru, D.M., Tautan, M.N.: Assessing the relationship between surface levels of PM2.5 and PM10 particulate matter impact on COVID-19 in Milan, Italy. Sci. Total Environ. **738**, 139825 (2020)
12. Travaglio, M., Yu, Y., Popovic, R., Selley, L., Leal, N.S., Martins, L.M.: Links between air pollution and COVID-19 in England. Environ. Pollut. **268**, 115859 (2021)
13. Pozzer, A., Dominici, F., Haines, A., Witt, C., Münzel, T., Lelieveld, J.: Regional and global contributions of air pollution to risk of death from COVID-19. Cardiovasc. Res. **116**(14), 2247–2253 (2020)
14. Delhi pollution: Is air quality in the Indian capital now improving?' BBC News, 6 Nov 2019, https://www.bbc.com/news/world-asia-india-49729291. [Accessed 22 Nov 2020]
15. Rizwan, S., Nongkynrih, B., Gupta, S.K.: Air pollution in Delhi: Its magnitude and effects on health. Indian J. Community Med. **38**(1), 4–8 (2013)
16. Central Control Room for Air Quality Management. https://app.cpcbccr.com/ccr/#/caaqm-dashboard/caaqm-landing/data. [Accessed 22 Nov 2020]
17. COVID19-India API. https://api.covid19india.org/; 2020. [Accessed 22 Nov 2020]
18. Clay, K.: Pollution, infectious disease, and mortality: Evidence from the 1918 Spanish influenza pandemic. J. Econ. Hist. **78**, 1179–1209 (2018)

19. Cui, Y., Zhang, Z.F., Froines, J., Zhao, J., Wang, H., Yu, S.Z., Detels, R.: Air pollution and case fatality of SARS in the People's Republic of China: An ecologic study. Environ. Health **2**(1), 15 (2003)
20. Mele, M., Magazzino, C.: Pollution, economic growth, and COVID-19 deaths in India: A machine learning evidence. Environ. Sci. Pollut. Res. Int. **28**(3), 2669–2677 (2021)
21. Vasquez-Apestegui, B.V., Parras-Garrido, E., Tapia, V., Paz-Aparicio, V.M., Rojas, J.P., Sanchez-Ccoyllo, O.R., Gonzales, G.F.: Association between air pollution in Lima and the high incidence of COVID-19: Findings from a post hoc analysis. BMC Public Health **21**(1), 1161 (2021)
22. Richardson, S., Hirsch, J.S., Narasimhan, M., Crawford, J.M., McGinn, T., Davidson, K.W., The Northwell C-RC, Barnaby, D.P., Becker, L.B., Chelico, J.D., Cohen, S.L., Cookingham, J., Coppa, K., Diefenbach, M.A., Dominello, A.J., Duer-Hefele, J., Falzon, L., Gitlin, J., Hajizadeh, N., Harvin, T.G., Hirschwerk, D.A., Kim, E.J., Kozel, Z.M., Marrast, L.M., Mogavero, J.N., Osorio, G.A., Qiu, M., Zanos, T.P.: Presenting characteristics, comorbidities, and outcomes among 5700 patients hospitalized with COVID-19 in the New York City area. JAMA **323**(20), 2052–2059 (2020)
23. Miller, M.R.: Oxidative stress and the cardiovascular effects of air pollution. Free Radic. Biol. Med. **151**, 69–87 (2020)
24. Münzel, T., Gori, T., Al-Kindi, S., Deanfield, J., Lelieveld, J., Daiber, A., Rajagopalan, S.: Effects of gaseous and solid constituents of air pollution on endothelial function. Eur. Heart J. **39**(38), 3543–3550 (2018)
25. Deek, S.A.: Chronic exposure to air pollution implications on COVID-19 severity. Med. Hypotheses **145**, 110303 (2020)
26. Bourdrel, T., Annesi-Maesano, I., Alahmad, B., Maesano, C.N., Bind, M.A.: The impact of outdoor air pollution on COVID-19: A review of evidence from in vitro, animal, and human studies. Eur. Respir. Rev. **30**(159), 200242 (2021)
27. London, N.R., Lina, I., Ramanathan, M.: Aeroallergens, air pollutants, and chronic rhinitis and rhinosinusitis. World J. Otorhinolaryngol Head Neck Surg. **4**(3), 209–215 (2018)
28. Setti, L., Passarini, F., De Gennaro, G., Barbieri, P., Licen, S., Perrone, M.G., Piazzalunga, A., Borelli, M., Palmisani, J., Di Gilio, A., Rizzo, E., Colao, A., Piscitelli, P., Miani, A.: Potential role of particulate matter in the spreading of COVID-19 in Northern Italy: First observational study based on initial epidemic diffusion. BMJ Open **10**(9), 039338 (2020)

# Artificial Neural Network Based Synthesis of 12-Lead ECG Signal from Three Predictor Leads

**Ato Kapfo, Sumit Datta, Samarendra Dandapat, and Prabin Kumar Bora**

**Abstract** In clinical practice, continuous recording and monitoring of the standard 12-lead electrocardiogram (ECG) is often not feasible. The emerging technology and advancement to record the ECG signal without the help of the medical expert's in-home care or ambulatory conditions with minimal complexity have become more common in recent times. We aim to devise a model to obtain the 12-lead ECG from a reduced number of leads to reduce the intricacy and enhance patient comfort and care. We propose a discrete wavelet transform (DWT) based artificial neural network (ANN) model that transforms a 3-lead ECG into a standard 12-lead ECG without losing diagnostic information. Prominent distortion measures, namely, correlation coefficient, $R^2$ statistics, and wavelet energy diagnostic distortion (WEDD) are employed to evaluate the quality of the synthesis by the proposed model. The performance of the suggested model is compared with the antecedent models. The experimental result shows that the proposed technique can successfully synthesize the standard 12-lead ECG from the reduced lead sets.

**Keywords** Electrocardiogram · ECG synthesis · Discrete wavelet transform · Artificial neural network · Diagnostic information

---

A. Kapfo · S. Datta (✉) · S. Dandapat · P. K. Bora
Department of Electronics and Electrical Engineering, Indian Institute of Technology Guwahati, Guwahati 781039, India
e-mail: sumit89@iitg.ac.in; sumit.datta@duk.ac.in

A. Kapfo
e-mail: ato.kapfo@iitg.ac.in

S. Dandapat
e-mail: samaren@iitg.ac.in

P. K. Bora
e-mail: prabin@iitg.ac.in

S. Datta
School of Electronic Systems and Automation, Kerala University of Digital Sciences, Innovation and Technology, Trivandrum, India



D. Gupta et al. (eds.), *Pattern Recognition and Data Analysis with Applications*, Lecture Notes in Electrical Engineering 888,
https://doi.org/10.1007/978-981-19-1520-8_51

# 1 Introduction

The graphical representation of the electrical signals of the human heart through electrodes is the Electrocardiogram (ECG). In clinical practice, the ECG is used as a standard tool to assess cardiac activities and diagnosed the myocardium condition even after a century of its invention due to its non-invasive, cost-effective, and reliability [1]. Commonly, to acquire the standard 12-lead ECG, ten electrodes are positioned on a specific location on the human body's surface, as shown in Fig. 1. However, the utilization of multiple electrodes for monitoring and recording is not suitable in some cases. These include remote healthcare, continuous monitoring, personalized health care, ambulatory monitoring, etc. The emerging technology and advancement in the development of miniature and wireless ECG with minimal complexity have gained popularity in recent times [2]. In continuous 12-lead ECG monitoring, the ten electrodes placed in predefined positions on the body surface may cause unwanted noise owing to muscle activity and electrode artifacts due to the patient's movement. On the other hand, if we reduce the number of leads for recording ECG, this may make the diagnosis difficult. This problem can be addressed by recording fewer leads and synthesizing the remaining leads using the acquired leads. It is possible because the information about the heart vector can be collected from these few leads and then use to derive other leads by utilizing the inter-lead and the intra-lead relationship among the different ECG leads. It will also enhance the patient's comfort due to reduced leads over multiple electrodes for monitoring as well as recording. Therefore, synthesis of the standard 12-lead ECG from the independent reduced lead sets is desirable.

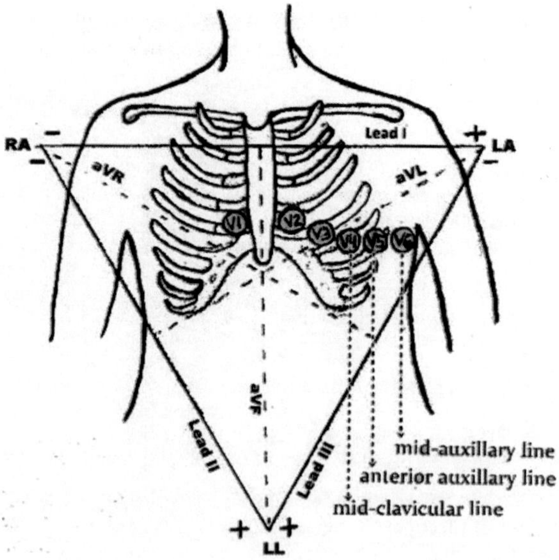


**Fig. 1.** 12-lead ECG signal acquisition

The standard 12-lead ECG consists of six frontal leads (I, II, III, aVR, aVL, and aVF) and six precordial leads (V1–V6). The frontal leads are not independent of each other, and hence, the leads III, aVR, aVL, and aVF can be mathematically derived from lead I and lead II [1]. According to volume conduction theory, developed by Burger and van Milaan [3], the human body is a 3-dimensional, uneven structure, and have volume conduction. This theory relied on the hypothesis that by projecting a heart vector on the lead vector in 3-D space, the voltage difference at any point on the surface of the torso can be determined [4]. This principle can facilitate to reconstruct the 12-lead ECG system by capturing the essential features corresponding to the heart vector from the reduced-lead sets. Most of the established work in existing literature utilized lead I, lead II, and one lead from the precordial leads as a predictor lead set.

Various methods are proposed and studied to synthesize the standard 12-lead ECG from the minimal lead sets in the past decade. Several studies adopted linear transformations to achieve the aim of deriving 12-lead ECG from its subset [4, 5]. Utilizing linear models to derive 12-lead ECG by EASI lead system using four electrodes and the Mason-Likar system are reported in [6] and [7], respectively. In literature, numerous linear regression models are also proposed to synthesize the 12-lead ECG from its subset [8, 9]. Tsouri et al. [10] propose an adaptive method of synthesizing a 12-lead ECG using independent component analysis (ICA) from the two sets of three leads. Maheshwari et al. [11], employed the principal component analysis (PCA) to obtain 12-lead ECG from a reduced 3-lead system.

## 2 Method

The proposed approach to synthesize the standard 12-lead ECG involves several stages, as depicted in a schematic representation in Fig. 2. In this study, lead I, lead

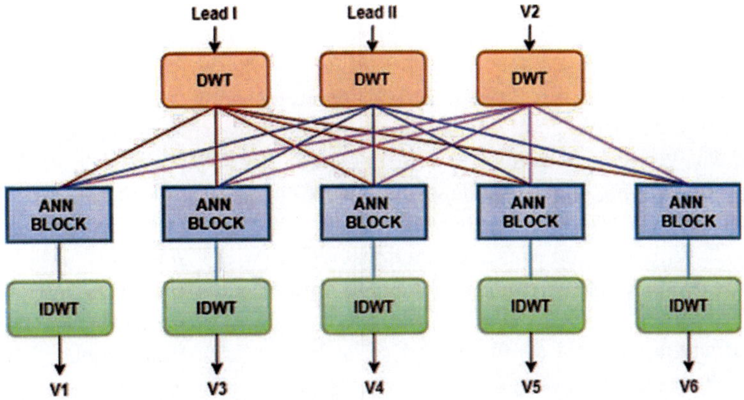


**Fig. 2** Block diagram of the proposed method

II, and precordial lead V2 are chosen as the predictor lead set as they achieved the good performance described in [9, 12]. In this work, the input three predictor leads (I, II, V2) are input to DWT multi-resolution block. The DWT decomposed the input ECG signal up to six levels. The approximation coefficients and the details coefficients obtained from the DWT are fed to the ANN. The multi-resolution technique has the ability to resolve the ECG signal into various time–frequency resolution called the subbands. These subbands contain important attributes of the ECG signal. When the individual subband of the decomposed ECG signal is given as an input to an ensemble of ANN algorithm, it is expected that the network may learn better than the whole frame as an input signal. ANN is proven to be an effective model to characterize an ECG signal due to its ability to map complex and non-linear problems. The outcome from each ANN is fused and fed to the inverse discrete wavelet transform (IDWT). The output of the IDWT is the derived ECG signal, namely, V1, V3, V4, V5, and V6. The detailed description of the proposed method is presented in the following subsections.

## 2.1 *Preprocessing*

Initially, the 1 kHz ECG signal is downsampled to 500 Hz to reduce the computational complexity while preserving significant information. ECG signal is corrupted with diverse categories of noise, namely, the powerline noise, the high-frequency electromyogram noise, and the low-frequency baseline wandering drift. The notch filter is used to remove the 50 Hz powerline interference noise. To get rid of the high-frequency noise and baseline wandering, the ECG signal is subjected to DWT and decomposed up to level 10. The approximation subband, detail 1, and detail 2 subband are subtracted from the ECG signal.

## 2.2 *DWT of ECG Signal*

Wavelets are minute waves, compactly supported having an inherently bounded period of time with good localization in both time and frequency domains. The DWT grossly segments the three predictor lead ECG signals at different resolutions by decomposing the signals into coarse approximation and detail subband. DWT employs two sets of functions, called scaling functions $\O_{L,m}(n) = 2^{-L/2}\O\left(2^{-L}n - m\right)$ and wavelet functions $\psi_{L,m}(n) = 2^{-k/2}\psi\left(2^{-m}n - l\right)$. The multiresolution decomposition technique is implemented by the dyadic wavelet transform gives $L + 1$ subbands. After decomposition of $i^{\text{th}}$ ECG lead, we have approximation subband coefficient $cA^{i}_{L,m}$ at level $L$ and details subbands coefficients, $cD^{i}_{l,m}$ at level $l$, where $l = 1, 2, \ldots, L$. The approximation and the various detail subband coefficients are obtained by the inner product of the ECG signal with the scaling function and wavelet function, respectively. The input ECG signal $x^i(n)$ are evaluated as $cA^i_{L,m} = \langle x^i(n), \emptyset_{L,m}(n)\rangle$ for approximation wavelet coefficient and $cD^i_{l,m} = \langle x^i(n), \psi_{l,m}(n)\rangle$ for detail wavelet coefficient. Six-level wavelet decomposition is performed in this study. When the three predictor leads are decomposed with the same level of decomposition and mother wavelet, it produced the same number of coefficients in different subbands. The wavelet coefficients acquired from the $L$-level decomposition are organized in $L+1$ subband matrices. The rows of the matrix portray the coefficients, and the column corresponds to the ECG predictor leads. The approximation subband matrix can be presented as $A_L = \left[cA^1_{L,m}, cA^2_{L,m}, cA^{V2}_{L,m}\right]$. The details of matrices are $D_l = \left[cD^1_{l,m}, cD^2_{l,m}, cD^{V2}_{l,m}\right]$, where m is the subband coefficient and superscript 1, 2 and V2 denote lead I, lead II, and lead V2, respectively. The approximation and details subband matrices are given as an input to the neural network algorithm.

## 2.3 *Artificial Neural Network Architecture Design*

ANNs are widely used to approximate non-linear biomedical signals with high generality. To synthesize the independent ECG signals i.e. V1, V3, V4, V5, and V6 from the predictor (I, II, V2) 12-lead ECG subset, we employed a set of multi-layer feed-forward ANNs which applied single input layer, single hidden layer, and single output layer trained back-propagation method. The non-linear relationship between input and output layers of the model is derived by the different layers, weights, and biases. The weights and biases determine the neuron's output with the activation function. Back-propagation computes the derivatives of the activation functions in each successive neuron to obtain the optimal weights to generate the best outcome. The log sigmoid function is employed as an activation function. In this work, 5 ANN blocks are employed to derive the five targeted leads, as shown in Fig. 2. Each ANN block consists of seven individual ANN. The input to the first ANN is the approximation subband coefficients of the three predictor lead (I, II, V2), and the input of the second ANN are the coefficients of the detail 6th subband and so on, as shown in Fig. 3. The

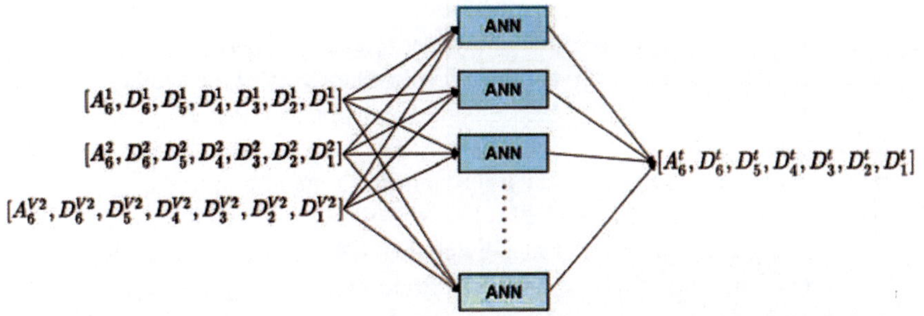


**Fig. 3** Frame of ANN model

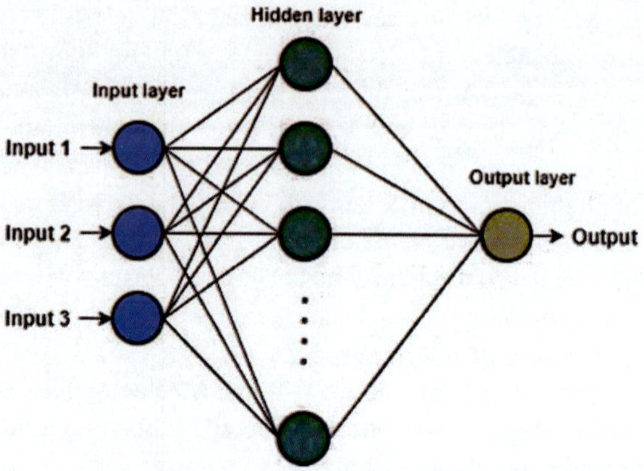


**Fig. 4** ANN architecture

outputs of the ANNs are the coefficients of the targeted leads. Each ANN consists of three input neurons (subband coefficients for each input), ten hidden neurons, and an output layer as depicted in Fig. 4. The output of ANNs are the approximation and the details coefficients of the target lead $A_6^t$, $D_6^t$, $D_5^t$, $D_4^t$, $D_3^t$, $D_2^t$, $D_1^t$ where $t$ is the target lead. This is fed as an input to the IDWT to obtain the synthesized ECG lead.

## 3 Results and Discussion

In this work, the Physikalisch-Technische Bundesanstalt (PTB) diagnostic ECG database is utilized to assess the proposed method. The database comprises 290 subjects with a total record of 549 from healthy volunteers and patients with various cardiovascular diseases assembled. Each individual record contains standard 12-leads along with three vectorcardiogram orthogonal leads. The 15-leads are recorded simultaneously and digitized at 1 kHz, with a resolution of 16 bit over a range of ±16.384 mV. Here, 16,000 samples are used to train the model, and 10,000 samples are used to test the derived lead sets.

The synthesized signal is compared with the actual signal for assessing the proposed method. Overall performance is evaluated by standard measures, namely, correlation coefficient $r$ and coefficient of determination $R^2$ between the synthesized signals and corresponding originals. The diagnostic quality of the synthesized leads is assessed by wavelet energy-based diagnostic distortion (WEDD) measure [13]. The result of the proposed method is established from the average obtained value from the performance metrics of the five synthesized precordial leads.

Figures 5 and 6 show the original (blue) signal and the synthesized (red) signal of the derived precordial leads of healthy control (HC) signal and pathology signal of myocardial infarction (MI), respectively. From Figs. 5 and 6, we can notice a fine resemblance between the original and the synthesized leads. It is visible from the figures that the synthesized leads preserve the shape feature. It is also observed from

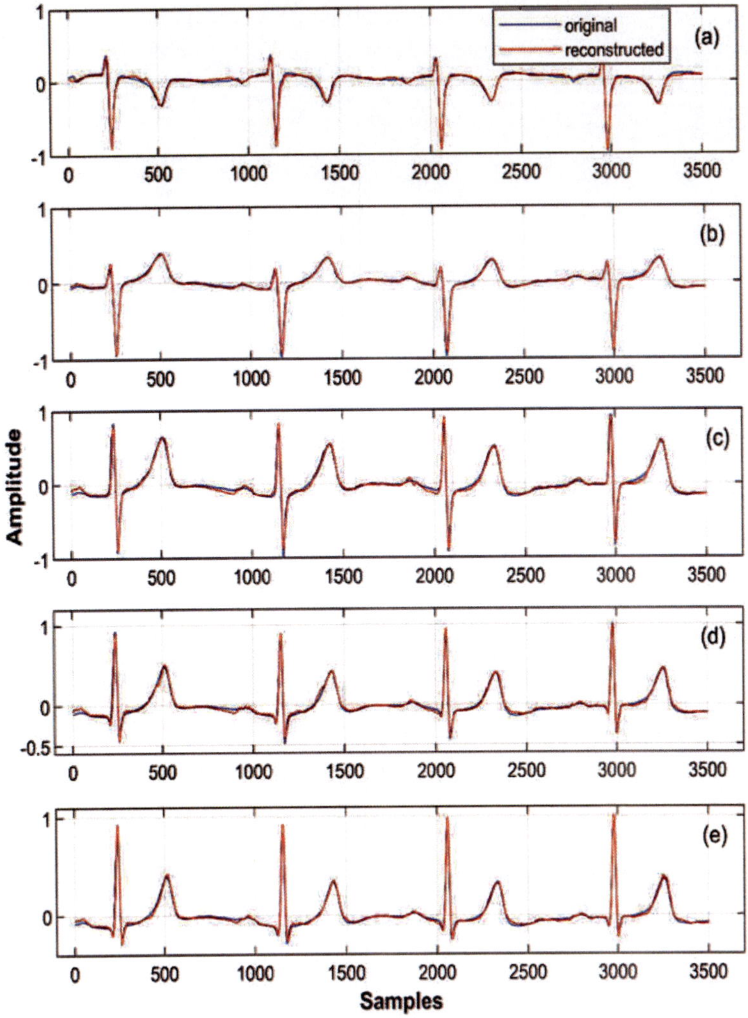


**Fig. 5** Original (blue) signal and synthesized (red) signal of precordial leads for HC. **a** V1, **b** V3, **c** V4, **d** V5, **e** V6

the figure that QRS-complex preserves the original shape and amplitude features. In Fig. 6, we observe the shape and amplitude features are preserved very well. So, it is apparent that the suggested model can synthesize the pathological signal of MI accurately.

Table 1 shows the performance assessment of the proposed model. The lead V2 is used in the predictor lead set. Hence the experimental result is shown in Table 1 for five target leads, namely, V1, V3, V4, V5, and V6. The WEDD value is low, while the correlation coefficient and $R^2$ values are high. The higher average correlation

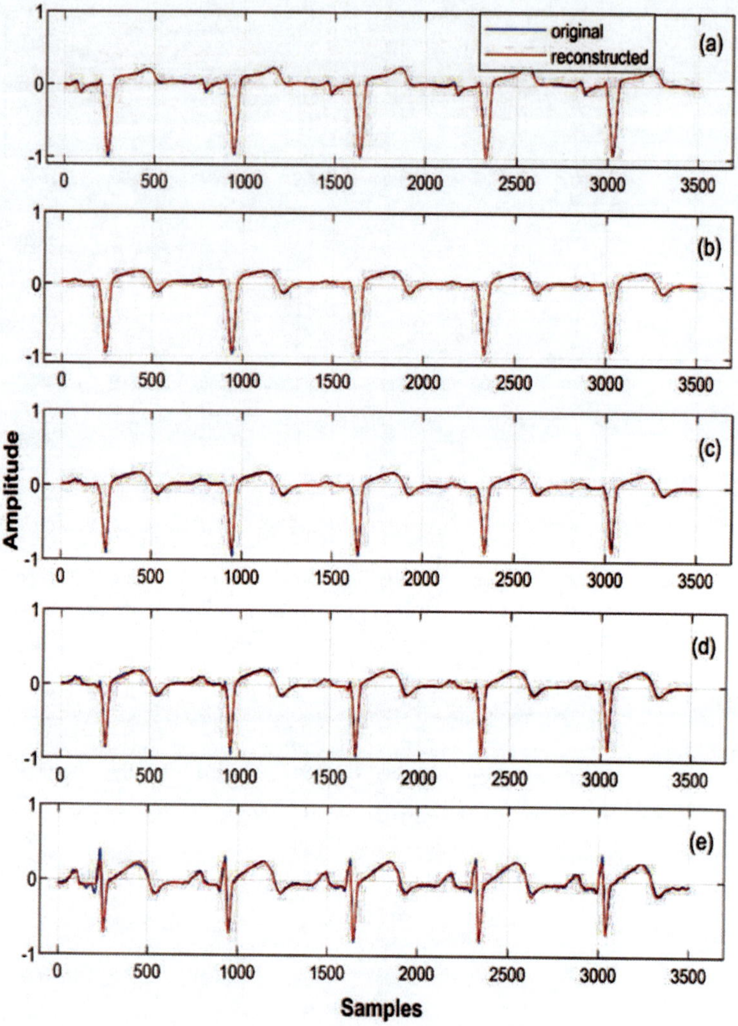


**Fig. 6** Original (blue) signal and synthesized (red) signal of precordial leads for MI. **a** V1, **b** V3, **c** V4, **d** V5, **e** V6

**Table 1** Performance evaluation of the proposed method

| | $R$ | $R^2$ (%) | WEDD (%) |
|---|---|---|---|
| V1 | 0.98 | 95.21 | 18.34 |
| V3 | 0.98 | 95.65 | 14.04 |
| V4 | 0.97 | 92.56 | 21.11 |
| V5 | 0.97 | 93.00 | 20.63 |
| V6 | 0.98 | 94.24 | 19.27 |
| Average | 0.98 | 94.13 | 18.68 |

**Table 2** Summary and performance comparison with the previously established method

| Author | Number of leads | Technique | Performance | | |
|---|---|---|---|---|---|
| | | | $r$ | $R^2$ (%) | WEDD (%) |
| Nelwan et al. [8] | 3 | LinR | 0.96 | *88.61* | 26.03 |
| Tsouri et al. [10] | 3 | ICA | 0.93 | 86.72 | 26.99 |
| Maheshwari et al. [11] | 8 | PCA | 0.95 | *90.21* | 16.07 |
| Kaewfoongrunsi et al. [14] | 4 | SVR | 0.97 | – | 26.64 |
| Atoui et al. [12] | 3 | ANN | 0.98 | – | 15.01 |
| Nallikuzhy et al. [9] | 3 | LinR | 0.98 | *94.23* | 18.97 |
| Proposed method | 3 | DWT and ANN | 0.98 | *94.13* | 18.68 |

value of 0.98 indicates that the synthesized leads bear a good resemblance with their original. The average $R^2$ value of 94.13% shows that the proposed model achieved high-quality synthesized ECG signal for the set of data that has been evaluated. The WEDD obtained an average value of 18.68% which indicates the synthesized leads preserve the significant diagnostic information. Taking account of the individual precordial lead, lead V3 achieves the best result in terms of $R^2$ and WEDD with a value of 95.65%, and 14.04% respectively. It is apparent from the Table that based on the performance metrics, the proposed method has the potential to synthesize the 12-lead ECG with minimal synthesis error. The good performance obtained by the proposed system may be because the ANN learns better signal characteristics from a particular subset of the decomposed frequency bands than the whole frame of the ECG signal.

Table 2 summarizes the comparison of the various techniques to synthesize the 12-lead ECG from different reduced lead sets. Maheshwari et al. [10] employed PCA and conducted a study using eight leads in their work. In [14], a support vector regression model was used for deriving the 12-lead ECG using 4-leads. The rest of the works reported in the table use 3-leads to synthesize the 12-leads ECG. Besides, the result obtained by the proposed model is compared with the previously established work to derive the standard 12-leads. In Table 2, we observe that the proposed model achieves better or comparable results than linear synthesis methods. The correlation value of the model has the same value as the study done by Nallikuzhy et al. [9] and Atoui et al. [12] which is 0.98. A non-linear ANN method developed by Atoui et al. [12] yields a WEDD value of 15.01%, which shows better performance than the proposed method, which is 18.68%. However, they have used ANN committees of 50 individual ANN in their study while we used only 5 ANN blocks. As a result, the computational complexity in the case of the proposed method is significantly less. Compared to different algorithms, it is apparent that the proposed model achieves superior performance than most of the antecedent methods.

## 4 Conclusion

This study investigates and evaluates a patient-specific approach to synthesize a standard 12-lead ECG based on a non-linear ANN model. The proposed technique's main advantage is that it learns from the non-identical frequency bands of the decomposed DWT to synthesize the five targeted precordial leads. Another advantage is, it yields improved performance than the linear synthesized models. The assessment of the proposed approach using three standard metrics: correlation coefficient, $R^2$ statistics, and WEDD bears good quality between the original and derived leads. The performance evaluation results shows that the proposed method has the promising capability to synthesize the standard 12-lead ECG without losing diagnostic information and yield improved performance over most of the previously established works.

## References


1. Goldberger, A.L., Goldberger, Z.D., Shvilkin, A.: Clinical electrocardiography: a simplified approach. Elsevier Health Sci. (2017)
2. Serhani, M.A., El Kassabi, H.T., Ismail, H., Nujum Navaz, A.: ECG Monitoring systems: Review, architecture, processes, and key challenges. Sensors **20**(6), 1796 (2020)
3. Burger, H.C., Milaan, J.B.V.: Heart-vector and leads, Br. Heart J. **8**(3), 157–161 (1946)
4. Tomašić, I., Trobec, R.: Electrocardiographic systems with reduced numbers of leads—synthesis of the 12-lead ECG. IEEE Rev. Biomed. Eng. **7**, 126–142 (2014)
5. Feild, D.Q., Zhou, S.H., Helfenbein, E.D., Gregg, R.E., Lindauer, J.M.: Technical challenges and future directions in lead reconstruction for reduced-lead systems. J. Electrocardiol. **41**(6), 466–473 (2008)
6. Dower, G.E., Yakush, A., Nazzal, S.B., Jutzy, R.V., Ruiz, C.E.: Deriving the 12-lead electrocardiogram from four (easi) electrodes. J. Electrocardiol. **21**, S182–S187 (1988)
7. Man, S.C., Maan, A.C., Kim, E., Draisma, H.H., Schalij, M.J., van der Wall, E.E., Swenne, C.A.: synthesis of standard 12-lead electrocardiograms from 12-lead electrocardiograms recorded with the mason-likar electrode configuration. J. Electrocardiol. **41**(3), 211–219 (2008)
8. Nelwan, S.P., Kors, J.A., Meij, S.H., van Bemmel, J.H., Simoons, M.L.: Reconstruction of the 12-lead electrocardiogram from reduced lead sets. J. Electrocardiol. **37**(1), 11–18 (2004)
9. Nallikuzhy, J.J., Dandapat, S.: Spatial enhancement of ECG using diagnostic similarity score based lead selective multi-scale linear model. Comput. Biol. Med. **85**, 53–62 (2017)
10. Tsouri, G., Ostertag, M.: Patient-specific 12-lead ecg reconstruction from sparse electrodes using independent component analysis. IEEE J. Biomed. Health Inform. **18**(2), 476–548 (2014)
11. Maheshwari, S., Acharyya, A., Schiariti, M., Puddu, P.E.: Personalized reduced 3-lead system formation methodology for remote health monitoring applications and reconstruction of standard 12-lead system. Transl. Cardiol., Int. Arch. Med. **8**(62), 1–15 (2015)
12. Atoui, H., Fayn, J., Rubel, P.: A novel neural-network model for deriving standard 12-lead ECGs from serial three-lead ECGs: Application to self-care. IEEE Trans. Inf. Technol. Biomed. **14**(3), 883–890 (2010)
13. Manikandan, M.S., Dandapat, S.: Wavelet energy based diagnostic distortion measure for ECG. Biomed. Signal Process. Control **2**(2), 80–96 (2007)
14. Kaewfoongrungsi, P., Hormdee, D.: Support vector regression-based synthesis of 12-lead ecg system from the standard 5 electrode system using lead v1. KKU Eng. J. **43**(S3), 494–498 (2016)

# Understanding Quantum Computing Through Drunken Walks

Sujit Biswas and Rajat S. Goswami

**Abstract** Quantum random walks have caught the attention of quantum information theorists in recent years. Classical walks have been used to solve or design several highly efficient randomized algorithms, and they are also used in many quantum algorithms, but quantum walks provide an exponential speedup over classical walks since they can solve some oracle problems. For several practical problems, such as the element distinctness problem, the triangle finding problem, and evaluating NAND trees, quantum walks provide polynomial speedups over classical algorithms. In this paper, we propose a new quantum random walk representation based on the drunken walk and the classical random walk. This current portrayal would make it easier for people to comprehend the potential of quantum computing. It will also be shown how to solve a large deviation analysis through quantum walks. This paper includes an explanation of how an inebriated person might locate his friend in a bar after leaving the restroom, as well as a comparison of quantum walks and classical walks.

**Keywords** Quantum computing · Quantum walk · Random walk · Classical walk · QASM

## 1 Introduction

Quantum computation is the most significant advancement in computing after, indeed, computing. While our universe is made up of quantum information, we interpret it as classical information. That is to say, there is a lot going on at small scales that we can't see from our eyes. Our minds are designed to think of Sabertooth cats, not Schrodinger's cats, so we evolved to interpret classical information, not quantum information. We can conveniently encode our classical knowledge with zeros and ones, but what about the additional information that makes up our universe? Is it possible to filter knowledge using the quantum essence of reality? Yes, let's talk

S. Biswas (✉) · R. S. Goswami
National Institute of Technology Arunachal Pradesh, Yupia 791112, India
e-mail: sujitbiswas.phd@nitap.ac.in



D. Gupta et al. (eds.), *Pattern Recognition and Data Analysis with Applications*,
Lecture Notes in Electrical Engineering 888,
https://doi.org/10.1007/978-981-19-1520-8_52

**Fig. 1** Drawing a 3D object on a 2D sheet

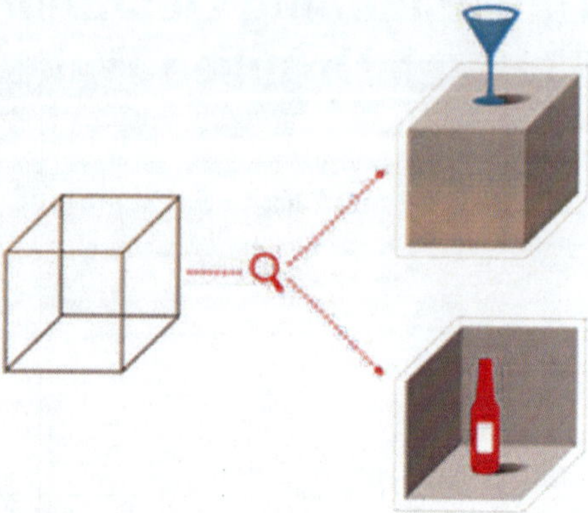

about the potential of quantum computation and then start writing some quantum code.

Understanding that, while many of the principles seem counterintuitive, the classical cosmos we know and love is merely a shadow of the quantum fabric of reality is the starting point for investigating quantum computing. Part of becoming comfortable with the quantum is to become comfortable with the limitations of our own perception. This limitation is analogous to drawing a 3D object on a 2D sheet of paper. Take a look at the (see Fig. 1). It can represent either a box (we can illustrate this with a glass on top) or it can be inverted into a corner (we can put the bottle inside to make it flip to a corner in our brains).

We are forced to see either one or the other and never both. We can change them back and forth, but because we are stuck in a two dimensional representation we can only see one or the other. Two dimensions is not enough for a perfect representation of a three dimensional object. Similarly the world of classical information in its simplest encoding is represented in bits, 0 s and 1 s. These are not enough though to describe the quantum world. In the quantum world, we need quantum bits, or qubits, to describe our information. Like putting the drink on top of the box or in the corner, we can make a measurement that will force our qubit to tell us a classical bit, but there is more information there that we can take advantage of.

Quantum computers will use the rest of the information to achieve more computational power. This will be transformative with applications in pharmacy, green new materials, logistics, finance, big data and more. For example, quantum computing will be better at calculating the energy of molecules because this is a fundamentally quantum problem. So if we can imagine an industry that deals with molecules, we can imagine an application of quantum computing. Often people want to know if quantum computers will be faster, and while they will be able to do computations faster, it is not because they are doing the same thing with more cycles. Instead quantum computers take advantage of a fundamentally different way of processing information. We'll walk through an example that demonstrates the potential of quantum computation to gain a sense of the underlying distinction.

## 2 Quantum Walks Using Coins

One of the most fundamental quantum translations of random walks will be our starting point: discrete-time quantum walks that occur in both discrete space and discrete time. To describe their development, an iterative implementation of a selected unitary operator would be employed, with each step advancing the walk by one step.

The concept of a discrete-time quantum walk was first proposed in Ref. [1]. The authors looked at the system's spatial evolution whose internal spin-1/2 state is denoted by the unitary—$U = \exp(-iSzP\delta/\hbar)$.

The operators $P$ and $S_z$ represent the particle's momentum and $z$-component of spin, respectively. The initial state $|\psi\ (\times 0) > (c+|\uparrow> + c\text{-}\ |\downarrow>)$ transforms into the state, is influenced by $U$.

$$|\Psi>= c-|\psi(x0--\delta) > |\downarrow> + c+|\psi(x0+\delta) > |\uparrow> \tag{1}$$

where $|\psi(x) >$ refers to the particle's wave function centered at position $x$.

The calculation is only done once in quantum walks, at the conclusion of the evolution. Quantum superposition and associations that occur during evolution are destroyed by a repeated measuring mechanism [1]. As a result, we'll use the definitions of quantum walks used in later sources [2, 3], with the calculation taking place only at the conclusion of the experiment.

## 3 A Quantum Drunk

We are going to think about the drunken walk. In the classical drunken walk (sometimes called the random walk) we have a drunk who is leaving the restroom and trying to find his friend at the bar (see Fig. 2).

Everyone basically looks the same at this hipster bar and he has had one too many so he is approaching a random person seated at the bar. When he discovers that the first person he has bothered is not his friend, he will randomly go to the next stool, either to the left or to the right. We can simulate our drunken walker by flipping a coin and saying heads he will go right, tails he will go left.

**Fig. 2** Moving probability of a quantum drunks

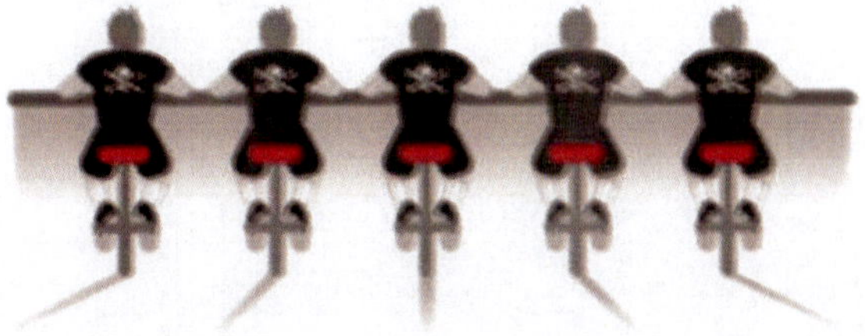

The next person is also wrong and his memory is short, so they will move on to either the left or right with equal probability (see Fig. 3). This will go on until security is called to throw out the drunkard.

The security team loves physics, so they decide to keep a tally of where they finally catch up to the drunk each time. Here is what security finds (see Fig. 4):

The shape is a bell curve (see Fig. 5) and the interesting feature of the bell curve is that the spread of the middle (the most likely place to find the drunk) is the square root of the number of steps [4] the drunken walker takes. When the drunk tries nine barstools, the spread of the bell curve is three; security can likely find them within three barstools of where they started. When the drunk makes 100 attempts, security will find the most likely within the ten closest stools to where they started. These statistics help security know where they are most likely to find the drunken walker, who is somewhere near the center.

Now the security team has a model they can use to keep up with the classical drunks (see Fig. 5), but unfortunately at this bar there are also quantum drunks (see Fig. 6). Whereas the classical drunk is a simple coin flip for each direction, for the quantum drunk the coin is quantum, and can be in a superposition of heads and tails

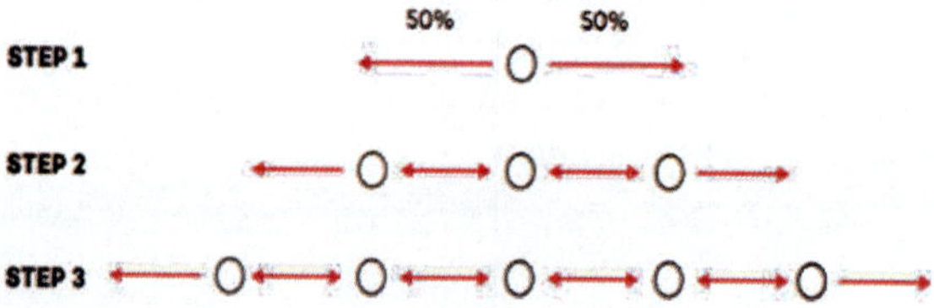


**Fig. 3** Moving probability

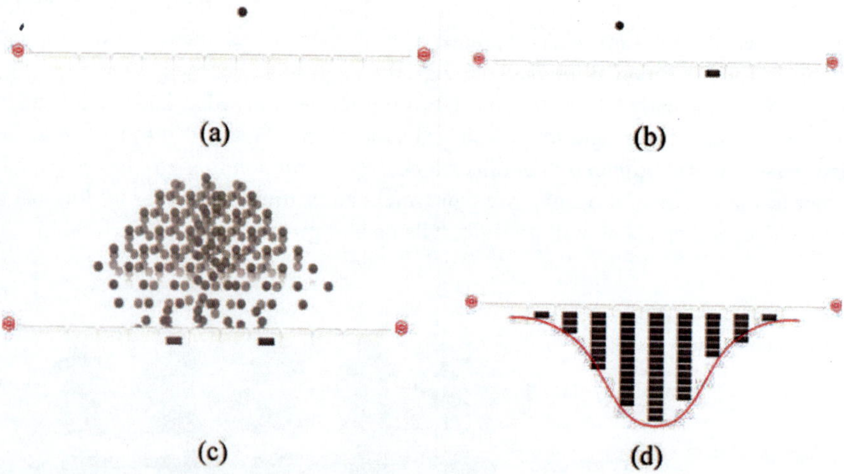


**Fig. 4** Classical walks always find in the center

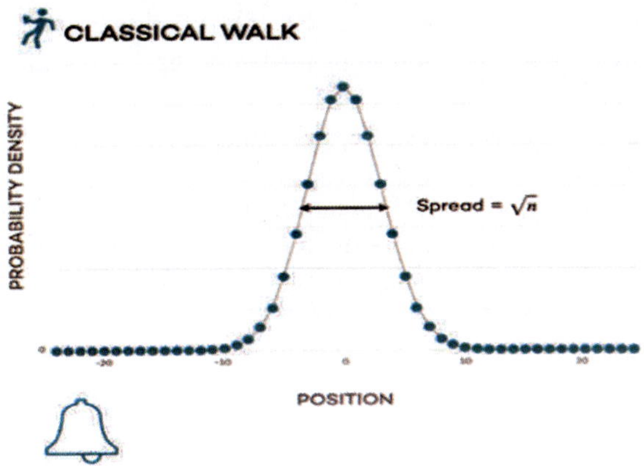


**Fig. 5** Graphical position of classical walks

**Fig. 6** Quantum drunks

at the same time. The quantum drunk is following a path that is a superposition of left and right at each bar stool.

Superposition is one of the fundamental concepts in quantum mechanics and one of the tools that differentiates quantum information and classical information.

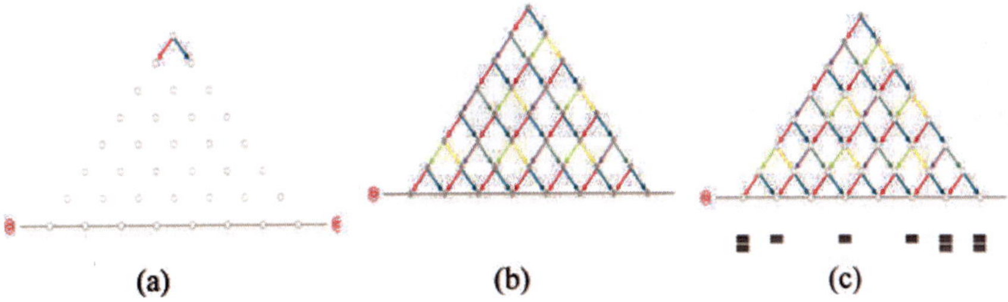

**Fig. 7** Distribution positions of the quantum drunk

When the security looks at the distribution of positions where the quantum drunk (see Fig. 7) is found, they will find a very different result from the classical drunk (see Fig. 9).

As opposed to a smooth bell curve distribution, they will find the "fangs" distribution shown below (see Fig. 8):

What is going on? Where is the quantum drunk? Why would the peaks of the distribution be on the outside? Why are there areas inside that are very low probability and others that are higher? The quantum drunk has new properties.

The drunkard tends to be farther away, and it is less likely to be close to the center. Certain paths are less probable because of interference, while some are more probable. The overall spread is much different too. Instead of the spread being related to the square root, the spread is related linearly to the number or steps. A quantum

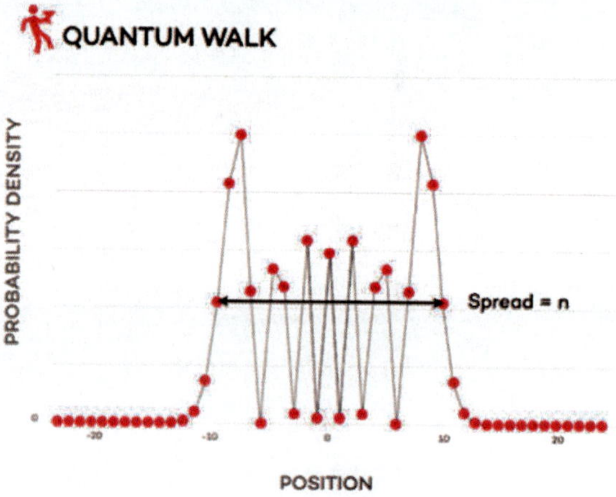


**Fig. 8** Graphical distribution position of the quantum walk

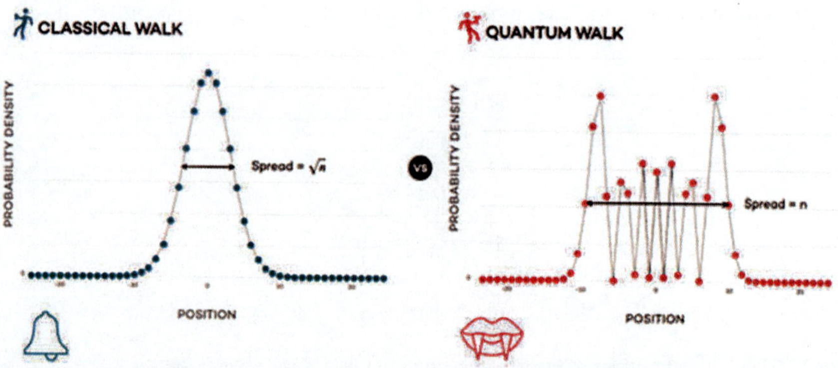


**Fig. 9** Quantum walk versus classical walk

drunk taking ten steps will most likely be found on the outside of a ten barstool spread, as wide as the distribution for a classical drunk taking 100 steps.

So how can we use this to our advantage? Is there some problem we can solve better with quantum drunks than with classical drunks? Well I'm glad you asked, because yes there is! To see this we are going to put the drunks to the task of solving a labyrinth. We are choosing a specific labyrinth that will illustrate the power of the quantum drunk. In this problem we have a tree structure that is mirrored and then stuck together.

On the left is the entrance to the labyrinth and on the right is the exit (see Fig. 10). We want to see how well our drunk walkers can find the exit. Remember that the classical drunk is going to be flipping a coin at every node, whereas the quantum drunk is creating a superposition of every path at each node. The drunks tend to get stuck in the random connections in the middle, taking more time to find their way out.

Since quantum drunks are more spread out, they can escape being stuck easier. This is why quantum drunks find the exit faster than classical drunks.

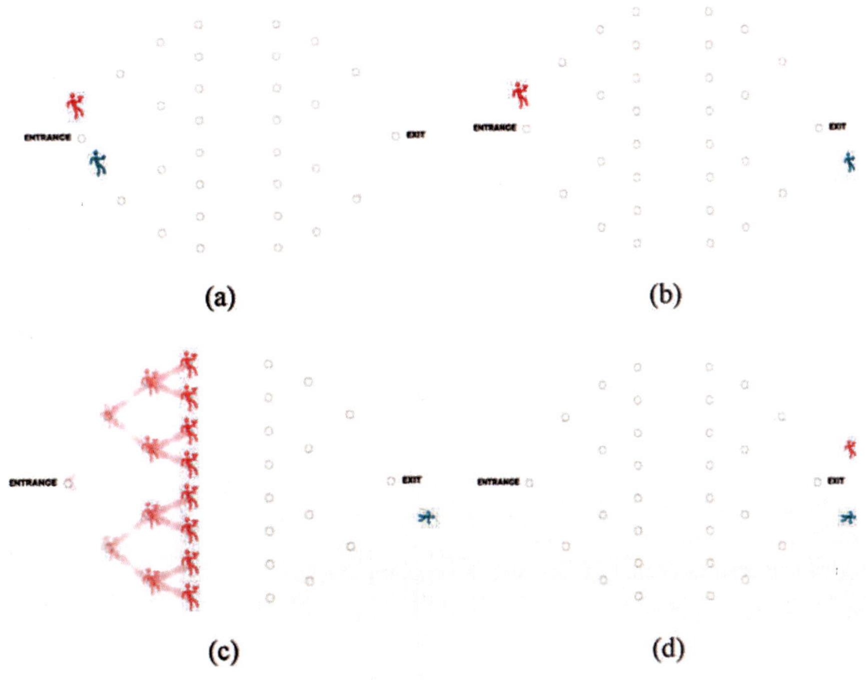


**Fig. 10** Moving direction of quantum drunks and classical drunks

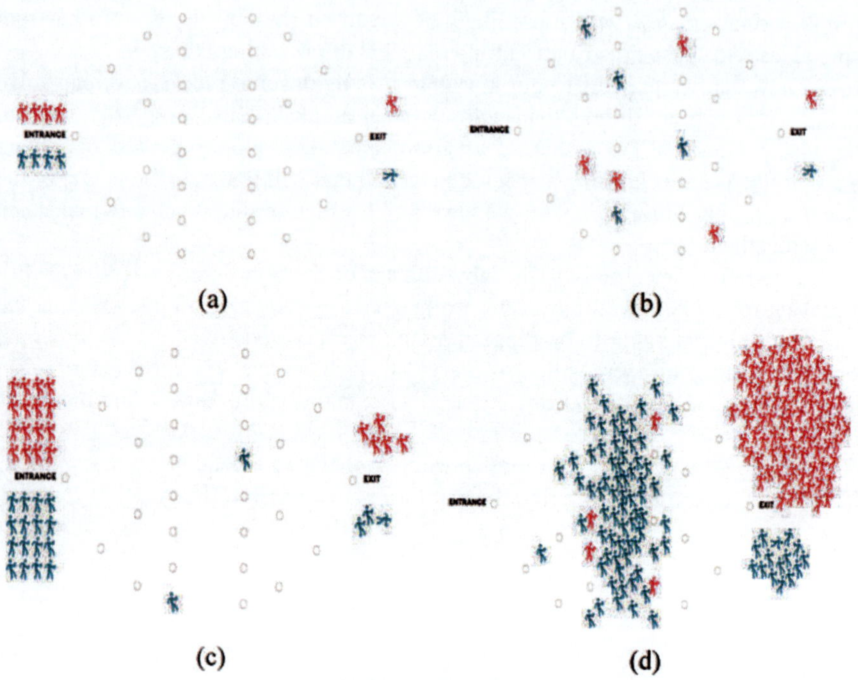


**Fig. 11** Quantum drunks find exit faster than classical drunks

For this problem as we send more and more drunks through, the quantum ones are going to make it through **exponentially** more than the classical ones (see Fig. 11).

## 4 Get Started Programming with Quantum Computers

**A Quantum Walk on Four Nodes**:

Imagine a square graph (see Fig. 12) with four nodes 00, 01, 10 and 11 and we to want start our journey from node 00.

This is a simple code for one step of a quantum walk.

```
OPENQASM 2.0;
include "qelib1.inc";
        //Initialization variables
qreg q[3];      //This creates three qubits.
        //The first two qubits, 0 and 1, representing the network of 4 nodes.
                //The last qubit, 2, represents the quantum coin.
creg c[2];          //This creates two classical bits to store the outcome of
measurements.
```

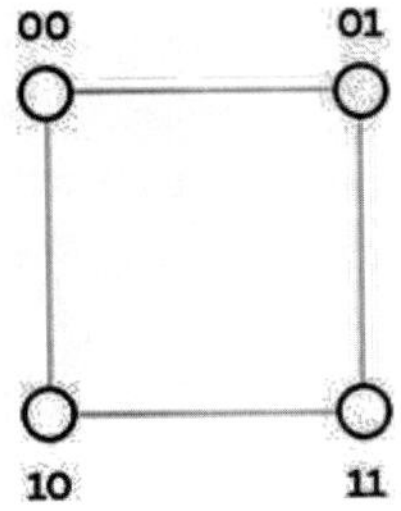


**Fig. 12** A Graph with four nodes

```
            //Preparing the quantum coin.
    h q[2];            //Rotate quantum coin into a superposition.
    barrier q[0], q[1], q[2];    //Barriers are a way break circuit portions into sections.
    //Quantum walk step
    cx q[2], q[0];    //Controlled-NOT gate on qubit 0 conditioned on the coin.
    x q[2];           //Bit-Flip (or x rotation) of the coin.
    cx q[2], q[1];    //Controlled-Not gate on qubit 1conditioned on the coin.
    barrier q[0], q[1], q[2]    //Barrier for readability of the quantum circuit.
    //This concludes one quantum walk step.
    //What would you have to do to have more steps on the walk?
    //Measurements
    //To finish, we measure the whole network to determine where the quantum walker is.
    measure q[0] -> c[0];        //Measure qubit 0, store the outcome in the bit 0.
    measure q[1] -> c[1];        // Measure qubit 1, store the outcome in the bit 1.
    //After the measurements, we can read out where the walker is.
```

## 5 Result Analysis

Imagine flipping a coin. If it is heads take a step to the left. If it is tails, take a step to the right. Repeat this process a few times. This is an example of a classical random walk. According to our histogram, we'll most likely end up in a range of the size of the standard deviation, which in this case is $\sqrt{N}$ where $N$ is the number of steps we took. Now, what if we used a qubit in a superposition instead of a coin to decide if we should take a step to the left or a step to the right. This is called a quantum walk. At each step, we enter a superposition of the previous step. Only when we finally make a measurement do we find that the histogram looks very different from the classical walk. Now there is a greater chance to be further from the center. In fact, the standard deviation of our new histogram is simply $N$, a quadratic increase. This increase is the power of quantum computing.

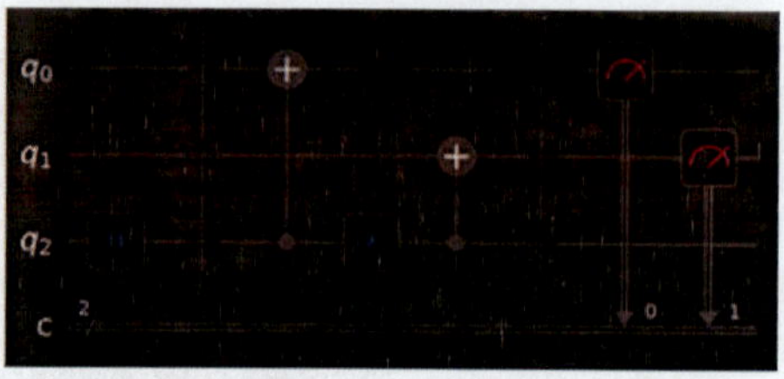


**Fig. 13** Circuit diagram

**Circuit Diagram**: Fig. 13, the first two qubits, q0 and q1, representing the network of four nodes: 00, 01, 10 and 11. We'll begin our walk on node 00. The last qubit, q2, represents the quantum coin.

**Circuit Resources**: In our example we used one Hadamard gate, one Pouli-X gate two Controlled-NOT gate, two barriers and two measurement (Fig. 14) form a one step of quantum walk.

**Histogram**: Fig. 15 is showing the histogram of our result.

From Fig. 15 we can see our step which has been starting from node 00, finally reach to node 01. This is an example of one step of quantum walk. Furthermore we can use the same code and changed our Pouli-X and Hadamard gate positions to get different result. We can also increase our qubit to make it different. We can change number of steps.

**Fig. 14** Gate count

**Fig. 15** Histogram of result

## 6 Conclusion

This is the power of quantum computing. Even though this is a simple example, all quantum algorithms work the same way: by exploiting the quantum spread in clever ways that fit the structure of a problem. There are many applications for quantum algorithms, so it is an exciting time to start exploring quantum programming. In the near term, the best applications are the design of pharmaceuticals and the engineering of new materials. Many of these chemistry applications are fundamentally quantum mechanical. This is because figuring out the energies of electrons for different molecules is more efficient using a quantum computer. Optimization problems are another area where quantum computing will have an impact in the not-too-distant future. This class of logistics problems include storage optimization or the distribution of goods, such as vaccines. Risk management for finance can be tackled using similar algorithms. Further afield are the technologies to build a quantum Internet that will replace some of our cryptographic systems, to ensure privacy and security.

## References

1. Aharonov, Y., Davidovich, L., Zagury, N.: Quantum random walks. Phys. Rev. A **48**, 1687 (1993). https://doi.org/10.1103/PhysRevA.48.1687
2. Aharonov, D., Ambainis, A., Kempe, J., Vazirani, U.: Quantum walks on graphs. STOC '01. In: Proceedings of the Thirty-Third Annual ACM Symposium on Theory of Computing, pp. 50–59 (2001). https://doi.org/10.1145/380752.380758
3. Ambainis, A., Bach, E., Nayak, A., Vishwanath, A., Watrous, J.: One-dimensional quantum walks. STOC '01. In: Proceedings of the Thirty-Third Annual ACM Symposium on Theory of Computing, pp. 37–49 (2001). https://doi.org/10.1145/380752.380757
4. Kempe, J.: Quantum random walks. An introductory overview. Contemp. Phys. **44**, 307–327 (2009)
5. Chowdhury, A.N., Somma, R.D.: Quantum algorithms for gibbs sampling and hitting time estimation. Quant. Inf. Comp. **17**(1/2), 0041–0064 (2017)
6. Montanaro, A.: Quantum speedup of Monte Carlo methods. In: Proceedings of the Royal Society A: Mathematical, Physical and Engineering Sciences (2015). https://doi.org/10.1098/rspa.2015.0301. ISSN: 1364-5021
7. Dhahri, A., Ko, K.C., Yoo, J.H.: Quantum Markov chains associated with open quantum random walks. J. Stat. Phys. **176**(5), 1272–1295 (2019)
8. Ambainis, A.: Quantum walk algorithm for element distinctness. SIAM J. Comput. **37**(1), 210–239 (2007). https://doi.org/10.1137/S0097539705447311
9. Magniez, F., Nayak, A., Roland, J., Santha, M.: Search via quantum walk. STOC '07: In: Proceedings of the Thirty-Ninth Annual ACM Symposium on Theory of Computing, pp. 575–584 (2007). https://doi.org/10.1145/1250790.1250874
10. Childs, M.A., Goldstone, J.: Spatial search by quantum walk. Phys. Rev. A **70**(2), 022314 (2004). https://doi.org/10.1103/PhysRevA.70.022314
11. Montanaro, A.: Quantum-walk speedup of backtracking algorithms. Theor. Comput. **14**, 1–14 (2018)
12. Sansoni, L., Sciarrino, F., Vallone, G., Mataloni, P., Crespi, A., Ramponi, R., Osellame, R.: Two-particle bosonic-fermionic quantum walk via integrated photonics. Phys. Rev. Lett. **108**(1) (2012). https://doi.org/10.1103/PhysRevLett.108.010502

13. Rakovszky, T., Asboth, K.J.: Localization, delocalization, and topological phase transitions in the one-dimensional split-step quantum walk. Phys. Rev. A **92**(5) (2015). https://doi.org/10.1103/PhysRevA.92.052311
14. Acasiete, F., Agostini, P.F., Moqadam, K.J., Portugal, R.: Implementation of quantum walks on IBM quantum computers. Quantum Inf. Proc. 19. https://doi.org/10.1007/s11128020-02938-5 (2020)
15. Krovi, H.: Symmetry in quantum walks. Quantum Phys. (quant-ph). arXiv:0711.1694 (2007)
16. Orthey, C.A., Amorim, M.P.E.: On the spreading of quantum walks starting from local and delocalized states. Quantum Phys. (quant-ph). arXiv:1706.06257 (2017)
17. Balu, R., Castillo, D., Siopsis, G.: Physical realization of topological quantum walks on IBM-q and beyond. Quantum Sci. Technol. **3**, 035001 (2018)
18. Moll, N., Barkoutsos, P., Bishop, L.S., Chow, M.J., Cross, A., Egger, J.D., Filipp, S., Fuhrer, A., Gambetta, M.J., Ganzhorn, M., et al.: Quantum optimization using variational algorithms on near-term quantum devices. Quantum Sci. Technol. **3**, 030503 (2018)

# Survey on Recent Malware Detection Techniques for IoT

Sangeeta Kakati, Debasish Chouhan, Amitava Nag, and Subir Panja

**Abstract** Malware has been growing at a rapid rate in recent times, and studying detection techniques has become a vital step. There are a varieties of malware each of which has its intended application and purpose. With the abundance of malware on the Internet, malware detection is critical because it serves as an alert system of Internet security. Since these two platforms are the most widely used nowadays, the problem of mobile malware detection on the IoT and Android has gotten a lot of attention. For example, the BrickerBot virus discovered in 2017 in the IoT platform tries to disable IoT devices that have been improperly configured. Bots, DDoS attacks and other vulnerabilities frequently compromise the IoT network. As a result, it is pivotal to conduct a survey on the strategies employed to detect these intrusions. According to reports, malware accounts for roughly 90% of all IoT breaches. Among these, the android devices are the worst affected. We have provided a comprehensive survey of some recent malware attacks in the IoT, as well as the most prominent malware detection techniques that uses machine learning such as static, dynamic and hybrid analysis, along with blockchain and convolutional neural network-based approaches, in this paper. We began with a broad overview of malware before moving on to IoT malware. The detecting mechanisms will be discussed in the subsequent sections of the paper. Finally, we gave a thorough characterization of existing works using these strategies.

**Keywords** IoT security · Malware · Android · Blockchain · Machine learning · CNN

S. Kakati (✉) · D. Chouhan
Indian Institute of Information Technology, Guwahati, India
e-mail: sangeeta.kakati@iiitg.ac.in

A. Nag · S. Panja
Central Institute of Technology, Kokrajhar, India



D. Gupta et al. (eds.), *Pattern Recognition and Data Analysis with Applications*, Lecture Notes in Electrical Engineering 888,
https://doi.org/10.1007/978-981-19-1520-8_53

# 1 Introduction

"Malware is dangerous, be very careful!". The Internet of things is one of the most interesting and fastest developing fields of the Internet and with 5G beginning to appear, the Internet of things or IoT for short, tipped to become even bigger and even more prevalent.

A malware is a collaborative name given to a variant of malicious codes in the aim of damaging data or gaining unauthorized access over a system. Malware analysis is an emerging need for IoT since the data from the IoT devices are shared directly over the Internet, which arises the need to secure the network for holding the large amount of traffic. When devices are built in a constrained environment they face security bugs and breaches. The main aim is to provide genuineness to the system by detecting cyber-attacks. IoT is not immune to hacking and DDoS attacks, Gateways are being compromised, sensitive information can be exploited.

Year 2017 reports says that the number surpassed nearly 8.4 billion devices of IoT which further exceeded 9.2 billion in 2019. It is in the year 2020 that the IoT connected devices worldwide has reached nearly 18 billions. Surveys resulted that the year 2022 will experience a growth of up to 22 billion IoT devices.

With a greater number of devices integrating with IoT, the attack risks on the system also increases. With every new system developed there also arises a threat model. Since the data from the devices are shared directly over the Internet, we are in the need to secure the network for holding the large amount of traffic. IoT is more vulnerable to malicious codes and a good source for security invasion by malware attackers because of, (i) Poor password: Since the end users of a IoT device is some common people and most devices have its default passwords unchanged hence attackers find it easier to compromise it. (ii) Multiple sources of data: IoT which have myriads of data and the network which receives this data from multiple sources needs to perform the streaming of this data saving it from malware attacks. This itself is a huge task to follow. (iii) The backdoor feature: A backdoor can be used for legid reasons as well as for unauthorized use too. It allows to gain the security measures of a device and improvise as a root user. The backdoor can even be an intentional manufacture feature used for troubleshooting but it becomes an easier way for malware attackers to evade this backdoor system. (iv) Security Deficit: The risk of software vulnerabilities that comes with a user who is unaware of the security flaws that may rise in a system. Since manufacturers wants a device to be more user friendly in the end user interface, therefore often the security features are compromised.

The learning-based Android malware detection system for IoT devices has increasingly improved as machine learning algorithms have progressed. These learning-based detection models, on the other hand, frequently sensitive to adversarial data. To aid in the security analysis of these learning-based malware detection systems for IoT devices, a framework for automated testing is required. There is a huge growth of android operators and in the coming decades, it is forecast to expand with introduction to 5G. Most of the application of IoT runs in an android operating system (OS). Some of the main IoT devices using android OS are smartphones, smart TV, camera, cars, smart watches and many more. According to GlobalStats, 71.81% of mobile operating system runs on android. There is a 40% increase in cyber-attacks in android since 2016 as reported by Avast. Android being open source strives the need for robust malware detection techniques.

To sum up, following are the key contributions of this paper:

- First, we provide a concise overview of the risks posed by malware, including why IoT is particularly vulnerable.
- Second, we explored why malware based on Android IoT is the most particularly at risk toward exploitation.
- Third, we undertake a detailed characterization of IoT malware detection methods, namely, static, dynamic and hybrid; also including a discussion of blockchain and CNN-based identification approaches.

The rest of this paper is laid out as follows. The related work on malware detection is reviewed in Sect. 2. Section 3 discusses Android IoT malware and its causes. Section 4 examines the identification methods focused on static, dynamic, hybrid, blockchain, and CNN in particular and displays the findings of some of the most recent detection techniques. Finally, Sect. 5 brings the paper to a close.

## 2 Related Works

Many of the IoT devices that are currently being distributed are inherently unstable. Developers discuss how obsolete Linux firmware can expose telnet ports and other vulnerabilities. The proliferation of vulnerable IoT devices opens up a slew of new cyber security attack vectors, the most notable of which are botnet attacks, in which quite a few compromised IoT devices are linked together. The Mirai botnet attack, which was at the time the largest DDoS attack ever registered, is the most prominent example among them. Several authors have mentioned these attacks in their works [1, 2]. The concept of malwares in distributed denial of service IoT was discussed in [3].

The threats and privacy measures of IoT based on 8 features (Ubiquitous, Interdependence, Constrained, etc.) is being analyzed in [4] where the authors presented research challenges from each of the eight features. One of the most widely used methods for detecting malware is the use of control flow graph [5], where the authors analyzed the robustness of detection using CGF with deep learning. Machine learning and deep learning strategies due to their practical performance are positively used in detection systems. Mohaisen et al. [6] presented a design to automatically detect samples based on components like behavior analysis of malware and automated labeling.

A collation of deep learning techniques is discussed in [7] where the authors differentiated three convolutional neural network viz. CNN on byte sequences, picture elements and assembly sequences. Xiao et al. [8] discussed a deep learning system for detecting malware in IoT environments that is integrated into a cloud platform. Behavior graph represents the programs, and the cloud platform further converts these features to binary vectors. Liu et al. [9] presented an android malware detection system which works by adding some deviations into the malware but not deviating the functionality of the malware.

Operation code is another feature which is actively used in malware detection [10]. It also states that graph-based methods are more advantageous than non-graph-based when it comes to dealing with unseen malwares, thus can precisely detect intrusions in the network. Blockchain smart contracts can be used to provide privacy in IoT and cloud environments [11]. It is specially useful on distributed intrusion detection models. Reaching a single trusted model using blockchain proves to be effective in detecting anomalies [12]. The wide range of applications in healthcare, agriculture, smart home, etc. and multiple platform integration makes IoT a more security deprived technology.

## 3 IoT Mobile Malware

Mobile phones have become an integral part of our everyday routine and their use has tremendously increased over the last couple of years. It is expected that the trajectory will remain increasing as mobile devices are becoming ubiquitous given this fact the number of malicious mobile applications is tremendously increasing. Android security and the emergence of security threats continues to be a major concern. In general, android is not only a day-to-day thing, but also a framework made up of three key components: computer hardware, android OS and android runtime. The android device hardware block encompasses a broad variety of hardware configurations on which Android will run, including smartphones, tablets, and other mobile devices. The banking Trojan family is the next one, which is a type of mobile malware that targets mobile banking services for monetary benefit. These Trojans are delivered as legitimate software with the added perk of intercepting user credentials or one-time passwords. Online smartphone ransomware is malicious program that prevents victims from completely using their computers before a ransom is paid in digital currency such as bitcoin. The definition of how malware gets to the end user can be generally acknowledged by looking at various delivery channels, the simplest one is app store distribution where the Trojans are uploaded to the app store in large number to take advantage of the download volume.

# 4 Malware Detection Methods for IoT

## 4.1 Using Machine Learning Classifier

Many methods for detecting IoT and android IoT malware using machine learning techniques have been suggested. Static analysis, dynamic analysis and hybrid analysis, which incorporates both static and dynamic analysis, are the three types of IoT and android malware detection methods.

**Static Analysis**: Static analysis is the extraction of data from a malware that is at rest. It is a low-cost method of detecting malware. We have discussed recent IoT and android malware detection and classification methods based on static features like control flow graph (CFG), file header, operation code (opcode), strings with various machine learning classifiers such as support vector machine (SVM), decision tree, random forest (RF), K-nearest neighbor (KNN), Naive Bayes, etc. This technique will not detect malware that employs code obfuscation.

*Works undertaken in static analysis (CFG-Based)*: CFG is a directed graph that shows all the possible pathways which can be traversed during the execution of a program. A CFG has a node for every basic block and an edge for each possible control transfer between blocks. CFG can be characterized by features like total nodes and edges, density, shortest path, centrality. Alasmary et al. [13] have used CFG to test about 6000 IoT malware samples with 99% detection accuracy. Phu et al. [14], Yamaguchi et al. [15] uses CFG for efficient malware detection technique. Experiments [16] shows that CFGs are much effective in differentiating IoT malware and android malware. A test result of sample with 2874 IoT and 201 android malware binaries shows android malware tend to have more number of nodes and edges and high density than IoT malware.

*Works undertaken in static analysis (ELF-Based)*: Executable and Linkable Format (ELF) is the standard format for storing all Linux executables. ELF file header contains different information and can give very good insight of IoT malware. ELF files provide information about the instruction set architecture for which the code in a relocatable, executable or shared object file is intended. Shahzad and Farooq [17] used a rule-based and decision tree classifier for IoT malware detection, which is based on 383 features extracted from the ELF file header including section headers, symbol sections and program headers, and has a 99% detection accuracy.

*Works undertaken in static analysis (Opcode-Based)*: Opcode is a common type of feature retrieved via static analysis and widely used for IoT malware detection. It is a single machine instruction that specifies the operation to be performed. The basic actions of a program are represented by the opcode sequences retrieved from disassembled executable files. Based on opcode sequences [18, 19] can detect IoT malware effectively using various ML classifiers. Dovom et al. [20] uses opcode and deploys fuzzy pattern tree method for IoT malware detection.

*Works undertaken in static analysis (Strings-Based)*: String-based analysis from the executable files employs indicators such as commands, payloads and other data for IoT malware detection. Within an executable file, each printable string could extract valuable information such as IP address associated, URL to link to and so on. Lee et al. [22] extracts useful features from printable strings from ELF binaries and uses Recurrent Networks (RN), SVM and KNN to get 98% accuracy. Torabi et al. [23] uses string-based analysis to group IoT malwares with similar characteristics.

*Malware detection in Android IoT*: For android malware, static analysis is done on the Android Package (APK) files rather than installing and running it. In static approach reverse engineering is performed on the APK file to decompile it and analyze benign patterns. Recent research of static analysis are mostly based on Application Program Interface (API) call and permissioned-based analysis. As permission is the most effective feature for an attacker to launch an attack. After getting permission an android malware can install itself on the device. Whereas API calls are used to interact with the different programs of an android framework to achieve functionality.

Tao et al. [24] have discussed permission-related APIs to discover hidden patterns of android malware and classify using random forest classifier with F1 score of 98.24%. SEDMDroid [25] uses Multi-layer perceptron and SVM atop of permission, sensitive APIs and other static features of android malware and it can achieve accuracy up to 89.07%. Huang et al. [26] have discussed malware detection technique using different permissions from the application. RanDroid [27] system deploys SVM, Naive Bayes, decision tree and random forest classifiers to detect and classify android malware. RanDroid uses both vulnerable API calls and permissions gained along with some key features and it can get 97.7% of classification accuracy. DroidAPIMiner [28] is another proposed model for extracting malware behavior captured at API level using KNN classifier. CogramDroid [29] is an android malware detection method employing opcode n-grams with 96.22% accuracy rate.

**Dynamic Analysis**: Dynamic analysis is used to observe real-time behavior of the application to discover malicious patterns. Dynamic analysis is the process of evaluating a sample by executing it in a controlled environment and monitoring its activities, interactions and impact on the system. System calls and API calls analysis and control flow analysis are the main methods used for dynamic analysis. Dynamic monitoring tools like Process Hacker and Process Monitor are used for inspecting process attributes and system interaction. Wireshark is used to capture network traffic. However, dynamic analysis is time consuming and resource intensive.

*Works undertaken in dynamic analysis*: A dynamic approach on system calls and their relations with fuzzy logic for malware detection is used in [30]. DAIMD [31] model performs dynamic analysis to extract behaviors related to memory, network, virtual file system and system calls from the IoT malware and uses CNN for classification. In [32] API calls and other key information are extracted to establish dependency chain which could describe the behavior of IoT malware based on similarity comparison.

In this approach android emulators are used to extract runtime behaviors of the features like API calls. The DynaLog framework [33] is a tool to extract and log high-level behaviors of API calls and system call information which could be used to assess the traits of the android malware. DroidCat [34] uses dynamic analysis based on API calls for classification. SDAC [35] evaluates API call sequences as the input vector for training neural network to detect any malicious behavior. DREBIN [36] is a light weight mobile-based dynamic android malware detection tool. It uses APIs and also hardware components and permission information as the feature for detection. DroidScope [37], DroidRanger [38] are some other dynamic analysis tools.

**Hybrid Analysis**: Hybrid analysis incorporates both static analysis and dynamic analysis. It overcomes the limitations of both static and dynamic analysis. It explores by examining malware code's signature and continue by combining it with other behavioral pattern factors to improve malware analysis, although this approach is much time consuming and costly.

*Works undertaken in hybrid analysis*: Shijo and Salim [39] proposed a hybrid technique to detect and classify IoT malware based on printable strings information static analysis and API calls from dynamic analysis. It shows a detection rate of 98.7%. Ma et al. [40] approach uses variety of features including static import routines and dynamic call functions with different ML classifiers. Hybrid analysis for android malware [41, 42], first investigates applications to be run on a device prior to their installation and then monitor their activities during runtime and determines the pattern using machine learning classifiers for any malicious activity. SAMADroid [43] is a 3-level hybrid android malware detection method with low resource consumption.

We can summarize the most recent studies on static, dynamic and hybrid analysis in Table 1.

## 4.2 *Using Blockchain Technology*

Blockchain being a distributed network enables secure communication between devices thus decreases the risk to cyber threats. The Internet layer, the ledger layer, and the application layer are the metrics for detecting malware in IoT devices using blockchain. The TCP/IP infrastructure could be seen in the Internet layer. The blockchain layer records the file with details about the malware. Consensus rules follows malware detection through some ML trained model. Finally, we have the network of IoT devices running through the P2P network. The application layer shows the exploitation of the malware in the applications where it is been used. The most efficient way of using IoT with blockchain is the installation of chips in the sensors and devices used in that particular IoT system.

*Works undertaken*: Blockchain is facilitating the security needs of IoT by identifying the malicious participant and keeping records which cannot be tampered [44]. A consortium-based blockchain was discussed in [45] where the test members share a consortium chain and a public chain is shared by the users. Smart contract-based detection where the smart contracts feature of blockchain can be used. The contract

**Table 1** Static, dynamic and hybrid analysis using ML classifiers

| Method | Contribution | Feature | ML classifier | Accuracy (%) | Malware class |
|---|---|---|---|---|---|
| Static | [13] | CFG | LR, SVM, RF, CNN | 99.6 | Benign, Gafgyt, Mirai, Tsunami |
| | [14] | CFG | SVM | 99.3 | IoT malware detection |
| | [15] | CFG | CNN | – | Code vulnerabilities |
| | [16] | CFG | – | – | IoT and Android malware |
| | [17] | ELF | Rule-based, DT, bio-inspired | 99 | Detect malicious executables |
| | [18] | Opcode | KNN, SVM, MLP, RF, DT | 99 | Polymorphic |
| Static | [19] | Opcode | CNN | 96 | Malware detection |
| | [20] | Opcode | Fuzzy pattern tree | 99.8 | IoT malware detection and categorization |
| | [21] | ELF, Opcode | SVM, CNN, ANN | 98 | Mirai, Tsunami |
| | [22] | String-based | RF, KNN, SVM | 98 | Mirai, Tsunami, Hajime, Dofloo, Bashlite, Xorddos, Android |
| | [23] | String-based | Similarity analysis | – | Mirai, Gafgyt, Tsunami, other |
| | [24] | API | RF | 98.24 | Android |
| | [25] | Permission, API | SVM | 94.9 | Android |
| | [26] | Permission | DT, SVM, NB, AdaBoost | 81 | Android |
| | [27] | Permission, API call | SVM, DT, RF, NB | 97.7 | Android |
| | [28] | API | KNN | 99 | Android |
| | [29] | Opcode | n-gram | 96.2 | Android |
| Dynamic | [30] | SEF | Fuzzy logic | – | IoT |
| | [31] | Memory, Network, System call, VFS, Process | CNN | 99.28 | IoT |
| | [32] | API call | Similarity | – | IoT |

(continued)

**Table 1** (continued)

| Method | Contribution | Feature | ML classifier | Accuracy (%) | Malware class |
|---|---|---|---|---|---|
| | [34] | API call | RF | 97 | Android |
| | [35] | API | Distance | 97.49 | Android |
| | [36] | Permission, API call, other | SVM | 93 | Android |
| | [37] | API | – | – | Android |
| | [38] | Permission, other | Heuristic | – | Android |
| Hybrid | [39] | String, API call | SVM | 98.7 | IoT |
| | [40] | PE header, System call | SVM, NB, Classification tree | – | IoT |
| | [41] | Opcode, System call and hardware info | LR, NB, J48 | 100 | Android Ransomware |
| | [42] | API, permission, System call | SVM | 93–99 | Android |
| | [43] | API, Permission, Hardware, System call | SVM | – | Android |

itself checks and gives information of the malware and stores all the information of the APKs which joined newly. Using blockchain as a malware detector in IoT is not an assured way since IoT contains a large network of devices and more numbers of end users, hence security in the multiple layers has to be prioritized [46]. The integration of blockchain into IoT and IIoT is discussed in [47] where it has been stated that creating a resilient and scalable blockchain-based security system for both moderate servers and low-powered IoT computers is a difficult task. Smart contracts are recently applied in many fields of IoT for instance, in data certification. Smart contract is a peer-to-peer-based network that can execute its own. It contains the conditions of an agreement among peers. It eliminates the need of a third party to keep record and validate a system. Hu et al. [48] employs a peer-to-peer file sharing strategy to distribute various versions of device firmware and minimize the likelihood of DDoS attacks in IoT systems.

### 4.3 Using Convolutional Neural Network

CNN is one of the most active machine learning techniques for predicting IoT and android malwares. It has the feature to detect malwares that are hidden inside benign malwares hence leaving no loophole in detecting. In graphic representation, CNN works better than any other methods of detecting IoT malwares since it converts the binary malware into 8-bit vector and further the 8-bit vector to an image. The final step is to build the CNN predictive model.

*Works undertaken*: Li et al. [49] presented a CNN-based approach for IoT malware detection which included two layers, namely, the convolutional layer for reducing image sizes and activation layer ReLU to introduce non linearity. They used samples for Trojans, worms and backdoors. This method had an accuracy of 98.57% on IoT malwares. Jeon et al. [31] presented an additional feature preprocessing phase in comparison to the traditional IoT malware detection systems. The model is used in a cloud environment with virtually embedded systems. They integrated the feature selection and classification phase using ZFNet model of CNN. Detecting malign behavior of IoT nodes is as important as detecting the entire network [50]. It can be implemented in IoT applications with 5G connectivity.

Taking into consideration the growth of android IoT malwares, Ren et al. [51] has presented an end-to-end-based android malware detection method. This method can reach an accuracy up to 95.8%. Vu and Jung [52] uses images for construction of an adjacency matrix to fit into the CNN for malware detection and has average 98.26% detection rate.

We can summarize recent contributions on IoT malware detection techniques using blockchain and CNN in Table 2.

**Table 2** IoT malware detection based on blockchain, CNN

| Method | Contribution | Feature | Accuracy (%) | Malware class |
|---|---|---|---|---|
| Blockchain | [44–48] | – | – | IoT |
| CNN | [13, 15, 19, 31] | CFG, Opcode Memory, Network, System call, VFS, Process | (Refer to Table 1) | IoT |
| | [49] | Grayscale Image | 98.57 | IoT |
| | [50] | Image-based | 92 | Mirai, Gafgyt |
| | [51] | Bytecode sequence | 95.8 | Android |
| | [52] | Image-based | 98.26 | Android |
| | [53] | PSI graph | 92 | IoT Botnet |
| ML and blockchain | [54] | API, Opcode, other | – | Android |

## 5 Conclusion

Malware has the power to even outstand factory resets under some conditions as such, (1) Factory backup location gets infected or is the source of infection, (2) Malware is aware that factory reset is done and can intercept the process (depends on devices), (3) Malware is spreaded through the local network and it infects the device just after the reset.

After reviewing a series of papers, we came to the conclusion that both known and unknown malware seemed to be more effectively detected using CNN and string-based approaches. Non-ML-based methods, such as the Blockchain, are slightly less successful at detecting malware than ML-based techniques. Learning algorithms could well track IoT system behavior and divide it into groups to clearly distinguish between objective and subjective IoT acts. Also in terms of detecting unknown malicious codes, graph-based methods outperform non-graph-based methods, regardless of the malware's complexity. However, these methods fail to detect some proportion of malwares.

There has been a lot of research in order to get a one-way solution for detecting and keeping away the malwares from a cutting-edge technology like IoT. But due to the ever evolving malwares being detected every day, there does not exist any proper methodology which could save the IoT world from malwares. IoT has made lives of the human being straightforward and comfortable. Whereas on the other hand it also increases the treat for security and safety. The IoT devices running on android OS are most vulnerable to these malware attacks. So careful deliberation needs be made, while providing the details on the Internet platform. More study is needed in order to arrive at a sustainable conclusion for recompensing the issues caused by malwares.

## References

1. Deogirikar, J., Vidhate, A.: Security attacks in IoT: a survey. In: I-SMAC, pp. 32–37. IEEE (2017)
2. Shah, S., Simnani, S.S.A., Banday, M.T.: A study of security attacks on internet of things and its possible solutions. In: ICACE, pp. 203–209. IEEE (2018)
3. Vishwakarma, R., Jain, A.K.: A survey of DDoS attacking techniques and defence mechanisms in the IoT network. Telecom. Syst. **73**(1), 3–25 (2020)
4. Zhou, W., Jia, Y., Peng, A., Zhang, Y., Liu, P.: The effect of IoT new features on security and privacy: new threats, existing solutions, and challenges yet to be solved. IEEE Internet Things J. **6**(2), 1606–1616 (2018)
5. Abusnaina, A., Khormali, A., Alasmary, H., Park, J., Anwar, A., Mohaisen, A.: Adversarial learning attacks on graph-based IoT malware detection systems. In: ICDCS, pp. 1296–1305. IEEE (2019)
6. Mohaisen, A., Alrawi, O., Mohaisen, M.: AMAL: high-fidelity, behavior-based automated malware analysis and classification. Comput. Secur. **52**, 251–266 (2015)
7. Nguyen, K.D.T., Tuan, T.M., Le, S.H., Viet, A.P., Ogawa, M., Le Minh, N.: Comparison of three deep learning-based approaches for IoT malware detection. In: KSE, 382–388. IEEE (2018)

8. Xiao, F., Lin, Z., Sun, Y., Ma, Y.: Malware detection based on deep learning of behavior graphs. Math. Prob. Eng. (2019)
9. Liu, X., Du, X., Zhang, X., Zhu, Q., Wang, H., Guizani, M.: Adversarial samples on android malware detection systems for IoT systems. Sensors **19**(4), 974 (2019)
10. Ngo, Q.D., Nguyen, H.T., Le, V.H., Nguyen, D.H.: A survey of IoT malware and detection methods based on static features. ICT Express **6**(4), 280–286 (2020)
11. Alkadi, O., Moustafa, N., Turnbull, B., Choo, K.K.R.: A deep blockchain framework-enabled collaborative intrusion detection for protecting IoT and cloud networks. IEEE Internet Things J. **8**(12), 9463–9472 (2020)
12. Mirsky, Y., Golomb, T., Elovici, Y.: Lightweight collaborative anomaly detection for the IoT using blockchain. J. Parallel Distrib. Comput. **145**, 75–97 (2020)
13. Alasmary, H., Khormali, A., Anwar, A., Park, J., Choi, J., Abusnaina, A., Mohaisen, A.: Analyzing and detecting emerging internet of things malware: a graph-based approach. IEEE Internet Things J. **6**(5), 8977–8988 (2019)
14. Phu, T.N., Hoang, L., Toan, N.N., Dai Tho, N., Binh, N.N.: C500-CFG: A novel algorithm to extract control flow-based features for IoT malware detection. In: ISCIT, pp. 568–573. IEEE (2019)
15. Yamaguchi, F., Golde, N., Arp, D., Rieck, K.: Modeling and discovering vulnerabilities with code property graphs, pp. 590–604. IEEE (2014)
16. Alasmary, H., Anwar, A., Park, J., Choi, J., Nyang, D., Mohaisen, A.: Graph-based comparison of IoT and android malware, pp. 259–272. Springer, Cham (2018)
17. Shahzad, F., Farooq, M.: ELF-miner: using structural knowledge and data mining methods to detect new (Linux) malicious executables. Knowl. Inf. Syst. **30**(3), 589–612 (2012)
18. Darabian, H., Dehghantanha, A., Hashemi, S., Homayoun, S., Choo, K.K.R.: An opcode-based technique for polymorphic Internet of Things malware detection. Concurrency Comput. Pract. Experience **32**(6), e5173 (2020)
19. Jeon, S., Moon, J.: Malware-detection method with a convolutional recurrent neural network using opcode sequences. Inf. Sci. **535**, 1–15 (2020)
20. Dovom, E.M., Azmoodeh, A., Dehghantanha, A., Newton, D.E., Parizi, R.M., Karimipour, H.: Fuzzy pattern tree for edge malware detection and categorization in IoT. J. Syst. Archit. **97**, 1–7 (2019)
21. Tien, C.W., Chen, S.W., Ban, T., Kuo, S.Y.: Machine learning framework to analyze IoT malware using elf and opcode features. Digit. Threats: Res. Pract. **1**(1), 1–19 (2020)
22. Lee, Y.T., Ban, T., Wan, T.L., Cheng, S.M., Isawa, R., Takahashi, T., Inoue, D.: Cross platform IoT-malware family classification based on printable strings. In: TrustCom, pp. 775–784. IEEE (2020)
23. Torabi, S., Dib, M., Bou-Harb, E., Assi, C., Debbabi, M.: A strings-based similarity analysis approach for characterizing IoT malware and inferring their underlying relationships. IEEE Netw. Lett. (2021)
24. Tao, G., Zheng, Z., Guo, Z., Lyu, M.R.: MalPat: mining patterns of malicious and benign Android apps via permission-related APIs. IEEE Trans. Reliab. **67**(1), 355–369 (2017)
25. Zhu, H., Li, Y., Li, R., Li, J., You, Z.H., Song, H.: SEDMDroid: an enhanced stacking ensemble of deep learning framework for android malware detection. IEEE Trans. Netw. Sci. Eng. (2020)
26. Huang, C.Y., Tsai, Y.T., Hsu, C.H.: Performance evaluation on permission-based detection for android malware. In: Advances in Intelligent Systems and Application, vol. 2, pp. 111–120. Springer (2013)
27. Koli, J.D.: RanDroid: Android malware detection using random machine learning classifiers. In: ICSESP, pp. 1–6. IEEE (2018)
28. Aafer, Y., Du, W., Yin, H.: DroidAPIMiner: mining API-level features for robust malware detection in android, pp. 86–103. Springer, Cham (2013)
29. Bhat, P., Dutta, K.: CogramDroid—an approach towards malware detection in Android using opcode n-grams. Concurrency Comput. Pract. Experience, e6332 (2021)
30. Bernardi, M.L., Cimitile, M., Martinelli, F., Mercaldo, F.: A fuzzy-based process mining approach for dynamic malware detection. In: FUZZ-IEEE, pp. 1–8 (2017)

31. Jeon, J., Park, J.H., Jeong, Y.S.: Dynamic analysis for IoT malware detection with convolution neural network model. IEEE Access **8**, 96899–96911 (2020)
32. Liang, G., Pang, J., Dai, C.: A behavior-based malware variant classification technique. Int. J. Inf. Educ. Technol. **6**(4), 291 (2016)
33. Alzaylaee, M.K., Yerima, S.Y., Sezer, S.: DynaLog: an automated dynamic analysis framework for characterizing android applications. In: Cyber Security, pp. 1–8. IEEE (2016)
34. Cai, H., Meng, N., Ryder, B., Yao, D.: Droidcat: effective android malware detection and categorization via app-level profiling. IEEE **14**(6), 1455–1470 (2018)
35. Xu, J., Li, Y., Deng, R., Xu, K.: SDAC: a slow-aging solution for Android malware detection using semantic distance based API clustering. IEEE Trans. Dependable Secure Comput. (2020)
36. Arp, D., Spreitzenbarth, M., Hubner, M., Gascon, H., Rieck, K., Siemens, C.E.R.T.: DREBIN: effective and explainable detection of android malware in your pocket. NDSS, 23–26 (2014)
37. Yan, L.K., Yin, H.: DroidScope: seamlessly reconstructing the {OS} and Dalvik semantic views for dynamic android malware analysis. In: USENIX Security, pp. 569–584 (2012)
38. Zhou, Y., Wang, Z., Zhou, W., Jiang, X.: Hey, you, get off of my market: detecting malicious apps in official and alternative android markets. NDSS **25**, 50–52 (2012)
39. Shijo, P.V., Salim, A.J.P.C.S.: Integrated static and dynamic analysis for malware detection. Proc. Comput. Sci. **46**, 804–811 (2015)
40. Ma, X., Biao, Q., Yang, W., Jiang, J.: Using multi-features to reduce false positive in malware classification. IEEE, 361–365 (2016)
41. Ferrante, A., Malek, M., Martinelli, F., Mercaldo, F., Milosevic, J.: Extinguishing ransomware-a hybrid approach to android ransomware detection. In: FPS, pp. 242–258. Springer (2017)
42. Liu, Y., Zhang, Y., Li, H., Chen, X.: A hybrid malware detecting scheme for mobile Android applications. In: ICCE, pp. 155–156. IEEE (2016)
43. Arshad, S., Shah, M.A., Wahid, A., Mehmood, A., Song, H., Yu, H.: SAMADroid: a novel 3-level hybrid malware detection model for android operating system. IEEE Access **6**, 4321–4339 (2018)
44. Yang, W., Aghasian, E., Garg, S., Herbert, D., Disiuta, L., Kang, B.: A survey on blockchain-based internet service architecture: requirements, challenges, trends, and future. IEEE Access **7**, 75845–75872 (2019)
45. Gu, J., Sun, B., Du, X., Wang, J., Zhuang, Y., Wang, Z.: Consortium blockchain-based malware detection in mobile devices. IEEE Access **6**, 12118–12128 (2018)
46. Waheed, N., He, X., Ikram, M., Usman, M., Hashmi, S.S., Usman, M.: Security and privacy in IoT using machine learning and blockchain: threats and countermeasures. ACM Comput. Surv. (CSUR) **53**(6), 1–37 (2020)
47. Sengupta, J., Ruj, S., Bit, S.D.: A comprehensive survey on attacks, security issues and blockchain solutions for IoT and IIoT. J. Network Comput. Appl. **149**, 102481 (2020)
48. Hu, J.W., Yeh, L.Y., Liao, S.W., Yang, C.S.: Autonomous and malware-proof blockchain-based firmware update platform with efficient batch verification for Internet of Things devices. Comput. Secur. **86**, 238–252 (2019)
49. Li, Q., Mi, J., Li, W., Wang, J., Cheng, M.: CNN-based malware variants detection method for Internet of Things. IEEE Internet Things J. (2021)
50. Zaza, A.M., Kharroub, S.K., Abualsaud, K.: Lightweight IoT malware detection solution using CNN classification. In: 5GWF, pp. 212–217. IEEE (2020)
51. Ren, Z., Wu, H., Ning, Q., Hussain, I., Chen, B.: End-to-end malware detection for android IoT devices using deep learning. Ad Hoc Netw. **101**, 102098 (2020)
52. Vu, L.N., Jung, S.: AdMat: a CNN-on-matrix approach to android malware detection and classification. IEEE Access **9**, 39680–39694 (2021)
53. Nguyen, H.T., Ngo, Q.D., Le, V.H.: IoT botnet detection approach based on PSI graph and DGCNN classifier. In: ICICSP, pp. 118–122. IEEE (2018)
54. Kumar, R., Zhang, X., Wang, W., Khan, R.U., Kumar, J., Sharif, A.: A multimodal malware detection technique for Android IoT devices using various features. IEEE access **7**, 64411–64430 (2019)

# Enhancing Accuracy of Symptom-Based Disease Prediction Using Ensemble Techniques and Feature Selection

**Abhijeet Chavan, Atharva Dixit, Gaurav Mandke, and Vaibhav Khatavkar**

**Abstract** Machine Learning has, over the years, played a crucial role in shaping human lifestyle and simplifying innumerous tasks. Biomedical and healthcare domains consist of a huge volume of data, which is present in a relatively unstructured form. The current disease prediction system displays the probable diseases upon entering the symptoms, to the user. However, the disadvantage with this approach is, the user can only enter his symptoms first, forcing the system to take just those into consideration, without actually asking any follow-up questions, and analysis of the input data is based just on a supervised learning algorithm. The system also fails to take into account the symptom severity. This paper proposes to build up on such a system by enhancing its accuracy using ensemble learning and/or feature selection on supervised learning algorithms.

**Keywords** Machine learning · Supervised learning · Prediction · Accuracy · Ensemble learning · Feature selection

## 1 Introduction

Searching for a possible infection based on the symptoms directly on the web is to be best avoided. The reason being, online sources vary widely in terms of credibility and information. The proposed system, with respect to application point of view, will create a generalized application with a wide disease database, which takes in the initial user input symptom(s), asks follow-up questions which shows other possible co-occurring symptoms with the ones the user initially enters. Once the user feels all the relevant data has been entered, the system shows the top five possible diseases the user could be suffering from, in decreasing order of their probabilities. This paper aims focuses on research point of view of this application. In order to arrive at the maximum possible accuracy for a particular supervised learning model, the project aims to apply ensemble learning methods as well as feature selection methods.

A. Chavan · A. Dixit · G. Mandke (✉) · V. Khatavkar
College of Engineering, Pune (COEP), Pune, India
e-mail: gauravsmm007@gmail.com



D. Gupta et al. (eds.), *Pattern Recognition and Data Analysis with Applications*, Lecture Notes in Electrical Engineering 888,
https://doi.org/10.1007/978-981-19-1520-8_54

Ensemble learning makes use of a diverse set of supervised learning models, instead of a single model, in order to improve the net system performance for prediction purposes. For instance, in the case of a decision trees, rather than relying on a single decision tree and hoping the system chooses the correct path at a given split from a node to its children, it is much more efficient to build a final predictor that has the ability to calculate the features to be used by amalgamating multiple decision trees. Random Forest Classifier is the ensemble learning technique for executing mentioned system, wherein each tree will be split on different features and such trees are finally averaged to generate the final model. Feature Selection process is used to reduce the number of input variables/features that are necessary to build a predictive model. Certain features that are irrelevant in the system, lowering the model efficiency, adding to the computational cost must be removed. Systematic experimentation is needed to come up with the best suited supervised feature set.

## 2 Literature Review

The authors in [1], use datasets from Scopus and Pubmed containing 48 articles, which were used for more than 1 type of supervised ML algorithms for disease prediction. Random Forest displayed highest accuracy of 53%. In [2], the benchmark dataset and SEER dataset are utilized for analysis to conclude that data mining procedures in all the medical services applications give an accuracy of 97.77% for cancer detection (malignant) and 70% for assessing the achievement pace of IVF therapy. Sharmila et al. [3], using the UCI ML repository show that Fuzzy Neural Network gives the highest accuracy score (over Decision Tree, Fuzzy Logic) to help the diagnosis of liver disease. Singh and Kumar [4] use heart disease symptoms dataset from UCI to describe the techniques used for predicting the risk factor of heart disease. KNN, DT, SVM and LR are trained and tested for accuracy. Caponetto [5], uses MNIST dataset to describe hyperparameter optimization used for enhancing the accuracy of the heart disease prediction. Random search was time-efficient and found better models. Lowd and Domingos [6] uses 47 datasets from the UCI repository to conclude that Naive Bayes models show high accuracy compared to other Bayesian networks. Gao et al. [7], apply bagging and boosting to classify heart disease along with KNN, DT, NB, and RF. Utilization of the heart disease dataset for training and subsequent evaluation of the models is carried out. It is concluded, with the aid of multiple Feature Extraction algorithms, that bagging along with Decision Tree and PCA show high accuracy. In [8], Priya et al. present a hybrid classifier model that uses logarithmic regression with an accuracy of 95% in predicting diseases. Modified Artificial Plant Optimization (MAPO) is used for optimal feature selector, along with fingertip video dataset. Proposed model using KNN and CNN can predict with the accuracy of 95%. Jackins et al. [9], have used correlation coefficient and confusion matrix to predict disease classification, and conclude that the random forest model was best suited for training datasets, as well as real-time data. Classification results show an improved performance over the existing results. Latha and Jeeva [10], make use of the Cleveland heart dataset from UCI, and test ensembling methods like boosting, bagging, stacking and majority voting on the test dataset, to conclude that majority voting provides the best jump in accuracy, whereas bagging and boosting work well on weak classifiers.

In this paper, the methodology of comparing supervised ML algorithms for disease prediction mentioned in [1] is used as initial base work and this is further extended to comparing models for the same using ensemble techniques based on [10]. However, this paper discusses to predict different types of diseases based on variety of symptoms as features unlike that in [10] where the prediction is limited to heart disease and corresponding signs and symptoms.

## 3 Proposed Methodology

The paper starts by initially comparing the accuracy of certain supervised machine learning algorithms used for analysing the dataset—these include Decision Trees, K-Nearest Neighbours, Logistic Regression, Support Vector Machines, Multinomial Naive Bayes, and Multilayer Perceptron Classifier. Then, the cross-validation score is calculated to conclude an efficient model. The system prompts the user to enter the symptoms based on which model predicts disease with the highest probability and scores.

Initially, User Input Processing is undertaken, during which the stop words from the query list are removed, tokenization is done and subsequent lemmatization of the tokens is taken as the input for the Query Expansion stage. Here, each list element is appended with its synonyms, which are obtained using the Wordnet dictionary in Python and the thesaurus.com website.

In the Symptom Selection stage, the related symptoms in the dataset are explored using the expanded symptom query. Each symptom in a dataset is divided into tokens, and each token is tested for its existence in an expanded query to find similar symptoms. A similarity score is determined based on this, and if the symptom's score exceeds the threshold rating, it counts as being identical to the user's symptom and is recommended to the user. Based on the particular user input, top co-occurring symptoms are shown till the user wants to stop.

Different machine learning models that accept the user input symptoms in vector form are then able to display a list of the top probable diseases, in decreasing order of their probabilities. Among these models, on applying ensemble methods, feature selection can give a cross-validation score as an ideal parameter to identify the most efficient approach for the problem statement (Fig. 1).

# 4 Materials and Methods

## *4.1 Dataset Description*

To start with, several datasets pertaining to the disease-symptom domain were gathered from Kaggle for the analysis phase. One such raw dataset contained 132 symptoms as columns (features) and 4920 rows (binary format of the combination of diseases and their different symptoms). This, however, generated vague predictions for even simple systems, fueling the need to obtain a dataset which could generate a satisfactory result upon training.

The dataset for use in this experiment was formed by scraping data from the National Health Portal (NHP) of India, using BeautifulSoup library, which fetches HTML code to filter out the HTML tags, in order to label each disease. For the corresponding symptoms, Wikipedia was scraped using the respective disease as input, by fetching the required page HTML code. All this data is further stored in a CSV file after preprocessing is done. This leads to a dictionary creation, with diseases being the keys and symptoms being the corresponding values.

**Fig. 1** Methodology

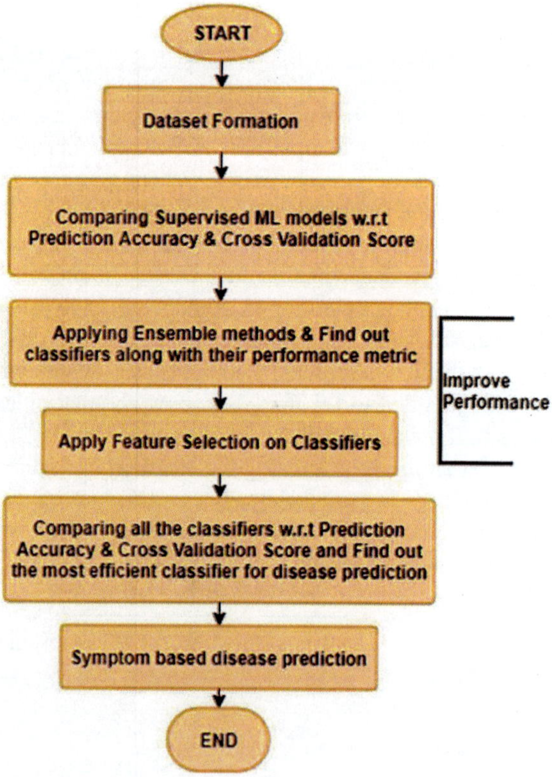

### 4.2 Classification Algorithms

Classification is a supervised learning approach, in which a model is trained upon by a certain set of data, and its predictive outcome efficiency is evaluated by testing it upon another dataset. The dataset to be used is divided into a training dataset, which trains the individual classifiers, and a testing dataset for target prediction. This paper aims to propose a method to diagnose several diseases using several classifiers, each of whose working is as follows:

1. **Multinomial Naive Bayes (MNB)**: It works better on discrete features and the multinomial distribution usually requires integer counts. Features are pertaining to symptoms and we need to maintain the count of distinct symptoms (which are not correlated).
2. **Decision Tree (DT)**: This is a supervised learning algorithm where a certain parameter is used to continuously split the input data at the decision nodes, in order to predict the class of a given categorical variable, based on the leaf nodes of the tree that are assigned a class label.
3. **Random Forest Classifier (RF)**: This is a supervised learning algorithm that combines multiple decision trees based on any given random sample within the training dataset, to generate a "forest". Using several such forests, the desired final result is obtained upon using Majority Voting. It is capable of handling missing values. However, it's major drawback is overfitting, which can be handled using parameter tuning appropriately.
4. **Multilayer Perceptron (MLP)**: This algorithm uses backpropagation as a supervised learning technique, and belongs to the class of Feedforward artificial Neural Network (ANN). Due to this, it has a single input layer and a single output layer, with multiple hidden layers in between them, where the training can be carried out by adjusting the weights and biases relative, but not limited to, Root Mean Squared Error (RMSE).
5. **Logistic Regression (LR)**: This model is used often in tasks that require binary classification. The classifier is derived from the sigmoid/logistic function.
6. **K-Nearest Neighbors (KNN)**: This supervised learning algorithm focuses on the classification tasks by using majority vote to its neighbours. The underlying principle behind a new data point classification is similarity, dependent on the number '$k$', whose optimum value can be found by running the algorithm multiple times, till a peak in accuracy (the least number of errors) is obtained.
7. **Support Vector Machines (SVM)**: This supervised learning algorithm takes in the data points from the training dataset, and generates a hyperplane in an $N$-dimensional space (where $N$ = number of features). Newly entered data points can then be classified into their required plane. The goal is to maximize as much as possible, the distance between the hyperplane and the data points, and thus a hinge loss function is used.

### 4.3 *Ensemble Techniques*

1. **Hard Voting**: This ensemble learning method, enables the combination of several predictions of the base learners, where each base learner is a voter, and each class is a contender. In order to choose a winner, it takes votes into account.
2. **Stacking**: This uses a meta-learning algorithm to combine the predictions from more than one base machine learning algorithms. It allows the integration of several high-performance models to make better predictions compared to a single model.
3. **Bagging**: Bootstrap-Aggregation, is an ensemble technique in which several independent models are fit together in order to obtain a cumulative ensemble model with a lower variance. This is done by building multiple bootstrap samples, each of which acts as a almost separate independent dataset drawn from the initial dataset. Then, for each of these samples, a weak classifier is fitted and combined to average their outputs.
4. **Soft Voting**: Voting Scheme in which the predicted probabilities for class labels are added together and the class label with the highest total probability is predicted. Once the average of all these weighted probabilities is taken, the target label with the highest value wins the vote.
5. **Boosting**: This generates strong learners from weak learners. It is an ensemble meta-algorithm for minimising bias and variance in supervised learning.

## 5 Experimentation and Results

Classifiers like MNB, DT w.r.t. cross-validation score, are weaker as compared to other classifiers. Since ensemble is a known technique for improving classification accuracy, the meta classification algorithms are used to assess the weak learners. Ensembling is done using 5 different techniques: bagging, boosting, soft voting, hard voting and stacking, with the outcomes analysed. The classification models' efficiency is evaluated using fivefold cross validation. When the dataset is classified using individual classifiers, the results obtained are as shown in Tables 1 and 2.

### 5.1 *Ensemble Techniques*

1. **Bagging**: Accuracy rates of DT, LR, MLP, KNN and MNB lie in the range of 83.94–91.29%, whereas the cross-validation accuracy ranges from 83.60 to 89.19%. The DT classifier exhibits best accuracy of 91.29%, whereas the cross-validation accuracy of LR is the highest (89.19%). Bagging has been seen to increase accuracy as well as cross-validation score by upto 3–4% (Figs. 2 and 3).

**Table 1** Model versus accuracy

| Model | Accuracy (%) |
|---|---|
| DT | 91.29 |
| LR | 90.72 |
| KNN | 91.29 |
| SVM | 90.05 |
| MNB | 83.94 |
| RF | 90.05 |
| MLP | 90.72 |

**Table 2** Model versus cross validation score

| Model | Score (%) |
|---|---|
| DT | 83.60 |
| LR | 89.19 |
| KNN | 87.03 |
| SVM | 88.62 |
| MNB | 84.50 |
| RF | 87.13 |
| MLP | 86.77 |

2. **Boosting**: As a part of Boosting, only DT was used for experimentation. The Adaboost model was useful in order to increase accuracy as well as cross-validation score. XGB classifier was also tested for classification, but there was no increase in the accuracy.
3. **Majority Voting**: Majority voting incorporated several classifiers to increase their accuracy. For our dataset, the proposed method revealed that classifiers were weak with low accuracy. The following is the increasing order of models based on average accuracy (normal + cross validation): MNB, DT, MLP, RF, KNN, SVM and LR.
   We can consider 4 strong models to form an ensemble with the remaining 3 weak models each. Note: Random Forests model is an ensemble of Decision Tree. However, it can be included in the combination to increase accuracy.

   Ensemble 1: LR + SVM + KNN + MNB
   Ensemble 2: LR + SVM + KNN + MLP
   Ensemble 3: LR + SVM + KNN + RF
   Ensemble 4: LR + SVM + KNN + DT

   Following a method similar in [10]: It can be seen from Fig. 4 that accuracy significantly increases when majority voting is used for an ensemble of strong classifiers and weak classifiers (MLP, RF, DT, MNB). Ensembling the latter with the strong classifiers: LR, SVM, KNN improves accuracy by 7.92%, and cross-validation score by 4.23%. Using the same strong classifier set to ensemble MLP improves the accuracy by 0.23%, and cross-validation score by 1.9%. Ensembling Random Forest using the same methodology improves the accuracy by 2.14%, and cross-validation score by 1.45%. In case of ensembling Decision Tree, the accuracy improves by 0.68%, and cross-validation score by 4.69%.

4. **Soft Voting**: Here each individual classifier generated a probability value for a certain data point to show if it is a part of a particular target class and one with the highest value won the vote. It can be inferred from Fig. 5 that the accuracy for weak classifiers increases more significantly, compared to hard voting (majority voting). However, the jump in the cross-validation score is smaller as compared to the one in case of majority voting. MNB ensembled with strong classifiers such as Logistic Regression, Support Vector Machine, and K-Nearest Neighbours, increased accuracy by 8.59% and the cross-validation score by 3.32%. Using the same strong classifier collection to ensemble Multilayer Perceptron increased accuracy by 2.27% and the cross-validation score by 1.04%. Random Forest ensembled with the same set increased accuracy by 2.37% and the cross-validation score by 0.97%. Ensembling Decision Tree with the set of strong classifiers improves the accuracy by 1.02%, and cross-validation score by 2.66%.
5. **Stacking**: Here there are 2 layer estimators: 1st baseline model layer for predicting the output of the testing dataset, and the 2nd meta-classifier layer generates further predictions based on the first layer output. In this paper, LR is used as a meta classifier. KNN, SVM, RF, MLP and DT combinations of these classifiers are used as base learners. Stacking base learners KNN, SVM and RF with LR generated an accuracy of 89.82%, and cross-validation score of 84.32%. Stacking base learners KNN, SVM, RF, MLP and DT with LR generated an accuracy of 90.27%, and cross-validation score of 87.09%. Stacking base learners KNN, SVM and MLP with LR generated an accuracy of 90.27%, and cross-validation score of 86.29%. Exception: whatever used in meta classifier can be used in base learner also, to observe the enhancement in accuracy. Using KNN, SVM, RF, MLP, LR as base learners, maintaining LR as a meta classifier as well, gave an accuracy of 90.38%, and cross-validation score of 86.89% (Fig. 6).

## 5.2 *Feature Selection*

Feature selection involves selecting the best features for improving the performance in terms of accuracy and time complexity. Performing preliminary feature selection methods of dropping constant features using Variance Threshold method, removes features having low variance. Following the primary approach for removing features setting the threshold value to zero, the training set was fit into the same. This resulted in three columns which had zero variance, leading to them being dropped. How-

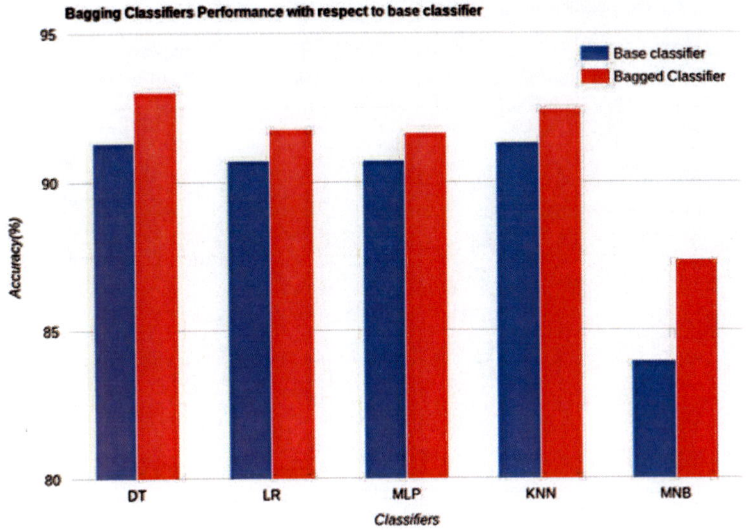


**Fig. 2** Bagging classifier performance w.r.t base classifier

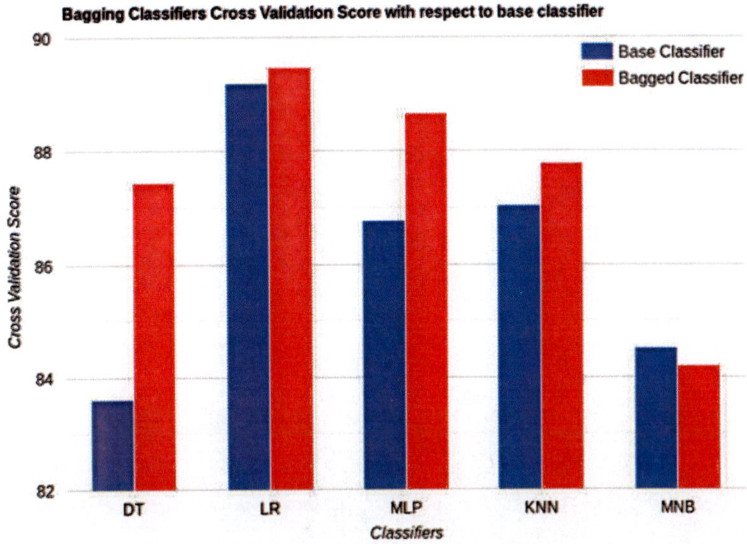


**Fig. 3** Bagging classifier performance w.r.t base classifier cross validation

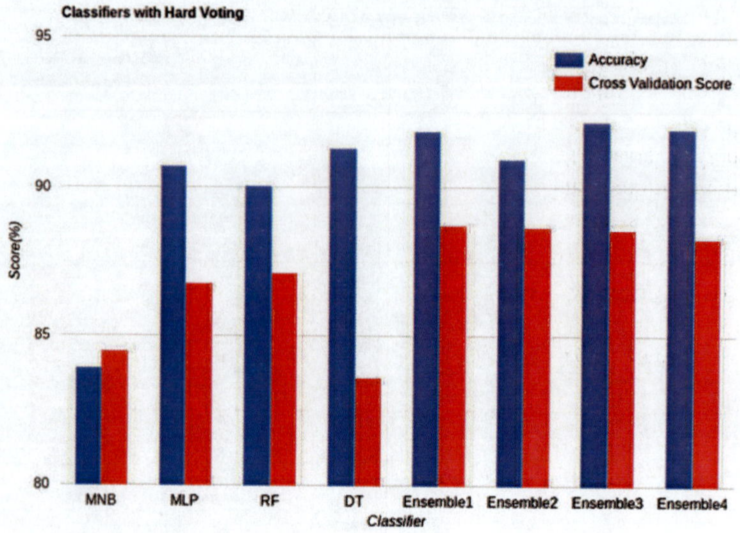


**Fig. 4** Classifier with hard voting

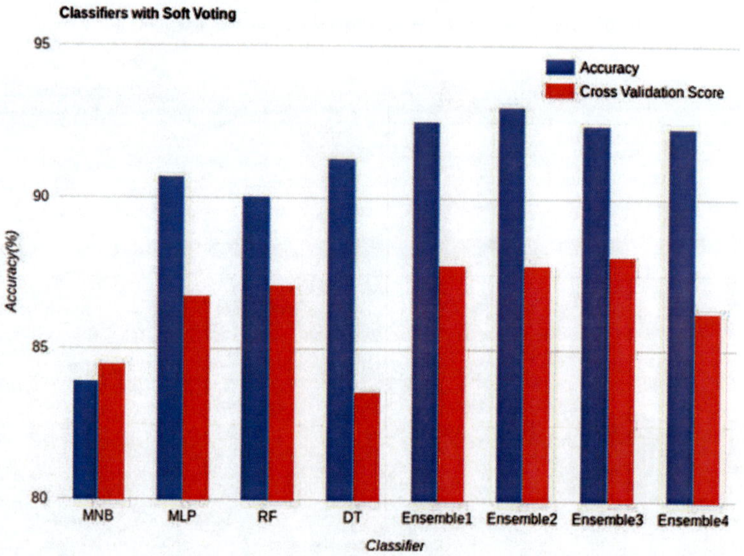


**Fig. 5** Classifier with soft voting

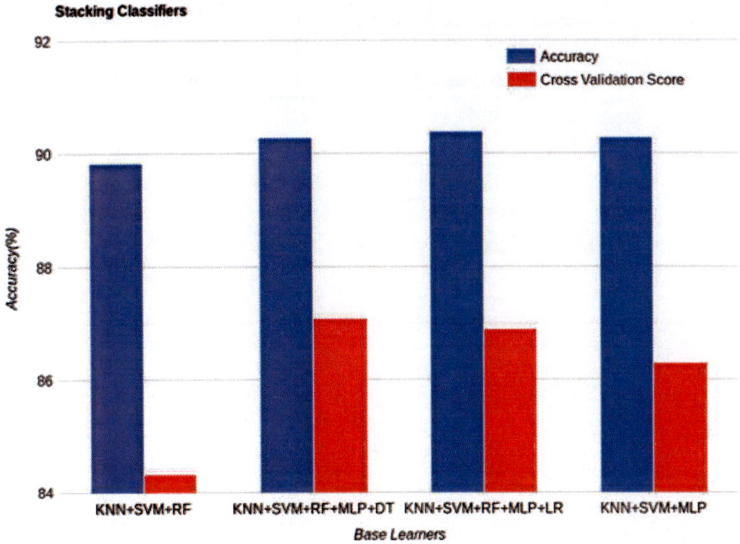


**Fig. 6** Stacking

ever, variance threshold using Pearson Correlation, we generated a heatmap using correlation matrix, from which it could be inferred that there were zero correlation features. This led to the combination of this feature selection method with the best performing ensemble classifier, in accordance with the above parameters. The highest increase of 0.18% in cross-validation score was observed in the case of soft voting ensemble of LR + SVM + KNN + MLP, when feature selection was applied to all the ensemble classifiers. However, the highest cross validation score was obtained for bagged-logistic regression when the obtained feature set was used which was 89.50%. The results of ensemble classifiers combined with feature set obtained are mentioned in Tables 3, 4 and 5 respectively.

## 6 Conclusion

The dataset plays an extremely important role in training the model. If we have a number of diseases associated with a particular set of symptoms, it helps keep the system domain sufficiently wide. But at the same time, it should not be skewed or sparse in nature. It can be observed that Hard Voting works well with the weak classifier set, as they show the highest '%' increase w.r.t accuracy as well as cross-validation score. But, the highest cross-validation score is seen with Bagged LR model, making it the most viable classifier. The Feature Set enhances accuracy further.

**Table 3** Comparison of cross-validation score (%) for bagged classifiers before and after feature selection

| Bagged model | Before FS (%) | After FS (%) |
|---|---|---|
| LR | 89.47 | 89.50 |
| DT | 87.44 | 87.52 |
| MLP | 88.66 | 88.75 |
| KNN | 87.78 | 87.65 |
| MNB | 84.19 | 84.25 |

**Table 4** Comparison of cross-validation score (%) for hard voting ensembles before and after feature selection

| Hard voting ensemble | Before FS (%) | After FS (%) |
|---|---|---|
| LR + SVM + KNN + MNB | 88.73 | 88.73 |
| LR + SVM + KNN + MLP | 88.67 | 88.73 |
| LR + SVM + KNN + RF | 88.58 | 88.64 |
| LR + SVM + KNN + DT | 88.29 | 88.25 |

**Table 5** Comparison of cross-validation score (%) for soft voting ensembles before and after feature selection

| Soft voting ensemble | Before FS (%) | After FS (%) |
|---|---|---|
| LR + SVM + KNN + MNB | 87.82 | 87.88 |
| LR + SVM + KNN + MLP | 87.81 | 87.98 |
| LR + SVM + KNN + RF | 88.10 | 88.15 |
| LR + SVM + KNN + DT | 86.26 | 86.23 |

# References


1. Khan, A., Hossain, M., Uddin, S.: Comparing different supervised machine learning algorithms for disease prediction. BMC Med. Inform. Decis. Mak
2. Ranjani, V., Durairaj, M.: Data mining applications in healthcare sector: a study. Int. J. Sci. Technol. Res. **2**, 29–35 (2013)
3. Sharmila, S.L., Dharuman, C., Venkatesan, P.: Disease classification using machine learning algorithms—a comparative study. Int. J. Pure Appl. Math. **114**, 1–10 (2017)
4. Singh, A., Kumar, R.: Heart disease prediction using machine learning algorithms. In: 2020 International Conference on Electrical and Electronics Engineering (ICE3), pp. 452–457 (2020)
5. Caponetto, G.: Random search vs grid search for hyperparameter optimization (2019)
6. Lowd, D., Domingos, P.: Naive Bayes models for probability estimation, pp. 529–536 (2005)

7. Gao, X.-Y., et al.: Improving the accuracy for analyzing heart diseases prediction based on the ensemble method (2021)
8. Priya, L., et al.: A novel intelligent diagnosis and disease prediction algorithm in green cloud using machine learning approach. J. Green Eng. **10**, 3421–3433 (2020)
9. Jackins, V., et al.: AI-based smart prediction of clinical disease using random forest classifier and Naive Bayes. J. Super Comput
10. Latha, C.B.C., Jeeva, S.C.: Improving the accuracy of prediction of heart disease risk based on ensemble classification techniques

# Pruning for Compression of Visual Pattern Recognition Networks: A Survey from Deep Neural Networks Perspective

**Seema A. Bhalgaonkar, Mousami V. Munot, and Alwin D. Anuse**

**Abstract** Visual Pattern Recognition Networks (VPRN) delivers high performance using deep neural networks. With the advancements in deep neural networks VPR network has gained wide popularity. Continuous advancements will be nurtured with the availability of big data and enormous computing powers. However, such DNN based VPRN models are plunged with computational complexities, intense memory requirements, huge energy expenses which impedes its deployment in resource constrained, strict latency required environments such as edgeAI. For instance, the VGG-16 model needs 500 MB of storage space, has 138 million parameters and involves 15.5 billion Floating Point Operations (FLOPs) to classify a single image with a 32 bit floating point addition that consumes 0.9 pJ. Such overheads demand for compression of VPRN models without impairing its performance. Various compression methods are reported in literature. This survey paper presents a survey on pruning, a popular compression technique for DNN as applied for VPRN. This paper presents comprehensive survey, comparison and points further research direction.

**Keywords** Visual pattern recognition · Deep neural networks · Compression techniques · Pruning

## 1 Introduction

A Visual Pattern Recognition (VPR) is description or recognition of measurements of visual or image data [1]. This VPR techniques has wide applicability in various AI applications such as Computer Vision [2], Edge AI [3], Medical Imaging [4], Industrial Automation [5] to name a few. VPR techniques can be classified as traditional machine learning algorithm based and deep learning based techniques [6].

S. A. Bhalgaonkar (✉) · M. V. Munot
SCTR's Pune Institute of Computer Technology, SPPU, Pune 411046, India
e-mail: seema.bhalgaonkar@gmail.com

A. D. Anuse
School of ECE, Dr. Vishwanath Karad MIT-WPU, Pune 411038, India

 
D. Gupta et al. (eds.), *Pattern Recognition and Data Analysis with Applications*, Lecture Notes in Electrical Engineering 888,
https://doi.org/10.1007/978-981-19-1520-8_55

**Table 1** Neural network architectures and their parameters

| Model | No. of parameters (M) | Model size (MB) | FLOPS |
|---|---|---|---|
| Alexnet (2012) | 60 | 240 | 724 M |
| VGG 16 (2014) | 138 | 500 | 15.5 |
| Resnet152 | 6.4 | 244 | 2 B |
| Inception-V 4 (2016) | 26.6 | 163 | 5.7 G |
| DenseNet (2017) | 28.6 | 230 | 8 B |
| EfficientNet-B5 (2019) | 30 | 118 | 9.9 B |
| HRNetV2-W48 (2020) | 65.9 | NA | 747.3 G |

*NA* Not Available

Currently, Deep learning (DL) is a cutting edge technique with its extraordinary performance supported by availability of high capability computational resources (Graphical Processing Units), advanced network technologies (cloud computing) and availability of Big data [7]. DL excels its counterpart with its self-learning capabilities without being explicitly programmed and avoids need of manual feature extraction [8].

While DL approaches have marked incredible success in all applications they are used; DL architecture leads to several complexities. These include tremendous increase in networks size due to billions of hyper parameters, higher computational complexities involving intensive mathematical operations, higher memory footprints, inference, training time and more power requirement [9] demanding model compression. For instance, the VGG-16 model needs 500 MB of storage space with 138 million parameters and involves 15.5 billion Floating Point Operations (FLOPs) to classify a single image. Energy consumption largely depends upon memory access operations for example a 32 bit floating point addition consumes 0.9 pJ [10].

Table 1 shows the network complexities involved in such hyper parameterized pertained Deep Neural Network (DNN) models with their year of introduction [11]. The necessity for compression of DNN based VPR Networks is to make it sizable and resource efficient while maintaining its performance as shown in Fig. 1.

**Fig. 1** Dense and pruned network

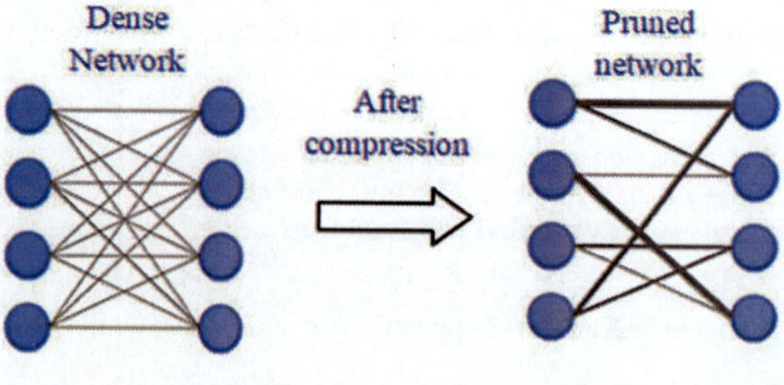

Various compression techniques are explored by the researchers to trade off between performance versus computational complexities and resource requirements of VPRN models [12]. Numerous techniques are proposed and published in last few years and it is essential to conduct its comprehensive survey to present overview of available techniques, their challenges and possible future opportunities of research.

This paper presents preliminary review of recent reported research on VPRN compression techniques. The paper is organized in various sections such as Sect. 2: Background; Sect. 3: Pruning Techniques; Sect. 4: Discussion and conclusion.

## 2 Background

Many DNN architectures have been implemented as a result of the growing success of DL in various applications. The architectures like Multi Layer Perceptron (MLP), Convolutional Neural Networks (CNN), Recurrent Neural Networks (RNN), and Generative Adversarial Network (GAN) have been a popular choice among researchers due to their superior performance even in highly complex datasets. Among these, CNN is most suitable for visual recognition application due to its distinctive capability to automatically learn features from given image dataset and its inherent architecture reduces the size of images using convolutional layers and pooling layers [2]. Although DNN based VPRN delivers superior performance with cutting-edge accuracy, it requires extremely deep networks involving millions of parameters and huge computational operations. Various compression techniques are explored by researchers to resolve these issues and have been successful to attain the same. These techniques can be divided into broad categories based on review of existing literature as: Pruning, Quantization, Knowledge Distillation, Low Rank Approximation and Hybrid Techniques.

**Pruning**: Pruning is removal of redundant parameters and connections in network while keeping retaining highly important parameters and aims to reduce computational and storage complexities.

**Quantization**: Quantization is based on representation of weights and activations using less number of bits to reduce number of Multiply-and-Accumulate (MAC) operations required and thus can reduce the size of trained Neural Networks (NN). Several quantization methods on NN have been devised by researchers for various application and have been successful in reducing size and MACs [13–15].

**Knowledge Distillation (KD)**: It is based on teacher-student architecture in which a small student model learns from a large teacher model. Various KD techniques have been proposed by researchers and are used for compression of model used in various complex visual applications in last few years [16–18].

**Low-rank Matrix and Tensor Decomposition**: Low-rank approximation is based on decomposition of high-dimensional weight tensors into the lower rank approximations. The convolution kernels can be thought of as a three-dimensional tensor.

These tensors can be decomposed in lower rank matrix and thus results in dimension reduction [19, 20]. Such technique has been successfully experimented by various researchers [21–23] for various applications and performs well for layer wise compression compared to global compression but is computationally expensive and needs higher model retraining time to achieve same performance as that of the original model.

**Combinational/Hybrid Approaches**: Combinations of above methods to strive for better compression performance is also experimented by few researchers. Han and Dally [10] have used pruning on the network for removal of connections followed by quantization of the weights to enforce weight sharing, finally application of Huffman coding resulted in network compression by 49× and 39× without loss of accuracy hence opens the doors for further research on similar lines.

This paper is focused on review of recent pruning techniques used for VPRN, since it is the most popular technique. Pruning has been successful on various well-known CNNs architectures such as Imagenet, Resnet, InceptionV3 as well as a variety of general image datasets such as CIFAR10, MNIST etc.

## 3 Pruning Techniques

Pruning is a techniques to remove redundant parameters of the network which are least important and does not affect performance of the network when removed. Pruning not only helps in reducing network size but also helps in reducing over-fitting [24]. The following criteria can be used to broadly categorize pruning techniques:

- Selection or removal of redundant parameters
- Type of redundant parameter
- Technique of removal.

A standard pruning procedure includes four stages as shown in Fig. 2. Firstly a CNN is trained with a large-scale dataset to learn all the features of input image, followed by pruning of model based on some criteria, retraining the pruned network to regain accuracy, and lastly fine-tuning the retrained network using a small dataset from the target application. Usually pruning used to achieve one of the objectives such as reducing size of network, MAC operations or memory requirements.

**Training a large network → Pruning → Retraining → Fine Tuning**

**Fig. 2** A typical pruning work flow

## 3.1 Categories of Pruning

### Redundant Parameter Removal Based (Structured and Unstructured)

*Unstructured Pruning*: Unstructured pruning eliminates redundant parameters without following a particular geometry and works particularly on removing individual parameter like weight.

Pruning of the weights eliminates unnecessary connections that take significant portion of the computations required during execution. This is achieved by setting weights to zero and not actual pruning to maintain the architecture consistency [25, 26].

Such experimentation by Shrinivas et al. [26] have used similarity between neuron as basis for its removal. The system removes the similar weights. Suppose $W_1$, $W_2$, ... $\in R_d$ are vectors of weights and $a_1, a_2, \ldots \in R$ are scalar weights in the next layer, $X \in R_d$ is the input, with the bias and $h(\cdot)$ is a non-linearity function. The output is given by

$$z = a_1 h\left(W_1^T X\right) + a_2 h\left(W_2^T X\right) + a_3 h\left(W_3^T X\right) + \cdots + a_n h\left(W_n^T X\right)$$

If weights are equal i.e. $W_1 = W_2$, it means $h(W^T{}_1 X) = h(W^T{}_2 X)$. In such cases $W_2$ by can be replaced by $W_1$, resulting in equation

$$z = (a_1 + a_2) h\left(W_1^T X\right) + 0.h\left(W_2^T X\right) + a_3 h\left(W_3^T X\right) + \cdots + a_n h\left(W_n^T X\right)$$

So by a simply changing co-efficient a1 to $a1 + a2$, $W_2$ by can be replaced by $W_1$. This technique have proved to be successful with little loss in accuracy from 99.06 to 97.99% and resulted in 87.45% compression rate. However, it is applicable only to fully connected layers.

Various other approaches for determining the weight zeroing criterion proposed by researchers are iterative thresholding selection [27], Huffman code [28], regularization [15, 29] which also proved to be successful. While weight selection criterion like thresholding, L norms, and even few novel techniques are successfuly experimented by [30, 31].

These unstructured pruning criterions suffers from possibility of incorrect weight pruning. Moreover it needs to store information for sparse weights and may take longer training times and hence needs further improvement.

*Structured Pruning*: Structured pruning usually follows a geometric structure and works on removing groups of consecutive parameters like entire neurons, filters, or channels [32, 33]. As shown in Fig. 3, structured pruning can be done at various levels of network to prune weights, nodes, channels, kernels, or filters.

Channel pruning removes all incoming and outgoing weights from an input feature map. Such approaches are successfully implemented by [34–36]. He et al. [34] used LASSO regression based redundant channel selection on VGG-16 with 5× speed-up with only slight increase in error. Authors have implemented an iterative two-step

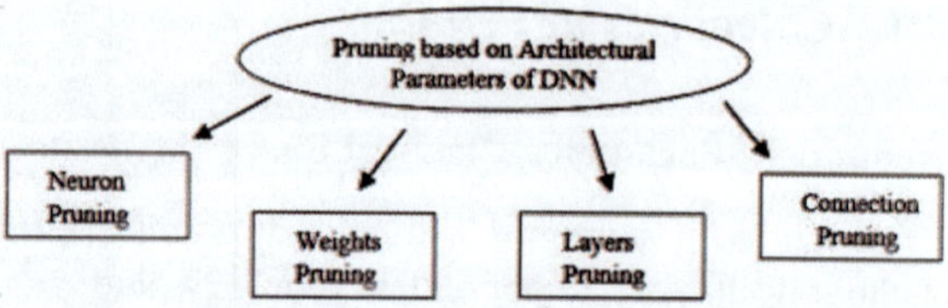


**Fig. 3** Pruning based on architectural parameters

algorithm. First step selects most representative channels and later reconstructs the outputs with remaining channels. As approach works on redundancy with specific formulation for LASSO Regression it has reduced inference-time. Another approach by Liu and Wu [35] is a global pruning based on mean gradient for convolutional kernels reducing FLOPS by 5.64% on VGG16 and CIFAR-10 dataset. The channels having flat gradient of loss function are removed hierarchically. This technique is limited to convolutional layers and can be further explored on other layers.

Experimentation by Han et al. [36] based on Variational Automatic Channel Pruning Algorithm utilizes structure optimization. It uses weights generator to produce weights for pruned CNN. Then optimal pruning structure is obtained by scaling channels with truncated factorized log-uniform prior and log-normal posterior. Although the approach had a success with compression ratio 34.60×, further work can be explored on other sparse prior and posterior channel scaling techniques.

Channel pruning has to be performed critically as removal of channel from one layer highly affects the inputs to next layers.

Kernel level pruning removes the entire kernel [37–39], while intra-kernel pruning removes weights inside a kernel [40–42].

These structured pruning techniques needs less training time than unstructured pruning as it works on removal of group of parameters at a time. However, entire block removal may affect the accuracy and hence the performance of the network. This demands critical balance between performance and accuracy.

The elimination of a node, along with all of its in-going connections, is known as node pruning. It lowers computational costs, and is more suitable to hardware and software optimization [43, 44]. Ben-Guigui et al.[43] removed node based on regularization techniques and when used along with a weight decay regularizer resultedin 50% node removal in an MLP for MNIST dataset. Another node pruning approach devised by Tianxing et al. [44] works on pruning the nodes in hidden layers depending on its importance calculated by the average L1-norm of the all incoming weights. After sorting these nodes, less importance valued nodes are removed. This method reduces complexity by 37.9% but is dependent on the model redundancy. Pruning either simplify CNN architectures or minimizes FLOPS directly [34, 39, 40, 45, 46]. Depending upon the pruning objective and the model's flexibility, level of pruning can be adjusted during the pruning process.

**Parameter Selection Criteria Based**: Pruning of a particular parameter in the network is typically based on its scoring criteria. This has tremendously wide range from simplest absolute values to highly sophisticated importance criterion such as contributions to network activation, or gradients etc.

*Absolute value/thresholding* by [47–49] prunes the parameters below a certain threshold. The network is then fine-tuned, and the process is repeated until the network's accuracy reaches optimum level. Experimentation by Han et al. [24] which is based on thresholding has shown encouraging results on LeNet-300-100Ref and Lenet-5 network with MNIST dataset. It reduces weights by 12× with same accuracy.

*Gradient descent based techniques by* [50, 51] selects the channels, based upon classification and reconstruction losses on intermediate layers. This helps not only in channel selection but also increases discriminative power. By measuring the mean gradient of feature maps in each layer, a hierarchical global pruning strategy was suggested by [35]. This is applied for the layers with similar sensitivity. Proposed method reported encouraging results with 2.48% reduction in FLOPs and parameters with slight decrease in accuracy. However, maximum accuracy is achieved for 70% reduction and hence can be further improved on higher compression ratio.

*Tailer series expansion* by You et al. [38] and Molchanov et al.[52] has been used to estimate the effect of weight perturbations on the loss function. Molchanov compared two variants—first and second order Taylor expansion and found that both variants provide promising results. First-order criterion is substantially faster to compute but is slightly less accurate and hence provides the scope for further developments.

*Regularization based pruning* [43, 53] has been used recently as it offers advantages like reduced over fitting and weight values. Experimentation on MNIST dataset using VGG 16 proved to be better performing than node and weight pruning techniques.

*Regression* by [34, 54, 55] explores sparsity in a network by using novel techniques like structured regularization which works on finding correlations between two consecutive layers for channel pruning. This technique does not affect accuracy greatly and is able to preserve the important features. Experimentation using various regression techniques like LASSO is carried out till now and can be further explored for optimized performance.

*Cosine similarity* by Roy Choudry et al. and Shao et al. [56, 57] details its use to identify similar filters with experimentation on MLP and CNN for the CIFAR-10 dataset. The system performs well in terms of accuracy however; system is highly sensitive to the similarity threshold setting which needs to be carefully selected otherwise resulting into merging of dissimilar filters.

**Iterations Based Pruning**: Typically pruning is iterated till the network learns with precise number of parameters yielding optimum accuracy. Pruning followed by retraining is one iteration; the minimum number of connections could be found after several iterations [24]. Most of the techniques needs number of iterations so as to fine tune the network to the original networks performance and hence suffers from longer training time. This issue of attaining original accuracy with minimum iterations is the crux of pruning. Although attempted by very few researchers [58] it can be explored further. Recently emerging pruning and optimization methods have proved better performing in terms of accuracy as compared to earlier heuristic methods.

## 4 Comparison of Pruning Techniques

Several network compression techniques have been explored in recent years that use pruning strategies and have been successful with outstanding results. These techniques are so diverse and have been applied to address varying parameters of network that it becomes highly difficult to bring them on single base line and compare. Experimentation reported works on varied DNN models, datasets, and aims to improve various independent performance parameters. This makes it extremely difficult to apply one standard benchmark or matrix to uniformly evaluate all pruning techniques.

Moreover, majority of experimentation conducted and results presented till date lack in clear analysis by virtue of commonality among all experimentation. A study by Davis Blalock et al. [59] shows that comparison of pruning techniques is largely confined due to various parameters such as lack of uniformity in architecture-dataset combinations, absence of comparison of result with previous or other methods, and use of inconsistent metric parameters. The various performance parameters which can be application specific is the way of comparing pruning techniques.

Pruning as whole can be used to improve one of the disadvantages associated with DNN such as number of parameters, FLOPS, size, training time, memory, energy consumption etc. Reported literature results shows that pruning is simple, effective and is quite successful in compressing the network without any accuracy loss.

The unstructured pruning techniques like neuron pruning is aims at removing redundant neurons while structured method works on removing groups of elements like entire layer or channel which directly reduce the feature map width and reduces model size. This method is effective, but it involves difficulty because removing channels can drastically alter the input of the next layer. Pruning a neuron results in more accuracy loss compared to a weight pruning.

Some advanced pruning techniques which are recently proposed by Mingwen et al. [56] have shown remarkable results. Mingwen have explored dynamic pruning depending upon the situation of each layer. Their technique has three steps, firstly use of cosine similarity to find similarity between the filters or feature maps of the same layer. Secondly prune filters and feature maps having higher similarity, and setting up dynamic pruning rate depending upon the sensitivity set. This sensitivity rate depends on the mean value and distance between the channel and the measuring centre, redundancy being inversely proportional to this distance. Experimentation results shows 52.70% compression ratio on CIFAR-10 which is quite promising.

Table 2 shows a comparison of various parameter pruning techniques, compression rate and accuracy offered after pruning and Table 3 shows the comparison of methods based on parameter selection. Most of the research work is carried out aiming in trading off between these performance parameters.

**Table 2** Comparison of parameter removal based techniques

| Pruning element | Model | Dataset | Base accuracy | Accuracy (%) | Compression (%) |
|---|---|---|---|---|---|
| Neuron Pruning [25] | LeNet-like | MNIST | – | 94.18 | 93.47 |
| Kernel [36] | VGG16 | CIFAR10 | 93.49 | 93.12 | 98.33 |
| Kernel [34] | ResNet | CIFAR-10 | 93.1 | 92.72 | 60 |
| Intra Kernel [37] | VGG | CIFAR-10 | 79.04 | 74.04 | 30 |
| Intra Kernel [37] | ResNet | CIFAR-10 | 93.1 | 92.81 | 50 |
| Node [39] | VGG | CIFAR-10 | 79.04 | 72.35 | 60 |

**Table 3** Comparison of parameter selection criteria based techniques

| Pruning criteria | Model | Dataset | Base accuracy | Accuracy (%) | Compression (%) |
|---|---|---|---|---|---|
| Thresholding [44] | MobileNet V2 | ImageNet | 79.8 | 79.1 | 26.4 |
| Gradient Descend [27] | VGG16 | CIFAR100 | 88 | 77 | 80 |
| Tailor Series Expansion [34] | ResNet | CIFAR10 | 93.1 | 92.8 | 54 |
| Regularization [49] | AlexNet | CIFAR10 | 93.54 | 90.91 | 2x |
| Regression [32] | ResNet 56 | CIFAR10 | 93 | 92.8 | 2x |
| Cosine Similarity [52] | ResNet 56 | CIFAR10 | 93 | 92.87 | 52.70 |

## 5 Discussion and Conclusion

With the current trend of DNN increasing exponentially, keeping track of new developments in the field of network compression has become utmost important. Performance of compression techniques can be evaluated by various evaluation parameters such as accuracy, model size, and number of FLOPS, memory footprint, inference latency, and training time. Majority of the compression techniques are applicable on VPRN using CNN as they are used extensively in image based tasks since last few decades.

Pruning for compression of such networks has been a popular choice among researchers since its introduction in early 90s as it addresses both issues size reduction and over fitting of a model. Recently, plenty of pruning techniques have emerged in an attempt to strike a balance between various evaluation matrices of the network.

- The continuous uprising of incredibly huge amount of related works and divergent reported approaches indicates that pruning techniques have not reached to its optimum performance and still is in progress and hence leaves the scope for future developments.

- Most of the reported pruning techniques explored by research community, claims outstanding performance. However, it is extremely challenging to conform its superiority over other as reported techniques works on differing methodologies, network architectures, datasets, and uses different performance matrix. Hence standard datasets, networks, metrics, and experimental practices are needed to facilitate additional experimentation and comparison.
- Structured pruning techniques outperform with their counterpart with superior performance and lower storage overheads, however these techniques are sensitive and prone to lose important features of input. Hence, structured techniques that maintain accuracy and remain robust with feature retention needs further experimentation.
- On trained models, pruning can be used to produce amazing results. However, they can easily lead to a sparse model, which is not always hardware efficient, and hence needs a further experimentation on techniques that strike a balance between size reduction and computational efficiency.
- A few approaches with possible combination of pruning with other compression techniques have been reported in literature. From a practical perspective, a thorough experimentation of such hybrid approach on a variety of architectures with varying datasets is likely to be extremely useful.
- Pruning suffers from longer training time due the number of iterations that are needed to reach to its optimum performance level. This necessitates the exploration of novel retraining methods as well as setting iteration count so as to reduce overall system development time to sustain in the competitive technology.

Our further research aims to explore these pruning techniques for VPRN. Although our study focused on pruning techniques, may also be applied to other compression techniques.

**Acknowledgements** Authors are immensely thankful to Dr. Y. Ravinder, and Dr. Surajkumar Saw for their valuable inputs and continuous support for this research. Authors extend their heartfelt thanks to Dr. Vibha Vyas for her motivation and guidance towards completion of this paper.

## References

1. Prandi, D., Gauthier, J.P.: Pattern recognition. In: SpringerBriefs Math, pp. 53–76 (2018). https://doi.org/10.1007/978-3-319-78482-3_5
2. Voulodimos, A., Doulamis, N., Doulamis, A., Protopapadakis, E.: Deep learning for computer vision: a brief review. Comput. Intell. Neurosci. (2018)
3. Lin, J.C., Lin, W., Cohn, Y., Gan, J., Han, C.S.: MCUNet: Tiny Deep Learning on IoT Devices, pp. 1–15 (2020). arXiv, no. NeurIPS
4. Lim, G., Hsu, W., Li Lee, M., Ting, D.S.W., Wong, T.Y.: Technical and Clinical Challenges of A.I. in Retinal Image Analysis. Elsevier Ltd. (2019)

5. Qin, X.Y., Wang, Z., Gang, M., Jun, D., Pai, G., Jun, W., Hongguang, P. et al.: A cable fault recognition method based on a deep belief network. Comput. Electr. Eng. **71**(July), 452–464 (2018)
6. Bye, S.J., Adams, A.: Neural network paradigm for visual pattern recognition. IEEE Conf. Publ. **372**, 11–15 (1993)
7. Xie, J.R., Huang, F., Xie, T., Liu, R., Wang, J., Liu, C.: A survey of machine learning techniques applied to software defined networking (SDN): research issues and challenges. IEEE Commun. Surv. Tutorials **21**(1), 393–430 (2019)
8. Ghods, A., Cook, D.: A survey of deep network techniques all classifiers can adopt. Data Min. Knowl. Discov. **35**(1), 46–87 (2021)
9. Dargan, S., Kumar, M., Ayyagari, M.R., Kumar, G.: A survey of deep learning and its applications: a new paradigm to machine learning. Arch. Comput. Methods Eng. **27**(4), 1071–1092 (2020)
10. Han, S.M., Dally, H.: Deep compression: compressing deep neural networks with pruning, trained quantization and Huffman coding. In: 4th International Conference on Learning Representations, ICLR 2016—Conference Track Proceedings, pp. 1–14 (2016)
11. Canziani, A., Paszke, A., Culurciello, E.: An Analysis of Deep Neural Network Models for Practical Applications, pp. 1–7 (2016). http://arxiv.org/abs/1605.07678
12. Neill, J.: An Overview of Neural Network Compression, pp. 1–73 (2020)
13. Kwasniewska, A.S., Ozga, M., Wolfe, M., Das, J., Zajac, A., Ruminski, A., Rad, A.: Deep learning optimization for edge devices: analysis of training quantization parameters. In: IECON Proceedings (Industrial Electronics Conference), pp. 96–101 (2019)
14. Park, E.A., Yoo, J.: Weighted-entropy-based quantization for deep neural networks. In: Proceedings, 30th IEEE Conference on Computer Vision and Pattern Recognition, CVPR 2017, vol. 2017, pp. 7197–7205 (2017)
15. Deng, B.L., Li, G., Han, S., Shi, L., Xie, Y.: Model compression and hardware acceleration for neural networks: a comprehensive survey. Proc. IEEE **108**(4), 485–532 (2020)
16. Alkhulaifi, A., Alsahli, F., Ahmad, I.: Knowledge Distillation in Deep Learning and Its Applications, arXiv (2020)
17. Ho, T.K.K., Gwak, J.: Utilizing knowledge distillation in deep learning for classification of chest X-ray abnormalities. IEEE Access **8**, 160749–160761 (2020)
18. Ni, H., Shen, J., Yuan, C.: Enhanced knowledge distillation for face recognition. In: Proceedings of 2019 IEEE International Conference on Parallel & Distributed Processing with Applications, Big Data & Cloud Computing, Sustainable Computing & Communications, Social Computing & Networking ISPA/BDCloud/SustainCom/SocialCom 2019, pp. 1441–1444 (2019)
19. Swaminathan, S., Garg, D., Kannan, R., Andres, F.: Sparse low rank factorization for deep neural network compression. Neurocomputing **398**, 185–196 (2020)
20. Yu, X., Liu, T., Wang, X., Tao, D.: On compressing deep models by low rank and sparse decomposition. In: Proceedings, 30th IEEE Conference on Computer Vision and Pattern Recognition, CVPR 2017, vol. 2017, pp. 67–76 (2017)
21. Sainath, T., Kingsbury, B., Sindhwani, V., Arisoy, E., Ramabhadran, B.: Low-rank matrix factorization for Deep Neural Network training with high-dimensional output targets. In: ICASSP, IEEE International Conference on Acoustics, Speech and Signal Processing—Proceedings, pp 6655–6659 (2013)
22. Tai, C., Xiao, T., Zhang, Y., Wang, X., Weinan, E.: Convolutional neural networks with low-rank regularization. In: 4th International Conference on Learning Representations, ICLR 2016—Conference Track Proceedings, vol. 1, no. 2014, pp. 1–11 (2016)
23. Bejani, M., Ghatee, M.: Adaptive low-rank factorization to regularize shallow and deep neural networks, no. 1, pp. 1–11 (2020)
24. Han, S., Pool, J., Tran, J., Dally, W.: Learning both weights and connections for efficient neural networks. Adv. Neural Inf. Process. Syst. **2015**, 1135–1143 (2015)
25. Xu, S., Huang, A., Chen, L., Zhang, B.: Convolutional neural network pruning: a survey. Chinese Control Conf. CCC 2020, 7458–7463 (2020)

26. Srinivas, S., Babu, R.: Data-Free Parameter Pruning for Deep Neural Networks, pp. 31.1–31.12 (2015). arXiv:1507.06149v1
27. Guo, Y., Yao, A., Chen, Y.: Dynamic network surgery for efficient DNNs. Adv. Neural Inf. Process. Syst. 1387–1395 (2016)
28. Zhu, M., Gupta, S.: To Prune, Or Not to Prune: Exploring the Efficacy of Pruning for Model Compression, arXiv (2017)
29. Huang, C., Chen, J., Wu, J.: Learning sparse neural networks through mixture-distributed regularization. Proceedings of the IEEE Computer Society Conference on Computer Vision and Pattern Recognition, vol. 2020, no. 1, pp. 2968–2977 (2020)
30. Cheng, Y., Yu, F., Feris, R., Kumar, S., Choudhary, A., Chang, S.: An exploration of parameter redundancy in deep networks with circulant projections. Proc. IEEE Int. Conf. Comput. Vis. **2015**(1), 2857–2865 (2015)
31. Sakshi Kumar, R.: A computationally efficient weight pruning algorithm for artificial neural network classifiers. Arab. J. Sci. Eng. **43**(12), 6787–6799 (2018)
32. Wang, H.Q., Zhang, C., Fu, Y.: Emerging Paradigms of Neural Network Pruning (2021). http://arxiv.org/abs/2103.06460
33. Vadera, S., Ameen, S.: Methods for Pruning Deep Neural Networks, pp. 1–36 (2019)
34. He, Y., Zhang, X., Sun, J.: Channel pruning for accelerating very deep neural networks. Proc. IEEE Int. Conf. Comput. Vis. **2017**, 1398–1406 (2017)
35. Liu, C., Wu, H.: Channel pruning based on mean gradient for accelerating. Convolut. Neural Netw. Signal Process. **156**, 84–91 (2019)
36. Han, S., Zhan, Y., Liu, X.: Variational automatic channel pruning algorithm based on structure optimization for convolutional neural networks. J. Internet Technol. **22**(2), 339–351 (2021)
37. Li, S., Hanson, E., Li, H., Chen, Y.: PENNI: Pruned Kernel Sharing for Efficient CNN Inference. arXiv (2020)
38. You, Z., Yan, K., Ye, J., Ma, M., Wang, P.: Gate decorator: Global filter pruning method for accelerating deep convolutional neural networks, NeurIPS, pp. 1–12 (2019). arXiv
39. Pasandi, M., Hajabdollahi, M., Karimi, N., Samavi, S.: Modeling of Pruning Techniques for Deep Neural Networks Simplification (2020). arXiv
40. Anwar, V., Hwang, K., Sung, W.: Structured pruning of deep convolutional neural networks. ACM J. Emerg. Technol. Comput. Syst. **13**(3), 1–18 (2017)
41. Chen, Y., Li, C., Gong, L., Wen, X., Zhang, Y., Shi, W.: A deep neural network compression algorithm based on knowledge transfer for edge devices. Comput. Commun. **163**(August), 186–194 (2020)
42. Salehinejad, H., Valaee, S.: Pruning of Convolutional Neural Networks Using Ising Energy Model, pp. 1–5 (2021). Available: http://arxiv.org/abs/2102.05437
43. Ben-Guigui, Y., Goldberger, J., Riklin-Raviv, T.: The Role of Regularization in Shaping Weight and Node Pruning Dependency and Dynamics, pp. 1–13 (2020). arXiv
44. He, T., Fan, Y., Qian, Y., Tan, T., Yu, K.: Reshaping deep neural network for fast decoding by node-pruning. In: 2014 IEEE International Conference on Acoustics, Speech and Signal Processing (ICASSP) (2014)
45. Pouyanfar, S., Sadiq, S., Yan, Y., Tian, H., Tao, Y., Reyes, M., Shyu, M., Chen, S., Iyengar, S.: A survey on deep learning: algorithms, techniques, and applications. ACM Comput. Surv. **51**, 5 (2018)
46. Zhou, H., Alvarez, J., Porikli, F.: Less is more: towards compact CNNs supplementary material. Eur. Conf. Comput. Vis. 662–677 (2016)
47. Hoefler, T., Alistarh, D., Ben-Nun, T., Dryden, N., Peste, A.: Sparsity in Deep Learning: Pruning and Growth for Efficient Inference and Training In Neural Networks (2021). http://arxiv.org/abs/2102.00554
48. Azarian, K., Bhalgat, Y., Lee, J., Blankevoort, T.: Learned Threshold Pruning, pp. 1–12 (2020). arXiv
49. Kusupati, A., Ramanujan, V., Somani, R., Wortsman, M., Jain, P., Kakade, S., Farhadi, A.: Soft threshold weight reparameterization for learnable sparsity. In: 37th International Conference on Machine Learning, ICML 2020, Vol. Part F168147–8, pp. 5500–5511 (2020)

50. Tian, Q., Arbel, T., Clark, J.: Task dependent deep LDA pruning of neural networks. Comput. Vis. Image Underst. **203**(2020), 103154 (2021)
51. Yeom, S., Seegerer, P., Lapuschkin, S., Binder, A., Wiedemann, S., Müller, K., Samek, W.: Pruning by explaining: a novel criterion for deep neural network pruning. Pattern Recognit. **115**, 107899 (2021)
52. Molchanov, P., Mallya, A., Tyree, S., Frosio, I., Kautz, J.: Importance estimation for neural network pruning. Proc. IEEE Comput. Soc. Conf. Comput. Vis. Pattern Recognit. **2019**(11256), 11256–11264 (2019)
53. Paupamah, K., James, S., Klein, R.: Quantisation and pruning for neural network compression and regularization. In: 2020 International SAUPEC/RobMech/PRASA Conference SAUPEC/RobMech/PRASA 2020, pp. 1–6 (2020)
54. Li, J., Qi, Q., Wang, J., Ge, C., Li, Y., Yue, Z., Sun, H.: OICSR: Out-in-channel sparsity regularization for compact deep neural networks. In: Proceedings of the IEEE Computer Society Conference on Computer Vision and Pattern Recognition, vol. 2019, pp. 7039–7048 (2019)
55. Zhao, C., Ni, B., Zhang, J., Zhao, Q., Zhang, W., Tian, Q.: Variational convolutional neural network pruning. Proceedings of the IEEE Computer Society Conference on Computer Vision and Pattern Recognition, vol. 2019, pp. 2775–2784 (2019). https://doi.org/10.1109/CVPR.2019.00289
56. Shao, M., Dai, J., Kuang, J., Meng, D.: A dynamic CNN pruning method based on matrix similarity. Signal, Image Video Process **15**(2), 381–389 (2021)
57. Roychowdhury, A., Sharma, P., Learned-Miller, E.: Reducing duplicate filters in deep neural networks, NIPS work. Deep Learn. Bridg. Theory Pract. (2017)
58. Hu, P., Peng, X., Zhu, H., Aly, M., Lin, J.: OPQ: compressing deep neural networks with one-shot pruning-quantization. In: Thirty-Fifth AAAI Conference on Artificial Intelligence, AAAI, pp. 7780–7788 (2020)
59. Blalock, D., Ortiz, J., Frankle, J., Guttag, J.: What is the State of Neural Network Pruning? (2020). http://arxiv.org/abs/2003.03033

# Performance Comparison of Classification Models for Identification of Breast Lesions in Ultrasound Images

**A. Prabhakara Rao, G. Prasanna Kumar, and Rakesh Ranjan**

**Abstract** Globally, breast cancer is the most common disease among women. A region endures from damage through any disease then the region is known as lesion. It is important to differentiate different types of breast lesions for proper treatment. Therefore, there is a significant impetus for the researchers in the development of computer-aided diagnostic (CAD) system that can assist clinicians in breast lesion diagnosis. This work presents the performance comparison of different renowned classifiers in terms of accuracy with which they identify the type of breast lesion in ultrasound (US) images of the given dataset. The CAD system is developed to assist the clinicians as a second opinion tool in identifying the type of breast lesion. In this system, ultrasound images are taken as inputs and the region of interest (ROI) for each US image is marked according to the shape of abnormality. Different texture features are extracted from the US images and further these images are classified into binary classes of malignant and benign using different classifiers. The performance of these classifiers is compared and it is observed that the law's mask texture features of dimension 5 provided a maximum classification of 97.4% than other feature extraction methods applicable for the classification of two-class breast lesions.

**Keywords** Breast cancer · Ultrasound imaging · CAD system · Texture features · Machine learning

## 1 Introduction

Globally, breast cancer is the foremost reason of death in women specifically in urban areas and the age group of 20–59 [1, 2]. In the year 2020, around 2.26 million new

A. Prabhakara Rao · G. Prasanna Kumar
Department of ECE, Vishnu Institute of Technology, Bhimavaram, AP 534202, India

R. Ranjan (✉)
Department of ECE, National Institute of Technology Patna, Bihar 800005, India
e-mail: rakesh.ec19@nitp.ac.in



D. Gupta et al. (eds.), *Pattern Recognition and Data Analysis with Applications*, Lecture Notes in Electrical Engineering 888,
https://doi.org/10.1007/978-981-19-1520-8_56

breast cancer cases were found worldwide, and 685,000 women lost their life due to this [3]. In the past 10 years from 2011 to 2020, a surge of 52% in breast cancer has been reported globally by the world health organization [4]. Being the biocide form of cancer that usually starts in the lobules supplying milk to the ducts or in the ducts that carry the milk. In general, every woman should regularly check and report if there are any changes in the appearance and feel of their breasts [5, 6]. However, the American cancer society does not recommend physical examination of the breast by self or a clinical professional as an effective screening mechanism. There are various commonly available medical imaging modalities that include breast exams by the physician (in the initial stage), X-ray, Ultrasonography, Magnetic Resonance Imagining (MRI), Biopsy etc. [7]. The biopsy is the way in which a sample of tissue from apparent abnormality is taken out for the analysis resulting in unbearable pain to the patient [8]. To reduce redundant biopsies, the most frequent methods prescribed are mammography, Ultrasound (US) imaging. Mammography has high false reports low conspicuity and noisy nature of images. US imaging offers non-radioactive, non-invasive, real-time, low cost and deep penetration as compared to X-ray mammography. Among all the clinical breast screening tests, US imaging is the proven choice for diagnosis of dense breast tissues specifically for younger women [5, 7].

The characterization of breast lesions based on their appearance is clinically important because the invasive nature causes a significant increase in the risk of breast cancer [9]. The radiologists predict the breast lesions after examining the ultrasounds, but this visual examination is highly subjective [10]. The classification of breast lesions is challenging in many cases, and it is seen as a formidable task even for experienced radiologists [11]. As a result, there is a strong push among various researchers to build computer aided diagnostic (CAD) systems capable of distinguishing between breast lesions. Many researchers have developed various CAD schemes to identify breast lesions into benign and malignant categories [11, 12]. These CAD tools assist the clinicians in the diagnosis and also help in reducing the rate of biopsies. Alveranga et al. [13] explored 7 morphological characteristics on US images to distinguish between benign and malignant breast tumours and achieved a classification accuracy of 84%. According to Wan's research group [14], the low-rank matrix utilizing feature selection approach will boost classification outcomes through detection and proper choice of critical features. The preparation datasets for the benign and malignant cases include 92 benign and 172 malignant cases, respectively, while the research datasets include 21 benign and 36 malignant cases. Shi and his fellow members had developed a CAD system [15], in which breast tumours were classified using texture characteristics derived from US images using a fuzzy support vector machine (SVM) classifier. Gomez et al. [16] extracted 22 textural features from 436 US images and they combined the histogram, grey level co-occurrence matrix (GLCM), and grey level run length matrix (GLRLM) to differentiate the type of lesion and obtained a classification accuracy of 92.83% on 5500 images of prostate cancer. Huang et al. [17] classification SVM classification on 118 breast US images using 19 morphological features and obtained an accuracy of 90.9%. Wu et al. [18] performed SVM classification on 210 US images of breast lesions by combining auto-covariance texture feature with morphological features and achieved an accuracy of 92.86%. Later on [19], they applied SVM genetic classifier on the same database and achieved an accuracy of 96.14%. Alvaranga et al. [20] performed fisher linear discriminant analysis (FLDA) on 246 US images and achieved an accuracy of 85.37%. Few more researchers had attempted to address to detect and classify the breast lesions from US images and succeeded up to some extents [7, 8]. In this work, to curtail the unresolved issues in the characterization of lesions into the benign and malignant classes, different machine learning classifiers are implemented using the statistical and signal processing based features to enhance the accuracy of breast lesions detection. The remainder of this paper is organized as follows: Sect. 2 explains the proposed methodology for the implementation of CAD system. Results are summarized in Sect. 3 and finally, the last section is presented with the conclusions.

## 2 Proposed Methodology

In this work, a CAD system design is proposed to classify different breast lesions using the underlying feature characteristics by taking into account the impact of ROI. Figure 1 depicts the structure of the proposed CAD system used in this study. The CAD structure consists of four blocks: (a) ROI selection block, (b) feature extraction block, (c) classification block and (d) decision block. The min–max normalization method is used to normalize each feature set. The normalized feature vector is then split into portions: training and testing [21]. The contribution of the four classifiers—k-nearest neighbour (KNN), probabilistic neural network (PNN), support vector machine (SVM) and smooth-SVM (SSVM) were assessed in the classification module. Features are extracted based on the texture in the feature extraction

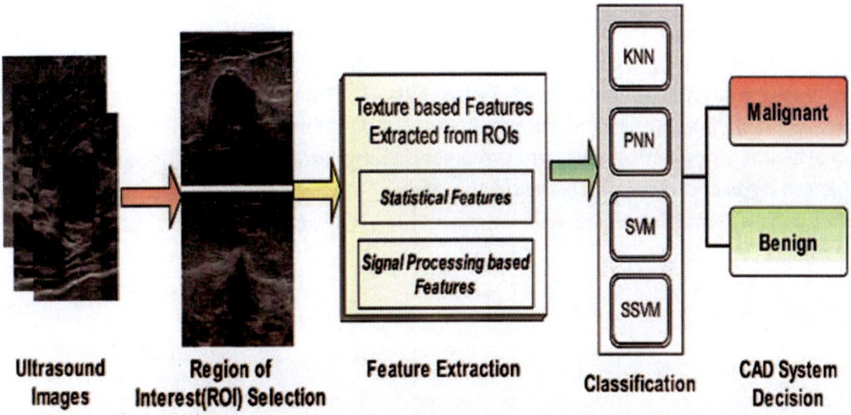


Fig. 1 Framework of the proposed CAD system for binary classification of breast lesions

module, which is classified into binary classes of malignant and benign using different classifiers to achieve better accuracy.

## 2.1 Data Collection

The CAD system proposed in this work is tested using benchmarked database obtained from the publicly available online repository Ultrasound Cases Info. developed by the Gelderse vallei hospital in collaboration with Hitachi medical systems, Europe [22]. This database comprises of a total of 167 US images of both the left and right breast. These images are categorized into three classes—primary benign (51 images), primary malignant cases (74 images) and secondary malignant cases (42 images) [22]. The database doesn't include the dataset of biopsy cases and the cases having the colour Doppler effect present. The CAD system has been implemented using two breast lesion classes, the dataset of primary benign is taken for the benign case and the cases of primary and secondary malignant are combined for the cases of malignant. To counter the issue of unbalanced data, batch-wise training is done in which the equal number of samples belong to individual classes have been taken in every iteration. To minimize the impact of bias, batch-wise training plays a significant role. The considered dataset is bifurcated into two-third of the dataset as train data set and the residual data as test data set.

## 2.2 Selection of Regions of Interest (ROIs)

The size of ROI is deliberately chosen to provide a sufficient population of pixels so that texture functionality can be computed. The skilled radiologist identifies and marks the abnormality in the US image, which is then segmented using the software Image J [23]. This software assists in loading the image, identifying the infected location, and segmenting it. A rectangular bounding box is used to enclose the segmented region adjoining the boundaries of abnormality. The benign type of cancer has a wider than taller shape, intense and uniform hyper echogenicity, a thin consistent capsule in an oval shape consisting of two to three gentle lobulations while malignant type has a marked hypo echogenicity, acoustic shadowing, microlobulation, an extension of the duct with wider form and angular margins [24]. The selection of ROI is represented in Fig. 2.

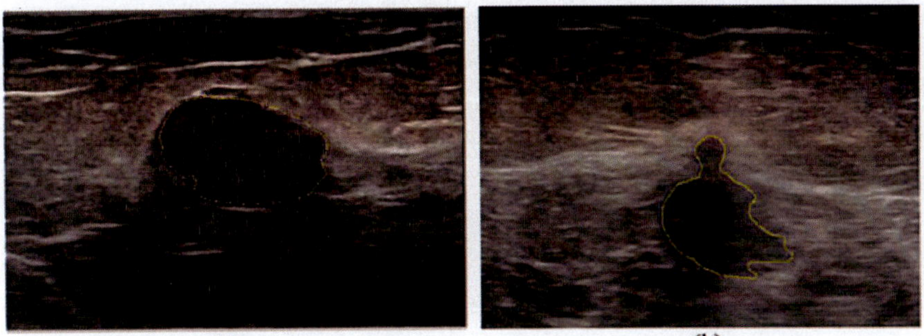


**Fig. 2** **a** Benign marked region, **b** Malignant marked regions

## 2.3 *Feature Extraction*

In this work, the characteristics that are visually extractable and non-extractable are transformed into mathematical descriptors. The textural characteristics consisting of statistical and signal processing features are comprised of such mathematical descriptors.

a. ***Statistical features***

Textural features of an image are extracted using statistical approaches which depend on the grey level intensity distribution of the pixels in the image. Regardless of the number of pixels used to compute the texture features, these approaches are categorized into first order, second order and other statistics. The first-order statistics are extracted from the grey level intensity histograms of the image. For each ROI, six characteristics are evaluated using standard mathematical equations as mentioned in Table 1 [25].

Second-order statistics include the computations with the GLCM [26]. GLCM describes the regular occurrence of pairs of pixels with different grey levels in an image with different dimensions distributed in opposite directions, such as 0°, 45°, 90° and 135°. All GLCM features considered in this implemented work are described and mentioned in Table 1. Other statistical features such as GLDS (Grey Level Difference Statistics) which is based on the co-occurrence of a pixel pair that have the same difference in grey levels are also considered in this work. Table 1 is describing all the statistical features applied in the CAD system used in this analysis.

b. ***Signal processing based features***

In the signal processing based features, law's mask texture analysis has been adopted in which convolution masks of small size are used as filters and these filters are convolved with ROIs to get the enhanced texture characteristics. Averaging, edge detection, spot detection, wave detection and ripple detection on these filters are performed to extract textural features. Mean, standard deviation, entropy, kurtosis

**Table 1** Mathematical formulas for statistical features applied in the CAD system

| Statistical features | Mathematical formulas |
|---|---|
| First-order statistics | $\text{Mean}(\overline{x}) = \frac{1}{N}\sum_{i=1}^{N} i.x_i$ where $x_i$ is the individual pixel values and $N$ is total number of pixels |
| | $\text{Std. deviation}(\sigma) = \frac{\sum_{i=1}^{N}(x_i-\overline{x})^2}{\sqrt{(N-1)}}$ |
| | $\text{Third\_Moment}(\mu_3) = \frac{\sum_{i=1}^{N}(x_i-\overline{x})^3}{N\sigma^3}$ |
| | $\text{uniformity} = \sum_{i=1}^{N} p(i)^2$ |
| | $\text{Entropy} = -\sum_{i=1}^{N} p(x_i).\log_2 p(x_i)$ |
| | $\text{Smoothness} = 1 - \frac{1}{(1+\sigma^2)}$ |
| GLCM features | $\text{Angular\_moment} = \sum_{i,j=1}^{N} P_{i,j}^2$ |
| | $\text{Contrast} = \sum_{i,j=1}^{N} P_{i,j}(i-j)^2$ |
| | $\text{Correlation} = \sum_{i,j=1}^{N} P_{i,j}\left[\frac{(i-\mu_i)(j-\mu_j)}{\sigma_i\sigma_j}\right]$ |
| | $\text{Variance} = \sum_{i,j=1}^{N} P_{i,j}(i-\mu_i)^2$ |
| | $\text{Homogenity} = \sum_{i,j}\frac{P_{i,j}}{1+(i-j)^2}$ |
| | Some other GLCM features such as sum entropy, difference entropy, difference variance |
| GLDS features | $\text{Energy} = \sqrt{(P_{i,j}^2)}$ |
| | Some other GLDS features such as contrast, homogeneity, entropy and mean |

and skewness from each ROI are computed using Law's masks of lengths 3, 5, 7 and 9 [27].

### c. *Classification*

The method of grouping testing samples appropriately is known as classification, which can be supervised or unsupervised. If the classes for the training sets have already been defined, classification is supervised; otherwise, classification is unsupervised. As mentioned earlier, this work includes four classifiers namely k-NN, PNN, SVM and SSVM [28]. The k-NN classifier is based on the concept of inferring an undefined instance's class from its neighbours. The k-NN classifier's classification efficiency is determined by the value of k. PNN is a supervised Bayesian based feed-forward neural network used for assessing the unknown class instances. SVM is the supervised machine learning method implemented using LibSVM library [29]. SVM is associated with the traditional quadratic programme, therefore to unconstraint smoothing unconstrained optimization reformulation SSVM classifier is used. SSVM works on the SVM related problems by smoothing the unconstrained optimization of the reformulation for the pattern classification.

After classification, CAD system makes the decision whether the breast lesions are malignant or benign. Various CAD system designs have the potential to assist the radiologists in routing medical practice as a second opinion tool for the classification of lesions in ultrasound images in cases where clear subjective discrimination is difficult. In light of this fact, a CAD system design employing the texture analysis techniques of feature extraction and feature classification have been proposed and analyzed in the present work for US breast lesions classification into two classes - benign and malignant. Some impressive results can be found in the further section of this article.

## 3 Results

A CAD system architecture for breast lesion classification is proposed in this work, which is evaluated through experimental trials. In the experiment, the classification accuracy of a texture features vector (TFV) incorporating various statistical and signal processing-based features is tested for breast lesion classification into benign and malignant classes using variously defined classifiers. According to Table 2, the average classification accuracy for statistical features using k-NN, PNN, SVM, and SSVM classifiers are 83.1, 80.5, 79.2 and 76.6%, respectively. The best individual class accuracy of 57.1% is achieved for a benign class using an SVM classifier while the best individual class accuracy of 100% is achieved for a malignant case using SSVM classifier. Maximum values are marked bold.

The description of texture feature vectors of signal processing methods is described in Table 3. In the signal processing based texture features described in Table 4, the average classification accuracy of 91.4, 88.6, 91.8 and 96.8% have been obtained using k-NN, PNN, SVM and SSVM classifiers respectively over the defined texture feature vectors. It is observed from experimental results that SSVM performs apparently better than all the others classifiers and shows the maximum classification of 97.4% for law's mask texture features of dimension 5.

**Table 2** Performance analysis of statistical features using different classifiers

| Classifiers | Overall classification accuracy [OCA (%)] | Individual class accuracy [$ICA_B$ (%)] | Individual class accuracy [$ICA_M$ (%)] |
|---|---|---|---|
| k-NN | **83.1** | **52.3** | 94.6 |
| PNN | 80.5 | **52.3** | 91 |
| SVM | 79.2 | 57.1 | 87.5 |
| SSVM | 76.6 | 44.2 | **100** |
| ***Mean*** | 79.8 | 51.5 | 93.3 |

*Note* $A_B$: accuracy for benign class, and $A_M$: accuracy for malignant class

**Table 3** Description of texture feature vectors of signal processing methods

| Features | Law's mask length | No. of statistical features | No. of rotation invariant texture images (TRI) | Total extracted features |
|---|---|---|---|---|
| TFV1 | 3 | 5 | 6 | $5 \times 06 = 30$ |
| TFV2 | 5 | 5 | 15 | $5 \times 15 = 75$ |
| TFV3 | 7 | 5 | 6 | $5 \times 06 = 30$ |
| TFV4 | 9 | 5 | 15 | $5 \times 15 = 75$ |

## 4 Conclusion

The proposed CAD system can assist the radiologists in routine medical practice in decision making as a second opinion tool for clear discrimination of benign and malignant breast lesions in US images. This could help to minimize the need for unnecessary biopsies. The signal processing based law's mask features are the most efficient texture features which take into account the textural changes exhibited by benign and malignant lesion when fed to SSVM classifier for predicting the labels unknown testing instances of US images. The law's mask texture features of dimension 5 is computed to achieve the maximum classification of 97.4% than other texture features. The improvement in individual class accuracy will enhance the applicability and reliability of the proposed system for clinical diagnosis would be the most priority future task in this work. The scope of this method is not limited to breast lesions only, it can be applied in the classification of other lesions which would be the prime focus in future works.

**Table 4** Classification performance of signal processing based features using different classifiers

| Classifiers | TVF1 | | | TVF2 | | | TVF3 | | | TVF4 | | |
|---|---|---|---|---|---|---|---|---|---|---|---|---|
| | OCA (%) | $ICA_B$ (%) | $ICA_M$ (%) | OCA (%) | $ICA_B$ (%) | $ICA_M$ (%) | OCA (%) | $ICA_B$ (%) | $ICA_M$ (%) | OCA (%) | $ICA_B$ (%) | $ICA_M$ (%) |
| k-NN | 92.2 | 76.1 | 98.2 | 90.2 | 71.4 | **100** | 89.6 | **61.9** | **100** | 93.5 | **76.1** | **100** |
| PNN | 88.3 | 57.1 | **100** | 89.6 | 61.9 | **100** | 85.7 | 47.6 | **100** | 90.9 | 66.6 | **100** |
| SVM | 89.6 | 66.6 | 98.2 | 92.2 | **90.4** | **100** | 92.0 | 57.1 | 98.2 | 93.5 | **76.1** | **100** |
| SSVM | **96.1** | **85.7** | **100** | **97.4** | 71.4 | **100** | **96.7** | 47.6 | 96.4 | **96.8** | 63.3 | **100** |
| ***Mean*** | 91.6 | 71.4 | 99.1 | 92.4 | 73.8 | 100 | 91.0 | 53.6 | 98.7 | 93.7 | 70.6 | 100 |

## References


1. Alanko, J., Tanner, M., Vanninen, R., Auvinen, A., Isola, J.: Triple-negative and HER2-positive breast cancers found by mammography screening show excellent prognosis. Breast Cancer Res. Treat. **187**, 267–274 (2021). https://doi.org/10.1007/s10549-020-06060-z
2. Heer, E., Harper, A., Escandor, N., Sung, H., McCormack, V., Fidler-Benaoudia, M.M.: Global burden and trends in premenopausal and postmenopausal breast cancer: a population-based study. Lancet Glob. Heal. **8**, e1027–e1037 (2020). https://doi.org/10.1016/S2214-109X(20)30215-1
3. Cancer. https://www.who.int/news-room/fact-sheets/detail/cancer
4. Ginsburg, O., Yip, C., Brooks, et. al.: Breast cancer early detection: a phased approach to implementation. Cancer. **126**, 2379–2393 (2020). https://doi.org/10.1002/cncr.32887
5. Version, D.: Breast Cancer in Young Women Aspects of Heredity and Contralateral Disease (2021)
6. Rao, A.P., Bokde, N., Sinha, S.: Photoacoustic imaging for management of breast cancer: a literature review and future perspectives. Appl. Sci. **10** (2020). https://doi.org/10.3390/app10030767
7. Elouassif, B., Idri, A., Hosni, M., Abran, A.: Classification techniques in breast cancer diagnosis: a systematic literature review. Comput. Methods Biomechan. Biomed. Eng. Imag. Vis. (2021). https://doi.org/10.1080/21681163.2020.1811159
8. Sree, S.V.: Breast imaging: a survey. World J. Clin. Oncol. **2**, 171 (2011). https://doi.org/10.5306/wjco.v2.i4.171
9. Zheng, Y., Jiang, Z., Xie, F., Zhang, H., Ma, Y., Shi, H., Zhao, Y.: Feature extraction from histopathological images based on nucleus-guided convolutional neural network for breast lesion classification. Pattern Recognit. **71**, 14–25 (2017). https://doi.org/10.1016/j.patcog.2017.05.010
10. Zhou, J., Zhang, Y., Chang, K.T., Lee, K.E., Wang, O., Li, J., Lin, Y., Pan, Z., Chang, P., Chow, D., Wang, M., Su, M.Y.: Diagnosis of benign and malignant breast lesions on DCE-MRI by using radiomics and deep learning with consideration of peritumor tissue. J. Magn. Reson. Imaging. **51**, 798–809 (2020). https://doi.org/10.1002/jmri.26981
11. Zahoor, S., Lali, I.U., Khan, M.A., Javed, K., Mehmood, W.: Breast cancer detection and classification using traditional computer vision techniques: a comprehensive review. Curr. Med. Imaging Former. Curr. Med. Imaging Rev. **16**, 1187–1200 (2020). https://doi.org/10.2174/1573405616666200406110547
12. Ragab, D.A., Attallah, O., Sharkas, M., Ren, J., Marshall, S.: A framework for breast cancer classification using Multi-DCNNs. Comput. Biol. Med. **131**, 104245 (2021). https://doi.org/10.1016/j.compbiomed.2021.104245
13. Alvarenga, A.V., Infantosi, A.F.C., Pereira, W.C.A., Azevedo, C.M.: Assessing the performance of morphological parameters in distinguishing breast tumors on ultrasound images. Med. Eng. Phys. **32**, 49–56 (2010). https://doi.org/10.1016/j.medengphy.2009.10.007
14. Wan, T., Liao, R., Qin, Z.: A robust feature selection approach using low rank matrices for breast tumors in ultrasonic images. In: 18th IEEE International Conference on Image Processing, pp. 1645–1648 (2011)
15. Shi, X., Cheng, H.D., Hu, L., Ju, W., Tian, J.: Detection and classification of masses in breast ultrasound images. Digit. Signal Process. A Rev. J. **20**, 824–836 (2010). https://doi.org/10.1016/j.dsp.2009.10.010
16. Gómez, W., Pereira, W.C.A., Infantosi, A.F.C.: Analysis of co-occurrence texture statistics as a function of gray-level quantization for classifying breast ultrasound. IEEE Trans. Med. Imaging. **31**, 1889–1899 (2012). https://doi.org/10.1109/TMI.2012.2206398
17. Huang, Y.L., Chen, D.R., Jiang, Y.R., Kuo, S.J., Wu, H.K., Moon, W.K.: Computer-aided diagnosis using morphological features for classifying breast lesions on ultrasound. Ultrasound Obstet. Gynecol. **32**, 565–572 (2008). https://doi.org/10.1002/uog.5205

18. Wu, W.J., Moon, W.K.: Ultrasound breast tumor image computer-aided diagnosis with texture and morphological features. Acad. Radiol. **15**, 873–880 (2008). https://doi.org/10.1016/j.acra.2008.01.010
19. Wu, W.J., Lin, S.W., Moon, W.K.: Combining support vector machine with genetic algorithm to classify ultrasound breast tumor images. Comput. Med. Imaging Graph. **36**, 627–633 (2012). https://doi.org/10.1016/j.compmedimag.2012.07.004
20. Alvarenga, A.V., Infantosi, A.F.C., Pereira, W.C.A., Azevedo, C.M.: Assessing the combined performance of texture and morphological parameters in distinguishing breast tumors in ultrasound images. Med. Phys. **39**, 7350–7358 (2012). https://doi.org/10.1118/1.4766268
21. Ranjan, R., Sahana, B.C.: An efficient facial feature extraction method based supervised classification model for human facial emotion identification. In: 2019 IEEE 19th International Symposium on Signal Processing and Information Technology, ISSPIT 2019. Institute of Electrical and Electronics Engineers Inc. (2019). https://doi.org/10.1109/ISSPIT47144.2019.9001839
22. Breast and Axilla I Ultrasound Cases. https://www.ultrasoundcases.info/cases/breast-and-axilla/. Last accessed 7 May 2021
23. Rasband, W.S.: ImageJ: image processing and analysis in Java. ascl. ascl:1206.013 (2012)
24. Kornecki, A.: Current status of breast ultrasound. Canad. Assoc. Radiol. J. (2011). https://doi.org/10.1016/j.carj.2010.07.006
25. Varga, D.: No-reference image quality assessment with global statistical features. J. Imaging. **7**, 29 (2021). https://doi.org/10.3390/jimaging7020029
26. Chekouo, T., Mohammed, S., Rao, A.: A Bayesian 2D functional linear model for gray-level co-occurrence matrices in texture analysis of lower grade gliomas. NeuroImage Clin. **28**, 102437 (2020). https://doi.org/10.1016/j.nicl.2020.102437
27. Qi, Y., Yang, Z., Lei, J., Lian, J., Liu, J., Feng, W., Ma, Y.: Morph_SPCNN model and its application in breast density segmentation. Multimed. Tools Appl. **80**, 2821–2845 (2021). https://doi.org/10.1007/s11042-020-09796-4
28. Nanaa, A., Akkus, Z., Lee, W.Y., Pantanowitz, L., Salama, M.E.: Machine learning and augmented human intelligence use in histomorphology for haematolymphoid disorders. Natl. Lib. Med. (2021). https://doi.org/10.1016/j.pathol.2020.12.004
29. Borkar, A., Sinha, S., Dhengre, N., Chinni, B., Dogra, V., Rao, N.: Diagnosis of prostate cancer with support vector machine using multiwavelength photoacoustic images. In: Advances in Intelligent Systems and Computing. pp. 247–254. Springer Science and Business Media Deutschland GmbH (2020). https://doi.org/10.1007/978-981-32-9088-4_21

# Analysis of Randomization-Based Approaches for Autism Spectrum Disorder

**Umesh Gupta, Deepak Gupta, and Umang Agarwal**

**Abstract** Autism spectrum disorder (ASD) is a severe neurodevelopmental disorder that affects an individual's sensory activity, social interaction, and cognitive abilities. In the mental illnesses ASD disorder, the problem starts in infancy and affects more habits as the age progresses, progressing to adolescence and adulthood, also known as a behavioural disorder. Everything nowadays is moving towards automated software, which is relevant not just in terms of time efficiency but also in terms of cost-effectiveness. As a result, there is a pressing need in the healthcare sector to incorporate machine learning to reap the greatest benefits. Over the last two decades, randomization-based approaches such as extreme learning machines (ELM) and random vector functional link (RVFL) have gained popularity amongst researchers due to their better generalization performance. In this work, the classification capability of the ELM and RVFL models with different activation functions is investigated to estimate the autism spectrum disorder in a human population of children, adolescents, and adults on the grounds of publicly accessible UCI datasets. The attainment of the randomization-based approaches is determined using various quality measures such as accuracy, precision, recall, negative predictive value, rate of misclassification, $F_1$-measure, G-mean, and Matthew's correlation coefficient. From the numerical experiment results, one can show that the generalization capability of RVFL using hardlim, hyperbolic tangent function, and sigmoidal activation function is superior to ELM based on several quality measures.

**Keywords** Extreme learning machine · Random vector function link · ASD disorder · Autism traits · Randomization-based approaches

U. Gupta · D. Gupta (✉)
National Institute of Technology Arunachal Pradesh, Jote, Arunachal Pradesh, India
e-mail: deepakjnu85@gmail.com; deepak@nitap.ac.in

U. Agarwal
Hills College of Teacher Education, Naharlagun, Arunachal Pradesh, India



D. Gupta et al. (eds.), *Pattern Recognition and Data Analysis with Applications*, Lecture Notes in Electrical Engineering 888,
https://doi.org/10.1007/978-981-19-1520-8_57

# 1 Introduction

Nowadays, the problem of autism spectrum disorder (ASD) is considered a very momentous brain developmental disorder which is not only restraining the person's communication and social behaviours but also hinders the learning skill in their natural growth [1–5]. ASD is a very serious issue in the human population which is seriously ignored for a very long time due to a lack of awareness and knowledge about this neurodevelopmental disorder amongst people and society but now the adverse effect of ASD is visible in society [3]. However, most researchers believe that it is linked to the genes of human beings but it can be realized using behavioural science. Early detection of this neuro disorder condition can make a significant difference in the person's life as well as the human beings of the entire community. In the research literature, there are several ASD diagnosis tools are available such as AQ, ADI-R, SRS, SCQ, Q-CHAT, ADOS-R, DSM-III-R, CHAT, CARS, red flags [6], and many more which are further classified into clinical and non-clinical tools. Even, the detection of ASD may be performed at any period of the lifecycle but its manifestation is generally noticeable in the early age of the child (mainly the first five years) and it grows as the age increase. It will become a difficult task to detect ASD in adults and teenagers, because there are many chances to suffer from many other mental diseases such as ADHD, anxiousness, epilepsy, depression, and many more [1, 5]. For most of the children who are suffering from this type of mental disorder, their behaviours like imaginative ability, learning ability, personal interaction ability, repetitive behaviours, social interactions, following strict life action plan and difficulty with concentrations, etc., seem to be changing very rapidly as well as their sensory traits may be badly affected like smelling, tasting, hearing, talking, visual impairment, etc. According to WHO [8], 1 out of 160 children is suffering from ASD disorder globally, and this number will gradually rise in the future. If early detection and prevention measures for ASD are not taken now, it will expand to a large number of people.

Several computational intelligence approaches such as machine learning and deep learning work as an automated and efficient solutions for mental and health-related problems due to their variability and applicability [2–4]. It is not only saving the monetary value of the person but also helpful in early-stage medication to their family and society. There are many machine learning approaches that have been pertained for the ASD disorder, named support vector machine (SVM) [9–11], support vector regression [12, 13], alternating decision tree (ADTree), rule classifier, fuzzy logic, artificial neural networks, logistic regression, random forest, Naive Bayes, k-nearest neighbours, and convolutional neural network [14]. The diagnosis and study of ASD are a conventional binary classification problem where a model is performed on the already predicted ASD and no-ASD class. There are very few numbers of literature reviews on the usage of machine learning algorithms on psychiatric disorders. For example, Cruz and Wishart [15] explored the machine learning approach for cancer diagnosis, whilst Gupta et al. [14] have experimented with 15 machine learning approaches for the prediction of diabetes. Li et al. [16] have performed a certain experiment using machine learning approaches and with the help of imitation, given a technique to identify the autistic adults on AQ-adolescent dataset having 40 kinematic constraints. Vaishali and Sasikala [17] have considered the swarm intelligence-based binary firefly algorithm on UCI autistic datasets for distinguishing the ASD and no-ASD cases with a range of accuracy 92–97%. To lessen the autism screening time and rapid diagnosis of autism traits, an alternating decision tree (ADTree) has been implemented using the ADI-R approach. Some researchers have used "red flags" to screening autistic traits and experimented with machine learning approaches. Support vector machine is one of the popular binary classifiers which have been considered for the same purpose [18, 19]. Related to this work, one can follow the advanced variants of the machine learning approach [19, 21, 22]. Here, the purpose of this paper is to analyze an effective and efficient autism prediction model by implementing randomization-based approaches [23, 24] on the autism spectrum disorder datasets. Also, the numerical experiment is used to illustrate the usefulness of randomization-based methods such as extreme learning machines [23] and random vector functional link network [24] in terms of various performance measures like accuracy, precision, recall, negative predictive value, rate of misclassification, F1-measure, G-mean and Matthew's correlation coefficient [25], and CPU training cost. The contribution of this work is

1. The classification potential of the ELM and RVFL models with various activation functions is explored in this research to quantify the autism spectrum disorder in the human population of infants, teenagers, and adults.
2. To validate the efficiency of randomization-based approaches, several quality metrics have been considered including accuracy, precision, recall, negative predictive value, rate of misclassification, F1-measure, G-mean, and Matthew's correlation coefficient on UCI ASD datasets.

In Sect. 2, we have introduced the formulation of randomization-based approaches in detail. Also, the autism spectrum disorder-related datasets with their attributes, features, and class labels are described in Sect. 3. In Sect. 4, numerical results are deliberated as the performance analysis between the two approaches ELM and RVFL using 6 activation functions. Ultimately, in Sect. 5, we summarize our study and make suggestions for new work.

## 2 Randomization-Based Approaches

In this work, suppose, we have an input vector $\mathbf{a}$ and its $L_2$-norm is represented as $||\mathbf{a}||$. Here, one's vector and identity matrix are written as $\mathbf{e}$ and $\mathbf{I}$, respectively.

## 2.1 Extreme Learning Machine (ELM)

There is a prominent model for classification named as an extreme learning machine that is associated with a single-layer feed-forward network (SLFNs) structure [23]. For each given input dataset $\mathbf{a}_\ell = (\mathbf{a}_{\ell 1}, \ldots, \mathbf{a}_{\ell p})^t \in \Re^p$ and its outcome $\mathbf{y}_\ell \in \Re$, the below mentioned (1) will satisfy over training as

$$\mathbf{y}_\ell = \sum_{r=1}^{M} \mu_r S(\eta_r, \lambda_r, \mathbf{a}_\ell) \text{ for } \ell = 1, \ldots, q, \tag{1}$$

where $M$ denotes the symbol of present hidden nodes on the hidden layer; $S(\eta_r, \lambda_r, \mathbf{a}_\ell)$ is the non-linear function that is the outcome of the $r$th enhancement node with the considered $\mathbf{a}_\ell$ input dataset; $\psi = (\psi_1, \ldots, \psi_M)^t \in \Re^M$ is the outcome weight vector which is connecting the hidden node to the outcome node. $\eta_r = (\eta_{r1}, \ldots, \eta_{rp})^t \in R^n$ and $\lambda_r \in \Re$ are the arbitrarily allocated input weight vector and the bias which are linked to the input layer to the $r$th hidden node.

Further, one can write the Eq. (1) in this way

$$\mathbf{H}\boldsymbol{\psi} = \mathbf{y}, \tag{2}$$

where $\mathbf{y} = (y_1, \ldots, y_q)^t \in \Re^q$ is the vector of possible outcomes. Here,

$$\mathbf{H} = \begin{bmatrix} S(\eta_1, \lambda_1, a_1) \ldots S(\eta_M, \lambda_M, a_1) \\ . \quad \ldots \quad . \\ S(\eta_1, \lambda_1, a_q) \ldots S(\eta_M, \lambda_M, x_q) \end{bmatrix}_{q \times M}, \tag{3}$$

To obtain the value of $\psi \in \Re^M$, (2) is required to be its minimum norm least square solution in this way

$$\boldsymbol{\psi} = \mathbf{H}^\dagger \mathbf{y}, \tag{4}$$

where the $(.)^\dagger$ is the symbolic representation of Moore–Penrose generalized inverse [20].

At last, the final classification function $f(.)$ is required (5) for any sample $\mathbf{a}_s \in \Re^p$ as

$$f(a_s) = h(\mathbf{a}_s) \cdot \boldsymbol{\psi}. \tag{5}$$

### 2.2 *Random Vector Functional Link Network (RVFL)*

RVFL [24] is also a famous single-layer feed-forward network (SLFNs) that arbitrarily assigns the weights to the hidden node and sticks them. Here, $\boldsymbol{\gamma} = (\gamma_1, \ldots, \gamma_r)^t$ and $r = M + p$ be the weight vector to the outcome neuron. The outcome when the activation function $h_s(\mathbf{a}_k) = S(\boldsymbol{\eta}_s, \lambda, \mathbf{a}_k)$ for $s = 1, \ldots, M$ and $k = 1, \ldots, q$ of the $s$th hidden layer neuron w. r. to the $k$th sample. $R$ is the $q \times p$ dimensional matrix of training examples. Further, tuning is not required.

The mathematical expression for RVFL is written as

$$\min \alpha ||\boldsymbol{\gamma}||^2 + || - \ \mathbf{W}\boldsymbol{\gamma} + \mathbf{y}||^2, \tag{6}$$

where $\mathbf{W} = [\mathbf{H} \quad \mathbf{R}]$.

By differentiation (6) w. r. to $\boldsymbol{\gamma}$, one can find the solution through (7) as follows,

$$\boldsymbol{\gamma} = (\mathbf{W^t W} + \alpha \mathbf{I})^{-1} \mathbf{W^t y}, \tag{7}$$

At last, For RVFL, the final classifier of any sample is attained as

$$f(\mathbf{a}_s) = sign\ ([h(\mathbf{a}_s) \quad \mathbf{a}_s]\boldsymbol{\gamma}), \tag{8}$$

where $\mathbf{h}(\mathbf{x}) = [h_1(x) \ \ldots \ h_M(x)]$.

In the randomization-based algorithms like ELM and RVFL, many activation functions are considered such as multiquadric (M), sigmoidal(S), sine, swish, hyperbolic tangent function (tanh), hard limit transfer function (hardlim), and many more. The definition of these activation functions is written [23, 24] as

1. For multiquadric: $f(\mathbf{a}) = \sqrt{(||\mathbf{a} - \eta||^2 + \lambda^2)}$,
2. For sigmoidal: $f(\mathbf{a}) = \frac{1}{1+e^{-\ \mathbf{a}\eta + \lambda}}$,
3. For sine: $f(\mathbf{a}) = \sin(\mathbf{a})$,
4. For swish: $f(\mathbf{a}) = \mathbf{a} \times \left(\frac{1}{1+e^{-\mathbf{a}}}\right)$,
5. For tanh: $f(\mathbf{a}) = \frac{e^{\mathbf{a}} - e^{-\mathbf{a}}}{e^{\mathbf{a}} + e^{-\mathbf{a}}}$,
6. For hardlim: $f(\mathbf{a}) = 1$, if $a \geq 0$ otherwise $f(\mathbf{a}) = -1$.

## 3 Autism Spectrum Disorder Datasets Description

In this work, autism spectrum disorder datasets of children, adolescent, and adult are used which is collected from UCI data repositories [7] that contains the three type of data for ASD screening and further apply on both the randomization-based approach to classifying the psychiatric disorders named ASD in the human population at any age. Hence, the ASD dataset contains 20 numbers common attributes like age, sex, nationality have jaundice or not at birth, knowledge of pervasive development disorders, country name, etc., and 292, 98, 704 numbers of samples for children, adolescent, and adult, respectively. The class labels are ASD (1) and no-ASD (0). The attribute type of these ASD datasets is binary, categorical, and continuous. ELM [23] and RVFL [24] approaches are applied to this ASD dataset [7].

## 4 Numerical Results

In this numerical result, the credibility and applicability of the randomization-based algorithms named such as ELM [23] and RVFL [24] are validated on standard ASD real-world datasets [7] for binary classification problems. To perform the numerical experiments, we arrange some resources in such a manner as one CPU, having 1 TB HD, i5 Intel(R) processor with 3.20 GHz, 8 GB memory, Windows 10 O.S., and for simulation MATLAB2019a software. Further, the six popular activations function is chosen under RBF hidden nodes. Input weights and biases were examined arbitrarily at the start. But considered estimations will not be altered for the entire test. The optimal parameters $M$ and $\alpha$ are selected from the range of $\{\ 20,\ 50,\ 100,\ 200,\ 500,\ 1000\ \}$ and $\{10^{-5}, 10^{-4}, \ldots,\ 10^{4},\ 10^{5}\}$, respectively, using tenfold cross-validation. To assess the fulfilment of the randomization-based approaches, we have gauged several quality measures such as negative predictive value, rate of misclassification, recall, accuracy, precision, $F_1$-measure, geometric-mean, and Matthew's correlation coefficient [14, 25] for three datasets. The definition of these quality measures is written as Table 1.

In the ASD dataset, the ratio of training to testing is considered in the 30:70. The attainment analysis is being inspected by using several quality measures where the appearance of ASD traits is advised to be "positive" class, and the inadequacy of ASD traits is assumed to be "negative" class. The comparative arrangement of the randomization-based approaches of ELM to RVFL based on accuracy on the ASD publicly available UCI real-world datasets is presented in Table 2, respectively.

**Table 1** Various evaluation quality measures

| Quality measures | Definition |
|---|---|
| Accuracy | $(\xi_{TP} + \xi_{TN})/(\xi_{TP} + \xi_{TN} + \xi_{FP} + \xi_{FN})$ |
| Precision (Pr) | $\xi_{TP}/(\xi_{TP} + \xi_{FP})$ |
| Recall (Re) | $\xi_{TP}/(\xi_{TP} + \xi_{FN})$ |
| Negative predictive value (NPV) | $\xi_{TN}/(\xi_{TN} + \xi_{FN})$ |
| Rate of misclassification (RME) | $(\xi_{FP} + \xi_{FN})/(\xi_{TP} + \xi_{TN} + \xi_{FP} + \xi_{FN})$ |
| $F_1$-measure | $2 \times (\text{Pr} \times \text{Re})/(\text{Pr} + \text{Re})$ |
| Geometric-mean | $\sqrt{(\text{Pr} \times \text{Re}}$ |
| Matthews's correlation coefficient (MCC) | $\frac{\xi_{TP}\times\xi_{TN}-\xi_{FP}\times\xi_{FN}}{\sqrt{(\xi_{TP}+\xi_{FP})(\xi_{TP}+\xi_{FN})(\xi_{TN}+\xi_{FP})(\xi_{TN}+\xi_{FN})}}$ |

Where TP denotes as $\xi_{TP}$; TN denotes as $\xi_{TN}$; FP denotes as $\xi_{FP}$; and FN denotes as $\xi_{FN}$

**Table 2** The comparative analysis of randomization-based models ELM and RVFL with different activation functions on ASD datasets for children, adolescents, and adults

| ASD dataset/models | Autism-child-data (88 × 20, 204 × 20) | Autism-adolescent-data (30 × 20, 68 × 20) | Autism-adult-data (212 × 20, 492 × 20) |
|---|---|---|---|
| ELM (hardlim)<br>Accuracy ± S.D.<br>($M$)<br>Time | 49.5098 ± 0<br>(1000)<br>0.64013 | 60.2941 ± 0<br>(1000)<br>0.23224 | 74.3902 ± 0<br>(1000)<br>1.11386 |
| ELM (sine)<br>Accuracy ± S.D.<br>($M$)<br>Time | 68.9216 ± 2.78594<br>(1000)<br>0.15729 | 63.8235 ± 2.46076<br>(1000)<br>0.05004 | 66.0163 ± 5.05563<br>(1000)<br>0.50473 |
| ELM (swish)<br>Accuracy ± S.D.<br>($M$)<br>Time | 90.4902 ± 0.55891<br>(20)<br>0.00354 | 73.8235 ± 2.82873<br>(500)<br>0.03354 | 92.0732 ± 0.14372<br>(20)<br>0.00746 |
| ELM (tanh)<br>Accuracy ± S.D.<br>($M$)<br>Time | 57.6471 ± 5.25219<br>(1000)<br>0.13693 | 71.4706 ± 3.2219<br>(50)<br>0.00281 | 73.9837 ± 0.28744<br>(20)<br>0.00652 |
| ELM (multiquard)<br>Accuracy ± S.D.<br>($M$)<br>Time | 84.3137 ± 0.60037<br>(20)<br>0.00525 | 66.1765 ± 1.8011<br>(1000)<br>0.08697 | 93.6585 ± 0.26501<br>(20)<br>0.01521 |
| ELM (sigmoidal)<br>Accuracy ± S.D.<br>($M$)<br>Time | 59.2157 ± 2.23026<br>(500)<br>0.07168 | 74.1176 ± 6.02762<br>(500)<br>0.02373 | 72.8862 ± 1.62982<br>(50)<br>0.0664 |
| RVFL (hardlim)<br>Accuracy ± S.D.<br>($M, \alpha$)<br>Time | **94.6078 ± 0**<br>(1000,$10^{-2}$)<br>0.00181 | **82.3529 ± 0**<br>(1000,$10^{0}$)<br>0.00149 | **97.1545 ± 0**<br>(100,$10^{-2}$)<br>0.00187 |
| RVFL (sine)<br>Accuracy ± S.D.<br>($M, \alpha$)<br>Time | 84.2157 ± 1.81439<br>(20,$10^{1}$)<br>0.0032 | 75.2941 ± 3.18816<br>(20,$10^{0}$)<br>0.00535 | 92.1138 ± 0.46349<br>(20,$10^{2}$)<br>0.00472 |
| RVFL (swish)<br>Accuracy ± S.D.<br>($M, \alpha$)<br>Time | 92.1569 ± 1.03986<br>(500,$10^{3}$)<br>0.00175 | 81.1765 ± 1.91741<br>(100,$10^{-1}$)<br>0.0007 | 94.2276 ± 0.42147<br>(1000,$10^{2}$)<br>0.00574 |
| RVFL (tanh)<br>Accuracy ± S.D.<br>($M, \alpha$)<br>Time | **94.6078 ± 0**<br>(1000,$10^{-2}$)<br>0.00423 | **82.3529 ± 0**<br>(1000,$10^{0}$)<br>0.00242 | **97.1545 ± 0**<br>(100,$10^{-2}$)<br>0.00161 |

(continued)

**Table 2** (continued)

| ASD dataset/models | Autism-child-data (88 × 20, 204 × 20) | Autism-adolescent-data (30 × 20, 68 × 20) | Autism-adult-data (212 × 20, 492 × 20) |
|---|---|---|---|
| RVFL (multiquard)<br>Accuracy ± S.D.<br>$(M, \alpha)$<br>Time | 93.6275 ± 0<br>$(1000,10^{-2})$<br>0.09594 | 78.2353 ± 0.65767<br>$(500,10^{0})$<br>0.01164 | 97.0325 ± 0.18179<br>$(1000,10^{-2})$<br>0.18504 |
| RVFL (sigmoidal)<br>Accuracy ± S.D.<br>$(M, \alpha)$<br>Time | **94.6078 ± 0**<br>$(1000,10^{-2})$<br>0.00794 | **82.3529 ± 0**<br>$(1000,10^{0})$<br>0.00039 | **97.1545 ± 0**<br>$(100,10^{-2})$<br>0.01095 |

Here, time in sec
Bold defines the best result

One can see from Table 2 that RVFL using hyperbolic tangent function (tanh), hard limit transfer function (hardlim), and sigmoidal RBF node is achieved better classification accuracy than ELM for all three datasets. We can easily conclude from Table 2 that the learning cost is determined by the considered hidden layer nodes. If the value of the hidden node is less, then the computational cost will be minimum. The approach RVFL takes comparable or minimum time in contrast to ELM. For more understanding, we have plotted the bar graph of the accuracy rank of all 12 approaches with three ASD datasets in Fig. 1. We have determined other quality measures named precision, recall, negative predictive value, rate of misclassification, F1-measure, G-mean, and Matthew's correlation coefficient [14, 25] for both the models as shown in Table 3 for autism-child-data, Table 4 for autism-adolescent-data, and Table 5 for autism-adult-aata, respectively.

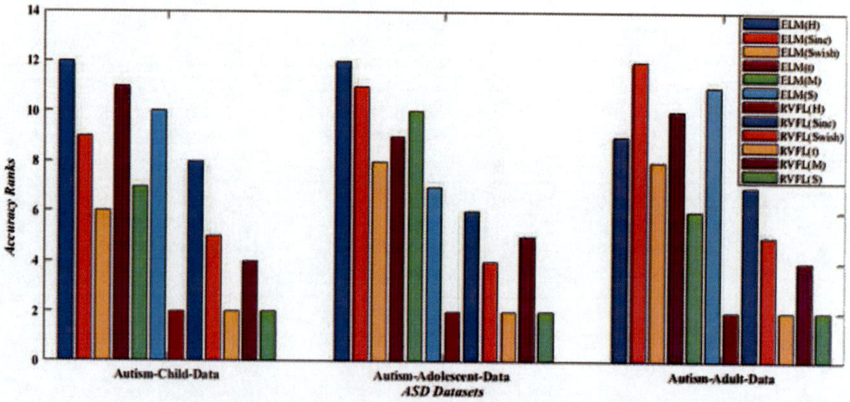


**Fig. 1** Accuracy ranks of the randomization-based model's ELM and RVFL with hardlim (H), sine, swish, tanh (t), multiquadric (M), sigmoidal (S) activation functions on ASD datasets

**Table 3** The performance analysis of randomization-based models ELM and RVFL with different activation functions on ASD datasets named autism-child-data in terms of various quality measures

| Measures/models | Precision | Recall | NPV | RME | $F_1$-measure | G-mean | MCC |
|---|---|---|---|---|---|---|---|
| ELM (hardlim) | 0.7064 | 0 | 0.4951 | 0.5049 | 0.655 | 0.8123 | 0.7769 |
| ELM (sine) | 0.6757 | 0.7398 | 0.7072 | 0.3108 | 0.7059 | 0.7068 | 0.3801 |
| ELM (swish) | 0.9067 | 0.9049 | 0.9033 | 0.0951 | 0.9057 | 0.9058 | 0.8099 |
| ELM (tanh) | 0.5692 | 0.6466 | 0.5884 | 0.4235 | 0.6046 | 0.6063 | 0.1545 |
| ELM (multiquard) | 0.8464 | 0.8427 | 0.8404 | 0.1569 | 0.8444 | 0.8445 | 0.6865 |
| ELM (sigmoidal) | 0.5855 | 0.6718 | 0.6092 | 0.4078 | 0.6224 | 0.6255 | 0.1884 |
| RVFL (hardlim) | **0.9259** | **0.9709** | **0.9688** | **0.0539** | **0.9479** | **0.9481** | **0.8932** |
| RVFL (sine) | 0.8341 | 0.8583 | 0.8516 | 0.1578 | 0.8458 | 0.846 | 0.6849 |
| RVFL (swish) | 0.945 | 0.8971 | 0.9008 | 0.0784 | 0.9201 | 0.9206 | 0.8447 |
| RVFL (tanh) | **0.9259** | **0.9709** | **0.9688** | **0.0539** | **0.9479** | **0.9481** | **0.8932** |
| RVFL (multiquard) | 0.9167 | 0.9612 | 0.9583 | 0.0637 | 0.9384 | 0.9387 | 0.8735 |
| RVFL (sigmoidal) | **0.9259** | **0.9709** | **0.9688** | **0.0539** | **0.9479** | **0.9481** | **0.8932** |

Bold defines the best result

**Table 4** The performance analysis ELM and RVFL with different activation functions on ASD datasets named autism-adolescent-data in terms of various quality measures

| Measures/models | Precision | Recall | NPV | RME | $F_1$-measure | G-mean | MCC |
|---|---|---|---|---|---|---|---|
| ELM (hardlim) | 0.6029 | **1** | 0.7149 | 0.3971 | 0.7523 | 0.7765 | 0.4159 |
| ELM (sine) | 0.6771 | 0.7659 | 0.5548 | 0.3618 | 0.7186 | 0.72 | 0.2208 |
| ELM (swish) | 0.7756 | 0.8049 | 0.6977 | 0.2618 | 0.7856 | 0.7879 | 0.4567 |
| ELM (tanh) | 0.6865 | 0.9756 | **0.9114** | 0.2853 | 0.8052 | 0.818 | 0.4154 |
| ELM (multiquard) | 0.7888 | 0.6 | 0.5542 | 0.3382 | 0.6815 | 0.6879 | 0.3492 |
| ELM (sigmoidal) | 0.719 | 0.9415 | 0.8251 | 0.2588 | 0.815 | 0.8226 | 0.453 |
| RVFL (hardlim) | 0.9143 | 0.7805 | 0.7273 | **0.1765** | **0.8421** | **0.8447** | **0.6553** |
| RVFL (sine) | 0.8831 | 0.6829 | 0.6439 | 0.2471 | 0.7678 | 0.7753 | 0.5345 |
| RVFL (swish) | **0.9325** | 0.7415 | 0.7011 | 0.1882 | 0.8259 | 0.8314 | 0.6467 |
| RVFL (tanh) | 0.9143 | 0.7805 | 0.7273 | **0.1765** | **0.8421** | **0.8447** | **0.6553** |
| RVFL (multiquard) | 0.9068 | 0.7122 | 0.6705 | 0.2176 | 0.7978 | 0.8036 | 0.5891 |
| RVFL (sigmoidal) | 0.9143 | 0.7805 | 0.7273 | **0.1765** | **0.8421** | **0.8447** | **0.6553** |

Bold defines the best result

We have also drawn Figs. 2, 3, and 4 corresponding to Tables 3, 4, and 5, respectively. In Table 4 for the autism-adolescent-data, the recall value of ELM (multiquadric) is better than other as well as negative predictive value is also better in the case of ELM (tanh). Tables 3, 4, 5 and Figs. 2, 3, 4 summarize the same results as Table 2 that RVFL using a hyperbolic tangent function (tanh), hard limit transfer function (hardlim), and sigmoidal activation function is far better rather than ELM. However,

**Table 5** The performance analysis of ELM and RVFL with different activation functions on ASD datasets named autism-adult-data in terms of various quality measures

| Measures/models | Precision | Recall | NPV | RME | $F_1$-measure | G-mean | MCC |
|---|---|---|---|---|---|---|---|
| ELM (hardlim) | 0.8513 | 0 | 0.7439 | 0.2561 | 0.5361 | 0.8345 | 0.5648 |
| ELM (sine) | 0.4047 | 0.6762 | 0.8538 | 0.3398 | 0.5068 | 0.5229 | 0.2924 |
| ELM (swish) | 0.8705 | 0.8111 | 0.9365 | 0.0793 | 0.8398 | 0.8403 | 0.7881 |
| ELM (tanh) | 0.7846 | 0 | 0.7429 | 0.2602 | 0.7356 | 0.7845 | 0.6983 |
| ELM (multiquard) | 0.9135 | 0.8317 | 0.9439 | 0.0634 | 0.8704 | 0.8715 | 0.8304 |
| ELM (sigmoidal) | 0.3211 | 0.0603 | 0.7478 | 0.2711 | 0.1004 | 0.1376 | 0.0368 |
| RVFL (hardlim) | **0.9746** | **0.9127** | **0.9706** | **0.0285** | **0.9426** | **0.9431** | **0.9246** |
| RVFL (sine) | 0.8601 | 0.827 | 0.9413 | 0.0789 | 0.8429 | 0.8432 | 0.7908 |
| RVFL (swish) | 0.918 | 0.8508 | 0.9499 | 0.0577 | 0.8831 | 0.8837 | 0.8459 |
| RVFL (tanh) | **0.9746** | **0.9127** | **0.9706** | **0.0285** | **0.9426** | **0.9431** | **0.9246** |
| RVFL (multiquard) | 0.9712 | 0.9111 | 0.97 | 0.0297 | 0.9402 | 0.9407 | 0.9213 |
| RVFL (sigmoidal) | **0.9746** | **0.9127** | **0.9706** | **0.0285** | **0.9426** | **0.9431** | **0.9246** |

Bold defines the best result

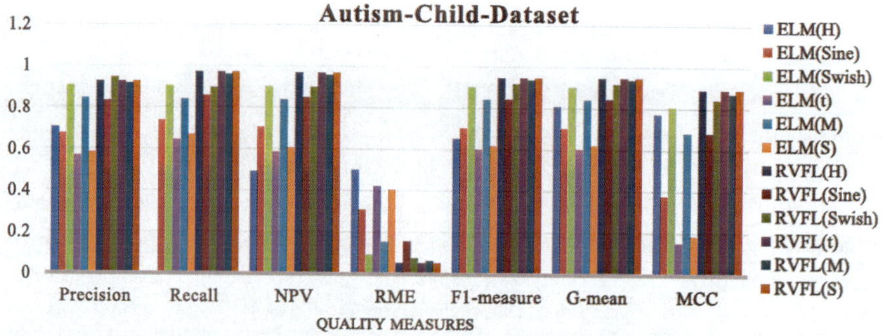


**Fig. 2** Comparative analysis of the ELM and RVFL with different activation functions on autism-child-datasets on various quality measures

overall RVFL with three activation functions is showing better performance in all datasets. The sigmoidal function is non-linear and continuously differentiable (0, 1) whilst construction of neural network-based model. The tanh function is a scaled-up or altered variant of the sigmoid function, although it still has the vanishing gradient problem.

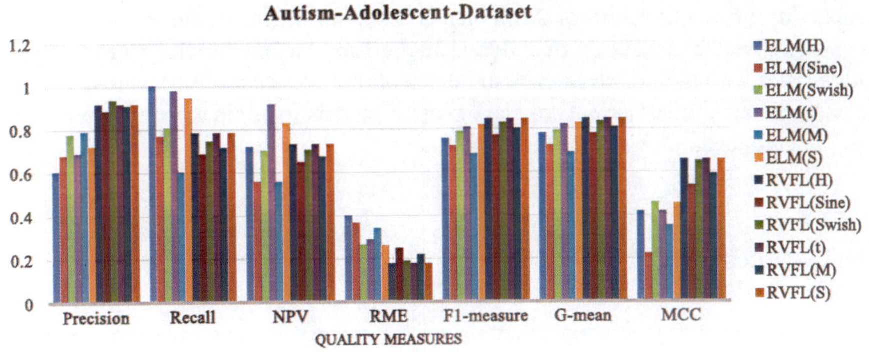


**Fig. 3** Comparative analysis of the ELM and RVFL with various activation functions on autism-adolescents-datasets on various quality measures

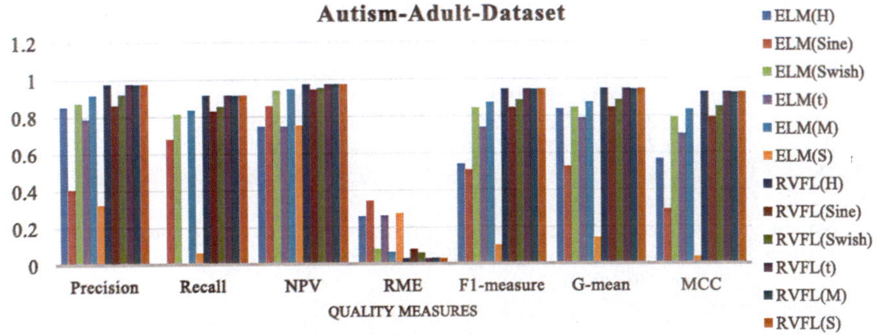


**Fig. 4** Comparative analysis of the ELM and RVFL with different activation functions on autism-adult-datasets on various quality measures

## 5 Conclusion and Future Work

Here, the classification efficacy of two randomization-based approaches, i.e., ELM and RVFL has experimented in the autism spectrum disorder dataset at all age groups. Theoretically, RVFL is comparable in speed compared to ELM. But, for a tiny dataset, it does not show any significance in the training time of ELM and RVFL approaches. Therefore, in results of the generalization of the approaches, we emphasized based on the quality measures like accuracy, precision, recall, negative predictive value, rate of misclassification, $F_1$-measure, G-mean, and Matthew's correlation coefficient. ELM and RVFL are compared with each other using different variants of activation functions to verify the classification performance. It could be derived from the above-computed results that accuracy, precision, recall, negative predictive value, $F_1$-measure, G-mean, and Matthew's correlation coefficient of ELM model are lower than the RVFL model which supports its generalization. It could be noted that the RME of RVFL is less compared to ELM. This shows that the RVFL has the superior capability of classification of ASD mental disorder. Since, in this work, we have considered the three datasets of children, adolescents, and adults which are confined. Large data with high dimensions will be one of the future aspects as well as we can also implement other variants of classification models to attain a better solution for ASD neurodevelopment disorder in the future.

## References


1. Thabtah, F.: Autism spectrum disorder screening: machine learning adaptation and DSM-5 fulfillment. In: Proceedings of the 1st International Conference on Medical and health Informatics, pp. 1–6. ACM (2017)
2. Hyde, K.K., Novack, M.N., LaHaye, N., Parlett-Pelleriti, C., Anden, R., Dixon, D. R., Linstead, E.: Applications of supervised machine learning in autism spectrum disorder research: a review. Rev. J. Autism Dev. Disord. **6**(2), 128–146 (2019)
3. Omar, K.S., Mondal, P., Khan, N.S., Rizvi, M.R.K., Islam, M.N.: A machine learning approach to predict autism spectrum disorder. In: International Conference on Electrical, Computer and Communication Engineering (ECCE), pp. 1–6. IEEE (2019)
4. Raj, S., Masood, S.: Analysis and detection of autism spectrum disorder using machine learning techniques. In: International Conference on Computational Intelligence and Data Science. Procedia Computer Science, 167, pp. 994–1004. ELSEVIER (2020)
5. Thabtah, F.: Machine learning in autistic spectrum disorder behavioral research: a review and ways forward. Inform. Health Soc. Care **44**(3), 278–297 (2019)
6. Barbaro, J., Dissanayake, C.: Autism spectrum disorders in infancy and toddlerhood: a review of the evidence on early signs, early identification tools, and early diagnosis. J. Dev. Behav. Pediatr. **30**(5), 447–459 (2009)
7. Dua, D., Graff, C.: In: Irvine, C.A.(ed.) UCI Machine Learning Repository. University of California, School of Information and Computer Science (2019). http://archive.ics.uci.edu/ml. Accessed 1 Jan 2021
8. W.H.O. https://www.who.int/news-room/fact-sheets/detail/autism-spectrum-disorders. Accessed 1 Mar 2021
9. Hearst, M.A., Dumais, S.T., Osuna, E., Platt, J., Scholkopf, B.: Support vector machines. IEEE Intel. Syst. Appl. **13**(4), 18–28 (1998)
10. Gupta, U., Gupta, D.: Lagrangian twin-bounded support vector machine based on L2-norm. In: Kalita, J., Balas, V., Borah, S., Pradhan, R. (eds.) Recent developments in machine learning and data analytics. Advances in Intelligent Systems and Computing, vol. 740, pp. 431–444. Springer, Singapore (2018)
11. Gupta, U., Gupta, D., Prasad, M.: Kernel target alignment based fuzzy least square twin bounded support vector machine. In: 2018 IEEE Symposium Series on Computational Intelligence (SSCI), pp. 228–235. IEEE (2018)
12. Gupta, U., Gupta, D.: Least squares large margin distribution machine for regression. Appl. Intel. 1–36 (2021)
13. Smola, A.J., Schölkopf, B.: A tutorial on support vector regression. Statist. Comput. **14**(3), 199–222 (2004)
14. Gupta, D., Choudhury, A., Gupta, U., Singh, P., Prasad, M.: Computational approach to clinical diagnosis of diabetes disease: a comparative study. Multimed. Tools Appl. 1–26 (2021)
15. Cruz, J.A., Wishart, D.S.: Applications of machine learning in cancer prediction and prognosis. Cancer Inform. **2**, 59–78 (2006)
16. Li, B., Sharma, A., Meng, J., Purushwalkam, S., Gowen, E.: Applying machine learning to identify autistic adults using imitation: an exploratory study. PLoS ONE **12**(8), 1–19 (2017)

17. Vaishali, R., Sasikala, R.: A machine learning based approach to classify autism with optimum behaviour sets. Int. J. Eng. Technol. **7**, 1–6 (2018)
18. Hazarika, B.B., Gupta, D., Berlin, M.: Modeling suspended sediment load in a river using extreme learning machine and twin support vector regression with wavelet conjunction. Environ. Earth Sci. **79**, 1–15 (2020)
19. Gupta, U., Gupta, D.: Regularized based implicit Lagrangian twin extreme learning machine in primal for pattern classification. Int. J. Mach. Learn. Cybern. 1–32 (2021)
20. Rao, C. R., Mitra, S.K.: Further contributions to the theory of generalized inverse of matrices and its applications. Sankhyā: The Indian J. Statist. Ser.A. 289–300 (1971)
21. Lai, M., Lee, J., Chiu, S., Charm, J., So, W.Y., Yuen, F.P., Kwok, C., Tsoi, J., Lin, Y., Zee, B.: A machine learning approach for retinal images analysis as an objective screening method for children with autism spectrum disorder. EClinicalMedicine. **28**, 100588 (2020)
22. Thabtah, F., David, P.: A new machine learning model based on induction of rules for autism detection. Health Inform. J. **26**(1), 264–286 (2020)
23. Huang, G.B., Zhu, Q.Y., Siew, C.K.: Extreme learning machine: theory and applications. Neurocomputing **70**(1–3), 489–501 (2006)
24. Pao, Y.H., Phillips, S.M., Sobajic, D.J.: Neural-net computing and the intelligent control of systems. Int. J. Control **56**(2), 263–289 (1992)
25. Gupta, U., Meher, P.: Statistical analysis of target tracking algorithms in thermal imagery. In: Mallick, P., Balas, V., Bhoi, A., Chae, G.S. (eds.) Cognitive Informatics and Soft Computing. Advances in Intelligent Systems and Computing, vol. 1040, pp. 635–646. Springer, Singapore (2020)

# A Twin Kernel Ridge Regression Classifier for Binary Classification

Barenya Bikash Hazarika, Deepak Gupta, and Parashjyoti Borah

**Abstract** Kernel ridge regression (KRR) is a popular machine learning technique for tasks related to both regression and classification. To improve the generalization ability of the KRR model, this paper suggests a twin KRR model for binary classification. The twin KRR (TKRR) solves two sets of linear equations rather than solving quadratic programming problems (QPPs) unlike support vector machine (SVM) and twin SVM (TWSVM). The conventional KRR learns only one large hyperplane while the TKRR learns two small hyperplanes. The proposed TKRR is novel, fast, and intuitive. Experimental simulations have been carried out on an artificial and a few interesting real-world datasets. The obtained classification accuracies of TKRR are statistically compared with SVM, TWSVM, and KRR models. The experimental outcomes demonstrate the efficacy and usability of the proposed TKRR model.

**Keywords** Machine learning · Kernel ridge regression · Binary classification · Least squares

## 1 Introduction

### *1.1 A Subsection Sample*

The support vector machine (SVM) is among the most popular machine learning (ML) models which follows the statistical learning theory [1]. Cortes and Vapnik [2] suggested the SVM for classification tasks, and it has emerged to be a very influential supervised model. SVM is based on the structural risk minimization (SRM) rule, which decreases the occurrence of risk during the training process and increases the generalization capability. Because of its improved performance, SVM

B. Bikash Hazarika · D. Gupta (✉)
National Institute of Technology Arunachal Pradesh, Jote, Arunachal Pradesh 791112, India
e-mail: deepakjnu85@gmail.com; deepak@nitap.ac.in

P. Borah
Indian Institute of Information Technology, Guwahati 781039, India



D. Gupta et al. (eds.), *Pattern Recognition and Data Analysis with Applications*, Lecture Notes in Electrical Engineering 888,
https://doi.org/10.1007/978-981-19-1520-8_58

is a commonly used classification model in several fields like pattern classification [2], class imbalance learning [3], and many more [4–6]. But one of the major problems with traditional SVM is seeking a solution to the complex QPP. Hence, it has a high-computational cost. To decrease the computational cost of SVM, Jayadeva et al. [7] came up with the idea of twin SVM (TWSVM). TWSVM solves two smaller QPPs as opposed to solving a bigger one. Theoretically, TWSVMs computational speed is quadrupled compared to conventional SVM model. Due to its fast training as well as high-generalization performance, the TWSVM has been successfully implemented in several fields like pattern classification [8], medical data classification [9], and so on.

The single layer feed forward networks are widely implemented for solving tasks related to both classification and regression [10]. The randomized version of feed forward network also known as the random vector functional link networks (RVFL) [11] has gained tremendous attention among researchers because of its fast training and high-generalization ability. An extensive study on the randomization-based feed forward networks has been carried out in [12]. Few recent RVFL-based studies can be explored in [13, 14]. Extreme learning machine [15–18] is another popular machine learning model that has been popularly used in wide areas of applications. Kernel ridge regression (KRR) [19]-based approaches have recently gained quite a lot of attention due to their non-iterative learning approach. KRR extends the ridge regression model for solving nonlinear problems. Although it was originally suggested for regression but it shows promising classification ability almost similar to SVM. Following that, a number of articles were published that used similar methodology as KRR but used other names and failed to reference the original KRR. Recently, a co-trained KRR was suggested by [20] using the "perturb and combine" strategy which trains two KRRs jointly. In this work, to improve the generalization ability of KRR, we suggest a novel twin KRR for classification. TKRR solves two systems of linear equations to obtain its optimum solution. The prime objectives of this paper are

1. To propose a novel twin KRR model in order to enhance the classification ability of KRR.
2. To statistically analyze the performance of TKRR with state-of-the-art SVM, TWSVM, and the KRR models.

In Sect. 2, the related works are addressed. Section 3 elaborates the proposed model. In Sect. 4, numerical simulation is discussed. Finally, the conclusion is demonstrated with future aspects in Sect. 5.

## 2 Related Works

### 2.1 The SVM Model

SVM [1] finds a hyperplane as $\phi(\mathbf{x})^t\mathbf{w} + \mathbf{e}b = 0$, $\phi(\mathbf{x})$ is maps the feature of $\mathbf{x}$. Consider $Y_i \in \{-1,\ 1\}$ indicates the class labels of the training matrix $\mathbf{X}$ of order $m \times n$, where $i = 1,\ 2,\ \ldots,\ m$. The unknowns $\mathbf{w}$ and $b$ are calculated by obtaining the solution of the optimization problem:

$$\begin{aligned} &\min \frac{1}{2}\|\mathbf{w}\|^2 + C\mathbf{e}^t\xi, \\ &\text{s.t.,}\ \ Y(\phi(\mathbf{x})^t\mathbf{w} + \mathbf{e}b) \geq \mathbf{e} - \xi,\ \xi \geq 0 \end{aligned} \tag{1}$$

where $\mathbf{e}$ is the one's vector and $\xi$ indicates the slack variable, respectively. $C \geq 0$ indicates the tradeoff parameter.

The dual problem of (1) can be achieved by introducing the Lagrangian multiplier, $\lambda \geq 0$ to (1) and further solving it by using the Karush–Kuhn–Tucker (KKT) condition:

$$\begin{aligned} &\min \frac{1}{2}\lambda^t\mathbf{Y}K(\mathbf{x}, \mathbf{x}^t)Y\lambda - \mathbf{e}^t\lambda, \\ &\text{s.t.,}\ \ 0 \leq \lambda \leq C\mathbf{e},\ \ \mathbf{y}^t\lambda = 0, \end{aligned} \tag{2}$$

where $\mathbf{Y} = \text{diag}(\mathbf{y})$ and $K(\mathbf{x}, \mathbf{x}^t)$ indicate the nonlinear kernel function so that$K(\mathbf{x}, \mathbf{x}) = \phi(\mathbf{x}).\phi(\mathbf{x}) = \phi(\mathbf{x})^t\phi(\mathbf{x})$. $\mathbf{w}$ and $b$ are determined by solving the QPP of (2) with respect to $\lambda$. For a new sample, $\mathbf{x} \in \Re^n$ the SVM classifier may be determined as

$$f(x) = \text{sign}\ (\phi(\mathbf{x})^t w + b). \tag{3}$$

### 2.2 The TWSVM Model

TWSVM [7] seeks for two non-parallel hyperplanes as $f_1(\mathbf{x}) = K(\mathbf{x^t}, \mathbf{T^t})\mathbf{w_1} + \mathbf{b_1} = 0$ and $f_2(x) = K(\mathbf{x^t}, \mathbf{T^t})\mathbf{w_2} + \mathbf{b_2} = 0$. Let two training matrices $\mathbf{X_1}$ and $\mathbf{X}_2$ are of $p \times n$ and $q \times n$ sizes, respectively. The unknown variables $\mathbf{w_1}$, $\mathbf{w}_2$ and $b_1,\ b_2$ are determined by solving two optimization problems as

$$\begin{aligned} &\min \frac{1}{2}\left\|K(\mathbf{X_1}, \mathbf{T}^t)\mathbf{w_1} + \mathbf{e}b_1\right\|^2 + C_1\mathbf{e}^t\xi_1, \\ &\text{s.t.,}\ -(K(\mathbf{X_2}, \mathbf{T}^t).\mathbf{w_1} + \mathbf{e}b_1) \geq \mathbf{e} - \xi_1,\ \xi_1 \geq 0 \end{aligned} \tag{4}$$

and

$$\min \frac{1}{2}\left\| K(\mathbf{X_2}, \mathbf{T}^t)\mathbf{w_2} + \mathbf{e}b_2 \right\|^2 + C_2\mathbf{e}^t\xi_2,$$
$$\text{s.t.,}\ (K(\mathbf{X_1}, T^t).\mathbf{w_2} + \mathbf{e}b_2) \geq \mathbf{e} - \xi_2,\ \ \xi_2 \geq 0, \tag{5}$$

where $\xi_1$ and $\xi_2$ are the slack variables. $C_1, C_2 > 0$ represents the penalty parameters, and $K(.)$ represents the kernel function. Also, consider $\mathbf{T} = [\mathbf{X_1};\ \mathbf{X_2}]$..

To solve the primal problems of (4) and (5), their dual can be determined by using the Lagrangian multipliers and solving it by using the KKT condition as

$$\min\ \frac{1}{2}\lambda_1^t\mathbf{M}(\mathbf{L^tL})^{-1}\mathbf{M^t}\lambda_1 - \mathbf{e}^t\lambda_1,$$
$$\text{s.t.,}\ 0 \leq \lambda_1 \leq C_1\mathbf{e}, \tag{6}$$

and

$$\min\ \frac{1}{2}\lambda_2^t\mathbf{L}(\mathbf{M^tM})^{-1}\mathbf{L}^t\lambda_2 - \mathbf{e}^t\lambda_2,$$
$$\text{s.t.,}\ 0 \leq \lambda_2 \leq C_2\mathbf{e}, \tag{7}$$

where $\lambda_1 \geq 0$ and $\lambda_2 \geq 0$ are Lagrangian multipliers. Also, $\mathbf{L} = [K(\mathbf{X_1}, \mathbf{T^t})\quad \mathbf{e}]$ and $\mathbf{M} = [K(\mathbf{X_2}, \mathbf{T}^t)\quad \mathbf{e}]$. For a new example, $\mathbf{x} \in R^n$ the TWSVM classifier may be expressed as

$$f(\mathbf{x}) = \arg\min\left|K(\mathbf{x^t}, \mathbf{X^t})\mathbf{w_i} + b_i\right|,\ i = 1, 2\ . \tag{8}$$

## 2.3 The KRR Model

The primal of KRR [19] may be presented by the following problem of optimization as shown below:

$$\min\ \frac{C}{2}\|\mathbf{w}\|^2 + \frac{1}{2}\|\xi\|^2,$$
$$\text{s.t.,}\ \ y - \phi(\mathbf{x})^t\mathbf{w} = \xi, \tag{9}$$

where $\mathbf{e}$ is the one's vector and $\xi$ indicates the slack vector.

The Lagrangian of (9) may be formulated as

$$L(\mathbf{w}, \xi; \lambda) = \frac{C}{2}\|\mathbf{w}\|^2 + \frac{1}{2}\|\xi\|^2 - \alpha^t\left\{\mathbf{y} - \phi(\mathbf{x})^t\mathbf{w} - \xi\right\}, \tag{10}$$

where $\lambda$ is the Lagrangian multiplier. Further applying the KKT condition, the dual form may be obtained as

$$L = -\frac{1}{2C}\lambda^t \phi(\mathbf{x})\phi(\mathbf{x})^t \lambda - \frac{1}{2}\lambda^t \mathbf{I}\lambda - \mathbf{y}^t \lambda, \tag{11}$$

where $\mathbf{I}$ is an identity matrix. After differentiating Eq. (11) with respect to $\lambda$ and further equating it to zero, we obtain

$$\begin{aligned} \frac{\partial L}{\partial w} &= \frac{1}{C}\phi(\mathbf{x})\phi(\mathbf{x})^t \lambda + \mathbf{I}\lambda + \mathbf{y} = 0, \\ \lambda &= -\left(\frac{1}{C}\phi(\mathbf{x})\phi(\mathbf{x})^t + \mathbf{I}\right)^{-1} \mathbf{y}, \end{aligned} \tag{12}$$

For a new input example, $\mathbf{x} \in R^N$ the KRR decision function may be generated as

$$f(\mathbf{x}) = \text{sign}\left\{-\frac{1}{C}\phi(\mathbf{x})^t \lambda\right\}. \tag{13}$$

## 3 The Proposed TKRR Model

TKRR is a binary classifier that solves two systems of linear equation. Now let us consider a binary problem of classification, where $m_1$ data points belong to class $+1$ and $m_2$ data points belongs to class $-1$ in $n-$ dimensional space $n \in \Re^n$. Let $\mathbf{A} \in R^{m_1 \times n}$ and $\mathbf{B} \in R^{m_2 \times n}$ represent the positive and negative classes, respectively. $\mathbf{w_1}, \mathbf{w_2}$ are the unknown variables. Given a binary classification problem, the primal problems of the proposed TKRR may be expressed as

$$\begin{aligned} &\min \frac{C_1}{2}\|\mathbf{w_1}\|^2 + \frac{1}{2}\|\mathbf{A}\mathbf{w_1}\|^2 + \frac{C_3}{2}\xi^t \xi, \\ &\text{s.t.}, -B\mathbf{w_1} = \mathbf{e} - \xi, \end{aligned} \tag{14}$$

and

$$\begin{aligned} &\min \frac{C_2}{2}\|\mathbf{w_2}\|^2 + \frac{1}{2}\|\mathbf{B}\mathbf{w_2}\|^2 + \frac{C_4}{2}\psi^t \psi, \\ &\text{s.t.}, \ A\mathbf{w_2} = \mathbf{e} - \psi, \end{aligned} \tag{15}$$

where $C_1, \ C_2, \ C_3$, and $C_4$ are user-defined parameters. $\mathbf{e}$ is the one's vector. $\xi$ and $\psi$ indicate the slack variables. Now replace the values of $\xi$ and $\psi$ in (14) and (15), we get

$$L_1 = \frac{C_1}{2}\|\mathbf{w_1}\|^2 + \frac{1}{2}\|\mathbf{Aw_1}\|^2 - \frac{C_3}{2}\|\mathbf{e} + \mathbf{Bw_1}\|^2, \tag{16}$$

and

$$L_2 = \frac{C_2}{2}\|\mathbf{w_2}\|^2 + \frac{1}{2}\|\mathbf{Bw_2}\|^2 + \frac{C_4}{2}\|\mathbf{e} - \mathbf{Aw_2}\|^2. \tag{17}$$

Now, differentiating (16) and (17) with respect to $\mathbf{w_1}$ and $\mathbf{w_2}$, respectively, and further equating it to zero, we get

$$\frac{\partial L_1}{\partial w_1} = C_1\mathbf{w_1} + \mathbf{A^tAw_1} - C_3\mathbf{B^t}(\mathbf{e} + \mathbf{Bw_1}) = 0, \tag{18}$$

and

$$\frac{\partial L_2}{\partial w_2} = C_2\mathbf{w_2} + \mathbf{B^tBw_2} + C_4\mathbf{A^t}(\mathbf{e} - \mathbf{Aw_2}) = 0. \tag{19}$$

The solutions of (18) and (19) can be derived to be

$$\mathbf{w_1} = -\left(\frac{C_1}{C_3}\mathbf{I} + \frac{1}{C_3}\mathbf{A^tA} + \mathbf{B^tB}\right)^{-1}\mathbf{B^te}, \tag{20}$$

and

$$\mathbf{w_2} = \left(\frac{C_2}{C_4}\mathbf{I} + \frac{1}{C4}\mathbf{B^tB} + \mathbf{A^tA}\right)^{-1}\mathbf{A^te}. \tag{21}$$

The TKRR classifier for any unknown point can be generated as

$$f(\mathbf{x}) = \arg\min\left|K(\mathbf{x}^t, \mathbf{X}^t)\mathbf{w}_i\right|,\ i = 1, 2\ .$$

## 4 Numerical Experiments

The experiments have been performed using MATLAB software on a computer with 32 gigabytes of random access memory, Intel Core i7 embedded with 3.20 Gh clock speed. The Gaussian RBF kernel is selected in the experiments which can be expressed as $k(x_l, x_m) = -\exp(-\mu\,||x_l - x_m||^2)$, where $x_l, x_m$ denotes any data point. The choice of parameter is very essential to get the optimum result for any model. In this work, we chose $C$ and $\mu$ from $\{10^{-5}, ..., 10^5\}$ and $\{2^{-5}, ..., 2^5\}$, respectively, for the reported models along with the proposed TKRR. The normalization is performed by considering, $\overline{x}_{lr} = \frac{x_{lr} - x_r^{\min}}{x_r^{\max} - x_r^{\min}}$, where $x_r^{\min} = \min\limits_{l=1,..,m}(x_{lr})$ and $x_r^{\max}$

$= \max_{r=1,..,m} (x_{lr})$ are the extreme and least values, respectively, of the $r^{th}$ attribute of all input samples $x_l$. $\overline{x}_{lr}$ is the normalized outcome of $x_{lr}$.

### 4.1 Simulation on an Artificial Dataset

The artificially created 2D Ripley dataset [21] has a total of 1250 data points of which 250 data points are used to train, and 1000 samples are used to test. From Fig. 1, it is possible to visualize the different classifiers obtained from the Ripley dataset. Table 1 displays the classification accuracy of Ripley's dataset using SVM, TWSVM, KRR, and TKRR along with their optimum parameters and training period.

One can observe that the proposed TKRR shows the best classification accuracy among the classifiers.

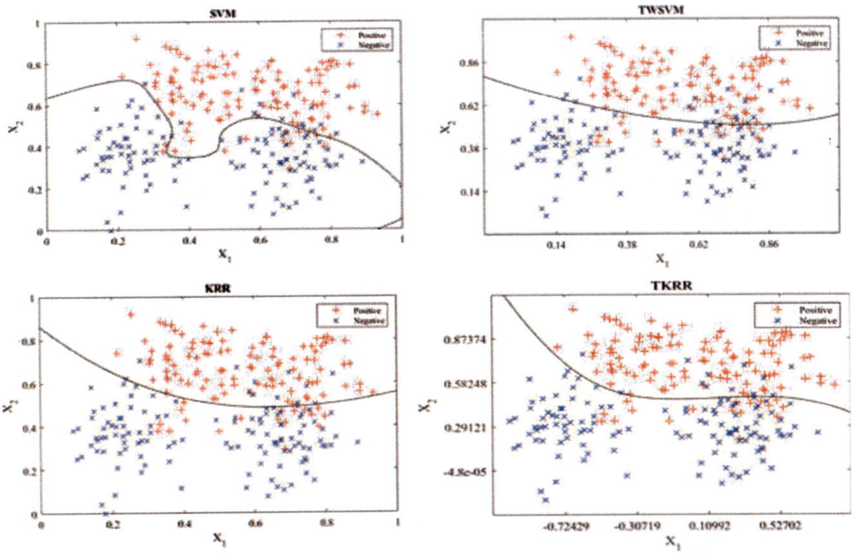


**Fig. 1** Classifiers obtained on the artificially created Ripley dataset

**Table 1** Accuracies obtained on the artificial Ripley dataset (best result is in boldface)

| Dataset (Train × test × attributes) | SVM $(C, \mu)$ Time (s) | TWSVM $(C_1 = C_2, \mu)$ Time (s) | KRR $(C, \mu)$ Time (s) | TKRR $(C_1 = C_2, C_3 = C_4, \mu)$ Time (s) |
|---|---|---|---|---|
| Ripley (250 × 1000 × 2) | 89.6<br>$(10^{-2}, 2^{-2})$<br>0.01924 | 89.1<br>$(10^{-5}, 2^{1})$<br>0.03753 | 90<br>$(10^{-5}, 2^{3})$<br>0.00391 | **90.5**<br>$(10^{-5}, 10^{0}, 2^{1})$<br>0.0179 |

## 4.2 Simulation on a Few Real-World Datasets

Datasets of different types and sizes are used for experiments. These datasets are collected from the UCI ML data repository [22] and KEEL repository [23]. The information about the datasets is presented in Table 2.

Classification accuracies with their ranks, optimal parameters, and training time (sec.) are exhibited in Table 3. It is noticeable from Table 3 that TKRR illustrates comparable or better generalization ability compared to SVM, TWSVM, and KRR. Additionally, it can be seen that TKRR shows low-computational cost compared to SVM and TWSVM which indicates the computational efficiency of the TKRR model. The time graph that is shown in Fig. 2 confirms the same where the training time of SVM, TWSVM, and TKRR is shown. It is further observable from Table 3 that the TKRR model has the lowest mean rank.

**Table 2** Information of the datasets

| Dataset | #Total samples | #Training samples | #Testing samples | #Attributes |
|---|---|---|---|---|
| Abalone9-18 | 731 | 439 | 292 | 7 |
| Australian | 690 | 414 | 276 | 14 |
| Dermatology | 358 | 215 | 143 | 34 |
| Ecoli-0-1_vs_2-3-5 | 244 | 147 | 97 | 7 |
| Ecoli-0-1_vs_5 | 240 | 144 | 96 | 6 |
| Ecoli-0-1-4-7_vs_5-6 | 332 | 200 | 132 | 6 |
| Ecoli-0-2-6-7_vs_3-5 | 224 | 135 | 89 | 7 |
| Ecoli-0-4-6_vs_5 | 203 | 122 | 81 | 6 |
| Glass | 214 | 129 | 85 | 9 |
| Glass-0-4_vs_5 | 92 | 56 | 36 | 9 |
| Iris | 150 | 90 | 60 | 4 |
| Ndc2k | 2200 | 1320 | 880 | 32 |
| New-thyroid1 | 215 | 129 | 86 | 5 |
| Seeds | 210 | 126 | 84 | 7 |
| Shuttle-6_vs_2-3 | 230 | 138 | 92 | 9 |
| Titanic | 2201 | 1321 | 880 | 3 |
| Yeast-0-5-6-7-9_vs_4 | 528 | 317 | 211 | 8 |
| Yeast1 | 2968 | 1781 | 1187 | 8 |
| Yeast3 | 1484 | 891 | 593 | 8 |
| 03subcl5-600-5-0-BI | 600 | 360 | 240 | 2 |

**Table 3** Accuracies and ranks obtained on the real-world datasets (best results are in boldface)

| Dataset | SVM (Rank)<br>$(C, \mu)$ Time (s) | TWSVM (Rank)<br>$(C_1 = C_2, \mu)$<br>Time (s) | KRR (Rank)<br>$(C, \mu)$<br>Time (s) | TKRR (Rank)<br>$(C_1 = C_2, C_3 = C_4, \mu)$ Time (s) |
|---|---|---|---|---|
| Abalone9-18 | 97.5945 (2.5)<br>$(10^3, 2^5)$<br>0.52146 | 97.5945 (2.5)<br>$(10^{-1}, 2^0)$<br>0.09557 | 96.5753 (4)<br>$(10^0, 2^{-2})$<br>0.01029 | **97.6027 (1)**<br>$(10^{-1}, 10^{-5}, 2^0)$<br>0.04471 |
| Australian | 84 (2.5)<br>$(10^{-1}, 2^2)$<br>0.05375 | 84 (2.5)<br>$(10^{-5}, 2^{-1})$<br>0.05661 | 83.6957 (4)<br>$(10^1, 2^1)$<br>0.013939 | **85.1449 (1)**<br>$(10^{-3}, 10^{-3}, 2^3)$<br>0.05269 |
| Dermatology | 99.2958 (4)<br>$(10^{-1},2^0)$<br>0.03636 | **100** (2)<br>$(10^{-5}, 2^0)$<br>0.03387 | **100** (2)<br>$(10^{-5}, 2^{-1})$<br>0.01004 | **100 (2)**<br>$(10^{-5}, 10^{-3}, 2^5)$<br>0.03332 |
| Ecoli-0-1_vs_2-3-5 | 93.75 (3)<br>$(10^1, 2^{-1})$<br>0.06669 | 92.7083 (4)<br>$(10^{-1}, 2^2)$<br>0.01752 | 93.8144 (2)<br>$(10^0, 2^{-1})$<br>0.00469 | **95.8763 (1)**<br>$(10^{-4}, 10^{-5}, 2^5)$<br>0.00775 |
| Ecoli-0-1_vs_5 | 96.8421 (3.5)<br>$(10^0, 2^{-1})$<br>0.05586 | 96.8421 (3.5)<br>$(10^0, 2^0)$<br>0.03229 | **96.875** (1.5)<br>$(10^{-1}, 2^{-1})$<br>0.00225 | **96.875 (1.5)**<br>$(10^{-5}, 10^{-5}, 2^0)$<br>0.0096 |
| Ecoli-0-1-4-7_vs_5-6 | 98.4733 (2)<br>$(10^0, 2^{-1})$<br>0.11443 | 97.7099 (4)<br>$(10^1, 2^0)$<br>0.02766 | 97.7273 (3)<br>$(10^0, 2^{-1})$<br>0.00420 | **98.4848 (1)**<br>$(10^{-3}, 10^{-5}, 2^4)$<br>0.01493 |
| Ecoli-0-2-6-7_vs_3-5 | **100** (1)<br>$(10^0, 2^{-1})$<br>0.05718 | 96.5909 (4)<br>$(10^{-3}, 2^{-1})$<br>0.02618 | 96.6292 (2.5)<br>$(10^{-3}, 2^0)$<br>0.00132 | 96.6292 (2.5)<br>$(10^{-2}, 10^5, 2^4)$<br>0.0069 |
| Ecoli-0-4-6_vs_5 | 96.25 (2.5)<br>$(10^0, 2^{-1})$<br>0.04093 | 96.25 (2.5)<br>$(10^{-1}, 2^{-1})$<br>0.02307 | 95.0617 (4)<br>$(10^{-5}, 2^{-5})$<br>0.00186 | **96.2963 (1)**<br>$(10^{-5}, 10^{-5}, 2^4)$<br>0.00554 |
| Glass | 78.5714 (4)<br>$(10^1, 2^{-3})$<br>0.04909 | 79.7619 (3)<br>$(10^{-5}, 2^{-3})$<br>0.03324 | 81.1765 (2)<br>$(10^{-1}, 2^{-3})$<br>0.00371 | **83.5294 (1)**<br>$(10^{-2}, 10^{-1}, 2^{-1})$<br>0.01303 |
| Glass-0-4_vs_5 | **100 (2)**<br>$(10^2, 2^0)$<br>0.01195 | **100 (2)**<br>$(10^0, 2^0)$<br>0.01743 | 97.2222 (4)<br>$(10^{-1}, 2^0)$<br>0.00034 | **100 (2)**<br>$(10^{-5},10^0, 2^3)$<br>0.00112 |
| Iris | **100 (2)**<br>$(10^{-2}, 2^{-1})$<br>0.02525 | 98.3051 (4)<br>$(10^{-5}, 2^{-3})$<br>0.01248 | **100 (2)**<br>$(10^{-5}, 2^{-5})$<br>0.00142 | **100 (2)**<br>$(10^{-5}, 10^{-5}, 2^{-5})$<br>0.00332 |
| Ndc2k | 96.3595 (4)<br>$(10^1, 2^{-1})$<br>0.80162 | 96.587 (3)<br>$(10^{-1}, 2^{-1})$<br>0.79199 | 96.7045 (2)<br>$(10^{-4}, 2^2)$<br>0.13552 | **97.0455 (1)**<br>$(10^{-5}, 10^{-5}, 2^3)$<br>0.47432 |
| New-thyroid1 | 98.8235 **(3)**<br>$(10^1, 2^{-1})$<br>0.04821 | 97.6471 **(4)**<br>$(10^{-5}, 2^{-2})$<br>0.01769 | **98.8372 (1.5)**<br>$(10^{-1}, 2^{-3})$<br>0.00115 | **98.8372 (1.5)**<br>$(10^{-5}, 10^{-4}, 2^2)$<br>0.00634 |

(continued)

**Table 3** (continued)

| Dataset | SVM (Rank) $(C, \mu)$ Time (s) | TWSVM (Rank) $(C_1 = C_2, \mu)$ Time (s) | KRR (Rank) $(C, \mu)$ Time (s) | TKRR (Rank) $(C_1 = C_2, C_3 = C_4, \mu)$ Time (s) |
|---|---|---|---|---|
| Seeds | 86.747 (4) $(10^0, 2^{-1})$ 0.04631 | 90.3614 (3) $(10^0, 2^1)$ 0.01792 | **92.8571 (1.5)** $(10^{-2}, 2^{-1})$ 0.00175 | **92.8571 (1.5)** $(10^{-5}, 10^0, 2^3)$ 0.00578 |
| Shuttle-6_vs_2-3 | **100 (2)** $(10^1, 2^{-1})$ 0.05799 | 98.9011 (4) $(10^{-4}, 2^{-1})$ 0.02959 | **100 (2)** $(10^{-3}, 2^1)$ 0.00123 | **100 (2)** $(10^{-5}, 10^{-5}, 2^5)$ 0.01047 |
| Titanic | 78.4983 (4) $(10^0, 2^{-1})$ 0.59380 | **79.7497 (1)** $(10^{-2}, 2^{-1})$ 0.64556 | 79.2045 (3) $(10^{-3}, 2^{-5})$ 0.60593 | 79.6591 (2) $(10^0, 10^1, 2^0)$ 0.47292 |
| Yeast-0-5-6-7-9_vs_4 | 90 (3.5) $(10^1, 2^{-2})$ 0.27344 | 90 (3.5) $(10^{-1}, 2^{-2})$ 0.04516 | 90.9953 (2) $(10^0, 2^{-2})$ 0.00677 | **91.9431 (1)** $(10^0, 10^0, 2^{-2})$ 0.02967 |
| Yeast1 | 99.5784 (2) $(10^0, 2^{-5})$ 1.3437 | 98.5666 (4) $(10^{-3}, 2^{-4})$ 1.4797 | 98.9048 (3) $(10^3, 2^{-5})$ 0.21819 | **100 (1)** $(10^{-4}, 10^{-5}, 2^{-4})$ 1.20761 |
| Yeast3 | 94.0878 (2) $(10^4, 2^2)$ 0.38104 | 93.9189 (3) $(10^{-2}, 2^{-1})$ 0.32531 | 92.5801 (4) $(10^3, 2^{-4})$ 0.05135 | **94.0978 (1)** $(10^0, 10^{-2}, 2^{-2})$ 0.21602 |
| 03subcl5-600-5-0-BI | 93.7238 (2) $(10^2, 2^{-4})$ 0.0355 | 91.6318 (4) $(10^{-5}, 2^{-5})$ 0.06756 | 93.3333 (3) $(10^0, 2^{-5})$ 0.00661 | **94.1667 (1)** $(10^{-1}, 10^{-1}, 2^5)$ 0.05419 |
| Mean | 94.1298 (2.775) | 93.8563 (3.275) | 94.1097 (2.55) | **94.9523 (1.4)** |

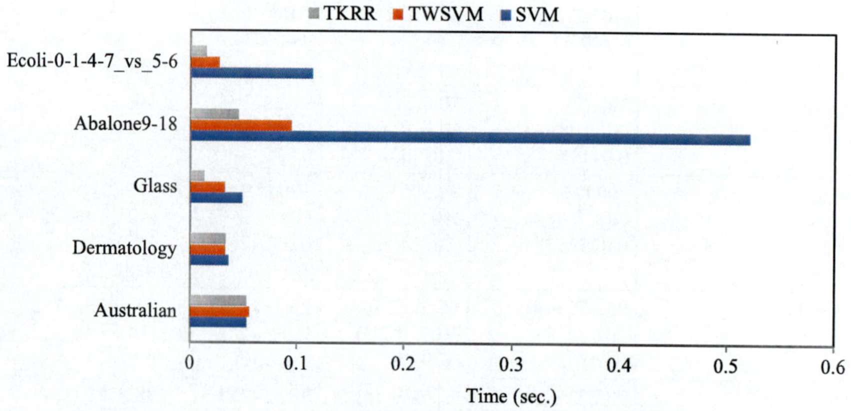


**Fig. 2** Training time (s) comparison among SVM, TWSVM, and TKRR

### 4.3 *Statistical Analysis*

The Friedman test analyzes performance of the models based on their average ranks. To form the Friedman test, the mean ranks of the 4 models over the 20 datasets are taken from Table 3.

From Table 3, we can generate the null hypothesis as

$$\chi_F^2 = \frac{12 \times 20}{4 \times 5}\left[2.775^2 + 3.275^2 + 2.55^2 + 1.4^2 - \frac{4 \times 5^2}{4}\right] = 22.665$$

$$F_F = \frac{(20-1) \times 22.665}{20 \times (4-1) - 22.665} = 11.5343$$

$F_F$ is distributed to $(4-1)$ and $(4-1) \times (20-1)$ degrees of freedom. The critical value $C_V$ for $F_F$ is 2.182 for $\alpha = 0.10$. Since $F_F > C_V$, we can reject the null hypothesis. The critical difference (CD) taking $p = 0.10$ can be computed as [24]:

$$\text{CD} = 2.78\sqrt{\frac{4 \times (4+1)}{6 \times 20}} = 1.1349$$

It is observable from the results that the difference between the mean rank of TKRR with SVM, TWSVM, and KRR is 1.375, 1.875, and 1.15, respectively, which are greater than the CD. Hence, TKRR shows improved classification performance compared to SVM, TWSVM, and KRR. It can be further noticed from Table 3 that the proposed TKRR shows the best results in 18 cases out of 20 which reveals the supremacy of TKRR over the other models.

## 5 Conclusion and Future Direction

This work suggests a novel KRR for classification called TKRR. The proposed TKRR is computationally efficient and intuitive. Numerical simulations have been performed on an artificial dataset and twenty benchmark datasets. Further statistical analysis has been carried out using the popular Friedman test with posthoc Nemenyi statistics. Experimental outcomes on different types of small and large-scale datasets reveal the effectiveness of the proposed TKRR. No external optimization toolbox is needed to solve the optimization problem of TKRR. However, the main limitations of the TKRR model are that it is sensitive to feature noise. In future, this drawback could be overcome by assigning the affinity and class probability-based fuzzy membership values to TKRR.

## References


1. Cortes, C., Vapnik, V.: Support-vector networks. Mach. Learn. **20**(3), 273–297 (1995)
2. Liu, L., Chu, M., Gong, R., Peng, Y.: Nonparallel support vector machine with large margin distribution for pattern classification. Pattern Recogn. **106**, 107374 (2020)
3. Hazarika, B.B., Gupta, D.: Density-weighted support vector machines for binary class imbalance learning. Neural Comput. Appl. **33**(9), 4243–4261 (2021)
4. Borah, P., Gupta, D.: Functional iterative approaches for solving support vector classification problems based on generalized Huber loss. Neural Comput. Appl. **32**(13), 1–21 (2019)
5. Hazarika, B.B., Gupta, D., Berlin, M.: A comparative analysis of artificial neural network and support vector regression for river suspended sediment load prediction. In Luhach, A.K., Kosa, J.A., Poonia, R.C., Gao, X.Z., & Singh, D. (eds.) Advances in Intelligent Systems and Computing, 1st International Conference on Sustainable Technologies for Computational Intelligence, Jaipur, August 2019. Proceedings of ICTSCI, pp. 339–349. Springer, Singapore (2020)
6. Gupta, U., Gupta, D., Prasad, M.: Kernel target alignment based fuzzy least square twin bounded support vector machine. In: Paper Presented at the IEEE Symposium Series on Computational Intelligence (SSCI), Canberra, Australia, 18–21 November 2018
7. Jayadeva, R.K., Chandra, S.: Twin support vector machines for pattern classification. IEEE Trans. Pattern Anal. Mach. Intel. **29**(5), 905–910 (2007)
8. Liu, L., Chu, M., Yang, Y., Gong, R.: Twin support vector machine based on adjustable large margin distribution for pattern classification. Int. J. Mach. Learn. Cybern. **11**(10), 2371–2389 (2020)
9. de Lima, M.D., Lima, J.D., Barbosa, R.M.: Medical data set classification using a new feature selection algorithm combined with twin-bounded support vector machine. Med. Biol. Eng. Comput. **58**(3), 519–528 (2020)
10. Zhang, L., Suganthan, P.N.: A comprehensive evaluation of random vector functional link networks. Inf. Sci. **367**, 1094–1105 (2016)
11. Suganthan, P.N., Katuwal, R.: On the origins of randomization-based feedforward neural networks. Appl. Soft Comput. **105**, 107239 (2021)
12. Pao, Y.H., Phillips, S.M., Sobajic, D.J.: Neural-net computing and the intelligent control of systems. Int. J. Control **56**(2), 263–289 (1992)
13. Shi, Q., Katuwal, R., Suganthan, P.N., Tanveer, M.: Random vector functional link neural network based ensemble deep learning. Pattern Recogn. **117**, 107978 (2021)
14. Hazarika, B.B., Gupta, D.: Modelling and forecasting of COVID-19 spread using wavelet-coupled random vector functional link networks. Appl. Soft Comput. **96**, 106626 (2020)
15. Huang, G.B., Zhu, Q.Y., Siew, C.K.: Extreme learning machine: theory and applications. Neurocomputing **70**(1–3), 489–501 (2006)
16. Zhu, Q.Y., Qin, A.K., Suganthan, P.N., Huang, G.B.: Evolutionary extreme learning machine. Pattern Recogn. **38**(10), 1759–1763 (2005)
17. Hazarika, B.B., Gupta, D., Berlin, M.: Modeling suspended sediment load in a river using extreme learning machine and twin support vector regression with wavelet conjunction. Environ. Earth Sci. **79**, 1–15 (2020)
18. Hazarika, B.B., Gupta, D., Berlin, M.: A coiflet LDMR and coiflet OB-ELM for river suspended sediment load prediction. Int. J. Environ. Sci. Technol. **18**, 2675–2692 (2021)
19. Saunders, C., Gammerman, A., Vovk, V.: Ridge Regression Learning Algorithm in Dual Variables, pp. 515–521 (1998)
20. Zhang, L., Suganthan, P.N.: Benchmarking ensemble classifiers with novel co-trained kernel ridge regression and random vector functional link ensembles [research frontier]. IEEE Comput. Intell. Mag. **12**(4), 61–72 (2017)
21. Ripley, B.D.: Pattern recognition and neural networks. Cambridge University Press (2007)
22. Blake, C.: UCI repository of machine learning databases (1998). http://www.ics.uci.edu/~mlearn/MLRepository.html. Accessed 14 Jan 2021

23. Alcalá-Fdez, J., Fernández, A., Luengo, J., Derrac, J., García, S., Sánchez, L., Herrera, F.: Keel data-mining software tool: data set repository, integration of algorithms and experimental analysis framework. J. Mult. Valued Logic Soft Comput. **17** (2011)
24. Demšar, J.: Statistical comparisons of classifiers over multiple data sets. J. Mach. Learn. Res. 7, 1–30 (2006)

# Performance Evaluation of CMOS Voltage-Controlled Oscillator for High-Frequency Communication System

**Abhijit Panigrahy, Abinash Patnaik, and Rajesh Kumar Patjoshi**

**Abstract** Voltage control oscillator (VCO) is one of the fundamental blocks in high-frequency communications systems. In recent scenario, Giga-Hertz band communication is one of the attractive features. Therefore, the demand for realization of a single chip transceiver is increasing; with the higher requirement of reduced size, lower cost and low power consumption. So, the designed VCO should be one of the essential considerations for high-frequency communication system. This paper contemplates the design and implementation of low power and delay voltage-controlled oscillators which targets to improve frequency generation performance in the field of high-frequency communication systems. The proposed models aim at providing better results in different sectors. All the circuits are validated by utilizing cadence virtuoso tool with United Microelectronics Corporation (UMC) 180 nm technology library. Eventually, these circuits are compared with respect to power, delay, area and frequency to justify the most suitable VCO for high-frequency communication system.

**Keywords** Voltage-controlled oscillator · Current starved oscillator · Op-amp-based VCO · UMC

## 1 Introduction

The primary objective of this paper is to design an ideal voltage-controlled oscillator to produce waveforms of variable frequency by altering the supplied voltage irrespective of the process variation. Voltage control oscillator (VCO) is one of the essential circuits for recent high-frequency communication system which generates high-frequency signal for communication system and act as a central block of radio frequency integrated circuit (RFIC) [1]. VCO can operate over a wide range of frequency based on the requirement such as multi-standard or multi-band radios.

A. Panigrahy (✉) · A. Patnaik · R. K. Patjoshi
Electronics and Communication Engineering, National Institute of Science and Technology, Berhampur, India
e-mail: abhijit316abhijit@gmail.com



D. Gupta et al. (eds.), *Pattern Recognition and Data Analysis with Applications*, Lecture Notes in Electrical Engineering 888,
https://doi.org/10.1007/978-981-19-1520-8_59

The proposed VCO architecture has lesser power utilization, a wider range of frequency tuning, less phase noise and a higher performance rate as compared to the existing VCO's. Moreover, this circuit aims at minimal delay and area consumption. Different types of VCO's were taken into consideration and studied. Schematics of various architectures were done followed by their simulation.

Voltage-controlled oscillator can be stated as an electronic device which generates oscillating frequency with an application of variable input voltage. The control voltage governs the oscillating frequency of the circuit [2]. VCO is an electronics structure which uses feedback circuit, resonant circuit and an amplification circuit for generating various signals at a particular frequency. As the VCO circuits operates over a wide range of frequency with minimal area and power consumption, it has become an important part of the modern communication systems [3]. They are also responsible for timing of digital systems and frequency translation in high-frequency systems [4]. In modern era, the super-charged systems like phase-locked loops (PLLs) is a popularly used circuit structure which is having various applications in data recovery, clock generation and frequency synthesis, where VCO is treated as the integral part of those high-performance PLLs [5]. PLL consists of a voltage-controlled oscillator, a loop filter, a charge pump, a low-pass filter and a phase detector. VCO affects the stability of the systems with respect to noise performance and power consumption [6]. For controlling the frequency with respect to control voltage and produce higher range of frequency, current starved oscillator structure is used [7]. VCO can also be implemented by using various delay cells and circuit structures [8]. In recent days, it is seen that the high-frequency circuit implemented using CMOS process is better in power and speed and also cost-effective as compared to that implemented using bipolar technique [9]. The quadrature output can be obtained from the two-stage differential VCO structure [10].

Here we designed different VCO structure like ring oscillator, current starved VCO, op-amp-based VCO and compared their performance with reference to power, speed and frequency tuning. Thus, we intend to design a suitable VCO which generates output signals in the required range so that they can be used further in PLL networks.

## 2 Overview of Voltage-Controlled Oscillator

This section will discuss the fundamentals of the design before going to the design and simulations. But before going to op-amp, the behavior of other oscillator circuits such as ring oscillator, current starved oscillator needs to be studied in detail. As in VCO, the frequency of the output voltage is determined by the input control voltage. Also, the power, speed and area factors should be considered as important parameters.

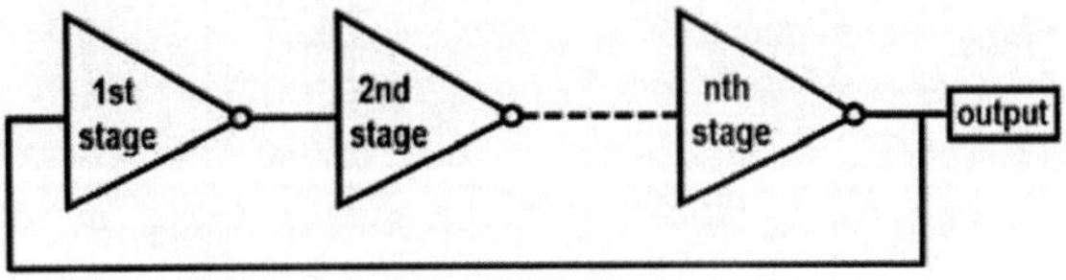


**Fig. 1** Basic structure of a *n*th stage ring VCO

## 2.1 *Ring Oscillator*

In today's time, oscillators play an vital role in most of the communication systems and optical devices. A ring oscillator is a feedback circuit consisting of odd number of series connected identical inverter stages forming a closed loop. The output fed to the input causes the anticipated oscillations. The circuit only needs power supply, and the oscillation starts on its own because of the feedback circuit. With the change in the number of inverter stages or with the change in supply voltage, the frequency of the oscillation can be further altered. It also produces quadrature phase and phase outputs when the numbers of delay cells used are even.

Inverter cell is the basic element of a ring oscillator circuit. Inverter cells are chosen wisely considering that both of them are identical and symmetric in a way that channel length of n and p transistors are same. All the inverter stages in a ring oscillator are connected in a series forming a loop. The circuit is designed carefully so that it meets the Barkhausen criteria of oscillation which insist that the circuit should produce a phase shift of $2\pi$ and have a unity voltage gain in order to perform oscillation. Figure 1 shows a basic structure of a nth stage ring oscillator, and for '$N$' number of stages, the oscillation frequency can be calculated as,

$$f = \frac{1}{2 * T * N} \tag{1}$$

where '$f$' is the oscillation frequency and '$T$' is the delay time.

## 2.2 *Current Starved Oscillator*

An oscillator produces waveforms of specific frequency and constant amplitude. In case of a ring oscillator, delays produced by each inverter stages determine its oscillation frequency but in case of current starved oscillator, by controlling the flow of current to the capacitive loads connected at each stage, delay of the circuit can be controlled. Figure 2 shows the basic structure of a current starved oscillator.

In a current starved VCO, the control voltage regulates the resistances of consisting pull-down and pull-up transistors of that circuit through current mirror technique. These wavering resistances control the availability of current to the capacitive loads for charging and discharging operations which determine the frequency of the oscillation. In this architecture, the output is calculated from the drains of $M1$ and $M2$

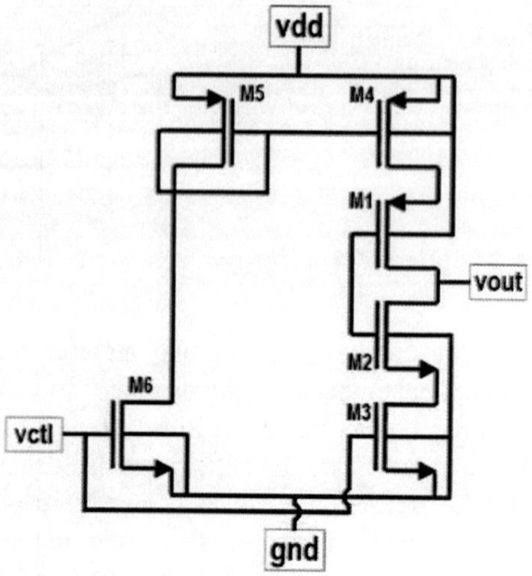


**Fig. 2** Basic structure of a current starved oscillator

transistors which act as inverters. The perk of using this configuration is a variation in the value of control voltage which can tune the oscillatory frequency over a wide range of values. The linearity and frequency of VCO is determined by varying the control input voltage ($V_{\text{ctrl}}$). The oscillation frequency for '$N$' stage current starved oscillator can be calculated as

$$f = \frac{I_D}{\text{VDD} * N * C_{\text{tot}}} \tag{2}$$

where '$I_{\text{D}}$' is the drain current, 'VDD' is the supply voltage and '$C_{\text{tot}}$' is the total capacitance.

## 2.3 *VCO Using Operational Amplifier*

Op-amp-based VCO can also be implemented by using the two-stage operational amplifier, where the first stage will maximize the voltage swing and the second stage will enhance the gain of the amplifier. Width-to-length ($W/L$) ratio of all the metal–oxide–semiconductor field-effect transistor (MOSFET) was calculated theoretically and the values were used for practical implementation. Figure 3 demonstrates various blocks associated in the architecture of a VCO.

From the simulated Bode plot, we can analyze that the two-stage op-amp has better phase stability. For designing a VCO, we need an integrator and Schmitt trigger, where the triangular wave will be generated by the integrator and Schmitt trigger will generate the square wave. The integrator and Schmitt trigger circuits can be designed using two-stage operational amplifier [11]. In operational amplifier-based VCO, by varying the input voltage, the oscillating frequency can be supervised.

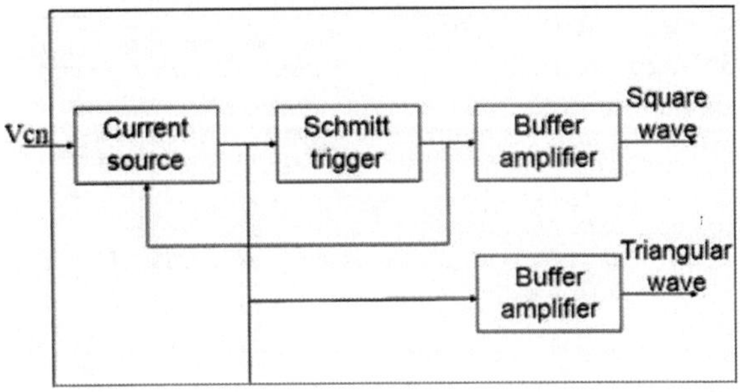


**Fig. 3** Block diagram of an op-amp VCO

# 3 Schematic Design and Implementation

## *3.1 Schematic Design of Ring Oscillator*

A five-stage ring oscillator has been designed in UMC 180 nm CMOS technology using cadence virtuoso tool. To design a five-stage ring oscillator, five inverter stages were cascaded where output of each stage were the inputs to their succeeding stage and the final output being connected to the input of circuit forming a feedback loop. For the oscillations to arise spontaneously, supply voltage and reset voltage were applied. The schematic of five-stage ring oscillator is shown in Fig. 4. Similarly, different stages of ring oscillators can be designed using the same stages of inverter stages. Table 1 represents the *W*/*L* ratios of the MOSFET's used in the circuit.

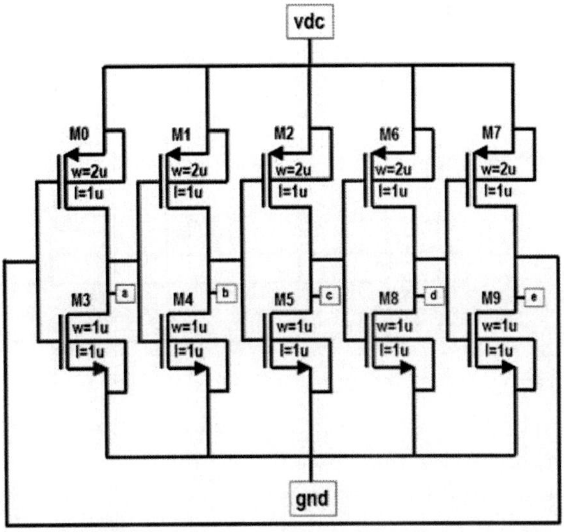


**Fig. 4** Schematic representation of five-stage ring oscillator

**Table 1** *W/L* ratio of ring oscillator

| Transistor number | (*W/L*) Ratio |
|---|---|
| $(W/L)_{0,1,2,6,7}$ | 2 |
| $(W/L)_{3,4,5,8,9}$ | 1 |

## 3.2 *Schematic Design of Current Starved Oscillator*

In a current starved oscillator, the control voltage alters the resistances of pull-down as well as pull-up transistors through a current mirror. Current starved oscillator is designed using ring oscillator as they are very much similar to each other. Schematic diagram of current starved oscillator has been illustrated in Fig. 5.

From the designed architecture, it can be observed that the $M0$ and $M3$ transistor act as an inverter, where as $M9$ and $M12$ transistors act as current sources and sink. The inverter transistors, i.e., $M0$ and $M3$, starve for the availability of current as the current sources limit the current available to them. The current flowing through transistors $M14$ and $M15$ is same as they are current-mirrored and are governed by the control voltage supplied at the input. Table 2 demonstrates the $W/L$ ratios of the transistors used in the circuit.

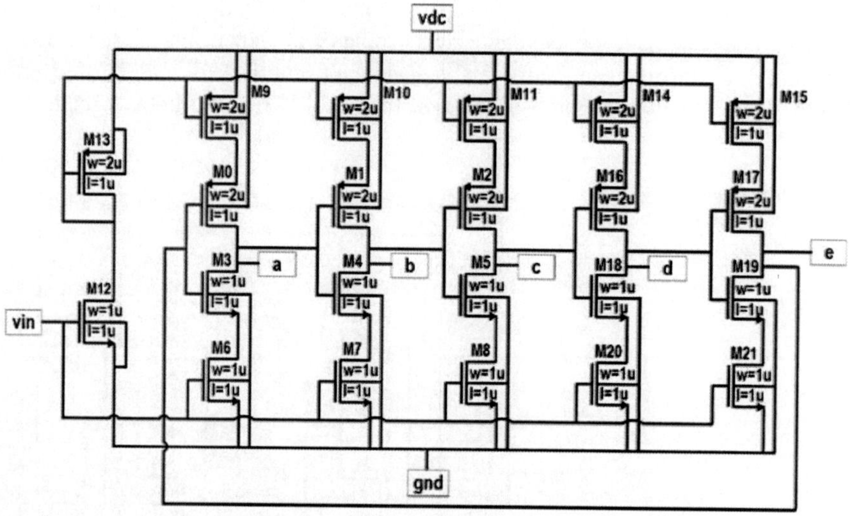


**Fig. 5** Schematic representation of current starved oscillator

**Table 2** *W/L* ratio of current starved oscillator

| Transistor number | (*W/L*) Ratio |
|---|---|
| $(W/L)_{0,1,2,9,10,11,13,14,15,16,17}$ | 2 |
| $(W/L)_{3,4,5,6,7,8,12,18,19,20,21}$ | 1 |

### *3.3 Schematic Design of Op-amp Based VCO*

Involvement of two-stage architecture in the design of a differential VCO makes it more balanced. Figure 6 shows the schematic representation of op-amp-based voltage-controlled oscillator. The various blocks are current source, Schmitt trigger and buffer amplifiers. Schmitt trigger is responsible for generating square wave signals. It is biased by a constant current source. Current source is used to bias primary Schmitt trigger. The capacitor voltage changes with the variation of current. With increase of current, the voltage increases and vice-versa. Voltage variation across the terminals of capacitor results in triangular waveforms. We can manipulate it using Schmitt trigger or comparators circuits, to generate square wave signals. The buffer amplifiers are optional and are used for impedance matching. To design the Schmitt trigger and integrators, we have to design a two-stage operational amplifier.

## 4 Simulation Results and Analysis

The design with optimization of ring oscillator, current starved oscillator and op-amp-based VCO is done using Cadence Virtuoso tool having 180 nm node and supply voltage of 1.8 V. Different analyses have been done such as transient analysis, delay analysis and power analysis to observe the circuit behavior in response to different parameters. Similarly, some process variation analyses have also been done to define the applicability of the circuit, which is needed because of the generation of different process variation constraints while manufacturing.

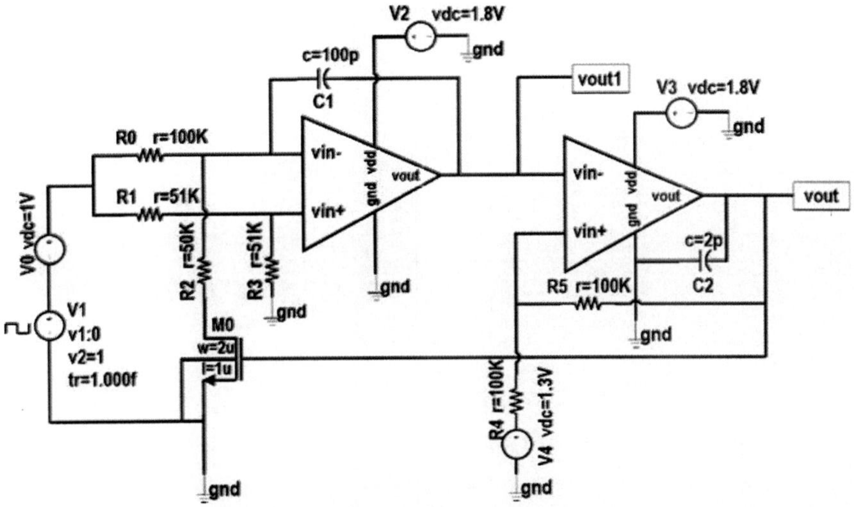


Fig. 6 Schematic representation of op-amp-based VCO

**Fig. 7** Transient analysis

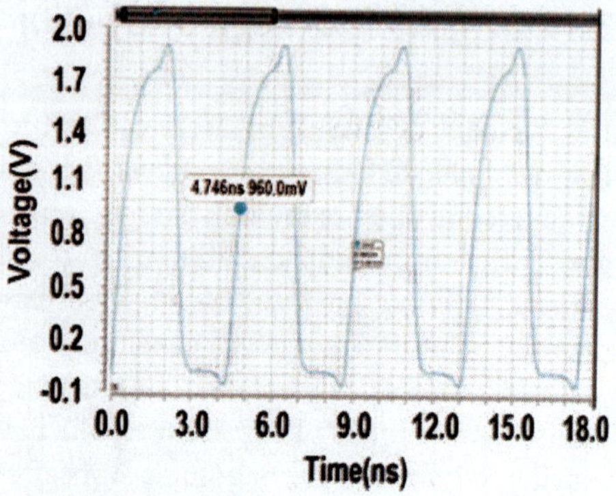


**Fig. 8** Delay analysis

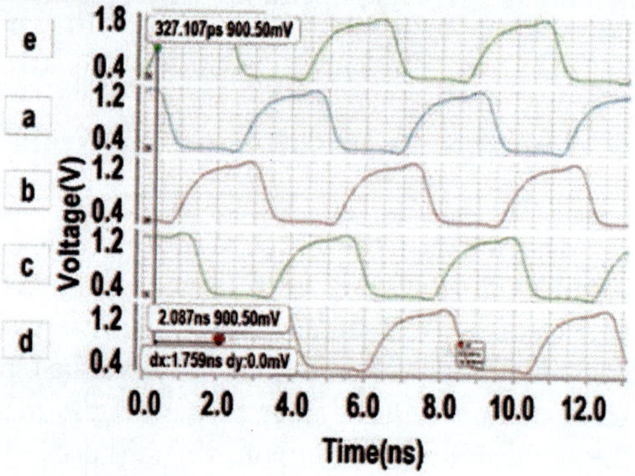


**Fig. 9** Power analysis

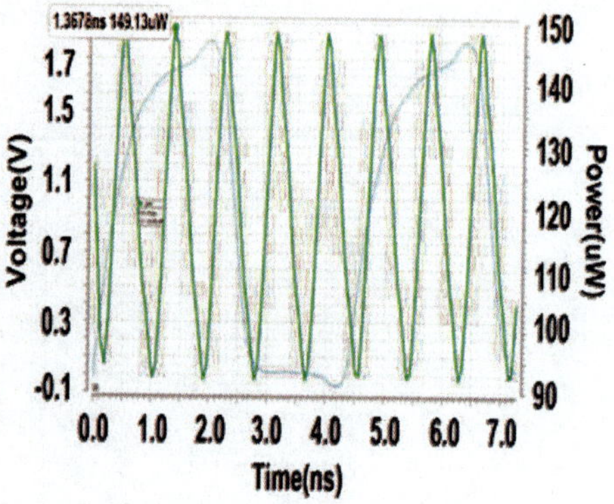

**Table 3** Summarizing the parameters of different types of ring oscillator

| Stages | Oscillation frequency (MHz) | Delay value (ns) | Power consumption (uW) |
|---|---|---|---|
| 3 | 479.3 | 0.896 | 134.20 |
| 5 | 210.97 | 1.76 | 149.13 |
| 7 | 163.46 | 2.64 | 149.21 |

## 4.1 Schematic Simulations of Five-Stage Ring Oscillator

Different simulations such as transient analysis, delay analysis and power analysis of the five-stage ring oscillator have been done and shown in Figs. 7, 8 and 9.

**Analysis report of Ring Oscillator**
Transient analysis is basically performed to illustrate the behavior of the waveform over a time period. Figure 7 shows the transient analysis of the five-stage ring oscillator. Delay analysis is required to calculate the oscillation frequency which is demonstrated in Fig. 8, and Fig. 9 presents the power analysis of the five-stage ring oscillator which is done to calculate the total power consumption of the whole circuit. Table 3 provides the oscillation frequency, delay value and the power consumption of three-stage, five-stage and seven-stage ring oscillators.

## 4.2 Schematic Simulation of Current Starved Oscillator

Transient analysis, delay analysis and power analysis of presented current starved oscillator are done, and simulation results are given below. Moreover, frequency-voltage characteristic of the oscillator is also being showed.

**Analysis Report of Current Starved Oscillator**
Figure 10 shows transient analysis of current starved oscillator, representing behavior of the circuit with respect to time domain. Figure 11 illustrates the delay analysis of current starved oscillator where the delay between the different inverter stage has been shown. Figure 12 provides the power analysis of the circuit which is observed as 132.297 uW at 1.1277 ns. Figure 13 gives a graphical representation of voltage vs frequency curve, which shows variation of the voltage from 0.4 to 1.6 V, respectively, which leads the variation in oscillation frequency form 1.24 to 108.60 MHz.

Fig. 10 Transient analysis

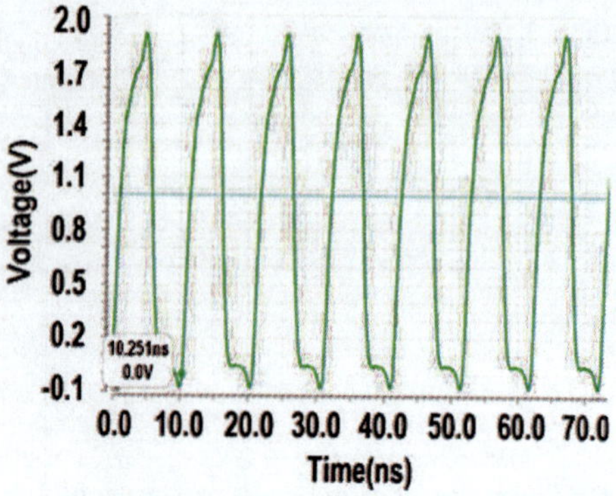


Fig. 11 Delay analysis

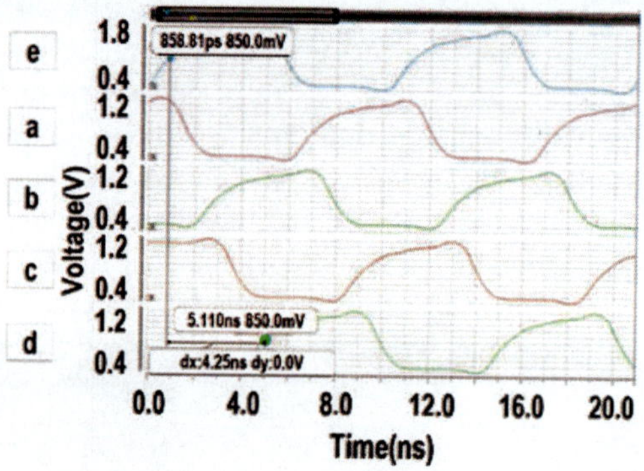


Fig. 12 Power analysis

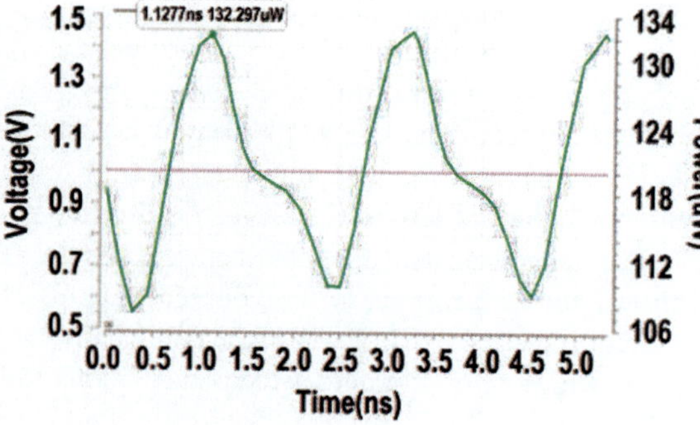

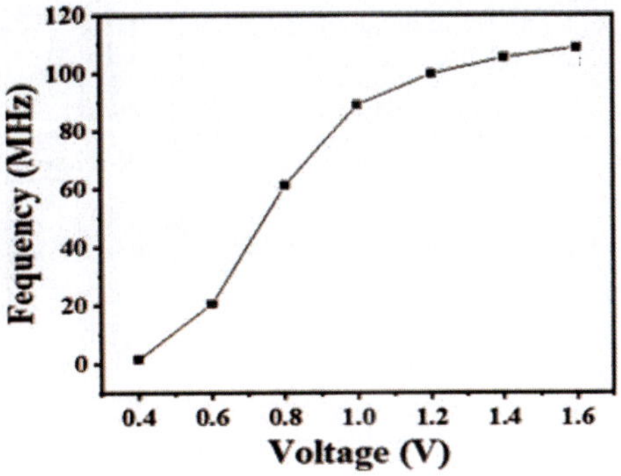


**Fig. 13** Graphical value of frequency-voltage

## 4.3 *Schematic Simulation of Op-amp VCO*

Process variation describes the feasibility of the circuit, which conveys about a substantial effect over analog circuits with regard to conceivable and computable variance of the output, and those results demand Monte Carlo and corner analysis.

**Analysis Report of Op-amp-Based VCO**

Figure 14 shows the transient analysis of op-amp-based VCO, where the time period is observed as 2.20981us. Figure 15 shows the Monte Carlo analysis, and Fig. 16 shows the corner analysis of the transient analysis. Figure 17 shows the power analysis of the circuit. Figure 18 shows the delay analysis of the triangular wave output, while delay analysis of the square wave output is shown by Fig. 19. Table 4 represents the variation of the oscillating frequencies with the variation in the input control voltage, and Table 5 ultimately gives a comparison of all the VCO structures that have been mentioned in this paper.

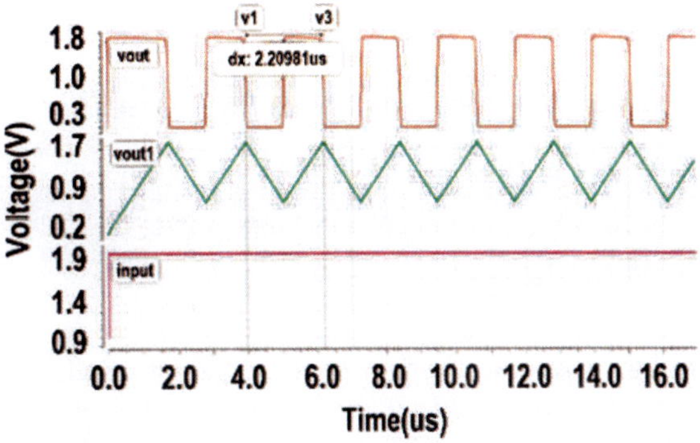


**Fig. 14** Transient analysis

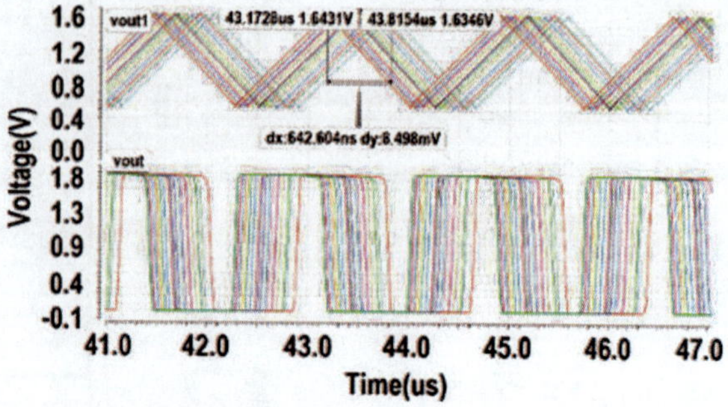


**Fig. 15** Transient analysis (Monte Carlo)

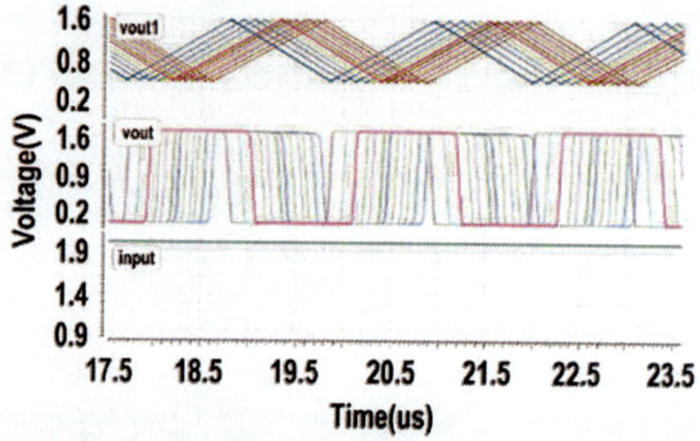


**Fig. 16** Transient analysis (corner)

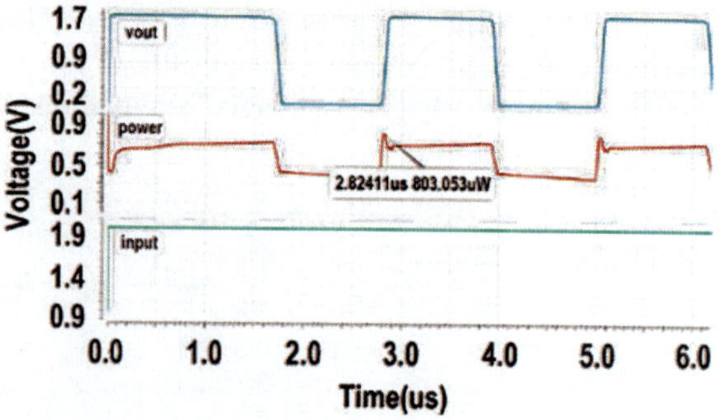


**Fig. 17** Power analysis

## 5 Conclusion

In this paper, at first, we studied different oscillator structures like ring oscillator, current starved oscillator and op-amp-based oscillator and verified their experimental result with the theoretical studies. We noted that in current starved oscillator the frequency is increased with increase in input voltage. As op-amp is a critical block in VCO design, so the values of the circuit components were calculated from the given design specifications, and the two-stage op-amp was designed. Further, we

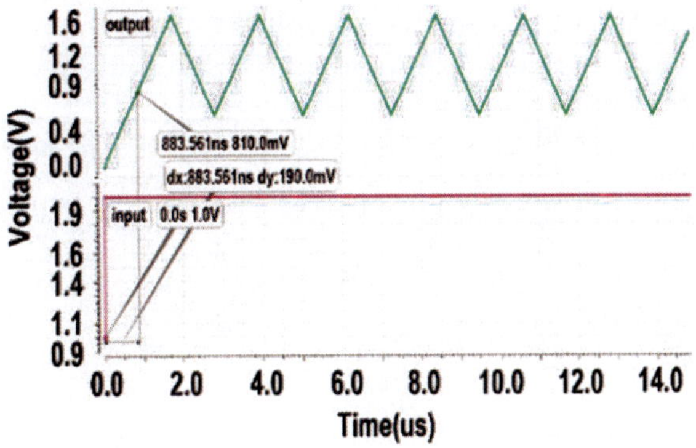


**Fig. 18** Delay analysis of a triangular wave

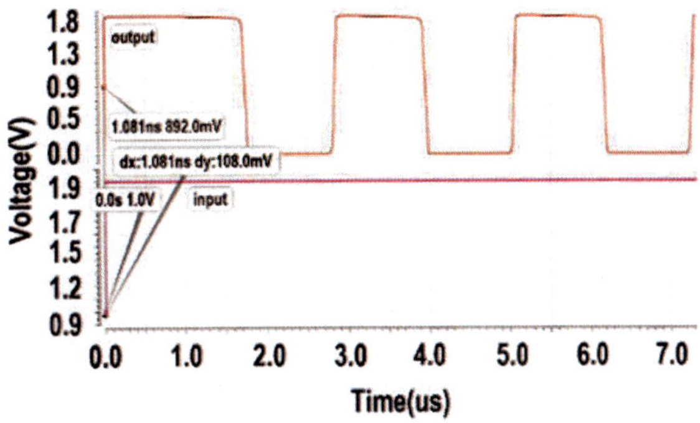


**Fig. 19** Delay analysis of a square wave

**Table 4** Representation of oscillation frequencies of an op-amp VCO

| Sl. No. | Input control voltage (V) | Oscillation frequency (kHz) |
|---|---|---|
| 1 | 0.6 | 378.78 |
| 2 | 0.8 | 415.80 |
| 3 | 1.0 | 460.82 |
| 4 | 1.2 | 495.04 |
| 5 | 1.4 | 540.54 |

designed a VCO using operational amplifier circuit which provides better result in terms of delay from other oscillator circuits. The behavior of this design was observed from the circuit analysis, in which we found that by increasing the control voltage the oscillation frequency is also increasing. This paper aimed at designing voltage-controlled oscillators having lower phase noise and consuming less power. VCO with low power consumption with good functionality would result in minimal power loss. Therefore, the discussed models can be act as great assets in the domain of communication because of its characteristics.

**Table 5** Comparison of various oscillator structures

| Types of oscillation structure | Frequency | Delay (ns) | Power (uW) | Comment |
|---|---|---|---|---|
| Op-amp-based VCO | 378–574 kHz | 1.05 | 803.05 | Low-frequency operation<br>Less delay<br>More complex design<br>More power dissipation |
| Current starved VCO | 1.24–108.60 MHz | 4.25 | 132.29 | High-frequency operation<br>More delay<br>Simple design<br>Very low-power dissipation |
| Ring oscillator | 210 MHz | 1.76 | 149.13 | High frequency (but no variation)<br>Less delay<br>Easy design<br>Low-power dissipation |

## References


1. Omar, F., Md Tawfiq Amin.: An ultra-low power area efficient voltage controlled oscillator based on tunable active inductor for wireless applications. Microprocessors Microsyst. 104291 (2021)
2. Sandhiya, S., Revathi, S., Vinothkumar, B.: Design of voltage controlled oscillator in 180nm cmos technology. Int. Res. J. Eng. Technol. (IRJET) **5**(3), 844–849 (2018)
3. Changzhi, Li., Jenshan, L.: A 1–9 GHz linear-wide-tuning-range quadrature ring oscillator in 130 nm CMOS for non—contact vital sign radar application. IEEE Microw. Wireless Compon. Lett. **20**(1), 34–36 (2010)
4. Halesh, M.R., Rasane, K.R., Rohini H.: Design and implementation of voltage control oscillator (VCO) using 180nm technology. In: International Conference on Advances in Computing, Communication and Control, vol. **125**, pp. 472–478. Springer, Berlin, Heidelberg (2011)
5. Madhumita, S., Sanjeev, M., Ali, Zoonubiya.: A study of different oscillator structures. Int. J. Innov. Res. Sci. Eng. Technol. **3**(5) (2014)
6. Saikee, C., Rajesh, M.: CMOS design and performance analysis of ring oscillator for different stages. Int. J. Eng. Trends. Technol (IJETT). **32**(5) (2016)
7. Prakash, R., Debiprasad, A., Ganapati, P.: A multiobjective optimization based fast and robust design methodology for low power and low phase noise current starved VCO. IEEE Trans. Semicond. Manuf. **27**(1), 43–50 (2014)
8. Aniket, P., Prajapati, P.: Analysis of Current Starved Voltage Controlled Oscillator using 45nm CMOS Technology. Int. J. Adv. Res. Electr. Electron. Instrum. Eng. **3**(3), 8076 (2014)
9. Shruti, S., Sharma, K., Ghosh, P.: Analysis and design of current starved ring VCO. In: International Conference on Electrical, Electronics, and Optimization Techniques (ICEEOT), pp. 3222–3227. IEEE (2016)
10. Shabbir Maeed, C., Shabbir, Z.: design of a wide band, low-noise differential voltage controlled ring oscillator in 90 nm CMOS process. National Acad. Sci. Lett **41**(1), pp. 47-52 (2018)
11. Abinash, P., Abhijit, P., Rajesh Kumar, P., Shasanka Sekhar, R.: Design and implementation of optimized parameter based operational amplifier for high speed analog signal processing. In: IEEE International Symposium on Sustainable Energy, Signal Processing and Cyber Security (iSSSC), pp. 1–6. IEEE (2020)

# Kernelized Random Vector Functional-Link Network

**Parashjyoti Borah, Deepak Gupta, and Sandeep Soumya Sekhar Mishra**

**Abstract** Random vector functional-link (RVFL) networks are closed-form solution-based methods that evaluate a linear function in the output layer. The input vectors are fed to the output layer using the direct links. In this study, similar to the kernel-based methods, the input space is mapped to a higher dimensional feature space before feeding them to the output layer of RVFL. Thus, the proposed method utilizes the benefits of the nonlinear kernel functions of the kernel-based methods and the nonlinear activation functions of the artificial neural networks. The proposed method achieves complete nonlinearity at the output layer. Experiments performed on real-world datasets establish efficacy of the proposed approach.

**Keywords** RVFL · KRR · Kernel-based methods · Activation functions

## 1 Introduction

Random vector functional-link networks [1], popularly referred to its corresponding acronym as RVFLs, are one of the popular supervised learning algorithms for regression and classification tasks. RVFL basically is a single hidden layer feedforward network (SLFN) where there exists one hidden layer (also called the enhancement nodes) between the input layer and the output layer. However, in addition to the links from the enhancement nodes to the output layer neuron(s), RVFL sets direct connections that establish links from the input layer to the output layer. In the simplest case, the output layer of RVFL consists of single neuron that finds linear output. Unlike

P. Borah (✉)
Department of Computer Science and Engineering, Indian Institute of Information Technology, Guwahati 781015, India
e-mail: parashjyoti@hotmail.com

D. Gupta
Department of Computer Science and Engineering, National Institute of Technology, Papum Pare, Arunachal Pradesh 791112, India

S. S. S. Mishra
Department of Computer Science and Engineering, Siksha 'O' Anusandhan, Bhubaneswar, India

 
D. Gupta et al. (eds.), *Pattern Recognition and Data Analysis with Applications*, Lecture Notes in Electrical Engineering 888,
https://doi.org/10.1007/978-981-19-1520-8_60

traditional artificial neural networks (ANNs) that obtain the solutions iteratively, RVFL is a closed-form solutions-based method that first assigns random weights to the links from input layer to the enhancement nodes and obtains the weights to the output layer neurons by solving a linear equation. Some latest RVFL-based studies can be explored in [2–4], and some advancements on RVFL are available at [5, 6].

Two other very popular algorithms that are similar to RVFL are kernel ridge regression (KRR) [7] and least squares support vector machine (LS-SVM) [8]. However, both KRR and LS-SVM have structural differences to RVFL which is graphically depicted in a later section of this paper. KRR first performs feature mapping from the input space into a feature space of higher dimension where it then evaluates the linear decision function. Similarly, LS-SVM in the nonlinear case first projects the input space to a higher dimensional feature space, and there it evaluates the decision function. The only architectural difference between KRR and LS-SVM is that, unlike KRR, LS-SVM introduces a bias term into its optimization problem, also called the intercept, that allows the decision boundary not to pass through the origin of the space. As the mapping function is not known sometimes and also turns out to be computationally costlier if known, KRR and LS-SVM obtain the solutions by using the kernel trick instead. The kernel trick replaces the dot product of two feature vectors by an appropriately chosen kernel function [7, 8]. Some latest KRR and LS-SVM-based approaches are available at [9–11]. In [2], authors have agreeably stated that replacing the matrix $\mathbf{DD}^t$ in RVFL solution by kernel matrix results in the KRR and therefore such kernelization of RVFL is meaningless. However, some alternative kernelization methods for RVFL are discussed in [12–14]. In [12], authors have proposed kernel-based random vector functional link (*K*-RVFL) where the activation function is replaced with kernel function in the hidden layer of RVFL. Thus, the hidden layer output matrix is the feature mapped kernel matrix instead of the activated outputs using the random weights of RVFL. In some sense, it can be visualized as KRR with dual kernels (linear|nonlinear). A diagrammatic representation of *K*-RVFL is presented in a later section. *K*-RVFL is further extended to multiclass classification in [13, 14].

In this paper, we propose a kernelized RVFL ($\text{RVFL}_{\text{ker}}$) that maps the input space into a higher dimensional feature space before feeding the input vectors to the output neuron through the direct link. The principal benefit of the proposed $\text{RVFL}_{\text{ker}}$ is that it can utilize the power of nonlinear activation functions of ANNs and nonlinear kernel functions of kernel-based methods like KRR to achieve complete nonlinearity at the output layer. It can be stated that activation functions of ANN map individual inputs for nonlinearity; i.e. the summed scalar output with weight and bias is passed to a nonlinear activation function. On the other hand, feature mapping function when kernelized maps the whole input space altogether to the higher dimensional feature space; i.e. the mapping depends upon other input samples. The proposed approach takes advantages of both the nonlinearities. We have diagrammatically shown the structural differences of $\text{RVFL}_{\text{ker}}$ from KRR, LS-SVM and K-RVFL, whereas $\text{RVFL}_{\text{ker}}$ retains the philosophy of RVFL. Experimental study on some real-world binary classification datasets establishes the efficacy of $\text{RVFL}_{\text{ker}}$.

## 2 RVFL Networks

All vectors are to be taken as column vectors. Let us consider, $\mathbf{x}_i \in \Re^n$ be an input vector and $\mathbf{X} = (\mathbf{x}_1, \ldots, \mathbf{x}_m)^t$ be the input data matrix of order $m \times n$. The output vector is considered to be $\mathbf{y} = (y_1, ...y_m)^t$, where $y_i \in \{+1, -1\}$ is the target output of the $i$th sample. Considering $\boldsymbol{\omega}_j \in \Re^n$ be the randomly initialized weight vector to the $j$th (for $j = 1, \ldots, l$) enhancement node, the weight matrix to the enhancement layer can be constructed as $\mathbf{W} = (\boldsymbol{\omega}_1, \ldots, \boldsymbol{\omega}_l)$, where $l$ is the number of hidden layer neurons, or in other words, $l$ is the number of enhancement nodes. The bias to the enhancement nodes is randomly initialized in the vector $\mathbf{b} \in \Re^l$. Selecting a suitable activation function $a(\mathbf{x}_i, \boldsymbol{\omega}_j, b_j)$ that finds the scalar output of the $j$th enhancement neuron to the input vector $\mathbf{x}_i$, the hessian matrix can be formed as $\mathbf{H} = (h(\mathbf{x}_1), \ldots, h(\mathbf{x}_m))^t$, where $h(\mathbf{x}_i) = (a(\mathbf{x}_i, \boldsymbol{\omega}_1, b_1), \ldots, a(\mathbf{x}_i, \boldsymbol{\omega}_l, b_l))^t$. Forming the augmented matrix $\mathbf{D} = [\mathbf{H} \quad \mathbf{X}]$ of size $(m \times (n+l))$ to be fed to the output layer and considering $\boldsymbol{\beta} \in \Re^{(n+l)}$ be the output layer weight vector, the optimization problem of RVFL with regularization can be defined as [2]

$$\min_{\boldsymbol{\beta}} \quad ||\mathbf{D}\boldsymbol{\beta} - \mathbf{y}||^2 + C||\boldsymbol{\beta}||^2. \tag{1}$$

With $\mathbf{I}$ as an identity matrix of appropriate dimension, the solution corresponding to the unconstrained optimization problem defined in Eq. (1) can be obtained in the dual space as

$$\boldsymbol{\beta} = \mathbf{D}^t(\mathbf{D}\mathbf{D}^t + C\mathbf{I})^{-1}\mathbf{y}. \tag{2}$$

## 3 Kernelized Random Vector Functional-Link Network (RVFL$_{\mathbf{ker}}$)

In the above section, we have seen that RVFL performs nonlinear transformations of the input vectors to form the hidden or enhancement layer before feeding them to the output layer. Nonlinearity can be achieved by applying a nonlinear activation function to the individual inputs, and no information of the input space is utilized in obtaining the enhancement node outputs. Along with the output from the enhancement nodes, the input vectors are linearly fed to the output layer. To achieve complete nonlinearity, in this work, we project the data samples from the input space to a feature space having higher number of dimensions as compared to the input space, and the feature mapped samples are then fed to the output layer through the direct links.

Let us redefine the matrix $\mathbf{D}$ with the feature mapped input matrix as below:

$$\mathbf{D} = [\mathbf{H} \quad \varphi(\mathbf{X})] \tag{3}$$

where $\varphi(\mathbf{x})$ is the feature mapping function, and thus, the matrix $\varphi(\mathbf{X})$ is defined as $\varphi(\mathbf{X}) = (\varphi(\mathbf{x}_1), \ldots, \varphi(\mathbf{x}_m))^{\mathrm{t}}$. For the solution of RVFL$_{\mathrm{ker}}$, Eq. (2) can be redefined with mapped features as below:

$$\boldsymbol{\beta} = [\mathbf{H} \quad \varphi(\mathbf{X})]^{\mathrm{t}}\left(\mathbf{H}\mathbf{H}^{\mathrm{t}} + \varphi(\mathbf{X})\varphi(\mathbf{X})^{\mathrm{t}} + C\mathbf{I}\right)^{-1}\mathbf{y}. \tag{4}$$

Now, we can apply the kernel trick to the solution for RVFL$_{\mathrm{ker}}$ expressed in Eq. (4). Considering $k(\bullet, *) = \varphi(\bullet) \cdot \varphi(*) = \varphi(\bullet)^{\mathrm{t}}\varphi(*)$ is an appropriately chosen kernel function, the Macer's kernel matrix is formed as

$$\mathbf{K} = \varphi(\mathbf{X})\varphi(\mathbf{X})^{\mathrm{t}} = \begin{bmatrix} k(\mathbf{x}_1, \mathbf{x}_1) & \cdots & k(\mathbf{x}_1, \mathbf{x}_m) \\ \vdots & \ddots & \vdots \\ k(\mathbf{x}_m, \mathbf{x}_1) & \cdots & k(\mathbf{x}_m, \mathbf{x}_m) \end{bmatrix} \tag{5}$$

Thus, we can rewrite Eq. (4) as

$$\boldsymbol{\beta} = [\mathbf{H} \quad \varphi(\mathbf{X})]^{\mathrm{t}}\left(\mathbf{H}\mathbf{H}^{\mathrm{t}} + \mathbf{K} + C\mathbf{I}\right)^{-1}\mathbf{y}. \tag{6}$$

And finally, the decision function for an unseen vector $\mathbf{x}$ can be expressed as below:

For regression:

$$f(\mathbf{x}) = \left(h(\mathbf{x})^{\mathrm{t}}\mathbf{H}^{\mathrm{t}} + k(\mathbf{x}^{\mathrm{t}}, \mathbf{X}^{\mathrm{t}})\right)\left(\mathbf{H}\mathbf{H}^{\mathrm{t}} + \mathbf{K} + C\mathbf{I}\right)^{-1}\mathbf{y}. \tag{7}$$

For classification:

$$f(\mathbf{x}) = \mathrm{sign}\Big(\left(h(\mathbf{x})^{\mathrm{t}}\mathbf{H}^{\mathrm{t}} + k(\mathbf{x}^{\mathrm{t}}, \mathbf{X}^{\mathrm{t}})\right)\left(\mathbf{H}\mathbf{H}^{t} + \mathbf{K} + C\mathbf{I}\right)^{-1}\mathbf{y}\Big) \tag{8}$$

It can be noted that the output neuron behaves like nonlinear classifiers such as KRR or nonlinear LS-SVM (in the presence of the bias term). However, unlike KRR or LS-SVM, RVFL$_{\mathrm{ker}}$ utilizes the additional features generated by the hidden layer (enhancement nodes) using some activation function. Moreover, the bias term $b$ can be introduced to the output neuron so that the output function does not always have to pass through the origin. A graphical idea of KRR [7], LS-SVM [8], RVFL [1, 2], $K$-RVFL [12] and RVFL$_{\mathrm{ker}}$ is presented in Fig. 1.

In Fig. 1c $\boldsymbol{\beta} = \begin{bmatrix} \boldsymbol{\beta}_{h(\mathbf{x})} \\ \boldsymbol{\beta}_{\mathbf{x}} \end{bmatrix}$, $\boldsymbol{\beta} = \begin{bmatrix} \boldsymbol{\beta}_{\phi(\mathbf{x})} \\ \boldsymbol{\beta}_{\mathbf{x}} \end{bmatrix}$ in Fig. 1d, and in Fig. 1e $\boldsymbol{\beta} = \begin{bmatrix} \boldsymbol{\beta}_{h(\mathbf{x})} \\ \boldsymbol{\beta}_{\phi(\mathbf{x})} \end{bmatrix}$. For Fig. 1c and e, $h(\mathbf{x}) = (h_1(\mathbf{x}), h_2(\mathbf{x}), \ldots, h_l(\mathbf{x}))^{t}$. It can be observed that RVFL$_{\mathrm{ker}}$ has significant structural difference from KRR [7], LS-SVM [8] and $K$-RVFL [12].

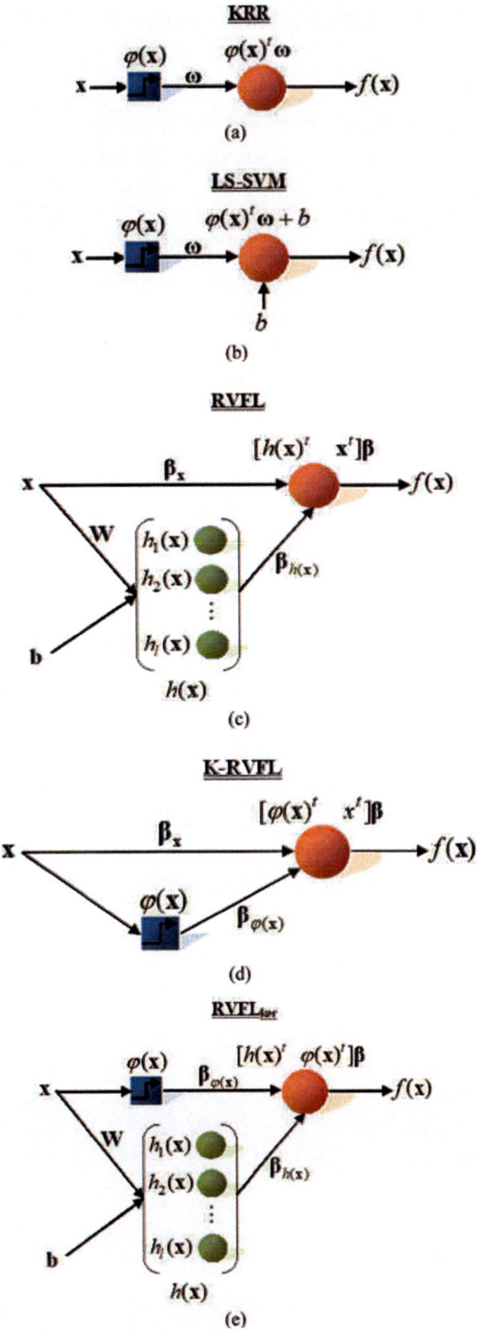


**Fig. 1** **a** KRR, **b** LS-SVM, **c** RVFL, **d** *K*-RVFL and **e** $\mathrm{RVFL}_{\mathrm{ker}}$

## 4 Experimental Study

In this section, we are presenting the experimental study carried out for binary classification problems and not regression problems. To test how $RVFL_{ker}$ performs on real-world problem-solving, numerical experiments are conducted on 11 real-world publicly available benchmark datasets. The collected datasets are accessible from the UCI machine learning repository [8]. All the experimental studies are carried out on a general-purpose PC with 8 GB RAM and Intel *i5* processor running Windows 10 OS. The MATLAB computational software is used to execute the algorithms. Datasets are normalized to [0, 1] before experiments are performed. The hold-out split strategy is applied for splitting the datasets into train, validation and test data in the ratio 5:2:3, respectively. In other worlds, the training set consists of 50% of the data samples, 20% are used for tuning the user-specified parameters, and 30% of the dataset constitute the test set. The results presented are obtained on the test dataset, i.e. the 30% of the total samples selected for testing. Results are compared with very similar classification algorithms, viz. KRR, LS-SVM (nonlinear), RVFL and *K*-RVFL. For the kernel-based methods, namely KRR, LS-SVM and *K*-RVFL, we use the popular Gaussian kernel. The sigmoid activation function is used for RVFL and $RVFL_{ker}$. The proposed method uses the Gaussian kernel for kernel mapping. Generalization performance on the selected binary classification datasets is presented in Table 1. It can be observed that $RVFL_{ker}$ could deliver better generalization performance in most of the datasets when compared with the other reported related methods.

**Table 1** Classification performance of KRR, LS-SVM, RVFL, *K*-RVFL and $RVFL_{ker}$ on UCI datasets. The best result is shown in bold

| Datasets | KRR | LS-SVM | RVFL | *K*-RVFL | $RVFL_{ker}$ |
|---|---|---|---|---|---|
| Australian credit | **89.372** | **89.372** | 86.8599 | 86.9565 | 88.1643 |
| Breast cancer Wisconsin | 94.6078 | 96.5686 | 95.9314 | 93.6275 | **97.3039** |
| Bupa or liver disorders | **69.9029** | 66.9903 | 60.9709 | 68.932 | 69.3204 |
| Cleveland | 77.5281 | 76.4045 | 78.8764 | 75.2809 | **79.6629** |
| German | 77 | 77 | **78.0667** | 77.6667 | 77.2333 |
| Heart-*c* | 70.7865 | 70.7865 | 69.2135 | 68.5393 | **74.0449** |
| Heart-stat | 77.7778 | 77.7778 | **85.3086** | 79.0123 | 80 |
| Monk1 | 94.5783 | 95.1807 | 83.4337 | 93.9759 | **95.6627** |
| Pima Indians diabetes | 74.3478 | 74.7826 | **75.0435** | 74.7826 | 74.8261 |
| WDBC | 98.2353 | **98.8235** | 97 | 97.6471 | **98.8235** |
| WPBC | 67.2414 | 74.1379 | 72.7586 | 77.5862 | **77.7586** |

## 5 Conclusion

The input vectors are directly fed to the output neuron of RVFL where the outputs are obtained using a linear decision function. However, the additionally generated feature outputs from the enhancement nodes may be obtained through nonlinear transformation using nonlinear activation function. $RVFL_{ker}$ performs feature mapping of the input vectors to a feature space with higher dimensions using feature mapping function before feeding them to the output neuron. Thus, $RVFL_{ker}$ could obtain nonlinearity in both, the direct feature mapped links as well as in the enhancement nodes. It is verified that most of the real-world datasets are nonlinearly separable in which case $RVFL_{ker}$ could achieve improved generalization performance as compared to RVFL. Moreover, generalization performance on real-world datasets is also compared with similar methods, viz. KRR, LS-SVM and K-RVFL, which also establishes efficacy of $RVFL_{ker}$. Future works may include extension of $RVFL_{ker}$ to multiclass classification and very large-scale datasets.

## References

1. Pao, Y.H., Phillips, S.M., Sobajic, D.J.: Neural-net computing and the intelligent control of systems. Int. J. Control **56**(2), 263–289 (1992)
2. Suganthan, P.N.: On non-iterative learning algorithms with closed-form solution. Appl. Soft Comput. **70**, 1078–1082 (2018)
3. Hazarika, B.B., Gupta, D.: Modelling and forecasting of COVID-19 spread using wavelet-coupled random vector functional link networks. Appl. Soft Comput. **96**, 106626 (2020)
4. Borah, P., Gupta, D.: Unconstrained convex minimization based implicit Lagrangian twin random vector functional-link networks for binary classification (ULTRVFLC). Appl. Soft Comput. **81**, 105534 (2019)
5. Zhang, P.B., Yang, Z.X.: A new learning paradigm for random vector functional-link network: RVFL+. Neural Netw. **122**, 94–105 (2020)
6. Shi, Q., Katuwal, R., Suganthan, P.N., Tanveer, M.: Random vector functional link neural network based ensemble deep learning. Pattern Recogn. **117**, 107978 (2021)
7. C. Saunders, A. Gammerman, V. Vovk, Ridge regression learning algorithm in dual variables, in *Proceedings of the 15th International Conference on Machine Learning, ICML*, pp. 515–521 (1998)
8. Suykens, J.A., Vandewalle, J.: Least squares support vector machine classifiers. Neural Process. Lett. **9**(3), 293–300 (1999)
9. H. Avron, M. Kapralov, C. Musco, C. Musco, A. Velingker, A. Zandieh, Random Fourier features for kernel ridge regression: approximation bounds and statistical guarantees. Int. Conf. Mach. Learn. 253–262 (2017), PMLR
10. P. Borah, D. Gupta, M. Prasad, Improved 2-norm based fuzzy least squares twin support vector machine, in *2018 IEEE Symposium Series on Computational Intelligence (SSCI)*, pp. 412–419, (IEEE, 2018)
11. P. Borah, D. Gupta, A two-norm squared fuzzy-based least squares twin parametric-margin support vector machine, In *Machine Intelligence and Signal Analysis*, pp. 119–134 (Springer, Singapore, 2019)
12. Xu, K.K., Li, H.X., Yang, H.D.: Kernel-based random vector functional-link network for fast learning of spatiotemporal dynamic processes. IEEE Trans. Syst. Man. Cybern. Syst. **49**(5), 1016–1026 (2017)

13. V. Chauhan, A. Tiwari, S. Arya, Multi-label classifier based on kernel random vector functional link network, in *2020 International Joint Conference on Neural Networks (IJCNN)*, pp. 1–70, (IEEE, 2020)
14. Nayak, D.R., Dash, R., Majhi, B., Acharya, U.R.: Application of fast curvelet tsallis entropy and kernel random vector functional link network for automated detection of multiclass brain abnormalities. Comput. Med. Imaging Graph. **77**, 101656 (2019)
15. D. Dua, C. Graff, UCI machine learning repository. University of California, School of Information and Computer Science, Irvine, CA (2019). http://archive.ics.uci.edu/ml

# Data Analytics Research in Nonprofit Organisations: A Bibliometric Analysis

**Idrees Alsolbi, Mengjia Wu, Yi Zhang, Siamak Tafavogh, Ashish Sinha, and Mukesh Prasad**

**Abstract** Profitable organisations that applied data analytics have obtained a double-digit improvement in reducing costs, predicting demands, and enhancing decision-making. However, in nonprofit organisations (NPOs), applying data analysis can interpret and discover more patterns of donors, volunteers, and forecasting future funds, gifts and grants. To uncover the usage of data analytics in different NPOs and understand its contribution, this article presents a bibliometric analysis of 2673 related publications to reveal the research landscape of data analytics applied in NPOs. Through a co-term analysis and scientific evolutionary pathways analysis, we profile the associations between data analysis techniques and NPOs and additionally identify the research topic changes in this field over time. The results yield us three major insights: (1) Robust and classic statistical methods-based data analysis techniques are dominantly prevalent in the NPOs field through all the time; (2) Healthcare and public affairs are two crucial sectors that involve data analytics to support decision-making and problem-solving; (3) Artificial Intelligence (AI)-based data analytics is a recently emerging trending, especially in the healthcare-related sector; however, it is still at an immature stage, and more efforts are needed to nourish its development.

**Keywords** Bibliometric analysis · Data analysis · Nonprofit organisations · Nongovernmental organisations

I. Alsolbi (✉) · M. Wu · Y. Zhang · M. Prasad
School of Computer Science, Australian Artificial Intelligence Institution, University of Technology Sydney, Sydney, Australia
e-mail: idna1257@gmail.com

S. Tafavogh
Department of Finance, Commonwealth Bank Health Society, Sydney, Australia

A. Sinha
Business School, University of Technology Sydney, Sydney, Australia



D. Gupta et al. (eds.), *Pattern Recognition and Data Analysis with Applications*, Lecture Notes in Electrical Engineering 888,
https://doi.org/10.1007/978-981-19-1520-8_61

## 1 Introduction

Data technology is evolving rapidly to address global needs. The professionals and practitioners from for-profit organisations are being urged to consider data analytics applications to maximise their potential and increase productivity. However, there are recent calls to apply such analytics on NPOs' activities such as analysing donor behaviours [1, 2]. NPOs differ from for-profit organisations as they are private, independent and self-generated funds [3]. Museums, colleges, libraries, research centres, health agencies, human welfare, human rights organisations, religious organisations, and charitable foundations are examples of NPOs. NPOs are expected to gather and analyse information from stakeholders to implement effective strategies for achieving their missions' objectives [4]. The role of data analytics in NPOs' activities is significant; it can assist such organisations in monitoring, evaluating, and determining barriers to their success, and can provide meaningful visualisations to support decision-makers. Kassen [5] doubted that governmental agencies and social community organisations can improve innovation by applying analytics on data. Using data verifiable to make decisions may be considered as a valuable strategy in the organisation [6]. Nevertheless, there are many concerns that NPOs do not benefit from the data analytics. Thus, NPOs collect data but less often use it to analyse certain activities such as analysing donor behaviours and predicting future donations [6].

NPOs face a number of significant challenges, including a shortage of technical skills, and financial and human resources to analyse their data [7]. NPOs face internal limitations in making the best use of data analytics and information technology including hardware, software, and technical skills [1]. Johnson [1] described the lack of technical skills amongst NPO employees as a major challenge. Moreover, NPOs fail to improve employees' skills and assist employees in sustaining a workforce [6]. Developing these skills remains essential for these organisations to enable the transformation of data-driven missions [8]. On the other hand, low budgets have led to an ignorance of data applications [1] for NPOs that not able to facilitate data-driven techniques [9].

Considering the opportunities and challenges of applying data in NPOs, it is crucial to avoid any gaps in the research and share the required knowledge. However, there is still need to present and report empirical studies, case studies, and experiments using data analytics in NPOs. To better investigate how data analytics is being applied in different NPOs, we conducted a bibliometric analysis to evaluate the contribution and the evolution of data analytics in NPOs scientific research. Bibliometrics, or Scientometrics, are quantitative methods used to measure and map the existing research in a scientific discipline [10]. The bibliometric analysis provides a comprehensive analysis of research trends and assesses science as a productive knowledge system [10]. Bibliometric analysis exposes the internal structure and development pattern of a specific research path or journal. It has been widely applied to various fields such as computers and information ethics [11].

This paper incorporates co-term analysis and scientific evolutionary pathways (SEP) to identify the associations between prevalent data analysing techniques and

**Table 1** Selection criteria

| Item | Restriction | Description |
|---|---|---|
| Search field | Titles, abstract, and keywords | The search for the data is only in three fields of each database: titles, abstracts, and keywords |
| Period | January 1973–January 2021 | For comprehensive coverage of literature |
| Language | English | Only English documents to be included for consistent processing and analysis |

various types of nonprofit organisations and identified the research topic changes over time. Through applying those analyses on 2673 research papers, this paper profiles the research landscape of this field and generates the following insights:

1. Robust and classic statistical methods-based data analysing techniques are dominantly prevalent in the NPO field through all the time;
2. Healthcare and public affairs are two crucial sectors that involve data analytics to support decision-making and problem-solving;
3. AI-based data analytics is a recently emerging trending, especially in the healthcare-related sector. However, it is still immature, and more efforts are needed to nourish its development.

The rest of this paper is structured as follows: Sect. 2 presents the materials and used methods, including collecting data, the searching databases, and the methodology used in this paper. Section 3 presents the main bibliometric analysis, the results and the research growth.

## 2 Materials and Methods

### 2.1 Data Collection and Pre-Processing

We chose three databases Web of Science,[1] Scopus[2] and ProQuest[3] to conduct the indexed data search for publications discussing data analytics in NPOs. The Web of Science (WoS), owned by Clarivate, is a well-recognised integrative platform of bibliometric data sources with a wide collection of scholarly journals, books and proceedings in the sciences and social sciences; its wide coverage and complete bibliographic information make it our primary choice for the data source. We also selected Scopus and ProQuest to complementarily obtain additional publications in the recent two decades [12]. We then applied multiple selection criteria to include precise publications that match research scope, as shown in Table 1. The three items that constitute

[1] https://www.webofscience.com/wos/woscc/basic-search.

[2] https://www.scopus.com/home.uri.

[3] https://www.proquest.com/.

**Table 2** Number change of publications during the process of collecting data

| Database | Raw results | After duplication removal | After manual assessment filtering |
|---|---|---|---|
| Scopus | 2089 | 3214 | 2673 |
| Web of science | 923 | | |
| ProQuest | 805 | | |

the selection criteria are: searching only in titles, abstracts, and keywords of each document, the period of publications is from January 1973 to January 2021, and only English written documents.

After that, we included a search string using major keywords and alternative synonyms for accurate results, using the Boolean operations (AND, OR). For example, on Scopus, TITLE-ABS-KEY ((“Data analy*” OR “Data-driven” OR “Analy*”) AND (“Nonprofi*” OR “Nonprofi*” OR “no*for*profit*”)). A detailed search string is shown in Appendix 1 to provide a full coverage of different types of data analysis in NPOs. The publications extracted (included bibliographic data, keywords, and citation data) after applying the selection criteria from each database.

Before moving on to the pre-processing data stage, we removed 603 duplications in our search results due to the database collection overlap. Then a manual assessment applied to exclude 541 publications that are not relevant to our research scope based on their titles and abstracts. Table 2 presents the number change of our search publications.

In the next stage, we conduct a data pre-processing procedure to obtain the inputs for co-occurrence and SEP analyses. Specifically, we exploited a natural language processing (NLP) function integrated in VantagePoint[4] to extract raw scientific terms from the titles and abstracts of the collected papers, further a term clumping process [13] was applied to the extracted terms to accomplish term cleaning and consolidation. The stepwise pre-processing results are given in Table 3.

## 2.2 *Co-term Analysis*

Co-term analysis is a classical topic analysis method in bibliometrics [11, 14]. It adopts the assumption that two terms that appear in the same context may share similar semantic meanings. The collection of such term co-occurrences can constitute a co-term network represented as $G = (V, E)$, where $V$ denotes the set of terms and $E$ denotes the co-occurrences between them. This paper adopts the terms selected in Step 7 of Table 3 and profile the associations between different types of NPOs and the prevalent data analysing techniques.

[4] VantagePoint is a commercial software used in text mining and particularly in science, technology and innovation text analysis. More details can be found on the website: https://www.thevantagepoint.com/.

**Table 3** Stepwise results of the term clumping process

| Step | Description | # Terms |
|---|---|---|
| 1 | Raw terms retrieved with NLP | 73,357 |
| 2 | Consolidated terms with the same stem, e.g. "information system" and "information systems" | 66,804 |
| 3 | Removed spelling variations, removed terms starting/ending with non-alphabetic characters, e.g. "Step 1" or "1.5 m/s", removed meaningless terms, e.g. pronouns, prepositions, and conjunctions | 55,604 |
| 4 | Removed general single-word terms, e.g. "information"[a] | 46,268 |
| 5 | Consolidated synonyms based on expert knowledge, e.g. "co-word analysis" and "word co-occurrence analysis" | 44,787 |
| 6 | Eliminated all terms occurring less than 2 times | 2349 |
| 7 | Manually select terms of data analysing methods and nonprofit organisations with top frequencies[a] | 10/17[b] |

[a] In this study, we aim to uncover the data analysing techniques associated with different types of nonprofit organisations; hence we manually pick high-frequency representative terms to indicate the two sets of concepts

[b] They are, respectively, 10 and 17 terms representing NPOs types and data analytic techniques

## 2.3 *Scientific Evolutionary Pathways*

Scientific evolutionary pathways (SEP) is a research topic tracking method developed by Zhang et al. [15]. It is used to identify the changes of research attention in time-labelled streaming documents. This basic assumption of this method is that scientific novelty derives from the accumulative changes and recombination of existing knowledge [16, 17]. We employed this method to identify the topic changes in at a macro level. The definitions and stepwise explanations of the SEP are stated below.

**Algorithm design**: Initially, the SEP represents every document with scientific terms (in our case, the terms are obtained from Step 6 in Table 3) and aligns all documents with a term-document matrix; in such case, every document can be represented as a term vector with the entire vocabulary as the feature space. By separating the documents into consecutive time slices, SEP measures the drifts of documents and assigns documents to topics generated in different years to indicate the topics and their changes.

**Concept definition**: A topic is defined as a collection of documents that share semantically similar research content, denoted as $T$. The centroid of a topic is represented by the mean vector of all corresponding document vectors in the topic, denoted as $c$. The radius of a topic is defined as the largest Euclidean distance of all the documents in the topic to the centroid, defined as $r$.

**Step 1**: Construct the document-term matrix and the documents into consecutive time slices.

**Step 2**: Group the first slice of articles to form an initial topic $T_0$ and label it with the top-frequency term, calculate its radius $r$ and centroid $c$, $T_0$ would constitute the starting topic of SEP.

**Step 3**: For every document $a$ entering in the second time slice, its Euclidean distance $E(a, c)$ with the centroid $c$ of $T_0$. If $\frac{E(a,c)-r}{r} < \sigma$ ($\sigma$ is a given threshold which is set as 0.1 by default), this document is newly added documents to $T_0$, otherwise it will be classified as a drifted article to $T_0$.

**Step 4**: Update the centroid and radius of $T_0$ with the newly added documents. For the other drifted documents, we will apply a K-means clustering algorithm to generate a new topic set $T_n$. All topics in $T_n$ is regarded as descendent topics of $T_0$.

**Step 5**: For all the following time slices, iterate Steps 3–4. But after the second slice, measure a document's similarity with all the existing topics and assign it to the most similar topic using Salton's cosine similarity [18], every new set of $T_n$ generated in Step 4 is also regarded as the corresponding descendant of its most similar topic. Also, the radius and centroids of all the topics are updated in Step 4.

More details of this method could be found in [15]. The final outcome of SEP is a visualising map with nodes representing the topics and directed edges linking those topics representing the descendent-predecessor relationships between connected topics.

## 3 Results

### 3.1 Co-term Analysis

By applying co-term analysis to the ten terms describing nonprofit organisations and seventeen terms indicating data analysis techniques, we produced a co-term map in Fig. 1 with the aid of Circos Table Viewer.[5] Figure 1. yields a bird's eye view of (1) the frequently mentioned data analysis methods in the field and the key types of NPOs, represented by the arc length of terms, and (2) how strong data analysis methods are related to different types of NPOs, represented by the width of ribbons connecting two terms.

Observing the arc length of those terms, we could identify that organisation management-related and statistics-based data analysing methods, such as *document analysis*, *knowledge management*, *data analysis*, *decision-making*, and *logistic regression*, are frequently seen in relevant NPOs publications. Intriguingly, data analysis methods in the computer science domain also appear in this map, including *decision support system*, *predictive model* and *machine learning*, even though they seem less prevalent than the top ones. The frequency ranking of those methods

[5] More details could be found at http://mkweb.bcgsc.ca/tableviewer/visualize/.

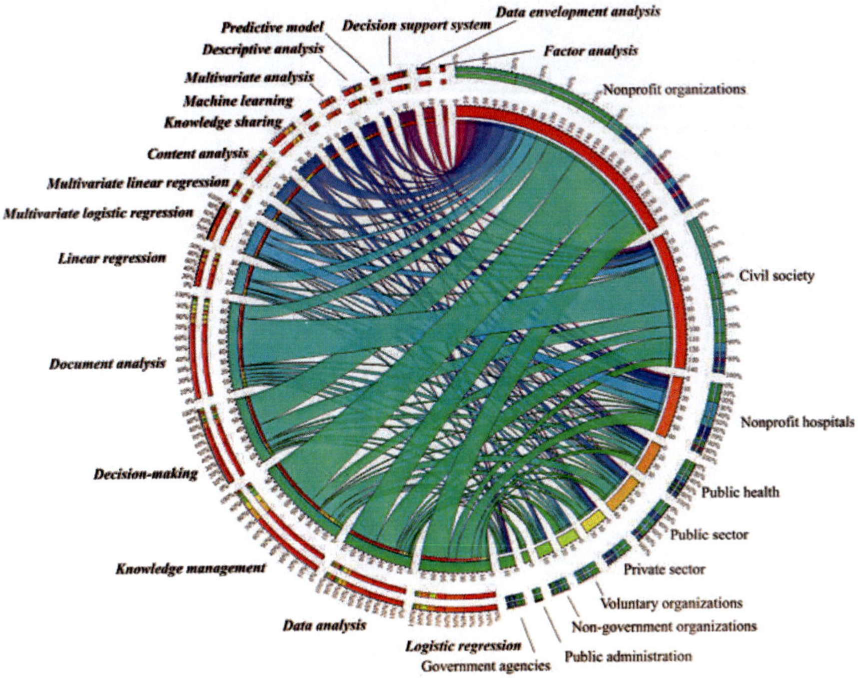


**Fig. 1** Co-term network visualisation

indicates that current NPOs still prefer to apply those robust, classical and organisation management-related methods in real-world implementations. Nevertheless, the utilisation of novel computer science-based methods is still an emerging trend.

Next, we focus on the linking strength of different NPOs data analysis method pairs. From the NPOs perspective, in healthcare-related NPOs such as *nonprofit hospitals* and *public health*, the dominating techniques are mostly derived from statistical analysis like *logistic regression*[6] and multivariate logistic regression, representing the typical data analysis methods in the medical domain. However, in other public affair related NPOs such as *public sector* or *public administration*, *knowledge management* seems to take a more crucial role in the associated studies. From the technique perspective, we could see *document analysis* is notably involved with civil society; in such case document analysis is widely used to identify the impacts and roles of civil society using multiple case studies [19–21]. Some AI-related techniques like *predictive model* and *machine learning* tend to have stronger associations with the healthcare domain including *nonprofit hospitals* and *public health*, indicating the emergence of those techniques in the healthcare sector.

[6] Although we noticed that multivariate logistic regression is a sub-method of logistic regression, we still consider it insightful to profile the two techniques from different granularities. This criterion applies to other technical terms that may overlap with each other as well.

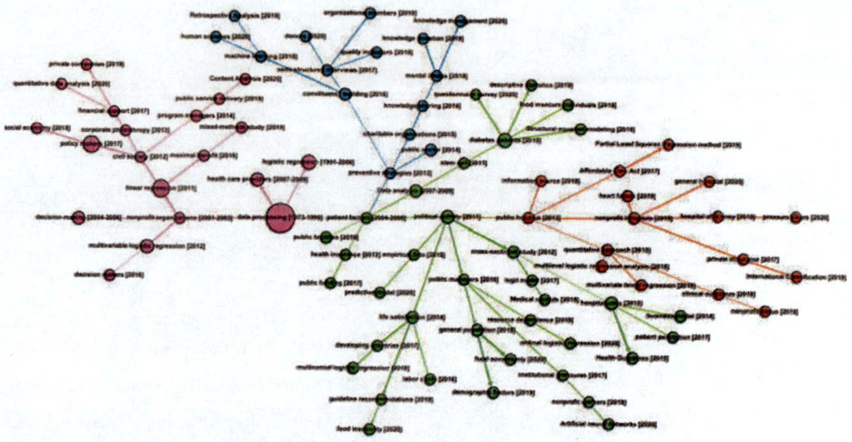

**Fig. 2** The evolutionary patterns of research topics

## 3.2 Evolutionary Relationship Identification

We generated a topic evolutionary with 87 nodes and 86 edges by applying SEP to the extracted terms. Each node represents a research topic, and each directed edge represents the predecessor-descendant relationship between the connected nodes, time labels in the brackets indicate when the topic was proposed. With the aid of Gephi [22], we applied a community detection algorithm to the SEP network and partitioned it into five topic communities with different colours, as shown in Fig. 2. The topic communities could, respectively, be concluded as *#1 data processing (pink), #2 public health and health management (green), #3 hospital management (orange)* and *#4 public affair and knowledge management (blue).*

The four communities represent the evolving patterns and divergence of research attention in this field over time. Tracing back to 1973, the initial topic data processing *[1973–1990]* indicates the emergence of data utilisation needs and data analysis means in NPOs at the early stage. This topic derives a pink topic community *#1 data processing*, in which we could observe topics focusing on classical and traditional statistical methods *(linear regression [2011], logistic regression [1991–2000])* and classical organisational roles *(programme managers [2014], policy makers [2017])*. Even the topics at the end of those branches, which represent the recent changes, are still related to basically organisational or data analysis concepts such as *decision-making [2004–2006], decision-makers [2016]*, and *quantitative data analysis [2020]*. Hence, we summarise that this community profiles a fundamental and traditional pathway of how the data analysis methods are applied in NPOs.

Derived from community *#1*, community *#2 public health* and *health management* has a clear emphasis on the healthcare sector. Some of those topics reveal the healthcare data sources like *empirical data [2015], medical records [2018]*, whilst some others reflect realistic problems that using data analytics to solve in the medical domain such as *stem cell [2011], diabetes patients [2015], life satisfaction [2014], labour costs [2016], guideline recommendations [2019].* The rest topics mostly refer to specific data analysis methods. Interestingly, except for the data analysis methods identified in community #1, a few AI-related topics like predictive model *[2020]* and artificial neural networks *[2020]* emerge from which we could have a glimpse of artificial intelligence implementation for data analysis in the healthcare domain. However, the limited amount of such topics also indicates this trend is still in its infancy.

Community *#3* typically inherits the healthcare attribute of community *#2* and focuses on a more specific domain of hospital management. Topics in this community cover hospital attributes *(public hospital [2013], nonprofit hospitals [2016], nonprofit status [2019],* and *general hospital [2020]),* evaluating metrics *(hospital efficiency [2018], clinical outcomes [2018],* and *pressure ulcers [2020]*[7]), policy-related issues (*private insurance [2017]* and *Affordable Care Act [2017]*), and data analysis methods used on hospital management issues. From the technical perspective, data analysing techniques applied in this topic community are still highly coupling with the fundamental ones in community #1.

Lastly, community #2 additionally derives another community #4 public affair and knowledge management. Different from community #2, community #4 expands the scope of topics to a wider public affair sector. Politics and charity-related topics (*charitable organisations [2015], constituency building [2016],* and *donors [2020]*) are frequently seen in this community, accoupling with a branch emphasising knowledge management concerns in organisations (*knowledge sharing [2016], knowledge donation [2019],* and *knowledge management [2020]*). The appearance of another AI-related topic, machine learning *[2018],* indicates AI's application has also started to emerge in those public sectors.

Evolutionary pathways span time born topics that could be seen in each cluster, meaning that certain publications inspiration is being disrupted by new awareness at various levels. The ability to track the evolutionary paths of scientific subjects is the SEP's key advantage; however, the graph of the SEP's visual routines can only show changes in topics. As *nonprofit organisations* were set to be used from 2001 to 2003 in data processing, it appears twice recently as alternative terms charitable organisations and nonprofit hospitals in 2015 and 2016, respectively. This indicates the excellent representative to present the evolutionary pathway in a dynamic way of disturbing data and feature space. Such findings will help: to address more data and issues for specific subjects and areas; provide enough statistical information to trace precise evolutionary pathways; and assist in the understanding of the SEP's visual routines.

[7] Pressure ulcer incidence rate is an important evaluation metric for clinical nursing quality.

## 4 Limitations of This Study

There are also some limitations in our current study. From the methodological perspective, co-term and SEP analyses are based on the correlation between semantic similarities of terms. Yet, they neglect the causality or sentiment associations (positive or negative) between terms, limiting us to interpret the deeper reasons of the term relationships and evolutionary patterns. Another limitation is that relevant research papers allocated beyond the scope of this study were excluded, although they have a high rate of keyword occurrence. For example, several studies related to medical research containing medical and health research keywords seem irrelevant to NPOs. The selection of keywords during the search stage was changed several times to ensure all the found documents have lied in the scope of applying data analytics in NPOs. In future studies, we aim to involve sentiment analysis and causality inference techniques to improve our methodology in order to provide a deeper data-driven interpretation of such results. Also, comprehensive databases will be added to ensure the full coverage of the literature.

## 5 Conclusions

This paper is a one-of-a-kind compilation of bibliometric research and recent advances in the field of data analytics in NPOs. This study conducted co-term and SEP analyses on 2673 documents to reveal the NPO-data analysis technique relationships and topic evolutionary patterns from data analytics research in NPOs. The results from co-term and SEP maps complement each other and together yield the following insights: (1) classic statistics-based data analysing techniques are dominantly applied in the NPO field through all the time; (2) Healthcare and public affairs are two significant sectors that involve data analytics to support decision-making and problem-solving; (3) AI-based data analytics is an emerging trending in this field, especially in the healthcare-related sector, however, it is still at an immature stage, and more efforts are needed to nourish its development. Findings in this study could benefit (1) Researchers with evidence to conduct longitudinal analyses and investigate the status quo of data analytics in NPOs, and (2) Researchers with empirical insights in recognising the potential of data analytics and implementing data analytics in some sectors such as public sectors, governmental agencies, and private–public organisations.

## Appendix 1

| Field tags | Search string |
|---|---|
| Scopus: TITLE-ABS-KEY | TITLE-ABS-KEY ("Data analy*" OR "Data-driven" OR "Predictive Analy*" OR "Analy*" OR "Big Data" OR "large*scale data" OR "Open Data" OR "natur* language process*" OR "NLP" OR "Machine Transla*" OR "lexical analys*" OR "Information extract*" OR "knowledge Graph" OR "Feature Select*" OR "Natur* language generat*" OR "NLG" OR "Natur* language interact*" OR "mode identif*" OR "virtual personal assistant" OR "Text to Speech" OR "sentiment analys*" OR "data mine*" OR "text mine*" OR "document mine*" OR "linguistic mine*" OR "data analys*" OR "text analys*" OR "document analys*" OR "linguistic analys*" OR "data Process*" OR "text Process*" OR "document Process*" OR "linguistic Process*" OR "Text Classif*" OR "Text Cluster*" OR "SYNTACTIC ANALYS*" OR "automatic summarise" OR "information filter*" OR "Expert System" OR "Decision Support System" OR "Model-based" OR "intelligen* system" OR "multi-agent system" OR "knowledge Management" OR "knowledge Represent*" OR "Semantic Net*" OR "Predicate logic" OR "knowledge engineer*" OR "Decision Tree" OR "Linear Regression" OR "BP Neural Network" OR "neural comput*" OR "Artificial Neural Network" OR "Bayesian Classification" OR "Support Vector Machine" OR "Logistic Regression" OR "Spectral Clustering" OR "Dimensionality Reduction" OR "Classification Accuracy" OR "Fuzzy Clustering" OR "Association Rules" OR "Combined Training" OR "Deep Learning" OR "Machine Learning" OR "Reinforcement Learning" OR "depth learning")<br>AND<br>TITLE-ABS-KEY ("Nonprofi*" OR "Non*Profi*" OR "no*for*profit*" OR "Charit*" OR "third sector" OR "giving dector" OR "voluntary sector*" OR "voluntary sector*" OR "civil society giving" OR "nongovernmental" OR "organisation philanthropy" OR "social capital") |

(continued)

(continued)

| Field tags | Search string |
| --- | --- |
| WoS: the search field changed three times using different tags which are: (TI = Title, AB = Abstract<br>AK = Author Keywords) | TI = ("Data analy*" OR "Data-driven" OR "Predictive Analy*" OR "Analy*" OR "Big Data" OR "large*scale data" OR "Open Data" OR "natur* language process*" OR "NLP" OR "Machine Transla*" OR "lexical analys*" OR "Information extract*" OR "knowledge Graph" OR "Feature Select*" OR "Natur* language generat*" OR "NLG" OR "Natur* language interact*" OR "mode identif*" OR "virtual personal assistant" OR "Text to Speech" OR "sentiment analys*" OR "data mine*" OR "text mine*" OR "document mine*" OR "linguistic mine*" OR "data analys*" OR "text analys*" OR "document analys*" OR "linguistic analys*" OR "data Process*" OR "text Process*" OR "document Process*" OR "linguistic Process*" OR "Text Classif*" OR "Text Cluster*" OR "SYNTACTIC ANALYS*" OR "automatic summarise" OR "information filter*" OR "Expert System" OR "Decision Support System" OR "Model-based" OR "intelligen* system" OR "multi-agent system" OR "knowledge Management" OR "knowledge Represent*" OR "Semantic Net*" OR "Predicate logic" OR "knowledge engineer*" OR "Decision Tree" OR "Linear Regression" OR "BP Neural Network" OR "neural comput*" OR "Artificial Neural Network" OR "Bayesian Classification" OR "Support Vector Machine" OR "Logistic Regression" OR "Spectral Clustering" OR "Dimensionality Reduction" OR "Classification Accuracy" OR "Fuzzy Clustering" OR "Association Rules" OR "Combined Training" OR "Deep Learning" OR "Machine Learning" OR "Reinforcement Learning" OR "depth learning")<br>AND<br>TI = ("Nonprofi*" OR "Non*Profi*" OR "no*for*profit*" OR "Charit*" OR "third sector" OR "giving dector" OR "voluntary sector*" OR "voluntary sector*" OR "civil society giving" OR "nongovernmental" OR "organisation philanthropy" OR "social capital") |

## References


1. Johnson, M.P.: Data, analytics and community-based organizations: transforming data to decisions for community development. ISJLP **11**, 49 (2015)
2. Mayer, L.H.: The promises and perils of using big data to regulate nonprofits. Wash. Law Rev. **94**(3), 1281–1336 (2019)
3. Anheier, H.K.: Nonprofit Organizations Theory, Management, Policy. Routledge Taylor & Francis Group (2005)
4. Mahmoud, M.A., Yusif, B.: Market orientation, learning orientation, and the performance of nonprofit organisations (NPOs). Int. J. Product. Perform. Manag. **61**, 624–652 (2012)
5. Kassen, M.: Adopting and managing open data: stakeholder perspectives, challenges and policy recommendations. Aslib J. Inf. Manag. **70**, 518–537 (2018)
6. Maxwell, N.L., Rotz, D., Garcia, C.: Data and decision making: same organization, different perceptions; different organizations, different perceptions. Am. J. Eval. **37**(4), 463–485 (2016)
7. Hou, Y., Wang, D.: Hacking with NPOs: collaborative analytics and broker roles in civic data hackathons. Proc. ACM Human-Comput. Inter. **1**(2), 1–16 (2017)
8. Shah, N., Irani, Z., Sharif, A.M.: Big data in an HR context: exploring organizational change readiness, employee attitudes and behaviors. J. Bus. Res. **70**, 366–378 (2017)
9. Bopp, C., Harmon, C., Voida, A.: Disempowered by data: nonprofits, social enterprises, and the consequences of data-driven work. In: 2017 ACM SIGCHI Conference on Human Factors in Computing Systems, pp. 3608–3619. Association for Computing Machinery, Denver, USA (2017)
10. Trivedi, G.: Visualization and scientometric mapping of global agriculture big data research. Libr. Philos. Pract. (2019)
11. Wang, X., Xu, Z., Su, S.-F., Zhou, W.: A comprehensive bibliometric analysis of uncertain group decision making from 1980 to 2019. Inf. Sci. **547**, 328–353 (2021)
12. Falagas, M.E., Pitsouni, E.I., Malietzis, G.A., Pappas, G.: Comparison of PubMed, Scopus, Web of science, and Google scholar: strengths and weaknesses. FASEB J. **22**(2), 338–342 (2008)
13. Zhang, Y., Porter, A.L., Hu, Z., Guo, Y., Newman, N.C.: "Term clumping" for technical intelligence: a case study on dye-sensitized solar cells. Technol. Forecast. Soc. Chang. **85**, 26–39 (2014)
14. Callon, M., Courtial, J.-P., Turner, W.A., Bauin, S.: From translations to problematic networks: an introduction to co-word analysis. Soc. Sci. Inf. **22**(2), 191–235 (1983)
15. Zhang, Y., Zhang, G., Zhu, D., Lu, J.: Scientific evolutionary pathways: identifying and visualizing relationships for scientific topics. J. Am. Soc. Inf. Sci. **68**(8), 1925–1939 (2017)
16. Fleming, L.: Recombinant uncertainty in technological search. Manage. Sci. **47**(1), 117–132 (2001)
17. Fleming, L., Sorenson, O.: Science as a map in technological search. Strateg. Manag. J. **25**, 909–928 (2004)
18. Salton, G., Harman, D.: Information retrieval. In: Encyclopedia of Computer Science, pp. 858–863. John Wiley and Sons Ltd. (2003)
19. Anaf, J., Baum, F.E., Fisher, M., Harris, E., Friel, S.: Assessing the health impact of transnational corporations: a case study on McDonald's Australia. Glob. Health **13**(1), 7 (2017)
20. Powell, L.J., Wittman, H.: Farm to school in British Columbia: mobilizing food literacy for food sovereignty. Agric. Hum. Values **35**(1), 193–206 (2018)
21. Secco, L., Favero, M., Masiero, M., Pettenella, D.M.: Failures of political decentralization in promoting network governance in the forest sector: observations from Italy. Land Use Policy **62**, 79–100 (2017)
22. Mathieu, B., Sebastien, H., Mathieu, J.: Gephi: An open source software for exploring and manipulating networks. In: Third International AAAI Conference on Weblogs and Social Media, pp. 361–362. Association for the Advancement of Artificial Intelligence (2009)

# Analyzing Donors Behaviors in Nonprofit Organizations: A Design Science Research Framework

**Idrees Alsolbi, Renu Agarwarl, Bhuva Narayan, Gnana Bharathy, Mahendra Samarawickrama, Siamak Tafavogh, and Mukesh Prasad**

**Abstract** In nonprofit organizations (NPOs), analyzing donor behavior remains critical and challenging due to internal and external factors, such as family, political, and environmental issues. Machine learning (ML) techniques are very promising to provide the solutions to analyze the customer behaviors and churns issues of many different organizations. However, it remains a challenge on how best to build and design an intelligent decision support system for analyzing donor behaviors in NPOs. This paper applies the underlying guidance from information systems by utilizing a design science research framework to create an artificial intelligence (AI)-enabled decision support system to analyze donors behaviors more effectively and efficiently. The framework aims to provide a theoretical foundation for creating generalized design principles and design features for designing an intelligent decision support system. It also presents the capabilities of data analytics and ML techniques to understand donors behaviors by exploring the external factors that affect donors' decision-making.

**Keywords** Donors behaviors · Decision support system · Design science · Nonprofit organization

---

I. Alsolbi (✉) · M. Prasad
School of Computer Science, Australian Artificial Intelligence Institute, University of Technology Sydney, Sydney, Australia
e-mail: idna1257@gmail.com

R. Agarwarl
Business School, University of Technology Sydney, Sydney, Australia

B. Narayan
School of Communication, University of Technology Sydney, Sydney, Australia

G. Bharathy
Information Systems and Modelling School, University of Technology Sydney, Sydney, Australia

M. Samarawickrama
Australian Red Cross, Sydney, Australia

S. Tafavogh
Department of Finance, Commonwealth Bank Health Society, Sydney, Australia



D. Gupta et al. (eds.), *Pattern Recognition and Data Analysis with Applications*, Lecture Notes in Electrical Engineering 888,
https://doi.org/10.1007/978-981-19-1520-8_62

# 1 Introduction

Nonprofit, also known as a nonprofit corporation, not-for-profit agency, or nonprofit institution, differs from the businesses and industries as they are private, independent, self-governing institutions and control their practices and goals [1]. Such institutions are museums, schools, universities, research institutions, human services, health organizations, human rights organizations, religious centers and organizations, and charitable foundations [1]. Mahmoud and Yusif [2] describe NPOs in business to meet individuals' and beneficiaries' requirements. NPOs have important social purposes, which contrast with an entity that operates as a business to generate a profit for its owners in a traditional market [3]. Anheier [1] mentioned that NPO's goals cover individual activities and the values and motivations that drive people to engage in activities to benefit society, the environment, and cultural heritage through charities, philanthropy, volunteering, and giving. NPOs' funding and income sources vary; in Australia, 49.1% of NPOs' income is self-generated, the government contributes 33.5%, and only 9.5% comes from public donations [3]. Notably, Australia's percentage of public donations is greater than Germany, France, and New Zealand but less than the USA [3]. Given this high NPO's funding across different countries, NPOs can significantly influence society by attracting donors (who provide monies/funds) and volunteers (who give their time) and establishing strong relationships with clients to pursue their NPOs' interests.

Donors support the goals of NPOs in different ways, such as giving money, gifts, time for volunteering, and using their experience in various events in many different ways such as playing music, singing, and photography. Private donations represent a significant factor in funding NPOs in the USA, which annually contribute to more than 10% of the Gross Domestic Product [4]. Dietz and Keller [5] reported that individuals donate to NPOs because of their deep passion or beliefs of NPOs' needs which attracted around $260 billion. It is believed that certain factors impact peoples' intentions toward donating, such as income, educational level, and previous giving history [4]. Today's NPOs focus is not only on gaining donations but also on knowing donors habits, leading NPOs to authentically interact with their donors and how they resonate with them [6]. One of the essential behaviors is the retuning or intention to donate for a second time. Only 19% of donors donate for the second time, which is a major concern for NPOs [6]. However, Sargeant and Jay [7] mentioned that targeting appropriate donors to charities and improving communications remain critical for NPOs.

Given this backdrop, understanding the fundamentals of donors is crucial [8]. Certain behaviors include donors intentions to donate either time or money, donor frequency (returning), donor engagement, donor communications, and volunteering engagement that require a deeper understanding of technologies and capabilities of data science and ML techniques. By analyzing the behaviors using ML techniques, NPOs improve the chances of increasing their current financial support and interaction with outgoing donors for potential opportunities for repeat donation activity [9].

This paper has two main contributions: first, to create a new design science theory of designing an artifact for analyzing donor behaviors and second, to design an artifact (an AI-enabled decision support system) to analyze donors behaviors in NPOs. This paper's remainder provides a literature review first, then introduces the design science approach, followed by coverage of the research framework, the collection and analysis of data, and finally, the research contribution and expected results.

## 2 Literature Review

### *2.1 Decision Support System*

Decision support system (DSS) became a common interest for many researchers since the last few decades in various fields such as information systems (IS), mathematics, and economics [10]. Decision support is the main component of IS research which evolved in improving and managing the decision-making process [11]. DSS is not based on combining all the ongoing alternatives but on choosing the right one based on priorities and goals [10]. DSS has been transformed from being traditional to intelligent-based systems, where AI, ML, cloud computing, and networking are the main reasons for this transformation [10]. These technologies become required when designing a DSS to ensure sustainability, high productivity, and advantages [10]. The intelligent DSS includes knowledge-driven, documents-driven, data-driven, and communication-driven DSS [12]. In addition, an intelligent DSS involves AI techniques to support decision-making, counted as "intelligent" [13].

Moreover, any DSS built based on ML is referred to as intelligent or AI-enabled DSS [14]. The term AI started in the 1950s and led to many AI-enabled systems [15]. ML techniques play a significant role in describing and predicting donations and donor behaviors in this context. ML can help NPOs to handle their current donors more effectively or utilize their existing assets. ML techniques have been used widely in various sciences in different disciplines for organizing the data, extracting helpful information, and recognizing patterns through supervised (i.e., classification) and unsupervised (i.e., clustering) algorithms [16]. For example, classifications can assign benefactions to predefined classes, whereas clusters find any relationships and hidden information without predefined classes [17].

### *2.2 Donors Behaviors*

There are some factors, including behaviors that affect donors' ability to donate funds or volunteer ability to spend their time, such as attitudes, norms, perceived behavioral control, subjective norms, past behaviors, and moral norms [8]. Farrokhvar et al. [4] illustrated some influential factors on donors' behaviors toward donating comprised donors' education level, sex, age, population, household income, and ethnicity. Using these factors, different ML models (support vector regression, multiple linear regression, artificial neural networks) were generated to estimate future charitable giving accurately from donors. The results recommend that educational level, population, and previous giving amount are independent variables and significant. Similarly, a multinomial logistic model in [18] was developed to investigate if multidonations individuals are different from a single donor or non-donors. Shehu et al. [18] used various predictors: geographical, health-related, psychographics, and sociodemographic variables to generate useful insights of donors behaviors. The results show useful insights into the donor engagement and retention techniques of NPOs along with donor recognized profile characteristics. However, none of the above studies [4, 8, 18] attempted to design a DSS for analyzing donors' behaviors in NPOs. Hence, considering a smart DSS for analyzing donors behaviors remains an essential gap for NPOs.

## 2.3 *DSS in NPOs*

Decision-making in NPOs has shown effectiveness in managing decisions [19]. However, decision-making in NPOs faces obstacles due to data growth, which provides more opportunities to manage the data [20]. Most of the data from NPOs are unstructured, which is very challenging to understand the hidden information and find some relationship [21]. Also, it is claimed that managing information in NPOs is a challenging task [21]. There are major challenges for NPOs, such as the lack of technical skills [22] and financial sources [23] for applying data analysis. Hence, if the data is not well collected and organized, NPOs will not benefit from the available data to draw insights and conclusions [19]. Managers may use performance data to gain useful insights into the organization's strengths and weaknesses, providing them the knowledge they need to make informed decisions [24]. Most importantly, developing a DSS for managing NPOs activities is crucial [25].

Nevertheless, the literature shows that no research has focused on designing AI-enabled DSS for analyzing donors behaviors in NPOs. The current literature lacks experimental and theoretical foundations for designing AI-enabled DSS in NPOs to analyze donors' behaviors. Designing an intelligent system is complex, which requires special characteristics such as autonomy, self-learning, and user interactions [16, 17]. All these characteristics distinguish the AI-enabled DSS from the traditional DSS. Moreover, DSS research recently requires more improvements on its relevance and quality [14]. We tackle these issues and research gaps by (1) developing a conceptual AI-enabled DSS design relying on the driven knowledge from theoretical sources and (2) creating an artifact according to this design to analyze donors behaviors. Thus, we found that Design Science Research (DSR) can overcome the complexity of designing DSS in NPOs. DSR has been an essential approach in DSS research because industry and profession may be involved in intellectually relevant ventures by design science studies [14]. Moreover, Arnott and Pervan [14] claimed that researchers are searching for assistance with preparing and implementing their DSR projects. In our research project context, DSR proposes a dialog between abstract theoretical knowledge and practical knowledge.

## 3 Design Science Research

Design science is creating artifacts and scientific studies to solve a particular problem [26]. DSR is a scientific problem-solving methodology developed specifically for the IS domain. The DSR has three aims: first, a nominal method model for design research in science, second, a mental model for the presentation, and third, evaluation of design research in IS [27]. The DSR represents a well-established process in the field of IS to create an artifact seeking to expand the barriers and limitations between people and organizations [28]. Artifacts are defined as constructs consisting of software, hardware, systems, or models [28]. The artifact must be creative, more productive, or useful in solving a previously unresolved problem or solving a known problem [28]. In the context of implemented software or algorithms, the artifact may range from simple instantiations to more efforts in the context of final design theories [28]. This research project will construct a design theory for designing an artifact (an AI-enabled DSS) to analyze donors behaviors using ML techniques. This artifact aims to help NPOs' managers make better decisions on future marketing, fundraising management, and other NPOs missions. The design theory will explain the artifact's functions, attributes, and features [29]. The design theory also provides prescriptions on how our AI-enabled DSS is designed and constructed.

### *3.1 Research Framework*

DSR seeks to bridge the gap between implementation and theory [30]. One of the advantages of the design science approach is an incremental and iterative process [28], which requires conducting at least three iterations [16, 31]. Thus, we realized that the design science research framework presented by Peffers et al. [27] suits our research project aims. This framework is selected because it has three iterations, a communication stage with scholars via publications and a design theory. Also, the iterative cycles imply constant reflection and abstraction [32], which we assume are necessary foundations for developing a design theory and artifact.

The framework combines the common stages of DSR approaches presented in the literature [28, 29, 33]. It involves three iterations and consists of six stages/phases, starting from identifying the problem, illustrating the solution's objectives, designing and developing the artifact, displaying the artifact's viability, assessing the artifact, and reporting the results through communication with scholars and professionals

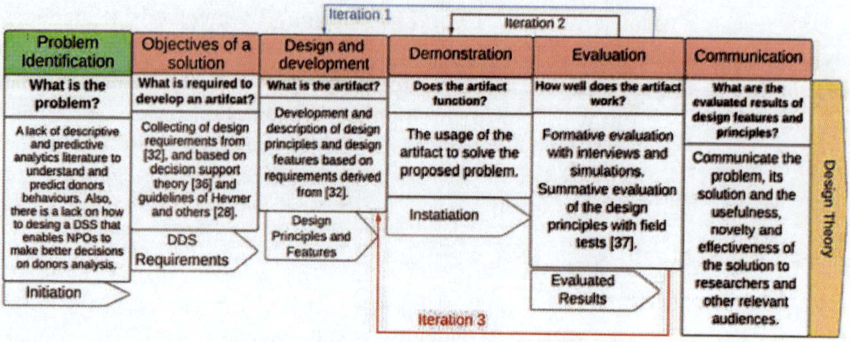


**Fig. 1** Proposed framework for designing AI-enabled DSS. Adapted from Peffers et al. [27]

via publications. Each stage produces an output used in the following stage. The proposed framework has six phases, as shown in Fig. 1, and outlined as follows:

**Phase 1: Problem Identification**

This phase identifies a research problem and the importance of solving the proposed problem. There are attempts to predict donors behaviors using ML techniques such as [4, 34]. However, we found a lack of descriptive and predictive analytics literature to understand and predict donors attitudes toward helping, donating, and giving to the NPOs, especially in the context of donating money and volunteering time. A DSS is developed by Barzanti et al. [25] to rank donors using a fuzzy method to predict the targeted campaign. Although this study is useful for our problem initiation, it lacks in developing guidelines for designing a DSS. The above studies [4, 25, 34] focus on domain-specific explanations that show the capabilities of some ML techniques to analyze donors behaviors. Notably, they are less focused on design knowledge that guides creating an artifact for a better decision-making process in NPOs.

To expand the awareness of the research problem, we conducted two informal interviews with experts from NPOs during this stage. During interviews, we asked the experts (1) to describe the process of donors behaviors analysis, (2) state the challenges they face to design such DSS that helps in describing and predicting donors behaviors, and (3) explicate the potentials of creating a design theory that guides the process of designing AI-enabled DSS. We noted all valuable insights from the interviews. For example, experts mentioned that descriptive and predictive analytics assist NPOs in making better decisions to increase the efficiency and performance of NPOs and understand the influential factors on donations. Furthermore, these analytics can be functioned and generated through a decision support system. At this stage, the interviews helped identifying the problem and increasing the awareness of creating a design theory of an artifact to analyze donors behaviors.

**Phase 2: Objectives of Solutions**

This stage elicits the intended artifact requirements and determines the main functionalities of the desired DSS. For designing the artifact, the initial requirements for

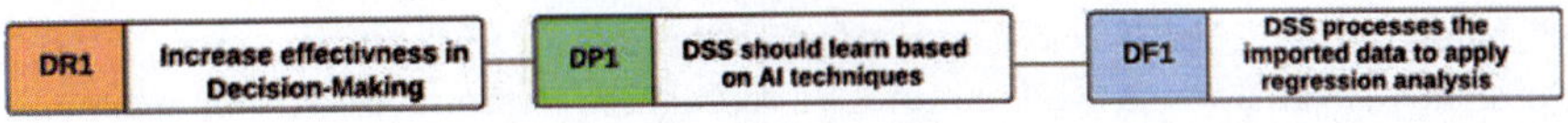


**Fig. 2** Preliminary conceptual map of DRs, DPs, and DFs

creating an artifact as a product are defined based on the design requirements of Meth et al. [32], the decision support theory [35], and DSR guidelines by Hevner et al. [28]. The approach of Meth et al. [32] is chosen because it applies the fundamentals of developing the decision support theory of Silver [35] and his decisional guidance. Also, the guidelines of creating an artifact suggested by Hevner et al. [28] are followed to ensure that a constructed artifact is scientific in its method and effects. Finally, the collected and generated requirements are evaluated before Iteration 1 through interviews with decision-making and data science experts in NPOs.

**Phase 3: Design and Development**

To meet the design requirements (DRs), this framework adopts Meth et al. [32] model to derive a collection of Design Principles (DPs) and Design Features (DFs). DPs can be a statement that tells what the artifact should do [16], and DFs are unique artifact capabilities to fulfill DPs [32]. The DPs will be derived based on the DRs formulated in the model of Meth et al. [32]. Therefore, the DFs will be formulated to satisfy the design principles [32]. The DPs and DFs will be mapped into an conceptualized artifact to present to experts for a formative evaluation in Iteration 1. Each DP can be mapped to one or more relevant DF. The mapping of DPs to design DFs supports evaluating the artifact [32]. Figure 2 shows an example of mapping a DR according to DP and DF.

**Phase 4: Demonstration**

The artifact aims to support aggregated, high-level data and better understand the donor behaviors data. This demonstration phase builds descriptive and predictive models that predict and describe donations and donor behaviors. The models are generated using ML techniques, which can be supervised or unsupervised methods. Supervised methods make predictions or classifications based on the given labeled input data. Unsupervised methods draw inferences and hidden relationships from unlabeled data. We are applying different techniques of supervised and unsupervised ML techniques. The purpose of this stage is to generate an instantiation to solve the proposed problem. Our intended instantiation is a complete design theory to design an artifact to analyze donors behaviors. Also, the artifact is intended to be an AI-enabled DSS that offers suggestions and presents useful descriptions and predictions of donors behaviors.

**Phase 5: Evaluation**

This phase uses evaluation of the framework introduced by Venable et al. [36], which has two types of evaluations, formative and summative. The assessment will be for the AI-enabled DSS and the design theory with relevant DRs, DPs, and DFs. Formative evaluation is involved in creating empirically validated explanations that provide a foundation for effective action to enhance the evaluated features or results [36]. Summative evaluation is used to provide a foundation to produce common meanings of the evaluation in a different context. The evaluation stage will run three iterations:

**Iteration 1**: DRs, DPs, and DFs will be evaluated to ensure their relevance to our research aims and objectives. We will conduct semi-interviews with NPOs decision-makers, data scientists, and managers. Those experts are involved during interviews to conduct a formative evaluation. The results of this iteration lead to apply any changes or suggestions on DRs, DPs, and DFs.

**Iteration 2**: DRs DPs, and DFs will be visible and conceptualized before starting this iteration. Then, descriptive and predictive models will be completed and fully functioning. For the evaluation, validation techniques associated with ML techniques will be applied, such as K-fold cross-validation, to ensure these models' effectiveness. K-fold cross-validation is a common technique to evaluate the models and estimate errors among practitioners [37].

**Iteration 3**: The evaluation before this iteration aims to ensure the success of DRs, DPs, and DFs. The designed AI-enabled DSS to analyze donors' behaviors will be tested by NPOs decision-makers, data scientists, and managers. We will then interview them for their feedback and apply any final suggestions or changes for this iteration. Following that, we will develop a functional front-end, a back-end, to a web-based DSS for analyzing donors behaviors using ML techniques. Finally, the web-based decision support system will meet all the DRs, DPs, and DFs to analyze donors' behaviors. The output of this iteration is to finalize the design theory by combining evaluation results and results of the developed AI-enabled decision support system.

**Phase 6: Communication**

We will create a complete expository instantiation of the design theory to be published in the communication stage. The design theory will be created based on the design theory profile by Gregor and Jones [38]. Thus, the design theory consists of a prescription on how the DSS is developed. Moreover, the results of the evaluation stage will be demonstrated to the researchers and experts in NPOs. The communication stage aims to communicate the effectiveness, usefulness, and novelty of the solution based on the evaluation stage analysis.

## 4 Data Collection and Analysis

The evaluation stage involves semi-structured interviews with experts (before Iteration 1 and Iteration 3), as illustrated in Fig. 1. The interviews generate data that will be analyzed using MAXQDA. MAXQDA is a software application to analyze qualitative data (such as interviews) and organize the information into categories or groups [39]. The analysis will drop insights for constructing the artifact and expound the design theory's implementation. The data source to feed the analytics models (predictive and descriptive) will be obtained from one or more NPOs. We will focus on certain variables in the data, such as (level of education, income, age, gender, living area, and ethnicity). The created models will be presented to experts during interviews in Iteration 3 to build insights on the accuracy, sufficiency, and visualization of the results.

## 5 Research Contribution and Expected Results

This research intends to design an AI-enabled DSS for analyzing donor behaviors in NPOs. This DSS will create descriptive and predictive models that intensively analyze donors behaviors and provide meaningful information for NPOs. The research contributes to the academic literature practically by implementing ML algorithms on donor behaviors. Therefore, the decision-making process may add value proposals for NPOs missions or improve internal data processing efficiency and effectiveness [24]. We also intend to introduce a design theory to design the smart DSS contributing to the IS literature. The theory of design is intended to collect the theoretical foundations of designing an AI-enabled DSS to analyze donors behaviors in NPOs.

## 6 Conclusion

Data analytics can transform NPOs into data-driven if appropriate analytical models, frameworks, and empirical studies to support and manage resources. One major gap is the lack of literature on designing an intelligent decision support system to analyze donor behaviors toward donating and volunteering. Donor behaviors vary due to various impacts such as income, level of education, sex, etc. Understanding these behaviors and the influencing factors on donors is critical for making decisions in NPOs. Thus, we have presented a design science framework to provide theoretical bases for designing an AI-enabled DSS to analyze donors behaviors. To support the development of the AI-enabled DSS for NPOs, we will (1) derive a theoretical design of a DSS for analyzing donor behaviors and (2) build an artifact (AI-enabled DSS to analyze donors behaviors). This research intends to demonstrate that AI-enabled DSS based on the design science approach can be used and adopted among the global NPOs. The DPs and DFs can also systematically help practitioners deploy the descriptive and predictive models on donor behavior data for further actions.

## References

1. Anheier, H.K.: Nonprofit Organizations Theory, Management, Policy. Routledge Taylor &

Francis Group (2005)
2. Mahmoud, M.A., Yusif, B.: Market orientation, learning orientation, and the performance of nonprofit organisations (NPOs). Int. J. Product. Perform. Manag. **61**, 624–652 (2012)
3. Productivity Commission: Contribution of the Not for Profit Sector. Australia, Canberra (2010)
4. Farrokhvar, L., Ansari, A., Kamali, B.: Predictive models for charitable giving using machine learning techniques. PLoS ONE **13**(10), 1–14 (2018)
5. Dietz, R., Keller, B.: A Deep Dive Into Donor Behaviors And Attitudes. Abila: Ablia (2016)
6. Te, N.: Study helps you better understand donor behaviors (2019)
7. Sargeant, A., Jay, E.: Fundraising Management: Analysis, Planning and Practice, 3rd edn. Routledge Taylor & Francis Group London and New York (2014)
8. Li, C., Wu, Y.: Understanding voluntary intentions within the theories of self-determination and planned behavior. J. Nonprofit Public Sect. Mark. **31**(4), 378–389 (2019)
9. Dunford, L.: To give or not to give: using an extended theory of planned behavior to predict charitable giving intent to international aid charities. University of Minnesota (2016)
10. Zeebaree, M., Aqel, M.: A comparison study between intelligent decision support systems and decision support systems. ISC Int'l J. Inf. Secur. **11**(3), 187–194 (2019)
11. Arnott, D., Pervan, G.: A critical analysis of decision support systems research revisited: the rise of design science. J. Inf. Technol. **29**, 269–293 (2014)
12. Power, D.: Decision support systems: a historical overview. In: Handbook on Decision Support Systems 1 Basic Themes, pp. 121–140. Springer (2008)
13. Burstein, F., Holsapple, C.W.: Handbook on Decision Support Systems 1: Basic Themes. London Ltd., Berlin Germany, Springer-Verlag (2008)
14. Arnott, D., Pervan, G.: Design science in decision support systems research: an assessment using the Hevner, March, Park, and Ram guidelines. J. Assoc. Inf. Syst. **13**, 923–949 (2012)
15. Rzepka, C., Berger, B.: User interaction with AI-enabled systems: a systematic review of IS research. In: Thirty Ninth International Conference on Information Systems. San Francisco, USA (2018)
16. Rhyn, M., Blohm, I.: Combining collective and articial intelligence: towards a design theory for decision support in crowdsfuncding. In: Twenty-Fifth European Conference on Information Systems (ECIS). Guimarães, Portugal (2017)
17. Rhyn, M., Leicht, N., Blohm, I., Leimeister, J.M.: Opening the black box: how to design intelligent decision support systems for crowdsourcing. In: 15th International Conference on Wirtschaftsinformatik. Potsdam, Germany (2020)
18. Shehu, E., Langmaack, A.C., Felchle, E., Clement, M.: Profiling donors of blood, money, and time: a simultaneous comparison of the German population. Nonprofit Manag. Leadersh. **25**(3), 269–295 (2015)
19. Maxwell, N.L., Rotz, D., Garcia, C.: Data and decision making: same organization, different perceptions; different organizations, different perceptions. Am. J. Eval. **37**(4), 463–485 (2016)
20. Fredriksson, C.: Big data creating new knowledge as support in decision-making: practical examples of big data use and consequences of using big data as decision support. J. Decis. Syst. **27**(1), 1–18 (2018)
21. Bopp, C., Harmon, C., Voida, A.: Disempowered by data: nonprofits, social enterprises, and the consequences of data-driven work. In: 2017 ACM SIGCHI Conference on Human Factors in Computing Systems, pp. 3608–3619. Association for Computing Machinery, Denver, USA (2017)
22. Hou, Y., Wang, D.: Hacking with NPOs: collaborative analytics and broker roles in civic data hackathons. Proc. ACM Human-Comput. Interact. **1**(2), 1–16 (2017)
23. Hackler, D., Saxton, G.D.: The strategic use of information technology by nonprofit organizations: increasing capacity and untapped potential. Public Adm. Rev. **67**(3), 474–487 (2007)
24. LeRoux, K., Wright, N.S.: Does performance measurement improve strategic decision making? Findings from a national survey of nonprofit social service agencies. Nonprofit Volunt. Sect. Q. **39**(4), 571–587 (2010)

25. Barzanti, L., Giove, S., Pezzi, A.: A decision support system for non profit organizations. In: Fuzzy Logic and Soft Computing Applications, pp. 270–280. Springer International Publishing, Cham (2017)
26. Johannesson, P., Perjons, E.: An Introduction to Design Science (2014)
27. Peffers, K., Tuunanen, T., Rothenberger, M., Chatterjee, S.: A design science research methodology for information systems research. J. Manag. Inf. Syst. **24**(3), 45–77 (2007)
28. Hevner, R.A., March, S., Park, J., Ram, S.: Design science in information systems research. Manag. Inf. Syst. Q. **28**, 75 (2004)
29. Walls, J.G., Widmeyer, G.R., El Sawy, O.A.: Building an information system design theory for vigilant EIS. Inf. Syst. Res. **3**(1), 36–59 (1992)
30. Holmstrom, J., Ketokivi, M., Hameri, A.: Bridging practice and theory: a design science approach. Decis. Sci. **40**(1), 65–87 (2009)
31. Rhyn, M., Blohm, I.: A machine learning approach for classifying textual data in crowdsourcing. In: 13th International Conference on Wirtschaftsinformatik (WI). St. Gallen, Switzerland (2017)
32. Meth, H., Mueller, B., Maedche, A.: Designing a requirement mining system. J. Assoc. Inf. Syst. **16**(9), 799–837 (2015)
33. Vaishnavi, V., Kuechler, B., Petter, S.: Design Science Research in Information Systems, pp. 1–62 (2019)
34. Korolov, R., Peabody, J., Lavoie, A., Das, S., Magdon-Ismail, M., Wallace, W.: Predicting charitable donations using social media. Soc. Netw. Anal. Min. **6**(1) (2016)
35. Silver, M.S.: Decisional guidance for computer-based decision support. MIS Q. **15**(1), 105–122 (1991)
36. Venable, J., Pries-Heje, J., Baskerville, R.: FEDS: a framework for evaluation in design science research. Eur. J. Inf. Syst. **25**(1), 77–89 (2016)
37. Anguita, D., Ghelardoni, L., Ghio, A., Oneto, L., Ridella, S.: The 'K' in K-fold cross validation. In: European Symposium on Artificial Neural Networks (ESANN), pp. 441–446. Bruges, Belgium (2012)
38. Gregor, S., Jones, D.: The anatomy of a design theory. J. Assoc. Inf. Syst. **8**(5), 312–335 (2007)
39. Fontanella, C.: The best 10 qualitative data analysis software in 2021, in hubspot (2021)

# Intelligent Health Care System Using Modified Feature Selection Algorithm

Rajalakshmi Shenbaga Moorthy and P. Pabitha

**Abstract** Machine learning has become predominately bringing optimal decision for any kind of decision support system. Medical field is the one where it requires optimal prediction as the incorrect result may leads to worst decision. In recent days, IoT devices generate vast amount of medical data. Not all the data generated are relevant. Some of the features may be irrelevant. Irrelevant features may debase the achievement of the classifier. Thus, feature selection is essential to select optimal subset of features before applying machine learning model. The main objective is to build a novel feature selector algorithm to build intelligent health care system where the accuracy of the prediction is maximum with minimum error rate. A novel algorithm called modified binary sine cosine algorithm (MBSCA) is built to select optimal set of features from the real-world datasets. The problem with conventional binary sine cosine algorithm (BSCA) like premature convergence and poor exploitation, in addition to the position of the best agent, the other agents like second-best agent and third-best agent like beta agent and delta agent is used to exploit the solution space with the goal to pick optimal solution. The proposed MBSCA is evaluated in the real-world medical datasets and compared with genetic algorithm and BSCA. From the experimental results, it is noticed that accuracy of the preferred method is maximum than the other algorithms.

**Keywords** Health care system · Binary sine cosine algorithm · Modified binary sine cosine algorithm · Feature selection

---

R. S. Moorthy (✉)
Sri Ramachandra Faculty of Engineering and Technology, SRIHER, Chennai, India
e-mail: srajiresearch@gmail.com

P. Pabitha
Anna University, MIT Campus, Chennai, India

 
D. Gupta et al. (eds.), *Pattern Recognition and Data Analysis with Applications*, Lecture Notes in Electrical Engineering 888,
https://doi.org/10.1007/978-981-19-1520-8_63

# 1 Introduction

The Internet, which is the wonderful invention of human, had dominance in various fields such as business, education, health care, science, government, computer science, electrical, electronics, network, communication [1]. In this today's technological era, the field of computer science, electronics and communication is fused together to give birth to new technical field called "Internet of Things (IoT)" [2]. The IoT is the notable advancement of Internet and brings changes in life of everyone day by day either knowingly or unknowingly. The IoT aims to collect and distribute data using which one can do analytics and turn the raw data into potential valuable insights. The emergence of IoT enables the researchers to devise a positive solution for pervasive health care application. A person at remote can get reliable solution arbitrarily at low cost. The rapid growth of three fields such as cloud computing, mobile computing, and wearable devices transforms tradition of health care to smarter pervasive health care through IoT devices [3]. Here comes the answer for the question, why there is a need for intelligent pervasive health care system. Nowadays, every activity happens via network, which was earlier happened as human to human interaction. Some of the unhealthy habits which are of common prevalence due to the technology growth are: (i) food at the door step, (ii) lack of physical exercise which includes usage of lift and escalators, (iii) unhealthy diet patterns, (iv) changes in the sleeping patterns. Of course, all these changes bring a change in the human body and make ill one fine day. Going to hospital daily to monitor the changes in the body is quite impossible and also incurs huge cost. Thus, when a person wears a sensor, the sensor sends the vital signs to the health care professional at remote and thereby getting the reliable treatment at once without compromising quality of service. Also in this machine world, there is no one to look after the elderly person or sick person at home. When they wore sensors, the doctors at remote can monitor the status. If any sudden discrepancies happen, treatment will be given to them at once [4]. The three main layers required for pervasive health care system are (i) body area network (BAN), (ii) local area network (LAN)/wireless communication, and (iii) cloud architecture [5]. Body area network consists of sensors that senses the body and produces heterogeneous medical data. The middle layer is responsible for transferring data generated in bottom layer BAN to the top layer cloud and also made the data available to the health care professional. The top layer cloud computing intends to provide platform for data storage, processing, and provisioning of optimal insights. Sensors and actuators are the building blocks of IoT devices. Sensors are responsible to sense the environment and continuously generate vast amount of data. Such vast quantity of data is often entitled as big data [6]. Data analytics which analyzes the big data using machine learning algorithm aims to provide intelligent health care services [7]. While designing the architecture for pervasive health care system, the following challenges have to be addressed.

- The health care data gathered has vast number of features of which some are irrelevant and redundant which degrade the efficiency of the machine learning algorithm.

- Thus, selecting optimal features is essential to accurately predict the disease.

The algorithm proposed to analyze the health care data should able to deliver the insight in timely and reliable manner. The question of why there is a need of machine learning for health care arises when building intelligent pervasive health care system. Machine learning intends to provide analytical and processing methodology for storing or reforming cognition in intelligent systems and in specific, applying any learning algorithms to extract knowledge from data [8]. The learning algorithms are of supervised, semi-supervised, unsupervised, and reinforcement depending upon the class label [9]. As the data grows, it is difficult to do analytics on hand. When a model is built using the samples gathered from patients coming to the hospital, the model must be able to tell whether a new person will be subjected to a disease or not. Thus, prediction of future is done using machine learning algorithm which is essential to health care as all is about to save life of an individual. In this work, designing intelligent pervasive health care system for optimal provisioning of analytics as a service takes the advantage of intelligence from machine learning and optimization from metaheuristic algorithm to overcome the above-mentioned challenges.

## 2 Related Works

Automated feature selection algorithm based on unlabeled observations was designed to select features from health care record. A sparse regression model was then used to build a predictor model for rheumatoid arthritis [10]. Recursive feature selection algorithm with support vector machine as a classifier had been used to pick the relevant features from the data gathered from IoT devices. The method was evaluated with Wisconsin Breast Cancer dataset, and SVM kernel achieved 99% classification accuracy [11]. Cardiovascular disease was detected accurately with 99.05% accuracy using the classifier random forest bagging method. The feature subset was selected using relief and least absolute shrinkage and selection operator (LASSO) [12]. To efficiently identify ocular diseases, a feature selection mechanism called binary particle swarm optimization was used as the clinical data may have redundant features. Having selected the essential features, SVM classifier was used to optimally predict the presence or absence of ocular disease [13]. Correlation-based associated feature choosing algorithm was used to select the feature subset for the medical datasets such as breast cancer, heart, and diabetes taken from UCI repository. Having selected the feature subset, the classifiers such as neural network, decision tree, and SVM were fed with the selected features which produced accuracy as 77.23% for breast cancer dataset [14]. Particle swarm optimization-based feature selection algorithm had been used to select the relevant features from the real dataset where the fitness is considered to be the SVM classifier [15]. Decision tree algorithm CART is used to diagnose breast cancer, and all the filter-based feature chosen methods were utilized to select the feature subset, and the selected feature subset was fed to CART for diagnosing the breast cancer [16]. Opposition-based crow search algorithm was used to select the features to improve the performance of deep learning for diagnosing lung cancer and Alzheimer's disease [17]. Hybridization of particle swarm optimization and gray wolf optimization was used for selecting the set of features, and the performance was evaluated on twenty datasets [18].

## 3 Proposed System

The medical data is taken as input which is fed to the modified binary sine cosine algorithm to choose the relevant features that are optimal. Having selected the optimal relevant features, they are given as input to the classifier K-NN for optimal prediction which is represented in Fig. 1. Though binary sine cosine algorithm provides good level of exploration to find optimal feature subset, it is contaminated with shortcomings such as premature convergence, low level of exploitation and inability to switch between exploration and exploitation, and thus trapping in local optimal solution. In order to overcome the above challenges, improvements are integrated in the sine cosine algorithm by not only considering the best agent but also considering the next two best positions of the agent. Algorithm 1 represents the working of the proposed modified binary sine cosine algorithm for selecting the relevant features to maximize the accuracy of the classifier 1-NN where $K = 1$ in K-NN. Maximizing the accuracy of K-NN represents minimizing the error rate which is considered as the fitness function for evaluating the agent in MBSCA. The fitness function is specified in Eq. (1).

$$\mathbf{Min}\ \boldsymbol{F}(\boldsymbol{A}_i) \leftarrow \mathbf{Error\_Rate} \tag{1}$$

Each agent $\boldsymbol{A}_i$ is having $\boldsymbol{N}$ dimensions where each dimension $\boldsymbol{d}_i$ is either 0 or 1. Here, 0 represents deselection of the feature and 1 represents selection of the feature. At each iteration, the fitness value is computed using Eq. (1). The current iteration is given as $\boldsymbol{t}$, and the maximum iterations required for convergence are given as **MAXT**.

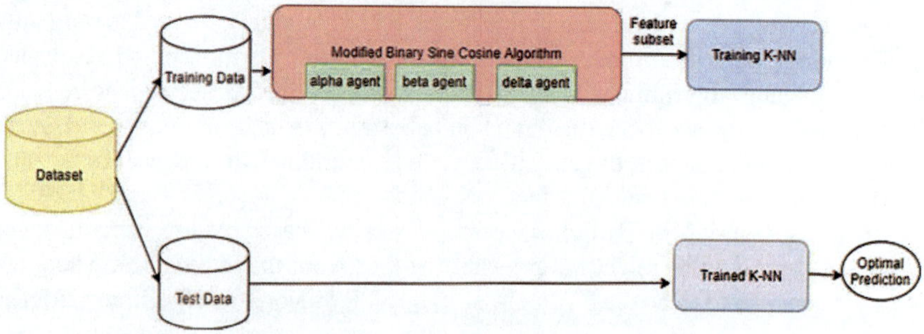


**Fig. 1** Flow diagram of proposed system

The position of the best agent is named as alpha agent indicated as $A_{\mathbf{alpha}}$, the second-best agent is named as beta agent indicated as $\boldsymbol{A}_{\mathbf{beta}}$, and third-best agent is named as delta agent indicated as $\boldsymbol{A}_{\mathbf{delta}}$. In order to have good level of exploitation as like exploration, two other agents, beta agent and delta agent, are chosen in addition to the best agent alpha. The variable $\boldsymbol{r}_1$ is computed using Eq. (2).

$$r_1 = r_1 - t\frac{r_1}{\mathrm{MAXT}} \tag{2}$$

Initially, the value of $r_1$ is set as 2, and in each iteration, it is gradually decreased using Eq. (2). When the value of $r_1$ is high, it paves the way for exploration, and as it approaches closer to 0, the agent does exploitation. The random variable $r_2$ is used to check whether the agent is moving toward the alpha agent. The random variable $r_3$ assigns weight to alpha agent. If value of $r_3$ is greater than $(1 - r_3)$, then agents will move toward the alpha agent, else the agent will move toward the beta agent and delta agent. $(1 - r_3)$ is the weight assigned to the beta agent and delta agent. The computation of position of the agent is based on Eqs. (3) and (4) based on the value of $r_4$. If the random variable $r_4$ is less than 0.5, then the position of the agent will be computed using Eq. (3), else the position of the agent will be computed using Eq. (4).

$$\begin{aligned} A_i^{t,d} \leftarrow A_i^{t-1,d} + r_1 * \sin(r_2) + \left| r_3 * A_{\text{alpha}}^{t,d} - A_i^{t,d} \right| \\ + \left| (1 - r_3) * A_{\text{beta}}^{t,d} - A_i^{t,d} \right| + \left| (1 - r_3) * A_{\text{delta}}^{t,d} - A_i^{t,d} \right| \end{aligned} \tag{3}$$

$$\begin{aligned} A_i^{t,d} \leftarrow A_i^{t-1,d} + r_1 * \cos(r_2) + \left| r_3 * A_{\text{alpha}}^{t,d} - A_i^{t,d} \right| \\ + \left| (1 - r_3) * A_{\text{beta}}^{t,d} - A_i^{t,d} \right| + \left| (1 - r_3) * A_{\text{delta}}^{t,d} - A_i^{t,d} \right| \end{aligned} \tag{4}$$

Sigmoid activation function is used to convert the position generated using Eqs. (3) and (4) to 0 or 1, and it is specified in Eqs. (5) and (6).

$$\sigma\left(A_i^{t,d}\right) \leftarrow \frac{1}{1 + \mathrm{e}^{-A_i^{t,d}}} \tag{5}$$

$$A_i^{t,d} \leftarrow \left\{ 1 \text{ if rand} < \sigma\left(A_i^{t,d}\right), 0 \text{ else} \right. \tag{6}$$

| Algorithm 1: Modified Binary Sine Cosine Algorithm | |
|---|---|
| Input: | Dimension set $D \leftarrow \{d_1, d_2, \ldots, d_N\}$<br>Maximum number of iterations $MAXT$<br>Nearest Neighbor $K = 1$<br>Number of Agents $\|A\|$ |
| Output: | Reduced set of dimensions $RSD \leftarrow \{d_1, d_2, \ldots, d_n\}$ where $n \leq N$ |

(continued)

(continued)

| | |
|---|---|
| | **for each** agent $A_i \in A$ |
| | $A \leftarrow Generate\_N\_Random\ Agents \in (0, 1)$ |
| | $A^{alpha} \leftarrow \infty$ |
| | $A^{beta} \leftarrow \infty$ |
| | $A^{delta} \leftarrow \infty$ |
| | **end for** |
| | **while** $t \leq MAXT$ |
| | **for each** Agent $A_i \in A$ |
| | Compute Fitness $F_{A_i}$ using Eq. (1) |
| | **end for** |
| | $A^{alpha} \leftarrow A_i \mid FirstMin\{ \mathrm{F}_{\mathrm{A_i}} \}$ |
| | $A^{beta} \leftarrow A_i \mid SecondMin\{ \mathrm{F}_{\mathrm{A_i}} \}$ |
| | $A^{delta} \leftarrow A_i \mid ThirdMin\{ \mathrm{F}_{\mathrm{A_i}} \}$ |
| | **for each** Agent $A_i \in A$ |
| | **for each** dimension $d_i$ |
| | Compute $r_1$ using Eq. (2) |
| | Generate $r_2 \leftarrow rand(\,)$ |
| | Generate $r_3 \leftarrow r\text{and}(\,)$ |
| | Generate $r_4 \in r\text{and}(0, 1)$ |
| | **if** $r_4 < 0.5$ |
| | Compute next position using Eq. (3) |
| | **else** |
| | Compute next position using Eq. (4) |
| | **end if** |
| | **end for** |
| | **end for** |
| | **end while** |
| | $RSD \leftarrow A^{alpha}$ |
| | return $RSD$ |

## 4 Experimental Results

The designed MBSCA algorithm is executed in Python, and the outcomes are evaluated with standard benchmarking datasets taken from UCI repository [19] represented in Table 1. Both breast cancer and heart datasets are considered as small-scale datasets as the number of features is less than 20. Also, the datasets are viewed

**Table 1** Details of dataset

| Dataset | #Instances | #Features | #Class |
|---|---|---|---|
| Breast cancer | 287 | 10 | 2 |
| Heart | 303 | 13 | 2 |

**Table 2** Features selected by different algorithms

| Dataset\algorithms | GA | BSCA | MBSCA |
|---|---|---|---|
| Breast cancer | 1, 6, 8, 9 | 3, 4, 5, 6 | 4, 5, 6 |
| Heart | 2, 3, 7, 8, 9, 10, 11, 12 | 2, 7, 9, 10, 13 | 2, 3, 8 |

as binary classification problem, as the number of class labels is two showing the existence of disease and non-existence of disease.

### 4.1 Comparison of Features Chosen

The features selected by various algorithms are represented in Table 2. The value in each cell represents the feature number chosen by different algorithms. From Table 2, it is conspicuous that MBSCA selects minimum features that other algorithms which not only save the storage space but also aimed to improve the accuracy and which in turn minimize the error rate.

### 4.2 Comparison of Dimensionality Reduction Ratio

Dimensionality reduction ratio DRR is computed as proportion of number of dimensions that are deselected given as $|DD|$ to the total number of dimensions in dataset given as $N$ which is represented in Eq. (7). From Table 3, it is evident that the devised MBSCA achieves maximum dimensionality reduction ratio than other algorithms. For the breast cancer dataset, the dimensionality reduction ratio of MBSCA is 17% better than GA and BSCA. Similarly for the heart dataset, MBSCA achieves 83.33% and 22.22% better accuracy than GA and BSCA, respectively.

$$\text{DRR} \leftarrow \frac{|DD|}{N} \tag{7}$$

**Table 3** Dimensionality reduction ratio

| Dataset\algorithms | GA | BSCA | MBSCA |
|---|---|---|---|
| Breast cancer | 0.6 | 0.6 | 0.7 |
| Heart | 0.43 | 0.64 | 0.79 |

## 4.3 Comparison of Accuracy

Accuracy is represented as the proportion of correctly classified samples to the population in the dataset which is specified in Eq. (8). Figure 2 represents the comparison of accuracy of different algorithms, and the same has been represented in Table 4. MBSCA improves accuracy by 16.57% than for breast cancer dataset considering all the features. The accuracy of MBSCA is 5.65% and 4.07% better than GA and BSCA conjointly for breast cancer dataset. As the proposed MBSCA considers the position of second-best and third-best agent, the algorithm does good level of exploitation which resulted in maximum accuracy.

$$\mathbf{Accuracy} = \frac{\mathbf{TP} + \mathbf{TN}}{\mathbf{TP} + \mathbf{TN} + \mathbf{FP} + \mathbf{FN}} \tag{8}$$

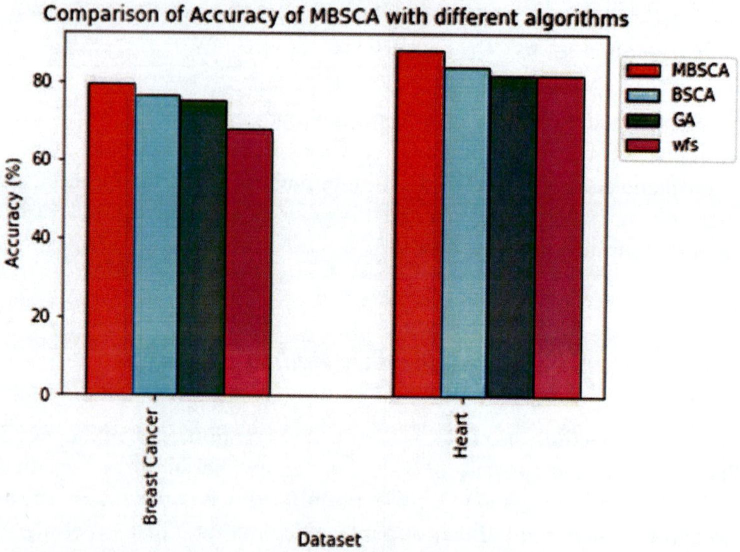


**Fig. 2** Comparison of accuracy of features selected by GA, BSCA, MBSCA algorithms and in the absence of feature selection

**Table 4** Comparison of accuracy

| Dataset\algorithms | Without feature selection | GA | BSCA | MBSCA |
|---|---|---|---|---|
| Breast cancer | 68.13 | 75.17 | 76.31 | 79.42 |
| Heart | 81.94 | 82.15 | 84.17 | 88.32 |

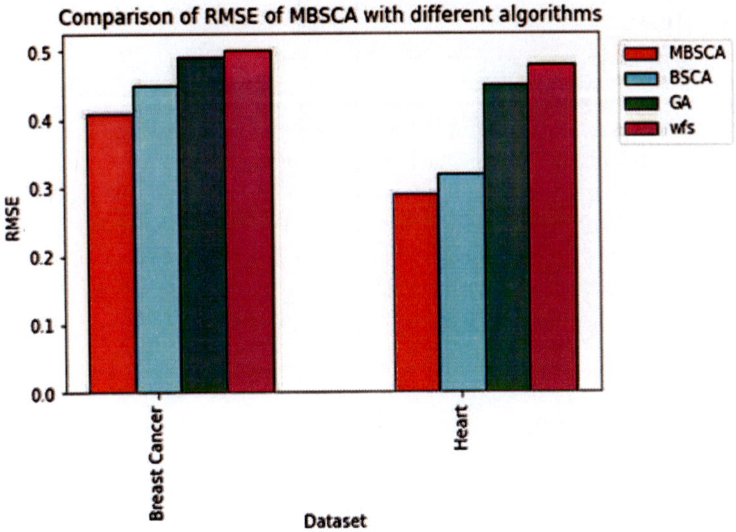


**Fig. 3** Comparison of RMSE of features selected by GA, BSCA, MBSCA algorithms and in the absence of feature selection

## 4.4 Comparison of Root Mean Square Error

Root mean square error is termed to find the error in misclassifying the instances. Equation (9) represents the computation of RMSE. Figure 3 represents the comparison of RMSE of with feature selection and without feature selection, and the same has been represented in Table 5. From Fig. 3, it is evident that RMSE is minimum for the proposed MBSCA algorithm for both the datasets. Also, for heart dataset, MBSCA algorithm reduces RMSE by 39.58% than without feature selection. The reduction in RMSE is 35.56% and 9.38% for GA and BSCA than MBSCA.

$$\mathrm{RMSE} = \sqrt{\frac{\sum_{i=1}^{|D|}\left(y_i - \widehat{y_i}\right)}{|D|}} \tag{9}$$

**Table 5** Comparison of root mean square error

| Dataset\algorithms | Without feature selection | GA | BSCA | MBSCA |
|---|---|---|---|---|
| Breast cancer | 0.5 | 0.49 | 0.45 | 0.41 |
| Heart | 0.48 | 0.45 | 0.32 | 0.29 |

## 5 Conclusion

Feature selection desires to choose the greatest set of features from the medical data. Accuracy is the prime concern for building intelligent pervasive health care system. Building classifier model using machine learning algorithm using all the features from the data gathered may lead to degrade the performance of the model. Thus, a metaheuristic algorithm called binary sine cosine algorithm is used with second- and third-best agent to do exploitation as equal as exploration thereby avoiding premature convergence to stuck in local optimal solution. The key findings of the research work include:

- Modified binary sine cosine algorithm was used for optimal selection of feature subset
- The introduction of beta agent and delta agent paved the way for exploitation, and alpha agent paved the way for exploration.

Having selected the optimal feature subset, the selected features can be fed into machine learning model for enhancing accuracy.

## References


1. Dave, E.: The Internet of Things how the next evolution of the internet is changing everything (2011)
2. Kulkarni, A., Sathe, S.: Healthcare applications of the Internet of Things: a review. Int. J. Comput. Sci. Inf. Technol. **5**(5), 6229–6232 (2014)
3. Alqahtani, F.H.: The application of the Internet of Things in healthcare. Int. J. Comput. Appl. **180**(18), 19–23 (2018)
4. Chen, M., Ma, Y., Li, Y., Wu, D., Zhang, Y., Youn, C.H.: Wearable 2.0: enabling human-cloud integration in next generation healthcare systems. IEEE Commun. Mag. **55**(1), 54–61 (2017)
5. Firouzi, F., Rahmani, A.M., Mankodiya, K., Badaroglu, M., Merrett, G.V., Wong, P., Farahani, B.: Internet-of-Things and big data for smarter healthcare: from device to architecture, applications and analytics (2018)
6. Manogaran, G., Thota, C., Lopez, D., Sundarasekar, R.: Big data security intelligence for healthcare industry 4.0. In: Cybersecurity for Industry 4.0, pp. 103–126. Springer, Cham (2017)
7. Ma, X., Wang, Z., Zhou, S., Wen, H., Zhang, Y.: Intelligent healthcare systems assisted by data analytics and mobile computing. Wirel. Commun. Mobile Comput. (2018)
8. Magoulas, G.D., Prentza, A.: Machine learning in medical applications. In: Advanced Course on Artificial Intelligence, pp. 300–307. Springer, Berlin, Heidelberg (1999)
9. Dietterich, T.G.: Ensemble methods in machine learning. In: International Workshop on Multiple Classifier Systems, pp. 1–15. Springer, Berlin, Heidelberg (2000)
10. Gronsbell, J., Minnier, J., Yu, S., Liao, K., Cai, T.: Automated feature selection of predictors in electronic medical records data. Biometrics **75**(1), 268–277 (2019)
11. Memon, M.H., Li, J.P., Haq, A.U., Memon, M.H., Zhou, W.: Breast cancer detection in the IoT health environment using modified recursive feature selection. Wirel. Commun. Mobile Comput. (2019)
12. Ghosh, P., Azam, S., Jonkman, M., Karim, A., Shamrat, F.J., Ignatious, E., Shultana, S., Beeravolu, A.R., De, B.F.: Efficient prediction of cardiovascular disease using machine learning algorithms with relief and LASSO feature selection techniques. IEEE Access **9**, 19304–19326 (2021)

13. Keerthiveena, B., Esakkirajan, S., Subudhi, B.N., Veerakumar, T.: A hybrid BPSO-SVM for feature selection and classification of ocular health. IET Image Proc. **15**(2), 542–555 (2021)
14. Rajeswari, K., Vaithiyanathan, V., Pede, S.V.: Feature selection for classification in medical data mining. Int. J. Emerg. Trends Technol. Comput. Sci. (IJETTCS) **2**(2), 492–497 (2013)
15. Chung, J.T., Chuang, L.Y., Yang, J.Y., Yang, C.H.: Feature selection using PSO-SVM. Int. J. Comput. Sci. **33**(1) (2007)
16. Lavanya, D., Rani, D.K.: Analysis of feature selection with classification: breast cancer datasets. Indian J. Comput. Sci. Eng. (IJCSE) **2**(5), 756–763 (2011)
17. Raj, R.J., Shobana, S.J., Pustokhina, I.V., Pustokhin, D.A., Gupta, D., Shankar, K.: Optimal feature selection-based medical image classification using deep learning model in internet of medical things. IEEE Access **8**, 58006–58017 (2020)
18. El-Hasnony, I.M., Barakat, S.I., Elhoseny, M., Mostafa, R.R.: Improved feature selection model for big data analytics. IEEE Access **8**, 66989–67004 (2020)
19. Dua, D., Graff, C.: UCI machine learning repository. University of California, School of Information and Computer Science, Irvine, CA (2019). http://archive.ics.uci.edu/ml

# Design of High Step-up Interleaved Boost Converter-fed Fuel Cell-Based Electric Vehicle System with Neural Network Controller

**M. Murali, Shaik Rafi Kiran, CH Hussaian Basha, S. Khaja Khizar, and P. M. Preethi Raj**

**Abstract** At present, fuel cell-based power generation system is playing a major role in all conventional and distribution power supply systems because of its advantages are less greenhouse gas emissions, high reliability, and less environmental pollution. The fuel cell power generation systems give nonlinear behavior characteristics. As a result, the overall system performance is reduced. In addition, the fuel cell harvest power is reduced extensively. Here, an artificial intelligence-based neural network technique is recommended to extract and improve the output power of the fuel stack system. The attractive features of neural network controller are less training time, ability to solve all complex and nonlinear problems, faster response, and serial data processing. The fuel cell production current is very high. So, the entire system switching and transmission losses are improved. In order to reduce the losses, an interleaved three-phase boost converter topology is used to step-up the fuel stack voltage.

**Keywords** Boost converter · Duty cycle · Fuel cell · Less converter output current · Neural network

M. Murali · P. M. Preethi Raj
Department of EEE, K.S.R.M College of Engineering, Kadapa, Andhra Pradesh, India
e-mail: murali@ksrmce.ac.in

P. M. Preethi Raj
e-mail: preethirajpandu7@gmail.com

CH Hussaian Basha (✉) · S. Khaja Khizar
CRI Laboratory, K.S.R.M College of Engineering, Kadapa, Andhra Pradesh, India
e-mail: sbasha238@gmail.com

S. Khaja Khizar
e-mail: khizar@ksrmce.ac.in

S. Rafi Kiran
Sri Venkateshwara Engineering College, Tirupati, Andhra Pradesh, India



D. Gupta et al. (eds.), *Pattern Recognition and Data Analysis with Applications*,
Lecture Notes in Electrical Engineering 888,
https://doi.org/10.1007/978-981-19-1520-8_64

## 1 Introduction

From the last few years, the usage of conventional energy sources is reduced excessively because of their own disadvantages. The classification of conventional power generation sources is coal, natural gas, petroleum oil, and nuclear. The coal is used in thermal power generation systems in order to convert the steam energy in to electrical energy. The merits of coal power generation systems are high reliability, good affordability, and high abundance, moderate safety [1]. But, it affects the environmental conditions by releasing high greenhouse gas emissions. In addition, the thermal power plants produce the million tons of wastage.

The drawbacks of the thermal systems are overcome by utilizing the natural gas-based power stations [2]. In this plant, the gas turbine rotates based on the principle of bray ton cycle. Here, the compressed natural gas is mixed with burned coal with a constant pressure. The resultant pressurized hot gas is converted to electrical energy by using different types of turbines. The attractive features of gas power stations are environmental friendly nature, and it starts burning neatly when compared to other fuels. In addition, this power plants are highly reliable. The disadvantage of gas power station is less efficiency. The oil power plants are used in most of the places to limit the disadvantages of gas power plants [3]. Here, the oil is burned to generate the heat water. The heated water is converted to steam to generate the electricity. The features of oil power systems are cheap in cost and easier to store. The only demerit is environmental pollution.

The nuclear power stations are used to overcome the limitations of other conventional power systems. Here, the uranium atoms are splitting to generate the steam, and it is supplied to the steam turbine to produce electricity. The most desirable merit of nuclear power generation is less environmental pollution when compared to thermal. However, the drawbacks of above conventional power sources are overcome by using the nonconventional energy sources which are solar, tidal, and wind sources [4].

At present, fuel cell power generation is the major focus in electrical vehicle charging when compared to the solar and wind power plants. The attractive merits of fuel cell stack system are less emission and high efficiency. In addition, it does not consist of no rotating parts. So, the noise pollution in the PV is less, and the corresponding heating and conduction losses are absent [5]. From the literature review, plenty of fuel stack topologies are available which are expended for diverse automotive industries applications. The classification of fuel cell stacks is proton exchange membrane fuel cell (PEMFC), direct methanol fuel cell (DMFC), and zinc-air fuel cell (ZFC) [6]. Among all of that, the PEMFC is applied in most of the distribution power systems and automotive industry. The merits of PEMFC stack are fast starting, high power density, and very simple construction. In addition to that, the size of the PEMFC is less. As a result, the catchment area of the PEMFC-based power generation device is reduced [7].

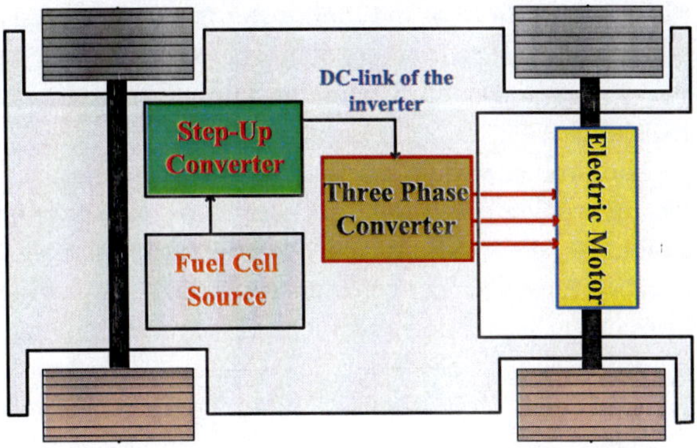


**Fig. 1** Schematic representation of PEMFC-fed electric vehicle system [7]

The fuel cell stack gives nonlinear voltage against current characteristics, and its output power varies continuously based on its operating temperature. In addition to that, the output power is majorly contingent on the layer of aquatic content, pressurized oxygen and hydrogen. The basic representation of converter topology is illustrated in Fig. 1. In article [8], a peak power point tracing technique is applied in fuel cell stack-based electrical vehicle configuration to find out the operating point of the fuel stack.

The classification of conventional power tracking methodologies is Perturb and Observe (P&O) [9], hill climb (HC) [10], and incremental conductance (IC) [11]. The P&O controller traces the MPP depending on variation of step size on nonlinear characteristics. The variation of present power with respect to the previous power gives positive result, then the perturbation step size is increased in an ascending manner. Otherwise, the step size is reduced in a descending manner. The disadvantages of P&O controller are high converter output voltage distortions, high transient oscillations, and less accuracy in MPP tracking.

The limitations of P&O method are overcome by using the hill climb technique. In this hill climb technique, the reference voltage is selected as a fuel cell maximum output voltage. The reference voltage value is compared with the instantaneous voltage. Based on the comparative results, the working point of fuel stack coincides with the peak power point on *V–I* curve. The demerit of hill climb technique is limited by using the IC technique. In this IC technique, the conductance of the fuel cell is compared with the previously existed conductance to improve the output of fuel cell. The comparison result gives positive sign, then the operating point of fuel cell is left side of the *V–I* curve. Otherwise, it is assumed to be right side of the voltage against current curve. However, the above techniques are not useful for the exact finding of MPP. To compensate the above drawbacks, in this work, a multilayer perceptron neural network is proposed.

From the previous study, the production voltage of the fuel stack is very small which is boosted by using the different types of DC–DC boost converters [12]. The classification of high step-up boost converters is isolated and non-isolated boost converters. The isolated boost converters required an external transformer to improve the voltage profile of the fuel power generation systems. In addition, it requires an additional rectifier for the isolated boost converter systems [13]. So, the system size plus cost is improved. The major limitation of isolated converter is voltage stress on the switch.

The problems of insulated converter circuits are limited by applying the non-isolated converter system. The basic boost converter topology is used in many applications because of its attractive features are less design cost, high reliable, more efficient, and less operating power losses. But, it is having a drawback of less magnetic overcurrent protection [14]. The disadvantage of basic boost converter is limited by using the Cuk converter. The Cuk converter topology is applied to stabilize the fuel cell output voltage. The limitations of this converter are high design and implementation complexity. In addition, it requires more number of power semiconductor devices. As a result, the overall system size is improved [15]. To overcome the drawbacks of all the above converters, in this work, an interleaved DC–DC converter is applied to the fuel cell topology to improve its output voltage.

## 2 Design and Analysis of PEM Fuel Cell

The PEMFC is an electromechanical device which is used to change the chemical energy into power energy, and it supplies to the distribution systems. The working of membrane-based fuel cell topology is given in Fig. 2.

From Fig. 2, it is clearly indicated that the fuel stack consists of three major blocks which are anode, electrolyte, and cathode. The polymer electrolyte membrane is used for the implementation of electrolyte of the fuel cell stack. From Fig. 2, the oxygen ($O_2$) and hydrogen ($H_2$) are the inputs to the proposed fuel system, and it is humidified and compressed with the support of humidifier and air pressure compressor. At the

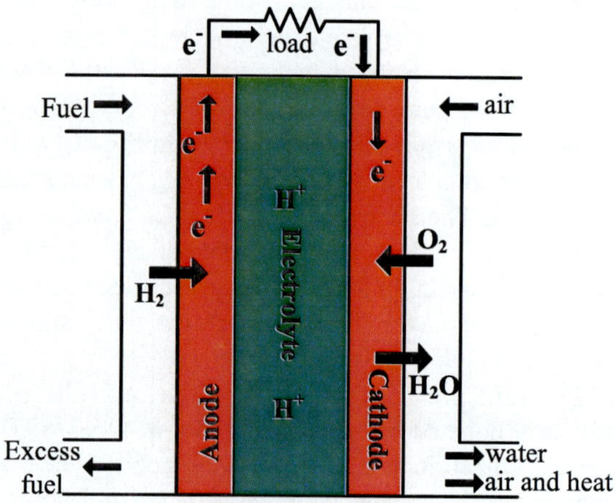


**Fig. 2** Block diagram of the fuel cell stack working condition [6]

anode side, the hydrogen is diverted in to electrons and protons. The electrons are shifted to from anode to the external circuit to supply the electrical power supply. Similarly, the protons are shifted from one side to cathode side with the membrane layer, and it reacts with the oxygen in order to generate the heat and water.

From Fig. 2, the chemical reaction of the total PEMFC stack is derived as,

$$H_2 \rightarrow 2H^+ + 2e^- \tag{1}$$

$$2H^+ + 2e^- + \frac{1}{2}O_2 \rightarrow H_2O \tag{2}$$

$$H_2 + \frac{1}{2}O_2 \rightarrow H_2O + \text{Energy} \tag{3}$$

where $H_2$ is the fuel cell generated hydrogen and $O_2$ is the oxygen generated from the cathode layer. The term $H_2O$ is the unused water which is the production of the cathode. The fuel cell generated output voltage is derived as,

$$V_{out} = N * V_{FC} \tag{4}$$

From Eq. (4), $V_{out}$ is the fuel stack total production voltage and VFC is the each cell output voltage. Similarly, $N$ is the total cells in stack. The each cell stack is derived as,

$$V_{FC} = E_{Oct} - V_{Oh} - V_{Act} - V_{Con} \tag{5}$$

The term $E_{Oct}$ is the thermodynamic potential, and $V_{Oh}$ is the ohmic polarization loss. Similarly, the $V_{Act}$ is the active polarization loss, and $V_{Con}$ is concentrated polarization loss. The detailed design parameters of fuel cell stack are given in Table 1, and its $V$–$I$ and $P$–$I$ characteristics are shown in Figs. 3a and b.

From Figs. 3a and b, it is clearly observed that the maximum peak to peak to voltage and currents of fuel cell stack are 24.23 V and 52 A. Here, it clearly says that the operating point of fuel cell stack varies continuously based on its operating temperature. For continuous MPP tracking, an MPPT controller is used.

## 3 Design of Perceptron Neural Network MPPT Controller

From the literature review, the neural networks are nothing but the series of algorithms that are endeavors to identify the relationship between different datasets through the human brain process. In addition, the neural networks working behavior is similar to the computing systems. Most of the time, neural networks are used for the pattern recognition, voice identification, language generation, character finding,

**Table 1** Design parameters of fuel cell stack system

| Parameters | Values |
|---|---|
| Power of the fuel cell at rated condition | 1.26 kW |
| Fuel cell voltage at high power position (VMPP) | 24 V |
| Fuel cell current at high power position (IMPP) | 52 A |
| Fuel cell open circuit condition voltage (VOC) | 42 V |
| Partial oxygen pressure of fuel cell | 1 bar |
| Partial hydrogen pressure of fuel cell | 1.5 bar |
| Cells used in the fuel cell stack (N) | 42 |
| Flowing air rate at nominal condition (Ipm) | 4615 |
| Constant of gases (R) | 84.092 [J $mol^{-1}.K^{-1}$] |
| Constant of faraday (F) | 95,432.218 [C.$mol^{-1}$] |
| Composition of oxidant | 21% |
| Composition of fuel rate | 99.95% |
| Hydrogen utilization of fuel cell | 99.92% |
| Oxygen utilization of fuel cell | 1.813% |

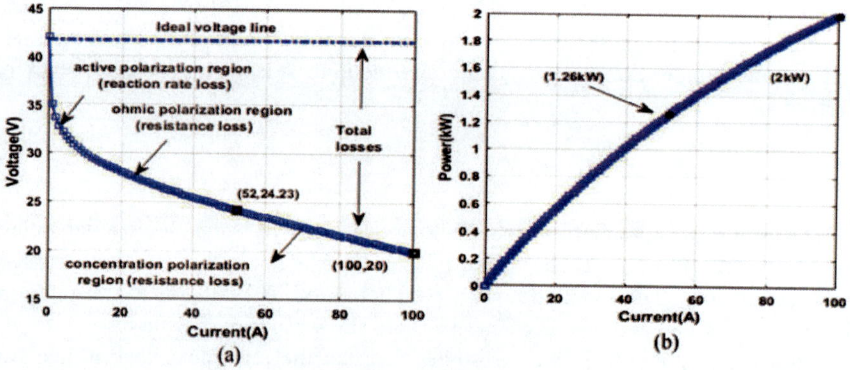


**Fig. 3** Fuel cell, **a** Voltage and current curve, and **b** Power and current curve

and multiple document summarization. In addition, it is a one of the machine learning tools which is used for the semantic parsing. The most attractive feature of neural network topology is used to solve all nonlinear problem solving application. As a result, the perceptron three-layer network is used in the fuel stack power generation system in order to find out the MPP at different atmospheric conditions. Also, the tracking speed of the proposed model is high when compared to the traditional techniques. However, the limitations of the neural networks are hardware dependency, unexplained functioning of network, difficulty of showing the problem to the network, and the duration of the neural network is unknown. The block diagram of the three-layer perceptron neural network structure is shown in Fig. 4.

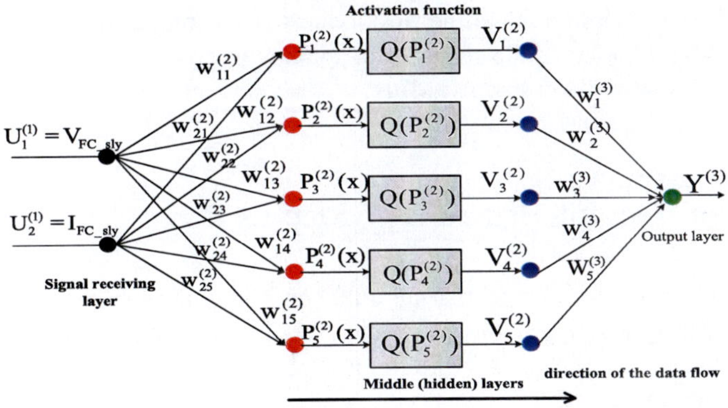


**Fig. 4** Multiple layer perceptron-based neural network-based power point tracing controller

The fuel cell output voltage and currents are given to the input layer neural network as shown in Fig. 4. From Fig. 4, it is clearly observed that the activation function is a nonlinear one, and its corresponding equations are derived as,

$$P_t^{(2)}(x) = \sum_{s=1}^{2} w_{ls}^{(2)} * U_l^1; \quad l = 1, 2, 3, 4, 5 \ldots x \tag{6}$$

$$V_l^{(2)}(x) = Q\left(P_s^{(2)}(x)\right) \tag{7}$$

$$Y^3(x) = \sum_{l=1}^{5} w_l^{(3)} * V_l^{(2)} \tag{8}$$

From Eq. (8), the entire network weights are monitored updated by exhausting the delta rule, and it is applied to the third layer in order to find out the total weights.

$$w_{ls}^{(2)} = w_{ls}^{(2)} + \Delta w_{ls} \tag{9}$$

$$w_l^{(3)} = w_{l3}^{(3)} + \Delta w_l \tag{10}$$

$$\Delta w_{ls} = u * \frac{\partial e}{\partial w_{ls}^{(2)}}, \text{ and } \Delta w_l = u * \frac{\partial e}{\partial w_l^{(3)}} \tag{11}$$

The neural network output layer error signal is given in Eq. (12), and it is used to determine optimum required duty of DC–DC converter.

$$\text{error} = \frac{1}{2}\left(Y_{\text{desired}} - Y^{(3)}\right)^2 \tag{12}$$

Finally, the perceptron neural network controller is compared with the adaptive P&O and VSS-IC methods in relations of maximum power extraction, efficiency, and steady-state oscillations across MPP.

## 4 Design of an Interleaved High Step-up Boost Converter

Basically, it has been observed that the fuel cell output current is very high. As a result, the direct interconnection of inverter to the fuel cell stack is not possible. So, the boost converter is integrated with the fuel cell-based power generation system. Here, the interleaved DC–DC converter topology is used for increasing the voltage profile of fuel cell system. The proposed interleaved converter circuit is illustrated in Fig. 5. From Fig. 5, it is observed that the circuit consists of three switches which are indicated as $Q_a$, $Q_b$, and $Q_c$. In addition, the proposed converter topology consists of three diodes which are $D_a$, $D_b$, and $D_c$. The utilized capacitors and inductors are indicated as $C_a$, $C_b$, and $C_c$ and $L_a$, $L_b$, and $L_c$. The currents flowing through the capacitors and inductors are represented as $I_{\mathrm{Ca}}$, $I_{\mathrm{Cb}}$, and $I_{\mathrm{Cc}}$ and $I_{\mathrm{La}}$, $I_{\mathrm{Lb}}$, and $I_{\mathrm{Lc}}$. In the first conduction mode of converter operation, the switches $Q_a$, $Q_b$, and $Q_c$ are in forward bias condition and the corresponding diodes work in opposite direction as shown in Table 2.

Similarly, in the second conduction state, the switch $Q_b$ is in working condition and the remaining switches are in OFF state. The corresponding diodes $D_a$ and $D_c$ are in conduction state and the diode $D_b$ in OFF state. In the third working state, the

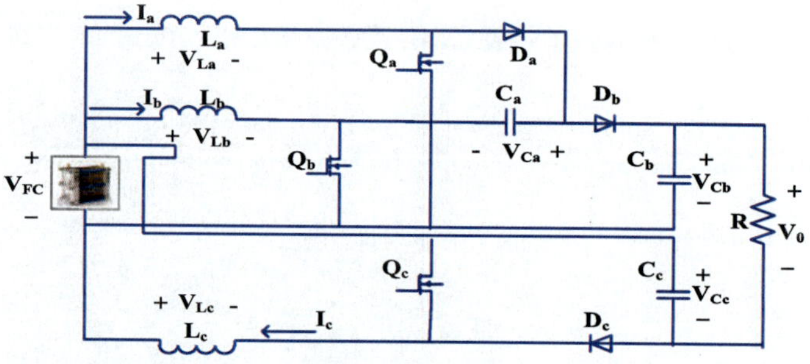


Fig. 5 Interleaved high voltage gain DC–DC converter

Table 2 Switching analysis of interleaved boost converter system

| Switch | $Q_a$ | $Q_b$ | $Q_c$ | $D_a$ | $D_b$ | $D_c$ |
|---|---|---|---|---|---|---|
| State-I | ON | ON | ON | OFF | OFF | OFF |
| State-II | OFF | ON | OFF | ON | OFF | ON |
| State-III | ON | OFF | ON | OFF | ON | OFF |

operating condition of switches is similar to the first condition. Finally, at last working state, the switch $S_b$ is reverse biased state and remaining switches are in working condition. The voltage transformation ratio of the DC–DC converter is determined by applying the voltage second balance concept transversely in the each capacitor and inductor. The voltage across the capacitors $C_a$, $C_b$, and $C_c$ is derived in terms of duty cycle and fuel cell output voltage.

$$\begin{cases} V_{Ca} = \frac{V_{FC}}{(1-\text{Duty})} \\ V_{Cb} = \frac{V_{FC}}{(1-\text{Duty})} + V_{Ca} \\ V_{Cc} = \frac{V_{FC}}{(1-\text{Duty})} \end{cases} \tag{13}$$

The total converter output voltage in terms of fuel cell output voltage and capacitor voltage is derived as,

$$V_{out} = V_{Ca} + V_{Cb} - V_{FC} \tag{14}$$

$$\frac{V_0}{V_{FC}} = \frac{2 + \text{Duty}}{1 - \text{Duty}} \tag{15}$$

The inductor and capacitor values are determined as,

$$L_a = L_b = L_c = L_{eq} = \frac{\text{Duty} * V_{FC}}{\Delta I * f_s} \tag{16}$$

$$C_a = \frac{V_{out}}{R f_s V_{Ca}} \tag{17}$$

$$C_b = C_c = C_{eq} = \frac{\text{Duty} * V_{out}}{R f_s V} \tag{18}$$

## 5 Simulation Results

The study of proposed fuel stack-fed interleaved boost converter topology has been done by using a MATLAB/Simulink window. Here, 1.26 kW rating fuel cell stack is used for the testing of boost converter performance. The input side capacitor is useful for stabilization of fuel cell output voltage. Similarly, the load side capacitor is used to obtain the continuous and smooth load voltage. The input and output inductor values are $L_a = 1.65$ mH, $L_b = 1.7$ mH, and $L_c = 1.72$ mH. The capacitors used for the step-up of the fuel cell output voltage are $C_a = 470$ μF, $C_b = 318$ μF, and $C_c = 330$ μF. Finally, the load resistor is $R = 46$ Ω.

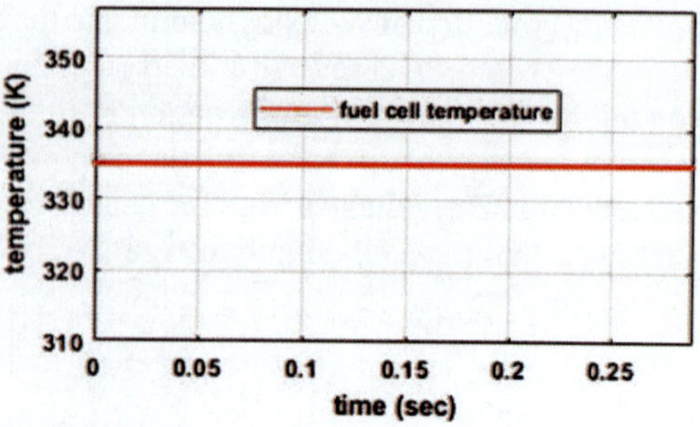


**Fig. 6** Fuel cell operated temperature in kelvin

### 5.1 *Static Temperature Condition of Fuel Cell Stack (335 K)*

At static temperature condition (see in Fig. 6), the P&O-based fuel cell and converter output voltages and currents are 26.43 V, 34 A, 197 V, and 4.3 A. So, the converter steps-down the current. So, the overall system losses are decreased excessively. The fuel cell and converter output voltage, power, and currents are given in Fig. 7. Similarly, the NN-based fuel cell and converter output voltage and currents are 26.62 V, 42 A, 218 V, and 4.72 A, respectively. From the above performance results, the NN-based power point tracing controller is giving high step-up voltage to reach the essential load request.

### 5.2 *Dynamic Temperature Condition of Fuel Cell Stack (335 K, 315 K, and 355 K)*

Similar to the static operating temperature condition of fuel cell stack, under dynamic condition (see in Fig. 8), the NN-based converter output voltage and currents at 315 K are 188 V and 4.1 A which are higher than the P&O-based controller. The fuel cell and converter output voltage, current, and powers are given in Fig. 9. Finally, at 355 K, the P&O-based fuel cell and converter output voltage and currents are 28 V, 35 A, 210 V, and 4.68 A, respectively. So, from the above observation, the neural network-based controller is giving good performance when compared to the conventional MPPT controller.

## 6 Conclusion

The proposed fuel cell stack-based high step-up DC–DC circuit is intended successfully by using the MATLAB/Simulink window. The merits of the proposed converter are high voltage conversion ratio, less distortion in converter output voltage, and wide input and output operation. In addition to that, it gives less voltage stress on power

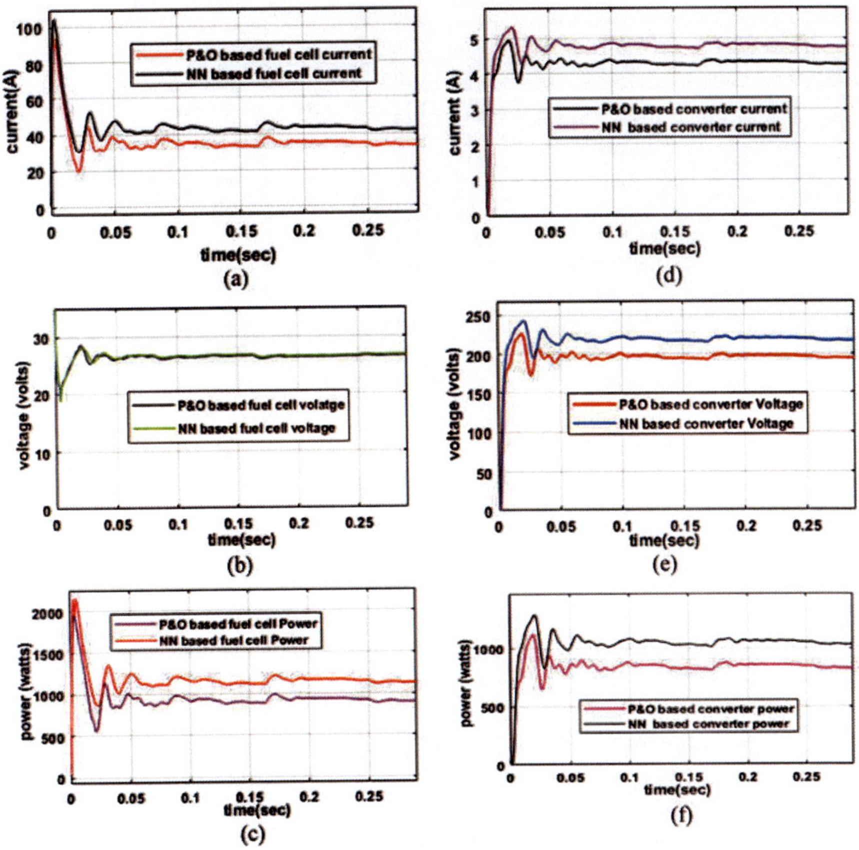


**Fig. 7** **a** Fuel cell output current, **b** Fuel cell output voltage, **c** Fuel cell output power, **d** Current of DC–DC converter, **e** Voltage of DC–DC converter, and **f** Power of DC–DC converter

**Fig. 8** Fuel cell temperature at dynamic condition

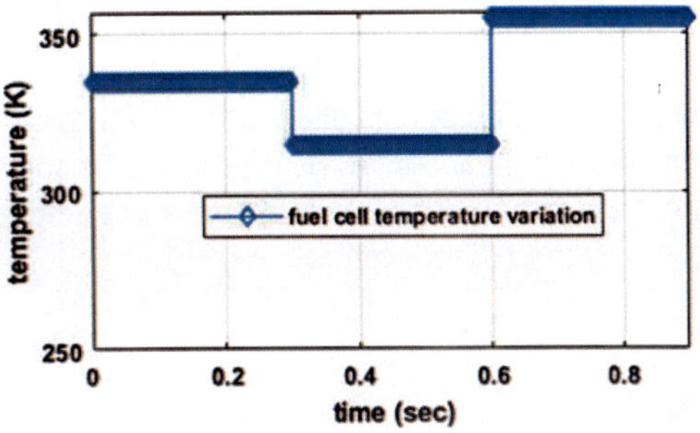

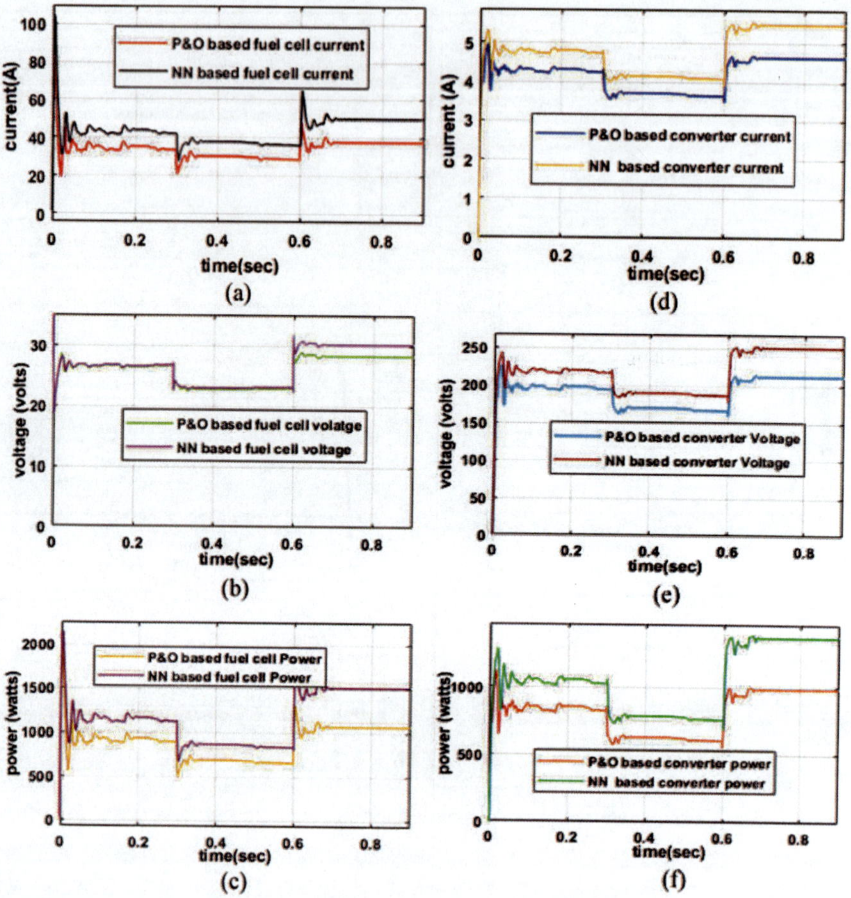


**Fig. 9** **a** Fuel cell output current, **b** Fuel cell output voltage, **c** Fuel cell output power, **d** Converter current, **e** Converter voltage, and **f** Converter power at dynamic temperature condition

semiconductor diodes. The proposed multilayer perceptron neural network controller is giving high accurate MPP position. In addition, it gives the optimum duty value to the boost converter. Finally, the soft computing MPPT controller design cost and complexity are reduced excessively.

## References

1. Fan, J.L., Shijie, W., Xian, Z., Lin, Y.: A comparison of the regional investment benefits of CCS retrofitting of coal-fired power plants and renewable power generation projects in China. Int. J. Greenhouse Gas Control **92**, 102858 (2020)
2. Ezekiel, J., Anozie, E., Benjamin, A.M., Martin, S.O.: Combining natural gas recovery and

$CO_2$-based geothermal energy extraction for electric power generation. Appl. Energy **269**, 115012 (2020)
3. Mahidin, S., Erdiwansyah, H., Hisbullah, H.A.P., Zhafran, M., Sidiq, M.A., Rinaldi, A., Fitria, B., Tarisma, R., Binda, Y.: Analysis of power from palm oil solid waste for biomass power plants: a case study in Aceh Province. Chemosphere **253**, 126714 (2020)
4. Hussaian, B.C.H., Rani, C.: Performance analysis of MPPT techniques for dynamic irradiation condition of solar PV. Int. J. Fuzzy Syst. **22**(8), 2577–2598 (2020)
5. Shen, D., Cheng, C.L., Peng, S.: Robust fuzzy model predictive control for energy management systems in fuel cell vehicles. Control. Eng. Pract. **98**, 104364 (2020)
6. Xueqin, L., Yinbo, W., Jie, L., Yangyang, Z., Chao, C., Peisong, W., Lingzheng, M.: Energy management of hybrid electric vehicles: a review of energy optimization of fuel cell hybrid power system based on genetic algorithm. Energy Convers. Manage. **205**, 112474 (2020)
7. Jun, S., Liang, X., Huawei, C., Zhengkai, T., Siew, H.C.: Partial flooding and its effect on the performance of a proton exchange membrane fuel cell. Energy Convers. Manage. **207**, 112537 (2020)
8. Derbeli, M., Oscar, B., Lassaad, S.: A robust maximum power point tracking control method for a PEM fuel cell power system. Appl. Sci. **8**(12), 2449 (2018)
9. Shashikant, Binod, S.: Comparison of SCA-optimized PID and P&O-based MPPT for an off-grid fuel cell system. In: Soft Computing in Data Analytics, pp. 51–58. Springer, Singapore (2019)
10. Mohamed, D., Oscar, B., Mohammed, Y.S., Cristian, N.: Real-time implementation of a new MPPT control method for a DC–DC boost converter used in a PEM fuel cell power system. Actuators **9**(4), 105 (2020)
11. Harrag, A., Hamza, B.: Novel neural network IC-based variable step size fuel cell MPPT controller: performance, efficiency and lifetime improvement. Int. J. Hydrogen Energy **42**(5), 3549–3563 (2017)
12. Shengrong, Z., Arnaud, G., Liangcai, X., Damien, P., Fei, G.: Extended state observer-based control of DC–DC converters for fuel cell application. IEEE Trans. Power Electron. **35**(9), 9925–9934 (2020)
13. Wang, H., Arnaud, G., Daniel, H.: A review of DC/DC converter-based electrochemical impedance spectroscopy for fuel cell electric vehicles. Renew. Energy **141**, 124–138 (2019)
14. Farhani, S., Amari, M., Marzougui, H., Bacha, F.: Analysis, modeling and implementation of an interleaved boost DC–DC converter for fuel cell used in electric vehicle. Int. J. Hydrogen Energy **42**(48), 28852–28864 (2017)
15. Chandrasekar, B., Chellammal, N., Bhargavi, N.: Non-isolated unidirectional three-port Cuk–Cuk converter for fuel cell/solar PV systems. J. Power Electron. **19**(5), 1278–1288 (2019)

# Design of Adaptive VSS-P&O-Based PSO Controller for PV-Based Electric Vehicle Application with Step-up Boost Converter

**CH Hussaian Basha, T. Mariprasath, M. Murali, C. N. Arpita, and Shaik Rafi Kiran**

**Abstract** At present, solar power generation system is the major concern in the most of the power distribution systems for satisfying future electrical energy demands. The solar photovoltaic (SPV) cell gives nonlinear output voltage characteristics. In addition, its output power varies continuously based on its different atmospheric irradiation intensity and temperature conditions. In this article, an adaptive variable step size Perturb & Observe (VSS-P&O)-based particle swarm optimization (PSO) controller is proposed as a maximum power point tracking (MPPT) controller for enhancing energy yield potential of SPV systems. Moreover, the efficiency of SPV system is low. Therefore, SPV output voltage has been significantly enhanced by the boost converter. The MATLAB/Simulink window is used to develop SPV with boost converter model which is used to investigate its performance at different atmospheric conditions.

**Keywords** Solar PV · Partial shading · MPPT · DC–DC converter · PSO · VSS

---

CH Hussaian Basha (✉) · T. Mariprasath
CRI Laboratory, K.S.R.M College of Engineering, Kadapa, Andhra Pradesh 516003, India
e-mail: sbasha238@gmail.com

T. Mariprasath
e-mail: ts.mariprasath@gmail.com

M. Murali · C. N. Arpita
Department of EEE, K.S.R.M College of Engineering, Kadapa, Andhra Pradesh 516003, India
e-mail: murali@ksrmce.ac.in

C. N. Arpita
e-mail: arpitha@ksrmce.ac.in

S. Rafi Kiran
Sri Venkateshwar Engineering College, Tirupati, Andhra Pradesh 517502, India
e-mail: rafikiran@gmail.com



D. Gupta et al. (eds.), *Pattern Recognition and Data Analysis with Applications*,
Lecture Notes in Electrical Engineering 888,
https://doi.org/10.1007/978-981-19-1520-8_65

## 1 Introduction

Traditional, fossil fuels-based power plants are dominating power sectors. Basically, the coal, oil, and natural gas are represented as fossil fuels. The fossil fuels occur due to the dead plants which are available within the earth. By using advance technology, the fossil fuels are extracted from the earth. Now, day by day, most of the non-renewable energy sources are reducing drastically. In addition to that, it produces highly inflammable gases. As result, the environmental pollution takes place [1]. The drawbacks and less availability of fossil fuels are compensated by using the natural resources such as solar, wind, hydro. These natural resources are called as renewable energy resource.

From the literature review, the wind power plants use the wind in order to generate the mechanical power by utilizing the wind turbines. In such plant, mechanical energy has been transformed into useful electrical energy by suitable electrical generators. The features of wind power are less impact on environmental conditions, high robust, and popular sustainable energy. In addition to that, the operating cost of wind power plants is very less and efficient utilization of land space [2]. The demerits of wind plants are high intermittent, moderate impact on atmospheric conditions, more visual pollution, and high noise. The drawbacks of wind power systems are overcome by utilizing the hydropower generation stations. In hydropower stations, the fast flow of water runs the generators in order to convert the kinetic energy into electrical power. The hydraulic energy is a clean source of energy, and it is having the features of high reliable and less emissions. In addition, it acts as a back-up power at the time of higher disruptions [3]. The drawbacks of this power generation system are expensive to build, limited reservoirs, and impact on fish.

The limitations of hydropower stations are overcome by using the geothermal stations. The geothermal power generation is high environmental friendly nature and clean source of energy [4]. The merits of these plants are non-pollutant, generate less wastage, and low maintenance when compared to the thermal power plants. In addition to that, it is extracted without burning of coal and oils. The major disadvantages of these plants are high initial operating cost, suitable only for particular regions, high sustainability issues, and required high operating temperature. The demerits of above power generation stations are overcome by applying the solar power plants [5].

From the above study, it shows either wind or hydro power plants harm the environments especially for living organism. When compared to all other RER, SPV-based electrical energy extraction paid more attention due to enormous potential. From the sun, huge irradiations fall on the earth surface. It is transferred into useful electrical energy by using photovoltaic cell. It is made from two types of crystalline materials such as P- and N-type semiconductors. When P-type semiconductor material received photon from sun, it exhibits a photon to the N-type semiconductor materials. This causes the release of the electron form N-type semiconductor, respectively. The each PV cell generation voltage is in between 0.7 and 0.8 which is not useful for the high-power electric vehicle application. Therefore, a number of PV cells are interconnected results of solar irradiations, and incident area has been increased which results high energy. Depending on the requirement, SPV cells are either connected series or connected parallel. Series connection offers high voltage, and high current is obtained from parallel connection, respectively [6, 7].

However, SPV power plant required high installation cost. From the literature review, there are different types of solar manufacturing technologies which are mono, polycrystalline, and thin film semiconductor manufacturing technologies. In these technologies, thin film is the most popular and high efficient PV cell topology because of that reason, most of research scholars are working on thin film PV cell technology [8].

Therefore, it is necessary to carry out research on high-energy extraction from SPV technology. So, a device is incorporated with existing PV system, such as MPPT. It is enhancing energy yield potential of SPV systems, respectively. Customarily, Perturb & Observation-based MPPT is popularly used due to simplicity and so costwise cheaper [9]. It is tracking MPP by using difference in two parameters such as present power and previous power. When the relative result gives the positive value, then the perturbation has been done in identical direction. Otherwise, it starts perturb the reverse direction, whereas limitation of the method is vast climatic condition and partial shaded condition, and its performance is very low. Incremental conductance (IC)-based MPPT controller [10].

In this IC MPPT method, the peak power of PV is obtained by comparing the instantaneous conductance with the past conductance which is stored in memory. The three comparative resultant rates of conductance are having the greater than zero value, then it varies in forward direction. Otherwise, it varies in opposite direction in order to find out the optimum MPP position [11]. The performance of INC is comparable that of P&O. The disadvantage of IC controller is high implementation cost. Few research article uses fraction open circuit voltage and current method for control of the duty cycle of boost converter. These are the approximated power point tracing techniques which are used where the accuracy of the MPP tracking is not necessary. But, these techniques are having the capability of high tracing speed of MPP. In addition, the implementation and understanding of the fractional open circuit voltage MPPT technique are very easy [12].

The parameters of PV output voltage and powers consist of ripples which are used as input signals to the ripple correlation control-based MPPT technique [13]. In this controller, the current and voltage consist of mutual relationship, respectively. This is applying to the boost converter for finding the operating point of PV. Perhaps, the switching frequency has been used to adjust duty cycle of the boost converter with the state flow technique, which is reported on [14]. The attractive features of state flow controller are easy to implement, high accuracy, and fast convergence. Added with, it is able to ready the fluctuation under steady-state and transient condition, respectively. Though, limitation of the method is high noise ratio of signal, in this work, an adaptive VSS-P&O-based PSO controller is proposed for enhancing energy yield potential of SPV system. It can suppress all the demerits caused by the traditional MPPT technique, respectively.

The PV output voltage has been step-up by the boost converter. Among all, the voltage enhancement rate strongly depends on the critical design methodology of boost converter, respectively. This research area has vast scope; from the literature, it was found that two types of boost converter are vital role in the boost converter design such as non-isolated and isolated [15]. The isolated boost converters are designed by using an additional transformer. Therefore, DC–DC converter (DC–DC) has isolation between input and output. The disadvantages of isolated DC–DC converters are overcome by using a non-isolated boost converter [16]. In this work, a conventional non-isolated boost converter is used to improve the voltage gain of the SPV.

## 2 Modeling of Three-diode-based SPV Cell

From the study found, various PV configurations are used which are classified according to the number diode such as single-, double-, and triple-diode PV models. The single-diode PV cell is designed by connecting a resistor in parallel with the diode, and this is the most commonly used circuit topology [17]. The merits are construction wise simple and so its characteristics are understandable. The major drawback of single-diode circuit topology is it gives inaccurate current versus voltage characteristics. Hence, it is not suitable for examining the overall system not feasible. However, two-diode model is presented in the article [18] to analyze the performance of PV system-based electrical vehicle charging system.

The overall parameters required for the design of two-diode model-based PV cell are seven which are obtained by using different optimization technique. The drawbacks of double-diode model PV cell are less fill factor, moderate efficiency, and less output power when compared to remaining PV cells. Here, a triple-diode model has been used to develop a new SPV system which is shown in Fig. 1. From Fig. 1, the PV cell output currents are derived as,

$$I_0 = I_{\mathrm{PV}} - I_{\mathrm{Da}} - I_{\mathrm{Db}} - I_{\mathrm{Dc}} - I_{\mathrm{sh}} \tag{1}$$

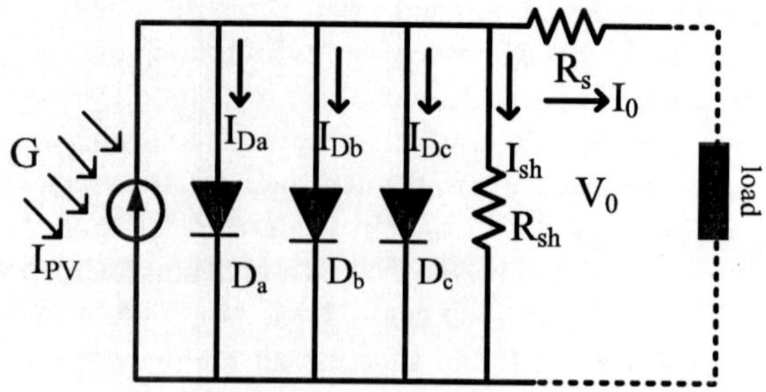


**Fig. 1** Triple-diode circuit-based solar PV cell

$$I_0 = I_{PV} - i_{0r\_1}\left(e^{\frac{q(V_0+I_0*R_s)}{\eta_1 K*T}} - 1\right) - i_{0r\_2}\left(e^{\frac{q(V_0+I_0*R_s)}{\eta_2 K*T}} - 1\right) - I_a \quad (2)$$

$$I_a = i_{0r\_3}\left(e^{\frac{q(V_0+I_0*R_s)}{\eta_3 K*T}} - 1\right) + \frac{V_0 + I_0 R_s}{R_{sh}} \quad (3)$$

$$I_{0r\_1} = I_{0r\_2} = I_{0r\_3} = I_{on}(\frac{T}{T_N})^3 e^{\frac{qEg}{nk}\left(\frac{1}{T_N}-\frac{1}{T}\right)} \quad (4)$$

$$I_{on} = I_{on\_1} = I_{on\_2} = I_{on\_3} = \frac{I_{sc\_n}}{e^{\left(\frac{Voc\_n}{\eta V_{Tn}}\right)}} \quad (5)$$

From Eq. (2), it is clearly observed that the solar cell current is mainly depending on its series and parallel resistances [19]. In addition, its output power varies at different operating temperature and partial shading conditions. Partial shading condition (PSC) is a phenomenon which directly affects the efficiency of SPV system since it has nonlinear characteristics. The primary causes of partial shading are clouds, tree, and nearby building shadow, respectively. The PSC causes adverse effect such as shaded cells consume power, therefore, it acts like a load. Due to power consumption, a hotspot is occurred on PV. It drastically reduced the efficiency of PV.

Consequently, a diode has been connected at reverse bias across the series-connected solar cell called as bypass diode. During normal operating condition, it offers high resistance to the flow of current, and under shading condition it allows current such it acts as a forward bias a shown in Fig. 2a. From Fig. 2a, it is clearly observed that the falling insolation on each PV cell is same. As a result, the overall

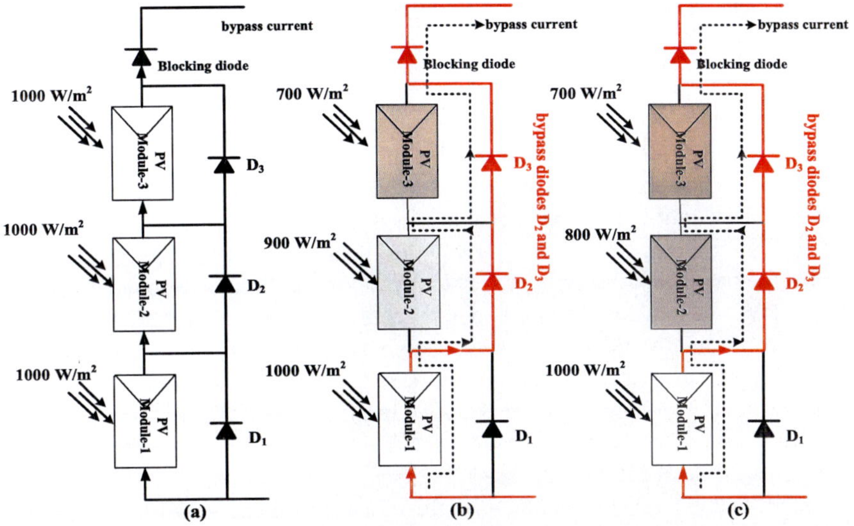


**Fig. 2** Shading phenomena on different solar PV modules, **a** Uniform, **b** PSC-1, and **c** PSC-2

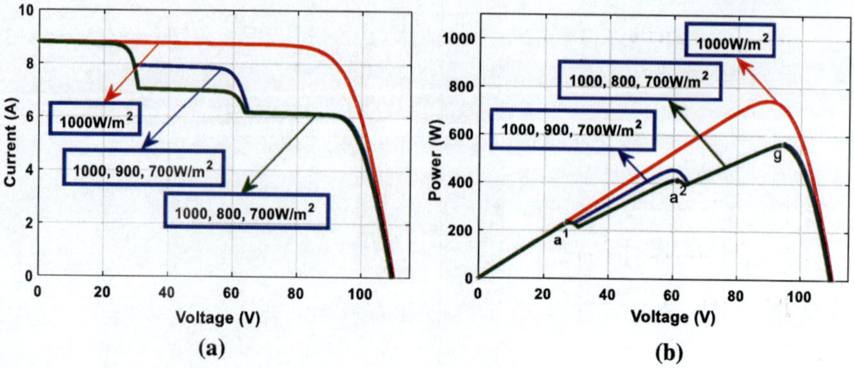


**Fig. 3** PV system, **a** *I–V* curve, and **b** *P–V* curve

system extracted power is more. Under second shading condition, the incident insolation on each PV module is different which is illustrated in Figs. 2b and c. The maximum extracted power under 1000 W/m$^2$, 1000 W/m$^2$, and 1000 W/m$^2$ is 750 W, and its corresponding maximum PV current and voltage are 7.97 A and 94 V. Under PSC-1, the obtained output power of PV is 576 W and its corresponding current and voltages are 6.198 A, and 93 V. Similarly, the PV power at second shading condition is 569 W and its corresponding voltage and current are 91.8 V and 6.19 A. The nonlinear *I–V* and *P–V* curves at different shading conditions are shown in Figs. 3a and b.

## 3 Design of an Adaptive VSS-P&O-based PSO Controller

The nonlinear power versus voltage curves gives multiple local MPPs and one required global MPP. The required effective point of SPV is traced by using the proposed hybrid MPPT technique. In this article [20], a variable step size P&O controller is used for the quadratic boost converter duty cycle adjustment. The controller is working effectively at uniform irradiation intensity values. In addition, its starting step size is very high. As a result, the tracing speed increases. Finally, the step is reduced in order to obtain the exact position of the operating point of the SPV, whereas this method is not suitable for dynamic irradiation operating conditions. Similarly, the particle swarm optimization controller is used in article [21] to improve the voltage magnitude of the supply system. In this PSO technique, the initializations of particles have not done properly, then it traces any one of the local MPPs. Also, the required number of iterations is more in PSO technique, then convergence time of MPP is increased.

To reduce the number of iterations in the PSO controller, a VSS-P&O controller is included to obtain the fast convergence speed of MPP which is shown in Fig. 4. Subsequently, a comparative analysis has been made between conventional PSO and

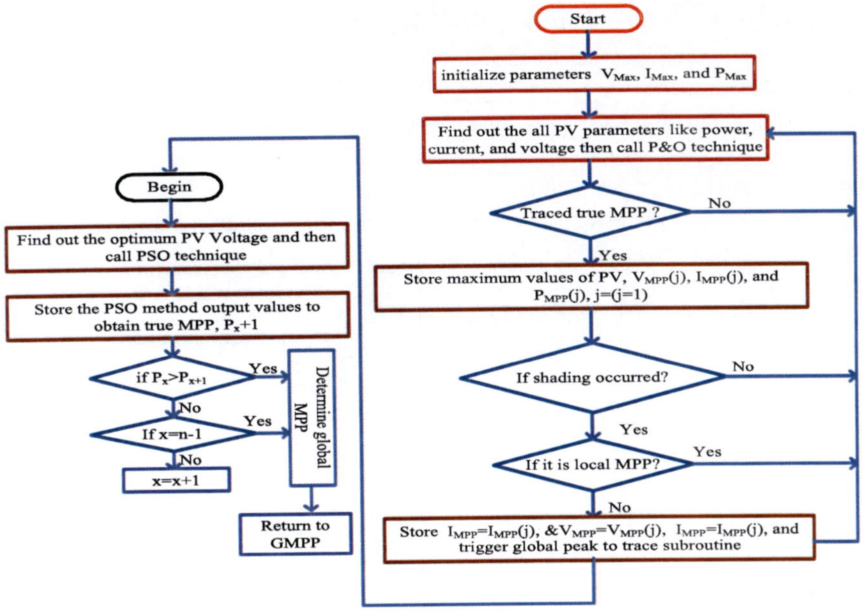


**Fig. 4** Proposed an adaptive VSS-P&O-based PSO MPPT controller

adaptive cuckoo search (ACS) power point tracking controllers. From Fig. 4, it is clearly observed that at starting variable step, P&O controller operates until the PV working point near to the true MPP. After the MPP reaching the global peak point, then the PSO starts to work in order to optimum the transient behavior of converter output voltage. In this hybrid PSO technique, the particle position and duty of high voltage gain boost converter are updated by applying Eqs. (6) and (7).

$$y^{k+1} = y_j^k + V_j^{k+1} \tag{6}$$

$$V^{k+1} = wV_j^k + c_{q1}r_{q1}\left(P_{\text{best\_}j} - y_j^k\right) + c_{q2}r_{q2}\left(G_{\text{best\_}j} - y_j^k\right) \tag{7}$$

To reduce the number of iterations in the PSO controller, a VSS-P&O controller is included to obtain the fast convergence speed of MPP which is shown in Fig. 4. The proposed controller is compared with conventional PSO and adaptive cuckoo search (ACS) power point tracking controllers. From Fig. 4, it is clearly observed that at starting variable step, P&O controller operates until the PV working point near to the true MPP. After the MPP reaching the global peak point, then the PSO starts to work in order to optimum the transient behavior of converter output voltage. In this hybrid PSO technique, the particle position and duty of high voltage gain boost converter are updated by applying Eqs. (6) and (7).

## 4 Design of High Step-up Boost Converter

From the literature review, it is found that to satisfy required energy demands, lot of DC–DC converters are used to adjust the voltage profile of SPV. Here, a conventional step-up converter is used to supply the continuous power from source to load. The block diagram of the boost converter is shown in Fig. 5a. The advantages of basic boost converter are high flexibility, more reliability, and less design cost when compared to the switched capacitor converters and two-phase interleaved converters. The proposed converter works in two genres of operations which are forward conduction state (see in Fig. 5b) and reverse biased state (see in Fig. 5c).

From Fig. 5b, the switch '*S*' working time is indicated as $DT_s$ and the diode '*D*' is in reverse working condition. In the switch forward condition, the input PV supply voltage is stored in inductor $L$ and it is represented as $V_L$. The output capacitor

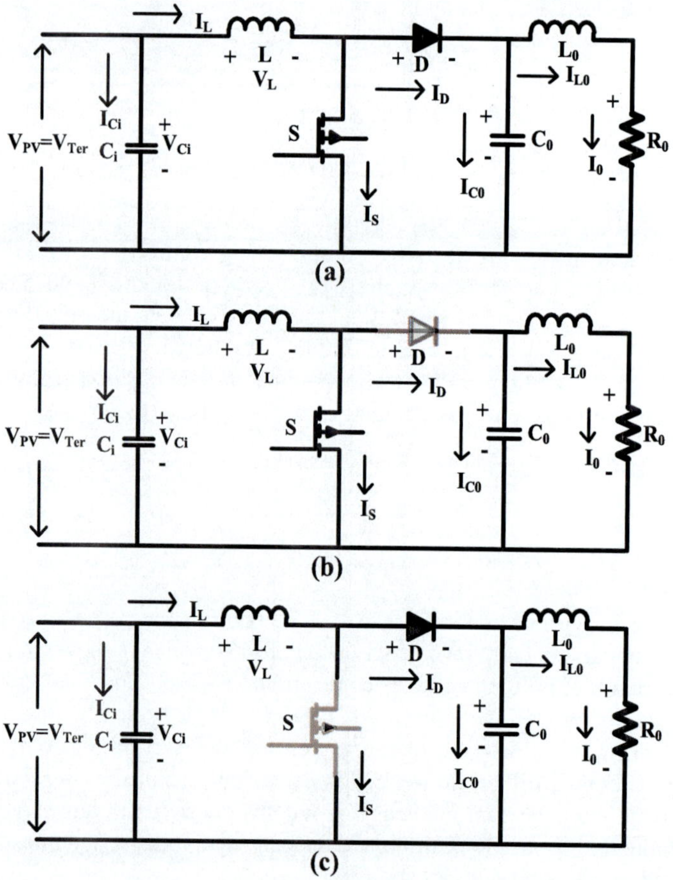


**Fig. 5** **a** Topology, **b** Switch ON condition, and **c** Switch OFF condition

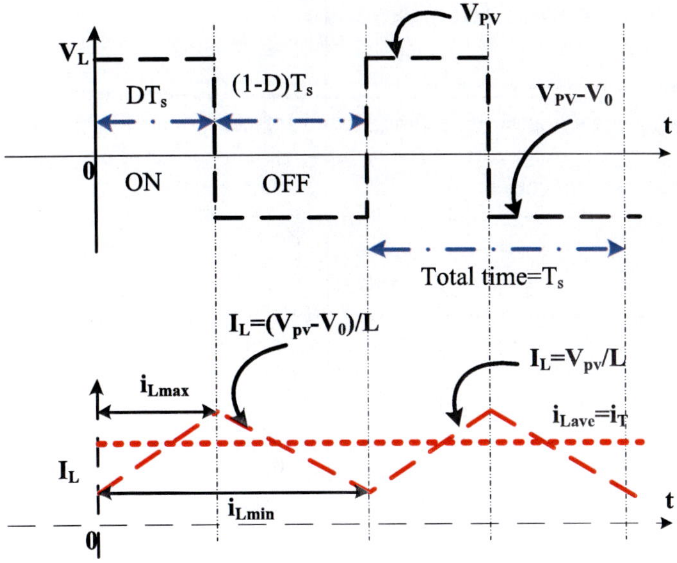


**Fig.6** Waveforms of high step-up boost converter

current ICO flows through the load. Similarly, in reverse blocking state of switch, the supply power flows from supply to load which is given in Fig. 5c. The performance waveforms of boost converter are shown in Fig. 6.

The output voltage gain ($V_O$) and duty cycle ($D$) of the converter are derived as,

$$DV_{PV}T_S + (1 - D) * (V_{PV} - V_0)T_S = 0 \tag{8}$$

$$-I_0DT_S + (1 - D) * (I_{PV} - I_0)T_S = 0 \tag{9}$$

Based on the above Eqs. (8) and (9), the converter output parameters such as voltage and current are determined as,

$$V_0 = V_{PV}/1 - D, \text{ and } I_0 = I_{PV}(1 - D) \tag{10}$$

$$V_0/I_0 = R_0, \text{ and } V_{PV}/I_{PV} = R_{PV} \tag{11}$$

From Eqs. (10), and (11), it is clearly indicated that the equivalent resistance across the PV module is controlled by varying the duty of boost converter.

# 5 Analysis of Simulation Results

Here, SPV is designed based on the three-diode PV. The parameters required for the mathematical modeling of triple-diode-based PV cell are open circuit voltage ($V_{oc}$ = 36.79 V), short circuit current ($I_{sc}$ = 8.84 A), cells per module ($N_{cell}$ = 60), PV cell generated output current ($I_{PV}$ = 8.84 A), shunt resistance ($R_{sh}$ = 315 Ω), series resistance ($R_s$ = 0.3 Ω), and diodes ideality factors ($\eta_1 = 2$, $\eta_2 = 0.8$, $\eta_3 = 0.94$, and $\eta_4 = 0.9$). Similarly, the PV side capacitor is equal to $C_i = 12$ µF, and it is used to maintain the constant PV output voltage.

Here, a triple-diode circuit-based solar PV cell model is used to design the solar array. The parameters required for the mathematical modeling of triple-diode-based PV cell are open circuit voltage ($V_{oc}$ = 36.79 V), short circuit current ($I_{sc}$ = 8.84 A), cells per module ($N_{cell}$ = 60), PV cell generated output current ($I_{PV}$ = 8.84 A), shunt resistance ($R_{sh}$ = 315 Ω), series resistance ($R_s$ = 0.3 Ω), and diodes ideality factors ($\eta_1 = 2$, $\eta_2 = 0.8$, $\eta_3 = 0.94$, and $\eta_4 = 0.9$). Similarly, the PV side capacitor is equal to $C_i = 12$ µF, and it is used to maintain the constant PV output voltage.

## 5.1 *First Shading Condition of Solar PV (1000 W/m$^2$, 900 W/m$^2$, 700 W/m$^2$)*

As we disused previously, the solar PV module nonlinear curves consist of different MPPs and its output supply reduced due its shading effect. In this condition, the falling irradiations on three PV modules in the each series string are 1000 W/m$^2$, 900 W/m$^2$, and 700 W/m$^2$. The PV module nonlinear *P–V*, power versus current, and *I–V* curves are given in Figs. 7a, b, and c. Here, the proposed MPPT controller is compared with the other power point tracing controllers in terms of tracking efficiency, MPP oscillations, number of iterations required, and steady-state settling time, etc. From Fig. 7a, the required PV system power is equal to 576 W. The convergence speed of VSS-P&O with PSO is very high when compared to PSO and ACS which is illustrated in Figs. 7d, e, and f.

At starting, the three MPPT techniques search alternatively on entire power versus voltage curve to find the required duty value for the DC–DC converter. For starting search, the initialized duty cycles are 0.12, 0.23, 0.67, and 0.85. After completing the entire first iteration, the duty cycles are adjusted in the second iteration which are identified as 0.19, 0.34, 0.92, and 0.98, respectively. The performance parameters of three nature-inspired optimization techniques are given in Table 1. Here, from Table 1, the steady-state settling time of VSS-P&O with PSO controller is 0.4 s which is less than the other two MPPT techniques.

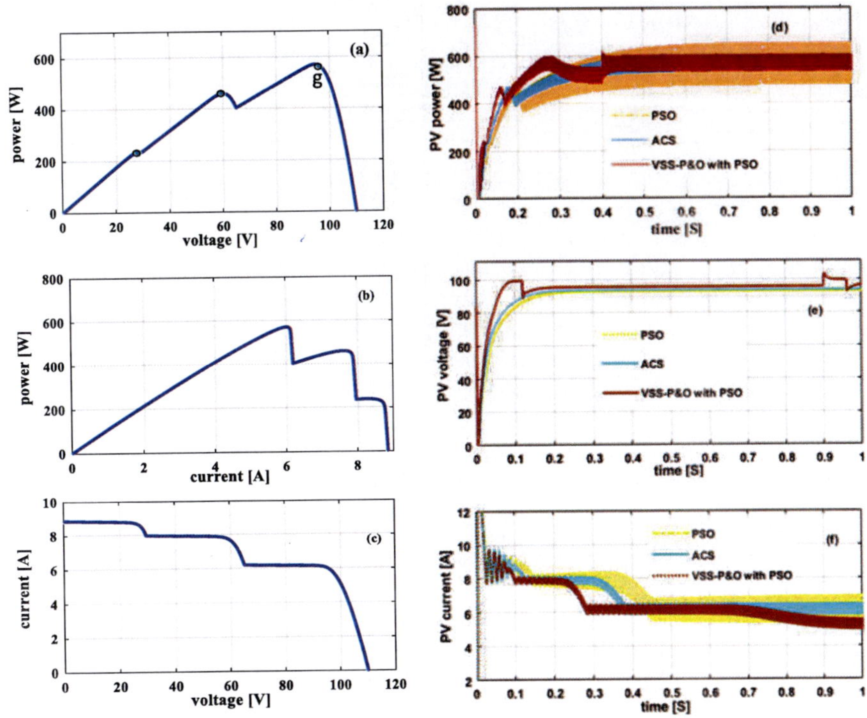


**Fig. 7** **a** *P–V*, **b** *P–I*, **c** *I–V* curves, **d** PV power, **e** PV voltage, and **f** PV current at the first PSC

## 5.2 *Second Shading Condition of Solar PV (1000 W/m$^2$, 800 W/m$^2$, 700 W/m$^2$)*

Similar to the first shading effect on PV modules, under second partial shading effect, the nonlinear *P–V*, *P–I*, and current versus voltage curves are given in Figs. 8a, b, and c. In this second condition, the required maximum extracted power of solar PV is 569 W. From Figs. 8d, e, and f, it is observed that the tracking speed and accuracy of VSS-P&O-based PSO are comparable to that of PSO, ACS MPPT techniques, respectively. The extracted power of PV by using PSO is moderate, while it offers high oscillation near MPP. As a result, it required lager number of iteration to tracking MPP in PSO. The detailed performance parameters of proposed MPPT technique for PV-fed DC–DC converter are given in Table 1. In this condition, the PSO power point finding controller extracts the maximum voltage and current when associated to the ACS. In addition to that, the ACS controller gives less tracking speed which is equal to 0.31 s.

**Table 1** Results analysis parameters of PSO, ACS, and VSS-P&O with PSO

| Parameters | PSO | ACS | VSS-P&O with PSO |
|---|---|---|---|
| *At first partial shading effect on solar PV modules* | | | |
| DC-link capacitor voltage | 93.832 | 94.99 | 102.55 |
| PV power (W) | 568.31 | 569.901 | 581.00 |
| PV current (A) | 6.058 | 5.99 | 5.680 |
| Global power (W) | 584 | 584 | 584 |
| Tracking efficiency (%) | 97.313 | 97.585 | 99.615 |
| Number of iterations | None | None | 11.00 |
| Steady-state settling time (s) | 0.4511 | 0.50 | 0.400 |
| Distortion in waveforms | Moderate | Moderate | Less |
| Tracking speed (s) | 0.200 | 0.280 | 0.1502 |
| Duty cycle | 0.7 | 0.4 | 0.6 |
| *At second partial shading effect on solar PV modules* | | | |
| DC-link capacitor voltage | 86.534 | 89.623 | 100.015 |
| PV power (W) | 560.97 | 569.901 | 579.471 |
| PV current (A) | 6.48 | 6.358 | 5.79384 |
| Global power (W) | 582.00 | 582.00 | 582.00 |
| Tracking efficiency (%) | 96.386 | 97.921 | 99.565 |
| Number of iterations | None | None | 9.00 |
| Steady-state settling time (s) | 0.34 | 0.450 | 0.300 |
| Distortion in waveforms | Moderate | Moderate | Less |
| Tracking speed (s) | 0.400 | 0.31 | 0.1241 |
| Duty cycle | 0.5 | 0.7 | 0.4 |

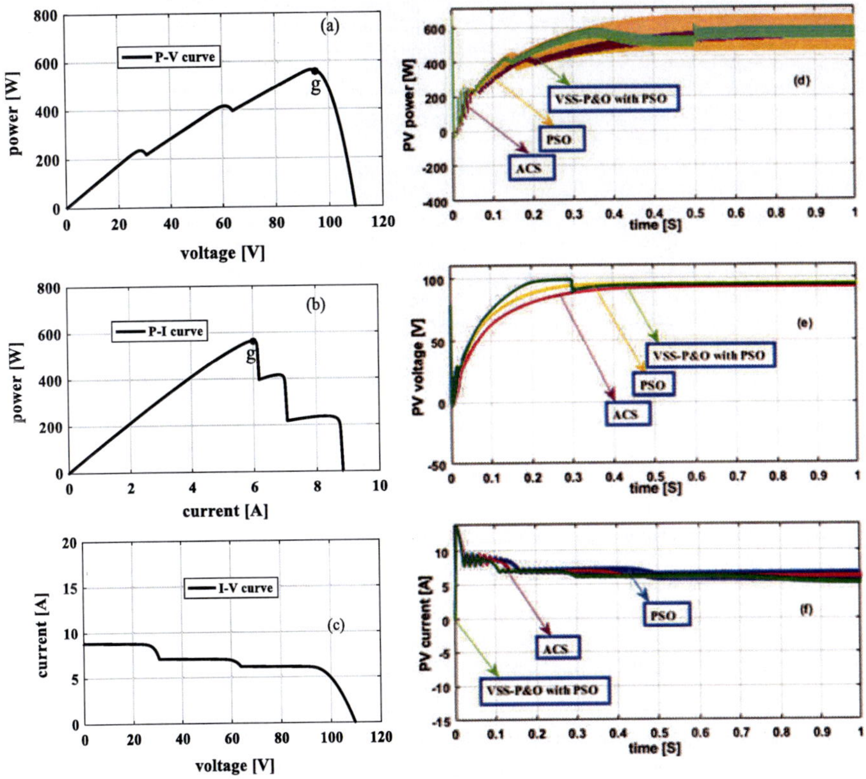


**Fig. 8** **a** *P–V*, **b** *P–I*, **c** *I–V*, **d** PV power, **e** PV voltage, and **f** PV current at the second PSC

## 6 Conclusion

The proposed variable step size considering P&O controller-based PSO power point tracking controller is implemented successfully with the other conventional PSO and ACS techniques by using MATLAB/Simulink window. The performance analysis of proposed controller has been done in terms of MPP tracking efficiency, iterations required to obtain optimum duty cycle value, and oscillations across MPP, etc. From the simulative performance parameters, it has been concluded that the proposed hybrid MPPT controller is tracing the MPP with high accuracy and less steady-state oscillations. In addition, the convergence speed of the hybrid controller is very high. Along with the hybrid MPPT, the boost converter is giving the constant output power with less distortions.

**Acknowledgements** We would like to thank the management of K.S.R.M. College of Engineering (Autonomous) for providing all the facilities to carry out this research work.

## References

1. Artigao, E., Vigueras-Rodriguez, A., Honrubia-Escribano, A., Martin-Martinez, S., Gomez-Lazaro, E.: Wind resource and wind power generation assessment for education in engineering. MDPI Sustain. **13**(5), 1–27 (2021)
2. Mohsin, M., Kamran, H.W., Nawaz, M.A., Hussain, M.S., Dahri, A.S.: Assessing the impact of transition from nonrenewable to renewable energy consumption on economic growth-environmental nexus from developing Asian economies. J. Environ. Manage. **284**, 111999 (2021)
3. He, Z., Wang, C., Wang, Y., Wei, B., Zhou, J., Zhang, H., Qin, H.: Dynamic programming with successive approximation and relaxation strategy for long-term joint power generation scheduling of large-scale hydropower station group. Energy **222**, 119960–119964 (2021)
4. Akar, S., Augustine, C., Kurup, P.: Global value chain and manufacturing analysis on geothermal power plant turbines. In: Thermodynamic Analysis and Optimization of Geothermal Power Plants, pp. 17–41. Elsevier (2021)
5. Hussaian Basha, CH., Rani, C., Brisilla, R.M., Odofin, S.: Simulation of metaheuristic intelligence MPPT techniques for solar PV under partial shading condition. In: Soft Computing for Problem Solving, pp. 773–785. Springer, Singapore (2020)
6. Hussaian Basha, C.H., Rani, C., Odofin, S.: A review on non-isolated inductor coupled DC–DC converter for photovoltaic grid-connected applications. Int. J. Renew. Energy Res. **7**, 1570–1585 (2017)
7. Hussaian Basha, C.H., Rani, C.: Different conventional and soft computing MPPT techniques for solar PV systems with high step-up boost converters: a comprehensive analysis. Energies **13**, 1–27 (2020)
8. Kumar, R.T., Ramakrishna, K.M., Sukumar, G.D.: A review on PV cells and nanocomposite-coated PV systems. Int. J. Energy Res. **42**(7), 2305–2319 (2018)
9. Alik, R., Jusoh, A.: Modified perturb and observe (P&O) with checking algorithm under various solar irradiation. Sol. Energy **148**, 128–139 (2017)
10. Mahmoud, K., Lehtonen, M., Darwish, M.M.F.: An efficient fuzzy-logic based variable-step incremental conductance MPPT method for grid-connected PV systems. IEEE Access **9**, 26420–26430 (2021)
11. Zand, S.J., Hsia, K.H., Eskandarian, N., Mobayen, S.: Improvement of self-predictive incremental conductance algorithm with the ability to detect dynamic conditions. Energies **14**, 1–14 (2021)
12. Nadeem, A., Sher, H.A., Murtaza, A.F., Ahmed, N.: Online current-sensorless estimator for PV open circuit voltage and short circuit current. Sol. Energy **213**, 198–210 (2021)
13. Singh, R., Yadav, R., Varshney, L., Sharma, S.: Analysis and comparison of PV array MPPT techniques to increase output power. In: International Conference on Advance Computing and Innovative Technologies in Engineering, pp. 1–8. IEEE, India (2021)
14. Roncero-Clemente, C.: Power-flow-based secondary control for autonomous droop-controlled ac nanogrids with peer-to-peer energy trading. IEEE Access **9**, 22339–22350 (2021)
15. Mumtaz, F., Yahaya, N.Z., Meraj, S.T., Singh, B.: Review on non-isolated DC-DC converters and their control techniques for renewable energy applications. Ain Shams Eng. J. 1–17 (2021)
16. Shahir, F.M., Babaei, E., Aberoumandazar, M.: New single-switch non-isolated boost DC–DC converter with free input current ripple. In: 12th Power Electronics, Drive Systems, and Technologies Conference, pp. 1–8. IEEE, Iran (2021)
17. Rasheed, M., Mohammed, O.Y., Shihab, S., Adili, A.A.: A comparative analysis of PV cell mathematical model. J. Phys.: Conf. Ser. 1–10 (2021)
18. Hussaian Basha, CH.: Mathematical design and analysis of photovoltaic cell using MATLAB/Simulink. In: Soft Computing for Problem Solving, pp. 711–726. Springer, Singapore (2019)
19. Obukhov, S.: Optimal performance of dynamic particle swarm optimization based maximum power trackers for stand-alone PV system under partial shading conditions. IEEE Access **8**, 20770–20785 (2020)

20. Chowdary, V.G., Sankar, V.U., Mathew, D., Hussaian Basha CH., Rani, C.: Hybrid fuzzy logic-based MPPT for wind energy conversion system. In: Soft Computing for Problem Solving, pp. 951–968. Springer, Singapore (2020)
21. Hussaian Basha, CH., Chowdary, V.G., Rani, C.: Design of SVPWM-based two-leg VSI for solar PV grid-connected systems. In: Soft Computing for Problem Solving, pp. 879–892. Springer, Singapore (2020)

# Spatiotemporal Data Compression on IoT Devices in Smart Irrigation System

Neha K. Nawandar and Vishal R. Satpute

**Abstract** With the advancement in technology and the evolution of the Internet of Things (IoT) era, the quantity of data generated by devices has increased tremendously. The IoT is an umbrella domain with enormously prolific applications like smart homes, waste management, agriculture, and retail. A huge amount of data is generated by these applications. This creates the need for data handling as well as management to be used for data processing. IoT applications are achieved by means of sensor-interfaced battery-powered device deployments. Low energy consumption, data handling, and fast performance are some of the key features an IoT device should possess. However, the limited memory of these devices is a concern as it is insufficient to handle the amount of data generated by the sensors. These concerns are targeted in the paper and applied to smart agriculture applications where real-time data fetched by sensors are compressed and saved onto the device, thus properly handling the data. The reconstructed data are obtained post-compression and decompression and are used for decision-making. It provides 100% accuracy, and compression has no ill effect on the data. Further, the energy efficacy of the work is compared, and it is found to be on an average ~47% efficient against the compared algorithms.

**Keywords** Spatiotemporal · Data compression · IoT · Smart farming · Adaptive irrigation

## 1 Introduction

Increasing advances in technology have led to a tremendous growth in the capabilities of Internet-enabled devices. This Internet connectivity makes communication and device operability possible with minimal effort. One such technology, Internet of

N. K. Nawandar (✉) · V. R. Satpute
Department of ECE, VNIT Nagpur, Nagpur, India
e-mail: nehanawandar@gmail.com

V. R. Satpute
e-mail: vrsatpute@ece.vnit.ac.in



D. Gupta et al. (eds.), *Pattern Recognition and Data Analysis with Applications*, Lecture Notes in Electrical Engineering 888,
https://doi.org/10.1007/978-981-19-1520-8_66

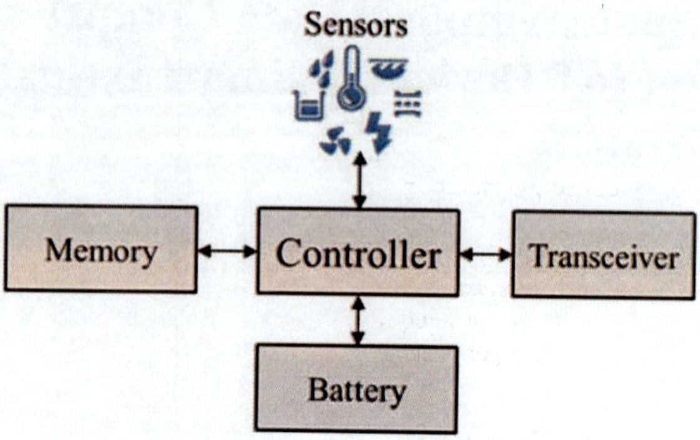


**Fig. 1** IoT system

Things (IoT) [1], has made human life much more comfortable and effortless. It extends its application domain to a wide horizon of applications such as: smart cities, ACs, and fans in homes and offices are controlled by the IoT devices [2], smart health: the devices with which patients send stats to their doctor without a technician actually measuring and sending it [3], smart agriculture: Smart systems manage irrigation [4], and machinery automatically harvests after end of growth cycle, etc. Based on their applications, IoT applications differ on most levels. However, one factor that is common in them is the device that gathers application-specific data and is used to initiate a decision.

Its basic functionalities include: presence of a controller with communication technology, sensors interfaced for getting real-world data, processing and decision-making capabilities, and power source and algorithm embedded for controlling as seen in Fig. 1. These devices work 24 $\times$ 7 on limited power. For efficacy, they are run on minimum energy consumption to ensure that the device runs on available power for maximum time.

In an IoT system, the sensors gather huge amount of data from the physical world which plays a major role in actuation. This collected data need to be stored for processing and for future requirements to allow precise decision-making based on past and present conditions. However, the amount of data generated by multiple sensors is huge and saving it is not memory effective. Instead, storing the data in its compressed form is required. There are various existing compression algorithms, but the main factor to be taken into consideration is that it has to be employed on a battery-operated device that already performs other functionalities. Compression in general is computationally costly, and an algorithm that uses minimal operations is beneficial. It also needs to be ensured that the algorithm does not come at the cost of same energy consumption as with normally transmitting the data. Instead, its inclusion must improve energy efficiency. It becomes evident that energy efficient and computationally effective algorithms are necessary for compression to be operated on an IoT device. On similar lines, this paper discusses the application of such energy-efficient compression algorithm on our IoT-based smart irrigation system.

Data compression has already been applied on various applications, some of which have been listed here. Lee and Kim [5] demonstrated a low-memory compression system to multimedia application. Smart meter is another place where compression has been employed, and Lee et al. [6] proposed a compressive sensing-based compression and authentication for a smart meter. It also finds suitability in wearable devices and Del Testa and Rossi [7] proposed a similar device that collects bio-signals and uses auto-encoders to compress the signals. Another medical-wearable device-related work in [8] shows a real-time QRS detector and compression architecture for ECG signals.

Apart from these applications, smart farming applications also require such algorithms due to the limited memory of IoT devices. Even so, implementation of data compression algorithms for smart farming applications still has solid research potential. Bhargava et al. [9] proposed a fog computing technique to address the memory issue in dairy farming. Vecchio et al. [10] proposed a lossless compression technique where environmental parameters were used. However, data compression algorithms on irrigation deployments have not yet blossomed. It is an opportunity which must be explored in the near future.

Our work aims to employ data compression algorithm on a battery-powered IoT device. Such an algorithm would prove beneficial as it will not only provide fast compression but also conserve energy which is one of our main concerns. In our work, we impose the algorithm presented in [11] on the generated sensor data for an automatic irrigation system deployment in the smart farming application domain. It is energy efficient, COordinate Rotation DIgital Computer-based (CORDIC) discrete cosine transform (DCT) that targets battery-operated portable devices. The work shows significant reduction in energy consumption, and it also maintains the SPAA trade-off, and its application can be seen in Sect. 2.1. The remaining paper is arranged as: Sect. 2 discusses the application of energy-efficient CORDIC algorithm to smart irrigation system, and Sect. 3 provides the experimental analysis and algorithm efficacy on the application. Section 4 concludes the paper, and references have been mentioned at the end.

## 2 CORDIC-Based Compression in Smart Irrigation System

This section discusses the CORDIC-based DCT algorithm and its application for data compression in a smart irrigation system. It starts with the algorithm followed by its application to IoT device in smart irrigation scenario.

### *2.1 Energy-Efficient CORDIC for Data Compression*

In this section, the energy-efficient data compression algorithm that is used in the work has been presented. Nawandar and Satpute [11] present a quality-tunable energy-efficient CORDIC-based algorithm that computes trigonometric values in real time. Such a design can prove useful to create a compression algorithm which is to be performed on battery-operated devices. To verify the functionality, the paper processes the algorithm on the sensor data, specifically temperature, humidity, and soil moisture. CORDIC provides cosine values of given input, and our aim is data compression where DCT has been used which requires computing some coefficients. These coefficients are the cosine values of some fixed inputs ranging from $\pi/16$ to $7\pi/16$, whose values are computed using the algorithm. These values are then used to form a DCT matrix which is finally used to compress the sensor data. In our work, we have employed this algorithm to compress our sensor data before saving and sending it. It is used because it ensures data compression with minute acceptable error in the reconstructed values.

CORDIC is capable of computing any mathematical operation using simply adders and shifters, thus consuming both less area and power. Our concern is power consumption, and CORDIC-based algorithm can help achieve maximum energy efficiency. However, the conventional CORDIC faces some drawbacks like data dependency and iterative nature, which has been overcome by [11] algorithm that also ensures more energy efficiency. This algorithm is preferred over the other available in literature.

The application of this algorithm can be seen further, where it is used to compress the data obtained using the sensors interfaced to an IoT device. This device is deployed in a test bed of an automatic irrigation system scenario. It is a well-known fact that agriculture is the backbone of the Indian economy where the livelihood of a large number of people is dependent on it. Agriculture faces certain threats and challenges such as: like pest and disease management, effective water utilization, data management and processing, and decision-making. These can be tackled by incorporating current technology and advancements in this field. So, here, we target application of technology to manage sensor data by using energy-efficient compression algorithm discussed in this section. Here, we present the application of data compression algorithm, whereas Sect. 3 presents its experimental analysis.

### 2.2 *Data Compression Algorithm in IoT Device for Smart Irrigation*

In this paper, sensor data related to an irrigation system are used as the test dataset for compression. The data are collected from sensors at regular intervals over a space which are compressed before storing and sending. Here, the spatiotemporal properties of the data are exploited for compression. Data collected at time instances are given by Eq. 1. It gives dataset of different time instances that are independent of each other, where, $t \rightarrow \{t_1{:}t_n\}$: the time instance at which data are gathered and $\{T, H, S\}$ are the temperature, humidity, and soil moisture at time $t$. Similarly, a set of spatial instances is given by Eq. 2, where $S \rightarrow \{S_1{:}S_n\}$: sample space that collects data at t time instant. Data are gathered at various time instances from multiple sensors placed at different locations and have redundancy in it, so the spatiotemporal properties can be utilized which make compression possible.

$$
\begin{aligned}
t_0 &: T_0, H_0, S_0 \\
t_1 &: T_1, H_1, S_1 \\
t_n &: T_n, H_n, S_n
\end{aligned}
\tag{1}
$$

$$
\begin{aligned}
S_0 &: [T, H, S]@t_0 \\
S_1 &: [T, H, S]@t_1 \\
S_n &: [T, H, S]@t_n
\end{aligned}
\tag{2}
$$

When DCT is performed on a data, it can be observed in the frequency domain that most of the energy is contained in first few coefficients only, whereas it is equally spread in the spatial domain. This is due to the energy compaction property of DCT which says that DCT packs critical information in fewer coefficients and requires less computational resources to transform the data. Here, DCT is applied to the sensor data, and Fig. 2 shows the energy contained by the DCT elements for temperature, humidity, and soil moisture data. From the figure, we can see that almost 90% of the total energy is contained in the first few elements only, thus providing an opportunity for performing compression. Algorithm 1 gives the pseudo-code which is implemented on our IoT device [12]. Here, the sensor data are used as the input which are compressed using quality tunable CORDIC to get the output in compressed form. The sensor data obtained in real time at a rate of 5 min are arranged in 2D form on which DCT is applied. The algorithm computes the DCT coefficients and forms an $8 \times 8$ matrix. It is used on the sensor-gathered data to compress it in batches of 64 sensor values. This $8 \times 8$ DCT transformed matrix is reduced to $4 \times 4$ which is used for reconstructing the data. This reconstructed data are used for final decision-making mechanism on the IoT device used in the irrigation system. The block diagram for DCT-based data compression technique on smart irrigation system can be seen from Fig. 3. The decision taken using the original as well as using the reconstructed data have been evaluated and compared for the reliability of the algorithm. The results have been mentioned in Sect. 3.

**Algorithm 1** Pseudo-code at the IoT device

```
Inputs: {T, H, S}
Output: compressed {T, H, S}
Variables: i, j, k, l, m
for setup do
    set variable values
    set θ: θ = {π/16:4π/16}
    for θ in range {π/16:4π/16} do
    compute coefficients
    end for
```

(continued)

(continued)

```
store a, b, ..., g coefficients
create 8 × 8 DCT matrix
end for
for continuous loop do
    fetch {T, H, S} @ t_i:t_{i+63}
    perform 2D-DCT and compression
    save the compressed {T, H, S}
end for
```

## 3 Experimental Analysis

This section discusses results obtained to prove the efficacy of algorithm usage in irrigation system scenario. To check its effectiveness, the algorithm has been initially modeled in MATLAB, and the results of an automatic irrigation application scenario have been obtained accordingly. After required results are obtained, the data compression algorithm is implemented on the device for the automatic irrigation system scenario. Temperature, humidity, and soil moisture data have been used by the algorithm which were obtained from sensors implanted at a test site that were controlled using our system [12]. The energy of CORDIC-based DCT algorithm has been compared with basic [13], scale-free [14], and look ahead [15] CORDIC architectures and mentioned in this section. The components used to create the device that runs the compression algorithm have been listed in Table 1.

After the data are obtained, it is used by the algorithm to check for the mean and variance in the data so that compression could be done accordingly. For this sample data values, the mean and variance in the original data could be seen in Fig. 4. These figures give the block-wise mean of temperature, humidity, and soil moisture values, where each block consists of 64 values of each data. These values are then compressed using the algorithm and reconstructed in order to see the algorithm efficiency. A similar analysis is done on the reconstructed data also.

Figure 5 shows the original and the reconstructed values for the sensor data. It can be observed from the figures that the reconstructed data follow the same pattern as that of the original uncompressed form, which proves that the algorithm maintains the data authenticity and is less prone to errors. It has been observed that the range of values of the sensor data $\{T, H\}$, i.e., the consecutively fetched values do not have significant difference between them. This on transform and compression does not lead to considerable error. (Here, $M$ is not included since its value is controlled by the irrigation system). The algorithm is used in the maximum energy-saving mode which introduces maximum amount of approximation in the data. The error

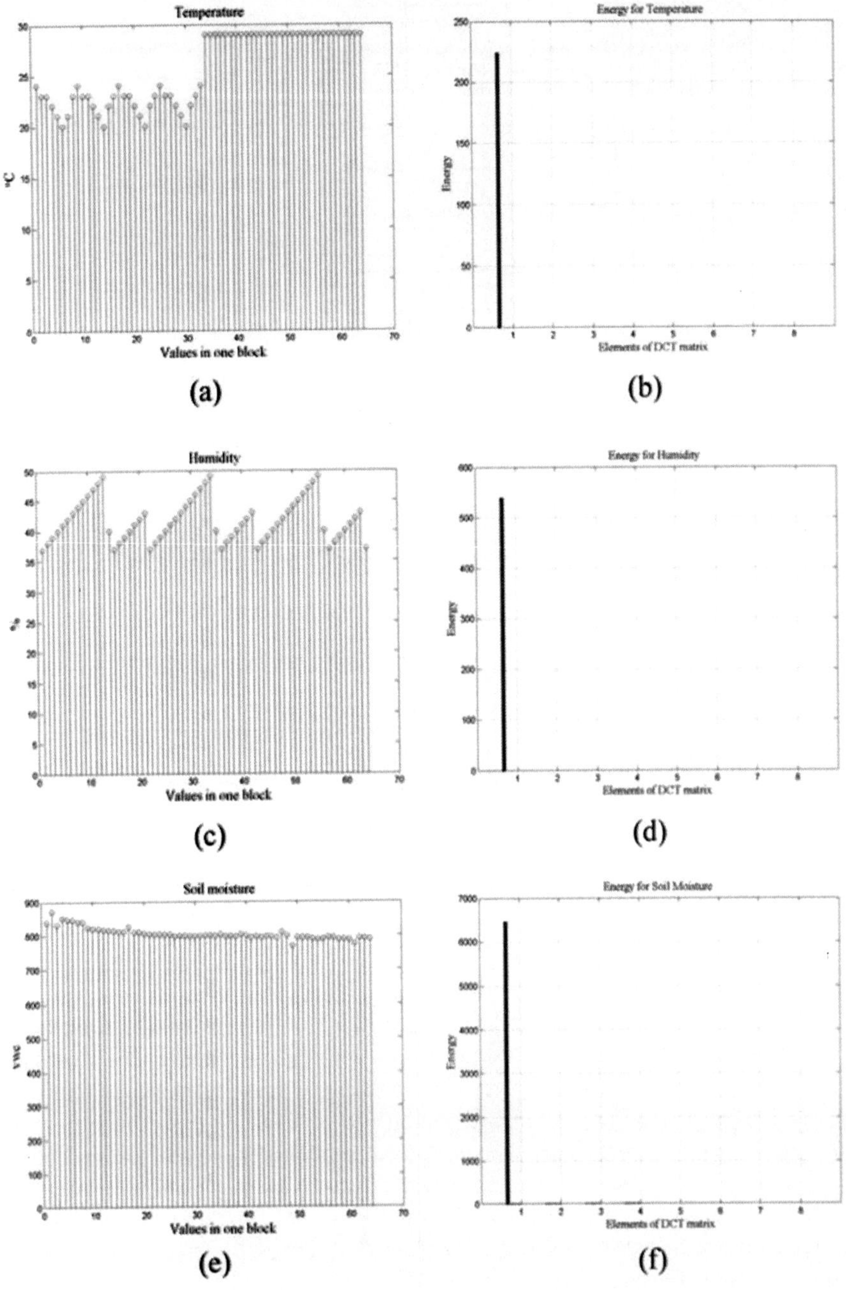


**Fig. 2** Temperature, humidity, and soil moisture values in one block and the corresponding energy contained in DCT operated on them (**a**, **b** temperature, **c**, **d** humidity and, **e**, **f** soil moisture)

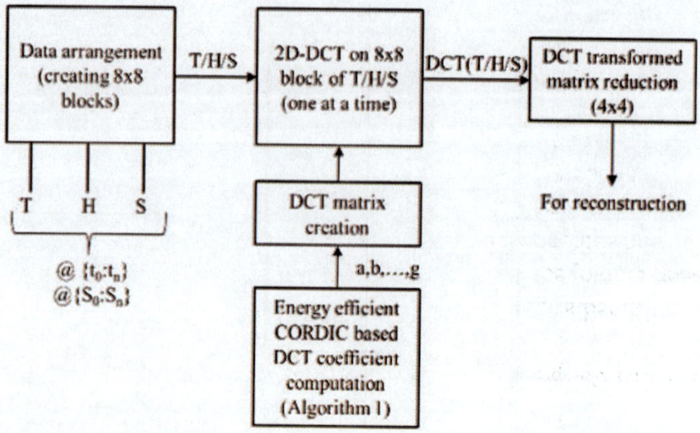


**Fig. 3** DCT-based data compression in our work

**Table 1** Components used

| Description | Model/type |
|---|---|
| Microcontroller board | Wemos D1 mini |
| Temperature and humidity sensors | DHT11 |
| Soil moisture sensors | RDL analog |
| Batteries | 6F22 |
| SD card (local/offline storage) | SanDisk 16 GB |
| Miscellaneous | Wires, connectors, cable, resistors, etc. |

reflected in the reconstructed values is obtained using this mode. Thus, the error seen in reconstructed data is the maximum possible error.

Apart from the mean and variance of sensor data and the reconstructed data, standard deviation, skewness, and kurtosis of the input as well as the reconstructed data have also been obtained, and their corresponding values can be seen from Table 2. Skewness gives the degree of distortion from the normal distribution. It is a measure of symmetry, and its value infers whether the data are symmetric or it lacks symmetry. A value between −0.5 and 0.5 implies that the data are fairly symmetrical; if it lies between −1 and −0.5, it means that the data are moderately skewed, whereas other values mean that the data are highly skewed. Kurtosis on the other hand is the measure of degree of distortion and a measure of outliers. It is used to know whether the dataset is free from outliers and if not, it gives a measure of the outlier presence. Data with kurtosis that fall beyond −3 or +3 mean outliers' presence, whereas other low values imply lack of outliers in the data. Thus, these parameters are important as they convey important information regarding our data. In light of this, before proceeding toward any processing on the data, it is beneficial to get this information about the data and proceed accordingly. The values for both the original dataset and the reconstructed

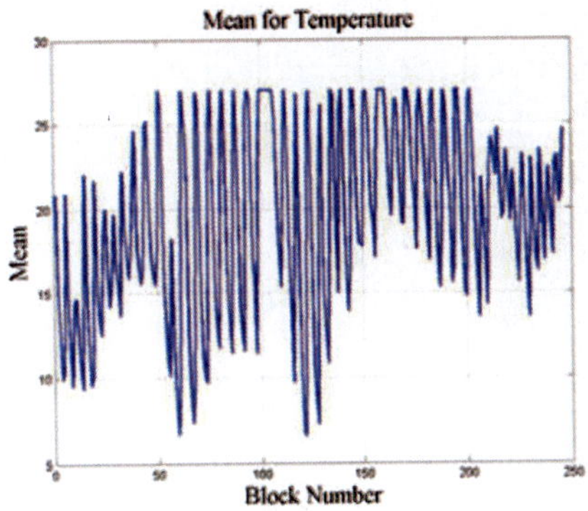


(a) Temperature mean

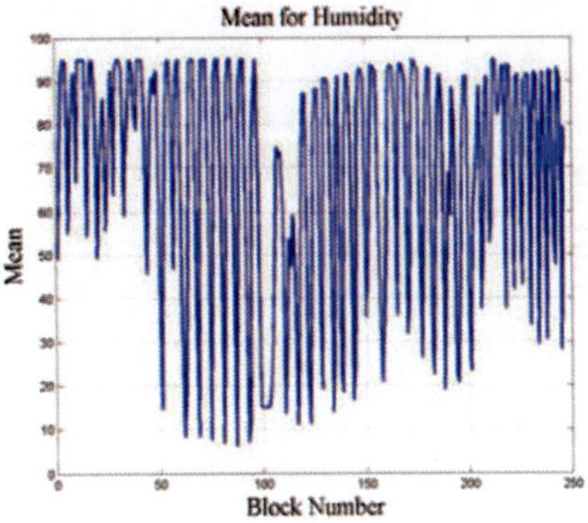


(b) Humidity mean

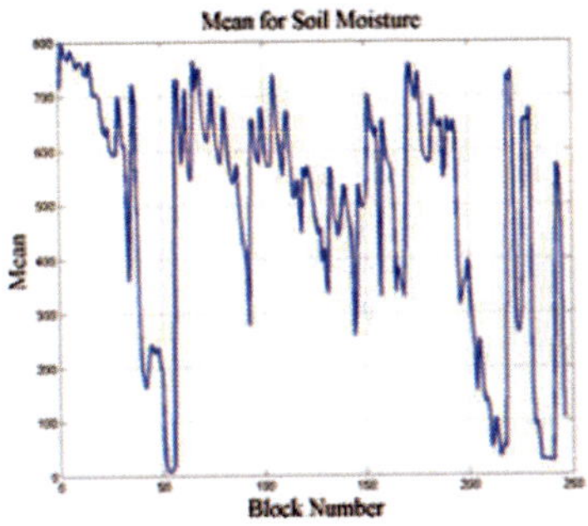


(c) Soil moisture mean

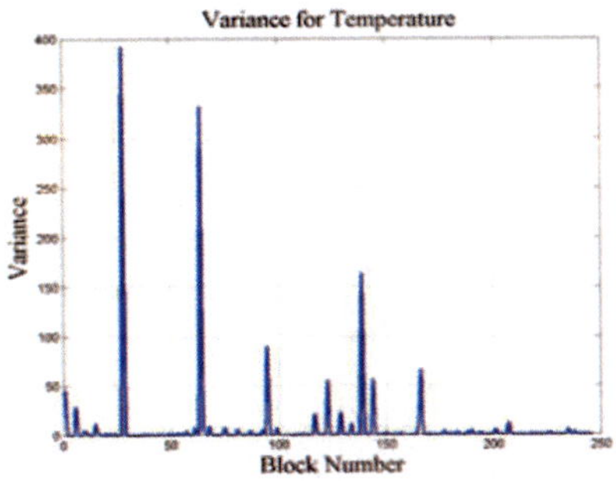


(d) Variance in temperature

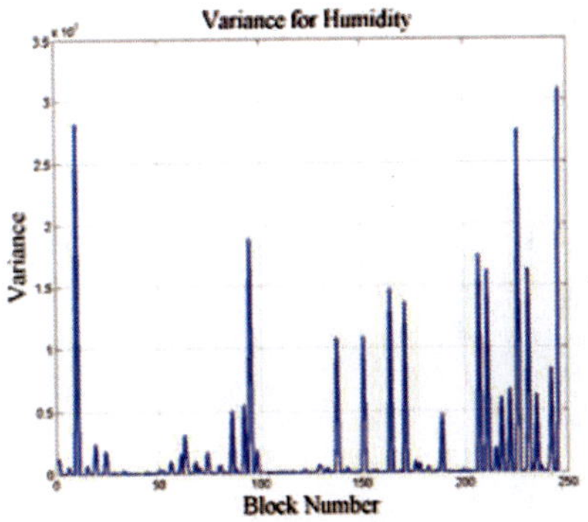


(e) Humidity variance

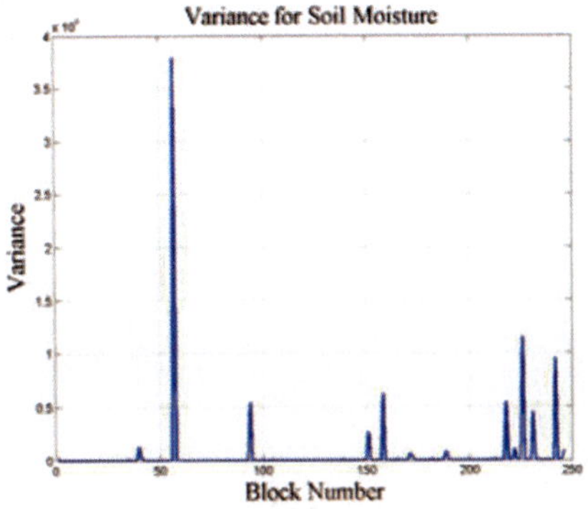


(f) Soil moisture variance

**Fig. 4** Block-wise mean and variance observed in original data values

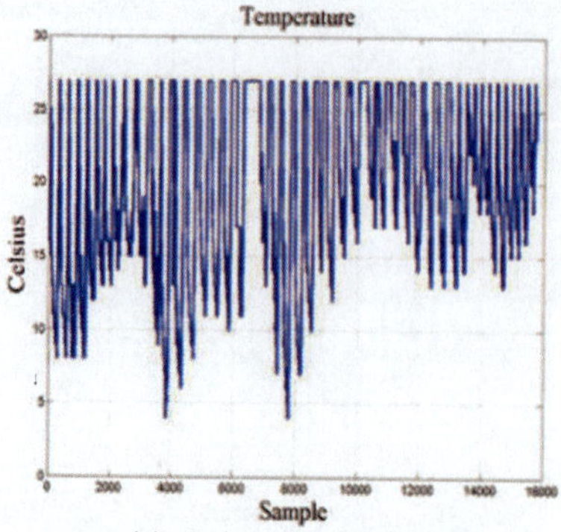


(a) Original temperature

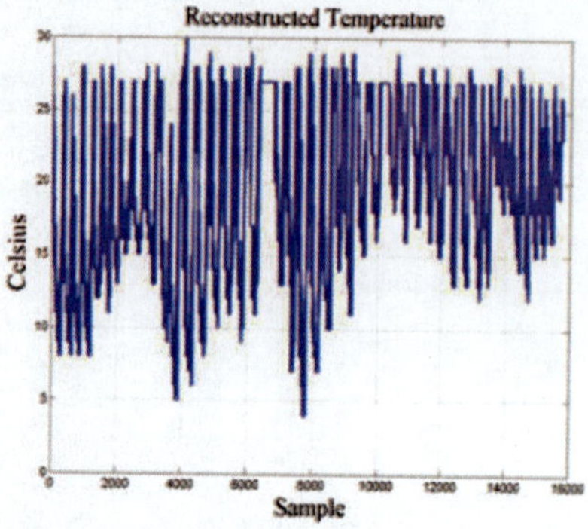


(b) Reconstructed temperature

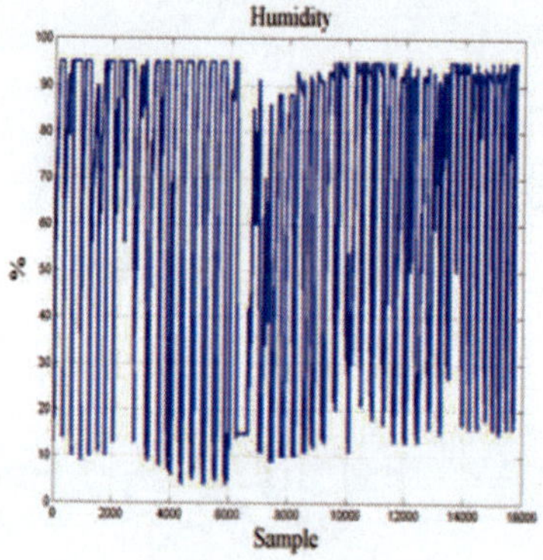


(c) Original humidity

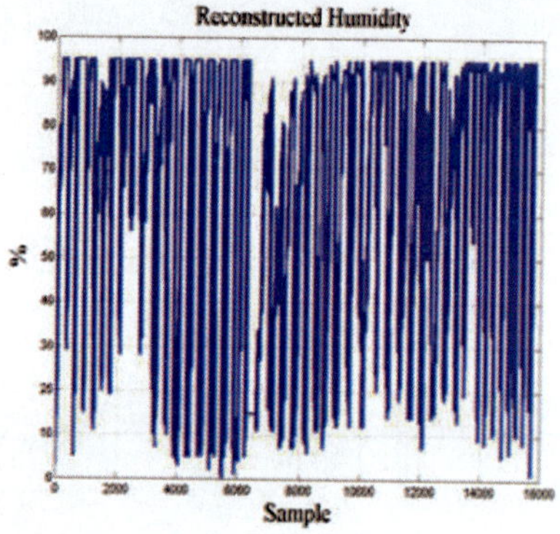


(d) Reconstructed humidity

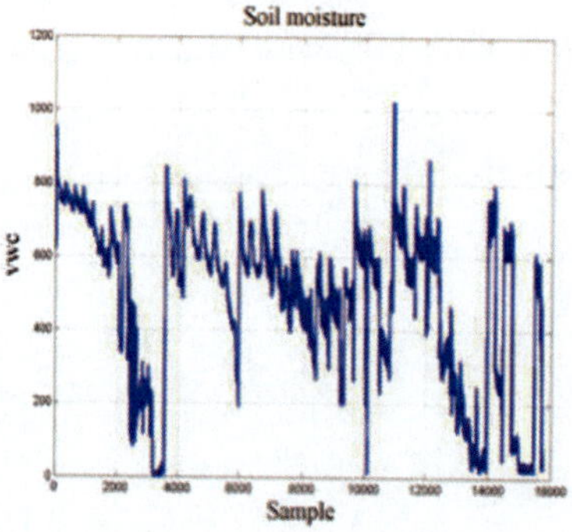


(e) Original soil moisture

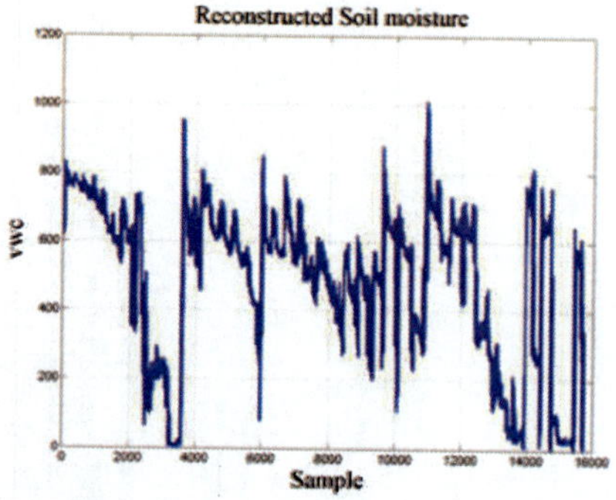


(f) Reconstructed soil moisture

**Fig. 5** Original and reconstructed temperature, humidity, and soil moisture data values

**Table 2** Standard deviation, skewness, and kurtosis of original and reconstructed data

| Parameter | | Temperature | Humidity | Soil moisture |
|---|---|---|---|---|
| Standard deviation | Original | 6.0703 | 31.5297 | 225.5850 |
| | Reconst | 6.0340 | 30.9883 | 223.7481 |
| Skewness | Original | −0.1415 | −0.7540 | −0.7513 |
| | Reconst | −0.1513 | −0.7434 | −0.7569 |
| Kurtosis | Original | 1.9050 | 1.9268 | 2.4594 |
| | Reconst | 1.9322 | 1.9591 | 2.4765 |

data can be seen from Table 2. It can be observed that the all the parameters of original and reconstructed data are almost similar. The value of skewness implies that the data is slightly skewed which is acceptable. Furthermore, from the value of kurtosis for temperature, humidity and soil moisture data it can be inferred that the data does not have any outliers. This means that the data consist of actual values fetched from the sensors, and no external additions have been made to the data.

The final outcome of an irrigation system is control of water supply to the crops. In such system that employs an IoT device, this decision is based on sensor values. The compression algorithm used here can be validated and proved authentic only if the decisions taken on original and reconstructed data are the same. This has been generated by the device, and Table 3 gives a comparison of the irrigation-related decisions taken on the original and reconstructed data values. It can be seen from the table that the decision taken on actual sensor data and the reconstructed data is the same. The decision taken on the reconstructed data is the same as that taken

**Table 3** Decision on original versus reconstructed data

| Inputs | | | Decision-based on[a] | |
|---|---|---|---|---|
| *T* | *H* | *S* | Original data | Reconstructed data |
| 20 | 20 | 300 | 0 | 0 |
| 24 | 37 | 390 | 0 | 0 |
| 27 | 34 | 485 | 0 | 0 |
| 27 | 37 | 620 | 0 | 0 |
| 32 | 27 | 320 | 0 | 0 |
| 39 | 20 | 555 | 0 | 0 |
| 40 | 29 | 170 | 1 | 1 |
| 27 | 38 | 250 | 1 | 1 |
| 21 | 40 | 155 | 1 | 1 |
| 27 | 37 | 350 | 1 | 1 |
| 27 | 36 | 150 | 1 | 1 |
| 39 | 20 | 250 | 1 | 1 |

[a] 0: motor off, 1: motor on

**Table 4** Energy consumption comparison

| CORDIC arch | Basic | Scalefree | Lookahead | Proposed |
|---|---|---|---|---|
| Energy consumption (nJ) | 80.1595 | 6.3569 | 12.1128 | 8.0488 |

using the original data, with nil error in the final decision. This means that the decision is maintained even after the data are compressed, thus making it reliable for use in decision-making in automatic irrigation systems. The comparison of energy consumption of different CORDIC architectures with the one implemented in this paper can be seen from Table 4. It can be observed that CORDIC algorithm used in our irrigation system outperforms existing basic, scale-free, and look ahead algorithms.

## 4 Conclusion and Future Scope

Irrespective of the application of IoT, huge amount of data is generated by the device which consists of useful information that needs to be stored. The limited memory of IoT devices makes it difficult to store the data offline in its original form. This calls for data compression mechanisms that can be implemented on the device before sending out the data for processing. Compression technique that is energy efficient is beneficial for a device which is battery operated. Work presented in this paper has considered one such application of smart agriculture, i.e., an irrigation system, where it uses the spatiotemporal properties of the data and an energy-efficient CORDIC-based DCT compression algorithm on the IoT device. This algorithm is 47% energy efficient when operated in least energy saving mode and 90% energy efficient than conventional CORDIC. It exploits the energy compaction property of DCT and the low-power needs of CORDIC, combines advantages of both algorithms and has applied it for data compression. Results obtained infer that compression is achieved at the cost of infinitesimal error in the reconstructed values. It is also free from outliers. As a result, the final decision outcome based on the reconstructed data is not compromised. This algorithm is not just specific to smart irrigation system, but can also it can be applied in any domain where enormous data are generated by the sensor. In future, more capability can be added by applying an outlier detection algorithm on the system. This almost removes the possibility of erroneous data due to malfunctioning of the sensors.

**Acknowledgements** The work proposed in this paper has been supported by Visvesvaraya a Ph.D. scheme, Ministry of Electronics and Information Technology, Government of India.

## References

1. Ashton, K.: That 'Internet of Things' thing. RFiD J. **22**(7), 97–114 (2009)
2. Jin, J., Gubbi, J., Marusic, S., Palaniswami, M.: An information framework for creating a smart city through internet of things. IEEE Internet Things J. **1**(2), 112–121 (2014)
3. Yang, G., Xie, L., Mantysalo, M., Zhou, X., Pang, Z., Da Xu, L., Kao-Walter, S., Chen, Q., Zheng, L.R.: A health-IoT platform based on the integration of intelligent packaging, unobtrusive bio-sensor, and intelligent medicine box. IEEE Trans. Ind. Inform. **10**(4), 2180–2191 (2014)
4. Gutierrez, J., Villa-Medina, J.F., Nieto-Garibay, A., Porta-G´andara, M.A.: Automated irrigation system using a wireless sensor network and GPRS module. IEEE Trans. Instrum. Measure. **63**(1), 166–176 (2014)
5. Lee, S.W., Kim, H.Y.: An energy-efficient low-memory image compression system for multimedia IoT products. EURASIP J. Image Video Process. **2018**(1), 87 (2018)
6. Lee, Y., Hwang, E., Choi, J.: A unified approach for compression and authentication of smart meter reading in Ami. IEEE Access **7**, 34383–34394 (2019)
7. Del Testa, D., Rossi, M.: Lightweight lossy compression of biometric patterns via denoising autoencoders. IEEE Signal Process. Lett. **22**(12), 2304–2308 (2015)
8. Tekeste, T., Saleh, H., Mohammad, B., Ismail, M.: Ultra-low power QRS detection and ECG compression architecture for IoT healthcare devices. IEEE Trans. Circuits Syst. I Regul. Pap. **66**(2), 669–679 (2018)
9. Bhargava, K., Ivanov, S., Donnelly, W., Kulatunga, C.: Using edge analytics to improve data collection in precision dairy farming. In: 2016 IEEE 41st Conference on Local Computer Networks Workshops (LCN Workshops), pp. 137–144. IEEE (2016)
10. Vecchio, M., Giaffreda, R., Marcelloni, F.: Adaptive lossless entropy compressors for tiny IoT devices. IEEE Trans. Wireless Commun. **13**(2), 1088–1100 (2014)
11. Nawandar, N.K., Satpute, V.R.: Energy efficient quality tunable CORDIC for DSP applications on battery operated portable devices. J. Circuits Syst. Comput. **27**(04), 1850051 (2018)
12. Nawandar, N.K., Satpute, V.R.: IoT based low cost and intelligent module for smart irrigation system. Comput. Electron. Agric. **162**, 979–990 (2019)
13. Volder, J.E.: The CORDIC trigonometric computing technique. IRE Trans. Electron. Comput. **3**, 330–334 (1959)
14. Aggarwal, S., Khare, K.: Hardware efficient architecture for generating Sine/Cosine waves. In: Proceedings of the 2012 25th International Conference on VLSI Design, pp. 57–61. IEEE Computer Society (2012)
15. Lee, M.W., Yoon, J.H., Park, J.: Reconfigurable CORDIC-based low-power DCT architecture based on data priority. IEEE Trans. Very Large Scale Integr. (VLSI) Syst. **22**(5), 1060–1068 (2013)

# Author Index




D. Gupta et al. (eds.), *Pattern Recognition and Data Analysis with Applications*,
Lecture Notes in Electrical Engineering 888,
https://doi.org/10.1007/978-981-19-1520-8